KU-163-813

Numerical Methods for Engineers and Scientists

Numerical Methods for Engineers and Scientists

Second Edition
Revised and Expanded

Joe D. Hoffman
Department of Mechanical Engineering
Purdue University
West Lafayette, Indiana

MARCEL DEKKER, INC.   NEW YORK · BASEL

The first edition of this book was published by McGraw-Hill, Inc. (New York, 1992).

ISBN: 0-8247-0443-6

This book is printed on acid-free paper.

Headquarters
Marcel Dekker, Inc.
270 Madison Avenue, New York, NY 10016
tel: 212-696-9000; fax: 212-685-4540

Eastern Hemisphere Distribution
Marcel Dekker AG
Hutgasse 4, Postfach 812, CH-4001 Basel, Switzerland
tel: 41-61-261-8482; fax: 41-61-261-8896

World Wide Web
http://www.dekker.com

The publisher offers discounts on this book when ordered in bulk quantities. For more information, write to Special Sales/Professional Marketing at the headquarters address above.

Copyright © 2001 by Marcel Dekker, Inc. All Rights Reserved.

Neither this book nor any part may be reproduced or transmitted in any form or by any means, electronic or mechanical, including photocopying, microfilming, and recording, or by any information storage and retrieval system, without permission in writing from the publisher.

Current printing (last digit)
10 9 8 7 6 5 4 3 2 1

PRINTED IN THE UNITED STATES OF AMERICA

To Cynthia Louise Hoffman

Preface

The second edition of this book contains several major improvements over the first edition. Some of these improvements involve format and presentation philosophy, and some of the changes involve old material which has been deleted and new material which has been added.

Each chapter begins with a chapter table of contents. The first figure carries a sketch of the application used as the example problem in the chapter. Section 1 of each chapter is an introduction to the chapter, which discusses the example application, the general subject matter of the chapter, special features, and solution approaches. The objectives of the chapter are presented, and the organization of the chapter is illustrated pictorially. Each chapter ends with a summary section, which presents a list of recommendations, dos and don'ts, and a list of what *you should be able to do after studying the chapter*. This list is actually an itemization of what the student should have learned from the chapter. It serves as a list of objectives, a study guide, and a review guide for the chapter.

Chapter 0, Introduction, has been added to give a thorough introduction to the book and to present several fundamental concepts of relevance to the entire book.

Chapters 1 to 6, which comprise Part I, Basic Tools of Numerical Analysis, have been expanded to include more approaches for solving problems. Discussions of pitfalls of selected algorithms have been added where appropriate. Part I is suitable for second-semester sophomores or first-semester juniors through beginning graduate students.

Chapters 7 and 8, which comprise Part II, Ordinary Differential Equations, have been rewritten to get to the methods for solving problems more quickly, with less emphasis on theory. A new section presenting extrapolation methods has been added in Chapter 7. All of the material has been rewritten to flow more smoothly with less repetition and less theoretical background. Part II is suitable for juniors through graduate students.

Chapters 9 to 15 of the first edition, which comprised Part III, Partial Differential Equations, has been shortened considerably to only four chapters in the present edition. Chapter 9 introduces elliptic partial differential equations. Chapter 10 introduces parabolic partial differential equations, and Chapter 11 introduces hyperbolic partial differential equations. These three chapters are a major condensation of the material in Part III of the first edition. The material has been revised to flow more smoothly with less emphasis on theoretical background. A new chapter, Chapter 12, The Finite Element Method, has been added to present an introduction to that important method of solving differential equations.

A new section, Programs, has been added to each chapter. This section presents several FORTRAN programs for implementing the algorithms developed in each chapter to solve the example application for that chapter. The application subroutines are written in a form similar to pseudocode to facilitate the implementation of the algorithms in other programming languages.

More examples and more problems have been added throughout the book.

The overall objective of the second edition is to improve the presentation format and material content of the first edition in a manner that not only maintains but enhances the usefullness and ease of use of the first edition.

Many people have contributed to the writing of this book. All of the people acknowledged in the Preface to the First Edition are again acknowledged, especially my loving wife, Cynthia Louise Hoffman. My many graduate students provided much help and feedback, especially Drs. D. Hofer, R. Harwood, R. Moore, and R. Stwalley. Thanks, guys. All of the figures were prepared by Mr. Mark Bass. Thanks, Mark. Once again, my expert word processing specialist, Ms. Janice Napier, devoted herself unsparingly to this second edition. Thank you, Janice. Finally, I would like to acknowledge my colleague, Mr. B. J. Clark, Executive Acquisitions Editor at Marcel Dekker, Inc., for his encouragement and support during the preparation of both editions of this book.

Joe D. Hoffman

Contents

Contents

0

Introduction

This Introduction contains a brief description of the objectives, approach, and organization of the book. The philosophy behind the Examples, Programs, and Problems is discussed. Several years' experience with the first edition of the book has identified several simple, but significant, concepts which are relevant throughout the book, but the place to include them is not clear. These concepts, which are presented in this Introduction, include the definitions of significant digits, precision, accuracy, and errors, and a discussion of number representation. A brief description of software packages and libraries is presented. Last, the Taylor series and the Taylor polynomial, which are indispensable in developing and understanding many numerical algorithms, are presented and discussed.

0.1 OBJECTIVE AND APPROACH

The objective of this book is to introduce the engineer and scientist to numerical methods which can be used to solve mathematical problems arising in engineering and science that cannot be solved by exact methods. With the general accessibility of high-speed digital computers, it is now possible to obtain rapid and accurate solutions to many complex problems that face the engineer and scientist.

The approach taken is as follows:

1. Introduce a type of problem.

1

2. Present sufficient background to understand the problem and possible methods of solution.
3. Develop one or more numerical methods for solving the problem.
4. Illustrate the numerical methods with examples.

In most cases, the numerical methods presented to solve a particular problem proceed from simple methods to complex methods, which in many cases parallels the chronological development of the methods. Some poor methods and some bad methods, as well as good methods, are presented for pedagogical reasons. Why one method does not work is almost as important as why another method does work.

0.2 ORGANIZATION OF THE BOOK

The material in the book is divided into three main parts:

I.   Basic Tools of Numerical Analysis
II. Ordinary Differential Equations
III. Partial Differential Equations

Part I considers many of the basic problems that arise in all branches of engineering and science. These problems include: solution of systems of linear algebraic equations, eigenproblems, solution of nonlinear equations, polynomial approximation and interpolation, numerical differentiation and difference formulas, and numerical integration. These topics are important both in their own right and as the foundation for Parts II and III.

Part II is devoted to the numerical solution of ordinary differential equations (ODEs). The general features of ODEs are discussed. The two classes of ODEs (i.e., initial-value ODEs and boundary-value ODEs) are introduced, and the two types of physical problems (i.e., propagation problems and equilibrium problems) are discussed. Numerous numerical methods for solving ODEs are presented.

Part III is devoted to the numerical solution of partial differential equations (PDEs). Some general features of PDEs are discussed. The three classes of PDEs (i.e., elliptic PDEs, parabolic PDEs, and hyperbolic PDEs) are introduced, and the two types of physical problems (i.e., equilibrium problems and propagation problems) are discussed. Several model PDEs are presented. Numerous numerical methods for solving the model PDEs are presented.

The material presented in this book is an introduction to numerical methods. Many practical problems can be solved by the methods presented here. Many other practical problems require other or more advanced numerical methods. Mastery of the material presented in this book will prepare engineers and scientists to solve many of their everyday problems, give them the insight to recognize when other methods are required, and give them the background to study other methods in other books and journals.

0.3 EXAMPLES

All of the numerical methods presented in this book are illustrated by applying them to solve an example problem. Each chapter has one or two example problems, which are solved by all of the methods presented in the chapter. This approach allows the analyst to compare various methods for the same problem, so accuracy, efficiency, robustness, and ease of application of the various methods can be evaluated.

Most of the example problems are rather simple and straightforward, thus allowing the special features of the various methods to be demonstrated clearly. All of the example problems have exact solutions, so the errors of the various methods can be compared. Each example problem begins with a reference to the problem to be solved, a description of the numerical method to be employed, details of the calculations for at least one application of the algorithm, and a summary of the remaining results. Some comments about the solution are presented at the end of the calculations in most cases.

0.4 PROGRAMS

Most numerical algorithms are generally expressed in the form of a computer program. This is especially true for algorithms that require a lot of computational effort and for algorithms that are applied many times. Several programming languages are available for preparing computer programs: FORTRAN, Basic, C, PASCAL, etc., and their variations, to name a few. Pseudocode, which is a set of instructions for implementing an algorithm expressed in conceptual form, is also quite popular. Pseudocode can be expressed in the detailed form of any specific programming language.

FORTRAN is one of the oldest programming languages. When carefully prepared, FORTRAN can approach pseudocode. Consequently, the programs presented in this book are written in simple FORTRAN. There are several vintages of FORTRAN: FORTRAN I, FORTRAN II, FORTRAN 66, 77, and 90. The programs presented in this book are compatible with FORTRAN 77 and 90.

Several programs are presented in each chapter for implementing the more prominent numerical algorithms presented in the chapter. Each program is applied to solve the example problem relevant to that chapter. The implementation of the numerical algorithm is contained within a completely self-contained *application subroutine* which can be used in other programs. These *application subroutines* are written as simply as possible so that conversion to other programming languages is as straightforward as possible. These subroutines can be used as they stand or easily modified for other applications.

Each *application subroutine* is accompanied by a *program main*. The variables employed in the *application subroutine* are defined by comment statements in *program main*. The numerical values of the variables are defined in *program main*, which then calls the *application subroutine* to solve the example problem and to print the solution. These main programs are not intended to be convertible to other programming languages. In some problems where a function of some type is part of the specification of the problem, that function is defined in a *function subprogram* which is called by the *application subroutine*.

FORTRAN compilers do not distinguish between uppercase and lowercase letters. FORTRAN programs are conventionally written in uppercase letters. However, in this book, all FORTRAN programs are written in lowercase letters.

0.5 PROBLEMS

Two types of problems are presented at the end of each chapter:

1. Exercise problems
2. Applied problems

Exercise problems are straightforward problems designed to give practice in the application of the numerical algorithms presented in each chapter. Exercise problems emphasize the mechanics of the methods.

Applied problems involve more applied engineering and scientific applications which require numerical solutions.

Many of the problems can be solved by hand calculation. A large number of the problems require a lot of computational effort. Those problems should be solved by writing a computer program to perform the calculations. Even in those cases, however, it is recommended that one or two passes through the algorithm be made by hand calculation to ensure that the analyst fully understands the details of the algorithm. These results also can be used to validate the computer program.

Answers to selected problems are presented in a section at the end of the book. All of the problems for which answers are given are denoted by an asterisk appearing with the corresponding problem number in the problem sections at the end of each chapter. The **Solutions Manual** contains the answers to nearly all of the problems.

0.6 SIGNIFICANT DIGITS, PRECISION, ACCURACY, ERRORS, AND NUMBER REPRESENTATION

Numerical calculations obviously involve the manipulation (i.e., addition, multiplication, etc.) of numbers. Numbers can be integers (e.g., 4, 17, -23, etc.), fractions (e.g., 1/2, $-2/3$, etc.), or an inifinite string of digits (e.g., $\pi = 3.1415926535\ldots$). When dealing with numerical values and numerical calculations, there are several concepts that must be considered:

1. Significant digits
2. Precision and accuracy
3. Errors
4. Number representation

These concepts are discussed briefly in this section.

Significant Digits

The **significant digits**, or figures, in a number are the digits of the number which are known to be correct. Engineering and scientific calculations generally begin with a set of data having a known number of significant digits. When these numbers are processed through a numerical algorithm, it is important to be able to estimate how many significant digits are present in the final computed result.

Precision and Accuracy

Precision refers to how closely a number represents the number it is representing. **Accuracy** refers to how closely a number agrees with the true value of the number it is representing.

Precision is governed by the number of digits being carried in the numerical calculations. Accuracy is governed by the errors in the numerical approximation. Precision and accuracy are quantified by the errors in a numerical calculation.

Errors

The **accuracy** of a numerical calculation is quantified by the **error** of the calculation. Several types of errors can occur in numerical calculations.

1. Errors in the parameters of the problem (assumed nonexistent).
2. Algebraic errors in the calculations (assumed nonexistent).
3. Iteration errors.
4. Approximation errors.
5. Roundoff errors.

Iteration error is the error in an iterative method that approaches the exact solution of an exact problem asymptotically. Iteration errors must decrease toward zero as the iterative process progresses. The iteration error itself may be used to determine the successive approximations to the exact solution. Iteration errors can be reduced to the limit of the computing device. The errors in the solution of a system of linear algebraic equations by the successive-over-relaxation (SOR) method presented in Section 1.5 are examples of this type of error.

Approximation error is the difference between the exact solution of an exact problem and the exact solution of an approximation of the exact problem. Approximation error can be reduced only by choosing a more accurate approximation of the exact problem. The error in the approximation of a function by a polynomial, as described in Chapter 4, is an example of this type of error. The error in the solution of a differential equation where the exact derivatives are replaced by algebraic difference approximations, which have truncation errors, is another example of this type of error.

Roundoff error is the error caused by the finite word length employed in the calculations. Roundoff error is more significant when small differences between large numbers are calculated. Most computers have either 32 bit or 64 bit word length, corresponding to approximately 7 or 13 significant decimal digits, respectively. Some computers have extended precision capability, which increases the number of bits to 128. Care must be exercised to ensure that enough significant digits are maintained in numerical calculations so that roundoff is not significant.

Number Representation

Numbers are represented in number systems. Any number of bases can be employed as the base of a number system, for example, the base 10 (i.e., decimal) system, the base 8 (i.e., octal) system, the base 2 (i.e., binary) system, etc. The base 10, or decimal, system is the most common system used for human communication. Digital computers use the base 2, or binary, system. In a digital computer, a binary number consists of a number of binary bits. The number of binary bits in a binary number determines the precision with which the binary number represents a decimal number. The most common size binary number is a 32 bit number, which can represent approximately seven digits of a decimal number. Some digital computers have 64 bit binary numbers, which can represent 13 to 14 decimal digits. In many engineering and scientific calculations, 32 bit arithmetic is adequate. However, in many other applications, 64 bit arithmetic is required. In a few special situations, 128 bit arithmetic may be required. On 32 bit computers, 64 bit arithmetic, or even 128 bit arithmetic, can be accomplished using software enhancements. Such calculations are called **double precision** or **quad precision**, respectively. Such software enhanced precision can require as much as 10 times the execution time of a single precision calculation.

Consequently, some care must be exercised when deciding whether or not higher precision arithmetic is required. All of the examples in this book are evaluated using 64 bit arithmetic to ensure that roundoff is not significant.

Except for integers and some fractions, all binary representations of decimal numbers are approximations, owing to the finite word length of binary numbers. Thus, some loss of precision in the binary representation of a decimal number is unavoidable. When binary numbers are combined in arithmetic operations such as addition, multiplication, etc., the true result is typically a longer binary number which cannot be represented exactly with the number of available bits in the binary number capability of the digital computer. Thus, the results are rounded off in the last available binary bit. This rounding off gives rise to **roundoff error**, which can accumulate as the number of calculations increases.

0.7 SOFTWARE PACKAGES AND LIBRARIES

Numerous commercial software packages and libraries are available for implementing the numerical solution of engineering and scientific problems. Two of the more versatile software packages are Mathcad and Matlab. These software packages, as well as several other packages and several libraries, are listed below with a brief description of each one and references to sources for the software packages and libraries.

A. Software Packages

Excel Excel is a spreadsheet developed by Microsoft, Inc., as part of Microsoft Office. It enables calculations to be performed on rows and columns of numbers. The calculations to be performed are specified for each column. When any number on the spreadsheet is changed, all of the calculations are updated. Excel contains several built-in numerical algorithms. It also includes the Visual Basic programming language and some plotting capability. Although its basic function is not numerical analysis, Excel can be used productively for many types of numerical problems. Microsoft, Inc. www.microsoft.com/office/Excel.

Macsyma Macsyma is the world's first artificial intelligence based math engine providing easy to use, powerful math software for both symbolic and numerical computing. Macsyma, Inc., 20 Academy St., Arlington, MA 02476-6412. (781) 646-4550, webmaster@macsyma.com, www.macsyma.com.

Maple Maple 6 is a technologically advanced computational system with both algorithms and numeric solvers. Maple 6 includes an extensive set of NAG (Numerical Algorithms Group) solvers for computational linear algebra. Waterloo Maple, Inc., 57 Erb Street W., Waterloo, Ontario, Canada N2L 6C2. (800) 267-6583, (519) 747-2373, info@maplesoft.com, www.maplesoft.com.

Mathematica Mathematica 4 is a comprehensive software package which performs both symbolic and numeric computations. It includes a flexible and intuitive programming language and comprehensive plotting capabilities. Wolfram Research, Inc., 100 Trade Center Drive, Champaign IL 61820-7237. (800) 965-3726, (217) 398-0700, info@wolfram.com, www.wolfram.com.

Mathcad Mathcad 8 provides a free-form interface which permits the integration of real math notation, graphs, and text within a single interactive worksheet. It includes statistical and data analysis functions, powerful solvers, advanced matrix manipulation, and the capability to create your own functions. Mathsoft, Inc., 101 Main Street, Cambridge, MA 02142-1521. (800) 628-4223, (617) 577-1017, info@mathsoft.com, www.mathcad.com.

Matlab Matlab is an integrated computing environment that combines numeric computation, advanced graphics and visualization, and a high-level programming language. It provides core mathematics and advanced graphics tools for data analysis, visualization, and algorithm and application development, with more than 500 mathematical, statistical, and engineering functions. The Mathworks, Inc., 3 Apple Hill Drive, Natick, MA 01760-2090. (508) 647-7000, info@mathworks.com, www.mathworks.com.

B. Libraries

GAMS GAMS (Guide to Available Mathematical Software) is a guide to over 9000 software modules contained in some 80 software packages at NIST (National Institute for Standards and Technology) and NETLIB. gams.nist.gov.

IMSL IMSL (International Mathematics and Statistical Library) is a comprehensive resource of more than 900 FORTRAN subroutines for use in general mathematics and statistical data analysis. Also available in C and Java. Visual Numerics, Inc., 1300 W. Sam Houston Parkway S., Suite 150, Houston TX 77042. (800) 364-8880, (713) 781-9260, info@houston.vni.com, www.vni.com.

LAPACK LAPACK is a library of FORTRAN 77 subroutines for solving linear algebra problems and eigenproblems. Individual subroutines can be obtained through NETLIB. The complete package can be obtained from NAG.

NAG NAG is a mathematical software library that contains over 1000 mathematical and statistical functions. Available in FORTRAN and C. NAG, Inc., 1400 Opus Place, Suite 200, Downers Grove, IL 60515-5702. (630) 971-2337, naginfo@nag.com, www.nag.com.

NETLIB NETLIB is a large collection of numerical libraries. netlib@research.att.com, netlib@ornl.gov, netlib@nec.no.

C. Numerical Recipes

Numerical Recipes is a book by William H. Press, Brian P. Flannery, Saul A. Teukolsky, and William T. Vetterling. It contains over 300 subroutines for numerical algorithms. Versions of the subroutines are available in FORTRAN, C, Pascal, and Basic. The source codes are available on disk. Cambridge University Press, 40 West 20th Street, New York, NY 10011. www.cup.org.

0.8   THE TAYLOR SERIES AND THE TAYLOR POLYNOMIAL

A power series in powers of x is a series of the form

$$\sum_{n=0}^{\infty} a_n x^n = a_0 + a_1 x + a_2 x^2 + \cdots \tag{0.1}$$

A power series in powers of $(x - x_0)$ is given by

$$\sum_{n=0}^{\infty} a_n (x - x_0)^n = a_0 + a_1 (x - x_0) + a_2 (x - x_0)^2 + \cdots \tag{0.2}$$

Within its radius of convergence, r, any continuous function, $f(x)$, can be represented exactly by a power series. Thus,

$$f(x) = \sum_{n=0}^{\infty} a_n(x - x_0)^n \tag{0.3}$$

is continuous for $(x_0 - r) < x < (x_0 + r)$.

A. Taylor Series in One Independent Variable

If the coefficients, a_n, in Eq. (0.3) are given by the rule:

$$a_0 = f(x_0),\, a_1 = \frac{1}{1!}f'(x_0),\, a_2 = \frac{1}{2!}f''(x_0),\, \ldots \tag{0.4}$$

then Eq. (0.3) becomes the **Taylor series** of $f(x)$ at $x = x_0$. Thus,

$$f(x) = f(x_0) + \frac{1}{1!}f'(x_0)(x - x_0) + \frac{1}{2!}f''(x_0)(x - x_0)^2 + \cdots \tag{0.5}$$

Equation (0.5) can be written in the simpler appearing form

$$f(x) = f_0 + f_0'\Delta x + \frac{1}{2}f_0''\Delta x^2 + \cdots + \frac{1}{n!}f_0^{(n)}\Delta x^n + \cdots \tag{0.6}$$

where $f_0 = f(x_0)$, $f^{(n)} = df^{(n)}/dx^n$, and $\Delta x = (x - x_0)$. Equation (0.6) can be written in the compact form

$$f(x) = \sum_{n=0}^{\infty} \frac{1}{n!}f_0^{(n)}(x - x_0)^n \tag{0.7}$$

When $x_0 = 0$, the Taylor series is known as the **Maclaurin series**. In that case, Eqs. (0.5) and (0.7) become

$$f(x) = f(0) + f'(0)x + \tfrac{1}{2}f''(0)x^2 + \cdots \tag{0.8}$$

$$f(x) = \sum_{n=0}^{\infty} \frac{1}{n!}f^{(n)}(0)x^n \tag{0.9}$$

It is, of course, impractical to evaluate an infinite Taylor series term by term. The Taylor series can be written as the finite Taylor series, also known as the *Taylor formula* or *Taylor polynomial* with remainder, as follows:

$$f(x) = f(x_0) + f'(x_0)(x - x_0) + \frac{1}{2}f''(x_0)(x - x_0)^2 + \cdots$$
$$+ \frac{1}{n!}f^{(n)}(x_0)(x - x_0)^n + R^{n+1} \tag{0.10}$$

where the term R^{n+1} is the **remainder term** given by

$$R^{n+1} = \frac{1}{(n+1)!}f^{(n+1)}(\xi)(x - x_0)^{n+1} \tag{0.11}$$

where ξ lies between x_0 and x. Equation (0.10) is quite useful in numerical analysis, where an approximation of $f(x)$ is obtained by truncating the remainder term.

B. Taylor Series in Two Independent Variables

Power series can also be written for functions of more than one independent variable. For a function of two independent variables, $f(x, y)$, the Taylor series of $f(x, y)$ at (x_0, y_0) is given by

$$
\begin{aligned}
f(x, y) = f_0 &+ \left.\frac{\partial f}{\partial x}\right|_0 (x - x_0) + \left.\frac{\partial f}{\partial y}\right|_0 (y - y_0) \\
&+ \frac{1}{2!}\left(\left.\frac{\partial^2 f}{\partial x^2}\right|_0 (x - x_0)^2 + 2\left.\frac{\partial^2 f}{\partial x \partial y}\right|_0 (x - x_0)(y - y_0) + \left.\frac{\partial^2 f}{\partial y^2}\right|_0 (y - y_0)^2 \right) + \cdots
\end{aligned}
$$

$$(0.12)$$

Equation (0.12) can be written in the general form

$$
f(x, y) = \sum_{n=0}^{\infty} \frac{1}{n!}\left((x - x_0)\frac{\partial}{\partial x} + (y - y_0)\frac{\partial}{\partial y} \right)^n f(x, y)|_0
\tag{0.13}
$$

where the term $(\cdots)^n$ is expanded by the binomial expansion and the resulting expansion operates on the function $f(x, y)$ and is evaluated at (x_0, y_0).

The Taylor formula with remainder for a function of two independent variables is obtained by evaluating the derivatives in the $(n + 1)$st term at the point (ξ, η), where (ξ, η) lies in the region between points (x_0, y_0) and (x, y).

I

Basic Tools of Numerical Analysis

Many different types of algebraic processes are required in engineering and science. These processes include the solution of systems of linear algebraic equations, the solution of eigenproblems, finding the roots of nonlinear equations, polynomial approximation and interpolation, numerical differentiation and difference formulas, and numerical integration. These topics are not only important in their own right, they lay the foundation for the solution of ordinary and partial differential equations, which are discussed in Parts II and III, respectively. Figure I.1 illustrates the types of problems considered in Part I.

The objective of Part I is to introduce and discuss the general features of each of these algebraic processes, which are the *basic tools of numerical analysis*.

I.1 SYSTEMS OF LINEAR ALGEBRAIC EQUATIONS

Systems of equations arise in all branches of engineering and science. These equations may be algebraic, transcendental (i.e., involving trigonometric, logarithmic, exponential, etc., functions), ordinary differential equations, or partial differential equations. The equations may be linear or nonlinear. Chapter 1 is devoted to the solution of *systems of linear algebraic equations* of the following form:

$$a_{11}x_1 + a_{12}x_2 + a_{13}x_3 + \cdots + a_{1n}x_n = b_1 \tag{I.1a}$$
$$a_{21}x_1 + a_{22}x_2 + a_{23}x_3 + \cdots + a_{2n}x_n = b_2 \tag{I.1b}$$
$$\cdots\cdots\cdots\cdots\cdots\cdots\cdots\cdots\cdots\cdots\cdots\cdots\cdots$$
$$a_{n1}x_1 + a_{n2}x_2 + a_{n3}x_3 + \cdots + a_{nn}x_n = b_n \tag{I.1n}$$

11

where x_j $(j = 1, 2, \ldots, n)$ denotes the unknown variables, $a_{i,j}$ $(i, j = 1, 2, \ldots, n)$ denotes the coefficients of the unknown variables, and b_i $(i = 1, 2, \ldots, n)$ denotes the nonhomogeneous terms. For the coefficients $a_{i,j}$, the first subscript i corresponds to equation i, and the second subscript j corresponds to variable x_j. The number of equations can range from two to hundreds, thousands, and even millions.

Systems of linear algebraic equations arise in many different problems, for example, (a) network problems (e.g., electrical networks), (b) fitting approximating functions (see Chapter 4), and (c) systems of finite difference equations that arise in the numerical solution of differential equations (see Chapters 7 to 12). The list is endless. Figure I.1a illustrates a static spring-mass system, whose static equilibrium configuration is governed by a system of linear algebraic equations. That system of equations is used throughout Chapter 1 as an example problem.

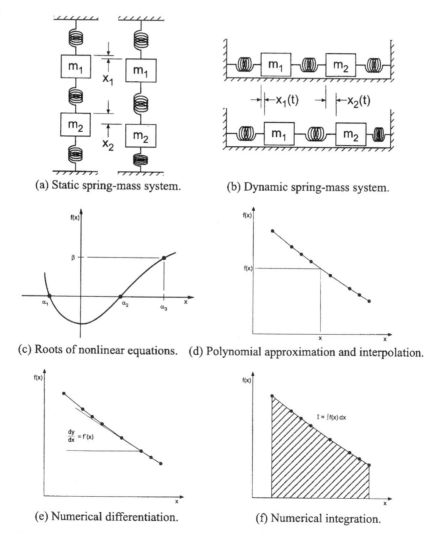

(a) Static spring-mass system.   (b) Dynamic spring-mass system.

(c) Roots of nonlinear equations.   (d) Polynomial approximation and interpolation.

(e) Numerical differentiation.   (f) Numerical integration.

Figure I.1 Basic tools of numerical analysis. (a) Static spring-mass system. (b) Dynamic spring-mass system. (c) Roots of nonlinear equations. (d) Polynomial approximation and interpolation. (e) Numerical differentiation. (f) Numerical integration.

Systems of linear algebraic equations can be expressed very conveniently in terms of matrix notation. Solution methods can be developed very compactly in terms of matrix notation. Consequently, the elementary properties of matrices and determinants are reviewed at the beginning of Chapter 1.

Two fundamentally different approaches can be used to solve systems of linear algebraic equations:

1. Direct methods
2. Iterative methods

Direct methods are systematic procedures based on algebraic elimination. Several direct elimination methods, for example, Gauss elimination, are presented in Chapter 1. Iterative methods obtain the solution asymptotically by an iterative procedure in which a trial solution is assumed, the trial solution is substituted into the system of equations to determine the mismatch, or error, and an improved solution is obtained from the mismatch data. Several iterative methods, for example, successive-over-relaxation (SOR), are presented in Chapter 1.

The notation, concepts, and procedures presented in Chapter 1 are used throughout the remainder of the book. A solid understanding of systems of linear algebraic equations is essential in numerical analysis.

I.2 EIGENPROBLEMS

Eigenproblems arise in the special case where a system of algebraic equations is homogeneous; that is, the nonhogeneous terms, b_i in Eq. (I.1), are all zero, and the coefficients contain an unspecified parameter, say λ. In general, when $b_i = 0$, the only solution to Eq. (I.1) is the trivial solution, $x_1 = x_2 = \cdots = x_n = 0$. However, when the coefficients $a_{i,j}$ contain an unspecified parameter, say λ, the value of that parameter can be chosen so that the system of equations is redundant, and an infinite number of solutions exist. The unspecified parameter λ is an *eigenvalue* of the system of equations. For example,

$$(a_{11} - \lambda)x_1 + a_{12}x_2 = 0 \tag{I.2a}$$

$$a_{21}x_1 + (a_{22} - \lambda)x_2 = 0 \tag{I.2b}$$

is a linear eigenproblem. The value (or values) of λ that make Eqs. (I.2a) and (I.2b) identical are the eigenvalues of Eqs. (I.2). In that case, the two equations are redundant, so the only unique solution is $x_1 = x_2 = 0$. However, an infinite number of solutions can be obtained by specifying either x_1 or x_2, then calculating the other from either of the two redundant equations. The set of values of x_1 and x_2 corresponding to a particular value of λ is an *eigenvector* of Eq. (I.2). Chapter 2 is devoted to the solution of eigenproblems.

Eigenproblems arise in the analysis of many physical systems. They arise in the analysis of the dynamic behavior of mechanical, electrical, fluid, thermal, and structural systems. They also arise in the analysis of control systems. Figure I.1b illustrates a dynamic spring-mass system, whose dynamic equilibrium configuration is governed by a system of homogeneous linear algebraic equations. That system of equations is used throughout Chapter 2 as an example problem. When the static equilibrium configuration of the system is disturbed and then allowed to vibrate freely, the system of masses will oscillate at special frequencies, which depend on the values of the masses and the spring constants. These special frequencies are the eigenvalues of the system. The relative values of x_1, x_2, etc. corresponding to each eigenvalue λ are the eigenvectors of the system.

The objectives of Chapter 2 are to introduce the general features of eigenproblems and to present several methods for solving eigenproblems. Eigenproblems are special problems of interest only in themselves. Consequently, an understanding of eigenproblems is not essential to the other concepts presented in this book.

I.3 ROOTS OF NONLINEAR EQUATIONS

Nonlinear equations arise in many physical problems. Finding their roots, or zeros, is a common problem. The problem can be stated as follows:

Given the continuous nonlinear function $f(x)$, find the value of $x = \alpha$ such that

$$f(\alpha) = 0$$

where α is the *root*, or zero, of the nonlinear equation. Figure I.1c illustrates the problem graphically. The function $f(x)$ may be an algebraic function, a transcendental function, the solution of a differential equation, or any nonlinear relationship between an input x and a response $f(x)$. Chapter 3 is devoted to the solution of nonlinear equations.

Nonlinear equations are solved by iterative methods. A trial solution is assumed, the trial solution is substituted into the nonlinear equation to determine the error, or mismatch, and the mismatch is used in some systematic manner to generate an improved estimate of the solution. Several methods for finding the roots of nonlinear equations are presented in Chapter 3. The workhorse methods of choice for solving nonlinear equations are Newton's method and the secant method. A detailed discussion of finding the roots of polynomials is presented. A brief introduction to the problems of solving systems of nonlinear equations is also presented.

Nonlinear equations occur throughout engineering and science. Nonlinear equations also arise in other areas of numerical analysis. For example, the shooting method for solving boundary-value ordinary differential equations, presented in Section 8.3, requires the solution of a nonlinear equation. Implicit methods for solving nonlinear differential equations yield nonlinear difference equations. The solution of such problems is discussed in Sections 7.11, 8.7, 9.11, 10.9, and 11.8. Consequently, a thorough understanding of methods for solving nonlinear equations is an essential requirement for the numerical analyst.

I.4 POLYNOMIAL APPROXIMATION AND INTERPOLATION

In many problems in engineering and science, the data under consideration are known only at discrete points, not as a continuous function. For example, as illustrated in Figure I.1d, the continuous function $f(x)$ may be known only at n discrete values of x:

$$y_i = y(x_i) \qquad (i = 1, 2, \ldots, n) \tag{I.3}$$

Values of the function at points other than the known discrete points may be needed (i.e., *interpolation*). The derivative of the function at some point may be needed (i.e., *differentiation*). The integral of the function over some range may be required (i.e., *integration*). These processes, for discrete data, are performed by fitting an approximating function to the set of discrete data and performing the desired processes on the approximating function. Many types of approximating functions can be used.

Because of their simplicity, ease of manipulation, and ease of evaluation, polynomials are an excellent choice for an approximating function. The general nth-degree polynomial is specified by

$$P_n(x) = a_0 + a_1 x + a_2 x^2 + \cdots + a_n x^n \tag{I.4}$$

Polynomials can be fit to a set of discrete data in two ways:

1. Exact fit
2. Approximate fit

An exact fit passes exactly through all the discrete data points. Direct fit polynomials, divided-difference polynomials, and Lagrange polynomials are presented in Chapter 4 for fitting nonequally spaced data or equally spaced data. Newton difference polynomials are presented for fitting equally spaced data. The least squares procedure is presented for determining approximate polynomial fits.

Figure I.1d illustrates the problem of interpolating within a set of discrete data. Procedures for interpolating within a set of discrete data are presented in Chapter 4.

Polynomial approximation is essential for interpolation, differentiation, and integration of sets of discrete data. A good understanding of polynomial approximation is a necessary requirement for the numerical analyst.

I.5   NUMERICAL DIFFERENTIATION AND DIFFERENCE FORMULAS

The evaluation of derivatives, a process known as *differentiation*, is required in many problems in engineering and science. Differentiation of the function $f(x)$ is denoted by

$$\frac{d}{dx}(f(x)) = f'(x) \tag{I.5}$$

The function $f(x)$ may be a known function or a set of discrete data. In general, known functions can be differentiated exactly. Differentiation of discrete data requires an approximate numerical procedure. Numerical differentiation formulas can be developed by fitting approximating functions (e.g., polynomials) to a set of discrete data and differentiating the approximating function. For polynomial approximating functions, this yields

$$\frac{d}{dx}(f(x)) = f'(x) \cong \frac{d}{dx}(P_n(x)) = P'_n(x) \tag{I.6}$$

Figure I.1e illustrates the problem of numerical differentiation of a set of discrete data. Numerical differentiation procedures are developed in Chapter 5.

The approximating polynomial may be fit exactly to a set of discrete data by the methods presented in Chapter 4, or fit approximately by the least squares procedure described in Chapter 4. Several numerical differentiation formulas based on differentiation of polynomials are presented in Chapter 5.

Numerical differentiation formulas also can be developed using Taylor series. This approach is quite useful for developing difference formulas for approximating exact derivatives in the numerical solution of differential equations. Section 5.5 presents a table of difference formulas for use in the solution of differential equations.

Numerical differentiation of discrete data is not required very often. However, the numerical solution of differential equations, which is the subject of Parts II and III, is one of the most important areas of numerical analysis. The use of difference formulas is essential in that application.

I.6 NUMERICAL INTEGRATION

The evaluation of integrals, a process known as *integration*, or quadrature, is required in many problems in engineering and science. Integration of the function $f(x)$ is denoted by

$$I = \int_a^b f(x)\, dx \tag{I.7}$$

The function $f(x)$ may be a known function or a set of discrete data. Some known functions have an exact integral. Many known functions, however, do not have an exact integral, and an approximate numerical procedure is required to evaluate Eq. (I.7). When a known function is to be integrated numerically, it must first be discretized. Integration of discrete data always requires an approximate numerical procedure. Numerical integration (quadrature) formulas can be developed by fitting approximating functions (e.g., polynomials) to a set of discrete data and integrating the approximating function. For polynomial approximating functions, this gives

$$I = \int_a^b f(x)\, dx \cong \int_a^b P_n(x)\, dx \tag{I.8}$$

Figure I.1f illustrates the problem of numerical integration of a set of discrete data. Numerical integration procedures are developed in Chapter 6.

The approximating function can be fit exactly to a set of discrete data by direct fit methods, or fit approximately by the least squares method. For unequally spaced data, direct fit polynomials can be used. For equally spaced data, the Newton forward-difference polynomials of different degrees can be integrated to yield the Newton-Cotes quadrature formulas. The most prominent of these are the trapezoid rule and Simpson's 1/3 rule. Romberg integration, which is a higher-order extrapolation of the trapezoid rule, is introduced. Adaptive integration, in which the range of integration is subdivided automatically until a specified accuracy is obtained, is presented. Gaussian quadrature, which achieves higher-order accuracy for integrating known functions by specifying the locations of the discrete points, is presented. The evaluation of multiple integrals is discussed.

Numerical integration of both known functions and discrete data is a common problem. The concepts involved in numerical integration lead directly to numerical methods for solving differential equations.

I.7 SUMMARY

Part I of this book is devoted to the *basic tools of numerical analysis*. These topics are important in their own right. In addition, they provide the foundation for the solution of ordinary and partial differential equations, which are discussed in Parts II and III, respectively. The material presented in Part I comprises the basic language of numerical analysis. Familiarity and mastery of this material is essential for the understanding and use of more advanced numerical methods.

1

Systems of Linear Algebraic Equations

1.1 INTRODUCTION

The static mechanical spring-mass system illustrated in Figure 1.1 consists of three masses m_1 to m_3, having weights W_1 to W_3, interconnected by five linear springs K_1 to K_5. In the configuration illustrated on the left, the three masses are supported by forces F_1 to F_3 equal to weights W_1 to W_3, respectively, so that the five springs are in a stable static equilibrium configuration. When the supporting forces F_1 to F_3 are removed, the masses move downward and reach a new static equilibrium configuration, denoted by x_1, x_2, and x_3, where x_1, x_2, and x_3 are measured from the original locations of the corresponding masses. Free-body diagrams of the three masses are presented at the bottom of Figure 1.1. Performing a static force balance on the three masses yields the following system of three linear algebraic equations:

$$(K_1 + K_2 + K_3)x_1 - K_2 x_2 - K_3 x_3 = W_1 \tag{1.1a}$$

$$-K_2 x_1 + (K_2 + K_4)x_2 - K_4 x_3 = W_2 \tag{1.1b}$$

$$-K_3 x_1 - K_4 x_2 + (K_3 + K_4 + K_5)x_3 = W_3 \tag{1.1c}$$

When values of K_1 to K_5 and W_1 to W_3 are specified, the equilibrium displacements x_1 to x_3 can be determined by solving Eq. (1.1).

The static mechanical spring-mass system illustrated in Figure 1.1 is used as the example problem in this chapter to illustrate methods for solving systems of linear

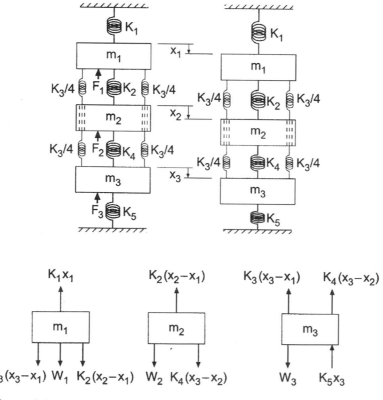

Figure 1.1   Static mechanical spring-mass system.

algebraic equations. For that purpose, let $K_1 = 40\,\text{N/cm}$, $K_2 = K_3 = K_4 = 20\,\text{N/cm}$, and $K_5 = 90\,\text{N/cm}$. Let $W_1 = W_2 = W_3 = 20\,\text{N}$. For these values, Eq. (1.1) becomes:

$$80x_1 - 20x_2 - 20x_3 = 20 \qquad\qquad (1.2\text{a})$$

$$-20x_1 + 40x_2 - 20x_3 = 20 \qquad\qquad (1.2\text{b})$$

$$-20x_1 - 20x_2 + 130x_3 = 20 \qquad\qquad (1.2\text{c})$$

The solution to Eq. (1.2) is $x_1 = 0.6\,\text{cm}$, $x_2 = 1.0\,\text{cm}$, and $x_3 = 0.4\,\text{cm}$, which can be verified by direct substitution.

Systems of equations arise in all branches of engineering and science. These equations may be algebraic, transcendental (i.e., involving trigonometric, logarithmic, exponential, etc. functions), ordinary differential equations, or partial differential equations. The equations may be linear or nonlinear. Chapter 1 is devoted to the solution of systems of linear algebraic equations of the following form:

$$a_{11}x_1 + a_{12}x_2 + a_{13}x_3 + \cdots + a_{1n}x_n = b_1 \qquad\qquad (1.3\text{a})$$

$$a_{21}x_1 + a_{22}x_2 + a_{23}x_3 + \cdots + a_{2n}x_n = b_2 \qquad\qquad (1.3\text{b})$$

$$\cdots\cdots\cdots\cdots\cdots\cdots\cdots\cdots\cdots\cdots\cdots\cdots\cdots$$

$$a_{n1}x_1 + a_{n2}x_2 + a_{n3}x_3 + \cdots + a_{nn}x_n = b_n \qquad\qquad (1.3\text{n})$$

where x_j ($j = 1, 2, \ldots, n$) denotes the unknown variables, $a_{i,j}$ ($i, j = 1, 2, \ldots, n$) denotes the constant coefficients of the unknown variables, and b_i ($i = 1, 2, \ldots, n$) denotes the nonhomogeneous terms. For the coefficients $a_{i,j}$, the first subscript, i, denotes equation i, and the second subscript, j, denotes variable x_j. The number of equations can range from two to hundreds, thousands, and even millions.

In the most general case, the number of variables is not required to be the same as the number of equations. However, in most practical problems, they are the same. That is the case considered in this chapter. Even when the number of variables is the same as the number of equations, several solution possibilities exist, as illustrated in Figure 1.2 for the following system of two linear algebraic equations:

$$a_{11}x_1 + a_{12}x_2 = b_1 \qquad\qquad (1.4\text{a})$$

$$a_{21}x_1 + a_{22}x_2 = b_2 \qquad\qquad (1.4\text{b})$$

The four solution possibilities are:

1. A unique solution (a consistent set of equations), as illustrated in Figure 1.2a
2. No solution (an inconsistent set of equations), as illustrated in Figure 1.2b
3. An infinite number of solutions (a redundant set of equations), as illustrated in Figure 1.2c
4. The trivial solution, $x_j = 0$ ($j = 1, 2, \ldots, n$), for a homogeneous set of equations, as illustrated in Figure 1.2d

Chapter 1 is concerned with the first case where a unique solution exists.

Systems of linear algebraic equations arise in many different types of problems, for example:

1. Network problems (e.g., electrical networks)
2. Fitting approximating functions (see Chapter 4)

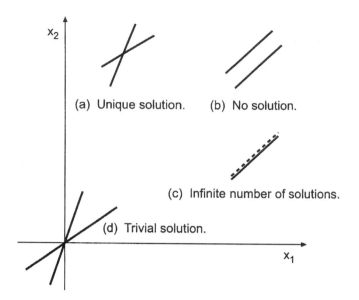

Figure 1.2   Solution of a system of two linear algebraic equations.

3. Systems of finite difference equations that arise in the numerical solution of differential equations (see Parts II and III)

The list is endless.

There are two fundamentally different approaches for solving systems of linear algebraic equations:

1. Direct elimination methods
2. Iterative methods

Direct elimination methods are systematic procedures based on algebraic elimination, which obtain the solution in a fixed number of operations. Examples of direct elimination methods are *Gauss elimination*, *Gauss-Jordan elimination*, the *matrix inverse method*, and *Doolittle LU factorization*. Iterative methods, on the other hand, obtain the solution asymptotically by an iterative procedure. A trial solution is assumed, the trial solution is substituted into the system of equations to determine the mismatch, or error, in the trial solution, and an improved solution is obtained from the mismatch data. Examples of iterative methods are *Jacobi iteration*, *Gauss-Seidel iteration*, and *successive-over-relaxation (SOR)*.

Although no absolutely rigid rules apply, direct elimination methods are generally used when one or more of the following conditions holds: (a) The number of equations is small (100 or less), (b) most of the coefficients in the equations are nonzero, (c) the system of equations is not diagonally dominant [see Eq. (1.15)], or (d) the system of equations is ill conditioned (see Section 1.6.2). Iterative methods are used when the number of equations is large and most of the coefficients are zero (i.e., a sparse matrix). Iterative methods generally diverge unless the system of equations is diagonally dominant [see Eq. (1.15)].

The organization of Chapter 1 is illustrated in Figure 1.3. Following the introductory material discussed in this section, the properties of matrices and determinants are reviewed. The presentation then splits into a discussion of direct elimination methods

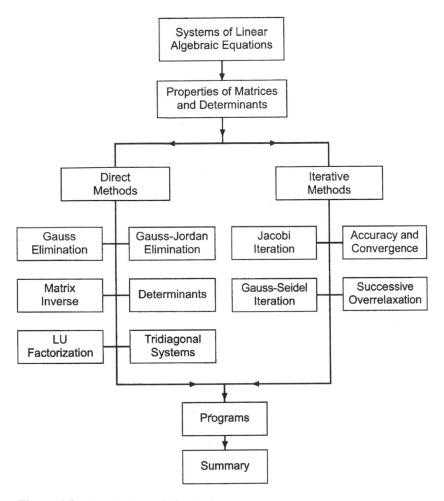

Figure 1.3 Organization of Chapter 1.

followed by a discussion of iterative methods. Several methods, both direct elimination and iterative, for solving systems of linear algebraic equations are presented in this chapter. Procedures for special problems, such as tridiagonal systems of equations, are presented. All these procedures are illustrated by examples. Although the methods apply to large systems of equations, they are illustrated by applying them to the small system of only three equations given by Eq. (1.2). After the presentation of the methods, three computer programs are presented for implementing the Gauss elimination method, the Thomas algorithm, and successive-over-relaxation (SOR). The chapter closes with a Summary, which discusses some philosophy to help you choose the right method for every problem and lists the things you should be able to do after studying Chapter 1.

1.2 PROPERTIES OF MATRICES AND DETERMINANTS

Systems of linear algebraic equations can be expressed very conveniently in terms of matrix notation. Solution methods for systems of linear algebraic equations can be developed very compactly using matrix algebra. Consequently, the elementary properties of matrices and determinants are presented in this section.

1.2.1. Matrix Definitions

A *matrix* is a rectangular array of elements (either numbers or symbols), which are arranged in orderly rows and columns. Each element of the matrix is distinct and separate. The location of an element in the matrix is important. Elements of a matrix are generally identified by a double subscripted lowercase letter, for example, $a_{i,j}$, where the first subscript i identifies the row of the matrix and the second subscript j identifies the column of the matrix. The size of a matrix is specified by the number of rows times the number of columns. A matrix with n rows and m columns is said to be an n by m, or $n \times m$, matrix. Matrices are generally represented by either a boldface capital letter, for example, $\mathbf{A}$, the general element enclosed in brackets, for example, $[a_{i,j}]$, or the full array of elements, as illustrated in Eq. (1.5):

$$\mathbf{A} = [a_{i,j}] = \begin{bmatrix} a_{11} & a_{12} & \cdots & \cdots & a_{1m} \\ a_{21} & a_{22} & \cdots & \cdots & a_{2m} \\ \cdots\cdots\cdots\cdots\cdots\cdots\cdots\cdots \\ a_{n1} & a_{n2} & \cdots & \cdots & a_{nm} \end{bmatrix} \qquad (i = 1, 2, \ldots, n; \ j = 1, 2, \ldots, m)$$

$$(1.5)$$

Comparing Eqs. (1.3) and (1.5) shows that the coefficients of a system of linear algebraic equations form the elements of an $n \times n$ matrix.

Equation (1.5) illustrates a convention used throughout this book for simplicity of appearance. When the general element $a_{i,j}$ is considered, the subscripts i and j are separated by a comma. When a specific element is specified, for example, a_{31}, the subscripts 3 and 1, which denote the element in row 3 and column 1, will not be separated by a comma, unless i or j is greater than 9. For example, a_{37} denotes the element in row 3 and column 7, whereas $a_{13,17}$ denotes the element in row 13 and column 17.

Vectors are a special type of matrix which has only one column or one row. Vectors are represented by either a boldface lowercase letter, for example, $\mathbf{x}$ or $\mathbf{y}$, the general element enclosed in brackets, for example, $[x_i]$ or $[y_i]$, or the full column or row of elements. A *column vector* is an $n \times 1$ matrix. Thus,

$$\mathbf{x} = [x_i] = \begin{bmatrix} x_1 \\ x_2 \\ \cdots \\ x_n \end{bmatrix} \qquad (i = 1, 2, \ldots, n) \tag{1.6a}$$

A *row vector* is a $1 \times n$ matrix. For example,

$$\mathbf{y} = [y_j] = [y_1 \quad y_2 \quad \cdots \quad y_n] \qquad (j = 1, 2, \ldots, n) \tag{1.6b}$$

Unit vectors, $\mathbf{i}$, are special vectors which have a magnitude of unity. Thus,

$$\|\mathbf{i}\| = (i_1^2 + i_2^2 + \cdots + i_n^2)^{1/2} = 1 \tag{1.7}$$

where the notation $\|\mathbf{i}\|$ denotes the length of vector $\mathbf{i}$. Orthogonal systems of unit vectors, in which all of the elements of each unit vector except one are zero, are used to define coordinate systems.

There are several special matrices of interest. A *square matrix* **S** is a matrix which has the same number of rows and columns, that is, $m = n$. For example,

$$\mathbf{S} = \begin{bmatrix} a_{11} & a_{12} & \cdots & a_{1n} \\ a_{21} & a_{22} & \cdots & a_{2n} \\ \multicolumn{4}{c}{\dotfill} \\ a_{n1} & a_{n2} & \cdots & a_{nn} \end{bmatrix} \tag{1.8}$$

is a square $n \times n$ matrix. Our interest will be devoted entirely to square matrices. The left-to-right downward-sloping line of elements from a_{11} to a_{nn} is called the *major diagonal* of the matrix. A *diagonal matrix* **D** is a square matrix with all elements equal to zero except the elements on the major diagonal. For example,

$$\mathbf{D} = \begin{bmatrix} a_{11} & 0 & 0 & 0 \\ 0 & a_{22} & 0 & 0 \\ 0 & 0 & a_{33} & 0 \\ 0 & 0 & 0 & a_{44} \end{bmatrix} \tag{1.9}$$

is a 4×4 diagonal matrix. The *identity matrix* **I** is a diagonal matrix with unity diagonal elements. The identity matrix is the matrix equivalent of the scalar number unity. The matrix

$$\mathbf{I} = \begin{bmatrix} 1 & 0 & 0 & 0 \\ 0 & 1 & 0 & 0 \\ 0 & 0 & 1 & 0 \\ 0 & 0 & 0 & 1 \end{bmatrix} \tag{1.10}$$

is the 4×4 identity matrix.

A *triangular matrix* is a square matrix in which all of the elements on one side of the major diagonal are zero. The remaining elements may be zero or nonzero. An *upper triangular matrix* **U** has all zero elements below the major diagonal. The matrix

$$\mathbf{U} = \begin{bmatrix} a_{11} & a_{12} & a_{13} & a_{14} \\ 0 & a_{22} & a_{23} & a_{24} \\ 0 & 0 & a_{33} & a_{34} \\ 0 & 0 & 0 & a_{44} \end{bmatrix} \tag{1.11}$$

is a 4×4 upper triangular matrix. A *lower triangular matrix* **L** has all zero elements above the major diagonal. The matrix

$$\mathbf{L} = \begin{bmatrix} a_{11} & 0 & 0 & 0 \\ a_{21} & a_{22} & 0 & 0 \\ a_{31} & a_{32} & a_{33} & 0 \\ a_{41} & a_{42} & a_{43} & a_{44} \end{bmatrix} \tag{1.12}$$

is a 4×4 lower triangular matrix.

A *tridiagonal matrix* **T** is a square matrix in which all of the elements not on the major diagonal and the two diagonals surrounding the major diagonal are zero. The elements on these three diagonals may or may not be zero. The matrix

$$\mathbf{T} = \begin{bmatrix} a_{11} & a_{12} & 0 & 0 & 0 \\ a_{21} & a_{22} & a_{23} & 0 & 0 \\ 0 & a_{32} & a_{33} & a_{34} & 0 \\ 0 & 0 & a_{43} & a_{44} & a_{45} \\ 0 & 0 & 0 & a_{54} & a_{55} \end{bmatrix} \qquad (1.13)$$

is a 5×5 tridiagonal matrix.

A *banded matrix* **B** has all zero elements except along particular diagonals. For example,

$$\mathbf{B} = \begin{bmatrix} a_{11} & a_{12} & 0 & a_{14} & 0 \\ a_{21} & a_{22} & a_{23} & 0 & a_{25} \\ 0 & a_{32} & a_{33} & a_{34} & 0 \\ a_{41} & 0 & a_{43} & a_{44} & a_{45} \\ 0 & a_{52} & 0 & a_{54} & a_{55} \end{bmatrix} \qquad (1.14)$$

is a 5×5 banded matrix.

The *transpose* of an $n \times m$ matrix **A** is the $m \times n$ matrix, $\mathbf{A}^T$, which has elements $a_{i,j}^T = a_{j,i}$. The transpose of a column vector, is a row vector and vice versa. *Symmetric* square matrices have identical corresponding elements on either side of the major diagonal. That is, $a_{i,j} = a_{j,i}$. In that case, $\mathbf{A} = \mathbf{A}^T$.

A *sparse matrix* is one in which most of the elements are zero. Most large matrices arising in the solution of ordinary and partial differential equations are sparse matrices.

A matrix is *diagonally dominant* if the absolute value of each element on the major diagonal is equal to, or larger than, the sum of the absolute values of all the other elements in that row, with the diagonal element being larger than the corresponding sum of the other elements for at least one row. Thus, diagonal dominance is defined as

$$|a_{i,i}| \geq \sum_{j=1, j \neq i}^{n} |a_{i,j}| \qquad (i = 1, \dots, n) \qquad (1.15)$$

with $>$ true for at least one row.

1.2.2. Matrix Algebra

Matrix algebra consists of *matrix addition*, *matrix subtraction*, and *matrix multiplication*. Matrix division is not defined. An analogous operation is accomplished using the matrix inverse.

Matrix addition and subtraction consist of adding or subtracting the corresponding elements of two matrices of equal size. Let **A** and **B** be two matrices of equal size. Then,

$$\mathbf{A} + \mathbf{B} = [a_{i,j}] + [b_{i,j}] = [a_{i,j} + b_{i,j}] = [c_{i,j}] = \mathbf{C} \qquad (1.16a)$$

$$\mathbf{A} - \mathbf{B} = [a_{i,j}] - [b_{i,j}] = [a_{i,j} - b_{i,j}] = [c_{i,j}] = \mathbf{C} \qquad (1.16b)$$

Unequal size matrices cannot be added or subtracted. Matrices of the same size are *associative* on addition. Thus,

$$\mathbf{A} + (\mathbf{B} + \mathbf{C}) = (\mathbf{A} + \mathbf{B}) + \mathbf{C} \qquad (1.17)$$

Matrices of the same size are *commutative* on addition. Thus,

$$\mathbf{A} + \mathbf{B} = \mathbf{B} + \mathbf{A} \tag{1.18}$$

Example 1.1.   Matrix addition.

Add the two 3×3 matrices $\mathbf{A}$ and $\mathbf{B}$ to obtain the 3×3 matrix $\mathbf{C}$, where

$$\mathbf{A} = \begin{bmatrix} 1 & 2 & 3 \\ 2 & 1 & 4 \\ 1 & 4 & 3 \end{bmatrix} \quad \text{and} \quad \mathbf{B} = \begin{bmatrix} 3 & 2 & 1 \\ -4 & 1 & 2 \\ 2 & 3 & -1 \end{bmatrix} \tag{1.19}$$

From Eq. (1.16a),

$$c_{i,j} = a_{i,j} + b_{i,j} \tag{1.20}$$

Thus, $c_{11} = a_{11} + b_{11} = 1 + 3 = 4$, $c_{12} = a_{12} + b_{12} = 2 + 2 = 4$, etc. The result is

$$\mathbf{A} + \mathbf{B} = \begin{bmatrix} (1+3) & (2+2) & (3+1) \\ (2-4) & (1+1) & (4+2) \\ (1+2) & (4+3) & (3-1) \end{bmatrix} = \begin{bmatrix} 4 & 4 & 4 \\ -2 & 2 & 6 \\ 3 & 7 & 2 \end{bmatrix} = \mathbf{C} \tag{1.21}$$

Matrix multiplication consists of row-element to column-element multiplication and summation of the resulting products. Multiplication of the two matrices $\mathbf{A}$ and $\mathbf{B}$ is defined only when the number of columns of matrix $\mathbf{A}$ is the same as the number of rows of matrix $\mathbf{B}$. Matrices that satisfy this condition are called *conformable* in the order $\mathbf{AB}$. Thus, if the size of matrix $\mathbf{A}$ is $n \times m$ and the size of matrix $\mathbf{B}$ is $m \times r$, then

$$\mathbf{AB} = [a_{i,j}][b_{i,j}] = [c_{i,j}] = \mathbf{C} \quad c_{i,j} = \sum_{k=1}^{m} a_{i,k}b_{k,j} \quad (i = 1, 2, \ldots, n, \; j = 1, 2, \ldots, r) \tag{1.22}$$

The size of matrix $\mathbf{C}$ is $n \times r$. Matrices that are not conformable cannot be multiplied.

It is easy to make errors when performing matrix multiplication by hand. It is helpful to trace across the rows of $\mathbf{A}$ with the left index finger while tracing down the columns of $\mathbf{B}$ with the right index finger, multiplying the corresponding elements, and summing the products. Matrix algebra is much better suited to computers than to humans.

Multiplication of the matrix $\mathbf{A}$ by the scalar α consists of multiplying each element of $\mathbf{A}$ by α. Thus,

$$\alpha \mathbf{A} = \alpha[a_{i,j}] = [\alpha a_{i,j}] = [b_{i,j}] = \mathbf{B} \tag{1.23}$$

Example 1.2.   Matrix multiplication.

Multiply the 3×3 matrix $\mathbf{A}$ and the 3×2 matrix $\mathbf{B}$ to obtain the 3×2 matrix $\mathbf{C}$, where

$$\mathbf{A} = \begin{bmatrix} 1 & 2 & 3 \\ 2 & 1 & 4 \\ 1 & 4 & 3 \end{bmatrix} \quad \text{and} \quad \mathbf{B} = \begin{bmatrix} 2 & 1 \\ 1 & 2 \\ 2 & 1 \end{bmatrix} \tag{1.24}$$

From Eq. (1.22),

$$c_{i,j} = \sum_{k=1}^{3} a_{i,k} b_{k,j} \qquad (i = 1, 2, 3, \ j = 1, 2) \tag{1.25}$$

Evaluating Eq. (1.25) yields

$$c_{11} = a_{11}b_{11} + a_{12}b_{21} + a_{13}b_{31} = (1)(2) + (2)(1) + (3)(2) = 10 \tag{1.26a}$$
$$c_{12} = a_{11}b_{12} + a_{12}b_{22} + a_{13}b_{32} = (1)(1) + (2)(2) + (3)(1) = 8 \tag{1.26b}$$
$$\cdots\cdots\cdots\cdots\cdots\cdots\cdots\cdots\cdots\cdots\cdots\cdots\cdots\cdots\cdots\cdots\cdots\cdots\cdots\cdots$$
$$c_{32} = a_{31}b_{12} + a_{32}b_{22} + a_{33}b_{32} = (1)(1) + (4)(2) + (3)(1) = 12 \tag{1.26c}$$

Thus,

$$\mathbf{C} = [c_{i,j}] = \begin{bmatrix} 10 & 8 \\ 13 & 8 \\ 12 & 12 \end{bmatrix} \tag{1.27}$$

Multiply the 3×2 matrix $\mathbf{C}$ by the scalar $\alpha = 2$ to obtain the 3×2 matrix $\mathbf{D}$. From Eq. (1.23), $d_{11} = \alpha c_{11} = (2)(10) = 20$, $d_{12} = \alpha c_{12} = (2)(8) = 16$, etc. The result is

$$\mathbf{D} = \alpha\mathbf{C} = 2\mathbf{C} = \begin{bmatrix} (2)(10) & (2)(8) \\ (2)(13) & (2)(8) \\ (2)(12) & (2)(12) \end{bmatrix} = \begin{bmatrix} 20 & 16 \\ 26 & 16 \\ 24 & 24 \end{bmatrix} \tag{1.28}$$

Matrices that are suitably conformable are *associative* on multiplication. Thus,

$$\mathbf{A}(\mathbf{BC}) = (\mathbf{AB})\mathbf{C} \tag{1.29}$$

Square matrices are *conformable* in either order. Thus, if $\mathbf{A}$ and $\mathbf{B}$ are $n \times n$ matrices,

$$\mathbf{AB} = \mathbf{C} \qquad \text{and} \qquad \mathbf{BA} = \mathbf{D} \tag{1.30}$$

where $\mathbf{C}$ and $\mathbf{D}$ are $n \times n$ matrices. However square matrices in general are not *commutative* on multiplication. That is, in general,

$$\mathbf{AB} \neq \mathbf{BA} \tag{1.31}$$

Matrices $\mathbf{A}$, $\mathbf{B}$, and $\mathbf{C}$ are *distributive* if $\mathbf{B}$ and $\mathbf{C}$ are the same size and $\mathbf{A}$ is *conformable* to $\mathbf{B}$ and $\mathbf{C}$. Thus,

$$\mathbf{A}(\mathbf{B} + \mathbf{C}) = \mathbf{AB} + \mathbf{AC} \tag{1.32}$$

Consider the two square matrices $\mathbf{A}$ and $\mathbf{B}$. Multiplying yields

$$\mathbf{AB} = \mathbf{C} \tag{1.33}$$

It might appear logical that the inverse operation of multiplication, that is, division, would give

$$\mathbf{A} = \mathbf{C}/\mathbf{B} \tag{1.34}$$

Unfortunately, matrix division is not defined. However, for square matrices, an analogous concept is provided by the matrix inverse.

Consider the two square matrices **A** and **B**. If $\mathbf{AB} = \mathbf{I}$, then **B** is the inverse of **A**, which is denoted as $\mathbf{A}^{-1}$. Matrix inverses *commute* on multiplication. Thus,

$$\mathbf{AA}^{-1} = \mathbf{A}^{-1}\mathbf{A} = \mathbf{I} \tag{1.35}$$

The operation desired by Eq. (1.34) can be accomplished using the matrix inverse. Thus, the inverse of the matrix multiplication specified by Eq. (1.33) is accomplished by matrix multiplication using the inverse matrix. Thus, the matrix equivalent of Eq. (1.34) is given by

$$\mathbf{A} = \mathbf{B}^{-1}\mathbf{C} \tag{1.36}$$

Procedures for evaluating the inverse of a square matrix are presented in Examples 1.12 and 1.16.

Matrix factorization refers to the representation of a matrix as the product of two other matrices. For example, a known matrix **A** can be represented as the product of two unknown matrices **B** and **C**. Thus,

$$\mathbf{A} = \mathbf{BC} \tag{1.37}$$

Factorization is not a unique process. There are, in general, an infinite number of matrices **B** and **C** whose product is **A**. A particularly useful factorization for square matrices is

$$\mathbf{A} = \mathbf{LU} \tag{1.38}$$

where **L** and **U** are lower and upper triangular matrices, respectively. The LU factorization method for solving systems of linear algebraic equations, which is presented in Section 1.4, is based on such a factorization.

A matrix can be *partitioned* by grouping the elements of the matrix into submatrices. These submatrices can then be treated as elements of a smaller matrix. To ensure that the operations of matrix algebra can be applied to the submatrices of two partitioned matrices, the partitioning is generally into square submatrices of equal size. Matrix partitioning is especially convenient when solving systems of algebraic equations that arise in the finite difference solution of systems of differential equations.

1.2.3. Systems of Linear Algebraic Equations

Systems of linear algebraic equations, such as Eq. (1.3), can be expressed very compactly in matrix notation. Thus, Eq. (1.3) can be written as the matrix equation

$$\boxed{\mathbf{Ax} = \mathbf{b}} \tag{1.39}$$

where

$$\mathbf{A} = \begin{bmatrix} a_{11} & a_{12} & \cdots & a_{1n} \\ a_{21} & a_{22} & \cdots & a_{2n} \\ \cdots & \cdots & \cdots & \cdots \\ a_{n1} & a_{n2} & \cdots & a_{nn} \end{bmatrix} \quad \mathbf{x} = \begin{bmatrix} x_1 \\ x_2 \\ \cdots \\ x_n \end{bmatrix} \quad \mathbf{b} = \begin{bmatrix} b_1 \\ b_2 \\ \cdots \\ b_n \end{bmatrix} \tag{1.40}$$

Equation (1.3) can also be written as

$$\sum_{j=1}^{n} a_{i,j}x_j = b_i \quad (i = 1, \ldots, n) \tag{1.41}$$

or equivalently as

$$a_{i,j}x_j = b_i \qquad (i,j=1,\dots,n) \tag{1.42}$$

where the *summation convention* holds, that is, the repeated index j in Eq. (1.42) is summed over its range, 1 to n. Equation (1.39) will be used throughout this book to represent a system of linear algebraic equations.

There are three so-called *row operations* that are useful when solving systems of linear algebraic equations. They are:

1. Any row (equation) may be multiplied by a constant (a process known as *scaling*).
2. The order of the rows (equations) may be interchanged (a process known as *pivoting*).
3. Any row (equation) can be replaced by a weighted linear combination of that row (equation) with any other row (equation) (a process known as *elimination*).

In the context of the solution of a system of linear algebraic equations, these three row operations clearly do not change the solution. The appearance of the system of equations is obviously changed by any of these row operations, but the solution is unaffected. When solving systems of linear algebraic equations expressed in matrix notation, these row operations apply to the rows of the matrices representing the system of linear algebraic equations.

1.2.4. Determinants

The term *determinant* of a square matrix $\mathbf{A}$, denoted $\det(\mathbf{A})$ or $|\mathbf{A}|$, refers to both the collection of the elements of the square matrix, enclosed in vertical lines, and the scalar value represented by that array. Thus,

$$\det(\mathbf{A}) = |\mathbf{A}| = \begin{vmatrix} a_{11} & a_{12} & \cdots & a_{1n} \\ a_{21} & a_{22} & \cdots & a_{2n} \\ \cdots & \cdots & \cdots & \cdots \\ a_{n1} & a_{n2} & \cdots & a_{nn} \end{vmatrix} \tag{1.43}$$

Only square matrices have determinants.

The scalar value of the determinant of a 2×2 matrix is the product of the elements on the major diagonal minus the product of the elements on the minor diagonal. Thus,

$$\det(\mathbf{A}) = |\mathbf{A}| = \begin{vmatrix} a_{11} & a_{12} \\ a_{21} & a_{22} \end{vmatrix} = a_{11}a_{22} - a_{21}a_{12} \tag{1.44}$$

The scalar value of the determinant of a 3×3 matrix is composed of the sum of six triple products which can be obtained from the augmented determinant:

$$\begin{vmatrix} a_{11} & a_{12} & a_{13} \\ a_{21} & a_{22} & a_{23} \\ a_{31} & a_{32} & a_{33} \end{vmatrix} \begin{matrix} a_{11} & a_{12} \\ a_{21} & a_{22} \\ a_{31} & a_{32} \end{matrix} \tag{1.45}$$

The 3×3 determinant is augmented by repeating the first two columns of the determinant on the right-hand side of the determinant. Three triple products are formed, starting with the elements of the first row multiplied by the two remaining elements on the right-downward-sloping diagonals. Three more triple products are formed, starting with the elements of the third row multiplied by the two remaining elements on the right-upward-sloping diagonals. The value of the determinant is the sum of the first three triple products minus the sum of the last three triple products. Thus,

$$
\det(\mathbf{A}) = |\mathbf{A}| = a_{11}a_{22}a_{33} + a_{12}a_{23}a_{31} + a_{13}a_{21}a_{32}
$$
$$
- a_{31}a_{22}a_{13} - a_{32}a_{23}a_{11} - a_{33}a_{21}a_{12} \tag{1.46}
$$

Example 1.3.   **Evaluation of a 3×3 determinant by the diagonal method.**

Let's evaluate the determinant of the coefficient matrix of Eq. (1.2) by the diagonal method. Thus,

$$
\mathbf{A} = \begin{bmatrix} 80 & -20 & -20 \\ -20 & 40 & -20 \\ -20 & -20 & 130 \end{bmatrix} \tag{1.47}
$$

The augmented determinant is

$$
\begin{vmatrix} 80 & -20 & -20 \\ -20 & 40 & -20 \\ -20 & -20 & 130 \end{vmatrix} \begin{matrix} 80 & -20 \\ -20 & 40 \\ -20 & -20 \end{matrix} \tag{1.48}
$$

Applying Eq. (1.46) yields

$$
\det(\mathbf{A}) = |\mathbf{A}| = (80)(40)(130) + (-20)(-20)(-20) + (-20)(-20)(-20)
$$
$$
- (-20)(40)(-20) - (-20)(-20)(80)
$$
$$
- (130)(-20)(-20) = 416{,}000 - 8{,}000 - 8{,}000
$$
$$
- 16{,}000 - 32{,}000 - 52{,}000 = 300{,}000 \tag{1.49}
$$

The diagonal method of evaluating determinants applies only to 2×2 and 3×3 determinants. It is incorrect for 4×4 or larger determinants. In general, the expansion of an $n \times n$ determinant is the sum of all possible products formed by choosing one and only one element from each row and each column of the determinant, with a plus or minus sign determined by the number of permutations of the row and column elements. One formal procedure for evaluating determinants is called *expansion by minors*, or the *method of cofactors*. In this procedure there are $n!$ products to be summed, where each product has n elements. Thus, the expansion of a 10×10 determinant requires the summation of 10! products ($10! = 3{,}628{,}800$), where each product involves 9 multiplications (the product of 10 elements). This is a total of 32,659,000 multiplications and 3,627,999 additions, not counting the work needed to keep track of the signs. Consequently, the evaluation of determinants by the method of cofactors is impractical, except for very small determinants.

Although the method of cofactors is not recommended for anything larger than a 4×4 determinant, it is useful to understand the concepts involved. The *minor* $M_{i,j}$ is the determinant of the $(n-1) \times (n-1)$ submatrix of the $n \times n$ matrix $\mathbf{A}$ obtained by deleting the ith row and the jth column. The cofactor $A_{i,j}$ associated with the minor $M_{i,j}$ is defined as

$$
A_{i,j} = (-1)^{i+j} M_{i,j} \tag{1.50}
$$

Using cofactors, the determinant of matrix $\mathbf{A}$ is the sum of the products of the elements of any row or column, multiplied by their corresponding cofactors. Thus, expanding across any fixed row i yields

$$\det(\mathbf{A}) = |\mathbf{A}| = \sum_{j=1}^{n} a_{i,j} A_{i,j} = \sum_{j=1}^{n} (-1)^{i+j} a_{i,j} M_{i,j} \tag{1.51}$$

Alternatively, expanding down any fixed column j yields

$$\det(\mathbf{A}) = |\mathbf{A}| = \sum_{i=1}^{n} a_{i,j} A_{i,j} = \sum_{i=1}^{n} (-1)^{i+j} a_{i,j} M_{i,j} \tag{1.52}$$

Each cofactor expansion reduces the order of the determinant by one, so there are n determinants of order $n - 1$ to evaluate. By repeated application, the cofactors are eventually reduced to 3×3 determinants which can be evaluated by the diagonal method. The amount of work can be reduced by choosing the expansion row or column with as many zeros as possible.

Example 1.4.   Evaluation of a 3×3 determinant by the cofactor method.

Let's rework Example 1.3 using the cofactor method. Recall Eq. (1.47):

$$\mathbf{A} = \begin{bmatrix} 80 & -20 & -20 \\ -20 & 40 & -20 \\ -20 & -20 & 130 \end{bmatrix} \tag{1.53}$$

Evaluate $|\mathbf{A}|$ by expanding across the first row. Thus,

$$|\mathbf{A}| = (80) \begin{vmatrix} 40 & -20 \\ -20 & 130 \end{vmatrix} - (-20) \begin{vmatrix} -20 & -20 \\ -20 & 130 \end{vmatrix} + (-20) \begin{vmatrix} -20 & 40 \\ -20 & -20 \end{vmatrix} \tag{1.54}$$

$$|\mathbf{A}| = 80(5200 + 400) - (-20)(-2600 + 400) + (-20)(400 + 800)$$

$$= 384000 - 60000 - 24000 = 300000 \tag{1.55}$$

If the value of the determinant of a matrix is zero, the matrix is said to be *singular*. A *nonsingular matrix* has a determinant that is nonzero. If any row or column of a matrix has all zero elements, that matrix is singular.

The determinant of a triangular matrix, either upper or lower triangular, is the product of the elements on the major diagonal. It is possible to transform any nonsingular matrix into a triangular matrix in such a way that the value of the determinant is either unchanged or changed in a well-defined way. That procedure is presented in Section 1.3.6. The value of the determinant can then be evaluated quite easily as the product of the elements on the major diagonal.

1.3   DIRECT ELIMINATION METHODS

There are a number of methods for the direct solution of systems of linear algebraic equations. One of the more well-known methods is Cramer's rule, which requires the evaluation of numerous determinants. Cramer's rule is highly inefficient, and thus not recommended. More efficient methods, based on the elimination concept, are recommended. Both Cramer's rule and elimination methods are presented in this section. After presenting *Cramer's rule*, the elimination concept is applied to develop Gauss elimination, Gauss-Jordan elimination, matrix inversion, and determinant evaluation. These concepts are extended to LU factorization and tridiagonal systems of equations in Sections 1.4 and 1.5, respectively.

1.3.1. Cramer's Rule

Although it is not an elimination method, *Cramer's rule* is a direct method for solving systems of linear algebraic equations. Consider the system of linear algebraic equations, $\mathbf{Ax} = \mathbf{b}$, which represents n equations. Cramer's rule states that the solution for x_j ($j = 1, \ldots, n$) is given by

$$x_j = \frac{\det(\mathbf{A}^j)}{\det(\mathbf{A})} \qquad (j = 1, \ldots, n) \tag{1.56}$$

where $\mathbf{A}^j$ is the $n \times n$ matrix obtained by replacing column j in matrix $\mathbf{A}$ by the column vector $\mathbf{b}$. For example, consider the system of two linear algebraic equations:

$$a_{11}x_1 + a_{12}x_2 = b_1 \tag{1.57a}$$

$$a_{21}x_1 + a_{22}x_2 = b_2 \tag{1.57b}$$

Applying Cramer's rule yields

$$x_1 = \frac{\begin{vmatrix} b_1 & a_{12} \\ b_2 & a_{22} \end{vmatrix}}{\begin{vmatrix} a_{11} & a_{12} \\ a_{21} & a_{22} \end{vmatrix}} \quad \text{and} \quad x_2 = \frac{\begin{vmatrix} a_{11} & b_1 \\ a_{21} & b_2 \end{vmatrix}}{\begin{vmatrix} a_{11} & a_{12} \\ a_{21} & a_{22} \end{vmatrix}} \tag{1.58}$$

The determinants in Eqs. (1.58) can be evaluated by the diagonal method described in Section 1.2.4.

For systems containing more than three equations, the diagonal method presented in Section 1.2.4 does not work. In such cases, the method of cofactors presented in Section 1.2.4 could be used. The number of multiplications and divisions N required by the method of cofactors is $N = (n-1)(n+1)!$. For a relatively small system of 10 equations (i.e., $n = 10$), $N = 360{,}000{,}000$, which is an enormous number of calculations. For $n = 100$, $N = 10^{157}$, which is obviously ridiculous. The preferred method for evaluating determinants is the elimination method presented in Section 1.3.6. The number of multiplications and divisions required by the elimination method is approximately $N = n^3 + n^2 - n$. Thus, for $n = 10$, $N = 1090$, and for $n = 100$, $N = 1{,}009{,}900$. Obviously, the elimination method is preferred.

Example 1.5.   Cramer's rule.

Let's illustrate Cramer's rule by solving Eq. (1.2). Thus,

$$80x_1 - 20x_2 - 20x_3 = 20 \tag{1.59a}$$

$$-20x_1 + 40x_2 - 20x_3 = 20 \tag{1.59b}$$

$$-20x_1 - 20x_2 + 130x_3 = 20 \tag{1.59c}$$

First, calculate det($\mathbf{A}$). From Example 1.4,

$$\det(\mathbf{A}) = \begin{vmatrix} 80 & -20 & -20 \\ -20 & 40 & -20 \\ -20 & -20 & 130 \end{vmatrix} = 300,000 \tag{1.60}$$

Next, calculate det($\mathbf{A}^1$), det($\mathbf{A}^2$), and det($\mathbf{A}^3$). For det($\mathbf{A}^1$),

$$\det(\mathbf{A}^1) = \begin{vmatrix} 20 & -20 & -20 \\ 20 & 40 & -20 \\ 20 & -20 & 130 \end{vmatrix} = 180,000 \tag{1.61}$$

In a similar manner, det($\mathbf{A}^2$) = 300,000 and det($\mathbf{A}^3$) = 120,000. Thus,

$$x_1 = \frac{\det(\mathbf{A}^1)}{\det(\mathbf{A})} = \frac{180,000}{300,000} = 0.60 \qquad x_2 = \frac{300,000}{300,000} = 1.00 \qquad x_3 = \frac{120,000}{300,000} = 0.40$$

$$\tag{1.62}$$

1.3.2. Elimination Methods

Elimination methods solve a system of linear algebraic equations by solving one equation, say the first equation, for one of the unknowns, say x_1, in terms of the remaining unknowns, x_2 to x_n, then substituting the expression for x_1 into the remaining $n-1$ equations to determine $n-1$ equations involving x_2 to x_n. This elimination procedure is performed $n-1$ times until the last step yields an equation involving only x_n. This process is called *elimination*.

The value of x_n can be calculated from the final equation in the elimination procedure. Then x_{n-1} can be calculated from modified equation $n-1$, which contains only x_n and x_{n-1}. Then x_{n-2} can be calculated from modified equation $n-2$, which contains only x_n, x_{n-1}, and x_{n-2}. This procedure is performed $n-1$ times to calculate x_{n-1} to x_1. This process is called *back substitution*.

1.3.2.1. Row Operations

The elimination process employs the row operations presented in Section 1.2.3, which are repeated below:

1. Any row (equation) may be multiplied by a constant (scaling).
2. The order of the rows (equations) may be interchanged (pivoting).
3. Any row (equation) can be replaced by a weighted linear combination of that row (equation) with any other row (equation) (elimination).

These row operations, which change the values of the elements of matrix $\mathbf{A}$ and $\mathbf{b}$, do not change the solution $\mathbf{x}$ to the system of equations.

The first row operation is used to scale the equations, if necessary. The second row operation is used to prevent divisions by zero and to reduce round-off errors. The third row operation is used to implement the systematic elimination process described above.

1.3.2.2. Elimination

Let's illustrate the elimination method by solving Eq. (1.2). Thus,

$$80x_1 - 20x_2 - 20x_3 = 20 \tag{1.63a}$$

$$-20x_1 + 40x_2 - 20x_3 = 20 \tag{1.63b}$$

$$-20x_1 - 20x_2 + 130x_3 = 20 \tag{1.63c}$$

Solve Eq. (1.63a) for x_1. Thus,

$$x_1 = [20 - (-20)x_2 - (-20)x_3]/80 \tag{1.64}$$

Substituting Eq. (1.64) into Eq. (1.63b) gives

$$-20\{[20 - (-20)x_2 - (-20)x_3]/80\} + 40x_2 - 20x_3 = 20 \tag{1.65}$$

which can be simplified to give

$$35x_2 - 25x_3 = 25 \tag{1.66}$$

Substituting Eq. (1.64) into Eq. (1.63c) gives

$$-20\{[20 - (-20)x_2 - (-20)x_3]/80\} - 20x_2 + 130x_3 = 20 \tag{1.67}$$

which can be simplified to give

$$-25x_2 + 125x_3 = 25 \tag{1.68}$$

Next solve Eq. (1.66) for x_2. Thus,

$$x_2 = [25 - (-25)x_3]/35 \tag{1.69}$$

Substituting Eq. (1.69) into Eq. (1.68) yields

$$-25\{[25 - (-25)x_3]/35\} + 125x_3 = 25 \tag{1.70}$$

which can be simplified to give

$$\tfrac{750}{7}x_3 = \tfrac{300}{7} \tag{1.71}$$

Thus, Eq. (1.63) has been reduced to the upper triangular system

$$80x_1 - 20x_2 - 20x_3 = 20 \tag{1.72a}$$

$$35x_2 - 25x_3 = 25 \tag{1.72b}$$

$$\tfrac{750}{7}x_3 = \tfrac{300}{7} \tag{1.72c}$$

which is equivalent to the original equation, Eq. (1.63). This completes the elimination process.

1.3.2.3. Back Substitution

The solution to Eq. (1.72) is accomplished easily by *back substitution*. Starting with Eq. (1.72c) and working backward yields

$$x_3 = 300/750 = 0.40 \tag{1.73a}$$

$$x_2 = [25 - (-25)(0.40)]/35 = 1.00 \tag{1.73b}$$

$$x_1 = [20 - (-20)(1.00) - (-20)(0.40)]/80 = 0.60 \tag{1.73c}$$

Example 1.6. Elimination.

Let's solve Eq. (1.2) by elimination. Recall Eq. (1.2):

$$80x_1 - 20x_2 - 20x_3 = 20 \tag{1.74a}$$

$$-20x_1 + 40x_2 - 20x_3 = 20 \tag{1.74b}$$

$$-20x_1 - 20x_2 + 130x_3 = 20 \tag{1.74c}$$

Elimination involves normalizing the equation above the element to be eliminated by the element immediately above the element to be eliminated, which is called the *pivot element*, multiplying the normalized equation by the element to be eliminated, and subtracting the result from the equation containing the element to be eliminated. This process systematically eliminates terms below the major diagonal, column by column, as illustrated below. The notation $R_i - (em)R_j$ next to the ith equation indicates that the ith equation is to be replaced by the ith equation minus em times the jth equation, where the elimination multiplier, em, is the quotient of the element to be eliminated and the pivot element.

For example, $R_2 - (-20/40)R_1$ beside Eq. (1.75.2) below means replace Eq. (1.75.2) by Eq. (1.75.2) $- (-20/40) \times$ Eq. (1.75.1). The elimination multiplier, em = $(-20/40)$, is chosen to eliminate the first coefficient in Eq. (1.75.2). All of the coefficients below the major diagonal in the first column are eliminated by linear combinations of each equation with the first equation. Thus,

$$\begin{bmatrix} 80x_1 - 20x_2 - 20x_3 = 20 \\ -20x_1 + 40x_2 - 20x_3 = 20 \\ -20x_1 - 20x_2 + 135x_3 = 20 \end{bmatrix} \begin{array}{l} \\ R_2 - (-20/80)R_1 \\ R_3 - (-20/80)R_1 \end{array} \qquad \begin{array}{l} (1.75.1) \\ (1.75.2) \\ (1.75.3) \end{array}$$

The result of this first elimination step is presented in Eq. (1.76), which also shows the elimination operation for the second elimination step. Next the coefficients below the major diagonal in the second column are eliminated by linear combinations with the second equation. Thus,

$$\begin{bmatrix} 80x_1 - 20x_2 - 20x_3 = 20 \\ 0x_1 + 35x_2 - 25x_3 = 25 \\ 0x_1 - 25x_2 + 125x_3 = 25 \end{bmatrix} \begin{array}{l} \\ \\ R_3 - (-25/35)R_2 \end{array} \qquad (1.76)$$

The result of the second elimination step is presented in Eq. (1.77):

$$\begin{bmatrix} 80x_1 - 20x_2 - 20x_3 = 20 \\ 0x_1 + 35x_2 - 25x_3 = 25 \\ 0x_1 + 0x_2 + 750/7x_3 = 300/7 \end{bmatrix} \qquad (1.77)$$

This process is continued until all the coefficients below the major diagonal are eliminated. In the present example with three equations, this process is now complete, and Eq. (1.77) is the final result. This is the process of elimination.

At this point, the last equation contains only one unknown, x_3 in the present example, which can be solved for. Using that result, the next to last equation can be solved for x_2. Using the results for x_3 and x_2, the first equation can be solved for x_1. This is the back substitution process. Thus,

$$x_3 = 300/750 = 0.40 \tag{1.78a}$$

$$x_2 = [25 - (-25)(0.40)]/35 = 1.00 \tag{1.78b}$$

$$x_1 = [20 - (-20)(1.00) - (-20)(0.40)/80 = 0.60 \tag{1.78c}$$

The extension of the elimination procedure to n equations is straightforward.

1.3.2.4. Simple Elimination

The elimination procedure illustrated in Example 1.6 involves manipulation of the coefficient matrix **A** and the nonhomogeneous vector **b**. Components of the **x** vector are fixed in their locations in the set of equations. As long as the columns are not interchanged, column j corresponds to x_j. Consequently, the x_j notation does not need to be carried throughout the operations. Only the numerical elements of **A** and **b** need to be considered. Thus, the elimination procedure can be simplified by augmenting the **A** matrix with the **b** vector and performing the row operations on the elements of the augmented **A** matrix to accomplish the elimination process, then performing the back substitution process to determine the solution vector. This simplified elimination procedure is illustrated in Example 1.7.

Example 1.7.　Simple elimination.

Let's rework Example 1.6 using simple elimination. From Example 1.6, the **A** matrix augmented by the **b** vector is

$$[A \mid b] = \begin{bmatrix} 80 & -20 & -20 \mid 20 \\ -20 & 40 & -20 \mid 20 \\ -20 & -20 & 130 \mid 20 \end{bmatrix} \tag{1.79}$$

Performing the row operations to accomplish the elimination process yields:

$$\begin{bmatrix} 80 & -20 & -20 \mid 20 \\ -20 & 40 & -20 \mid 20 \\ -20 & -20 & 130 \mid 20 \end{bmatrix} \begin{array}{l} \\ R_2 - (-20/80)R_1 \\ R_3 - (-20/80)R_1 \end{array} \tag{1.80}$$

$$\begin{bmatrix} 80 & -20 & -20 \mid 20 \\ 0 & 35 & -25 \mid 25 \\ 0 & -25 & 125 \mid 25 \end{bmatrix} R_3 - (-25/35)R_2 \tag{1.81}$$

$$\begin{bmatrix} 80 & -20 & -20 \mid 20 \\ 0 & 35 & -25 \mid 25 \\ 0 & 0 & 750/7 \mid 300/7 \end{bmatrix} \rightarrow \begin{array}{l} x_1 = [20 - (-20)(1.00) - (-20)(0.40)]/80 \\ \quad = 0.60 \\ x_2 = [25 - (-25)(0.4)]/35 = 1.00 \\ x_3 = 300/750 = 0.40 \end{array} \tag{1.82}$$

The back substitution step is presented beside the triangularized augmented **A** matrix.

1.3.2.5. Multiple b Vectors

If more than one **b** vector is to be considered, the **A** matrix is simply augmented by all of the **b** vectors simultaneously. The elimination process is then applied to the multiply augmented **A** matrix. Back substitution is then applied one column at a time to the modified **b** vectors. A more versatile procedure based on matrix factorization is presented in Section 1.4.

Example 1.8. Simple elimination for multiple b vectors.

Consider the system of equations presented in Example 1.7 with two **b** vectors, $\mathbf{b}_1^T = [20 \quad 20 \quad 20]$ and $\mathbf{b}_2^T = [20 \quad 10 \quad 20]$. The doubly augmented **A** matrix is

$$[\mathbf{A} \mid \mathbf{b}_1 \quad \mathbf{b}_2] = \begin{bmatrix} 80 & -20 & -20 \mid 20 \mid 20 \\ -20 & 40 & -20 \mid 20 \mid 10 \\ -20 & -20 & 130 \mid 20 \mid 20 \end{bmatrix} \tag{1.83}$$

Performing the elimination process yields

$$\begin{bmatrix} 80 & -20 & -20 & \mid & 20 & 20 \\ 0 & 35 & -25 & \mid & 25 & 15 \\ 0 & 0 & 750/7 & \mid & 300/7 & 250/7 \end{bmatrix} \tag{1.84}$$

Performing the back substitution process one column at a time yields

$$\mathbf{x}_1 = \begin{bmatrix} 0.60 \\ 1.00 \\ 0.40 \end{bmatrix} \quad \text{and} \quad \mathbf{x}_2 = \begin{bmatrix} 1/2 \\ 2/3 \\ 1/3 \end{bmatrix} \tag{1.85}$$

1.3.2.6. Pivoting

The element on the major diagonal is called the *pivot* element. The elimination procedure described so far fails immediately if the first pivot element a_{11} is zero. The procedure also fails if any subsequent pivot element $a_{i,i}$ is zero. Even though there may be no zeros on the major diagonal in the original matrix, the elimination process may create zeros on the major diagonal. The simple elimination procedure described so far must be modified to avoid zeros on the major diagonal. This result can be accomplished by rearranging the equations, by interchanging equations (rows) or variables (columns), before each elimination step to put the element of largest magnitude on the diagonal. This process is called *pivoting*. Interchanging both rows and columns is called *full pivoting*. Full pivoting is quite complicated, and thus it is rarely used. Interchanging only rows is called *partial pivoting*. Only partial pivoting is considered in this book.

Pivoting eliminates zeros in the pivot element locations during the elimination process. Pivoting also reduces round-off errors, since the pivot element is a divisor during the elimination process, and division by large numbers introduces smaller round-off errors than division by small numbers. When the procedure is repeated, round-off errors can compound. This problem becomes more severe as the number of equations is increased.

Example 1.9.   **Elimination with pivoting to avoid zero pivot elements.**

Use simple elimination with partial pivoting to solve the following system of linear algebraic equations, $\mathbf{Ax} = \mathbf{b}$:

$$
\begin{bmatrix} 0 & 2 & 1 \\ 4 & 1 & -1 \\ -2 & 3 & -3 \end{bmatrix} \begin{bmatrix} x_1 \\ x_2 \\ x_3 \end{bmatrix} = \begin{bmatrix} 5 \\ -3 \\ 5 \end{bmatrix}
\tag{1.86}
$$

Let's apply the elimination procedure by augmenting $\mathbf{A}$ with $\mathbf{b}$. The first pivot element is zero, so pivoting is required. The largest number (in magnitude) in the first column under the pivot element occurs in the second row. Thus, interchanging the first and second rows and evaluating the elimination multipliers yields

$$
\begin{bmatrix} 4 & 1 & -1 & | & -3 \\ 0 & 2 & 1 & | & 5 \\ -2 & 3 & -3 & | & 5 \end{bmatrix} \begin{array}{l} \\ R_2 - (0/4)R_1 \\ R_3 - (-2/4)R_1 \end{array}
\tag{1.87}
$$

Performing the elimination operations yields

$$
\begin{bmatrix} 4 & 1 & -1 & | & -3 \\ 0 & 2 & 1 & | & 5 \\ 0 & 7/2 & -7/2 & | & 7/2 \end{bmatrix}
\tag{1.88}
$$

Although the pivot element in the second row is not zero, it is not the largest element in the second column underneath the pivot element. Thus, pivoting is called for again. Note that pivoting is based only on the rows below the pivot element. The rows above the pivot element have already been through the elimination process. Using one of the rows above the pivot element would destroy the elimination already accomplished. Interchanging the second and third rows and evaluating the elimination multiplier yields

$$
\begin{bmatrix} 4 & 1 & -1 & | & -3 \\ 0 & 7/2 & -7/2 & | & 7/2 \\ 0 & 2 & 1 & | & 5 \end{bmatrix} \begin{array}{l} \\ \\ R_3 - (4/7)R_2 \end{array}
\tag{1.89}
$$

Performing the elimination operation yields

$$
\begin{bmatrix} 4 & 1 & -1 & | & -3 \\ 0 & 7/2 & -7/2 & | & 7/2 \\ 0 & 0 & 3 & | & 3 \end{bmatrix} \rightarrow \begin{array}{l} x_1 = -1 \\ x_2 = 2 \\ x_3 = 1 \end{array}
\tag{1.90}
$$

The back substitution results are presented beside the triangularized augmented $\mathbf{A}$ matrix.

1.3.2.7.   Scaling

The elimination process described so far can incur significant round-off errors when the magnitudes of the pivot elements are smaller than the magnitudes of the other elements in the equations containing the pivot elements. In such cases, scaling is employed to select the pivot elements. After pivoting, elimination is applied to the original equations. Scaling is employed only to select the pivot elements.

    Scaled pivoting is implemented as follows. Before elimination is applied to the first column, all of the elements in the first column are scaled (i.e., normalized) by the largest elements in the corresponding rows. Pivoting is implemented based on the scaled elements in the first column, and elimination is applied to obtain zero elements in the first column below the pivot element. Before elimination is applied to the second column, all of the elements from 2 to n in column 2 are scaled, pivoting is implemented, and elimination is applied to obtain zero elements in column 2 below the pivot element. The procedure is applied to the remaining rows 3 to $n-1$. Back substitution is then applied to obtain $\mathbf{x}$.

Example 1.10. Elimination with scaled pivoting to reduce round-off errors.

Let's investigate the advantage of scaling by solving the following linear system:

$$\begin{bmatrix} 3 & 2 & 105 \\ 2 & -3 & 103 \\ 1 & 1 & 3 \end{bmatrix} \mathbf{x} = \begin{bmatrix} 104 \\ 98 \\ 3 \end{bmatrix} \tag{1.91}$$

which has the exact solution $x_1 = -1.0$, $x_2 = 1.0$, and $x_3 = 1.0$. To accentuate the effects of round-off, carry only three significant figures in the calculations. For the first column, pivoting does not appear to be required. Thus, the augmented $\mathbf{A}$ matrix and the first set of row operations are given by

$$\begin{bmatrix} 3 & 2 & 105 \mid 104 \\ 2 & -3 & 103 \mid 98 \\ 1 & 1 & 3 \mid 3 \end{bmatrix} \begin{matrix} \\ R_2 - (0.667)R_1 \\ R_3 - (0.333)R_1 \end{matrix} \tag{1.92}$$

which gives

$$\begin{bmatrix} 3 & 2 & 105 \mid 104 \\ 0 & -4.33 & 33.0 \mid 28.6 \\ 0 & 0.334 & -32.0 \mid -31.6 \end{bmatrix} R_3 - (-0.0771)R_2 \tag{1.93}$$

Pivoting is not required for the second column. Performing the elimination indicated in Eq. (1.93) yields the triangularized matrix

$$\begin{bmatrix} 3 & 2 & 105 \mid 104 \\ 0 & -4.33 & 33.0 \mid 28.9 \\ 0 & 0 & -29.5 \mid -29.4 \end{bmatrix} \tag{1.94}$$

Performing back substitution yields $x_3 = 0.997$, $x_2 = 0.924$, and $x_1 = -0.844$, which does not agree very well with the exact solution $x_3 = 1.0$, $x_2 = 1.0$, and $x_1 = -1.0$. Round-off errors due to the three-digit precision have polluted the solution.

The effects of round-off can be reduced by scaling the equations before pivoting. Since scaling itself introduces round-off, it should be used only to determine if pivoting is required. All calculations should be made with the original unscaled equations.

Let's rework the problem using scaling to determine if pivoting is required. The first step in the elimination procedure eliminates all the elements in the first column under element a_{11}. Before performing that step, let's scale all the elements in column 1 by the largest element in each row. The result is

$$\mathbf{a}_1 = \begin{bmatrix} 3/105 \\ 2/103 \\ 1/3 \end{bmatrix} = \begin{bmatrix} 0.0286 \\ 0.0194 \\ 0.3333 \end{bmatrix} \tag{1.95}$$

where the notation $\mathbf{a}_1$ denotes the column vector consisting of the scaled elements from the first column of matrix $\mathbf{A}$. The third element of $\mathbf{a}_1$ is the largest element in $\mathbf{a}_1$, which indicates that rows 1 and 3 of matrix $\mathbf{A}$ should be interchanged. Thus, Eq. (1.91), with the elimination multipliers indicated, becomes

$$
\begin{bmatrix}
1 & 1 & 3 & | & 3 \\
2 & -3 & 103 & | & 98 \\
3 & 2 & 105 & | & 104
\end{bmatrix}
\begin{matrix}
\\
R_2 - (2/1)R_1 \\
R_3 - (3/1)R_1
\end{matrix}
\tag{1.96}
$$

Performing the elimination and indicating the next elimination multiplier yields

$$
\begin{bmatrix}
1 & 1 & 3 & | & 3 \\
0 & -5 & 97 & | & 92 \\
0 & -1 & 96 & | & 95
\end{bmatrix}
\begin{matrix}
\\
\\
R_3 - (1/5)R_2
\end{matrix}
\tag{1.97}
$$

Scaling the second and third elements of column 2 gives

$$
\mathbf{a}_2 =
\begin{bmatrix}
- \\
-5/97 \\
-1/96
\end{bmatrix}
=
\begin{bmatrix}
- \\
-0.0516 \\
-0.0104
\end{bmatrix}
\tag{1.98}
$$

Consequently, pivoting is not indicated. Performing the elimination indicated in Eq. (1.97) yields

$$
\begin{bmatrix}
1.0 & 1.0 & 3.0 & | & 3.0 \\
0.0 & -5.0 & 97.0 & | & 92.0 \\
0.0 & 0.0 & 76.6 & | & 76.6
\end{bmatrix}
\tag{1.99}
$$

Solving Eq. (1.99) by back substitution yields $x_1 = 1.00$, $x_2 = 1.00$, and $x_3 = -1.00$, which is the exact solution. Thus, scaling to determine the pivot element has eliminated the round-off error in this simple example.

1.3.3. Gauss Elimination

The elimination procedure described in the previous section, including scaled pivoting, is commonly called *Gauss elimination*. It is the most important and most useful direct elimination method for solving systems of linear algebraic equations. The Gauss Jordan method, the matrix inverse method, the LU factorization method, and the Thomas algorithm are all modifications or extensions of the Gauss elimination method. Pivoting is an essential element of Gauss elimination. In cases where all of the elements of the coefficient matrix $\mathbf{A}$ are the same order of magnitude, scaling is not necessary. However, pivoting to avoid zero pivot elements is always required. Scaled pivoting to decrease round-off errors, while very desirable in general, can be omitted at some risk to the accuracy of the solution. When performing Gauss elimination by hand, decisions about pivoting can be made on a case by case basis. When writing a general-purpose computer program to apply Gauss elimination to arbitrary systems of equations, however, scaled pivoting is an absolute necessity. Example 1.10 illustrates the complete Gauss elimination algorithm.

When solving large systems of linear algebraic equations on a computer, the pivoting step is generally implemented by simply keeping track of the order of the rows as they are interchanged without actually interchanging rows, a time-consuming and unnecessary operation. This is accomplished by using an order vector $\mathbf{o}$ whose elements denote the order in which the rows of the coefficient matrix $\mathbf{A}$ and the right-hand-side vector **b** are to be processed. When a row interchange is required, instead of actually interchanging the two rows of elements, the corresponding elements of the order vector are interchanged. The rows of the **A** matrix and the **b** vector are processed in the order indicated by the order vector **o** during both the elimination step and the back substitution step.

As an example, consider the second part of Example 1.10. The order vector has the initial value $\mathbf{o}^T = [1 \quad 2 \quad 3]$. After scaling, rows 1 and 3 are to be interchanged. Instead of actually interchanging these rows as done in Example 1.10, the corresponding elements of the order vector are changed to yield $\mathbf{o}^T = [3 \quad 2 \quad 1]$. The first elimination step then uses the third row to eliminate x_1 from the second and first rows. Pivoting is not required for the second elimination step, so the order vector is unchanged, and the second row is used to eliminate x_2 from the first row. Back substitution is then performed in the reverse order of the order vector, **o**, that is, in the order 1, 2, 3. This procedure saves computer time for large systems of equations, but at the expense of a slightly more complicated program.

The number of multiplications and divisions required for Gauss elimination is approximately $N = (n^3/3 - n/3)$ for matrix **A** and n^2 for each **b**. For $n = 10$, $N = 430$, and for $n = 100$, $N = 343,300$. This is a considerable reduction compared to Cramer's rule.

The Gauss elimination procedure, in a format suitable for programming on a computer, is summarized as follows:

1. Define the $n \times n$ coefficient matrix **A**, the $n \times 1$ column vector **b**, and the $n \times 1$ order vector **o**.
2. Starting with column 1, scale column k $(k = 1, 2, \ldots, n - 1)$ and search for the element of largest magnitude in column k and pivot (interchange rows) to put that coefficient into the $a_{k,k}$ pivot position. This step is actually accomplished by interchanging the corresponding elements of the $n \times 1$ order vector **o**.
3. For column k $(k = 1, 2, \ldots, n - 1)$, apply the elimination procedure to rows i $(i = k + 1, k + 2, \ldots, n)$ to create zeros in column k below the pivot element, $a_{k,k}$. Do not actually calculate the zeros in column k. In fact, storing the elimination multipliers, em $= (a_{i,k}/a_{k,k})$, in place of the eliminated elements, $a_{i,k}$, creates the Doolittle LU factorization presented in Section 1.4. Thus,

$$a_{i,j} = a_{i,j} - \left(\frac{a_{i,k}}{a_{k,k}}\right) a_{k,j} \qquad (i,j = k+1, k+2, \ldots, n) \qquad (1.100a)$$

$$b_i = b_i - \left(\frac{a_{i,k}}{a_{k,k}}\right) b_k \qquad (i = k+1, k+2, \ldots, n) \qquad (1.100b)$$

After step 3 is applied to all k columns, $(k = 1, 2, \ldots, n - 1)$, the original **A** matrix is upper triangular.

4. Solve for **x** using back substitution. If more than one **b** vector is present, solve for the corresponding **x** vectors one at a time. Thus,

$$x_n = \frac{b_n}{a_{n,n}} \qquad (1.101a)$$

$$x_i = \frac{b_i - \sum\limits_{j=i+1}^{n} a_{i,j} x_j}{a_{i,i}} \qquad (i = n-1, n-2, \ldots, 1) \qquad (1.101b)$$

1.3.4. Gauss-Jordan Elimination

Gauss-Jordan elimination is a variation of Gauss elimination in which the elements above the major diagonal are eliminated (made zero) as well as the elements below the major diagonal. The **A** matrix is transformed to a diagonal matrix. The rows are usually scaled to yield unity diagonal elements, which transforms the **A** matrix to the identity matrix, **I**. The transformed **b** vector is then the solution vector **x**. Gauss-Jordan elimination can be used for single or multiple b vectors.

The number of multiplications and divisions for Gauss-Jordan elimination is approximately $N = (n^3/2 - n/2) + n^2$, which is approximately 50 percent larger than for Gauss elimination. Consequently, Gauss elimination is preferred.

Example 1.11.   Gauss-Jordan elimination.

Let's rework Example 1.7 using simple Gauss-Jordan elimination, that is, elimination without pivoting. The augmented **A** matrix is [see Eq. (1.79)]

$$\begin{bmatrix} 80 & -20 & -20 \mid 20 \\ -20 & 40 & -20 \mid 20 \\ -20 & -20 & 130 \mid 20 \end{bmatrix} R_1/80 \tag{1.102}$$

Scaling row 1 to give $a_{11} = 1$ gives

$$\begin{bmatrix} 1 & -1/4 & -1/4 \mid 1/4 \\ -20 & 40 & -20 \mid 20 \\ -20 & -20 & 130 \mid 20 \end{bmatrix} \begin{matrix} \\ R_2 - (-20)R_1 \\ R_3 - (-20)R_1 \end{matrix} \tag{1.103}$$

Applying elimination below row 1 yields

$$\begin{bmatrix} 1 & -1/4 & -1/4 \mid 1/4 \\ 0 & 35 & -25 \mid 25 \\ 0 & -25 & 125 \mid 25 \end{bmatrix} R_2/35 \tag{1.104}$$

Scaling row 2 to give $a_{22} = 1$ gives

$$\begin{bmatrix} 1 & -1/4 & -1/4 \mid 1/4 \\ 0 & 1 & -5/7 \mid 5/7 \\ 0 & -25 & 125 \mid 25 \end{bmatrix} \begin{matrix} R_1 - (-1/4)R_2 \\ \\ R_3 - (-25)R_2 \end{matrix} \tag{1.105}$$

Applying elimination both above and below row 2 yields

$$\begin{bmatrix} 1 & 0 & -3/7 \mid 3/7 \\ 0 & 1 & -5/7 \mid 5/7 \\ 0 & 0 & 750/7 \mid 300/7 \end{bmatrix} R_3/(750/7) \tag{1.106}$$

Scaling row 3 to give $a_{33} = 1$ gives

$$\begin{bmatrix} 1 & 0 & -3/7 \mid 3/7 \\ 0 & 1 & -5/7 \mid 5/7 \\ 0 & 0 & 1 \mid 215 \end{bmatrix} \begin{matrix} R_1 - (-3/7)R_3 \\ R_2 - (-5/7)R_3 \\ \\ \end{matrix} \tag{1.107}$$

Applying elimination above row 3 completes the process.

$$\begin{bmatrix} 1 & 0 & 0 & | & 0.60 \\ 0 & 1 & 0 & | & 1.00 \\ 0 & 0 & 1 & | & 0.40 \end{bmatrix} \tag{1.108}$$

The $\mathbf{A}$ matrix has been transformed to the identity matrix $\mathbf{I}$ and the $\mathbf{b}$ vector has been transformed to the solution vector, $\mathbf{x}$. Thus, $\mathbf{x}^T = [0.60 \quad 1.00 \quad 0.40]$.

The inverse of a square matrix $\mathbf{A}$ is the matrix $\mathbf{A}^{-1}$ such that $\mathbf{AA}^{-1} = \mathbf{A}^{-1}\mathbf{A} = \mathbf{I}$. Gauss-Jordan elimination can be used to evaluate the inverse of matrix $\mathbf{A}$ by augmenting $\mathbf{A}$ with the identity matrix $\mathbf{I}$ and applying the Gauss-Jordan algorithm. The transformed $\mathbf{A}$ matrix is the identity matrix $\mathbf{I}$, and the transformed identity matrix is the matrix inverse, $\mathbf{A}^{-1}$. Thus, applying Gauss-Jordan elimination yields

$$\boxed{[\mathbf{A} \mid \mathbf{I}] \rightarrow [\mathbf{I} \mid \mathbf{A}^{-1}]} \tag{1.109}$$

The Gauss-Jordan elimination procedure, in a format suitable for programming on a computer, can be developed to solve Eq. (1.109) by modifying the Gauss elimination procedure presented in Section 1.3.C. Step 1 is changed to augment the $n \times n$ $\mathbf{A}$ matrix with the $n \times n$ identity matrix, $\mathbf{I}$. Steps 2 and 3 of the procedure are the same. Before performing Step 3, the pivot element is scaled to unity by dividing all elements in the row by the pivot element. Step 3 is expanded to perform elimination above the pivot element as well as below the pivot element. At the conclusion of step 3, the $\mathbf{A}$ matrix has been transformed to the identity matrix, $\mathbf{I}$, and the original identity matrix, $\mathbf{I}$, has been transformed to the matrix inverse, $\mathbf{A}^{-1}$.

Example 1.12.   Matrix inverse by Gauss-Jordan elimination.

Let's evaluate the inverse of matrix $\mathbf{A}$ presented in Example 1.7. First, augment matrix $\mathbf{A}$ with the identity matrix, $\mathbf{I}$. Thus,

$$[\mathbf{A} \mid \mathbf{I}] = \begin{bmatrix} 80 & -20 & -20 & | & 1 & 0 & 0 \\ -20 & 40 & -20 & | & 0 & 1 & 0 \\ -20 & -20 & 130 & | & 0 & 0 & 1 \end{bmatrix} \tag{1.110}$$

Performing Gauss-Jordan elimination transforms Eq. (1.110) to

$$\begin{bmatrix} 1 & 0 & 0 & | & 2/125 & 1/100 & 1/250 \\ 0 & 1 & 0 & | & 1/100 & 1/30 & 1/150 \\ 0 & 0 & 1 & | & 1/250 & 1/150 & 7/750 \end{bmatrix} \tag{1.111}$$

from which

$$\mathbf{A}^{-1} = \begin{bmatrix} 2/125 & 1/100 & 1/250 \\ 1/100 & 1/30 & 1/150 \\ 1/250 & 1/150 & 7/750 \end{bmatrix} = \begin{bmatrix} 0.016000 & 0.010000 & 0.004000 \\ 0.010000 & 0.033333 & 0.006667 \\ 0.004000 & 0.006667 & 0.009333 \end{bmatrix}$$

$$\tag{1.112}$$

Multiplying $\mathbf{A}$ times $\mathbf{A}^{-1}$ yields the identity matrix $\mathbf{I}$, thus verifying the computations.

1.3.5. The Matrix Inverse Method

Systems of linear algebraic equations can be solved using the matrix inverse, $\mathbf{A}^{-1}$. Consider the general system of linear algebraic equations:

$$\mathbf{Ax} = \mathbf{b} \tag{1.113}$$

Multiplying Eq. (1.113) by $\mathbf{A}^{-1}$ yields

$$\mathbf{A}^{-1}\mathbf{Ax} = \mathbf{Ix} = \mathbf{x} = \mathbf{A}^{-1}\mathbf{b} \tag{1.114}$$

from which

$$\boxed{\mathbf{x} = \mathbf{A}^{-1}\mathbf{b}} \tag{1.115}$$

Thus, when the matrix inverse $\mathbf{A}^{-1}$ of the coefficient matrix $\mathbf{A}$ is known, the solution vector $\mathbf{x}$ is simply the product of the matrix inverse $\mathbf{A}^{-1}$ and the right-hand-side vector $\mathbf{b}$. Not all matrices have inverses. Singular matrices, that is, matrices whose determinant is zero, do not have inverses. The corresponding system of equations does not have a unique solution.

Example 1.13.   The matrix inverse method.

Let's solve the linear system considered in Example 1.7 using the matrix inverse method. The matrix inverse $\mathbf{A}^{-1}$ of the coefficient matrix $\mathbf{A}$ for that linear system is evaluated in Example 1.12. Multiplying $\mathbf{A}^{-1}$ by the vector $\mathbf{b}$ from Example 1.7 gives

$$\mathbf{x} = \mathbf{A}^{-1}\mathbf{b} = \begin{bmatrix} 2/125 & 1/100 & 1/250 \\ 1/100 & 1/30 & 1/150 \\ 1/250 & 1/150 & 7/750 \end{bmatrix} \begin{bmatrix} 20 \\ 20 \\ 20 \end{bmatrix} \tag{1.116}$$

Performing the matrix multiplication yields

$$x_1 = (2/125)(20) + (1/100)(20) + (1/250)(20) = 0.60 \tag{1.117a}$$
$$x_2 = (1/100)(20) + (1/30)(20) + (1/150)(20) = 1.00 \tag{1.117b}$$
$$x_3 = (1/250)(20) + (1/150)(20) + (7/750)(20) = 0.04 \tag{1.117c}$$

Thus, $x^T = [0.60 \quad 1.00 \quad 0.40]$.

1.3.6. Determinants

The evaluation of determinants by the cofactor method is discussed in Section 1.2.4 and illustrated in Example 1.4. Approximately $N = (n-1)n!$ multiplications are required to evaluate the determinant of an $n \times n$ matrix by the cofactor method. For $n = 10$, $N = 32,659,000$. Evaluation of the determinants of large matrices by the cofactor method is prohibitively expensive, if not impossible. Fortunately, determinants can be evaluated much more efficiently by a variation of the elimination method.

First, consider the matrix **A** expressed in upper triangular form:

$$\mathbf{A} = \begin{bmatrix} a_{11} & a_{12} & a_{13} & \cdots & a_{1n} \\ 0 & a_{22} & a_{23} & \cdots & a_{2n} \\ 0 & 0 & a_{33} & \cdots & a_{3n} \\ \multicolumn{5}{c}{\dotfill} \\ 0 & 0 & 0 & \cdots & a_{nn} \end{bmatrix} \qquad (1.118)$$

Expanding the determinant of **A** by cofactors down the first column gives a_{11} times the $(n-1) \times (n-1)$ determinant having a_{22} as its first element in its first column, with the remaining elements in its first column all zero. Expanding that determinant by cofactors down its first column yields a_{22} times the $(n-2) \times (n-2)$ determinant having a_{33} as its first element in its first column with the remaining elements in its first column all zero. Continuing in this manner yields the result that the determinant of an upper triangular matrix (or a lower triangular matrix) is simply the product of the elements on the major diagonal. Thus,

$$\det(\mathbf{A}) = |\mathbf{A}| = \prod_{i=1}^{n} a_{i,i} \qquad (1.119)$$

where the $\prod$ notation denotes the product of the $a_{i,i}$. Thus,

$$\det(\mathbf{A}) = |\mathbf{A}| = a_{11}a_{22}\cdots a_{nn} \qquad (1.120)$$

This result suggests the use of elimination to triangularize a general square matrix, then to evaluate its determinant using Eq. (1.119). This procedure works exactly as stated if no pivoting is used. When pivoting is used, the value of the determinant is changed, but in a predictable manner, so elimination can also be used with pivoting to evaluate determinants. The row operations must be modified as follows to use elimination for the evaluation of determinants.

1. Multiplying a row by a constant multiplies the determinant by that constant.
2. Interchanging any two rows changes the sign of the determinant. Thus, an even number of row interchanges does not change the sign of the determinant, whereas an odd number of row interchanges does change the sign of the determinant.
3. Any row may be added to the multiple of any other row without changing the value of the determinant.

The modified elimination method based on the above row operations is an efficient way to evaluate the determinant of a matrix. The number of multiplications required is approximately $N = n^3 + n^2 - n$, which is orders and orders of magnitude less effort than the $N = (n-1)n!$ multiplications required by the cofactor method.

Example 1.14. Evaluation of a 3×3 determinant by the elimination method.

Let's rework Example 1.4 using the elimination method. Recall Eq. (1.53):

$$\mathbf{A} = \begin{bmatrix} 80 & -20 & -20 \\ -20 & 40 & -20 \\ -20 & -20 & 130 \end{bmatrix} \qquad (1.121)$$

From Example 1.7, after Gauss elimination, matrix $\mathbf{A}$ becomes

$$\begin{bmatrix} 80 & -20 & -20 \\ 0 & 35 & -25 \\ 0 & 0 & 750/7 \end{bmatrix} \tag{1.122}$$

There are no row interchanges or multiplications of the matrix by scalars in this example. Thus,

$$\det(\mathbf{A}) = |\mathbf{A}| = (80)(35)(750/7) = 300,000 \tag{1.123}$$

1.4 LU FACTORIZATION

Matrices (like scalars) can be *factored* into the product of two other matrices in an infinite number of ways. Thus,

$$\mathbf{A} = \mathbf{BC} \tag{1.124}$$

When $\mathbf{B}$ and $\mathbf{C}$ are lower triangular and upper triangular matrices, respectively, Eq. (1.124) becomes

$$\mathbf{A} = \mathbf{LU} \tag{1.125}$$

Specifying the diagonal elements of either $\mathbf{L}$ or $\mathbf{U}$ makes the factoring unique. The procedure based on unity elements on the major diagonal of $\mathbf{L}$ is called the *Doolittle method*. The procedure based on unity elements on the major diagonal of $\mathbf{U}$ is called the *Crout method*.

Matrix factoring can be used to reduce the work involved in Gauss elimination when multiple unknown $\mathbf{b}$ vectors are to be considered. In the Doolittle LU method, this is accomplished by defining the elimination multipliers, em, determined in the elimination step of Gauss elimination as the elements of the $\mathbf{L}$ matrix. The $\mathbf{U}$ matrix is defined as the upper triangular matrix determined by the elimination step of Gauss elimination. In this manner, multiple $\mathbf{b}$ vectors can be processed through the elimination step using the $\mathbf{L}$ matrix and through the back substitution step using the elements of the $\mathbf{U}$ matrix.

Consider the linear system, $\mathbf{Ax} = \mathbf{b}$. Let $\mathbf{A}$ be factored into the product $\mathbf{LU}$, as illustrated in Eq. (1.125). The linear system becomes

$$\mathbf{LUx} = \mathbf{b} \tag{1.126}$$

Multiplying Eq. (1.126) by $\mathbf{L}^{-1}$ gives

$$\mathbf{L}^{-1}\mathbf{LUx} = \mathbf{IUx} = \mathbf{Ux} = \mathbf{L}^{-1}\mathbf{b} \tag{1.127}$$

The last two terms in Eq. (1.127) give

$$\mathbf{Ux} = \mathbf{L}^{-1}\mathbf{b} \tag{1.128}$$

Define the vector $\mathbf{b}'$ as follows:

$$\mathbf{b}' = \mathbf{L}^{-1}\mathbf{b} \tag{1.129}$$

Multiplying Eq. (1.129) by $\mathbf{L}$ gives

$$\mathbf{Lb}' = \mathbf{LL}^{-1}\mathbf{b} = \mathbf{Ib} = \mathbf{b} \tag{1.130}$$

Equating the first and last terms in Eq. (1.130) yields

$$\boxed{\mathbf{Lb'} = \mathbf{b}}$$ (1.131)

Substituting Eq. (1.129) into Eq. (1.128) yields

$$\boxed{\mathbf{Ux} = \mathbf{b'}}$$ (1.132)

Equation (1.131) is used to transform the **b** vector into the **b'** vector, and Eq. (1.132) is used to determine the solution vector **x**. Since Eq. (1.131) is lower triangular, forward substitution (analogous to back substitution presented earlier) is used to solve for **b'**. Since Eq. (1.132) is upper triangular, back substitution is used to solve for **x**.

In the *Doolittle LU method*, the **U** matrix is the upper triangular matrix obtained by Gauss elimination. The **L** matrix is the lower triangular matrix containing the elimination multipliers, em, obtained in the Gauss elimination process as the elements below the diagonal, with unity elements on the major diagonal. Equation (1.131) applies the steps performed in the triangularization of **A** to **U** to the **b** vector to transform **b** to **b'**. Equation (1.132) is simply the back substitution step of the Gauss elimination method. Consequently, once **L** and **U** have been determined, any **b** vector can be considered at any later time, and the corresponding solution vector **x** can be obtained simply by solving Eqs. (1.131) and (1.132), in that order. The number of multiplicative operations required for each **b** vector is n^2.

Example 1.15. The Doolittle LU method.

Let's solve Example 1.7 using the Doolittle LU method. The first step is to determine the **L** and **U** matrices. The **U** matrix is simply the upper triangular matrix determined by the Gauss elimination procedure in Example 1.7. The **L** matrix is simply the record of the elimination multipliers, em, used to transform **A** to **U**. These multipliers are the numbers in parentheses in the row operations indicated in Eqs. (1.80) and (1.81) in Example 1.7. Thus, **L** and **U** are given by

$$\mathbf{L} = \begin{bmatrix} 1 & 0 & 0 \\ -1/4 & 1 & 0 \\ -1/4 & -5/7 & 1 \end{bmatrix} \quad \text{and} \quad \mathbf{U} = \begin{bmatrix} 80 & -20 & -20 \\ 0 & 35 & -25 \\ 0 & 0 & 750/7 \end{bmatrix}$$ (1.133)

Consider the first **b** vector from Example 1.8: $\mathbf{b}_1^T = [20 \quad 20 \quad 20]$. Equation (1.131) gives

$$\begin{bmatrix} 1 & 0 & 0 \\ -1/4 & 1 & 0 \\ -1/4 & -5/7 & 1 \end{bmatrix} \begin{bmatrix} b_1' \\ b_2' \\ b_3' \end{bmatrix} = \begin{bmatrix} 20 \\ 20 \\ 20 \end{bmatrix}$$ (1.134)

Performing forward substitution yields

$$b_1' = 20$$ (1.135a)

$$b_2' = 20 - (-1/4)(20) = 25$$ (1.135b)

$$b_3' = 20 - (-1/4)(20) - (-5/7)(25) = 300/7$$ (1.135c)

The $\mathbf{b}'$ vector is simply the transformed $\mathbf{b}$ vector determined in Eq. (1.82). Equation (1.132) gives

$$\begin{bmatrix} 80 & -20 & -20 \\ 0 & 35 & -25 \\ 0 & 0 & 750/7 \end{bmatrix} \begin{bmatrix} x_1 \\ x_2 \\ x_3 \end{bmatrix} = \begin{bmatrix} 20 \\ 25 \\ 300/7 \end{bmatrix} \tag{1.136}$$

Performing back substitution yields $x_1^T = [0.60 \quad 1.00 \quad 0.40]$. Repeating the process for $\mathbf{b}_2^T = [20 \quad 10 \quad 20]$ yields

$$\begin{bmatrix} 1 & 0 & 0 \\ -1/4 & 1 & 0 \\ -1/4 & -5/7 & 1 \end{bmatrix} \begin{bmatrix} b_1' \\ b_2' \\ b_3' \end{bmatrix} = \begin{bmatrix} 20 \\ 10 \\ 20 \end{bmatrix} \begin{matrix} b_1' = 20 \\ b_2' = 15 \\ b_3' = 250/7 \end{matrix} \tag{1.137}$$

$$\begin{bmatrix} 80 & -20 & -20 \\ 0 & 35 & -25 \\ 0 & 0 & 750/7 \end{bmatrix} \begin{bmatrix} x_1 \\ x_2 \\ x_3 \end{bmatrix} = \begin{bmatrix} 20 \\ 15 \\ 250/7 \end{bmatrix} \begin{matrix} x_1 = 1/2 \\ x_2 = 2/3 \\ x_3 = 1/3 \end{matrix} \tag{1.138}$$

When pivoting is used with **LU** factorization, it is necessary to keep track of the row order, for example, by an order vector $\mathbf{o}$. When the rows of $\mathbf{A}$ are interchanged during the elimination process, the corresponding elements of the order vector $\mathbf{o}$ are interchanged. When a new $\mathbf{b}$ vector is considered, it is processed in the order corresponding to the elements of the order vector $\mathbf{o}$.

The major advantage of LU factorization methods is their efficiency when multiple unknown $\mathbf{b}$ vectors must be considered. The number of multiplications and divisions required by the complete Gauss elimination method is $N = (n^3/3 - n/3) + n^2$. The forward substitution step required to solve $\mathbf{Lb}' = \mathbf{b}$ requires $N = n^2/2 - n/2$ multiplicative operations, and the back substitution step required to solve $\mathbf{Ux} = \mathbf{b}'$ requires $N = n^2/2 + n/2$ multiplicative operations. Thus, the total number of multiplicative operations required by LU factorization, after $\mathbf{L}$ and $\mathbf{U}$ have been determined, is n^2, which is much less work than required by Gauss elimination, especially for large systems.

The Doolittle LU method, in a format suitable for programming for a computer, is summarized as follows:

1. Perform steps 1, 2, and 3 of the Gauss elimination procedure presented in Section 1.3.3. Store the pivoting information in the order vector $\mathbf{o}$. Store the row elimination multipliers, em, in the locations of the eliminated elements. The results of this step are the $\mathbf{L}$ and $\mathbf{U}$ matrices.
2. Compute the $\mathbf{b}'$ vector in the order of the elements of the order vector $\mathbf{o}$ using forward substitution:

$$b_i' = b_i - \sum_{k=1}^{i-1} l_{i,k} b_k' \qquad (i = 2, 3, \ldots, n) \tag{1.139}$$

   where $l_{i,k}$ are the elements of the $\mathbf{L}$ matrix.
3. Compute the $\mathbf{x}$ vector using back substitution:

$$x_i = b_i' - \sum_{k=i+1}^{n} u_{i,k} x_k / u_{i,i} \qquad (i = n-1, n-2, \ldots, 1) \tag{1.140}$$

   where $u_{i,k}$ and $u_{i,i}$ are elements of the $\mathbf{U}$ matrix.

As a final application of LU factorization, it can be used to evaluate the inverse of matrix $\mathbf{A}$, that is, $\mathbf{A}^{-1}$. The matrix inverse is calculated in a column by column manner using unit vectors for the right-hand-side vector $\mathbf{b}$. Thus, if $\mathbf{b}_1^T = [1 \quad 0 \quad \cdots \quad 0]$, $\mathbf{x}_1$ will be the first column of $\mathbf{A}^{-1}$. The succeeding columns of $\mathbf{A}^{-1}$ are calculated by letting $\mathbf{b}_2^T = [0 \quad 1 \quad \cdots \quad 0]$, $\mathbf{b}_3^T = [0 \quad 0 \quad 1 \quad \cdots \quad 0]$, etc., and $\mathbf{b}_n^T = [0 \quad 0 \quad \cdots \quad 1]$. The number of multiplicative operations for each column is n^2. There are n columns, so the total number of multiplicative operations is n^3. The number of multiplicative operations required to determine $\mathbf{L}$ and $\mathbf{U}$ are $(n^3/3 - n/3)$. Thus, the total number of multiplicative operations required is $4n^3/3 - n/3$, which is smaller than the $3n^3/2 - n/2$ operations required by the Gauss-Jordan method.

Example 1.16. Matrix inverse by the Doolittle LU method.

Let's evaluate the inverse of matrix $\mathbf{A}$ presented in Example 1.7 by the Doolittle LU method:

$$\mathbf{A} = \begin{bmatrix} 80 & -20 & -20 \\ -20 & 40 & -20 \\ -20 & -20 & 130 \end{bmatrix} \tag{1.141}$$

Evaluate the $\mathbf{L}$ and $\mathbf{U}$ matrices by Doolittle LU factorization. Thus,

$$\mathbf{L} = \begin{bmatrix} 1 & 0 & 0 \\ -1/4 & 1 & 0 \\ -1/4 & -5/7 & 1 \end{bmatrix} \quad \text{and} \quad \mathbf{U} = \begin{bmatrix} 80 & -20 & -20 \\ 0 & 35 & -25 \\ 0 & 0 & 750/7 \end{bmatrix} \tag{1.142}$$

Let $\mathbf{b}_1^T = [1 \quad 0 \quad 0]$. Then, $\mathbf{Lb}_1' = \mathbf{b}_1$ gives

$$\begin{bmatrix} 1 & 0 & 0 \\ -1/4 & 1 & 0 \\ -1/4 & -5/7 & 1 \end{bmatrix} \begin{bmatrix} b_1' \\ b_2' \\ b_3' \end{bmatrix} = \begin{bmatrix} 1 \\ 0 \\ 0 \end{bmatrix} \rightarrow \mathbf{b}_1' = \begin{bmatrix} 1 \\ 1/4 \\ 3/7 \end{bmatrix} \tag{1.143a}$$

Solve $\mathbf{Ux} = \mathbf{b}_1'$ to determine $\mathbf{x}_1$. Thus,

$$\begin{bmatrix} 80 & -20 & -20 \\ 0 & 35 & -25 \\ 0 & 0 & 750/7 \end{bmatrix} \begin{bmatrix} x_1 \\ x_2 \\ x_3 \end{bmatrix} = \begin{bmatrix} 1 \\ 1/4 \\ 3/7 \end{bmatrix} \rightarrow \mathbf{x}_1 = \begin{bmatrix} 2/125 \\ 1/100 \\ 1/250 \end{bmatrix} \tag{1.143b}$$

where $\mathbf{x}_1$ is the first column of $\mathbf{A}^{-1}$. Letting $\mathbf{b}_2^T = [0 \quad 1 \quad 0]$ gives $\mathbf{x}_2^T = [1/100 \quad 1/30 \quad 1/150]$, and letting $\mathbf{b}_3^T = [0 \quad 0 \quad 1]$ gives $\mathbf{x}_3^T = [1/250 \quad 1/150 \quad 7/750]$. Thus, $\mathbf{A}^{-1}$ is given by

$$\mathbf{A}^{-1} = [\mathbf{x}_1 \quad \mathbf{x}_2 \quad \mathbf{x}_3] = \begin{bmatrix} 2/125 & 1/100 & 1/250 \\ 1/100 & 1/30 & 1/150 \\ 1/250 & 1/150 & 7/750 \end{bmatrix}$$

$$= \begin{bmatrix} 0.016000 & 0.010000 & 0.004000 \\ 0.01000 & 0.033333 & 0.006667 \\ 0.004000 & 0.006667 & 0.009333 \end{bmatrix} \tag{1.143c}$$

which is the same result obtained by Gauss-Jordan elimination in Example 1.12.

1.5 TRIDIAGONAL SYSTEMS OF EQUATIONS

When a large system of linear algebraic equations has a special pattern, such as a tridiagonal pattern, it is usually worthwhile to develop special methods for that unique pattern. There are a number of direct elimination methods for solving systems of linear algebraic equations which have special patterns in the coefficient matrix. These methods are generally very efficient in computer time and storage. Such methods should be considered when the coefficient matrix fits the required pattern, and when computer storage and/or execution time are important. One algorithm that deserves special attention is the algorithm for tridiagonal matrices, often referred to as the Thomas (1949) algorithm. Large tridiagonal systems arise naturally in a number of problems, especially in the numerical solution of differential equations by implicit methods. Consequently, the Thomas algorithm has found a large number of applications.

To derive the Thomas algorithm, let's apply the Gauss elimination procedure to a tridiagonal matrix $\mathbf{T}$, modifying the procedure to eliminate all unnecessary computations involving zeros. Consider the matrix equation:

$$\boxed{\mathbf{Tx} = \mathbf{b}} \tag{1.144}$$

where $\mathbf{T}$ is a tridiagonal matrix. Thus,

$$\mathbf{T} = \begin{bmatrix} a_{11} & a_{12} & 0 & 0 & 0 & \cdots & 0 & 0 & 0 \\ a_{21} & a_{22} & a_{23} & 0 & 0 & \cdots & 0 & 0 & 0 \\ 0 & a_{32} & a_{33} & a_{34} & 0 & \cdots & 0 & 0 & 0 \\ 0 & 0 & a_{43} & a_{44} & a_{45} & \cdots & 0 & 0 & 0 \\ \multicolumn{9}{c}{\cdots\cdots\cdots\cdots\cdots\cdots\cdots\cdots\cdots\cdots\cdots\cdots\cdots\cdots\cdots\cdots\cdots} \\ 0 & 0 & 0 & 0 & 0 & \cdots & a_{n-1,n-2} & a_{n-1,n-1} & a_{n-1,n} \\ 0 & 0 & 0 & 0 & 0 & \cdots & 0 & a_{n,n-1} & a_{n,n} \end{bmatrix} \tag{1.145}$$

Since all the elements of column 1 below row 2 are already zero, the only element to be eliminated in row 2 is a_{21}. Thus, replace row 2 by $R_2 - (a_{21}/a_{11})R_1$. Row 2 becomes

$$[0 \quad a_{22} - (a_{21}/a_{11})a_{12} \quad a_{23} \quad 0 \quad 0 \quad \cdots \quad 0 \quad 0 \quad 0] \tag{1.146}$$

Similarly, only a_{32} in column 2 must be eliminated from row 3, only a_{43} in column 3 must be eliminated from row 4, etc. The eliminated element itself does not need to be calculated. In fact, storing the elimination multipliers, $em = (a_{21}/a_{11})$, etc., in place of the eliminated elements allows this procedure to be used as an LU factorization method. Only the diagonal element in each row is affected by the elimination. Elimination in rows 2 to n is accomplished as follows:

$$a_{i,i} = a_{i,i} - (a_{i,i-1}/a_{i-1,i-1})a_{i-1,i} \qquad (i = 2, \ldots, n) \tag{1.147}$$

Thus, the elimination step involves only $2n$ multiplicative operations to place $\mathbf{T}$ in upper triangular form.

The elements of the $\mathbf{b}$ vector are also affected by the elimination process. The first element b_1 is unchanged. The second element b_2 becomes

$$b_2 = b_2 - (a_{21}/a_{11})b_1 \tag{1.148}$$

Subsequent elements of the $\mathbf{b}$ vector are changed in a similar manner. Processing the $\mathbf{b}$ vector requires only one multiplicative operation, since the elimination multiplier, em $= (a_{21}/a_{11})$, is already calculated. Thus, the total process of elimination, including the operation on the **b** vector, requires only $3n$ multiplicative operations.

The $n \times n$ tridiagonal matrix **T** can be stored as an $n \times 3$ matrix **A**$'$ since there is no need to store the zeros. The first column of matrix **A**$'$, elements $a'_{i,1}$, corresponds to the subdiagonal of matrix **T**, elements $a_{i,i-1}$. The second column of matrix **A**$'$, elements $a'_{i,2}$, corresponds to the diagonal elements of matrix **T**, elements $a_{i,i}$. The third column of matrix **A**$'$, elements $a'_{i,3}$, corresponds to the superdiagonal of matrix **T**, elements $a_{i,i+1}$. The elements $a'_{1,1}$ and $a'_{n,3}$ do not exist. Thus,

$$\mathbf{A}' = \begin{bmatrix} - & a'_{1,2} & a'_{1,3} \\ a'_{2,1} & a'_{2,2} & a'_{2,3} \\ a'_{3,1} & a'_{3,2} & a'_{3,3} \\ \cdots\cdots\cdots\cdots\cdots \\ a'_{n-1,1} & a'_{n-1,2} & a'_{n-1,3} \\ a'_{n,1} & a'_{n,2} & - \end{bmatrix} \tag{1.149}$$

When the elements of column 1 of matrix **A**$'$ are eliminated, that is, the elements $a'_{i,1}$, the elements of column 2 of matrix **A**$'$ become

$$a'_{1,2} = a'_{1,2} \tag{1.150a}$$
$$a'_{i,2} = a'_{i,2} - (a'_{i,1}/a'_{i-1,2})a'_{i-1,3} \qquad (i = 2, 3, \ldots, n) \tag{1.150b}$$

The **b** vector is modified as follows:

$$b_1 = b_1 \tag{1.151a}$$
$$b_i = b_i - (a'_{i,1}/a'_{i-1,2})b_{i-1} \qquad (i = 2, 3, \ldots, n) \tag{1.151b}$$

After $a'_{i,2}$ $(i = 2, 3, \ldots, n)$ and **b** are evaluated, the back substitution step is as follows:

$$x_n = b_n/a'_{n,2} \tag{1.152a}$$
$$x_i = (b_i - a'_{i,3}x_{i+1})/a'_{i,2} \qquad (i = n-1, n-2, \ldots, 1) \tag{1.152b}$$

Example 1.17.   The Thomas algorithm.

Let's solve the tridiagonal system of equations obtained in Example 8.4, Eq. (8.54). In that example, the finite difference equation

$$T_{i-1} - (2 + \alpha^2 \, \Delta x^2)T_i + T_{i+1} = 0 \tag{1.153}$$

is solved for $\alpha = 4.0$ and $\Delta x = 0.125$, for which $(2 + \alpha^2 \, \Delta x^2) = 2.25$, for $i = 2, \ldots, 8$, with $T_1 = 0.0$ and $T_9 = 100.0$. Writing Eq. (1.153) in the form of the $n \times 3$ matrix **A**$'$ (where the temperatures T_i of Example 8.4 correspond to the elements of the **x** vector) yields

$$\mathbf{A}' = \begin{bmatrix} - & -2.25 & 1.0 \\ 1.0 & -2.25 & 1.0 \\ 1.0 & -2.25 & 1.0 \\ 1.0 & -2.25 & 1.0 \\ 1.0 & -2.25 & 1.0 \\ 1.0 & -2.25 & 1.0 \\ 1.0 & -2.25 & - \end{bmatrix} \quad \text{and} \quad \mathbf{b} = \begin{bmatrix} 0.0 \\ 0.0 \\ 0.0 \\ 0.0 \\ 0.0 \\ 0.0 \\ -100.0 \end{bmatrix} \tag{1.154}$$

The major diagonal terms (the center column of the $\mathbf{A}'$ matrix) are transformed according to Eq. (1.150). Thus, $a'_{1,2} = -2.25$ and $a'_{2,2}$ is given by

$$a'_{2,2} = a'_{2,2} - (a'_{2,1}/a'_{1,2})a'_{1,3} = -2.25 - [1.0/(-2.25)](1.0) = -1.805556 \quad (1.155)$$

The remaining elements of column 2 are processed in the same manner. The $\mathbf{A}'$ matrix after elimination is presented in Eq. (1.157), where the elimination multipliers are presented in parentheses in column 1. The $\mathbf{b}$ vector is transformed according to Eq. (1.151). Thus, $b_1 = 0.0$, and

$$b'_2 = b_2 - (a'_{2,1}/a'_{1,2})b_1 = 0.0 - [1.0/(-2.25)](0.0) = 0.0 \quad (1.156)$$

The remaining elements of $\mathbf{b}$ are processed in the same manner. The results are presented in Eq. (1.157). For this particular $\mathbf{b}$ vector, where elements b_1 to b_{n-1} are all zero, the $\mathbf{b}$ vector does not change. This is certainly not the case in general. The final result is:

$$\mathbf{A}' = \begin{bmatrix} - & -2.250000 & 1.0 \\ (-0.444444) & -1.805556 & 1.0 \\ (-0.553846) & -1.696154 & 1.0 \\ (-0.589569) & -1.660431 & 1.0 \\ (-0.602253) & -1.647747 & 1.0 \\ (-0.606889) & -1.643111 & 1.0 \\ (-0.608602) & -1.641398 & - \end{bmatrix} \quad \text{and} \quad \mathbf{b}' = \begin{bmatrix} 0.0 \\ 0.0 \\ 0.0 \\ 0.0 \\ 0.0 \\ 0.0 \\ -100.0 \end{bmatrix} \quad (1.157)$$

The solution vector is computed using Eq. (1.152). Thus,

$$x_7 = b_7/a'_{7,2} = (-100)/(-1.641398) = 60.923667 \quad (1.158a)$$

$$x_6 = (b_6 - a'_{6,3}x_7)/a'_{6,2} = [0 - (1.0)(60.923667)]/(-1.643111)$$

$$= 37.078251 \quad (1.158b)$$

Processing the remaining rows yields the solution vector:

$$\mathbf{x} = \begin{bmatrix} 1.966751 \\ 4.425190 \\ 7.989926 \\ 13.552144 \\ 22.502398 \\ 37.078251 \\ 60.923667 \end{bmatrix} \quad (1.158c)$$

Equation (1.158c) is the solution presented in Table 8.9.

Pivoting destroys the tridiagonality of the system of linear algebraic equations, and thus cannot be used with the Thomas algorithm. Most large tridiagonal systems which represent real physical problems are diagonally dominant, so pivoting is not necessary.

The number of multiplicative operations required by the elimination step is $N = 2n - 3$ and the number of multiplicative operations required by the back substitution step is $N = 3n - 2$. Thus, the total number of multiplicative operations is $N = 5n - 4$ for the complete Thomas algorithm. If the $\mathbf{T}$ matrix is constant and multiple $\mathbf{b}$ vectors are to be considered, only the back substitution step is required once the $\mathbf{T}$ matrix has been factored into $\mathbf{L}$ and $\mathbf{U}$ matrices. In that case, $N = 3n - 2$ for subsequent $\mathbf{b}$ vectors. The advantages of the Thomas algorithm are quite apparent when compared with either the Gauss elimination method, for which $N = (n^3/3 - n/3) + n^2$, or the Doolittle LU method, for which $N = n^2 - n/2$, for each **b** vector after the first one. The Thomas algorithm, in a format suitable for programming for a computer, is summarized as follows:

1. Store the $n \times n$ tridiagonal matrix **T** in the $n \times 3$ matrix **A'**. The right-hand-side vector **b** is an $n \times 1$ column vector.
2. Compute the $a'_{i,2}$ terms from Eq. (1.150). Store the elimination multipliers, $em = a'_{i,1}/a'_{i-1,2}$, in place of $a'_{i,1}$.
3. Compute the b_i terms from Eq. (1.151).
4. Solve for x_i by back substitution using Eq. (1.152).

An extended form of the Thomas algorithm can be applied to *block tridiagonal matrices*, in which the elements of **T** are partitioned into submatrices having similar patterns. The solution procedure is analogous to that just presented for scalar elements, except that matrix operations are employed on the submatrix elements.

An algorithm similar to the Thomas algorithm can be developed for other special types of systems of linear algebraic equations. For example, a pentadiagonal system of linear algebraic equations is illustrated in Example 8.6.

1.6. PITFALLS OF ELIMINATION METHODS

All nonsingular systems of linear algebraic equations have a solution. In theory, the solution can always be obtained by Gauss elimination. However, there are two major pitfalls in the application of Gauss elimination (or its variations): (a) the presence of round-off errors, and (b) ill-conditioned systems. Those pitfalls are discussed in this section. The effects of round-off can be reduced by a procedure known as iterative improvement, which is presented at the end of this section.

1.6.1. Round-Off Errors

Round-off errors occur when exact infinite precision numbers are approximated by finite precision numbers. In most computers, single precision representation of numbers typically contains about 7 significant digits, double precision representation typically contains about 14 significant digits, and quad precision representation typically contains about 28 significant digits. The effects of round-off errors are illustrated in the following example.

Example 1.18. Effects of round-off errors.

Consider the following system of linear algebraic equations:

$$0.0003x_1 + 3x_2 = 1.0002 \tag{1.159a}$$

$$x_1 + x_2 = 1 \tag{1.159b}$$

Solve Eq. (1.159) by Gauss elimination. Thus,

$$\begin{bmatrix} 0.0003 & 3 \mid 1.0002 \\ 1 & 1 \mid 1 \end{bmatrix} R_2 - R_1/0.0003 \tag{1.160a}$$

$$\begin{bmatrix} 0.0003 & 3 \mid & 1.0002 \\ 0 & -9999 \mid & 1 - \dfrac{1.0002}{0.0003} \end{bmatrix} \tag{1.160b}$$

Table 1.1. Solution of Eq. (1.162)

Precision	x_2	x_1
3	0.333	3.33
4	0.3332	1.333
5	0.33333	0.70000
6	0.333333	0.670000
7	0.3333333	0.6670000
8	0.33333333	0.66670000

The exact solution of Eq. (1.160) is

$$x_2 = \frac{1 - \dfrac{1.0002}{0.0003}}{-9999} = \frac{\dfrac{0.0003 - 1.0002}{0.0003}}{-9999} = \frac{\dfrac{-0.9999}{0.0003}}{-9999} = \frac{1}{3} \quad (1.161a)$$

$$x_1 = \frac{1.0002 - 3x_2}{0.0003} = \frac{1.0002 - 3(1/3)}{0.0003} = \frac{0.0002}{0.0003} = \frac{2}{3} \quad (1.161b)$$

Let's solve Eq. (1.161) using finite precision arithmetic with two to eight significant figures. Thus,

$$x_2 = \frac{1 - \dfrac{1.0002}{0.0003}}{-9999} \quad \text{and} \quad x_1 = \frac{1.0002 - 3x_2}{0.0003} \quad (1.162)$$

The results are presented in Table 1.1. The algorithm is clearly performing very poorly.

Let's rework the problem by interchanging rows 1 and 2 in Eq. (1.159). Thus,

$$x_1 + x_2 = 1 \quad (1.163a)$$

$$0.0003x_1 + 3x_2 = 1.0002 \quad (1.163b)$$

Solve Eq. (1.163) by Gauss elimination. Thus,

$$\begin{bmatrix} 1 & 1 & | & 1 \\ 0.0003 & 3 & | & 1.0002 \end{bmatrix} R_2 - 0.0003R_1 \quad (1.164a)$$

$$\begin{bmatrix} 1 & 1 & | & 1 \\ 0 & 2.9997 & | & 0.9999 \end{bmatrix} \quad (1.164b)$$

$$x_2 = \frac{0.9999}{2.9997} \quad \text{and} \quad x_1 = 1 - x_2 \quad (1.164c)$$

Let's solve Eq. (1.164c) using finite precision arithmetic. The results are presented in Table 1.2. These results clearly demonstrate the benefits of pivoting.

Table 1.2. Solution of Eq. (1.164c)

Precision	x_2	x_1
3	0.333	0.667
4	0.3333	0.6667
5	0.33333	0.66667

Round-off errors can never be completely eliminated. However, they can be minimized by using high precision arithmetic and pivoting.

1.6.2. System Condition

All well-posed nonsingular numerical problems have an exact solution. In theory, the exact solution can always be obtained using fractions or infinite precision numbers (i.e., an infinite number of significant digits). However, all practical calculations are done with finite precision numbers which necessarily contain round-off errors. The presence of round-off errors alters the solution of the problem.

A *well-conditioned* problem is one in which a small change in any of the elements of the problem causes only a small change in the solution of the problem.

An *ill-conditioned* problem is one in which a small change in any of the elements of the problem causes a large change in the solution of the problem. Since ill-conditioned systems are extremely sensitive to small changes in the elements of the problem, they are also extremely sensitive to round-off errors.

Example 1.19. System condition.

Let's illustrate the behavior of an ill-conditioned system by the following problem:

$$x_1 + x_2 = 2 \tag{1.165a}$$
$$x_1 + 1.0001x_2 = 2.0001 \tag{1.165b}$$

Solve Eq. (1.165) by Gauss elimination. Thus,

$$\begin{bmatrix} 1 & 1 & | \ 2 \\ 1 & 1.0001 & | \ 2.0001 \end{bmatrix} R_2 - R_1 \tag{1.166a}$$

$$\begin{bmatrix} 1 & 1 & | \ 2 \\ 0 & 0.0001 & | \ 0.0001 \end{bmatrix} \tag{1.166b}$$

Solving Eq. (1.166b) yields $x_2 = 1$ and $x_1 = 1$.

Consider the following slightly modified form of Eq. (1.165) in which a_{22} is changed slightly from 1.0001 to 0.9999:

$$x_1 + x_2 = 2 \tag{1.167a}$$
$$x_1 + 0.9999x_2 = 2.0001 \tag{1.167b}$$

Solving Eq. (1.167) by Gauss elimination gives

$$\begin{bmatrix} 1 & 1 & | \ 2 \\ 1 & 0.9999 & | \ 2.0001 \end{bmatrix} R_2 - R_1 \tag{1.168a}$$

$$\begin{bmatrix} 1 & 1 & | \ 2 \\ 0 & -0.0001 & | \ 0.0001 \end{bmatrix} \tag{1.168b}$$

Solving Eq. (1.168b) yields $x_2 = -1$ and $x_1 = 3$, which is greatly different from the solution of Eq. (1.165).

Consider another slightly modified form of Eq. (1.165) in which b_2 is changed slightly from 2.0001 to 2:

$$x_1 + x_2 = 2 \tag{1.169a}$$
$$x_1 + 1.0001x_2 = 2 \tag{1.169b}$$

Solving Eq. (1.169) by Gauss elimination gives

$$\begin{bmatrix} 1 & 1 & | \ 2 \\ 1 & 1.0001 & | \ 2 \end{bmatrix} R_2 - R_1 \tag{1.170a}$$

$$\begin{bmatrix} 1 & 1 & | \ 2 \\ 0 & 0.0001 & | \ 0 \end{bmatrix} \tag{1.170b}$$

Solving Eq. (1.170) yields $x_2 = 0$ and $x_1 = 2$, which is greatly different from the solution of Eq. (1.165).

This problem illustrates that very small changes in any of the elements of **A** or **b** can cause extremely large changes in the solution, **x**. Such a system is ill-conditioned.

With infinite precision arithmetic, ill-conditioning is not a problem. However, with finite precision arithmetic, round-off errors effectively change the elements of **A** and **b** slightly, and if the system is ill-conditioned, large changes (i.e., errors) can occur in the solution. Assuming that scaled pivoting has been performed, the only possible remedy to ill-conditioning is to use higher precision arithmetic.

There are several ways to check a matrix **A** for ill-conditioning. If the magnitude of the determinant of the matrix is small, the matrix may be ill-conditioned. However, this is not a foolproof test. The inverse matrix $\mathbf{A}^{-1}$ can be calculated, and $\mathbf{AA}^{-1}$ can be computed and compared to **I**. Similarly, $(\mathbf{A}^{-1})^{-1}$ can be computed and compared to **A**. A close comparison in either case suggests that matrix **A** is well-conditioned. A poor comparison suggests that the matrix is ill-conditioned. Some of the elements of **A** and/or **b** can be changed slightly, and the solution repeated. If a drastically different solution is obtained, the matrix is probably ill-conditioned. None of these approaches is foolproof, and none give a quantitative measure of ill-conditioning. The surest way to detect ill-conditioning is to evaluate the condition number of the matrix, as discussed in the next subsection.

1.6.3. Norms and the Condition Number

The problems associated with an ill-conditioned system of linear algebraic equations are illustrated in the previous discussion. In the following discussion, ill-conditioning is quantified by the *condition number* of a matrix, which is defined in terms of the *norms* of the matrix and its inverse. Norms and the condition number are discussed in this section.

1.6.3.1. Norms

The measure of the magnitude of $\mathbf{A}$, $\mathbf{x}$, or $\mathbf{b}$ is called its *norm* and denoted by $\|\mathbf{A}\|$, $\|\mathbf{x}\|$, and $\|\mathbf{b}\|$, respectively. Norms have the following properties:

$$\|\mathbf{A}\| > 0 \tag{1.171a}$$

$$\|\mathbf{A}\| = 0 \qquad \text{only if } \mathbf{A} = \mathbf{0} \tag{1.171b}$$

$$\|k\mathbf{A}\| = |k|\|\mathbf{A}\| \tag{1.171c}$$

$$\|\mathbf{A} + \mathbf{B}\| \leq \|\mathbf{A}\| + \|\mathbf{B}\| \tag{1.171d}$$

$$\|\mathbf{AB}\| \leq \|\mathbf{A}\|\|\mathbf{B}\| \tag{1.171e}$$

The norm of a scalar is its absolute value. Thus, $\|k\| = |k|$. There are several definitions of the norm of a vector. Thus,

$$\|\mathbf{x}\|_1 = \sum |x_i| \qquad \text{Sum of magnitudes} \tag{1.172a}$$

$$\|\mathbf{x}\|_2 = \|\mathbf{x}\|_e = \left(\sum x_i^2\right)^{1/2} \qquad \text{Euclidean norm} \tag{1.172b}$$

$$\|\mathbf{x}\|_\infty = \max_{1 \leq i \leq n} |x_i| \qquad \text{Maximum magnitude norm} \tag{1.172c}$$

The *Euclidean* norm is the length of the vector in n-space.

In a similar manner, there are several definitions of the norm of a matrix. Thus,

$$\|\mathbf{A}\|_1 = \max_{1 \leq j \leq n} \sum_{i=1}^{n} |a_{i,j}| \qquad \text{Maximum column sum} \tag{1.173a}$$

$$\|\mathbf{A}\|_\infty = \max_{1 \leq i \leq n} \sum_{j=1}^{n} |a_{i,j}| \qquad \text{Maximum row sum} \tag{1.173b}$$

$$\|\mathbf{A}\|_2 = \min \lambda_i \qquad \text{(eigenvalue)} \qquad \text{Spectral norm} \tag{1.173c}$$

$$\|\mathbf{A}\|_e = \left(\sum_{i=1}^{n} \sum_{j=1}^{n} a_{i,j}^2\right)^{1/2} \qquad \text{Euclidean norm} \tag{1.173d}$$

1.6.3.2. Condition Number

The *condition number* of a system is a measure of the sensitivity of the system to small changes in any of its elements. Consider a system of linear algebraic equations:

$$\mathbf{Ax} = \mathbf{b} \tag{1.174}$$

For Eq. (1.174),

$$\|\mathbf{b}\| \leq \|\mathbf{A}\|\|\mathbf{x}\| \tag{1.175}$$

Consider a slightly modified form of Eq. (1.174) in which $\mathbf{b}$ is altered by $\delta\mathbf{b}$, which causes a change in the solution $\delta\mathbf{x}$. Thus,

$$\mathbf{A}(\mathbf{x} + \delta\mathbf{x}) = \mathbf{b} + \delta\mathbf{b} \tag{1.176}$$

Subtracting Eq. (1.174) from Eq. (1.176) gives

$$\mathbf{A}\,\delta\mathbf{x} = \delta\mathbf{b} \tag{1.177}$$

Solving Eq. (1.177) for $\delta\mathbf{x}$ gives

$$\delta\mathbf{x} = \mathbf{A}^{-1}\,\delta\mathbf{b} \tag{1.178}$$

For Eq. (1.178),

$$\|\delta\mathbf{x}\| \le \|\mathbf{A}^{-1}\|\|\delta\mathbf{b}\|$$ (1.179)

Multiplying the left-hand and right-hand sides of Eqs. (1.175) and (1.179) gives

$$\|\mathbf{b}\|\|\delta\mathbf{x}\| \le \|\mathbf{A}\|\|\mathbf{x}\|\|\mathbf{A}^{-1}\|\|\delta\mathbf{b}\|$$ (1.180)

Dividing Eqs. (1.180) by $\|\mathbf{b}\|\|\mathbf{x}\|$ yields

$$\frac{\|\delta\mathbf{x}\|}{\|\mathbf{x}\|} \le \|\mathbf{A}\|\|\mathbf{A}^{-1}\|\frac{\|\delta\mathbf{b}\|}{\|\mathbf{b}\|} = C(\mathbf{A})\frac{\|\delta\mathbf{b}\|}{\|\mathbf{b}\|}$$ (1.181)

where $C(\mathbf{A})$ is the *condition number* of matrix $\mathbf{A}$:

$$C(\mathbf{A}) = \|\mathbf{A}\|\|\mathbf{A}^{-1}\|$$ (1.182)

Equation (1.182) determines the sensitivity of the solution, $\|\delta\mathbf{x}\|/\|\mathbf{x}\|$, to changes in the vector $\mathbf{b}$, $\|\delta\mathbf{b}\|/\|\mathbf{b}\|$. The sensitivity is determined directly by the value of the condition number $C(\mathbf{A})$. Small values of $C(\mathbf{A})$, of the order of unity, show a small sensitivity of the solution to changes in $\mathbf{b}$. Such a problem is well-conditioned. Large values of $C(\mathbf{A})$ show a large sensitivity of the solution to changes in $\mathbf{b}$. Such a problem is ill-conditioned.

It can be shown by a similar analysis that perturbing the matrix $\mathbf{A}$ instead of the vector $\mathbf{b}$ gives

$$\frac{\|\delta\mathbf{x}\|}{\|\mathbf{x} + \delta\mathbf{x}\|} \le C(\mathbf{A})\frac{\|\delta\mathbf{A}\|}{\|\mathbf{A}\|}$$ (1.183)

The use of the condition number is illustrated in Example 1.20.

Example 1.20.   Norms and condition numbers.

Consider the coefficient matrix of Eq. (1.159):

$$\mathbf{A} = \begin{bmatrix} 0.0003 & 3 \\ 1 & 1 \end{bmatrix}$$ (1.184)

The Euclidian norm of matrix $\mathbf{A}$ is

$$\|\mathbf{A}\|_e = [(0.0003)^2 + 3^2 + 1^2 + 1^2]^{1/2} = 3.3166$$ (1.185)

The inverse of matrix $\mathbf{A}$ is

$$\mathbf{A}^{-1} = \begin{bmatrix} -\dfrac{1}{(0.0003)9{,}999} & \dfrac{10{,}000}{9{,}999} \\ \dfrac{1}{(0.0003)9{,}999} & -\dfrac{1}{9{,}999} \end{bmatrix}$$ (1.186)

The Euclidian norm of $\mathbf{A}^{-1}$ is $\|\mathbf{A}^{-1}\|_e = 1.1057$. Thus, the condition number of matrix $\mathbf{A}$ is

$$C(\mathbf{A}) = \|\mathbf{A}\|_e\|\mathbf{A}^{-1}\|_e = 3.3166(1.1057) = 3.6672$$ (1.187)

This relatively small condition number shows that matrix **A** is well-conditioned. As shown in Section 1.6.1, Eq. (1.159) is sensitive to the precision of the arithmetic (i.e., round-off effects), even though it is well-conditioned. This is a precision problem, not a condition problem.

Consider the coefficient matrix of Eq. (1.165):

$$\mathbf{A} = \begin{bmatrix} 1 & 1 \\ 1 & 1.0001 \end{bmatrix} \tag{1.188}$$

The Euclidean norm of matrix **A** is $\|\mathbf{A}\|_e = 2.00005$. The inverse of matrix **A** is

$$\mathbf{A}^{-1} = \begin{bmatrix} 10{,}001 & -10{,}000 \\ -10{,}000 & 10{,}000 \end{bmatrix} \tag{1.189}$$

The Euclidean norm of $\mathbf{A}^{-1}$ is $\|\mathbf{A}^{-1}\|_e = 20{,}000.5$. Thus, the condition number of matrix **A** is

$$C(\mathbf{A}) = \|\mathbf{A}\|_e \|\mathbf{A}^{-1}\|_e = (2.00005)20{,}000.5 = 40{,}002.0 \tag{1.190}$$

This large condition number shows that matrix **A** is ill-conditioned.

1.6.4. Iterative Improvement

In all direct elimination methods, the effects of round-off propagate as the solution progresses through the system of equations. The accumulated effect of round-off is *round-off error* in the computed values. Round-off errors in any calculation can be decreased by using higher precision (i.e., more significant digits) arithmetic. Round-off errors in direct elimination methods of solving systems of linear algebraic equations are minimized by using scaled pivoting. Further reduction in round-off errors can be achieved by a procedure known as *iterative improvement*.

Consider a system of linear algebraic equations:

$$\mathbf{Ax} = \mathbf{b} \tag{1.191}$$

Solving Eq. (1.191) by a direct elimination method yields $\tilde{\mathbf{x}}$, where $\tilde{\mathbf{x}}$ differs from the exact solution **x** by the error $\delta\mathbf{x}$, where $\tilde{\mathbf{x}} = \mathbf{x} + \delta\mathbf{x}$. Substituting $\tilde{\mathbf{x}}$ into Eq. (1.191) gives

$$\mathbf{A}\tilde{\mathbf{x}} = \mathbf{A}(\mathbf{x} + \delta\mathbf{x}) = \mathbf{Ax} + \mathbf{A}\,\delta\mathbf{x} = \mathbf{b} + \delta\mathbf{b} \tag{1.192}$$

From the first and last terms in Eq. (1.192), $\delta\mathbf{b}$, is given by

$$\delta\mathbf{b} = \mathbf{A}\tilde{\mathbf{x}} - \mathbf{b} \tag{1.193}$$

Subtracting $\mathbf{A}\tilde{\mathbf{x}} = \mathbf{A}(\mathbf{x} + \delta\mathbf{x}) = \mathbf{b} + \mathbf{A}\,\delta\mathbf{x}$ into Eq. (1.193) gives a system of linear algebraic equations for $\delta\mathbf{x}$. Thus,

$$\boxed{\mathbf{A}\,\delta\mathbf{x} = \delta\mathbf{b}} \tag{1.194}$$

Equation (1.194) can be solved for $\delta\mathbf{x}$, which can be added to $\tilde{\mathbf{x}}$ to give an improved approximation to **x**. The procedure can be repeated (i.e., iterated) if necessary. A convergence check on the value of $\delta\mathbf{x}$ can be used to determine if the procedure should be repeated. If the procedure is iterated, LU factorization should be used to reduce the computational effort since matrix $\mathbf{A}$ is constant. Equation (1.194) should be solved with higher precision than the precision used in the solution of Eq. (1.191).

1.7 ITERATIVE METHODS

For many large systems of linear algebraic equations, $\mathbf{Ax} = \mathbf{b}$, the coefficient matrix $\mathbf{A}$ is extremely sparse. That is, most of the elements of $\mathbf{A}$ are zero. If the matrix is diagonally dominant [see Eq. (1.15)], it is generally more efficient to solve such systems of linear algebraic equations by iterative methods than by direct elimination methods. Three iterative methods are presented in this section: *Jacobi iteration*, *Gauss-Seidel iteration*, and *successive-over-relaxation (SOR)*.

Iterative methods begin by assuming an initial solution vector $\mathbf{x}^{(0)}$. The initial solution vector is used to generate an improved solution vector $\mathbf{x}^{(1)}$ based on some strategy for reducing the difference between $\mathbf{x}^{(0)}$ and the actual solution vector $\mathbf{x}$. This procedure is repeated (i.e., iterated) to convergence. The procedure is convergent if each iteration produces approximations to the solution vector that approach the exact solution vector as the number of iterations increases.

Iterative methods do not converge for all sets of equations, nor for all possible arrangements of a particular set of equations. *Diagonal dominance* is a sufficient condition for convergence of Jacobi iteration, Gauss-Seidel iteration, and SOR, for any initial solution vector. Diagonal dominance is defined by Eq. (1.15). Some systems that are not diagonally dominant can be rearranged (i.e., by row interchanges) to make them diagonally dominant. Some systems that are not diagonally dominant may converge for certain initial solution vectors, but convergence is not assured. Iterative methods should not be used for systems of linear algebraic equations that cannot be made diagonally dominant.

When repeated application of an iterative method produces insignificant changes in the solution vector, the procedure should be terminated. In other words, the algorithm is repeated (iterated) until some specified convergence criterion is achieved. Convergence is achieved when some measure of the relative or absolute change in the solution vector is less than a specified convergence criterion. The number of iterations required to achieve convergence depends on:

1. The dominance of the diagonal coefficients. As the diagonal dominance increases, the number of iterations required to satisfy the convergence criterion decreases.
2. The method of iteration used.
3. The initial solution vector.
4. The convergence criterion specified.

1.7.1. The Jacobi Iteration Method

Consider the general system of linear algebraic equations, $\mathbf{Ax} = \mathbf{b}$, written in index notation:

$$\sum_{j=1}^{n} a_{i,j} x_j = b_i \qquad (i = 1, 2, \ldots, n) \tag{1.195}$$

In Jacobi iteration, each equation of the system is solved for the component of the solution vector associated with the diagonal element, that is, x_i. Thus,

$$x_i = \frac{1}{a_{i,i}} \left(b_i - \sum_{j=1}^{i-1} a_{i,j} x_j - \sum_{j=i+1}^{n} a_{i,j} x_j \right) \qquad (i = 1, 2, \ldots, n) \tag{1.196}$$

An initial solution vector $\mathbf{x}^{(0)}$ is chosen. The superscript in parentheses denotes the iteration number, with zero denoting the initial solution vector. The initial solution vector $\mathbf{x}^{(0)}$ is substituted into Eq. (1.196) to yield the first improved solution vector $\mathbf{x}^{(1)}$. Thus,

$$x_i^{(1)} = \frac{1}{a_{i,i}} \left(b_i - \sum_{j=1}^{i-1} a_{i,j} x_j^{(0)} - \sum_{j=i+1}^{n} a_{i,j} x_j^{(0)} \right) \qquad (i = 1, 2, \ldots, n) \tag{1.197}$$

This procedure is repeated (i.e., iterated) until some convergence criterion is satisfied. The Jacobi algorithm for the general iteration step (k) is:

$$x_i^{(k+1)} = \frac{1}{a_{i,i}} \left(b_i - \sum_{j=1}^{i-1} a_{i,j} x_j^{(k)} - \sum_{j=i+1}^{n} a_{i,j} x_j^{(k)} \right) \qquad (i = 1, 2, \ldots, n) \tag{1.198}$$

An equivalent, but more convenient, form of Eq. (1.198) can be obtained by adding and subtracting $x_i^{(k)}$ from the right-hand side of Eq. (1.198) to yield

$$x_i^{(k+1)} = x_i^{(k)} + \frac{1}{a_{i,i}} \left(b_i - \sum_{j=1}^{n} a_{i,j} x_j^{(k)} \right) \qquad (i = 1, 2, \ldots, n) \tag{1.199}$$

Equation (1.199) is generally written in the form

$$x_i^{(k+1)} = x_i^{(k)} + \frac{R_i^{(k)}}{a_{i,i}} \qquad (i = 1, 2, \ldots, n) \tag{1.200a}$$

$$R_i^{(k)} = b_i - \sum_{j=1}^{n} a_{i,j} x_j^{(k)} \qquad (i = 1, 2, \ldots, n) \tag{1.200b}$$

where the term $R_i^{(k)}$ is called the *residual* of equation i. The residuals $R_i^{(k)}$ are simply the net values of the equations evaluated for the approximate solution vector $\mathbf{x}^{(k)}$.

The Jacobi method is sometimes called the method of simultaneous iteration because all values of x_i are iterated simultaneously. That is, all values of $x_i^{(k+1)}$ depend only on the values of $x_i^{(k)}$. The order of processing the equations is immaterial.

Example 1.21.   The Jacobi iteration method.

To illustrate the Jacobi iteration method, let's solve the following system of linear algebraic equations:

$$\begin{bmatrix} 4 & -1 & 0 & 1 & 0 \\ -1 & 4 & -1 & 0 & 1 \\ 0 & -1 & 4 & -1 & 0 \\ 1 & 0 & -1 & 4 & -1 \\ 0 & 1 & 0 & -1 & 4 \end{bmatrix} \begin{bmatrix} x_1 \\ x_2 \\ x_3 \\ x_4 \\ x_5 \end{bmatrix} = \begin{bmatrix} 100 \\ 100 \\ 100 \\ 100 \\ 100 \end{bmatrix} \tag{1.201}$$

Table 1.3. Solution by the Jacobi Iteration Method

k	x_1	x_2	x_3	x_4	x_5
0	0.000000	0.000000	0.000000	0.000000	0.000000
1	25.000000	25.000000	25.000000	25.000000	25.000000
2	25.000000	31.250000	37.500000	31.250000	25.000000
3	25.000000	34.375000	40.625000	34.375000	25.000000
4	25.000000	35.156250	42.187500	35.156250	25.000000
5	25.000000	35.546875	42.578125	35.546875	25.000000
...					
16	25.000000	35.714284	42.857140	35.714284	25.000000
17	25.000000	35.714285	42.857142	35.714285	25.000000
18	25.000000	35.714285	42.857143	35.714285	25.000000

Equation (1.201), when expanded, becomes

$$4x_1 - x_2 + x_4 = 100 \tag{1.202.1}$$

$$-x_1 + 4x_2 - x_3 + x_5 = 100 \tag{1.202.2}$$

$$-x_2 + 4x_3 - x_4 = 100 \tag{1.202.3}$$

$$x_1 - x_3 + 4x_4 - x_5 = 100 \tag{1.202.4}$$

$$x_2 - x_4 + 4x_5 = 100 \tag{1.202.5}$$

Equation (1.202) can be rearranged to yield expressions for the residuals, R_i. Thus,

$$R_1 = 100 - 4x_1 + x_2 - x_4 \tag{1.203.1}$$

$$R_2 = 100 + x_1 - 4x_2 + x_3 - x_5 \tag{1.203.2}$$

$$R_3 = 100 + x_2 - 4x_3 + x_4 \tag{1.203.3}$$

$$R_4 = 100 - x_1 + x_3 - 4x_4 + x_5 \tag{1.203.4}$$

$$R_5 = 100 - x_2 + x_4 - 4x_5 \tag{1.203.5}$$

To initiate the solution, let $\mathbf{x}^{(0)T} = [0.0 \quad 0.0 \quad 0.0 \quad 0.0 \quad 0.0]$. Substituting these values into Eq. (1.203) gives $R_i^{(0)} = 100.0$ ($i = 1, \dots, 5$). Substituting these values into Eq. (1.200a) gives $x_1^{(1)} = x_2^{(1)} = x_3^{(1)} = x_4^{(1)} = x_5^{(1)} = 25.0$. The procedure is then repeated with these values to obtain $\mathbf{x}^{(2)}$, etc.

The first and subsequent iterations are summarized in Table 1.3. Due to the symmetry of the coefficient matrix $\mathbf{A}$ and the symmetry of the $\mathbf{b}$ vector, $x_1 = x_5$ and $x_2 = x_4$. The calculations were carried out on a 13-digit precision computer and iterated until all $|\Delta x_i|$ changed by less than 0.000001 between iterations, which required 18 iterations.

1.7.2. Accuracy and Convergence of Iterative Methods

All nonsingular systems of linear algebraic equations have an exact solution. In principle, when solved by direct methods, the exact solution can be obtained. However, all real calculations are performed with finite precision numbers, so round-off errors pollute the solution. Round-off errors can be minimized by pivoting, but even the most careful calculations are subject to the round-off characteristics of the computing device (i.e., hand computation, hand calculator, personal computer, work station, or mainframe computer).

Iterative methods are less susceptable to round-off errors than direct elimination methods for three reasons: (a) The system of equations is diagonally dominant, (b) the system of equations is typically sparse, and (c) each iteration through the system of equations is independent of the round-off errors of the previous iteration.

When solved by iterative methods, the exact solution of a system of linear algebraic equations is approached asymptotically as the number of iterations increases. When the number of iterations increases without bound, the numerical solution yields the exact solution within the round-off limit of the computing device. Such solutions are said to be correct to machine accuracy. In most practical solutions, machine accuracy is not required. Thus, the iterative process should be terminated when some type of accuracy criterion (or criteria) has been satisfied. In iterative methods, the term *accuracy* refers to the number of significant figures obtained in the calculations, and the term *convergence* refers to the point in the iterative process when the desired accuracy is obtained.

1.7.2.1. Accuracy

The *accuracy* of any approximate method is measured in terms of the *error* of the method. There are two ways to specify error: *absolute error* and *relative error*. Absolute error is defined as

$$\text{Absolute error} = \text{approximate value} - \text{exact value} \qquad (1.204)$$

and relative error is defined as

$$\text{Relative error} = \frac{\text{absolute error}}{\text{exact value}} \qquad (1.205)$$

Relative error can be stated directly or as a percentage.

Consider an iterative calculation for which the desired absolute error is ± 0.001. If the exact solution is 100.000, then the approximate value is 100.000 ± 0.001, which has five significant digits. However, if the exact solution is 0.001000, then the approximate value is 0.001000 ± 0.001, which has no significant digits. This example illustrates the danger of using absolute error as an accuracy criterion. When the magnitude of the exact solution is known, an absolute accuracy criterion can be specified to yield a specified number of significant digits in the approximate solution. Otherwise, a relative accuracy criterion is preferable.

Consider an iterative calculation for which the desired relative error is ± 0.00001. If the exact solution is 100.000, then the absolute error must be $100.000 \times (\pm 0.00001) = \pm 0.001$ to satisfy the relative error criterion. This yields five significant digits in the approximate value. If the exact solution is 0.001000, then the absolute error must be $0.001000 \times (\pm 0.00001) = \pm 0.00000001$ to satisfy the relative error criterion. This yields five significant digits in the approximate solution. A relative error criterion yields the same number of significant figures in the approximate value, regardless of the magnitude of the exact solution.

1.7.2.2. Convergence

Convergence of an iterative procedure is achieved when the desired accuracy criterion (or criteria) is satisfied. Convergence criteria can be specified in terms of absolute error or relative error. Since the exact solution is unknown, the error at any step in the iterative process is based on the change in the quantity being calculated from one step to the next. Thus, for the iterative solution of a system of linear algebraic equations, the error, $\Delta x_i = x_i^{(k+1)} - x_i^{\text{exact}}$, is approximated by $x_i^{(k+1)} - x_i^{(k)}$. The error can also be specified by the magnitudes of the residuals R_i. When the exact answer (or the exact answer to machine accuracy) is obtained, the residuals are all zero. At each step in the iterative procedure, some of the residuals may be near zero while others are still quite large. Therefore, care is needed to ensure that the desired accuracy of the complete system of equations is achieved.

Let ε be the magnitude of the convergence tolerance. Several convergence criteria are possible. For an absolute error criterion, the following choices are possible:

$$\left|(\Delta x_i)_{\max}\right| \le \varepsilon \qquad \sum_{i=1}^{n} \left|\Delta x_i\right| \le \varepsilon \qquad \text{or} \qquad \left[\sum_{i=1}^{n}(\Delta x_i)^2\right]^{1/2} \le \varepsilon \tag{1.206}$$

For a relative error criterion, the following choices are possible:

$$\left|\frac{(\Delta x_i)_{\max}}{x_i}\right| \le \varepsilon \qquad \sum_{i=1}^{n}\left|\frac{\Delta x_i}{x_i}\right| \le \varepsilon \qquad \text{or} \qquad \left[\sum_{i=1}^{n}\left(\frac{\Delta x_i}{x_i}\right)^2\right]^{1/2} \le \varepsilon \tag{1.207}$$

The concepts of accuracy and convergence discussed in this section apply to all iterative procedures, not just the iterative solution of a system of linear algebraic equations. They are relevant to the solution of eigenvalue problems (Chapter 2), to the solution of nonlinear equations (Chapter 3), etc.

1.7.3. The Gauss-Seidel Iteration Method

In the Jacobi method, all values of $\mathbf{x}^{(k+1)}$ are based on $\mathbf{x}^{(k)}$. The Gauss-Seidel method is similar to the Jacobi method, except that the most recently computed values of all x_i are used in all computations. In brief, as better values of x_i are obtained, use them immediately. Like the Jacobi method, the Gauss-Seidel method requires diagonal dominance to ensure convergence. The Gauss-Seidel algorithm is obtained from the Jacobi algorithm, Eq. (1.198), by using $x_j^{(k+1)}$ values in the summation from $j = 1$ to $i - 1$ (assuming the sweeps through the equations proceed from $i = 1$ to n). Thus,

$$x_i^{(k+1)} = \frac{1}{a_{i,i}}\left(b_i - \sum_{j=1}^{i-1} a_{i,j}x_j^{(k+1)} - \sum_{j=i+1}^{n} a_{i,j}x_j^{(k)}\right) \qquad (i = 1, 2, \ldots, n) \tag{1.208}$$

Equation (1.208) can be written in terms of the residuals R_i by adding and subtracting $x_i^{(k)}$ from the right-hand side of the equation and rearranging to yield

$$x_i^{(k+1)} = x_i^{(k)} + \frac{R_i^{(k)}}{a_{i,i}} \qquad (i = 1, 2, \ldots, n) \tag{1.209}$$

$$R_i^{(k)} = b_i - \sum_{j=1}^{i-1} a_{i,j}x_j^{(k+1)} - \sum_{j=i}^{n} a_{i,j}x_j^{(k)} \qquad (i = 1, 2, \ldots, n) \tag{1.210}$$

The Gauss-Seidel method is sometimes called the method of successive iteration because the most recent values of all x_i are used in all the calculations. Gauss-Seidel iteration generally converges faster than Jacobi iteration.

Table 1.4.   Solution by the Gauss-Seidel Iteration Method

k	x_1	x_2	x_3	x_4	x_5
0	0.000000	0.000000	0.000000	0.000000	
1	25.000000	31.250000	32.812500	26.953125	23.925781
2	26.074219	33.740234	40.173340	34.506226	25.191498
3	24.808502	34.947586	42.363453	35.686612	25.184757
4	24.815243	35.498485	42.796274	35.791447	25.073240
5	24.926760	35.662448	42.863474	35.752489	25.022510
...					
13	25.000002	35.714287	42.857142	35.714285	25.999999
14	25.000001	35.714286	42.857143	35.714285	25.000000
15	25.000000	35.714286	42.857143	35.714286	25.000000

Example 1.22.   The Gauss-Seidel iteration method.

Let's rework the problem presented in Example 1.21 using Gauss-Seidel iteration. The residuals are given by Eq. (1.210). Substituting the initial solution vector, $\mathbf{x}^{(0)T} = [0.0 \quad 0.0 \quad 0.0 \quad 0.0 \quad 0.0]$, into Eq. (1.210.1) gives $R_1^{(0)} = 100.0$. Substituting that result into Eq. (1.209.1) gives $x_1^{(1)} = 25.0$. Substituting $\mathbf{x}^T = [25.0 \quad 0.0 \quad 0.0 \quad 0.0 \quad 0.0]$ into Eq. (1.210.2) gives

$$R_2^{(1)} = (100.0 + 25.0) = 125.0 \qquad (1.211a)$$

Substituting this result into Eq. (1.209.2) yields

$$x_2^{(1)} = 0.0 + \frac{125.0}{4} = 31.25 \qquad (1.211b)$$

Continuing in this manner yields $R_3^{(1)} = 131.250$, $x_3^{(1)} = 32.81250$, $R_4^{(1)} = 107.81250$, $x_4^{(1)} = 26.953125$, $R_5^{(1)} = 95.703125$, and $x_5^{(1)} = 23.925781$.

The first and subsequent iterations are summarized in Table 1.4. The intermediate iterates are no longer symmetrical as they were in Example 1.21. The calculations were carried out on a 13-digit precision computer and iterated until all $|\Delta x_i|$ changed by less than 0.000001 between iterations, which required 15 iterations, which is three less than required by the Jacobi method in Example 1.21.

1.7.4.   The Successive-Over-Relaxation (SOR) Method

Iterative methods are frequently referred to as relaxation methods, since the iterative procedure can be viewed as relaxing $\mathbf{x}^{(0)}$ to the exact value $\mathbf{x}$. Historically, the method of relaxation, or just the term relaxation, refers to a specific procedure attributed to Southwell (1940). *Southwell's relaxation method* embodies two procedures for accelerating the convergence of the basic iteration scheme. First, the relaxation order is determined by visually searching for the residual of greatest magnitude, $|R_i|_{\max}$, and then relaxing the corresponding equation by calculating a new value of x_i so that $(R_i)_{\max} = 0.0$. This changes the other residuals that depend on x_i. As the other residuals are relaxed, the value of R_i moves away from zero. The procedure is applied repetitively until all the residuals satisfy the convergence criterion (or criteria).

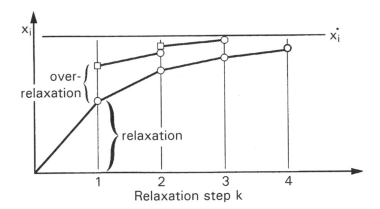

Figure 1.4 Over-relaxation.

Southwell observed that in many cases the changes in x_i from iteration to iteration were always in the same directions. Consequently, over-correcting (i.e., over-relaxing) the values of x_i by the right amount accelerates convergence. This procedure is illustrated in Figure 1.4.

Southwell's method is quite efficient for hand calculation. However, the search for the largest residual is inefficient for computer application, since it can take almost as much computer time to search for the largest residual as it does to make a complete pass through the iteration procedure. On the other hand, the over-relaxation concept is easy to implement on the computer and is very effective in accelerating the convergence rate of the Gauss-Seidel method.

The Gauss-Seidel method can be modified to include over-relaxation simply by multiplying the residual $R_i^{(k)}$ in Eq. (1.209), by the over-relaxation factor, ω. Thus, the *successive-over-relaxation method* is given by

$$x_i^{(k+1)} = x_i^{(k)} + \omega \frac{R_i^{(k)}}{a_{i,i}} \qquad (i = 1, 2, \ldots, n) \tag{1.212}$$

$$R_i^{(k)} = b_i - \sum_{j=1}^{i-1} a_{i,j} x_j^{(k+1)} - \sum_{j=i}^{n} a_{i,j} x_j^{(k)} \qquad (i = 1, 2, \ldots, n) \tag{1.213}$$

When $\omega = 1.0$, Eq. (1.212) yields the Gauss-Seidel method. When $1.0 < \omega < 2.0$, the system of equations is over-relaxed. Over-relaxation is appropriate for systems of linear algebraic equations. When $\omega < 1.0$, the system of equations is under-relaxed. Under-relaxation is appropriate when the Gauss-Seidel algorithm causes the solution vector to overshoot and move farther away from the exact solution. This behavior is generally associated with the iterative solution of systems of nonlinear algebraic equations. The iterative method diverges if $\omega \geq 2.0$. The relaxation factor does not change the final solution since it multiplies the residual R_i, which is zero when the final solution is reached. The major difficulty with the over-relaxation method is the determination of the best value for the over-relaxation factor, ω. Unfortunately, there is not a good general method for determining the optimum over-relaxation factor, ω_{opt}.

The optimum value of the over-relaxation factor ω_{opt} depends on the size of the system of equations (i.e., the number of equations) and the nature of the equations (i.e., the strength of the diagonal dominance, the structure of the coefficient matrix, etc.). As a general rule, larger values of ω_{opt} are associated with larger systems of equations. In Section 9.6, Eqs. (9.51) and (9.52), a procedure is described for estimating ω_{opt} for the system of equations obtained when solving the Laplace equation in a rectangular domain with Dirichlet boundary conditions. In general, one must resort to numerical experimentation to determine ω_{opt}. In spite of this inconvenience, it is almost always worthwhile to search for a near optimum value of ω if a system of equations is to be solved many times. In some problems, the computation time can be reduced by factors as large as 10 to 50. For serious calculations with a large number of equations, the potential is too great to ignore.

Example 1.23. The SOR method.

To illustrate the SOR method, let's rework the problem presented in Example 1.22 using $\omega = 1.10$. The residuals are given by Eq. (1.213). Substituting the initial solution vector, $\mathbf{x}^{(0)T} = [0.0 \quad 0.0 \quad 0.0 \quad 0.0 \quad 0.0]$, into Eq. (1.213.1) gives $R_1^{(0)} = 100.0$. Substituting that value into Eq. (1.212.1) with $\omega = 1.10$ gives

$$x_1^{(1)} = 0.0 + 1.10\frac{100.0}{4} = 27.500000 \tag{1.214a}$$

Substituting $\mathbf{x}^T = [27.50 \quad 0.0 \quad 0.0 \quad 0.0 \quad 0.0]$ into Eq. (1.213.2) gives

$$R_2^{(0)} = (100.0 + 27.50) = 127.50 \tag{1.214b}$$

Substituting this result into Eq. (1.212.2) gives

$$x_2^{(1)} = 0.0 + 1.10\frac{127.50}{4} = 35.062500 \tag{1.214c}$$

Continuing in this manner yields the results presented in Table 1.5.

The first and subsequent iterations are summarized in Table 1.5. The calculations were carried out on a 13-digit precision computer and iterated until all $|\Delta x_i|$ changed by less than 0.000001 between iterations, which required 13 iterations, which is 5 less than required by the Jacobi method and 2 less than required by the Gauss-Seidel method. The value of over-relaxation is modest in this example. Its value becomes more significant as the number of equations increases.

Table 1.5. Solution by the SOR Method

k	x_1	x_2	x_3	x_4	x_5
0	0.000000	0.000000	0.000000	0.000000	0.000000
1	27.500000	35.062500	37.142188	30.151602	26.149503
2	26.100497	34.194375	41.480925	35.905571	25.355629
3	24.419371	35.230346	42.914285	35.968342	25.167386
4	24.855114	35.692519	42.915308	35.790750	25.010375
5	24.987475	35.726188	42.875627	35.717992	24.996719
. . .	. . . .	. . . .	. . . .	. . . .	. . . .
11	24.999996	35.714285	42.857145	35.714287	25.000000
12	25.000000	35.714286	42.857143	35.714286	25.000000
13	25.000000	35.714286	42.857143	35.714286	25.000000

Table 1.6. Number of Iterations k as a Function of ω

ω	k	ω	k	ω	k
1.00	15	1.06	13	1.12	13
1.01	14	1.07	13	1.13	13
1.02	14	1.08	13	1.14	13
1.03	14	1.09	13	1.15	14
1.04	14	1.10	13		
1.05	13	1.11	13		

The optimum value of ω can be determined by experimentation. If a problem is to be worked only once, that procedure is not worthwhile. However, if a problem is to be worked many times with the same **A** matrix for many different **b** vectors, then a search for ω_{opt} may be worthwhile. Table 1.6 presents the results of such a search for the problem considered in Example 1.23. For this problem, $1.05 \leq \omega \leq 1.14$ yields the most efficient solution. Much more dramatic results are obtained for large systems of equations.

1.8. PROGRAMS

Four FORTRAN subroutines for solving systems of linear algebraic equations are presented in this section:

1. Simple Gauss elimination
2. Doolittle LU factorization
3. The Thomas algorithm
4. Successive-over-relaxation (SOR)

The basic computational algorithms are presented as completely self-contained subroutines suitable for use in other programs. Input data and output statements are contained in a main (or driver) program written specifically to illustrate the use of each subroutine.

1.8.1. Simple Gauss Elimination

The elimination step of simple Gauss elimination is based on Eq. (1.100). For each column k $(k = 1, 2, \ldots, n - 1)$,

$$a_{i,j} = a_{i,j} - (a_{i,k}/a_{k,k})a_{k,j} \qquad (i, j = k + 1, k + 2, \ldots, n) \tag{1.215a}$$

$$b_i = b_i - (a_{i,k}/a_{k,k})b_k \qquad (i = k + 1, k + 2, \ldots, n) \tag{1.215b}$$

The back substitution step is based on Eq. (1.101):

$$x_n = b_n/a_{n,n} \tag{1.216a}$$

$$x_i = \frac{b_i - \sum\limits_{j=i+1}^{n} a_{i,j}x_j}{a_{i,i}} \qquad (i = n - 1, n - 2, \ldots, 1) \tag{1.216b}$$

A FORTRAN subroutine, *subroutine gauss*, for solving these equations, without pivoting, is presented below. Note that the eliminated elements from matrix **A** have been replaced by the elimination multipliers, em, so *subroutine gauss* actually evaluates the **L** and **U** matrices needed for Doolittle LU factorization, which is presented in Section 1.8.2. *Program main* defines the data set and prints it, calls *subroutine gauss* to implement the solution, and prints the solution.

Program 1.1.   Simple Gauss elimination program.

```
      program main
c     main program to illustrate linear equation solvers
c     ndim  array dimension, ndim = 6 in this example
c     n        number of equations, n
c     a        coefficient matrix, A(i,j)
c     b        right-hand side vector, b(i)
c     x        solution vector, x(i)
      dimension a(6,6),b(6),x(6)
      data ndim,n / 6, 3 /
      data (a(i,1),i=1,3) /  80.0, -20.0, -20.0 /
      data (a(i,2),i=1,3) / -20.0,  40.0, -20.0 /
      data (a(i,3),i=1,3) / -20.0, -20.0, 130.0 /
      data (b(i),i=1,3) / 20.0, 20.0, 20.0 /
      write (6,1000)
      do i=1,n
         write (6,1010) i,(a(i,j),j=1,n),b(i)
      end do
      call gauss (ndim,n,a,b,x)
      write (6,1020)
      do i=1,n
         write (6,1010) i,(a(i,j),j=1,n),b(i),x(i)
      end do
      stop
 1000 format (' Simple Gauss elimination'/' '/' '/' A and b'/' ')
 1010 format (i2,7f12.6)
 1020 format (' '/' A, b, and x after elimination'/' ')
      end

      subroutine gauss (ndim,n,a,b,x)
c     simple gauss elimination
      dimension a(ndim,ndim),b(ndim),x(ndim)
c     forward elimination
      do k=1,n-1
         do i=k+1,n
            em=a(i,k)/a(k,k)
            a(i,k)=em
            b(i)=b(i)-em*b(k)
            do j=k+1,n
               a(i,j)=a(i,j)-em*a(k,j)
            end do
         end do
      end do
```

```
c      back substitution
       x(n)=b(n)/a(n,n)
       do i=n-1,1,-1
          x(i)=b(i)
          do j=n,i+1,-1
             x(i)=x(i)-a(i,j)*x(j)
          end do
          x(i)=x(i)/a(i,i)
       end do
       return
       end
```

The data set used to illustrate *subroutine gauss* is taken from Example 1.7. The output generated by the simple Gauss elimination program is presented below.

Output 1.1.   Solution by simple Gauss elimination.

```
Simple Gauss elimination

A and b

1    80.000000   -20.000000   -20.000000    20.000000
2   -20.000000    40.000000   -20.000000    20.000000
3   -20.000000   -20.000000   130.000000    20.000000

A, b, and x after elimination

1    80.000000   -20.000000   -20.000000    20.000000    0.600000
2    -0.250000    35.000000   -25.000000    25.000000    1.000000
3    -0.250000    -0.714286   107.142857    42.857143    0.400000
```

1.8.2.   Doolittle LU Factorization

Doolittle LU factorization is based on the LU factorization implicit in Gauss elimination. *Subroutine gauss* presented in Section 1.8.1 is modified to evaluate the **L** and **U** matrices simply by removing the line evaluating $b(i)$ from the first group of statements and entirely deleting the second group of statements, which evaluates the back substitution step. The modified subroutine is named *subroutine lufactor*.

A second subroutine, *subroutine solve*, based on steps 2 and 3 in the description of the Doolittle LU factorization method in Section 1.4, is required to process the **b** vector to the **b'** vector and to process the **b'** vector to the **x** vector. These steps are given by Eqs. (1.139) and (1.140):

$$b'_i = b_i - \sum_{k=1}^{i-1} l_{i,k} b'_k \qquad (i = 2, 3, \ldots, n) \tag{1.217a}$$

$$x_i = b'_i - \sum_{k=i+1}^{n} u_{i,k} x_k / u_{i,i} \qquad (i = n-1, n-2, \ldots, 1) \tag{1.217b}$$

FORTRAN subroutines for implementing Doolittle LU factorization are presented below. *Program main* defines the data set and prints it, calls *subroutine lufactor* to evaluate the **L** and **U** matrices, calls *subroutine solve* to implement the solution for a specified **b** vector, and prints the solution. *Program main* below shows only the statements which are different from the statements in *program main* in Section 1.8.1.

Program 1.2.   Doolittle LU factorization program.

```
      program main
c     main program to illustrate linear equation solvers
c     bp    b' vector, bp(i)
      dimension a(6,6),b(6),bp(6),x(6)
      call lufactor (ndim,n,a)
      write (6,1020)
      do i=1,n
         write (6,1010) i,(a(i,j),j=1,n)
      end do
      call solve (ndim,n,a,b,bp,x)
      write (6,1030)
      do i=1,n
         write (6,1010) i,b(i),bp(i),x(i)
      end do
      stop
 1000 format (' Doolittle LU factorization'/' '/' A and b'/' ')
 1010 format (i2,7f12.6)
 1020 format (' '/' L and U stored in A'/' ')
 1030 format (' '/' b, bprime, and x vectors'/' ')
      end


      subroutine lufactor (ndim,n,a)
c     Doolittle LU factorization, stores L and U in A
      dimension a(ndim,ndim)
      do k=1,n-1
         do i=k+1,n
            em=a(i,k)/a(k,k)
            a(i,k)=em
            do j=k+1,n
               a(i,j)=a(i,j)-em*a(k,j)
            end do
         end do
      end do
      return
      end


      subroutine solve (ndim,n,a,b,bp,x)
c     processes b to b' and b' to x
      dimension a(ndim,ndim),b(ndim),bp(ndim),x(ndim)
c     forward elimination step to calculate b'
      bp(1)=b(1)
      do i=2,n
         bp(i)=b(i)
         do j=1,i-1
            bp(i)=bp(i)-a(i,j)*bp(j)
         end do
      end do
```

```
c      back substitution step to calculate x
       x(n)=bp(n)/a(n,n)
       do i=n-1,1,-1
          x(i)=bp(i)
          do j=n,i+1,-1
             x(i)=x(i)-a(i,j)*x(j)
          end do
          x(i)=x(i)/a(i,i)
       end do
       return
       end
```

The data set used to illustrate *subroutines lufactor* and *solve* is taken from Example 1.15. The output generated by the Doolittle LU factorization program is presented below.

Output 1.2.   Solution by Doolittle LU factorization.

```
Doolittle LU factorization

A and b

1    80.000000   -20.000000   -20.000000
2   -20.000000    40.000000   -20.000000
3   -20.000000   -20.000000   130.000000

L and U matrices stored in A matrix

1    80.000000   -20.000000   -20.000000
2    -0.250000    35.000000   -25.000000
3    -0.250000    -0.714286   107.142857

b, bprime, and x vectors

1    20.000000    20.000000     0.600000
2    20.000000    25.000000     1.000000
3    20.000000    42.857143     0.400000
```

1.8.3.   The Thomas Algorithm

The elimination step of the Thomas algorithm is based on Eqs. (1.150) and (1.151):

$$a'_{1,2} = a'_{1,2} \tag{1.218a}$$

$$a'_{i,2} = a'_{i,2} - (a'_{i,1}/a'_{i-1,2})a'_{i-1,3} \qquad (i = 2, 3, \ldots, n) \tag{1.218b}$$

$$b_1 = b_1 \tag{1.218c}$$

$$b_i = b_i - (a'_{i,1}/a'_{i-1,2})b_{i-1} \qquad (i = 2, 3, \ldots, n) \tag{1.218d}$$

The back substitution step is based on Eq. (1.152):

$$x_n = b_n/a'_{n,2} \tag{1.219a}$$

$$x_i = (b_i - a'_{i,3}x_{i+1})/a'_{i,2} \qquad (i = n-1, n-2, \ldots, 1) \tag{1.219b}$$

A FORTRAN subroutine, *subroutine thomas*, for solving these equations is presented below. Note that the eliminated elements from matrix **A** have been replaced by the elimination multipliers, em, so *subroutine thomas* actually evaluates the **L** and **U** matrices needed for Doolittle LU factorization. *Program main* defines the data set and prints it, calls *subroutine thomas* to implement the solution, and prints the solution.

Program 1.3.   The Thomas algorithm program.

```
      program main
c     main program to illustrate linear equation solvers
c     ndim  array dimension, ndim = 9 in this example
c     n     number of equations, n
c     a     coefficient matrix, A(i,3)
c     b     right-hand side vector, b(i)
c     x     solution vector, x(i)
      dimension a(9,3),b(9),x(9)
      data ndim,n / 9, 7 /
      data (a(1,j),j=1,3) / 0.0, -2.25, 1.0 /
      data (a(2,j),j=1,3) / 1.0, -2.25, 1.0 /
      data (a(3,j),j=1,3) / 1.0, -2.25, 1.0 /
      data (a(4,j),j=1,3) / 1.0, -2.25, 1.0 /
      data (a(5,j),j=1,3) / 1.0, -2.25, 1.0 /
      data (a(6,j),j=1,3) / 1.0, -2.25, 1.0 /
      data (a(7,j),j=1,3) / 1.0, -2.25, 0.0 /
      data (b(i),i=1,7) / 0.0, 0.0, 0.0, 0.0, 0.0, 0.0, -100.0 /
      write (6,1000)
      do i=1,n
         write (6,1010) i,(a(i,j),j=1,3),b(i)
      end do
      call thomas (ndim,n,a,b,x)
      write (6,1020)
      do i=1,n
         write (6,1010) i,(a(i,j),j=1,3),b(i),x(i)
      end do
      stop
 1000 format (' The Thomas algorithm'/' '/' A and b'/' ')
 1010 format (i2,6f12.6)
 1020 format (' '/' A, b, and x after elimination'/' ')
      end

      subroutine thomas (ndim,n,a,b,x)
c     the Thomas algorithm for a tridiagonal system
      dimension a(ndim,3),b(ndim),x(ndim)
c     forward elimination
      do i=2,n
         em=a(i,1)/a(i-1,2)
         a(i,1)=em
         a(i,2)=a(i,2)-em*a(i-1,3)
         b(i)=b(i)-a(i,1)*b(i-1)
      end do
```

```
c       back substitution
        x(n)=b(n)/a(n,2)
        do i=n-1,1,-1
           x(i)=(b(i)-a(i,3)*x(i+1))/a(i,2)
        end do
        return
        end
```

The data set used to illustrate *subroutine thomas* is taken from Example 1.17. The output generated by the Thomas algorithm program is presented below.

Output 1.3.   Solution by the Thomas algorithm.

```
The Thomas algorithm

A and b

1      0.000000    -2.250000    1.000000    0.000000
2      1.000000    -2.250000    1.000000    0.000000
3      1.000000    -2.250000    1.000000    0.000000
4      1.000000    -2.250000    1.000000    0.000000
5      1.000000    -2.250000    1.000000    0.000000
6      1.000000    -2.250000    1.000000    0.000000
7      1.000000    -2.250000    0.000000 -100.000000

A, b, and x after elimination

1      0.000000    -2.250000    1.000000    0.000000    1.966751
2     -0.444444    -1.805556    1.000000    0.000000    4.425190
3     -0.553846    -1.696154    1.000000    0.000000    7.989926
4     -0.589569    -1.660431    1.000000    0.000000   13.552144
5     -0.602253    -1.647747    1.000000    0.000000   22.502398
6     -0.606889    -1.643111    1.000000    0.000000   37.078251
7     -0.608602    -1.641398    0.000000  100.000000   60.923067
```

1.8.4.   Successive-Over-Relaxation (SOR)

Successive-over-relaxation (SOR) is based on Eqs. (1.212) and (1.213):

$$x_i^{(k+1)} = x_i^{(k)} + \omega \frac{R_i^{(k)}}{a_{i,i}} \qquad (i = 1, 2, \ldots, n) \tag{1.220a}$$

$$R_i^{(k)} = b_i - \sum_{j=1}^{i-1} a_{i,j} x_j^{(k+1)} - \sum_{j=i}^{n} a_{i,j} x_j^{(k)} \qquad (i = 1, 2, \ldots, n) \tag{1.220b}$$

A FORTRAN subroutine, *subroutine sor*, for solving these equations is presented below. *Program main* defines the data set and prints it, calls *subroutine sor* to implement the solution, and prints the solution. Input variable *iw* is a flag for output of intermediate results. When $iw = 0$, no intermediate results are output. When $iw = 1$, intermediate results are output.

Program 1.4. Successive-over-relaxation (SOR) program.

```
      program main
c     main program to illustrate linear equation solvers
c     ndim  array dimension, ndim = 6 in this example
c     n     number of equations, n
c     a     coefficient matrix, A(i,j)
c     b     right-hand side vector, b(i)
c     x     solution vector, x(i)
c     iter  number of iterations allowed
c     tol   convergence tolerance
c     omega over-relaxation factor
c     iw    flag for intermediate output: 0 no, 1 yes
      dimension a(6,6),b(6),x(6)
      data ndim,n,iter,tol,omega,iw / 6,5,25,0.000001,1.0,1 /
      data (a(i,1),i=1,5) /  4.0, -1.0,  0.0,  1.0,  0.0 /
      data (a(i,2),i=1,5) / -1.0,  4.0, -1.0,  0.0,  1.0 /
      data (a(i,3),i=1,5) /  0.0, -1.0,  4.0, -1.0,  0.0 /
      data (a(i,4),i=1,5) /  1.0,  0.0, -1.0,  4.0, -1.0 /
      data (a(i,5),i=1,5) /  0.0,  1.0,  0.0, -1.0,  4.0 /
      data (b(i),i=1,5) / 100.0, 100.0, 100.0, 100.0, 100.0 /
      data (x(i),i=1,5) /  0.0,  0.0,  0.0,  0.0,  0.0 /
      write (6,1000)
      do i=1,n
         write (6,1010) i,(a(i,j),j=1,n),b(i)
      end do
      write (6,1020)
      it=0
      write (6,1010) it,(x(i),i=1,n)
      call sor (ndim,n,a,b,x,iter,tol,omega,iw,it)
      if (iw.eq.0) write (6,1010) it,(x(i),i=1,n)
      stop
 1000 format (' SOR iteration'/' '/' A and b'/' ')
 1010 format (i2,7f12.6)
 1020 format (' '/' i                       x(1) to x(n)'/' ')
      end


      subroutine sor (ndim,n,a,b,x,iter,tol,omega,iw,it)
c     sor iteration
      dimension a(ndim,ndim),b(ndim),x(ndim)
      do it=1,iter
         dxmax=0.0
         do i=1,n
            residual=b(i)
            do j=1,n
               residual=residual-a(i,j)*x(j)
            end do
            if (abs(residual).gt.dxmax) dxmax=abs(residual)
         x(i)=x(i)+omega*residual/a(i,i)
         end do
```

```
      if (iw.eq.1) write (6,1000) it,(x(i),i=1,n)
      if (dxmax.lt.tol) return
   end do
   write (6,1010)
   return
1000 format (i2,7f12.6)
1010 format (' '/' Solution failed to converge'/' ')
   end
```

The data set used to illustrate *subroutine sor* is taken from Example 1.23. The output generated by the SOR program is presented below.

Output 1.4.   Solution by successive-over-relaxation (SOR).

```
SOR iteration

A and b

1    4.000000   -1.000000    0.000000    1.000000    0.000000 100.000000
2   -1.000000    4.000000   -1.000000    0.000000    1.000000 100.000000
3    0.000000   -1.000000    4.000000   -1.000000    0.000000 100.000000
4    1.000000    0.000000   -1.000000    4.000000   -1.000000 100.000000
5    0.000000    1.000000    0.000000   -1.000000    4.000000 100.000000

i                  x(1) to x(n)

0    0.000000    0.000000    0.000000    0.000000    0.000000
1   25.000000   31.250000   32.812500   26.953125   23.925781
2   26.074219   33.740234   40.173340   34.506226   25.191498
3   24.808502   34.947586   42.363453   35.686612   25.184757
4   24.815243   35.498485   42.796274   35.791447   25.073240
    . . . . . . . . . . . . . . . . . . . . . . . . . . . . . . . .
15  25.000000   35.714286   42.857143   35.714286   25.000000
16  25.000000   35.714286   42.857143   35.714286   25.000000
```

1.8.5.   Packages for Systems of Linear Algebraic Equations

Numerous libraries and software packages are available for solving systems of linear algebraic equations. Many work stations and mainframe computers have such libraries attached to their operating systems. If not, libraries such as IMSL (International Mathematics and Statistics Library) or LINPACK (Argonne National Laboratory) can be added to the operating systems.

   Most commercial software packages contain solvers for systems of linear algebraic equations. Some of the more prominent packages are Matlab and Mathcad. The spreadsheet Excel can also be used to solve systems of equations. More sophisticated packages, such as Mathematica, Macsyma, and Maple, also contain linear equation solvers. Finally, the book *Numerical Recipes* (Press et al., 1989) contains several subroutines for solving systems of linear algebraic equations.

1.9 SUMMARY

The basic methods for solving systems of linear algebraic equations are presented in this chapter. Some general guidelines for selecting a method for solving systems of linear algebraic equations are given below.

- Direct elimination methods are preferred for small systems ($n \lesssim 50$ to 100) and systems with few zeros (nonsparse systems). Gauss elimination is the method of choice.
- For tridiagonal systems, the Thomas algorithm is the method of choice.
- LU factorization methods (e.g., the Doolittle method) are the methods of choice when more than one **b** vector must be considered.
- For large systems that are not diagonally dominant, the round-off errors can be large.
- Iterative methods are preferred for large, sparse matrices that are diagonally dominant. The SOR method is the method of choice. Numerical experimentation to find the optimum over-relaxation factor ω_{opt} is usually worthwhile if the system of equations is to be solved for many **b** vectors.

After studying Chapter 1, you should be able to:

1. Describe the general structure of a system of linear algebraic equations.
2. Explain the solution possibilities for a system of linear algebraic equations: (a) a unique solution, (b) no solution, (c) an infinite number of solutions, and (d) the trivial solution.
3. Understand the differences between direct elimination methods and iterative methods.
4. Understand the elementary properties of matrices and determinants.
5. Recognize the structures of square matrices, diagonal matrices, the identity matrix, upper and lower triangular matrices, tridiagonal matrices, banded matrices, and singular matrices.
6. Perform elementary matrix algebra.
7. Understand the concept of diagonal dominance.
8. Understand the concept of the inverse of a matrix.
9. Define the determinant of a matrix.
10. Express a system of linear algebraic equations in matrix form.
11. Understand matrix row operations.
12. Explain the general concept behind direct elimination methods.
13. Apply Cramer's rule to solve a system of linear algebraic equations.
14. Solve a system of linear algebraic equations by Gauss elimination.
15. Understand pivoting and scaling in Gauss elimination.
16. Evaluate 2×2 and 3×3 determinants by the diagonal method.
17. Evaluate a determinant by the cofactor method.
18. Evaluate a determinant by Gauss elimination.
19. Solve a system of linear algebraic equations by Gauss-Jordan elimination.
20. Determine the inverse of a matrix by Gauss-Jordan elimination.
21. Solve a system of linear algebraic equations by the matrix inverse method.
22. Explain the concept of matrix factorization.
23. Explain how Doolittle LU factorization is obtained by Gauss elimination.
24. Solve a system of linear algebraic equations by Doolittle LU factorization.
25. Determine the inverse of a matrix by Doolittle LU factorization.

26. Understand the special properties of a tridiagonal matrix.
27. Solve a tridiagonal system of linear algebraic equation by the Thomas algorithm.
28. Understand the concept of block tridiagonal systems of linear algebraic equations.
29. Understand the pitfalls of direct elimination methods.
30. Explain the effects of round-off on numerical algorithms.
31. Explain the concept of system condition.
32. Describe the effects of ill-conditioning.
33. Define the norm of a vector or matrix.
34. Define the condition number of a matrix.
35. Explain the significance of the condition number of a matrix.
36. Explain the general concept behind iterative methods.
37. Understand the structure and significance of a sparse matrix.
38. Explain the importance of diagonal dominance for iterative methods.
39. Solve a system of linear algebraic equations by Jacobi iteration.
40. Solve a system of linear algebraic equations by Gauss-Seidel iteration.
41. Solve a system of linear algebraic equations by successive-over relaxation (SOR).
42. Appreciate the importance of using the optimum overrelaxation factor, ω_{opt}.
43. Define the meaning and significance of the residual in an iterative method.
44. Explain accuracy of an approximate method.
45. Understand the difference between absolute error and relative error.
46. Understand convergence of an iterative method and convergence criteria.
47. Choose a method for solving a system of linear algebraic equations based on its size and structure.

EXERCISE PROBLEMS

Section 1.2.   Properties of Matrices and Determinants

The following four matrices are considered in this section:

$$A = \begin{bmatrix} 1 & 1 & 3 \\ 5 & 3 & 1 \\ 2 & 3 & 1 \end{bmatrix} \quad B = \begin{bmatrix} 2 & 3 & 5 \\ 3 & 1 & -2 \\ 1 & 3 & 4 \end{bmatrix} \quad C = \begin{bmatrix} 2 & 1 \\ 3 & 4 \\ 2 & 5 \end{bmatrix} \quad D = \begin{bmatrix} 1 & 1 & 1 \\ 1 & 3 & 3 \\ 1 & 3 & 5 \end{bmatrix}$$

1. Determine the following quantities, if defined: (a) $A + B$, (b) $B + A$, (c) $A + D$, (d) $A + C$, (e) $B + C$.
2. Determine the following quantities, if defined: (a) $A - B$, (b) $B - A$, (c) $A - D$, (d) $B - D$, (e) $A - C$, (f) $B - C$, (g) $C - B$, (h) $D - C$.
3. Determine the following quantities, if defined: (a) AC, (b) BC, (c) CA, (d) CB, (e) AD, (f) BD, (g) $C^T A$, (h) $C^T B$, (i) $C^T D$, (j) AA^T, (k) BB^T, (l) DD^T.
4. (a) Compute AB and BA and show that $AB \neq BA$. (b) Compute AD and DA and show that $AD \neq DA$. (c) Compute BD and DB and show that $BD \neq DB$.
5. Show that: (a) $(AB)^T = B^T A^T$ and (b) $AB = (Ab_1 \quad Ab_2 \quad Ab_3)$, where b_1, b_2, and b_3 are the columns of B.
6. Work Problem 5 for the general 3×3 matrices A and B.
7. Verify that (a) $A + (B + D) = (A + B) + D$ and (b) $A(BD) = (AB)D$.

8. Calculate the following determinants by the diagonal method, if defined:
    (a) det(**A**)   (b) det(**B**)   (c) det(**C**)   (d) det(**D**)   (e) det(**AB**)
    (f) det(**AD**)   (g) det(**BA**)   (h) det(**DA**)   (i) det(**CD**)   (j) det($\mathbf{C}^T\mathbf{A}$)
9. Work Problem 8 using the cofactor method.
10. Show that det(**A**) det(**B**) = det(**AB**).
11. Show that det(**A**) det(**D**) = det(**AD**).
12. Show that for the general 2×2 matrices **A** and **B**, det(**A**) det(**B**) = det(**AB**).

Section 1.3.   Direct Elimination Methods

Consider the following eight systems of linear algebraic equations, $\mathbf{Ax} = \mathbf{b}$:

$$\begin{array}{rl} -2x_1 + 3x_2 + x_3 = 9 \\ 3x_1 + 4x_2 - 5x_3 = 0 \quad\text{(A)} \\ x_1 - 2x_2 + x_3 = -4 \end{array} \qquad \begin{bmatrix} 1 & 1 & 3 \\ 5 & 3 & 1 \\ 2 & 3 & 1 \end{bmatrix}\begin{bmatrix} x \\ y \\ z \end{bmatrix} = \begin{bmatrix} 2 \\ 3 \\ -1 \end{bmatrix} \text{(B)}$$

$$\begin{array}{rl} x_1 + 3x_2 + 2x_3 - x_4 = 9 \\ 4x_1 + 2x_2 + 5x_3 + x_4 = 27 \\ 3x_1 - 3x_2 + 2x_3 + 4x_4 = 19 \quad\text{(C)} \\ -x_1 + 2x_2 - 3x_3 + 5x_4 = 14 \end{array} \qquad \begin{bmatrix} 3 & 1 & -1 & 3 \\ 2 & 1 & -2 & 0 \\ 0 & 3 & 2 & -2 \\ 1 & 1 & 1 & 5 \end{bmatrix}[x_i] = \begin{bmatrix} 4 \\ -1 \\ 4 \\ -2 \end{bmatrix} \text{(D)}$$

$$\begin{bmatrix} 1 & -2 & 1 \\ 2 & 1 & 2 \\ -1 & 1 & 3 \end{bmatrix}\begin{bmatrix} x_1 \\ x_2 \\ x_3 \end{bmatrix} = \begin{bmatrix} -1 \\ 3 \\ 8 \end{bmatrix} \text{(E)} \qquad \begin{bmatrix} 2 & 3 & 5 \\ 3 & 1 & -2 \\ 1 & 3 & 4 \end{bmatrix}\begin{bmatrix} x_1 \\ x_2 \\ x_3 \end{bmatrix} = \begin{bmatrix} 0 \\ -2 \\ -3 \end{bmatrix} \text{(F)}$$

$$\begin{bmatrix} 2 & -2 & 2 & 1 \\ 2 & -4 & 1 & 3 \\ -1 & 3 & -4 & 2 \\ 2 & 4 & 3 & -2 \end{bmatrix}\begin{bmatrix} x_1 \\ x_2 \\ x_3 \\ x_4 \end{bmatrix} = \begin{bmatrix} 7 \\ 10 \\ -14 \\ 1 \end{bmatrix} \text{(G)} \qquad \begin{bmatrix} 1 & 1 & 1 \\ 1 & 2 & 1 \\ 3 & 3 & 4 \end{bmatrix}\begin{bmatrix} x_1 \\ x_2 \\ x_3 \end{bmatrix} = \begin{bmatrix} 0 \\ -4 \\ 1 \end{bmatrix} \text{(H)}$$

Cramer's Rule

13. Solve Eq. (A) by Cramer's rule.
14. Solve Eq. (B) by Cramer's rule.
15. Solve Eq. (C) by Cramer's rule.
16. Solve Eq. (D) by Cramer's rule.
17. Solve Eq. (E) by Cramer's rule.
18. Solve Eq. (F) by Cramer's rule.
19. Solve Eq. (G) by Cramer's rule.
20. Solve Eq. (H) by Cramer's rule.

Gauss Elimination

21. Solve Eq. (A) by Gauss elimination without pivoting.
22. Solve Eq. (B) by Gauss elimination without pivoting.
23. Solve Eq. (C) by Gauss elimination without pivoting.
24. Solve Eq. (D) by Gauss elimination without pivoting.
25. Solve Eq. (E) by Gauss elimination without pivoting.
26. Solve Eq. (F) by Gauss elimination without pivoting.
27. Solve Eq. (G) by Gauss elimination without pivoting.
28. Solve Eq. (H) by Gauss elimination without pivoting.

Gauss-Jordan Elimination

29. Solve Eq. (A) by Gauss-Jordan elimination.
30. Solve Eq. (B) by Gauss-Jordan elimination.
31. Solve Eq. (C) by Gauss-Jordan elimination.
32. Solve Eq. (D) by Gauss-Jordan elimination.
33. Solve Eq. (E) by Gauss-Jordan elimination.
34. Solve Eq. (F) by Gauss-Jordan elimination.
35. Solve Eq. (G) by Gauss-Jordan elimination.
36. Solve Eq. (H) by Gauss-Jordan elimination.

The Matrix Inverse Method

37. Solve Eq. (A) using the matrix inverse method.
38. Solve Eq. (B) using the matrix inverse method.
39. Solve Eq. (C) using the matrix inverse method.
40. Solve Eq. (D) using the matrix inverse method.
41. Solve Eq. (E) using the matrix inverse method.
42. Solve Eq. (F) using the matrix inverse method.
43. Solve Eq. (G) using the matrix inverse method.
44. Solve Eq. (H) using the matrix inverse method.

Section 1.4. LU Factorization

45. Solve Eq. (A) by the Doolittle LU factorization method.
46. Solve Eq. (B) by the Doolittle LU factorization method.
47. Solve Eq. (C) by the Doolittle LU factorization method.
48. Solve Eq. (D) by the Doolittle LU factorization method.
49. Solve Eq. (E) by the Doolittle LU factorization method.
50. Solve Eq. (F) by the Doolittle LU factorization method.
51. Solve Eq. (G) by the Doolittle LU factorization method.
52. Solve Eq. (H) by the Doolittle LU factorization method.

Section 1.5. Tridiagonal Systems of Equations

Consider the following tridiagonal systems of linear algebraic equations:

$$
\begin{bmatrix} 2 & 1 & 0 & 0 \\ 1 & 2 & 1 & 0 \\ 0 & 1 & 2 & 1 \\ 0 & 0 & 1 & 2 \end{bmatrix}
\begin{bmatrix} x_1 \\ x_2 \\ x_3 \\ x_4 \end{bmatrix}
=
\begin{bmatrix} 4 \\ 8 \\ 12 \\ 11 \end{bmatrix} \text{ (I)}
\qquad
\begin{bmatrix} 3 & 2 & 0 & 0 \\ 2 & 3 & 2 & 0 \\ 0 & 2 & 3 & 2 \\ 0 & 0 & 2 & 3 \end{bmatrix}
\begin{bmatrix} x_1 \\ x_2 \\ x_3 \\ x_4 \end{bmatrix}
=
\begin{bmatrix} 12 \\ 17 \\ 14 \\ 7 \end{bmatrix} \text{ (J)}
$$

$$
\begin{bmatrix} -2 & 1 & 0 & 0 \\ 1 & -2 & 1 & 0 \\ 0 & 1 & -2 & 1 \\ 0 & 0 & 1 & -2 \end{bmatrix}
\begin{bmatrix} x_1 \\ x_2 \\ x_3 \\ x_4 \end{bmatrix}
=
\begin{bmatrix} 1 \\ 2 \\ -7 \\ -1 \end{bmatrix} \text{ (K)}
\qquad
\begin{bmatrix} -2 & 1 & 0 & 0 \\ 1 & -2 & 1 & 0 \\ 0 & 1 & -2 & 1 \\ 0 & 0 & 1 & -2 \end{bmatrix}
\begin{bmatrix} x_1 \\ x_2 \\ x_3 \\ x_4 \end{bmatrix}
=
\begin{bmatrix} 5 \\ 1 \\ 0 \\ 8 \end{bmatrix} \text{ (L)}
$$

$$
\begin{bmatrix} 4 & -1 & 0 & 0 \\ -1 & 4 & -1 & 0 \\ 0 & -1 & 4 & -1 \\ 0 & 0 & -1 & 4 \end{bmatrix}
[x_i] =
\begin{bmatrix} 150 \\ 200 \\ 150 \\ 100 \end{bmatrix} \text{ (M)}
\qquad
\begin{bmatrix} 2 & -1 & 0 & 0 \\ -1 & 2 & -1 & 0 \\ 0 & -1 & 2 & -1 \\ 0 & 0 & -1 & 2 \end{bmatrix}
\begin{bmatrix} x_1 \\ x_2 \\ x_3 \\ x_4 \end{bmatrix}
=
\begin{bmatrix} 5 \\ 1 \\ 0 \\ 8 \end{bmatrix} \text{ (N)}
$$

53. Solve Eq. (I) by the Thomas algorithm.
54. Solve Eq. (J) by the Thomas algorithm.
55. Solve Eq. (K) by the Thomas algorithm.
56. Solve Eq. (L) by the Thomas algorithm.
57. Solve Eq. (M) by the Thomas algorithm.
58. Solve Eq. (N) by the Thomas algorithm.

Section 1.7. Iterative Methods

Solve the following problems by iterative methods. Let $\mathbf{x}^{(0)T} = [0.0 \quad 0.0 \quad 0.0 \quad 0.0]$. For hand calculations, make at least five iterations. For computer solutions, iterate until six digits after the decimal place converges.

Jacobi Iteration

59. Solve Eq. (I) by Jacobi iteration.
60. Solve Eq. (K) by Jacobi iteration.
61. Solve Eq. (L) by Jacobi iteration.
62. Solve Eq. (M) by Jacobi iteration.
63. Solve Eq. (N) by Jacobi iteration.

Gauss-Seidel Iteration

64. Solve Eq. (I) by Gauss-Seidel iteration.
65. Solve Eq. (K) by Gauss-Seidel iteration.
66. Solve Eq. (L) by Gauss-Seidel iteration.
67. Solve Eq. (M) by Gauss-Seidel iteration.
68. Solve Eq. (N) by Gauss-Seidel iteration.

Successive Over-Relaxation

69. Solve Eq. (I) by the SOR method with $\omega = 1.27$.
70. Solve Eq. (K) by the SOR method with $\omega = 1.27$.
71. Solve Eq. (L) by the SOR method with $\omega = 1.27$.
72. Solve Eq. (M) by the SOR method with $\omega = 1.05$.
73. Solve Eq. (N) by the SOR method with $\omega = 1.25$.
74. Solve Eq. (I) by the SOR method for $1.25 \leq \omega \leq 1.35$ with $\Delta\omega = 0.01$.
75. Solve Eq. (K) by the SOR method for $1.25 \leq \omega \leq 1.35$ with $\Delta\omega = 0.01$.
76. Solve Eq. (L) by the SOR method for $1.25 \leq \omega \leq 1.35$ with $\Delta\omega = 0.01$.
77. Solve Eq. (M) by the SOR method for $1.00 \leq \omega = 1.10$ with $\Delta\omega = 0.01$.
78. Solve Eq. (N) by the SOR method for $1.25 \leq \omega = 1.35$ with $\Delta\omega = 0.01$.

Section 1.8. Programs

79. Implement the simple Gauss elimination program presented in Section 1.8.1. Check out the program using the given data.
80. Solve any of Eqs. (A) to (H) using the Gauss elimination program.
81. Implement the Doolittle LU factorization program presented in Section 1.8.2. Check out the program using the given data.
82. Solve any of Eqs. (A) to (H) using the Doolittle LU factorization program.
83. Implement the Thomas algorithm program presented in Section 1.8.3. Check out the program using the given data.
84. Solve any of Eqs. (I) to (N) using the Thomas algorithm program.
85. Implement the SOR program presented in Section 1.8.4. Check out the program using the given data.
86. Solve any of Eqs. (I) and (K) to (N) using the SOR program.

2

Eigenproblems

2.1 INTRODUCTION

Consider the dynamic mechanical spring-mass system illustrated in Figure 2.1. Applying Newton's second law of motion, $\sum F = m\ddot{x}$, to each individual mass gives

$$K_2(x_2 - x_1) + K_3(x_3 - x_1) - K_1 x_1 = m_1 \ddot{x}_1 \tag{2.1a}$$

$$-K_2(x_2 - x_1) + K_4(x_3 - x_2) = m_2 \ddot{x}_2 \tag{2.1b}$$

$$-K_3(x_3 - x_1) - K_4(x_3 - x_2) - K_5 x_3 = m_3 \ddot{x}_3 \tag{2.1c}$$

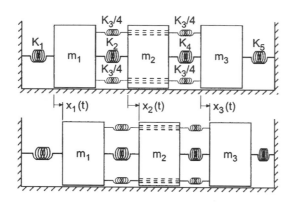

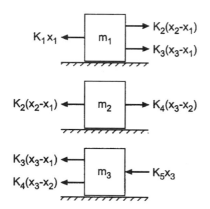

Figure 2.1 Dynamic mechanical spring-mass system.

Rearranging Eq. (2.1) yields:

$$-(K_1 + K_2 + K_3)x_1 + K_2 x_2 + K_3 x_3 = m_1 \ddot{x}_1 \tag{2.2a}$$

$$K_2 x_1 - (K_2 + K_4)x_2 + K_4 x_3 = m_2 \ddot{x}_2 \tag{2.2b}$$

$$K_3 x_1 + K_4 x_2 - (K_3 + K_4 + K_5)x_3 = m_3 \ddot{x}_3 \tag{2.2c}$$

For steady periodic motion at large time,

$$\mathbf{x}(t) = \mathbf{X} \sin(\omega t) \tag{2.3a}$$

where $\mathbf{x}(t)^T = [x_1(t) \quad x_2(t) \quad x_3(t)]$, $\mathbf{X}^T = [X_1 \quad X_2 \quad X_3]$ is the amplitude of the oscillation of the masses, and ω is the undamped natural frequency of the system. Differentiating Eq. (2.3a) gives

$$\frac{d\mathbf{x}}{dt} = \dot{\mathbf{x}} = \omega \mathbf{X} \cos(\omega t) \qquad \text{and} \qquad \frac{d^2\mathbf{x}}{dt^2} = \ddot{\mathbf{x}} = -\omega^2 \mathbf{X} \sin(\omega t) \tag{2.3b}$$

Substituting Eq. (2.3b) into Eq. (2.2) gives

$$-(K_1 + K_2 + K_3)X_1 + K_2 X_2 + K_3 X_3 = -m_1 \omega^2 X_1 \tag{2.4a}$$

$$K_2 X_1 - (K_2 + K_4)X_2 + K_4 X_3 = -m_2 \omega^2 X_2 \tag{2.4b}$$

$$K_3 X_1 + K_4 X_2 - (K_3 + K_4 + K_5)X_3 = -m_3 \omega^2 X_3 \tag{2.4c}$$

Rearranging Eq. (2.4) yields the *system equation*:

$$(K_1 + K_2 + K_3) - m_1\omega^2)X_1 - K_2X_2 - K_3X_3 = 0 \tag{2.5a}$$

$$-K_2X_1 + (K_2 + K_4 - m_2\omega^2)X_2 - K_4X_3 = 0 \tag{2.5b}$$

$$-K_3X_1 - K_4X_2 + (K_3 + K_4 + K_5 - m_3\omega^2)X_3 = 0 \tag{2.5c}$$

Let's nondimensionalize Eq. (2.5) using K_{ref} and m_{ref}. Thus, $\bar{m} = m/m_{\text{ref}}$ and $\bar{K} = K/K_{\text{ref}}$. Substituting these definitions into Eq. (2.5) and dividing by K_{ref} gives:

$$\left[\bar{K}_1 + \bar{K}_2 + \bar{K}_3 - \bar{m}_1\left(\frac{m_{\text{ref}}\omega^2}{K_{\text{ref}}}\right)\right]X_1 - \bar{K}_2X_2 - \bar{K}_3X_3 = 0 \tag{2.6a}$$

$$-\bar{K}_2X_1 + \left[\bar{K}_2 + \bar{K}_4 - \bar{m}_2\left(\frac{m_{\text{ref}}\omega^2}{K_{\text{ref}}}\right)\right]X_2 - \bar{K}_4X_3 = 0 \tag{2.6b}$$

$$-\bar{K}_3X_1 - \bar{K}_4X_2 + \left[\bar{K}_3 + \bar{K}_4 + \bar{K}_5 - \bar{m}_3\left(\frac{m_{\text{ref}}\omega^2}{K_{\text{ref}}}\right)\right]X_3 = 0 \tag{2.6c}$$

Define the parameter λ as follows:

$$\boxed{\lambda = \frac{m_{\text{ref}}\omega^2}{K_{\text{ref}}}} \tag{2.7}$$

Substituting Eq. (2.7) into Eq. (2.6) gives the *nondimensional system equation*:

$$(\bar{K}_1 + \bar{K}_2 + \bar{K}_3 - \bar{m}_1\lambda)X_1 - \bar{K}_2X_2 - \bar{K}_3X_3 = 0 \tag{2.8a}$$

$$-\bar{K}_2X_1 + (\bar{K}_2 + \bar{K}_4 - \bar{m}_2\lambda)X_2 - \bar{K}_4X_3 = 0 \tag{2.8b}$$

$$-\bar{K}_3X_1 - \bar{K}_4X_2 + (\bar{K}_3 + \bar{K}_4 + \bar{K}_5 - \bar{m}_3\lambda)X_3 = 0 \tag{2.8c}$$

Consider a specific system for which $K_1 = 40\,\text{N/cm}$, $K_2 = K_3 = K_4 = 20\,\text{N/cm}$, and $K_5 = 90\,\text{N/cm}$, and $m_1 = m_2 = m_3 = 2\,\text{kg}$. Let $K_{\text{ref}} = 10\,\text{N/cm}$ and $m_{\text{ref}} = 2\,\text{kg}$. For these values, Eq. (2.8) becomes:

$$\boxed{\begin{aligned}(8 - \lambda)X_1 - 2X_2 - 2X_3 &= 0 \\ -2X_1 + (4 - \lambda)X_2 - 2X_3 &= 0 \\ -2X_1 - 2X_2 + (13 - \lambda)X_3 &= 0\end{aligned}}\tag{2.9a, 2.9b, 2.9c}$$

Equation (2.9) is a system of three homogeneous linear algebraic equations. There are four unknowns: X_1, X_2, X_3, and λ (i.e., ω). Clearly unique values of the four unknowns cannot be determined by three equations. In fact, the only solution, other than the trivial solution $\mathbf{X} = 0$, depends on the special values of λ, called *eigenvalues*. Equation (2.9) is a classical *eigenproblem*. The values of λ that satisfy Eq. (2.9) are called *eigenvalues*. Unique values of $\mathbf{X}^T = [X_1 \quad X_2 \quad X_3]$ cannot be determined. However, for every value of λ, relative values of X_1, X_2, and X_3 can be determined. The corresponding values of $\mathbf{X}$ are called *eigenvectors*. The eigenvectors determine the *mode* of oscillation (i.e., the relative values of X_1, X_2, X_3). Equation (2.9) can be written as

$$\boxed{(\mathbf{A} - \lambda\mathbf{I})\mathbf{X} = 0} \tag{2.10}$$

where

$$\mathbf{A} = \begin{bmatrix} 8 & -2 & -2 \\ -2 & 4 & -2 \\ -2 & -2 & 13 \end{bmatrix} \tag{2.11}$$

Equation (2.10) is a system of homogeneous linear algebraic equations. Equation (2.10) is the classical form of an eigenproblem.

Chapter 1 is devoted to the solution of systems of nonhomogeneous linear algebraic equations:

$$\mathbf{Ax} = \mathbf{b} \tag{2.12}$$

As discussed in Section 1.1, Eq. (2.12) may have a unique solution (the case considered in Chapter 1), no solution, an infinite number of solutions, or the trivial solution, $\mathbf{x} = 0$, if the system of equations is homogeneous:

$$\mathbf{Ax} = \mathbf{0} \tag{2.13}$$

Chapter 2 is concerned with solutions to Eq. (2.13) other than $\mathbf{x} = 0$, which are possible if the coefficient matrix $\mathbf{A}$ contains an unknown parameter, called an *eigenvalue*.

Every nonsingular square $n \times n$ matrix $\mathbf{A}$ has a set of n eigenvalues λ_i $(i = 1, \ldots, n)$ and n eigenvectors $\mathbf{x}_i$ $(i = 1, \ldots, n)$ that satisfy the equation

$$\boxed{(\mathbf{A} - \lambda\mathbf{I})\mathbf{x} = 0 \qquad \text{or} \qquad \mathbf{Ax} = \lambda\mathbf{x}} \tag{2.14}$$

The eigenvalues may be real or complex and distinct or repeated. The elements of the corresponding eigenvectors $\mathbf{x}_i$ are not unique. However, their relative magnitudes can be determined.

Consider an eigenproblem specified by two homogeneous linear algebraic equations:

$$(a_{11} - \lambda)x_1 + a_{12}x_2 = 0 \tag{2.15a}$$

$$(a_{21}x_1 + (a_{22} - \lambda)x_2 = 0 \tag{2.15b}$$

Equation (2.15) represents two straight lines in the x_1x_2 plane, both passing through the origin $x_1 = x_2 = 0$. Rearranging Eq. (2.15) gives

$$x_2 = -\frac{(a_{11} - \lambda)}{a_{12}}x_1 = m_1x_1 \tag{2.16a}$$

$$x_2 = -\frac{a_{21}}{(a_{22} - \lambda)}x_1 = m_2x_1 \tag{2.16b}$$

where m_1 and m_2 are the slopes of the two straight lines. Figure 2.2 illustrates Eq. (2.16) in the x_1x_2 plane. Both straight lines pass through the origin where $x_1 = x_2 = 0$, which is the trivial solution. If the slopes m_1 and m_2 are different, there is no other solution, as illustrated in Figure 2.2a. However, if $m_1 = m_2 = m$, as illustrated in Figure 2.2b, then the two straight lines lie on top of each other, and there are an infinite number of solutions. For any value of x_1, there is a corresponding value of x_2. The ratio of x_2 to x_1 is specified by value of the slope m. The values of λ which make $m_1 = m_2 = m$ are called *eigenvalues*, and the solution vector $\mathbf{x}$ corresponding to λ is called an *eigenvector*. Problems involving eigenvalues and eigenvectors are called *eigenproblems*.

Eigenproblems arise in the analysis of many physical systems. They arise in the analysis of the dynamic behavior of mechanical, electrical, fluid, thermal, and structural

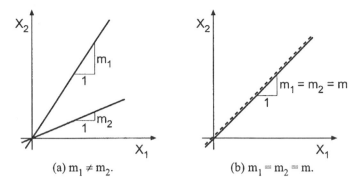

Figure 2.2 Graphical representation of Eq. (2.16).

systems. They also arise in the analysis of control systems. The objectives of this chapter are to introduce the general features of eigenproblems, to present several methods for solving simple eigenproblems, and to illustrate those methods by examples.

There are several methods for solving eigenproblems. Equation (2.14) can be solved *directly* by setting the determinant of $(\mathbf{A} - \lambda \mathbf{I})$ equal to zero and solving the resulting polynomial, which is called the *characteristic equation*, for λ. An iterative method, called the *power method*, is based on the repetitive matrix multiplication of an assumed eigenvector $\mathbf{x}$ by matrix $\mathbf{A}$, which eventually yields both λ and $\mathbf{x}$. The power method and the direct method are illustrated in this chapter. A more general and more powerful method, called the QR method, is based on more advanced concepts. The QR method is presented in Section 2.5. Serious students of eigenproblems should use the QR method.

The organization of Chapter 2 is illustrated in Figure 2.3. After a discussion of the general features of eigenproblems in this section, the mathematical characteristics of eigenproblems are discussed in Section 2.2. The power method and its variations are presented in Section 2.3. Section 2.4 presents the direct method. The most powerful method, the QR method, is developed in Section 2.5. The evaluation of eigenvectors is discussed in Section 2.6. A brief mention of other methods is then presented. Two programs for solving eigenproblems follow. The chapter closes with a Summary, which presents some general philosophy about solving eigenproblems, and lists the things you should be able to do after studying Chapter 2.

2.2 MATHEMATICAL CHARACTERISTICS OF EIGENPROBLEMS

The general features of eigenproblems are introduced in Section 2.1. The mathematical characteristics of eigenproblems are presented in this section.

Consider a system of nonhomogeneous linear algebraic equations:

$$\mathbf{Cx} = \mathbf{b} \tag{2.17}$$

Solving for $\mathbf{x}$ by Cramer's rule yields

$$x_j = \frac{\det(\mathbf{C}^j)}{\det(\mathbf{C})} \qquad (j = 1, \ldots, n) \tag{2.18}$$

where matrix $\mathbf{C}^j$ is matrix $\mathbf{C}$ with column j replaced by the vector $\mathbf{b}$. In general $\det(\mathbf{C}) \neq 0$, and unique values are found for x_j.

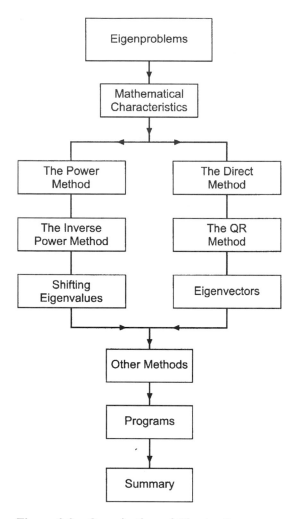

Figure 2.3   Organization of Chapter 2.

Consider a system of homogeneous linear algebraic equations:

$$Cx = 0 \tag{2.19}$$

Solving for **x** by Cramer's rule yields

$$x_j = \frac{\det(C^j)}{\det(C)} = \frac{0}{|C|} \qquad (j = 1, \ldots, n) \tag{2.20}$$

Therefore, $x = 0$ unless $\det(C) = 0$. In general, $\det(C) \neq 0$, and the only solution is the trivial solution, $x = 0$. For certain forms of **C** that involve an unspecified arbitrary scalar λ, the value of λ can be chosen to force $\det(C) = 0$, so that a solution other than the trivial solution, $x = 0$, is possible. In that case **x** is not unique, but relative values of x_j can be found.

Consider the coefficient matrix **C** to be of the form

$$C = A - \lambda B \tag{2.21}$$

where λ is an unspecified scalar. Then

$$\mathbf{Cx} = (\mathbf{A} - \lambda\mathbf{B})\mathbf{x} = 0 \qquad (2.22)$$

The values of λ are determined so that

$$\det(\mathbf{C}) = \det(\mathbf{A} - \lambda\mathbf{B}) = 0 \qquad (2.23)$$

The corresponding values of λ are the *eigenvalues*.

The homogeneous system of equations is generally written in the form

$$\mathbf{Ax} = \lambda\mathbf{Bx} \qquad (2.24)$$

In many problems $\mathbf{B} = \mathbf{I}$, and Eq. (2.24) becomes

$$\boxed{\mathbf{Ax} = \lambda\mathbf{x}} \qquad (2.25)$$

In problems where $\mathbf{B} \neq \mathbf{I}$, define the matrix $\bar{\mathbf{A}} = (\mathbf{B}^{-1}\mathbf{A})$. Then Eq. (2.24) becomes

$$\boxed{\bar{\mathbf{A}}\mathbf{x} = \lambda\mathbf{x}} \qquad (2.26)$$

which has the same form as Eq. (2.25). Equation (2.25) can be written in the alternate form

$$\boxed{(\mathbf{A} - \lambda\mathbf{I})\mathbf{x} = 0} \qquad (2.27)$$

which is the most common form of an eigenproblem statement.

The eigenvalues can be found by expanding $\det(\mathbf{A} - \lambda\mathbf{I}) = 0$ and finding the roots of the resulting nth-order polynomial, which is called the *characteristic equation*. This procedure is illustrated in the following discussion.

Consider the dynamic spring-mass problem specified by Eq. (2.9):

$$(\mathbf{A} - \lambda\mathbf{I})\mathbf{x} = \begin{bmatrix} (8 - \lambda) & -2 & -2 \\ -2 & (4 - \lambda) & -2 \\ -2 & -2 & (13 - \lambda) \end{bmatrix} \mathbf{x} = 0 \qquad (2.28)$$

The characteristic equation, $|\mathbf{A} - \lambda\mathbf{I}| = 0$, is

$$(8 - \lambda)\,[(4 - \lambda)(13 - \lambda) - 4] - (-2)\,[(-2)(13 - \lambda) - 4] + (-2)[4 + 2(4 - \lambda)] = 0 \qquad (2.29)$$

$$\lambda^3 - 25\lambda^2 + 176\lambda - 300 = 0 \qquad (2.30)$$

The eigenvalues are

$$\lambda = 13.870585, \quad 8.620434, \quad 2.508981 \qquad (2.31)$$

which can be demonstrated by direct substitution. From Eq. (2.7), the corresponding natural frequencies of oscillation, in terms of $f = 2\pi\omega$, where $\omega^2 = \lambda K_{ref}/m_{ref}$, are

$$f_1 = 2\pi\omega_1 = 2\pi\sqrt{\frac{\lambda_1 K_{ref}}{m_{ref}}} = 2\pi\sqrt{\frac{(13.870585)(10)}{2}} = 16.656 \text{ Hz} \qquad (2.32a)$$

$$f_2 = 2\pi\omega_2 = 2\pi\sqrt{\frac{\lambda_2 K_{ref}}{m_{ref}}} = 2\pi\sqrt{\frac{(8.620434)(10)}{2}} = 13.130 \text{ Hz} \qquad (2.32b)$$

$$f_3 = 2\pi\omega_3 = 2\pi\sqrt{\frac{\lambda_3 K_{ref}}{m_{ref}}} = 2\pi\sqrt{\frac{(2.508981)(10)}{2}} = 7.084 \text{ Hz} \qquad (2.32c)$$

where Hz = Hertz = 1.0 cycle/sec.

The eigenvectors corresponding to λ_1 to λ_3 are determined as follows. For each eigenvalue $\lambda_i (i = 1, 2, 3)$, find the amplitudes X_2 and X_3 relative to the amplitude X_1 by letting $X_1 = 1.0$. Any two of the three equations given by Eq. (2.9) can be used to solve for X_2 and X_3 with $X_1 = 1.0$. From Eqs. (2.9a) and (2.9c),

$$(8 - \lambda)X_1 - 2X_2 - 2X_3 = 0 \qquad (2.33a)$$
$$-2X_1 - 2X_2 + (13 - \lambda)X_3 = 0 \qquad (2.33b)$$

Solving Eqs. (2.33a) and (2.33b) for X_3 and substituting that result in Eq. (2.33a) yields

$$X_3 = \frac{(10 - \lambda)}{15} - \lambda \quad \text{and} \quad X_2 = \frac{(8 - \lambda)}{2} - X_3 \qquad (2.33c)$$

Substituting λ_1 to λ_3 into Eq. (2.33c) yields:

For $\lambda_1 = 13.870586$:

$$\mathbf{X}_1 = [1.000000 \quad 0.491779 \quad -3.427072] \qquad (2.34a)$$

For $\lambda_2 = 8.620434$:

$$\mathbf{X}_2 = [1.000000 \quad -0.526465 \quad 0.216247] \qquad (2.34b)$$

For $\lambda_3 = 2.508981$:

$$\mathbf{X}_3 = [1.000000 \quad 2.145797 \quad 0.599712] \qquad (2.34c)$$

The modes of oscillation corresponding to these results are illustrated in Figure 2.4.

In summary, eigenproblems arise from homogeneous systems of equations that contain an unspecified arbitrary parameter in the coefficients. The *characteristic equation* is determined by expanding the determinant

$$\det(\mathbf{A} - \lambda\mathbf{I}) = 0 \qquad (2.35)$$

which yields an nth-degree polynomial in λ. Solving the characteristic equation yields n eigenvalues λ_i ($i = 1, 2, \ldots, n$). The n eigenvectors $\mathbf{x}_i$ ($i = 1, 2, \ldots, n$), corresponding to the n eigenvalues λ_i ($i = 1, 2, \ldots, n$) are found by substituting the individual eigenvalues into the homogeneous system of equations, which is then solved for the eigenvectors.

In principle, the solution of eigenproblems is straightforward. In practice, when the size of the system of equations is very large, expanding the characteristic determinant to obtain the characteristic equation is difficult. Solving high-degree polynomials for the eigenvalues presents yet another difficult problem. Consequently, more straightforward

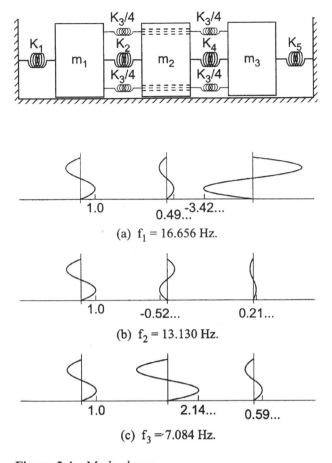

Figure 2.4   Mode shapes.

procedures for solving eigenproblems are desired. An iterative numerical procedure, called the *power method*, and its variations are presented in Section 2.3 to illustrate the numerical solution of eigenproblems. The direct method is presented in Section 2.4. The most general method, the QR method, is presented in Section 2.5.

2.3   THE POWER METHOD

Consider the linear eigenproblem:

$$\boxed{\mathbf{Ax} = \lambda \mathbf{x}} \tag{2.36}$$

The *power method* is based on repetitive multiplication of a trial eigenvector $\mathbf{x}^{(0)}$ by matrix $\mathbf{A}$ with a scaling of the resulting vector $\mathbf{y}$, so that the scaling factor approaches the largest eigenvalue λ and the scaled $\mathbf{y}$ vector approaches the corresponding eigenvector $\mathbf{x}$. The power method and several of its variations are presented in this section.

2.3.1. The Direct Power Method

When the largest (in absolute value) eigenvalue of $\mathbf{A}$ is distinct, its value can be found using an iterative technique called the *direct power method*. The procedure is as follows:

1. Assume a trial value $\mathbf{x}^{(0)}$ for the eigenvector $\mathbf{x}$. Choose one component of $\mathbf{x}$ to be unity. Designate that component as the *unity* component.
2. Perform the matrix multiplication:

$$\mathbf{A}\mathbf{x}^{(0)} = \mathbf{y}^{(1)} \tag{2.37}$$

3. Scale $\mathbf{y}^{(1)}$ so that the unity component remains unity:

$$\mathbf{y}^{(1)} = \lambda^{(1)}\mathbf{x}^{(1)} \tag{2.38}$$

4. Repeat steps 2 and 3 with $\mathbf{x} = \mathbf{x}^{(1)}$. Iterate to convergence. At convergence, the value λ is the largest (in absolute value) eigenvalue of $\mathbf{A}$, and the vector $\mathbf{x}$ is the corresponding eigenvector (scaled to unity on the unity component).

The general algorithm for the power method is as follows:

$$\boxed{\mathbf{A}\mathbf{x}^{(k)} = \mathbf{y}^{(k+1)} = \lambda^{(k+1)}\mathbf{x}^{(k+1)}} \tag{2.39}$$

When the iterations indicate that the unity component could be zero, a different unity component must be chosen. The method is slow to converge when the magnitudes (in absolute value) of the largest eigenvalues are nearly the same. When the largest eigenvalues are of equal magnitude, the power method, as described, fails.

Example 2.1. The direct power method.

Find the largest (in absolute value) eigenvalue and the corresponding eigenvector of the matrix given by Eq. (2.11):

$$\mathbf{A} = \begin{bmatrix} 8 & -2 & -2 \\ -2 & 4 & -2 \\ -2 & -2 & 13 \end{bmatrix} \tag{2.40}$$

Assume $\mathbf{x}^{(0)T} = [1.0 \quad 1.0 \quad 1.0]$. Scale the third component x_3 to unity. Then apply Eq. (2.39).

$$\mathbf{A}\mathbf{x}^{(0)} = \begin{bmatrix} 8 & -2 & -2 \\ -2 & 4 & -2 \\ -2 & -2 & 13 \end{bmatrix} \begin{bmatrix} 1.0 \\ 1.0 \\ 1.0 \end{bmatrix} = \begin{bmatrix} 4.00 \\ 0.00 \\ 9.00 \end{bmatrix} \quad \lambda^{(1)} = 9.00 \quad \mathbf{x}^{(1)} = \begin{bmatrix} 0.444444 \\ 0.000000 \\ 1.000000 \end{bmatrix} \tag{2.41}$$

$$\mathbf{A}\mathbf{x}^{(1)} = \begin{bmatrix} 8 & -2 & -2 \\ -2 & 4 & -2 \\ -2 & -2 & 13 \end{bmatrix} \begin{bmatrix} 0.444444 \\ 0.000000 \\ 1.000000 \end{bmatrix} = \begin{bmatrix} 1.555555 \\ -2.888888 \\ 12.111111 \end{bmatrix}$$

$$\lambda^{(2)} = 12.111111 \qquad \mathbf{x}^{(2)} = \begin{bmatrix} 0.128440 \\ -0.238532 \\ 1.000000 \end{bmatrix} \tag{2.42}$$

Table 2.1. The Power Method

k	λ	x_1	x_2	x_3
0		1.000000	1.000000	1.000000
1	9.000000	0.444444	0.000000	1.000000
2	12.111111	0.128440	-0.238532	1.000000
3	13.220183	-0.037474	-0.242887	1.000000
4	13.560722	-0.133770	-0.213602	1.000000
5	13.694744	-0.192991	-0.188895	1.000000
29	13.870583	-0.291793	-0.143499	1.000000
30	13.870584	-0.291794	-0.143499	1.000000

The results of the first two iterations presented above and subsequent iterations are presented in Table 2.1. These results were obtained on a 13-digit precision computer. The iterations were continued until λ changed by less than 0.000001 between iterations. The final solution for the largest eigenvalue, denoted as λ_1, and the corresponding eigenvector $\mathbf{x}_1$ is

$$\lambda_1 = 13.870584 \quad \text{and} \quad \mathbf{x}_1^T = [-0.291794] \quad -0.143499 \quad 1.000000] \quad (2.43)$$

This problem converged very slowly (30 iterations), which is a large number of iterations for a 3×3 matrix. A procedure for accelerating the convergence of a slowly converging eigenproblem is presented in Example 2.5.

2.3.2. Basis of the Power Method

The basis of the power method is as follows. Assume that $\mathbf{A}$ is an $n \times n$ nonsingular matrix having n eigenvalues, $\lambda_1, \lambda_2, \ldots, \lambda_n$, with n corresponding linearly independent eigenvectors, $\mathbf{x}_1, \mathbf{x}_2, \ldots, \mathbf{x}_n$. Assume further that $|\lambda_1| > |\lambda_2| > \cdots > |\lambda_n|$. Since the eigenvectors, $\mathbf{x}_i$ $(i = 1, 2, \ldots, n)$, are linearly independent (i.e., they span the n-dimensional space), any arbitrary vector $\mathbf{x}$ can be expressed as a linear combination of the eigenvectors. Thus,

$$\mathbf{x} = C_1\mathbf{x}_1 + C_2\mathbf{x}_2 + \cdots + C_n\mathbf{x}_n = \sum_{i=1}^{n} C_i\mathbf{x}_i \quad (2.44)$$

Multiplying both sides of Eq. (2.44) by $\mathbf{A}, \mathbf{A}^2, \ldots, \mathbf{A}^k$, etc., where the superscript denotes repetitive matrix multiplication, and recalling that $\mathbf{A}\mathbf{x}_i = \lambda_i\mathbf{x}_i$, yields

$$\mathbf{A}\mathbf{x} = \sum_{i=1}^{n} C_i\mathbf{A}\mathbf{x}_i = \sum_{i=1}^{n} C_i\lambda_i\mathbf{x}_i = \mathbf{y}^{(1)} \quad (2.45)$$

$$\mathbf{A}^2\mathbf{x} = \mathbf{A}\mathbf{y}^{(1)} = \sum_{i=1}^{n} C_i\lambda_i\mathbf{A}\mathbf{x}_i = \sum_{i=1}^{n} C_i\lambda_i^2\mathbf{x}_i = \mathbf{y}^{(2)} \quad (2.46)$$

$$\mathbf{A}^k\mathbf{x} = \mathbf{A}\mathbf{y}^{(k-1)} = \sum_{i=1}^{n} C_i\lambda_i^{k-1}\mathbf{A}\mathbf{x}_i = \sum_{i=1}^{n} C_i\lambda_i^k\mathbf{x}_i = \mathbf{y}^{(k)} \quad (2.47)$$

Factoring λ_1^k out of the next to last term in Eq. (2.47) yields

$$\mathbf{A}^k\mathbf{x} = \lambda_1^k \sum_{i=1}^{n} C_i \left(\frac{\lambda_i}{\lambda_1}\right)^k \mathbf{x}_i = \mathbf{y}^{(k)} \tag{2.48}$$

Since $|\lambda_1| > |\lambda_i|$ for $i = 2, 3, \ldots, n$, the ratios $(\lambda_i/\lambda_1)^k \to 0$ as $k \to \infty$, and Eq. (2.48) approaches the limit

$$\mathbf{A}^k\mathbf{x} = \lambda_1^k C_1 \mathbf{x}_1 = \mathbf{y}^{(k)} \tag{2.49}$$

Equation (2.49) approaches zero if $|\lambda_1| < 1$ and approaches infinity if $|\lambda_1| > 1$. Thus, Eq. (2.49) must be scaled between steps.

Scaling can be accomplished by scaling any component of vector $\mathbf{y}^{(k)}$ to unity at each step in the process. Choose the first component of vector $\mathbf{y}^{(k)}, y_1^{(k)}$, to be that component. Thus, $x_1 = 1.0$, and the first component of Eq. (2.49) is

$$y_1^{(k)} = \lambda_1^k C_1 \tag{2.50}$$

Applying Eq. (2.49) one more time (i.e., from k to $k + 1$) yields

$$y_1^{(k+1)} = \lambda_1^{k+1} C_1 \tag{2.51}$$

Taking the ratio of Eq. (2.51) to Eq. (2.50) gives

$$\frac{y_1^{(k+1)}}{y_1^{(k)}} = \frac{\lambda_1^{k+1} C_1}{\lambda_1^k C_1} = \lambda_1 \tag{2.52}$$

Thus, if $y_1^{(k)} = 1$, then $y_1^{(k+1)} = \lambda_1$. If $y_1^{(k+1)}$ is scaled by λ_1 so that $y_1^{(k+1)} = 1$, then $y_1^{(k+2)} = \lambda_1$, etc. Consequently, scaling a particular component of vector $\mathbf{y}$ each iteration essentially factors λ_1 out of vector $\mathbf{y}$, so that Eq. (2.49) converges to a finite value. In the limit as $k \to \infty$, the scaling factor approaches λ_1, and the scaled vector $\mathbf{y}$ approaches the eigenvector $\mathbf{x}_1$.

Several restrictions apply to the power method.

1. The largest eigenvalue must be distinct.
2. The n eigenvectors must be independent.
3. The initial guess $\mathbf{x}_i^{(0)}$ must contain some component of eigenvector $\mathbf{x}_i$, so that $C_i \neq 0$.
4. The convergence rate is proportional to the ratio

$$\frac{|\lambda_i|}{|\lambda_{i-1}|}$$

where λ_i is the largest (in magnitude) eigenvalue and λ_{i-1} is the second largest (in magnitude) eigenvalue.

2.3.3. The Inverse Power Method

When the smallest (in absolute value) eigenvalue of matrix $\mathbf{A}$ is distinct, its value can be found using a variation of the power method called the *inverse power method*. Essentially, this involves finding the largest (in magnitude) eigenvalue of the inverse matrix $\mathbf{A}^{-1}$, which is the smallest (in magnitude) eigenvalue of matrix $\mathbf{A}$. Recall the original eigenproblem:

$$\mathbf{A}\mathbf{x} = \lambda\mathbf{x} \tag{2.53}$$

Multiplying Eq. (2.53) by $\mathbf{A}^{-1}$ gives

$$\mathbf{A}^{-1}\mathbf{A}\mathbf{x} = \mathbf{I}\mathbf{x} = \mathbf{x} = \lambda\mathbf{A}^{-1}\mathbf{x} \tag{2.54}$$

Rearranging Eq. (2.54) yields an eigenproblem for $\mathbf{A}^{-1}$. Thus,

$$\boxed{\mathbf{A}^{-1}\mathbf{x} = \left(\frac{1}{\lambda}\right)\mathbf{x} = \lambda_{\text{inverse}}\mathbf{x}} \tag{2.55}$$

The eigenvalues of matrix $\mathbf{A}^{-1}$, that is, λ_{inverse}, are the reciprocals of the eigenvalues of matrix $\mathbf{A}$. The eigenvectors of matrix $\mathbf{A}^{-1}$ are the same as the eigenvectors of matrix $\mathbf{A}$. The power method can be used to find the largest (in absolute value) eigenvalue of matrix $\mathbf{A}^{-1}$, λ_{inverse}. The reciprocal of that eigenvalue is the smallest (in absolute value) eigenvalue of matrix $\mathbf{A}$.

In practice the LU method is used to solve the inverse eigenproblem instead of calculating the inverse matrix $\mathbf{A}^{-1}$. The power method applied to matrix $\mathbf{A}^{-1}$ is given by

$$\mathbf{A}^{-1}\mathbf{x}^{(k)} = \mathbf{y}^{(k+1)} \tag{2.56}$$

Multiplying Eq. (2.56) by $\mathbf{A}$ gives

$$\mathbf{A}\mathbf{A}^{-1}\mathbf{x}^{(k)} = \mathbf{I}\mathbf{x}^{(k)} = \mathbf{x}^{(k)} = \mathbf{A}\mathbf{y}^{(k+1)} \tag{2.57}$$

which can be written as

$$\boxed{\mathbf{A}\mathbf{y}^{(k+1)} = \mathbf{x}^{(k)}} \tag{2.58}$$

Equation (2.58) is in the standard form $\mathbf{A}\mathbf{x} = \mathbf{b}$, where $\mathbf{x} = \mathbf{y}^{(k+1)}$ and $\mathbf{b} = \mathbf{x}^{(k)}$. Thus, for a given $\mathbf{x}^{(k)}$, $\mathbf{y}^{(k+1)}$ can be found by the Doolittle LU method. The procedure is as follows:

1. Solve for $\mathbf{L}$ and $\mathbf{U}$ such that $\mathbf{L}\mathbf{U} = \mathbf{A}$ by the Doolittle LU method.
2. Assume $\mathbf{x}^{(0)}$. Designate a component of $\mathbf{x}$ to be unity.
3. Solve for $\mathbf{x}'$ by forward substitution using the equation

$$\mathbf{L}\mathbf{x}' = \mathbf{x}^{(0)} \tag{2.59}$$

4. Solve for $\mathbf{y}^{(1)}$ by back substitution using the equation

$$\mathbf{U}\mathbf{y}^{(1)} = \mathbf{x}' \tag{2.60}$$

5. Scale $\mathbf{y}^{(1)}$ so that the unity component is unity. Thus,

$$\mathbf{y}^{(1)} = \lambda_{\text{inverse}}^{(1)}\mathbf{x}^{(1)} \tag{2.61}$$

6. Repeat steps 3 to 5 with $\mathbf{x}^{(1)}$. Iterate to convergence. At convergence, $\lambda = 1/\lambda_{\text{inverse}}$, and $\mathbf{x}^{(k+1)}$ is the corresponding eigenvector.

The inverse power method algorithm is as follows:

$$\boxed{\begin{aligned} \mathbf{L}\mathbf{x}' &= \mathbf{x}^{(k)} \\ \mathbf{U}\mathbf{y}^{(k+1)} &= \mathbf{x}' \\ \mathbf{y}^{(k+1)} &= \lambda_{\text{inverse}}^{(k+1)}\mathbf{x}^{(k+1)} \end{aligned}} \quad\begin{aligned} &(2.62) \\ &(2.63) \\ &(2.64) \end{aligned}$$

Example 2.2. The inverse power method.

Find the smallest (in absolute value) eigenvalue and the corresponding eigenvector of the matrix given by Eq. (2.11):

$$\mathbf{A} = \begin{bmatrix} 8 & -2 & -2 \\ -2 & 4 & -2 \\ -2 & -2 & 13 \end{bmatrix} \tag{2.65}$$

Assume $\mathbf{x}^{(0)T} = [1.0 \quad 1.0 \quad 1.0]$. Scale the first component of $\mathbf{x}$ to unity. The first step is to solve for $\mathbf{L}$ and $\mathbf{U}$ by the Doolittle LU method. The results are

$$\mathbf{L} = \begin{bmatrix} 1 & 0 & 0 \\ -1/4 & 1 & 0 \\ -1/4 & -5/7 & 1 \end{bmatrix} \quad \text{and} \quad \mathbf{U} = \begin{bmatrix} 8 & -2 & -2 \\ 0 & 7/2 & -5/2 \\ 0 & 0 & 75/7 \end{bmatrix} \tag{2.66}$$

Solve for $\mathbf{x}'$ by forward substitution using $\mathbf{Lx}' = \mathbf{x}^{(0)}$.

$$\begin{bmatrix} 1 & 0 & 0 \\ -1/4 & 1 & 0 \\ -1/4 & -5/7 & 1 \end{bmatrix} \begin{bmatrix} x_1' \\ x_2' \\ x_3' \end{bmatrix} = \begin{bmatrix} 1.0 \\ 1.0 \\ 1.0 \end{bmatrix}$$

$$x_1' = 1.0$$
$$x_2' = 1.0 - (-1/4)(1.0) = 5/4$$
$$x_3' = 1.0 - (-1/4)(1.0) - (-5/7)(5/4) = 15/7 \tag{2.67}$$

Solve for $\mathbf{y}^{(1)}$ by back substitution using $\mathbf{Uy}^{(1)} = \mathbf{x}'$.

$$\begin{bmatrix} 8 & -2 & -2 \\ 0 & 7/2 & -5/2 \\ 0 & 0 & 75/7 \end{bmatrix} \begin{bmatrix} y_1^{(1)} \\ y_2^{(2)} \\ y_3^{(1)} \end{bmatrix} = \begin{bmatrix} 1.0 \\ 5/4 \\ 15/7 \end{bmatrix}$$

$$y_1^{(1)} = [1.0 - (-2.0)(0.5) - (-2)(0.2)]/8 = 0.30$$
$$y_2^{(1)} = [5/4 - (-5/2)(0.2)]/(7/2) = 0.50$$
$$y_3^{(1)} = (15/7)/(75/7) = 0.20 \tag{2.68}$$

Scale $\mathbf{y}^{(1)}$ so that the unity component is unity.

$$\mathbf{y}^{(1)} = \begin{bmatrix} 0.30 \\ 0.50 \\ 0.20 \end{bmatrix} \quad \lambda_{\text{inverse}}^{(1)} = 0.300000 \quad \mathbf{x}^{(1)} = \begin{bmatrix} 1.000000 \\ 1.666667 \\ 0.666667 \end{bmatrix} \tag{2.69}$$

  The results of the first iteration presented above and subsequent iterations are presented in Table 2.2. These results were obtained on a 13-digit precision computer. The iterations were continued until λ_{inverse} changed by less than 0.000001 between iterations.

  The final solution for the smallest eigenvalue λ_3 and the corresponding eigenvector $\mathbf{x}_3$ is

$$\lambda_3 = \frac{1}{\lambda_{\text{inverse}}} = \frac{1}{0.398568} = 2.508981 \quad \text{and}$$

$$\mathbf{x}_3^T = [1.000000 \quad 2.145797 \quad 0.599712] \tag{2.70}$$

Table 2.2. The Inverse Power Method

k	λ_{inverse}	x_1	x_2	x_3
0		1.000000	1.000000	1.000000
1	0.300000	1.000000	1.666667	0.666667
2	0.353333	1.000000	1.981132	0.603774
3	0.382264	1.000000	2.094439	0.597565
4	0.393346	1.000000	2.130396	0.598460
....	....			
12	0.398568	1.000000	2.145796	0.599712
13	0.398568	1.000000	2.145797	0.599712

2.3.4. The Shifted Power Method

The eigenvalues of a matrix $\mathbf{A}$ may be shifted by a scalar s by subtracting $s\mathbf{I}\mathbf{x} = s\mathbf{x}$ from both sides of the standard eigenproblem, $\mathbf{A}\mathbf{x} = \lambda\mathbf{x}$. Thus,

$$\mathbf{A}\mathbf{x} - s\mathbf{I}\mathbf{x} = \lambda\mathbf{x} - s\mathbf{x} \tag{2.71}$$

which yields

$$(\mathbf{A} - s\mathbf{I})\mathbf{x} = (\lambda - s)\mathbf{x} \tag{2.72}$$

which can be written as

$$\boxed{\mathbf{A}_{\text{shifted}}\mathbf{x} = \lambda_{\text{shifted}}\mathbf{x}} \tag{2.73}$$

where $\mathbf{A}_{\text{shifted}} = (\mathbf{A} - s\mathbf{I})$ is the shifted matrix and $\lambda_{\text{shifted}} = \lambda - s$ is the eigenvalue of the shifted matrix. Shifting a matrix $\mathbf{A}$ by a scalar, s, shifts the eigenvalues by s. Shifting a matrix by a scalar does not affect the eigenvectors. Shifting the eigenvalues of a matrix can be used to:

1. Find the *opposite extreme eigenvalue*, which is either the smallest (in absolute value) eigenvalue or the largest (in absolute value) eigenvalue of opposite sign
2. Find *intermediate eigenvalues*
3. *Accelerate convergence* for slowly converging eigenproblems

2.3.4.1. Shifting Eigenvalues to Find the Opposite Extreme Eigenvalue

Consider a matrix whose eigenvalues are all the same sign, for example 1, 2, 4, and 8. For this matrix, 8 is the largest (in absolute value) eigenvalue and 1 is the opposite extreme eigenvalue. Solve for the largest (in absolute value) eigenvalue, $\lambda_{\text{Largest}} = 8$, by the direct power method. Shifting the eigenvalues by $s = 8$ yields the shifted eigenvalues $-7, -6, -4$, and 0. Solve for the largest (in absolute value) eigenvalue of the shifted matrix, $\lambda_{\text{shifted,Largest}} = -7$, by the power method. Then $\lambda_{\text{Smallest}} = \lambda_{\text{shifted,Largest}} + 8 = -7 + 8 = 1$. This procedure yields the same eigenvalue as the inverse power method applied to the original matrix.

Consider a matrix whose eigenvalues are both positive and negative, for example, $-1, 2, 4$, and 8. For this matrix, 8 is the largest (in absolute value) eigenvalue and -1 is the opposite extreme eigenvalue. Solve for the largest (in absolute value) eigenvalue,

$\lambda_{Largest} = 8$, by the power method. Shifting the eigenvalues by $s = 8$ yields the shifted eigenvalues -9, -6, -4, and 0. Solve for the largest (in absolute value) eigenvalue of the shifted matrix, $\lambda_{shifted,Largest} = -9$, by the power method. Then $\lambda_{Largest,Negative} = \lambda_{shifted,Largest} + 8 = -9 + 8 = -1$.

Both of the cases described above are solved by shifting the matrix by the largest (in absolute value) eigenvalue and applying the direct power method to the shifted matrix. Generally speaking, it is not known a priori which result will be obtained. If all the eigenvalues of a matrix have the same sign, the smallest (in absolute value) eigenvalue will be obtained. If a matrix has both positive and negative eigenvalues, the largest eigenvalue of opposite sign will be obtained.

The above procedure is called the *shifted direct power method*. The procedure is as follows:

1. Solve for the largest (in absolute value) eigenvalue $\lambda_{Largest}$.
2. Shift the eigenvalues of matrix $\mathbf{A}$ by $s = \lambda_{Largest}$ to obtain the shifted matrix $\mathbf{A}_{shifted}$.
3. Solve for the eigenvalue $\lambda_{shifted}$ of the shifted matrix $\mathbf{A}_{shifted}$ by the direct power method.
4. Calculate the opposite extreme eigenvalue of matrix $\mathbf{A}$ by $\lambda = \lambda_{shifted} + s$.

Example 2.3. The shifted direct power method for opposite extreme eigenvalues.

Find the opposite extreme eigenvalue of matrix $\mathbf{A}$ by shifting the eigenvalues by $s = \lambda_{Largest} = 13.870584$. The original and shifted matrices are:

$$\mathbf{A} = \begin{bmatrix} 8 & -2 & -2 \\ -2 & 4 & -2 \\ -2 & -2 & 13 \end{bmatrix} \tag{2.74}$$

$$\mathbf{A}_{shifted} = \begin{bmatrix} (8 - 13.870584) & -2 & -2 \\ -2 & (4 - 13.870584) & -2 \\ -2 & -2 & (13 - 13.870584) \end{bmatrix}$$

$$= \begin{bmatrix} -5.870584 & -2.000000 & -2.000000 \\ -2.000000 & -9.870584 & -2.000000 \\ -2.000000 & -2.000000 & -0.870584 \end{bmatrix} \tag{2.75}$$

Assume $\mathbf{x}^{(0)T} = [1.0 \quad 1.0 \quad 1.0]$. Scale the second component to unity. Applying the power method to matrix $\mathbf{A}_{shifted}$ gives

$$\mathbf{A}_{shifted}\mathbf{x}^{(0)} = \begin{bmatrix} -5.870584 & -2.000000 & -2.000000 \\ -2.000000 & -9.870584 & -2.000000 \\ -2.000000 & -2.000000 & -0.870584 \end{bmatrix} \begin{bmatrix} 1.0 \\ 1.0 \\ 1.0 \end{bmatrix}$$

$$= \begin{bmatrix} -9.870584 \\ -13.870584 \\ -4.870584 \end{bmatrix} = \mathbf{y}^{(1)} \tag{2.76}$$

Table 2.3. Shifting Eigenvalues to Find the Opposite Extreme Eigenvalue

k	λ_{shifted}	x_1	x_2	x_3
0		1.000000	1.000000	1.000000
1	-13.870584	0.711620	1.000000	0.351145
2	-11.996114	0.573512	1.000000	0.310846
3	-11.639299	0.514510	1.000000	0.293629
4	-11.486864	0.488187	1.000000	0.285948
....	....			
19	-11.361604	0.466027	1.000000	0.279482
20	-11.361604	0.466027	1.000000	0.279482

Scaling the unity component of $\mathbf{y}^{(1)}$ to unity gives

$$\lambda_{\text{shifted}}^{(1)} = -13.870584 \qquad \text{and} \qquad \mathbf{x}^{(1)} = \begin{bmatrix} 0.711620 \\ 1.000000 \\ 0.351145 \end{bmatrix} \qquad (2.77)$$

The results of the first iteration presented above and subsequent iterations are presented in Table 2.3. These results were obtained on a 13-digit precision computer with an absolute convergence tolerance of 0.000001.

The largest (in magnitude) eigenvalue of $\mathbf{A}_{\text{shifted}}$ is $\lambda_{\text{shifted,Largest}} = -11.361604$. Thus, the opposite extreme eigenvalue of matrix $\mathbf{A}$ is

$$\lambda = \lambda_{\text{shifted,Largest}} + 13.870584 = -11.361604 + 13.870586 = 2.508980 \qquad (2.78)$$

Since this eigenvalue, $\lambda = 2.508980$, has the same sign as the largest (in absolute value) eigenvalue, $\lambda = 13.870584$, it is the smallest (in absolute value) eigenvalue of matrix $\mathbf{A}$, and all the eigenvalues of matrix $\mathbf{A}$ are positive.

2.3.4.2. Shifting Eigenvalues to Find Intermediate Eigenvalues

Intermediate eigenvalues λ_{Inter} lie between the largest eigenvalue and the smallest eigenvalue. Consider a matrix whose eigenvalues are 1, 2, 4, and 8. Solve for the largest (in absolute value) eigenvalue, $\lambda_{\text{Largest}} = 8$, by the power method and the smallest eigenvalue, $\lambda_{\text{Smallest}} = 1$, by the inverse power method. Two intermediate eigenvalues, $\lambda_{\text{Inter}} = 2$ and 4, remain to be determined. If λ_{Inter} is guessed to be $\lambda_{\text{Guess}} = 5$ and the eigenvalues are shifted by $s = 5$, the eigenvalues of the shifted matrix are $-4, -3, -1$, and 3. Applying the inverse power method to the shifted matrix gives $\lambda_{\text{shifted}} = -1$, from which $\lambda = \lambda_{\text{shifted}} + s = -1 + 5 = 4$. The power method is not an efficient method for finding intermediate eigenvalues. However, it can be used for that purpose.

The above procedure is called the *shifted inverse power method*. The procedure as follows:

1. Guess a value λ_{Guess} for the intermediate eigenvalue of the shifted matrix.
2. Shift the eigenvalues by $s = \lambda_{\text{Guess}}$ to obtain the shifted matrix $\mathbf{A}_{\text{shifted}}$.
3. Solve for the eigenvalue $\lambda_{\text{shifted,inverse}}$ of the inverse shifted matrix $\mathbf{A}_{\text{shifted}}^{-1}$ by the inverse power method applied to matrix $\mathbf{A}_{\text{shifted}}$.

4. Solve for $\lambda_{\text{shifted}} = 1/\lambda_{\text{shifted,inverse}}$.
5. Solve for the intermediate eigenvalue $\lambda_{\text{Inter}} = \lambda_{\text{shifted}} + s$.

Example 2.4.   The shifted inverse power method for intermediate eigenvalues.

Let's attempt to find an intermediate eigenvalue of matrix $\mathbf{A}$ by guessing its value, for example, $\lambda_{\text{Guess}} = 10.0$. The corresponding shifted matrix $\mathbf{A}_{\text{shifted}}$ is

$$\mathbf{A}_{\text{shifted}} = (\mathbf{A} - \lambda_{\text{Guess}}\mathbf{I}) = \begin{bmatrix} (8-10.0) & -2 & -2 \\ -2 & (4-10.0) & -2 \\ -2 & -2 & (13-10.0) \end{bmatrix}$$

$$= \begin{bmatrix} -2.0 & -2.0 & -2.0 \\ -2.0 & -6.0 & -2.0 \\ -2.0 & -2.0 & 3.0 \end{bmatrix} \tag{2.79}$$

Solving for $\mathbf{L}$ and $\mathbf{U}$ by the Doolittle LU method yields:

$$\mathbf{L} = \begin{bmatrix} 1.0 & 0.0 & 0.0 \\ 1.0 & 1.0 & 0.0 \\ 1.0 & 0.0 & 1.0 \end{bmatrix} \quad \text{and} \quad \mathbf{U} = \begin{bmatrix} -2.0 & -2.0 & -2.0 \\ 0.0 & -4.0 & 0.0 \\ 0.0 & 0.0 & 5.0 \end{bmatrix} \tag{2.80}$$

Assume $\mathbf{x}^{(0)T} = [1.0 \quad 1.0 \quad 1.0]$. Scale the first component of $\mathbf{x}$ to unity. Solve for $\mathbf{x}'$ by forward substitution using $\mathbf{L}\mathbf{x}' = \mathbf{x}^{(0)}$:

$$\begin{bmatrix} 1.0 & 0.0 & 0.0 \\ 1.0 & 1.0 & 0.0 \\ 1.0 & 0.0 & 1.0 \end{bmatrix} \begin{bmatrix} x_1' \\ x_2' \\ x_3' \end{bmatrix} = \begin{bmatrix} 1.0 \\ 1.0 \\ 1.0 \end{bmatrix} \tag{2.81}$$

which yields

$$x_1' = 1.0 \tag{2.82a}$$
$$x_2' = 1.0 - 1.0(1.0) = 0.0 \tag{2.82b}$$
$$x_3' = 1.0 - 1.0(1.0) - 1.0(0.0) = 0.0 \tag{2.82c}$$

Solve for $\mathbf{y}^{(1)}$ by back substitution using $\mathbf{U}\mathbf{y}^{(1)} = \mathbf{x}'$.

$$\begin{bmatrix} -2.0 & -2.0 & -2.0 \\ 0.0 & -4.0 & 0.0 \\ 0.0 & 0.0 & 5.0 \end{bmatrix} \begin{bmatrix} y_1^{(1)} \\ y_2^{(1)} \\ y_3^{(1)} \end{bmatrix} = \begin{bmatrix} 1.0 \\ 0.0 \\ 0.0 \end{bmatrix} \tag{2.83}$$

which yields

$$y_3^{(1)} = 0.0/(5.0) = 0.0 \tag{2.84a}$$
$$y_2^{(1)} = [0.0 - (0.0)(0.0)]/(-4.0) = 0.0 \tag{2.84b}$$
$$y_1^{(1)} = [1.0 - (-2.0)(0.0) - (-2.0)(0.0)]/(-2.0) = -0.50 \tag{2.84c}$$

Table 2.4. Shifting Eigenvalues to Find Intermediate Eigenvalues

k	$\lambda_{\text{shifted,inverse}}$	x_1	x_2	x_3
0		1.000000	1.000000	1.000000
1	-0.500000	1.000000	0.000000	0.000000
2	-0.550000	1.000000	-0.454545	0.363636
3	-0.736364	1.000000	-0.493827	0.172840
4	-0.708025	1.000000	-0.527463	0.233653
. . . . .				
14	-0.724865	1.000000	-0.526465	0.216248
15	-0.724866	1.000000	-0.526465	0.216247

Scale $\mathbf{y}^{(1)}$ so that the unity component is unity.

$$\mathbf{y}^{(1)} = \begin{bmatrix} -0.50 \\ 0.00 \\ 0.00 \end{bmatrix} \qquad \lambda^{(1)}_{\text{shifted,inverse}} = -0.50 \qquad \mathbf{x}^{(1)} = \begin{bmatrix} 1.00 \\ 0.00 \\ 0.00 \end{bmatrix} \qquad (2.85)$$

The first iteration and subsequent iterations are summarized in Table 2.4. These results were obtained on a 13-digit precision computer with an absolute convergence tolerance of 0.000001.

Thus, the largest (in absolute value) eigenvalue of matrix $\mathbf{A}_{\text{shifted}}^{-1}$ is $\lambda_{\text{shifted,inverse}} = -0.724866$. Consequently, the corresponding eigenvalue of matrix $\mathbf{A}_{\text{shifted}}$ is

$$\lambda_{\text{shifted}} = \frac{1}{\lambda_{\text{shifted,inverse}}} = \frac{1}{-0.724866} = -1.379566 \qquad (2.86)$$

Thus, the intermediate eigenvalue of matrix $\mathbf{A}$ is

$$\lambda_I = \lambda_{\text{shifted}} + s = -1.379566 + 10.000000 = 8.620434 \qquad (2.87)$$

and the corresponding eigenvector is $\mathbf{x}^T = [1.0 \quad -0.526465 \quad 0.216247]$.

2.3.4.3. Shifting Eigenvalues to Accelerate Convergence

The shifting eigenvalue concept can be used to accelerate the convergence of the power method for a slowly converging eigenproblem. When an estimate λ_{Est} of an eigenvalue of matrix $\mathbf{A}$ is known, for example, from several initial iterations of the direct power method, the eigenvalues can be shifted by this approximate value so that the shifted matrix has an eigenvalue near zero. This eigenvalue can then be found by the inverse power method.

The above procedure is called the *shifted inverse power method*. The procedure is as follows:

1. Obtain an estimate λ_{Est} of the eigenvalue λ, for example, by several applications of the direct power method.
2. Shift the eigenvalues by $s = \lambda_{\text{Est}}$ to obtain the shifted matrix, $\mathbf{A}_{\text{shifted}}$.
3. Solve for the eigenvalue $\lambda_{\text{shifted,inverse}}$ of the inverse shifted matrix $\mathbf{A}_{\text{shifted}}^{-1}$ by the inverse power method applied to matrix $\mathbf{A}_{\text{shifted}}$. Let the first guess for $\mathbf{x}$ be the value of $\mathbf{x}$ corresponding to λ_{Est}.

4. Solve for $\lambda_{\text{shifted}} = 1/\lambda_{\text{shifted,inverse}}$.
5. Solve for $\lambda = \lambda_{\text{shifted}} + s$.

Example 2.5.   The shifted inverse power method for accelerating convergence.

The first example of the power method, Example 2.1. converged very slowly since the two largest eigenvalues of matrix **A** are close together (i.e., 13.870584 and 8.620434). Convergence can be accelerated by using the results of an early iteration, say iteration 5, to shift the eigenvalues by the approximate eigenvalue, and then using the inverse power method on the shifted matrix to accelerate convergence. From Example 2.1, after 5 iterations, $\lambda^{(5)} = 13.694744$ and $\mathbf{x}^{(5)T} = [-0.192991 \quad -0.188895 \quad 1.000000]$. Thus, shift matrix **A** by $s = 13.694744$:

$$\mathbf{A}_{\text{shifted}} = (\mathbf{A} - s\mathbf{I}) = \begin{bmatrix} -5.694744 & -2.000000 & -2.00000 \\ -2.000000 & -9.694744 & -2.00000 \\ -2.000000 & -2.000000 & -0.694744 \end{bmatrix} \tag{2.88}$$

The corresponding **L** and **U** matrices are

$$\mathbf{L} = \begin{bmatrix} 1.000000 & 0.000000 & 0.000000 \\ 0.351201 & 1.000000 & 0.000000 \\ 0.351201 & 0.144300 & 1.000000 \end{bmatrix}$$

$$\mathbf{U} = \begin{bmatrix} -5.694744 & -2.000000 & -2.000000 \\ 0.000000 & -8.992342 & -1.297598 \\ 0.000000 & 0.000000 & 0.194902 \end{bmatrix} \tag{2.89}$$

Let $\mathbf{x}^{(0)T} = \mathbf{x}^{(5)T}$ and continue scaling the third component of **x** to unity. Applying the inverse power method to matrix $\mathbf{A}_{\text{shifted}}$ yields the results presented in Table 2.5. These results were obtained on a 13-digit precision computer with an absolute convergence tolerance of 0.000001. The eigenvalue λ_{shifted} of the shifted matrix $\mathbf{A}_{\text{shifted}}$ is

$$\lambda_{\text{shifted}} = \frac{1}{\lambda_{\text{shifted,inverse}}} = \frac{1}{5.686952} = 0.175841 \tag{2.90}$$

Table 2.5.   Shifting Eigenvalues to Accelerate Convergence

k	$\lambda_{\text{shifted,inverse}}$	x_1	x_2	x_3
0		-0.192991	-0.188895	1.000000
1	5.568216	-0.295286	-0.141881	1.000000
2	5.691139	-0.291674	-0.143554	1.000000
3	5.686807	-0.291799	-0.143496	1.000000
4	5.686957	-0.291794	-0.143498	1.000000
5	5.686952	-0.291794	-0.143498	1.000000
6	5.686952	-0.291794	-0.143498	1.000000

Thus, the eigenvalue λ of the original matrix $\mathbf{A}$ is

$$\lambda = \lambda_{\text{shifted}} + s = 0.175841 + 13.694744 = 13.870585 \tag{2.91}$$

This is the same result obtained in the first example with 30 iterations. The present solution required only 11 total iterations: 5 for the initial solution and 6 for the final solution.

2.3.5. Summary

In summary, the largest eigenvalue, $\lambda = 13.870584$, was found by the power method; the smallest eigenvalue, $\lambda = 2.508981$, was found by both the inverse power method and by shifting the eigenvalues by the largest eigenvalue; and the third (and intermediate) eigenvalue, $\lambda = 8.620434$, was found by shifting eigenvalues. The corresponding eigenvectors were also found. These results agree with the exact solution of this problem presented in Section 2.2.

2.4 THE DIRECT METHOD

The power method and its variations presented in Section 2.3 apply to linear eigenproblems of the form

$$\mathbf{A}\mathbf{x} = \lambda\mathbf{x} \tag{2.92}$$

Nonlinear eigenproblems of the form

$$\boxed{\mathbf{A}\mathbf{x} = \mathbf{B}(\lambda)\mathbf{x}} \tag{2.93}$$

where $\mathbf{B}(\lambda)$ is a nonlinear function of λ, cannot be solved by the power method. Linear eigenproblems and nonlinear eigenproblems both can be solved by a direct approach which involves finding the zeros of the characteristic equation directly.

For a linear eigenproblem, the characteristic equation is obtained from

$$\det(\mathbf{A} - \lambda\mathbf{I}) = 0 \tag{2.94}$$

Expanding Eq. (2.94), which can be time consuming for a large system, yields an nth-degree polynomial in λ. The roots of the characteristic polynomial can be determined by the methods presented in Section 3.5 for finding the roots of polynomials.

For a nonlinear eigenproblem, the characteristic equation is obtained from

$$\det[\mathbf{A} - \mathbf{B}(\lambda)] = 0 \tag{2.95}$$

Expanding Eq. (2.95) yields a nonlinear function of λ, which can be solved by the methods presented in Chapter 3.

An alternate approach for solving for the roots of the characteristic equation directly is to solve Eqs. (2.94) and (2.95) iteratively. This can be accomplished by applying the secant method, presented in Section 3.4.3, to Eqs. (2.94) and (2.95). Two initial approximations of λ are assumed, λ_0 and λ_1, the corresponding values of the characteristic determinant are computed, and these results are used to construct a linear relationship between λ and the value of the characteristic determinant. The solution of that linear relationship is taken as the next approximation to λ, and the procedure is repeated iteratively to convergence. Reasonable initial approximations are required, especially for nonlinear eigenproblems.

The direct method determines only the eigenvalues. The corresponding eigenvectors must be determined by substituting the eigenvalues into the system of equations and solving for the corresponding eigenvectors directly, or by applying the inverse power method one time as illustrated in Section 2.6.

Example 2.6. The direct method for a linear eigenproblem.

Let's find the largest eigenvalue of the matrix given by Eq. (2.11) by the direct method. Thus,

$$\mathbf{A} = \begin{bmatrix} 8 & -2 & -2 \\ -2 & 4 & -2 \\ -2 & -2 & 13 \end{bmatrix} \tag{2.96}$$

The characteristic determinant corresponding to Eq. (2.96) is

$$f(\lambda) = \det(\mathbf{A} - \lambda\mathbf{I}) = \begin{bmatrix} (8-\lambda) & -2 & -2 \\ -2 & (4-\lambda) & -2 \\ -2 & -2 & (13-\lambda) \end{bmatrix} = 0 \tag{2.97}$$

Equation (2.97) can be solved by the secant method presented in Section 3.4.3. Let $\lambda_0 = 15.0$ and $\lambda_1 = 13.0$. Thus,

$$f(\lambda_0) = f(15.0) = \begin{vmatrix} (8-15.0) & -2 & -2 \\ -2 & (4-15.0) & -2 \\ -2 & -2 & (13-15.0) \end{vmatrix} = -90.0 \quad (2.98a)$$

$$f(\lambda_1) = f(13.0) = \begin{vmatrix} (8-13.0) & -2 & -2 \\ -2 & (4-13.0) & -2 \\ -2 & -2 & (13-13.0) \end{vmatrix} = 40.0 \quad (2.98b)$$

The determinants in Eq. (2.98) were evaluated by Gauss elimination, as described in Section 1.3.6. Write the linear relationship between λ and $f(\lambda)$:

$$\frac{f(\lambda_1) - f(\lambda_0)}{\lambda_1 - \lambda_0} = \text{slope} = \frac{f(\lambda_2) - f(\lambda_1)}{\lambda_2 - \lambda_1} \tag{2.99}$$

where $f(\lambda_2) = 0$ is the desired solution. Thus,

$$\text{slope} = \frac{40.0 - (-90.0)}{13.0 - 15.0} = -65.0 \tag{2.100}$$

Solving Eq. (2.99) for λ_2 to give $f(\lambda_2) = 0.0$ gives

$$\lambda_2 = \lambda_1 - \frac{f(\lambda_1)}{\text{slope}} = 13.0 - \frac{40.0}{(-65.0)} = 13.615385 \tag{2.101}$$

The results of the first iteration presented above and subsequent iterations are presented in Table 2.6. The solution is $\lambda = 13.870585$. The solution is quite sensitive to the two initial guesses. These results were obtained on a 13-digit precision computer.

Table 2.6. The Direct Method for a Linear Eigenproblem

k	λ_k (deg)	$f(\lambda_k)$	$(\text{slope})_k$
0	15.000000	-90.000000	
1	13.000000	40.000000	-65.000000
2	13.615385	14.157487	-41.994083
3	13.952515	-4.999194	-56.822743
4	13.864536	0.360200	-60.916914
5	13.870449	0.008098	-59.547441
6	13.870585	-0.000014	-59.647887
7	13.870585	0.000000	

Example 2.6 presents the solution of a linear eigenproblem by the direct method. Nonlinear eigenproblems also can be solved by the direct method, as illustrated in Example 2.7.

Example 2.7. The direct method for a nonlinear eigenproblem.

Consider the nonlinear eigenproblem:

$$x_1 + 0.4x_2 = \sin(\lambda)\,x_1 \tag{2.102a}$$

$$0.2x_1 + x_2 = \cos(\lambda)\,x_2 \tag{2.102b}$$

The characteristic determinant corresponding to Eq. (2.102) is

$$f(\lambda) = \det[\mathbf{A} - \mathbf{B}(\lambda)] = \begin{vmatrix} [1 - \sin(\lambda)] & 0.4 \\ 0.2 & [1 - \cos(\lambda)] \end{vmatrix} = 0 \tag{2.103}$$

Let's solve Eq. (2.103) by the secant method. Let $\lambda_0 = 50.0$ deg and $\lambda_1 = 55.0$ deg. Thus,

$$f(\lambda_0) = \begin{vmatrix} [1 - \sin(50)] & 0.4 \\ 0.2 & [1 - \cos(50)] \end{vmatrix} = \begin{vmatrix} 0.233956 & 0.4 \\ 0.2 & 0.357212 \end{vmatrix} = 0.003572 \tag{2.104a}$$

$$f(\lambda_1) = \begin{vmatrix} [1 - \sin(55)] & 0.4 \\ 0.2 & [1 - \cos(55)] \end{vmatrix} = \begin{vmatrix} 0.180848 & 0.4 \\ 0.2 & 0.426424 \end{vmatrix} = -0.002882 \tag{2.104b}$$

Writing the linear relationship between λ and $f(\lambda)$ yields

$$\frac{f(\lambda_1) - f(\lambda_0)}{\lambda_1 - \lambda_0} = \text{slope} = \frac{f(\lambda_2) - f(\lambda_1)}{\lambda_2 - \lambda_1} \tag{2.105}$$

where $f(\lambda_2) = 0.0$ is the desired solution, Thus,

$$\text{Slope} = \frac{(-0.002882) - (-0.003572)}{55.0 - 50.0} = -0.001292 \tag{2.106}$$

Solving Eq. (2.105) for λ_2 to give $f(\lambda_2) = 0.0$ yields

$$\lambda_2 = \lambda_1 - \frac{f(\lambda_1)}{\text{slope}} = 55.0 - \frac{(-0.002882)}{(-0.001292)} = 52.767276 \tag{2.107}$$

Table 2.7. The Direct Method for a Nonlinear Eigenproblem

k	λ_k, deg	$f(\lambda_k)$	(Slope)$_k$
0	50.0	0.003572	
1	55.0	-0.002882	-0.001292
2	52.767276	0.000496	-0.001513
3	53.095189	0.000049	-0.001365
4	53.131096	-0.000001	

The results of the first iteration presented above and subsequent iterations are presented in Table 2.7. The solution is $\lambda = 53.131096$ deg. These results were obtained on a 13-digit precision computer and terminated when the change in $f(\lambda)$ between iterations was less than 0.000001.

2.5 THE QR METHOD

The power method presented in Section 2.3 finds individual eigenvalues, as does the direct method presented in Section 2.4. The QR method, on the other hand, finds all of the eigenvalues of a matrix simultaneously. The development of the QR method is presented by Strang (1988). The implementation of the QR method, without proof, is presented in this section.

Triangular matrices have their eigenvalues on the diagonal of the matrix. Consider the upper triangular matrix $\mathbf{U}$:

$$\mathbf{U} = \begin{bmatrix} u_{11} & u_{12} & u_{13} & \cdots & u_{1n} \\ 0 & u_{22} & u_{23} & \cdots & u_{2n} \\ 0 & 0 & u_{33} & \cdots & u_{3n} \\ \cdots & \cdots & \cdots & \cdots & \cdots \\ 0 & 0 & 0 & \cdots & u_{nn} \end{bmatrix} \tag{2.108}$$

The eigenproblem, $(\mathbf{U} - \lambda \mathbf{I})$, is given by

$$(\mathbf{U} - \lambda\mathbf{I}) = \begin{bmatrix} (u_{11}-\lambda) & u_{12} & u_{13} & \cdots & u_{1n} \\ 0 & (u_{22}-\lambda) & u_{23} & \cdots & u_{2n} \\ 0 & 0 & (u_{33}-\lambda) & \cdots & u_{3n} \\ \cdots & \cdots & \cdots & \cdots & \cdots \\ 0 & 0 & 0 & \cdots & (u_{nn}-\lambda) \end{bmatrix} \tag{2.109}$$

The characteristic polynomial, $|\mathbf{U} - \lambda\mathbf{I}|$, yields

$$(u_{11} - \lambda)(u_{22} - \lambda)(u_{33} - \lambda)\cdots(u_{nn} - \lambda) = 0 \tag{2.110}$$

The roots of Eq. (2.110) are the eigenvalues of matrix $\mathbf{U}$. Thus,

$$\lambda_i = u_{i,i} \qquad (i = 1, 2, \ldots, n) \tag{2.111}$$

The QR method use similarity transformations to transform matrix $\mathbf{A}$ into triangular form. A similarity transformation is defined as $\mathbf{A}' = \mathbf{M}^{-1}\mathbf{A}\mathbf{M}$. Matrices $\mathbf{A}$ and $\mathbf{A}'$ are said to be similar. The eigenvalues of similar matrices are identical, but the eigenvectors are different.

The *Gram-Schmidt process* starts with matrix $\mathbf{A}$, whose columns comprise the column vectors $\mathbf{a}_1, \mathbf{a}_2, \ldots, \mathbf{a}_n$, and constructs the matrix $\mathbf{Q}$, whose columns comprise a set of orthonormal vectors $\mathbf{q}_1, \mathbf{q}_2, \ldots, \mathbf{q}_n$. A set of orthonormal vectors is a set of mutually orthogonal unit vectors. The matrix that connects matrix $\mathbf{A}$ to matrix $\mathbf{Q}$ is the upper triangular matrix $\mathbf{R}$ whose elements are the vector products

$$r_{i,j} = \mathbf{q}_i^T \mathbf{a}_j \qquad (i, j = 1, 2, \ldots, n) \tag{2.112}$$

The result is the QR factorization:

$$\boxed{\mathbf{A} = \mathbf{QR}} \tag{2.113}$$

The QR process starts with the Gauss-Schmidt process, Eq. (2.113). That process is then reversed to give

$$\mathbf{A}' = \mathbf{RQ} \tag{2.114}$$

Matrices $\mathbf{A}$ and $\mathbf{A}'$ can be shown to be similar as follows. Premultiply Eq. (2.113) by $\mathbf{Q}^{-1}$ to obtain

$$\mathbf{Q}^{-1}\mathbf{A} = \mathbf{Q}^{-1}\mathbf{QR} = \mathbf{IR} = \mathbf{R} \tag{2.115}$$

Postmultiply Eq. (2.115) by $\mathbf{Q}$ to obtain

$$\mathbf{Q}^{-1}\mathbf{AQ} = \mathbf{RQ} = \mathbf{A}' \tag{2.116}$$

Equation (2.116) shows that matrices $\mathbf{A}$ and $\mathbf{A}'$ are similar, and thus have the same eigenvalues.

The steps in the Gram-Schmidt process are as follows. Start with the matrix $\mathbf{A}$ expressed as a set of column vectors:

$$\mathbf{A} = \begin{bmatrix} a_{11} & a_{12} & \cdots & a_{1n} \\ a_{21} & a_{22} & \cdots & a_{2n} \\ \cdots\cdots\cdots\cdots\cdots\cdots \\ a_{n1} & a_{n2} & \cdots & a_{nn} \end{bmatrix} = [\mathbf{a}_1 \quad \mathbf{a}_2 \quad \cdots \quad \mathbf{a}_n] \tag{2.117}$$

Assuming that the column vectors $\mathbf{a}_i$ $(i = 1, 2, \ldots, n)$ are linearly independent, they span the n-dimensional space. Thus, any arbitrary vector can be expressed as a linear combination of the column vectors $\mathbf{a}_i$ $(i = 1, 2, \ldots n)$. An orthonormal set of column vectors $\mathbf{q}_i$ $(i = 1, 2, \ldots, n)$ can be created from the column vectors $\mathbf{a}_i$ $(i = 1, 2, \ldots, n)$ by the following steps.

Choose $\mathbf{q}_1$ to have the direction of $\mathbf{a}_1$. Then normalize $\mathbf{a}_1$ to obtain $\mathbf{q}_1$:

$$\mathbf{q}_1 = \frac{\mathbf{a}_1}{\|\mathbf{a}_1\|} \tag{2.118a}$$

where $\|\mathbf{a}_1\|$ denotes the magnitude of $\mathbf{a}_1$:

$$\|\mathbf{a}_1\| = [a_{11}^2 + a_{12}^2 + \cdots + a_{n1}^2]^{1/2} \tag{2.118b}$$

To determine $\mathbf{q}_2$, first subtract the component of $\mathbf{a}_2$ in the direction of $\mathbf{q}_1$ to determine vector $\mathbf{a}_2'$, which is normal to $\mathbf{q}_1$. Thus,

$$\mathbf{a}_2' = \mathbf{a}_2 - (\mathbf{q}_1^T \mathbf{a}_2)\mathbf{q}_1 \tag{2.119a}$$

Choose $\mathbf{q}_2$ to have the direction of $\mathbf{a}_2'$. Then normalize $\mathbf{a}_2'$ to obtain $\mathbf{q}_2$:

$$\mathbf{q}_2 = \frac{\mathbf{a}_2'}{\|\mathbf{a}_2'\|} \tag{2.119b}$$

This process continues until a complete set of n orthonormal unit vectors is obtained. Let's evaluate one more orthonormal vector $\mathbf{q}_3$ to illustrate the process. To determine $\mathbf{q}_3$, first subtract the components of $\mathbf{a}_3$ in the directions of $\mathbf{q}_1$ and $\mathbf{q}_2$. Thus,

$$\mathbf{a}_3' = \mathbf{a}_3 - (\mathbf{q}_1^T \mathbf{a}_3)\mathbf{q}_1 - (\mathbf{q}_2^T \mathbf{a}_3)\mathbf{q}_2 \tag{2.120a}$$

Choose $\mathbf{q}_3$ to have the direction of $\mathbf{a}_3'$. Then normalize $\mathbf{a}_3'$ to obtain $\mathbf{q}_3$:

$$\mathbf{q}_3 = \frac{\mathbf{a}_3'}{\|\mathbf{a}_3'\|} \tag{2.120b}$$

The general expression for $\mathbf{a}_i'$ is

$$\mathbf{a}_i' = \mathbf{a}_i - \sum_{k=1}^{i-1}(\mathbf{q}_k^T \mathbf{a}_i)\mathbf{q}_k \qquad (i = 2, 3, \ldots, n) \tag{2.121}$$

and the general expression for $\mathbf{q}_i$ is

$$\mathbf{q}_i = \frac{\mathbf{a}_i'}{\|\mathbf{a}_i'\|} \qquad (i = 1, 2, \ldots, n) \tag{2.122}$$

The matrix $\mathbf{Q}$ is composed of the column vectors $\mathbf{q}_i$ $(i = 1, 2, \ldots, n)$. Thus,

$$\mathbf{Q} = [\mathbf{q}_1 \quad \mathbf{q}_2 \quad \cdots \quad \mathbf{q}_n] \tag{2.123}$$

The upper triangular matrix $\mathbf{R}$ is assembled from the elements computed in the evaluation of $\mathbf{Q}$. The diagonal elements of $\mathbf{R}$ are the magnitudes of the $\mathbf{a}_i'$ vectors:

$$r_{i,i} = \|\mathbf{a}_i'\| \qquad (i = 1, 2, \ldots, n) \tag{2.124}$$

The off-diagonal elements of $\mathbf{R}$ are the components of the $\mathbf{a}_i$ vectors which are subtracted from the $\mathbf{a}_i$ vectors in the evaluation of the $\mathbf{a}_i'$ vectors. Thus,

$$r_{i,j} = \mathbf{q}_i^T \mathbf{a}_j \qquad (i = 1, 2, \ldots, n, j = i + 1, \ldots, n) \tag{2.125}$$

The values of $r_{i,i}$ and $r_{i,j}$ are calculated during the evaluation of the orthonormal unit vectors $\mathbf{q}_i$. Thus, $\mathbf{R}$ is simply assembled from already calculated values. Thus,

$$\mathbf{R} = \begin{bmatrix} r_{11} & r_{12} & \cdots & r_{1n} \\ 0 & r_{22} & \cdots & r_{2n} \\ \cdots\cdots\cdots\cdots\cdots\cdots \\ 0 & 0 & \cdots & r_{nn} \end{bmatrix} \tag{2.126}$$

The first step in the QR process is to set $\mathbf{A}^{(0)} = \mathbf{A}$ and factor $\mathbf{A}^{(0)}$ by the Gram-Schmidt process into $\mathbf{Q}^{(0)}$ and $\mathbf{R}^{(0)}$. The next step is to reverse the factors $\mathbf{Q}^{(0)}$ and $\mathbf{R}^{(0)}$ to obtain

$$\mathbf{A}^{(1)} = \mathbf{R}^{(0)}\mathbf{Q}^{(0)} \tag{2.127}$$

$\mathbf{A}^{(1)}$ is similar to $\mathbf{A}$, so the eigenvalues are preserved. $\mathbf{A}^{(1)}$ is factored by the Gram-Schmidt process to obtain $\mathbf{Q}^{(1)}$ and $\mathbf{R}^{(1)}$, and the factors are reversed to obtain $\mathbf{A}^{(2)}$. Thus,

$$\mathbf{A}^{(2)} = \mathbf{R}^{(1)}\mathbf{Q}^{(1)} \tag{2.128}$$

The process is continued to determine $\mathbf{A}^{(3)}, \mathbf{A}^{(4)}, \ldots, \mathbf{A}^{(n)}$. When $\mathbf{A}^{(n)}$ approaches triangular form, within some tolerance, the eigenvalues of $\mathbf{A}$ are the diagonal elements. The process is as follows:

$$\mathbf{A}^{(k)} = \mathbf{Q}^{(k)}\mathbf{R}^{(k)} \tag{2.129}$$

$$\mathbf{A}^{(k+1)} = \mathbf{R}^{(k)}\mathbf{Q}^{(k)} \tag{2.130}$$

Equations (2.129) and (2.130) are the basic QR algorithm. Although it generally converges, it can be fairly slow. Two modifications are usually employed to increase its speed:

1. Preprocessing matrix $\mathbf{A}$ into a more nearly triangular form
2. Shifting the eigenvalues as the process proceeds

With these modifications, the QR algorithm is generally the preferred method for solving eigenproblems.

Example 2.8. The basic QR method.

Let's apply the QR method to find all the eigenvalues of matrix $\mathbf{A}$ given by Eq. (2.11) simultaneously. Recall:

$$\mathbf{A} = \begin{bmatrix} 8 & -2 & -2 \\ -2 & 4 & -2 \\ -2 & -2 & 13 \end{bmatrix} \tag{2.131}$$

The column vectors associated with matrix $\mathbf{A}$, $\mathbf{A} = [\mathbf{a}_1 \quad \mathbf{a}_2 \quad \mathbf{a}_3]$, are

$$\mathbf{a}_1 = \begin{bmatrix} 8 \\ -2 \\ -2 \end{bmatrix} \quad \mathbf{a}_2 = \begin{bmatrix} -2 \\ 4 \\ -2 \end{bmatrix} \quad \mathbf{a}_3 = \begin{bmatrix} -2 \\ -2 \\ 13 \end{bmatrix} \tag{2.132}$$

First let's solve for $\mathbf{q}_1$. Let $\mathbf{q}_1$ have the direction of $\mathbf{a}_1$, and divide by the magnitude of $\mathbf{a}_1$. Thus,

$$\|\mathbf{a}_1\| = [8^2 + (-2)^2 + (-2)^2]^{1/2} = 8.485281 \tag{2.133}$$

Solving for $\mathbf{q}_1 = \mathbf{a}_1/\|\mathbf{a}_1\|$ gives

$$\mathbf{q}_1^T = [0.942809 \quad -0.235702 \quad -0.235702] \tag{2.134}$$

Next let's solve for $\mathbf{q}_2$. First subtract the component of $\mathbf{a}_2$ in the direction of $\mathbf{q}_1$:

$$\mathbf{a}_2' = \mathbf{a}_2 - (\mathbf{q}_1^T\mathbf{a}_2)\mathbf{q}_1 \tag{2.135a}$$

$$\mathbf{a}_2' = \begin{bmatrix} -2 \\ 4 \\ -2 \end{bmatrix} - [0.942809 \quad -0.235702 \quad -0.235702] \begin{bmatrix} -2 \\ 4 \\ -2 \end{bmatrix} \begin{bmatrix} 0.942809 \\ -0.235702 \\ -0.235702 \end{bmatrix}$$

$$\tag{2.135b}$$

Performing the calculations gives $\mathbf{q}_1^T \mathbf{a}_2 = -2.357023$ and

$$\mathbf{a}_2' = \begin{bmatrix} -2 & -(-2.222222) \\ 4 & -(0.555555) \\ -2 & -(0.555555) \end{bmatrix} = \begin{bmatrix} 0.222222 \\ 3.444444 \\ -2.555557 \end{bmatrix} \tag{2.136}$$

The magnitude of $\mathbf{a}_2'$ is $\|\mathbf{a}_2'\| = 4.294700$. Thus, $\mathbf{q}_2 = \mathbf{a}_2'/\|\mathbf{a}_2'\|$ gives

$$\mathbf{q}_2 = [0.051743 \quad 0.802022 \quad -0.595049] \tag{2.137}$$

Finally let's solve for $\mathbf{q}_3$. First subtract the components of $\mathbf{a}_3$ in the directions of $\mathbf{q}_1$ and $\mathbf{q}_2$:

$$\mathbf{a}_3' = \mathbf{a}_3 - (\mathbf{q}_1^T \mathbf{a}_3)\mathbf{q}_1 - (\mathbf{q}_2^T \mathbf{a}_3)\mathbf{q}_2 \tag{2.138}$$

$$\mathbf{a}_3' = \begin{bmatrix} -2 \\ -2 \\ 13 \end{bmatrix} - [0.942809 \quad -0.235702 \quad -0.235702] \begin{bmatrix} -2 \\ -2 \\ 13 \end{bmatrix} \begin{bmatrix} 0.942809 \\ -0.235702 \\ -0.235702 \end{bmatrix}$$

$$-[0.051743 \quad 0.802022 \quad -0.595049] \begin{bmatrix} -2 \\ -2 \\ 13 \end{bmatrix} \begin{bmatrix} 0.051743 \\ 0.802022 \\ -0.595049 \end{bmatrix} \tag{2.139}$$

Performing the calculations gives $\mathbf{q}_1^T \mathbf{a}_3 = -4.478343$, $\mathbf{q}_2^T \mathbf{a}_3 = -9.443165$, and

$$\mathbf{a}_3' = \begin{bmatrix} -2 & -(-4.222222) & -(-0.488618) \\ -2 & -(1.055554) & -(-7.573626) \\ 13 & -(1.055554) & -(5.619146) \end{bmatrix} = \begin{bmatrix} 2.710840 \\ 4.518072 \\ 6.325300 \end{bmatrix} \tag{2.140}$$

The magnitude of $\mathbf{a}_3'$ is $\|\mathbf{a}_3'\| = 8.232319$. Thus, $\mathbf{q}_3 = \mathbf{a}_3'/\|\mathbf{a}_3'\|$ gives

$$\mathbf{q}_3' = [0.329293 \quad 0.548821 \quad 0.768350] \tag{2.141}$$

In summary, matrix $\mathbf{Q}^{(0)} = [\mathbf{q}_1 \quad \mathbf{q}_2 \quad \mathbf{q}_3]$ is given by

$$\mathbf{Q}^{(0)} = \begin{bmatrix} 0.942809 & 0.051743 & 0.329293 \\ -0.235702 & 0.802022 & 0.548821 \\ -0.235702 & -0.595049 & 0.768350 \end{bmatrix} \tag{2.142}$$

Matrix $\mathbf{R}^{(0)}$ is assembled from the elements computed in the calculation of matrix $\mathbf{Q}^{(0)}$. Thus, $r_{11} = \|\mathbf{a}_1'\| = 8.485281$, $r_{22} = \|\mathbf{a}_2'\| = 4.294700$, and $r_{33} = \|\mathbf{a}_3'\| = 8.232319$. The off-diagonal elements are $r_{12} = \mathbf{q}_1^T \mathbf{a}_2 = -2.357023$, $r_{13} = \mathbf{q}_1^T \mathbf{a}_3 = -4.478343$, and $r_{23} = \mathbf{q}_2^T \mathbf{a}_3 = -9.443165$. Thus, matrix $\mathbf{R}^{(0)}$ is given by

$$\mathbf{R}^{(0)} = \begin{bmatrix} 8.485281 & -2.357023 & -4.478343 \\ 0.000000 & 4.294700 & -9.443165 \\ 0.000000 & 0.000000 & 8.232819 \end{bmatrix} \tag{2.143}$$

It can be shown by matrix multiplication that $\mathbf{A}^{(0)} = \mathbf{Q}^{(0)}\mathbf{R}^{(0)}$.

Table 2.8. The Basic QR Method

k	λ_1	λ_2	λ_3
1	9.611111	9.063588	6.325301
2	10.743882	11.543169	2.712949
3	11.974170	10.508712	2.517118
4	12.929724	9.560916	2.509360
....	....		
19	13.870584	8.620435	2.508981
20	13.870585	8.620434	2.508981

The next step is to evaluate matrix $\mathbf{A}^{(1)} = \mathbf{R}^{(0)}\mathbf{Q}^{(0)}$. Thus,

$$\mathbf{A}^{(1)} = \begin{bmatrix} 8.485281 & -2.357023 & -4.478343 \\ 0.000000 & 4.294700 & -9.443165 \\ 0.000000 & 0.000000 & 8.232819 \end{bmatrix}$$

$$\times \begin{bmatrix} 0.942809 & 0.051743 & 0.329293 \\ -0.235702 & 0.802022 & 0.548821 \\ -0.235702 & -0.595049 & 0.768350 \end{bmatrix} \quad (2.144)$$

$$\mathbf{A}^{(1)} = \begin{bmatrix} 9.611111 & 1.213505 & -1.940376 \\ 1.213505 & 9.063588 & -4.898631 \\ -1.940376 & -4.898631 & 6.325301 \end{bmatrix} \quad (2.145)$$

The diagonal elements in matrix $\mathbf{A}^{(1)}$ are the first approximation to the eigenvalues of matrix $\mathbf{A}$. The results of the first iteration presented above and subsequent iterations are presented in Table 2.8. The final values of matrices $\mathbf{Q}, \mathbf{R}$, and $\mathbf{A}$ are given below:

$$\mathbf{Q}^{(19)} = \begin{bmatrix} 1.000000 & -0.000141 & 0.000000 \\ 0.000141 & 1.000000 & 0.000000 \\ 0.000000 & 0.000000 & 1.000000 \end{bmatrix} \quad (2.146)$$

$$\mathbf{R}^{(19)} = \begin{bmatrix} 13.870585 & 0.003171 & 0.000000 \\ 0.000000 & 8.620434 & 0.000000 \\ 0.000000 & 0.000000 & 2.508981 \end{bmatrix} \quad (2.147)$$

$$\mathbf{A}^{(20)} = \begin{bmatrix} 13.870585 & 0.001215 & 0.000000 \\ 0.001215 & 8.620434 & 0.000000 \\ 0.000000 & 0.000000 & 2.508981 \end{bmatrix} \quad (2.148)$$

The final values agree well with the values obtained by the power method and summarized at the end of Section 2.3. The QR method does not yield the corresponding eigenvectors. The eigenvector corresponding to each eigenvalue can be found by the inverse shifted power method presented in Section 2.6.

2.6 EIGENVECTORS

Some methods for solving eigenproblems, such as the power method, yield both the eigenvalues and the corresponding eigenvectors. Other methods, such as the direct method and the QR method, yield only the eigenvalues. In these cases, the corresponding eigenvectors can be evaluated by shifting the matrix by the eigenvalues and applying the inverse power method one time.

Example 2.9. Eigenvectors.

Let's apply this technique to evaluate the eigenvector x_1 corresponding to the largest eigenvalue of matrix A, $\lambda_1 = 13.870584$, which was obtained in Example 2.1. From that example,

$$A = \begin{bmatrix} 8 & -2 & -2 \\ -2 & 4 & -2 \\ -2 & -2 & 13 \end{bmatrix} \tag{2.149}$$

Shifting matrix A by $\lambda = 13.870584$, gives

$$A_{\text{shifted}} = (A - sI) = \begin{bmatrix} -5.870584 & -2.000000 & -2.000000 \\ -2.000000 & -9.870584 & -2.000000 \\ -2.000000 & -2.000000 & -0.870584 \end{bmatrix} \tag{2.150}$$

Applying the Doolittle LU method to A_{shifted} yields L and U:

$$L = \begin{bmatrix} 1.000000 & 0.000000 & 0.000000 \\ 0.340682 & 1.000000 & 0.000000 \\ 0.340682 & 0.143498 & 1.000000 \end{bmatrix}$$

$$U = \begin{bmatrix} -5.870584 & -2.000000 & -2.000000 \\ 0.000000 & -9.189221 & -1.318637 \\ 0.000000 & 0.000000 & 0.000001 \end{bmatrix} \tag{2.151}$$

Let the initial guess for $x^{(0)T} = [1.0 \quad 1.0 \quad 1.0]$. Solve for x' by forward substitution using $Lx' = x$.

$$\begin{bmatrix} 1.000000 & 0.000000 & 0.000000 \\ 0.340682 & 1.000000 & 0.000000 \\ 0.340682 & 0.143498 & 1.000000 \end{bmatrix} \begin{bmatrix} x_1' \\ x_2' \\ x_3' \end{bmatrix} = \begin{bmatrix} 1.0 \\ 1.0 \\ 1.0 \end{bmatrix} \rightarrow \begin{matrix} x_1' = 1.000000 \\ x_2' = 0.659318 \\ x_3' = 0.564707 \end{matrix}$$

$$\tag{2.152}$$

Solve for y by back substitution using $Uy = x'$.

$$\begin{bmatrix} -5.870584 & -2.000000 & -2.000000 \\ 0.000000 & -9.189221 & -1.318637 \\ 0.000000 & 0.000000 & 0.000001 \end{bmatrix} \begin{bmatrix} y_1 \\ y_2 \\ y_3 \end{bmatrix} = \begin{bmatrix} 1.000000 \\ 0.659318 \\ 0.564707 \end{bmatrix} \tag{2.153}$$

The solution, including scaling the first component to unity, is

$$y = \begin{bmatrix} -0.132662 \times 10^6 \\ -0.652404 \times 10^6 \\ 0.454642 \times 10^6 \end{bmatrix} \rightarrow 0.454642 \times 10^6 \begin{bmatrix} -0.291794 \\ -0.143498 \\ 1.000000 \end{bmatrix} \tag{2.154}$$

Thus, $\mathbf{x}^T = [-0.291794 \quad -0.143498 \quad 1.000000]$, which is identical to the value obtained by the direct power method in Example 2.1.

2.7 OTHER METHODS

The power method, including its variations, and the direct method are very inefficient when all the eigenvalues of a large matrix are desired. Several other methods are available in such cases. Most of these methods are based on a two-step procedure. In the first step, the original matrix is transformed to a simpler form that has the same eigenvalues as the original matrix. In the second step, iterative procedures are used to determine these eigenvalues. The best general purpose method is generally considered to be the *QR method*, which is presented in Section 2.5.

Most of the more powerful methods apply to special types of matrices. Many of them apply to symmetric matrices. Generally speaking, the original matrix is transformed into a simpler form that has the same eigenvalues. Iterative methods are then employed to evaluate the eigenvalues. More information on the subject can be found in Fadeev and Fadeeva (1963), Householder (1964), Wilkinson (1965), Steward (1973), Ralston and Rabinowitz (1978), and Press, Flannery, Teukolsky, and Vetterling (1989). Numerous compute programs for solving eigenproblems can be found in the IMSL (International Mathematical and Statistics Library) library and in the EISPACK program (Argonne National Laboratory). See Rice (1983) and Smith et al. (1976) for a discussion of these programs.

The Jacobi method transforms a symmetric matrix into a diagonal matrix. The off-diagonal elements are eliminated in a systematic manner. However, elimination of subsequent off-diagonal elements creates nonzero values in previously eliminated elements. Consequently, the transformation approaches a diagonal matrix iteratively. The Given method and the Householder method reduce a symmetric matrix to a tridiagonal matrix in a direct rather than an iterative manner. Consequently, they are more efficient than the Jacobi method. The resulting tridiagonal matrix can be expanded, and the corresponding characteristic equation can be solved for the eigenvalues by iterative techniques.

For more general matrices, the QR method is recommended. Due to its robustness, the QR method is generally the method of choice. See Wilkinson (1965) and Strang (1988) for a discussion of the QR method. The Householder method can be applied to nonsymmetrical matrices to reduce them to Hessenberg matrices, whose eigenvalues can then be found by the QR method.

Finally, deflation techniques can be employed for symmetric matrices. After the largest eigenvalue λ_1 of matrix $\mathbf{A}$ is found, for example, by the power method, a new matrix $\mathbf{B}$ is formed whose eigenvalues are the same as the eigenvalues of matrix $\mathbf{A}$, except that the largest eigenvalue λ_1 is replaced by zero in matrix $\mathbf{B}$. The power method can then be applied to matrix $\mathbf{B}$ to determine its largest eigenvalue, which is the second largest eigenvalue λ_2 of matrix $\mathbf{A}$. In principle, deflation can be applied repetitively to find all the eigenvalues of matrix $\mathbf{A}$. However, round-off errors generally pollute the results after a few deflations. The results obtained by deflation can be used to shift matrix $\mathbf{A}$ by the approximate eigenvalues, which are then solved for by the shifted inverse power method presented in Section 2.3.4 to find more accurate values.

2.8 PROGRAMS

Two FORTRAN subroutines for solving eigenproblems are presented in this section:

1. The power method
2. The inverse power method

The basic computational algorithms are presented as completely self-contained subroutines suitable for use in other programs. Input data and output statements are contained in a main (or driver) program written specifically to illustrate the use of each subroutine.

2.8.1. The Power Method

The *direct power method* evaluates the largest (in magnitude) eigenvalue of a matrix. The general algorithm for the power method is given by Eq. (2.39):

$$\mathbf{A}\mathbf{x}^{(k)} = \mathbf{y}^{(k+1)} = \lambda^{(k+1)}\mathbf{x}^{(k+1)} \tag{2.155}$$

A FORTRAN subroutine, *subroutine power*, for implementing the direct power method is presented below. *Subroutine power* performs the matrix multiplication, $\mathbf{A}\mathbf{x} = \mathbf{y}$, factors out the approximate eigenvalue λ to obtain the approximate eigenvector $\mathbf{x}$, checks for convergence, and returns or continues. After *iter* iterations, an error message is printed out and the iteration is terminated. *Program main* defines the data set and prints it, calls *subroutine power* to implement the solution, and prints the solution.

Program 2.1.   The direct power method program.

```
        program main
c       main program to illustrate eigenproblem solvers
c       ndim   array dimension, ndim = 6 in this example
c       n      number of equations, n
c       a      coefficient matrix, A(i,j)
c       x      eigenvector, x(i)
c       y      intermediate vector, y(i)
c       norm   specifies unity component of eigenvector
c       iter   number of iterations allowed
c       tol    convergence tolerance
c       shift amount by which eigenvalue is shifted
c       iw     intermediate results output flag:  0 no,  1 yes
        dimension a(6,6),x(6),y(6)
        data ndim,n,norm,iter,tol,shift,iw / 6,3,3,50,1.e-6,0.0,1/
c       data ndim,n,norm,iter,tol,shift,iw/6,3,3,50,1.e-6,13.870584,2/
        data (a(i,1),i=1,3) /  8.0, -2.0, -2.0 /
        data (a(i,2),i=1,3) / -2.0,  4.0, -2.0 /
        data (a(i,3),i=1,3) / -2.0, -2.0, 13.0 /
        data (x(i),i=1,3) /  1.0,  1.0,  1.0 /
        write (6,1000)
        do i=1,n
           write (6,1010) i,(a(i,j),j=1,n),x(i)
        end do
```

```
      if (shift.gt.0.0) then
         write (6,1005) shift
         do i=1,n
            a(i,i)=a(i,i)-shift
            write (6,1010) i,(a(i,j),j=1,n)
         end do
      end if
      call power (ndim,n,a,x,y,norm,iter,tol,shift,iw,k,ev2)
      write (6,1020)
      write (6,1010) k,ev2,(x(i),i=1,n)
      stop
 1000 format (' The power method'/' '/' A and x(0)'/' ')
 1005 format (' '/' A shifted by shift = ',f10.6/' ')
 1010 format (1x,i3,6f12.6)
 1020 format (' '/'   k     lambda and eigenvector components'/' ')
      end


      subroutine power (ndim,n,a,x,y,norm,iter,tol,shift,iw,k,ev2)
c     the direct power method
      dimension a(ndim,ndim),x(ndim),y(ndim)
      ev1=0.0
      if (iw.eq.1) write (6,1000)
      if (iw.eq.1) write (6,1010) k,ev1,(x(i),i=1,n)
      do k=1,iter
c     calculate y(i)
         do i=1,n
            y(i)=0.0
            do j=1,n
               y(i)=y(i)+a(i,j)*x(j)
            end do
         end do
c     calculate lambda and x(i)
         ev2=y(norm)
         if (abs(ev2).le.1.0e-3) then
            write (6,1020) ev2
            return
         else
            do i=1,n
               x(i)=y(i)/ev2
            end do
         end if
         if (iw.eq.1) write (6,1010) k,ev2,(x(i),i=1,n)
c     check for convergence
         if (abs(ev2-ev1).le.tol) then
            if (shift.ne.0.0) then
               ev1=ev2
               ev2=ev2+shift
               write (6,1040) ev1,ev2
            end if
            return
         else
            ev1=ev2
```

```
        end if
      end do
      write (6,1030)
      return
1000 format (' '/'   k      lambda and eigenvector components'/' ')
1010 format (1x,i3,6f12.6)
1020 format (' '/' lambda = ',e10.2,' approaching zero, stop')
1030 format (' '/' Iterations did not converge, stop')
1040 format (' '/' lambda shifted =',f12.6,' and lambda =',f12.6)
      end
```

The data set used to illustrate the use of *subroutine power* is taken from Example 2.1. The output generated by the power method program is presented below.

Output 2.1. Solution by the direct power method.

```
The power method

A and x(0)

   1     8.000000    -2.000000    -2.000000     1.000000
   2    -2.000000     4.000000    -2.000000     1.000000
   3    -2.000000    -2.000000    13.000000     1.000000


   k     lambda and eigenvector components

   0     0.000000     1.000000     1.000000     1.000000
   1     9.000000     0.444444     0.000000     1.000000
   2    12.111111     0.128440    -0.238532     1.000000
   3    13.220183    -0.037474    -0.242887     1.000000
   4    13.560722    -0.133770    -0.213602     1.000000
   5    13.694744    -0.192991    -0.188895     1.000000
  ..     ........     ........     ........     ........
  29    13.870583    -0.291793    -0.143499     1.000000
  30    13.870584    -0.291794    -0.143499     1.000000


   k     lambda and eigenvector components

  30    13.870584    -0.291794    -0.143499     1.000000
```

Subroutine power also implements the *shifted direct power method*. If the input variable, *shift*, is nonzero, matrix **A** is shifted by the value of *shift* before the direct power method is implemented. This implements the shifted direct power method. Example 2.3 illustrating the shifted direct power method can be solved by *subroutine power* simply by defining *norm* = 2 and *shift* = 13.870584 in the *data* statement. The *data* statement for this additional case is included in *program main* as a *comment* statement.

2.8.2. The Inverse Power Method

The *inverse power method* evaluates the largest (in magnitude) eigenvalue of the inverse matrix, $\mathbf{A}^{-1}$. The general equation for the inverse power method is given by

$$\mathbf{A}^{-1}\mathbf{x} = \lambda_{\text{inverse}}\mathbf{x} \tag{2.156}$$

This can be accomplished by evaluating $\mathbf{A}^{-1}$ by Gauss-Jordan elimination applied to the identity matrix $\mathbf{I}$ or by using the Doolittle LU factorization approach described in Section 2.3.3. Since a subroutine for Doolittle LU factorization is presented in Section 1.8.2, that approach is taken here. The general algorithm for the inverse power method based on the LU factorization approach is given by Eqs. (2.62) to (2.64):

$$\mathbf{L}\mathbf{x}' = \mathbf{x}^{(k)} \tag{2.157a}$$

$$\mathbf{U}\mathbf{y}^{(k+1)} = \mathbf{x}' \tag{2.157b}$$

$$\mathbf{y}^{(k+1)} = \lambda_{\text{inverse}}^{(k+1)}\mathbf{x}^{(k+1)} \tag{2.157c}$$

A FORTRAN subroutine, *subroutine invpower*, for implementing the inverse power method is presented below. *Program main* defines the data set and prints it, calls *subroutine invpower* to implement the inverse power method, and prints the solution. *Subroutine invpower* calls *subroutine lufactor* and *subroutine solve* from Section 1.8.2 to evaluate $\mathbf{L}$ and $\mathbf{U}$. This is indicated in *subroutine invpower* by including the subroutine declaration statements. The subroutines themselves must be included when *subroutine invpower* is to be executed. *Subroutine invpower* then evaluates $\mathbf{x}'$, $\mathbf{y}^{(k+1)}$, $\lambda_{\text{inverse}}^{(k+1)}$, and $\mathbf{x}^{(k+1)}$. Convergence of λ is checked, and the solution continues or returns. After *iter* iterations, an error message is printed and the solution is terminated. *Program main* in this section contains only the statements which are different from the statements in *program main* in Section 2.8.1.

Program 2.2. The inverse power method program.

```
      program main
c     main program to illustrate eigenproblem solvers
c     xp     intermediate solution vector
      dimension a(6,6),x(6),xp(6),y(6)
      data ndim,n,norm,iter,tol,shift,iw / 6,3,1,50,1.e-6,0.0,1 /
c     data ndim,n,norm,iter,tol,shift,iw / 6,3,1,50,1.e-6,10.0,1 /
c     data ndim,n,norm,iter,tol,shift,iw/6,3,3,50,1.e-6,13.694744,1/
c     data ndim,n,norm,iter,tol,shift,iw/6,3,3,1,1.e-6,13.870584,1/
      data (x(i),i=1,3) / 1.0, 1.0, 1.0 /
c     data (x(i),i=1,3) / -0.192991, -0.188895, 1.0 /
      call invpower (ndim,n,a,x,xp,y,norm,iter,tol,iw,shift,k,ev2)
 1000 format (' The inverse power method'/' '/' A and x(0)'/' ')
      end

      subroutine invpower (ndim,n,a,x,xp,y,norm,iter,tol,iw,shift,k,
     1 ev2)
c     the inverse power method.
      dimension a(ndim,ndim),x(ndim),xp(ndim),y(ndim)
```

```
c     perform the LU factorization
      call lufactor (ndim,n,a)
      if (iw.eq.1) then
         write (6,1000)
         do i=1,n
            write (6,1010) i,(a(i,j),j=1,n)
         end do
      end if
      if (iw.eq.1) write (6,1005) (x(i),i=1,n)
c     iteration loop
      do k=1,iter
         call solve (ndim,n,a,x,xp,y)
         ev2=y(norm)
         if (abs(ev2).le.1.0e-3) then
            write (6,1020) ev2
            return
         else
            do i=1,n
               x(i)=y(i)/ev2
            end do
         end if
         if (iw.eq.1) then
            write (6,1010) k,(xp(i),i=1,n)
            write (6,1015) (y(i),i=1,n)
            write (6,1015) (x(i),i=1,n),ev2
         end if
c        check for convergence
         if (abs(ev2-ev1).le.tol) then
            ev1=ev2
            ev2=1.0/ev2
            if (iw.eq.1) write (6,1040) ev1,ev2
            if (shift.ne.0.0) then
               ev1=ev2
               ev2=ev2+shift
               if (iw.eq.1) write (6,1050) ev1,ev2
            end if
            return
         else
            ev1=ev2
         end if
      end do
      if (iter.gt.1) write (6,1030)
      return
1000  format (' '/' L and U matrices stored in matrix A'/' ')
1005  format (' '/' row 1: k, xprime; row 2: y; row 3: x, ev2'
     1  /' '/4x,6f12.6)
1010  format (1x,i3,6f12.6)
1015  format (4x,6f12.6)
1020  format (' '/' ev2 = ',e10.2,' is approaching zero, stop')
1030  format (' '/' Iterations did not converge, stop')
1040  format (' '/' lambda inverse =',f12.6,' and lambda ='f12.6)
```

```
1050 format (' '/' lambda shifted =',f12.6,'  and lambda =',f12.6)
     end

     subroutine lufactor (ndim,n,a)
c    implements LU factorization and stores L and U in A
     end

     subroutine solve (ndim,n,a,b,bp,x)
c    process b to b' and b' to x
     end
```

The data set used to illustrate *subroutine invpower* is taken from Example 2.2. The output generated by the inverse power method program is presented below.

Output 2.2. Solution by the inverse power method.

```
The inverse power method

A and x(0)

  1    8.000000   -2.000000   -2.000000    1.000000
  2   -2.000000    4.000000   -2.000000    1.000000
  3   -2.000000   -2.000000   13.000000    1.000000

L and U matrices stored in matrix A

  1    8.000000   -2.000000   -2.000000
  2   -0.250000    3.500000   -2.500000
  3   -0.250000   -0.714286   10.714286

row 1: k, xprime; row 2: y; row 3: x, ev2

       1.000000    1.000000    1.000000
  1    1.000000    1.050000    2.142857
       0.300000    0.500000    0.200000
       1.000000    1.666667    0.666667    0.300000
  2    1.000000    1.916667    2.285714
       0.353333    0.700000    0.213333
       1.000000    1.981132    0.603774    0.353333
  3    1.000000    2.231132    2.447439
       0.382264    0.800629    0.228428
       1.000000    2.094439    0.597565    0.382264
 ..    ........    ........    ........    ........
 12    1.000000    2.395794    2.560994
       0.398568    0.855246    0.239026
       1.000000    2.145796    0.599712    0.398568

lambda inverse =    0.398568   and lambda =    2.508983

  k    lambda and eigenvector components

 12    2.508983    1.000000    2.145796    0.599712
```

Subroutine invpower also implements the *shifted inverse power method*. If the input variable, *shift*, is nonzero, matrix **A** is shifted by the value of *shift* before the inverse power method is implemented. This implements the shifted inverse power method. Example 2.4 illustrating the evaluation of an intermediate eigenvalue by the shifted inverse power method can be solved by *subroutine invpower* simply by defining *shift* = 10.0 in the *data* statement. Example 2.5 illustrating shifting eigenvalues to accelerate convergence by the shifted inverse power method can be solved by *subroutine invpower* by defining *norm* = 3 and *shift* = 13.694744 in the *data* statement and defining $x(i) = -0.192991$, $-0.188895, 1.0$. Example 2.9 illustrating the evaluation of the eigenvector corresponding to a known eigenvalue can be solved by *subroutine invpower* simply by defining *shift* = 13.870584 and *iter* = 1 in the *data* statement. The *data* statements for these additional cases are included in *program main* as *comment* statements.

2.8.3. Packages for Eigenproblems

Numerous libraries and software packages are available for solving eigenproblems. Many workstations and mainframe computers have such libraries attached to their operating systems. If not, libraries such as EISPACK can be added to the operating systems.

Many commercial software packages contain eigenproblem solvers. Some of the more prominent packages are Matlab and Mathcad. More sophisticated packages, such as ISML, Mathematica, Macsyma, and Maple, also contain eigenproblem solvers. Finally, the book *Numerical Recipes* [Pross et al. (1989)] contains subroutines and advice for solving eigenproblems.

2.9 SUMMARY

Some general guidelines for solving eigenproblems are summarized below.

- When only the largest and/or smallest eigenvalue of a matrix is required, the power method can be employed.
- Although it is rather inefficient, the power method can be used to solve for intermediate eigenvalues.
- The direct method is not a good method for solving linear eigenproblems. However, it can be used for solving nonlinear eigenproblems.
- For serious eigenproblems, the QR method is recommended.
- Eigenvectors corresponding to a known eigenvalue can be determined by one application of the shifted inverse power method.

After studying Chapter 2, you should be able to:

1. Explain the physical significance of an eigenproblem.
2. Explain the mathematical characteristics of an eigenproblem.
3. Explain the basis of the power method.
4. Solve for the largest (in absolute value) eigenvalue of a matrix by the power method.
5. Solve for the smallest (in absolute value) eigenvalue of a matrix by the inverse power method.
6. Solve for the opposite extreme eigenvalue of a matrix by shifting the eigenvalues of the matrix by the largest (in absolute value) eigenvalue and applying

the inverse power method to the shifted matrix. This procedure yields either the smallest (in absolute value) eigenvalue or the largest (in absolute value) eigenvalue of opposite sign.

7. Solve for an intermediate eigenvalue of a matrix by shifting the eigenvalues of the matrix by an estimate of the intermediate eigenvalue and applying the inverse power method to the shifted matrix.

8. Accelerate the convergence of an eigenproblem by shifting the eigenvalues of the matrix by an approximate value of the eigenvalue obtained by another method, such as the direct power method, and applying the inverse power method to the shifted matrix.

9. Solve for the eigenvalues of a linear or nonlinear eigenproblem by the direct method.

10. Solve for the eigenvalues of a matrix by the QR method.

11. Solve for the eigenvector corresponding to a known eigenvalue of a matrix by applying the inverse power method one time.

EXERCISE PROBLEMS

Consider the linear eigenproblem, $\mathbf{Ax} = \lambda\mathbf{x}$, for the matrices given below. Solve the problems presented below for the specified matrices. Carry at least six figures after the decimal place. Iterate until the values of λ change by less than three digits after the decimal place. Begin all problems with $\mathbf{x}^{(0)T} = [1.0 \quad 1.0 \quad \cdots \quad 1.0]$ unless otherwise specified. Show all the results for the first three iterations. Tabulate the results of subsequent iterations. Several of these problems require a large number of iterations.

$$\mathbf{A} = \begin{bmatrix} 2 & 1 \\ 3 & 4 \end{bmatrix} \quad \mathbf{B} = \begin{bmatrix} 3 & 2 \\ 3 & 4 \end{bmatrix} \quad \mathbf{C} = \begin{bmatrix} 2 & 3 \\ 1 & 4 \end{bmatrix}$$

$$\mathbf{D} = \begin{bmatrix} 1 & 1 & 2 \\ 2 & 1 & 1 \\ 1 & 1 & 3 \end{bmatrix} \quad \mathbf{E} = \begin{bmatrix} 1 & 1 & 2 \\ 2 & 1 & 3 \\ 1 & 1 & 1 \end{bmatrix} \quad \mathbf{F} = \begin{bmatrix} 2 & 1 & 2 \\ 1 & 1 & 3 \\ 1 & 1 & 1 \end{bmatrix}$$

$$\mathbf{G} = \begin{bmatrix} 1 & 1 & 1 & 2 \\ 2 & 1 & 1 & 1 \\ 3 & 2 & 1 & 2 \\ 2 & 1 & 1 & 4 \end{bmatrix} \quad \mathbf{H} = \begin{bmatrix} 1 & 2 & 1 & 2 \\ 2 & 1 & 1 & 1 \\ 3 & 2 & 1 & 2 \\ 2 & 1 & 1 & 4 \end{bmatrix}$$

2.2 Basic Characteristics of Eigenproblems

1. Solve for the eigenvalues of (a) matrix $\mathbf{A}$, (b) matrix $\mathbf{B}$, and (c) matrix $\mathbf{C}$ by expanding the determinant of $(\mathbf{A} - \lambda\mathbf{I})$ and solving the characteristic equation by the quadratic formula. Solve for the corresponding eigenvectors by substituting the eigenvalues into the equation $(\mathbf{A} - \lambda\mathbf{I})\mathbf{x} = 0$ and solving for $\mathbf{x}$. Let the first component of $\mathbf{x}$ be unity.

2. Solve for the eigenvalues of (a) matrix $\mathbf{D}$, (b) matrix $\mathbf{E}$, and (c) matrix $\mathbf{F}$ by expanding the determinant of $(\mathbf{A} - \lambda\mathbf{I})$ and solving the characteristic equation by Newton's method. Solve for the corresponding eigenvectors by substituting the eigenvalues into the equation $(\mathbf{A} - \lambda\mathbf{I})\mathbf{x} = 0$ and solving for $\mathbf{x}$. Let the first component of $\mathbf{x}$ be unity.

2.3 The Power Method

The Direct Power Method

3. Solve for the largest (in magnitude) eigenvalue of matrix **A** and the corresponding eigenvector **x** by the power method. (a) Let the first component of **x** be the unity component. (b) Let the second component of **x** be the unity component. (c) Show that the eigenvectors obtained in parts (a) an (b) are equivalent.

4. Solve for the largest (in magnitude) eigenvalue of matrix **A** and the corresponding eigenvector **x** by the power method with $x^{(0)T} = [1.0 \quad 0.0]$ and $[0.0 \quad 1.0]$. (a) For each $x^{(0)}$, let the first component of **x** be the unity component. (b) For each $x^{(0)}$, let the second component of **x** be the unity component.

5. Solve Problem 3 for matrix **B**.

6. Solve Problem 4 for matrix **B**.

7. Solve Problem 3 for matrix **C**.

8. Solve Problem 4 for matrix **C**.

9. Solve for the largest (in magnitude) eigenvalue of matrix **D** and the corresponding eigenvector **x** by the power method. (a) Let the first component of **x** be the unity component. (b) Let the second component of **x** be the unity component. (c) Let the third component of **x** be the unity component. (d) Show that the eigenvectors obtained in parts (a), (b), and (c) are equivalent.

10. Solve for the largest (in magnitude) eigenvalue of matrix **D** and the corresponding eigenvector **x** by the power method with $x^{(0)T} = [1.0 \quad 0.0 \quad 0.0]$, $[0.0 \quad 1.0 \quad 0.0]$, and $[0.0 \quad 0.0 \quad 1.0]$. (a) For each $x^{(0)}$, let the first component of **x** be the unity component. (b) For each $x^{(0)}$, let the second component of **x** be the unity component. (c) For each $x^{(0)}$, let the third component of **x** be the unity component.

11. Solve Problem 9 for matrix **E**.

12. Solve Problem 10 for matrix **E**.

13. Solve Problem 9 for matrix **F**.

14. Solve Problem 10 for matrix **F**.

15. Solve for the largest (in magnitude) eigenvalue of matrix **G** and the corresponding eigenvector **x** by the power method. (a) Let the first component of **x** be the unity component. (b) Let the second component of **x** be the unity component. (c) Let the third component of **x** be the unity component. (d) Let the fourth component of **x** be the unity component. (e) Show that the eigenvectors obtained in parts (a) to (d) are equivalent.

16. Solve for the largest (in magnitude) eigenvalue of matrix **G** and the corresponding eigenvector **x** by the power method with $x^{(0)T} =$ $[1.0 \quad 0.0 \quad 0.0 \quad 0.0]$, $[0.0 \quad 1.0 \quad 0.0 \quad 0.0]$, $[0.0 \quad 0.0 \quad 1.0 \quad 0.0]$, and $[0.0 \quad 0.0 \quad 0.0 \quad 1.0]$. (a) For each $x^{(0)}$, let the first component of **x** be the unity component. (b) For each $x^{(0)}$, let the second component of **x** be the unity component. (c) For each $x^{(0)}$, let the third component of **x** be the unity component. (d) For each $x^{(0)}$, let the fourth component of **x** be the unity component.

17. Solve Problem 15 for matrix **H**.

18. Solve Problem 16 for matrix **H**.

The Inverse Power Method

19. Solve for the smallest (in magnitude) eigenvalue of matrix **A** and the corresponding eigenvector **x** by the inverse power method using the matrix inverse. Use Gauss-Jordan elimination to find the matrix inverse. (a) Let the first component of **x** be the unity component. (b) Let the second component of **x** be the unity component. (c) Show that the eigenvectors obtained in parts (a) and (b) are equivalent.

20. Solve for the smallest (in magnitude) eigenvalue of matrix **A** and the corresponding eigenvector **x** by the inverse power method using the matrix inverse with $x^{(0)T} = [1.0 \quad 0.0]$ and $[0.0 \quad 1.0]$. (a) For each $x^{(0)}$, let the first component of **x** be the unity component. (b) For each $x^{(0)}$, let the second component of **x** be the unity component.

21. Solve Problem 19 for matrix **B**.
22. Solve Problem 20 for matrix **B**.
23. Solve Problem 19 for matrix **C**.
24. Solve Problem 20 for matrix **C**.

25. Solve for the smallest (in magnitude) eigenvalue of matrix **D** and the corresponding eigenvector **x** by the inverse power method using the matrix inverse. Use Gauss-Jordan elimination to find the matrix inverse. (a) Let the first component of **x** be the unity component. (b) Let the second component of **x** be the unity component. (c) Let the third component of **x** be the unity component. (d) Show that the eigenvectors obtained in parts (a), (b), and (c) are equivalent.

26. Solve for the smallest (in magnitude) eigenvalue of matrix **D** and the corresponding eigenvector **x** by the inverse power method using the matrix inverse with $x^{(0)T} = [1.0 \quad 0.0 \quad 0.0], [0.0 \quad 1.0 \quad 0.0]$, and $[0.0 \quad 0.0 \quad 1.0]$. (a) For each $x^{(0)}$, let the first component of **x** be the unity component. (b) For each $x^{(0)}$, let the second component of **x** be the unity component. (c) For each $x^{(0)}$, let the third component of **x** be the unity component.

27. Solve Problem 25 for matrix **E**.
28. Solve Problem 26 for matrix **E**.
29. Solve Problem 25 for matrix **F**.
30. Solve Problem 26 for matrix **F**.

31. Solve for the smallest (in magnitude) eigenvalue of matrix **G** and the corresponding eigenvector **x** by the inverse power method using the matrix inverse. Use Gauss-Jordan elimination to find the matrix inverse. (a) Let the first component of **x** be the unity component. (b) Let the second component of **x** be the unity component. (c) Let the third component of **x** be the unity component. (d) Let the fourth component of **x** be the unity component. (e) Show that the eigenvectors obtained in parts (a) to (d) are equivalent.

32. Solve for the smallest (in magnitude) eigenvalue of matrix **G** and the corresponding eigenvector **x** by the inverse power method using the matrix inverse with $x^{(0)T} = [1.0 \quad 0.0 \quad 0.0 \quad 0.0], [0.0 \quad 1.0 \quad 0.0 \quad 0.0], [0.0 \quad 0.0 \quad 1.0 \quad 0.0]$, and $[0.0 \quad 0.0 \quad 0.0 \quad 1.0]$. (a) For each $x^{(0)}$, let the first component of **x** be the unity component. (b) For each $x^{(0)}$, let the second component of **x** be the unity component. (c) For each $x^{(0)}$, let the third component of **x** be the unity component. (d) For each $x^{(0)}$, let the fourth component of **x** be the unity component.

33. Solve Problem 31 for matrix **H**.
34. Solve Problem 32 for matrix **H**.
35. Solve Problem 19 using Doolittle LU factorization.
36. Solve Problem 21 using Doolittle LU factorization.
37. Solve Problem 23 using Doolittle LU factorization.
38. Solve Problem 25 using Doolittle LU factorization.
39. Solve Problem 27 using Doolittle LU factorization.
40. Solve Problem 29 using Doolittle LU factorization.
41. Solve Problem 31 using Doolittle LU factorization.
42. Solve Problem 33 using Doolittle LU factorization.

Shifting Eigenvalues to Find the Opposite Extreme Eigenvalue

43. Solve for the smallest eigenvalue of matrix **A** and the corresponding eigen-vector **x** by shifting the eigenvalues by $s = 5.0$ and applying the shifted power method. Let the first component of **x** be the unity component.
44. Solve for the smallest eigenvalue of matrix **B** and the corresponding eigen-vector **x** by shifting the eigenvalues by $s = 6.0$ and applying the shifted power method. Let the first component of **x** be the unity component.
45. Solve for the smallest eigenvalue of matrix **C** and the corresponding eigen-vector **x** by shifting the eigenvalues by $s = 5.0$ and applying the shifted power method. Let the first component of **x** be the unity component.
46. Solve for the smallest eigenvalue of matrix **D** and the corresponding eigen-vector **x** by shifting the eigenvalues by $s = 4.5$ and applying the shifted power method. Let the first component of **x** be the unity component.
47. Solve for the smallest eigenvalue of matrix **E** and the corresponding eigen-vector **x** by shifting the eigenvalues by $s = 4.0$ and applying the shifted power method. Let the first component of **x** be the unity component.
48. Solve for the smallest eigenvalue of matrix **F** and the corresponding eigen-vector **x** by shifting the eigenvalues by $s = 4.0$ and applying the shifted power method. Let the first component of **x** be the unity component.
49. Solve for the smallest eigenvalue of matrix **G** and the corresponding eigen-vector **x** by shifting the eigenvalues by $s = 6.6$ and applying the shifted power method. Let the first component of **x** be the unity component.
50. Solve for the smallest eigenvalue of matrix **H** and the corresponding eigen-vector **x** by shifting the eigenvalues by $s = 6.8$ and applying the shifted power method. Let the first component of **x** be the unity component.

Shifting Eigenvalues to Find Intermediate Eigenvalues

51. The third eigenvalue of matrix **D** and the corresponding eigenvector **x** can be found in a trial and error manner by assuming a value for λ between the smallest (in absolute value) and largest (in absolute value) eigenvalues, shifting the matrix by that value, and applying the inverse power method to the shifted matrix. Solve for the third eigenvalue of matrix **D** by shifting by $s = 0.8$ and applying the shifted inverse power method using Doolittle LU factorization. Let the first component of **x** be the unity component.
52. Repeat Problem 51 for matrix **E** by shifting by $s = -0.4$.
53. Repeat Problem 51 for matrix **F** by shifting by $s = 0.6$.
54. The third and fourth eigenvalues of matrix **G** and the corresponding eigen-

vectors **x** can be found in a trial and error manner by assuming a value for λ between the smallest (in absolute value) and largest (in absolute value) eigenvalues, shifting the matrix by that value, and applying the shifted inverse power method to the shifted matrix. This procedure can be quite time consuming for large matrices. Solve for these two eigenvalues by shifting **G** by $s = 1.5$ and -0.5 and applying the shifted inverse power method using Doolittle LU factorization. Let the first component of **x** be the unity component.

55. Repeat Problem 54 for matrix **H** by shifting by $s = 1.7$ and -0.5.

Shifting Eigenvalues to Accelerate Convergence

The convergence rate of an eigenproblem can be accelerated by stopping the iterative procedure after a few iterations, shifting the approximate result back to determine an improved approximation of λ, shifting the original matrix by this improved approximation of λ, and continuing with the inverse power method.

56. Apply the above procedure to Problem 46. After 10 iterations in Problem 46, $\lambda_s^{(10)} = -4.722050$ and $\mathbf{x}^{(10)T} = [1.0 \quad -1.330367 \quad 0.047476]$.
57. Apply the above procedure to Problem 47. After 20 iterations in Problem 47, $\lambda_s^{(20)} = -4.683851$ and $\mathbf{x}^{(20)T} = [0.256981 \quad 1.0 \quad -0.732794]$.
58. Apply the above procedure to Problem 48. After 10 iterations in Problem 48, $\lambda_s^{(10)} = -4.397633$ and $\mathbf{x}^{(10)T} = [1.0 \quad 9.439458 \quad -5.961342]$.
59. Apply the above procedure to Problem 49. After 20 iterations in Problem 49, $\lambda_s^{(20)} = -7.388013$ and $\mathbf{x}^{(20)T} = [1.0 \quad -0.250521 \quad -1.385861 \quad -0.074527]$.
60. Apply the above procedure to Problem 50. After 20 iterations in Problem 50, $\lambda_s^{(20)} = -8.304477$ and $\mathbf{x}^{(20)T} = [1.0 \quad -1.249896 \quad 0.587978 \quad -0.270088]$.

2.4 The Direct Method

61. Solve for the largest eigenvalue of matrix **D** by the direct method using the secant method. Let $\lambda^{(0)} = 5.0$ and $\lambda^{(1)} = 4.0$.
62. Solve for the largest eigenvalue of matrix **E** by the direct method using the secant method. Let $\lambda^{(0)} = 3.0$ and $\lambda^{(1)} = 4.0$.
63. Solve for the largest eigenvalue of matrix **F** by the direct method using the secant method. Let $\lambda^{(0)} = 5.0$ and $\lambda^{(1)} = 4.0$.
64. Solve for the largest eigenvalue of matrix **G** by the direct method using the secant method. Let $\lambda^{(0)} = 7.0$ and $\lambda^{(1)} = 6.0$.
65. Solve for the largest eigenvalue of matrix **H** by the direct method using the secant method. Let $\lambda^{(0)} = 7.0$ and $\lambda^{(1)} = 6.0$.
66. Solve for the smallest eigenvalue of matrix **D** by the direct method using the secant method. Let $\lambda^{(0)} = 0.0$ and $\lambda^{(1)} = -0.5$.
67. Solve for the smallest eigenvalue of matrix **E** by the direct method using the secant method. Let $\lambda^{(0)} = -0.5$ and $\lambda^{(1)} = -1.0$.
68. Solve for the smallest eigenvalue of matrix **F** by the direct method using the secant method. Let $\lambda^{(0)} = -0.5$ and $\lambda^{(1)} = -1.0$.
69. Solve for the smallest eigenvalue of matrix **G** by the direct method using the secant method. Let $\lambda^{(0)} = -0.8$ and $\lambda^{(1)} = -1.0$.
70. Solve for the smallest eigenvalue of matrix **H** by the direct method using the secant method. Let $\lambda^{(0)} = -1.1$ and $\lambda^{(1)} = -1.5$.

2.5 The QR Method

71. Solve for the eigenvalues of matrix **A** by the QR method.
72. Solve for the eigenvalues of matrix **B** by the QR method.
73. Solve for the eigenvalues of matrix **C** by the QR method.
74. Solve for the eigenvalues of matrix **D** by the QR method.
75. Solve for the eigenvalues of matrix **E** by the QR method.
76. Solve for the eigenvalues of matrix **F** by the QR method.
77. Solve for the eigenvalues of matrix **G** by the QR method.
78. Solve for the eigenvalues of matrix **H** by the QR method.

2.6 Eigenvectors

79. Solve for the eigenvectors of matrix **A** corresponding to the eigenvalues found in Problem 71 by applying the shifted inverse power method one time. Let the first component of **x** be the unity component.
80. Solve for the eigenvectors of matrix **B** corresponding to the eigenvalues found in Problem 72 by applying the shifted inverse power method one time. Let the first component of **x** be the unity component.
81. Solve for the eigenvectors of matrix **C** corresponding to the eigenvalues found in Problem 73 by applying the shifted inverse power method one time. Let the first component of **x** be the unity component.
82. Solve for the eigenvectors of matrix **D** corresponding to the eigenvalues found in Problem 74 by applying the shifted inverse power method one time. Let the first component of **x** be the unity component.
83. Solve for the eigenvectors of matrix **E** corresponding to the eigenvalues found in Problem 75 by applying the shifted inverse power method one time. Let the first component of **x** be the unity component.
84. Solve for the eigenvectors of matrix **F** corresponding to the eigenvalues found in Problem 76 by applying the shifted inverse power method one time. Let the first component of **x** be the unity component.
85. Solve for the eigenvectors of matrix **G** corresponding to the eigenvalues found in Problem 77 by applying the shifted inverse power method one time. Let the first component of **x** be the unity component.
86. Solve for the eigenvectors of matrix **H** corresponding to the eigenvalues found in Problem 78 by applying the shifted inverse power method one time. Let the first component of **x** be the unity component.

2.8 Programs

87. Implement the direct power method program presented in Section 2.8.1. Check out the program using the given data set.
88. Solve any of Problems 3 to 18 using the direct power method program.
89. Check out the shifted direct power method of finding the opposite extreme eigenvalue using the data set specified by the *comment statements*.
90. Solve any of Problems 43 to 50 using the shifted direct power method program.
91. Implement the inverse power method program presented in Section 2.8.2. Check out the program using the given data set.

92. Solve any of Problems 19 to 34 using the inverse power method program.
93. Check out the shifted inverse power method for finding intermediate eigen-values using the data set specified by the *comment statements*.
94. Solve any of Problems 51 to 55 using the shifted inverse power method program.
95. Check out the shifted inverse power method for accelerating convergence using the data set specified by the *comment statements*.
96. Solve any of Problems 56 to 60 using the shifted inverse power method program.
97. Check out the shifted inverse power method program for evaluating eigen-vectors for a specified eigenvalue using the data set specified by the *comment statements*.
98. Solve any of Problems 79 to 86 using the shifted inverse power method program.

3

Nonlinear Equations

3.1 INTRODUCTION

Consider the four-bar linkage illustrated in Figure 3.1. The angle $\alpha = \theta_4 - \pi$ is the input to this mechanism, and the angle $\phi = \theta_2$ is the output. A relationship between α and ϕ can be obtained by writing the vector loop equation:

$$\vec{r}_2 + \vec{r}_3 + \vec{r}_4 - \vec{r}_1 = 0 \tag{3.1}$$

127

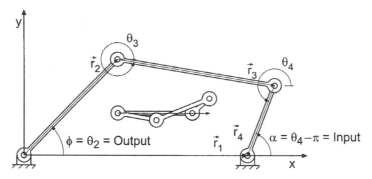

Figure 3.1 Four-bar linkage.

Let $\vec{r}_1$ lie along the x axis. Equation (3.1) can be written as two scalar equations, corresponding to the x and y components of the $\vec{r}$ vectors. Thus,

$$r_2 \cos(\theta_2) + r_3 \cos(\theta_3) + r_4 \cos(\theta_4) - r_1 = 0 \qquad (3.2a)$$

$$r_2 \sin(\theta_2) + r_3 \sin(\theta_3) + r_4 \sin(\theta_4) = 0 \qquad (3.2b)$$

Combining Eqs. (3.2a) and (3.2b), letting $\theta_2 = \phi$ and $\theta_4 = \alpha + \pi$, and simplifying yields Freudenstein's (1955) equation:

$$R_1 \cos(\alpha) - R_2 \cos(\phi) + R_3 - \cos(\alpha - \phi) = 0 \qquad (3.3)$$

where

$$R_1 = \frac{r_1}{r_2} \qquad R_2 = \frac{r_1}{r_4} \qquad R_3 = \frac{r_1^2 + r_2^2 + r_3^2 + r_4^2}{2 r_2 r_4} \qquad (3.4)$$

Consider the particular four-bar linkage specified by $r_1 = 10, r_2 = 6, r_3 = 8$, and $r_4 = 4$, which is illustrated in Figure 3.1. Thus, $R_1 = \frac{5}{3}$, $R_2 = \frac{5}{2}$, $R_3 = \frac{11}{6}$, and Eq. (3.3) becomes

$$\tfrac{5}{3} \cos(\alpha) - \tfrac{5}{2} \cos(\phi) + \tfrac{11}{6} - \cos(\alpha - \phi) = 0 \qquad (3.5)$$

The exact solution of Eq. (3.5) is tabulated in Table 3.1 and illustrated in Figure 3.2. Table 3.1 and Figure 3.2 correspond to the case where links 2, 3, and 4 are in the upper half-plane. This problem will be used throughout Chapter 3 to illustrate methods of solving for the roots of nonlinear equations. A mirror image solution is obtained for the case where links 2, 3, and 4 are in the lower half-plane. Another solution and its mirror image about the x axis are obtained if link 4 is in the upper half plane, link 2 is in the lower half-plane, and link 3 crosses the x axis, as illustrated by the small insert in Figure 3.1.

Table 3.1. Exact Solution of the Four-Bar Linkage Problem

α, deg	ϕ, deg	α, deg	ϕ, deg	α, deg	ϕ, deg
0.0	0.000000	70.0	54.887763	130.0	90.124080
10.0	8.069345	80.0	62.059980	140.0	92.823533
20.0	16.113229	90.0	68.888734	150.0	93.822497
30.0	24.104946	100.0	75.270873	160.0	92.734963
40.0	32.015180	110.0	81.069445	170.0	89.306031
50.0	39.810401	120.0	86.101495	180.0	83.620630
60.0	47.450827				

Many problems in engineering and science require the solution of a nonlinear equation. The problem can be stated as follows:

> Given the continuous nonlinear function $f(x)$,
> find the value $x = \alpha$ such that $f(\alpha) = 0$.

Figure 3.3 illustrates the problem graphically. The nonlinear equation, $f(x) = 0$, may be an algebraic equation (i.e., an equation involving $+$, $-$, $\times$, $/$, and radicals), a transcendental equation (i.e., an equation involving trigonometric, logarithmic, exponential, etc., functions), the solution of a differential equation, or any nonlinear relationship between an input x and an output $f(x)$.

There are two phases to finding the roots of a nonlinear equation: *bounding the root* and *refining the root* to the desired accuracy. Two general types of root-finding methods exist: *closed domain (bracketing) methods* which bracket the root in an ever-shrinking closed interval, and *open domain (nonbracketing) methods*. Several classical methods of both types are presented in this chapter. Polynomial root finding is considered as a special case. There are numerous pitfalls in finding the roots of nonlinear equations, which are discussed in some detail.

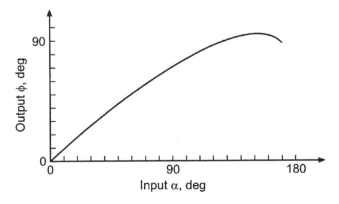

Figure 3.2 Exact solution of the four-bar linkage problem.

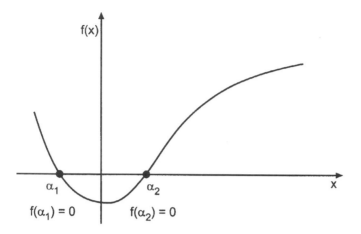

Figure 3.3 Solution of a nonlinear equation.

Figure 3.4 illustrates the organization of Chapter 3. After the introductory material presented in this section, some of the general features of root finding are discussed. The material then splits into a discussion of closed domain (bracketing) methods and open domain methods. Several special procedures applicable to polynomials are presented. After the presentation of the root finding methods, a section discussing some of the pitfalls of root finding and some other methods of root finding follows. A brief introduction to finding the roots of systems of nonlinear equations is presented. A section presenting several programs for solving nonlinear equations follows. The chapter closes with a Summary, which presents some philosophy to help you choose a specific method for a particular problem and lists the things you should be able to do after studying Chapter 3.

3.2 GENERAL FEATURES OF ROOT FINDING

Solving for the zeros of an equation, a process known as *root finding*, is one of the oldest problems in mathematics. Some general features of root finding are discussed in this section.

There are two distinct phases in finding the roots of a nonlinear equation: (1) *bounding the solution* and (2) *refining the solution*. These two phases are discussed in Sections 3.2.1 and 3.2.2, respectively. In general, nonlinear equations can behave in many different ways in the vicinity of a root. Typical behaviors are discussed in Section 3.2.3. Some general philosophy of root finding is discussed in Section 3.2.4.

3.2.1. Bounding the Solution

Bounding the solution involves finding a rough estimate of the solution that can be used as the initial approximation, or the starting point, in a systematic procedure that refines the solution to a specified tolerance in an efficient manner. If possible, the root should be bracketed between two points at which the value of the nonlinear function has opposite signs. Several possible bounding procedures are:

1. Graphing the function
2. Incremental search

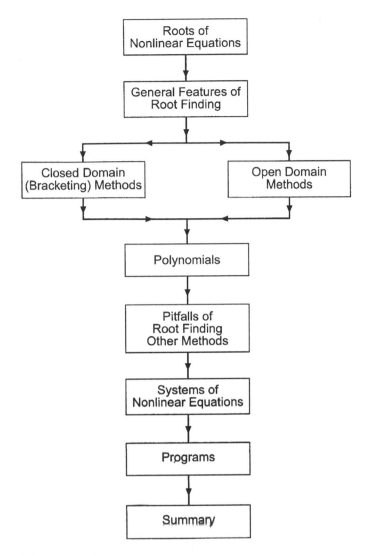

Figure 3.4 Organization of Chapter 3.

3. Past experience with the problem or a similar problem
4. Solution of a simplified approximate model
5. Previous solution in a sequence of solutions

Graphing the function involves plotting the nonlinear function over the range of interest. Many hand calculators have the capability to graph a function simply by defining the function and specifying the range of interest. Spreadsheets generally have graphing capability, as does software like Matlab and Mathcad. Very little effort is required. The resolution of the plots is generally not precise enough for an accurate result. However, the results are generally accurate enough to bound the solution. Plots of a nonlinear function display the general behavior of the nonlinear equation and permit the anticipation of problems.

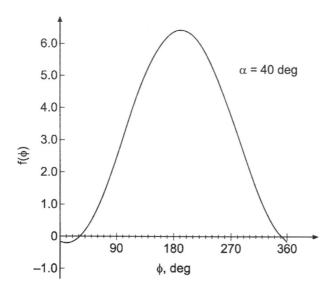

Figure 3.5   Graph of Eq. (3.5) for $\alpha = 40$ deg.

As an example of graphing a function to bound a root, consider the four-bar linkage problem presented in Section 3.1. Consider an input of $\alpha = 40$ deg. The graph of Eq. (3.5) with $\alpha = 40$ deg is presented in Figure 3.5. The graph shows that there are two roots of Eq. (3.5) when $\alpha = 40$ deg: one root between $\phi = 30$ deg and $\phi = 40$ deg, and one root between $\phi = 350$ (or -10) deg and $\phi = 360$ (or 0) deg.

An *incremental search* is conducted by starting at one end of the region of interest and evaluating the nonlinear function at small increments across the region. When the value of the function changes sign, it is assumed that a root lies in that interval. The two end points of the interval containing the root can be used as initial guesses for a refining method. If multiple roots are suspected, check for sign changes in the derivative of the function between the ends of the interval.

To illustrate an incremental search, let's evaluate Eq. (3.5) with $\alpha = 40$ deg for ϕ from 0 to 360 deg for $\Delta\phi = 10$ deg. The results are presented in Table 3.2. The same two roots identified by graphing the function are located.

Table 3.2.   Incremental Search for Eq. (3.5) with $\alpha = 40$ deg

ϕ, deg	$f(\phi)$	ϕ, deg	$f(\phi)$	ϕ, deg	$f(\phi)$	ϕ, deg	$f(\phi)$
0.0	−0.155970	100.0	3.044194	190.0	6.438119	280.0	3.175954
10.0	−0.217971	110.0	3.623104	200.0	6.398988	290.0	2.597044
20.0	−0.178850	120.0	4.186426	210.0	6.259945	300.0	2.033722
30.0	−0.039797	130.0	4.717043	220.0	6.025185	310.0	1.503105
40.0	0.194963	140.0	5.198833	230.0	5.701851	320.0	1.021315
50.0	0.518297	150.0	5.617158	240.0	5.299767	330.0	0.602990
60.0	0.920381	160.0	5.959306	250.0	4.831150	340.0	0.260843
70.0	1.388998	170.0	6.214881	260.0	4.310239	350.0	0.005267
80.0	1.909909	180.0	6.376119	270.0	3.752862	360.0	−0.155970
90.0	2.467286						

Whatever procedure is used to bound the solution, the initial approximation must be sufficiently close to the exact solution to ensure (a) that the systematic refinement procedure converges, and (b) that the solution converges to the desired root of the nonlinear equation.

3.2.2. Refining the Solution

Refining the solution involves determining the solution to a specified tolerance by an efficient systematic procedure. Several methods for refining the solution are:

1. Trial and error
2. Closed domain (bracketing) methods
3. Open domain methods

Trial and error methods simply guess the root, $x = \alpha$, evaluate $f(\alpha)$, and compare to zero. If $f(\alpha)$ is close enough to zero, quit. If not, guess another α, and continue until $f(\alpha)$ is close enough to zero. This approach is totally unacceptable.

Closed domain (bracketing) methods are methods that start with two values of x which bracket the root, $x = \alpha$, and systematically reduce the interval while keeping the root trapped within the interval. Two such methods are presented in Section 3.3:

1. Interval halving (bisection)
2. False position (regula falsi)

Bracketing methods are robust in that they are guaranteed to obtain a solution since the root is trapped in the closed interval. They can be slow to converge.

Open domain methods do not restrict the root to remain trapped in a closed interval. Consequently, they are not as robust as bracketing methods and can actually diverge. However, they use information about the nonlinear function itself to refine the estimates of the root. Thus, they are considerably more efficient than bracketing methods. Four open domain methods are presented in Section 3.4:

1. The fixed-point iteration method
2. Newton's method
3. The secant method
4. Muller's method

3.2.3. Behavior of Nonlinear Equations

Nonlinear equations can behave in various ways in the vicinity of a root. Algebraic and transcendental equations may have distinct (i.e., simple) real roots, repeated (i.e., multiple) real roots, or complex roots. Polynomials may have real or complex roots. If the polynomial coefficients are all real, complex roots occur in conjugate pairs. If the polynomial coefficients are complex, single complex roots can occur.

Figure 3.6 illustrates several distinct types of behavior of nonlinear equations in the vicinity of a root. Figure 3.6a illustrates the case of a single real root, which is called a *simple root*. Figure 3.6b illustrates a case where no real roots exist. Complex roots may

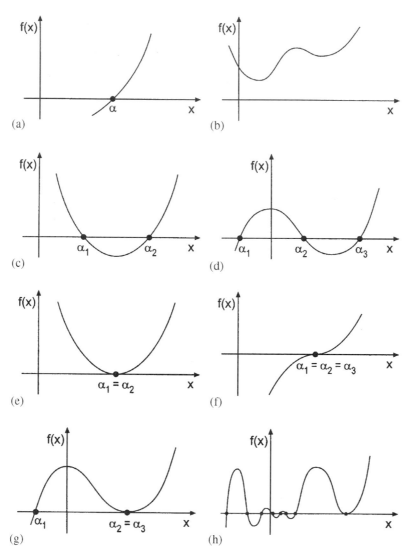

Figure 3.6   Solution behavior. (a) Simple root. (b) No real roots. (c) Two simple roots. (d) Three simple roots. (e) Two multiple roots. (f) Three multiple roots. (g) One simple and two multiple roots. (h) General case.

exist in such a case. Situations with two and three simple roots are illustrated in Figure 3.6c and d, respectively. Situations with two and three multiple roots are illustrated in Figure 3.6e and f, respectively. A situation with one simple root and two multiple roots is illustrated in Figure 3.6g. Lastly, Figure 3.6h illustrates the general case where any number of simple or multiple roots can exist.

Many problems in engineering and science involve a simple root, as illustrated in Figure 3.6a. Almost any root-finding method can find such a root if a reasonable initial approximation is furnished. In the other situations illustrated in Figure 3.6, extreme care may be required to find the desired roots.

3.2.4.   Some General Philosophy of Root Finding

There are numerous methods for finding the roots of a nonlinear equation. The roots have specific values, and the method used to find the roots does not affect the values of the roots. However, the method can determine whether or not the roots can be found and the amount of work required to find them. Some general philosophy of root finding is presented below.

1. Bounding methods should bracket a root, if possible.
2. Good initial approximations are extremely important.
3. Closed domain methods are more robust than open domain methods because they keep the root bracketed in a closed interval.
4. Open domain methods, when they converge, generally converge faster than closed domain methods.
5. For smoothly varying functions, most algorithms will always converge if the initial approximation is close enough. The rate of convergence of most algorithms can be determined in advance.
6. Many, if not most, problems in engineering and science are well behaved and straightforward. In such cases, a straightforward open domain method, such as Newton's method presented in Section 3.4.2 or the secant method presented in Section 3.4.3, can be applied without worrying about special cases and peculiar behavior. If problems arise during the solution, then the peculiarities of the nonlinear equation and the choice of solution method can be reevaluated.
7. When a problem is to be solved only once or a few times, the efficiency of the method is not of major concern. However, when a problem is to be solved many times, efficiency of the method is of major concern.
8. Polynomials can be solved by any of the methods for solving nonlinear equations. However, the special techniques applicable to polynomials should be considered.
9. If a nonlinear equation has complex roots, that must be anticipated when choosing a method.
10. Analyst's time versus computer time must be considered when selecting a method
11. Blanket generalizations about root-finding methods are generally not possible.

Root-finding algorithms should contain the following features:

1. An upper limit on the number of iterations.
2. If the method uses the derivative $f'(x)$, it should be monitored to ensure that it does not approach zero.
3. A convergence test for the change in the magnitude of the solution, $|x_{i+1} - x_i|$, or the magnitude of the nonlinear function, $|f(x_{i+1})|$, must be included.
4. When convergence is indicated, the final root estimate should be inserted into the nonlinear function $f(x)$ to guarantee that $f(x) = 0$ within the desired tolerance.

3.3   CLOSED DOMAIN (BRACKETING) METHODS

Two of the simplest methods for finding the roots of a nonlinear equation are:

1. Interval halving (bisection)
2. False position (regula falsi)

In these two methods, two estimates of the root which bracket the root must first be found by the bounding process. The root, $x = \alpha$, is bracketed by the two estimates. The objective is to locate the root to within a specified tolerance by a systematic procedure while keeping the root bracketed. Methods which keep the root bracketed during the refinement process are called *closed domain*, or *bracketing*, *methods*.

3.3.1. Interval Halving (Bisection)

One of the simplest methods for finding a root of a nonlinear equation is *interval halving* (also known as *bisection*). In this method, two estimates of the root, $x = a$ to the left of the root and $x = b$ to the right of the root, which bracket the root, must first be obtained, as illustrated in Figure 3.7, which illustrates the two possibilities with $f'(x) > 0$ and $f'(x) < 0$. The root, $x = \alpha$, obviously lies between a and b, that is, in the interval (a, b). The interval between a and b can be halved by averaging a and b. Thus, $c = (a + b)/2$. There are now two intervals: (a, c) and (c, b). The interval containing the root, $x = \alpha$, depends on the value of $f(c)$. If $f(a)f(c) < 0$, which is the case in Figure 3.7a, the root is in the interval (a, c). Thus, set $b = c$ and continue. If $f(a)f(c) > 0$, which is the case in Figure 3.7b, the root is in the interval (c, b). Thus, set $a = c$ and continue. If $f(a)f(c) = 0$, c is the root. Terminate the iteration. The algorithm is as follows:

$$c = \frac{a + b}{2} \tag{3.6}$$

If $f(a)f(c) < 0$: $a = a$ and $b = c$ (3.7a)

If $f(a)f(c) > 0$: $a = c$ and $b = b$ (3.7b)

Interval halving is an iterative procedure. The solution is not obtained directly by a single calculation. Each application of Eqs. (3.6) and (3.7) is an iteration. The iterations are continued until the size of the interval decreases below a prespecified tolerance ε_1, that is, $|b_i - a_i| \le \varepsilon_1$, or the value of $f(x)$ decreases below a prespecified tolerance ε_2, that is, $|f(c_i)| \le \varepsilon_2$, or both.

If a nonlinear equation, such as $f(x) = 1/(x - d)$ which has a singularity at $x = d$, is bracketed between a and b, interval halving will locate the discontinuity, $x = d$. A check on $|f(x)|$ as $x \to d$ would indicate that a discontinuity, not a root, is being found.

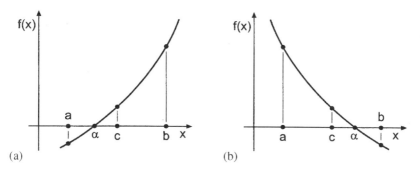

Figure 3.7   Interval halving (bisection).

Example 3.1. Interval halving (bisection).

Let's solve the four-bar linkage problem presented in Section 3.1 for an input of $\alpha = 40$ deg by interval halving. In calculations involving trigonometric functions, the angles must be expressed in radians. However, degrees (i.e., deg) are a more common unit of angular measure. Consequently, in all of the examples in this chapter, angles are expressed in degrees in the equations and in the tabular results, but the calculations are performed in radians. Recall Eq. (3.5) with $\alpha = 40.0$ deg:

$$f(\phi) = \tfrac{5}{3}\cos(40.0) - \tfrac{5}{2}\cos(\phi) + \tfrac{11}{6} - \cos(40.0 - \phi) = 0.0 \tag{3.8}$$

From the bounding procedure presented in Section 3.2, let $\phi_a = 30.0$ deg and $\phi_b = 40.0$ deg. From Eq. (3.8),

$$f(\phi_a) = f(30.0) = \tfrac{5}{3}\cos(40.0) - \tfrac{5}{2}\cos(30.0) + \tfrac{11}{6} - \cos(40.0 - 30.0)$$

$$= -0.03979719 \tag{3.9a}$$

$$f(\phi_b) = f(40.0) = \tfrac{5}{3}\cos(40.0) - \tfrac{5}{2}\cos(40.0) + \tfrac{11}{6} - \cos(40.0 - 40.0)$$

$$= 0.19496296 \tag{3.9b}$$

Thus, $\phi_a = 30.0$ deg and $\phi_b = 40.0$ deg bracket the solution. From Eq. (3.6),

$$\phi_c = \frac{\phi_a + \phi_b}{2} = \frac{30.0 + 40.0}{2} = 35.0 \text{ deg} \tag{3.10}$$

Substituting $\phi_c = 35.0$ deg into Eq. (3.8) yields

$$f(\phi_c) = f(35.0) = \tfrac{5}{3}\cos(40.0) - \tfrac{5}{2}\cos(35.0) + \tfrac{11}{6} - \cos(40.0 - 35.0) = 0.06599926 \tag{3.11}$$

Since $f(\phi_a)f(\phi_c) < 0$, $\phi_b = \phi_c$ for the next iteration and ϕ_a remains the same.

The solution is presented in Table 3.3. The convergence criterion is $|\phi_a - \phi_b| \leq 0.000001$ deg, which requires 24 iterations. Clearly, convergence is rather slow. The results presented in Table 3.3 were obtained on a 13-digit precision computer.

Table 3.3. Interval Halving (Bisection)

i	ϕ_a, deg	$f(\phi_a)$	ϕ_b, deg	$f(\phi_b)$	ϕ_c, deg	$f(\phi_c)$
1	30.0	−0.03979719	40.0	0.19496296	35.0	0.06599926
2	30.0	−0.03979719	35.0	0.06599926	32.50	0.01015060
3	30.0	−0.03979719	32.50	0.01015060	31.250	−0.01556712
4	31.250	−0.01556712	32.50	0.01015060	31.8750	−0.00289347
5	31.8750	−0.00289347	32.50	0.01015060	32.18750	0.00358236
6	31.8750	−0.00289347	32.18750	0.00358236	32.031250	0.00033288
7	31.8750	−0.00289347	32.031250	0.00033288	31.953125	−0.00128318
...	....	....	....	....	....	....
22	32.015176	−0.00000009	32.015181	0.00000000	32.015178	−0.00000004
23	32.015178	−0.00000004	32.015181	0.00000000	32.015179	−0.00000002
24	32.015179	−0.00000002	32.015181	0.00000000	32.015180	−0.00000001
	32.015180	−0.00000001				

The results in the table are rounded in the sixth digit after the decimal place. The final solution agrees with the exact solution presented in Table 3.1 to six digits after the decimal place.

The interval halving (bisection) method has several advantages:

1. The root is bracketed (i.e., trapped) within the bounds of the interval, so the method is guaranteed to converge.
2. The maximum error in the root is $|b_n - a_n|$.
3. The number of iterations n, and thus the number of function evaluations, required to reduce the initial interval, $(b_0 - a_0)$, to a specified interval, $(b_n - a_n)$, is given by

$$(b_n - a_n) = \frac{1}{2^n}(b_0 - a_0) \tag{3.12}$$

since each iteration reduces the interval size by a factor of 2. Thus, n is given by

$$n = \frac{1}{\log(2)} \log\left(\frac{b_0 - a_0}{b_n - a_n}\right) \tag{3.13}$$

The major disadvantage of the interval halving (bisection) method is that the solution converges slowly. That is, it can take a large number of iterations, and thus function evaluations, to reach the convergence criterion.

3.3.2. False Position (Regula Falsi)

The interval-halving (bisection) method brackets a root in the interval (a, b) and approximates the root as the midpoint of the interval. In the *false position (regula falsi) method*, the nonlinear function $f(x)$ is assumed to be a linear function $g(x)$ in the interval (a, b), and the root of the linear function $g(x)$, $x = c$, is taken as the next approximation of the root of the nonlinear function $f(x)$, $x = \alpha$. The process is illustrated graphically in Figure 3.8. This method is also called the *linear interpolation method*. The root of the linear function $g(x)$, that is, $x = c$, is not the root of the nonlinear function $f(x)$. It is a false position (in Latin, regula falsi), which gives the method its name. We now have two intervals, (a, c) and (c, b). As in the interval-halving (bisection) method, the interval containing the root of the nonlinear function $f(x)$ is retained, as described in Section 3.3.1, so the root remains bracketed.

The equation of the linear function $g(x)$ is

$$\frac{f(c) - f(b)}{c - b} = g'(x) \tag{3.14}$$

where $f(c) = 0$, and the slope of the linear function $g'(x)$ is given by

$$g'(x) = \frac{f(b) - f(a)}{b - a} \tag{3.15}$$

Solving Eq. (3.14) for the value of c which gives $f(c) = 0$ yields

$$\boxed{c = b - \frac{f(b)}{g'(x)}} \tag{3.16}$$

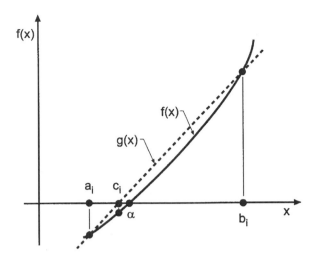

Figure 3.8 False position (regula falsi).

Note that $f(a)$ and a could have been used in Eqs. (3.14) and (3.16) instead of $f(b)$ and b. Equation (3.16) is applied repetitively until either one or both of the following two convergence criteria are satisfied:

$$|b - a| \le \varepsilon_1 \qquad \text{and/or} \qquad |f(c)| \le \varepsilon_2 \qquad (3.17)$$

Example 3.2. False position (regula falsi).

As an example of the false position (regula falsi) method, let's solve the four-bar linkage problem presented in Section 3.1. Recall Eq. (3.5) with $\alpha = 40.0$ deg:

$$f(\phi) = \tfrac{5}{3}\cos(40.0) - \tfrac{5}{2}\cos(\phi) + \tfrac{11}{6} - \cos(40.0 - \phi) = 0.0 \qquad (3.18)$$

Let $\phi_a = 30.0$ deg and $\phi_b = 40.0$ deg. From Eq. (3.18),

$$f(\phi_a) = f(30.0) = \tfrac{5}{3}\cos(40.0) - \tfrac{5}{2}\cos(30.0) + \tfrac{11}{6} - \cos(40.0 - 30.0)$$

$$= -0.03979719 \qquad (3.19a)$$

$$f(\phi_b) = f(40.0) = \tfrac{5}{3}\cos(40.0) - \tfrac{5}{2}\cos(40.0) + \tfrac{11}{6} - \cos(40.0 - 40.0)$$

$$= 0.19496296 \qquad (3.19b)$$

Thus, $\phi_a = 30.0$ deg and $\phi_b = 40.0$ deg bracket the solution. From Eq. (3.15),

$$g'(\phi_b) = \frac{0.19496296 - (-0.03979719)}{40.0 - 30.0} = 0.02347602 \qquad (3.20)$$

Substituting these results into Eq. (3.16) yields

$$\phi_c = \phi_b - \frac{f(\phi_b)}{g'(\phi_b)} = 40.0 - \frac{0.19496296}{0.02347602} = 31.695228 \text{ deg} \qquad (3.21)$$

Table 3.4. False Position (Regula Falsi)

i	ϕ_a, deg	$f(\phi_a)$	ϕ_b, deg	$f(\phi_b)$	ϕ_c, deg	$f(\phi_c)$
1	30.0	−0.03979719	40.0	0.19496296	31.695228	−0.00657688
2	31.695228	−0.00657688	40.0	0.19496296	31.966238	−0.00101233
3	31.966238	−0.00101233	40.0	0.19496296	32.007738	−0.00015410
4	32.007738	−0.00015410	40.0	0.19496296	32.014050	−0.00002342
5	32.014050	−0.00002342	40.0	0.19496296	32.015009	−0.00000356
6	32.015009	−0.00000356	40.0	0.19496296	32.015154	−0.00000054
7	32.015154	−0.00000054	40.0	0.19496296	32.015176	−0.00000008
8	32.015176	−0.00000008	40.0	0.19496296	32.015180	−0.00000001
9	32.015180	−0.00000001	40.0	0.19496296	32.015180	−0.00000000
	32.015180	0.00000000				

Substituting ϕ_c into Eq. (3.18) gives

$$f(\phi_c) = \tfrac{5}{3}\cos(40.0) - \tfrac{5}{2}\cos(31.695228) + \tfrac{11}{6} - \cos(40.0 - 31.695228) = -0.00657688$$

(3.22)

Since $f(\phi_a)f(\phi_c) > 0.0$, ϕ_a is set equal to ϕ_c and ϕ_b remains the same. This choice of ϕ_a and ϕ_b keeps the root bracketed.

These results and the results of subsequent iterations are presented in Table 3.4. The convergence criterion, $|\phi_b - \phi_a| \leq 0.000001$ deg, is satisfied on the ninth iteration. Notice that ϕ_b does not change in this example. The root is approached monotonically from the left. This type of behavior is common for the false position method.

3.3.3. Summary

Two closed domain (bracketing) methods for finding the roots of a nonlinear equation are presented in this section: interval halving (bisection) and false position (regula falsi). Both of these methods are guaranteed to converge because they keep the root bracketed within a continually shrinking closed interval. The interval-halving method gives an exact upper bound on the error of the solution, the interval size. It converges rather slowly. The false position method generally converges more rapidly than the interval halving method, but it does not give a bound on the error of the solution.

Both methods are quite robust, but converge slowly. The open domain methods presented in Section 3.4 are generally preferred because they converge much more rapidly. However, they do not keep the root bracketed, and thus, they may diverge. In such cases, the more slowly converging bracketing methods may be preferred.

3.4 OPEN DOMAIN METHODS

The interval halving (bisection) method and the false position (regula falsi) method presented in Section 3.3 converge slowly. More efficient methods for finding the roots of a nonlinear equation are desirable. Four such methods are presented in this section:

1. Fixed-point iteration
2. Newton's method

3. The secant method
4. Muller's method

These methods are called *open domain methods* since they are not required to keep the root bracketed in a closed domain during the refinement process.

Fixed-point iteration is not a reliable method and is not recommended for use. It is included simply for completeness since it is a well-known method. Muller's method is similar to the secant method. However, it is slightly more complicated, so the secant method is generally preferred. Newton's method and the secant method are two of the most efficient methods for refining the roots of a nonlinear equation.

3.4.1. Fixed-Point Iteration

The procedure known as *fixed-point iteration* involves solving the problem $f(x) = 0$ by rearranging $f(x)$ into the form $x = g(x)$, then finding $x = \alpha$ such that $\alpha = g(\alpha)$, which is equivalent to $f(\alpha) = 0$. The value of x such that $x = g(x)$ is called a *fixed point* of the relationship $x = g(x)$. Fixed-point iteration essentially solves two functions simultaneously: $x(x)$ and $g(x)$. The point of intersection of these two functions is the solution to $x = g(x)$, and thus to $f(x) = 0$. This process is illustrated in Figure 3.9.

Since $g(x)$ is also a nonlinear function, the solution must be obtained iteratively. An initial approximation to the solution x_1 must be determined by a bounding method. This value is substituted into the function $g(x)$ to determine the next approximation. The algorithm is as follows:

$$\boxed{x_{i+1} = g(x_i)} \tag{3.23}$$

The procedure is repeated (iterated) until a convergence criterion is satisfied. For example,

$$|x_{i+1} - x_i| \le \varepsilon_1 \quad \text{and/or} \quad |f(x_{i+1})| \le \varepsilon_2 \tag{3.24}$$

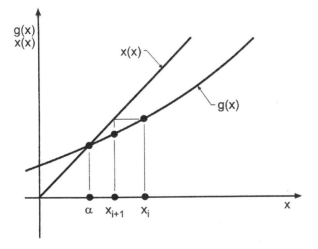

Figure 3.9 Fixed-point iteration.

Example 3.3. Fixed-point iteration.

Let's solve the four-bar linkage problem presented in Section 3.1 by fixed-point iteration. Recall Eq. (3.3):

$$f(\phi) = R_1 \cos(\alpha) - R_2 \cos(\phi) + R_3 - \cos(\alpha - \phi) = 0 \tag{3.25}$$

Equation (3.25) can be rearranged into the form $\phi = g(\phi)$ by separating the term $\cos(\alpha - \phi)$ and solving for ϕ. Thus,

$$\phi = \alpha - \cos^{-1}[R_1 \cos(\alpha) - R_2 \cos(\phi) + R_3] = \alpha - \cos^{-1}[u(\phi)] = g(\phi) \tag{3.26}$$

where

$$u(\phi) = R_1 \cos(\alpha) - R_2 \cos(\phi) + R_3 \tag{3.27}$$

The derivative of $g(\phi)$, that is, $g'(\phi)$, is of interest in the analysis of convergence presented at the end of this section. Recall

$$d(\cos^{-1}(u)) = -\frac{1}{\sqrt{1 - u^2}} du \tag{3.28}$$

Differentiating Eq. (3.26) gives

$$g'(\phi) = -\frac{1}{\sqrt{1 - u^2}} \frac{du}{d\phi} \tag{3.29}$$

which yields

$$g'(\phi) = \frac{R_2 \sin(\phi)}{\sqrt{1 - u^2}} \tag{3.30}$$

For the four-bar linkage problem presented in Section 3.1, $R_1 = \frac{5}{3}, R_2 = \frac{5}{2}$, and $R_3 = \frac{11}{6}$. Let's find the output ϕ for an input $\alpha = 40$ deg. Equations (3.27), (3.26), and (3.30) become

$$u(\phi_i) = \frac{5}{3} \cos(40.0) - \frac{5}{2} \cos(\phi_i) + \frac{11}{6} \tag{3.31}$$

$$\phi_{i+1} = g(\phi_i) = 40.0 - \cos^{-1}[u(\phi_i)] \tag{3.32}$$

$$g'(\phi_i) = \frac{5}{2} \frac{\sin(\phi_i)}{\sqrt{1 - [u(\phi_i)]^2}} \tag{3.33}$$

Let $\phi_1 = 30.0$ deg. Substituting $\phi_1 = 30.0$ deg into Eq. (3.25) gives $f(\phi_1) = -0.03979719$. Equations (3.31) to (3.33) give

$$u(30.0) = \frac{5}{3} \cos(40.0) - \frac{5}{2} \cos(30.0) + \frac{11}{6} = 0.945011 \tag{3.34}$$

$$\phi_2 = 40.0 - \cos^{-1}(0.945011) = 20.910798 \text{ deg} \tag{3.35}$$

$$g'(20.910798) = \frac{5}{2} \frac{\sin(20.910798)}{\sqrt{1 - (0.945011)^2}} = 2.728369 \tag{3.36}$$

Substituting $\phi_2 = 20.910798$ deg into Eq. (3.25) gives $f(\phi_2) = -0.17027956$. The entire procedure is now repeated with $\phi_2 = 20.910798$ deg.

Table 3.5. Fixed-Point Iteration for the First Form of $g(\phi)$

i	ϕ_i, deg	$f(\phi_i)$	$u(\phi_i)$	$g'(\phi_i)$	ϕ_{i+1}, deg	$f(\phi_{i+1})$
1	30.000000	−0.03979719	0.94501056	2.548110	20.910798	−0.17027956
2	20.910798	−0.17027956	0.77473100	0.940796	0.780651	−0.16442489
3	0.780651	−0.16442489	0.61030612	0.028665	−12.388360	0.05797824
4	−12.388360	0.05797824	0.66828436	−0.480654	−8.065211	−0.03348286
5	−8.065211	−0.03348286	0.63480150	−0.302628	−10.594736	0.01789195
...						
29	−9.747104	0.00000002	0.64616253	−0.369715	−9.747106	0.00000001
30	−9.747106	−0.00000001	0.64616254	−0.369715	−9.747106	−0.00000001
	−9.747105	−0.00000001				

These results and the results of 29 more iterations are summarized in Table 3.5. The solution has converged to the undesired root illustrated by the small insert inside the linkage illustrated in Figure 3.1. This configuration cannot be reached if the linkage is connected as illustrated in the large diagram. However, this configuration could be reached by reconnecting the linkage as illustrated in the small insert.

The results presented in Table 3.5 illustrate the major problem of open (nonbracketing) methods. Even though the desired root is bracketed by the two initial guesses, 30 deg and 40 deg, the method converged to a root outside that interval.

Let's rework the problem by rearranging Eq. (3.25) into another form of $\phi = g(\phi)$ by separating the term $R_2 \cos(\phi)$ and solving for ϕ. Thus,

$$\phi = \cos^{-1}\left\{\frac{1}{R_2}[R_1 \cos(\alpha) + R_3 - \cos(\alpha - \phi)]\right\} = \cos^{-1}[u(\phi)] = g(\phi) \quad (3.37)$$

where

$$u(\phi) = \frac{1}{R_2}[R_1 \cos(\alpha) + R_3 - \cos(\alpha - \phi)] \quad (3.38)$$

The derivative of $g(\phi)$ is

$$g'(\phi) = \frac{\sin(\alpha - \phi)}{R_2\sqrt{1 - u^2}} \quad (3.39)$$

For the four-bar linkage problem presented in Section 3.1, $R_1 = \frac{5}{3}, R_2 = \frac{5}{2}$, and $R_3 = \frac{11}{6}$. Let's find the output ϕ for the input $\alpha = 40$ deg. Equations (3.38), (3.37), and (3.39) become

$$u(\phi) = \frac{2}{5}[\frac{5}{3}\cos(40.0) + \frac{11}{6} - \cos(40.0 - \phi)] \quad (3.40)$$

$$\phi_{i+1} = g(\phi_i) = \cos^{-1}[u(\phi_i)] \quad (3.41)$$

$$g'(\phi_i) = \frac{2}{5}\frac{\sin(40.0 - \phi_i)}{\sqrt{1 - [u(\phi_i)]^2}} \quad (3.42)$$

Table 3.6. Fixed-Point Iteration for the Second Form of $g(\phi)$

i	ϕ_i, deg	$f(\phi_i)$	$u(\phi_i)$	$g'(\phi)$	ϕ_{i+1} deg	$f(\phi_{i+1})$
1	30.000000	-0.03979719	0.85010653	0.131899	31.776742	-0.00491050
2	31.776742	-0.00491050	0.84814233	0.107995	31.989810	-0.00052505
3	31.989810	-0.00052505	0.84793231	0.105148	32.012517	-0.00005515
4	32.012517	-0.00005515	0.84791025	0.104845	32.014901	-0.00000578
5	32.014901	-0.00000578	0.84790794	0.104814	32.015151	-0.00000061
6	32.015151	-0.00000061	0.84790769	0.104810	32.015177	-0.00000006
7	32.015177	-0.00000006	0.84790767	0.104810	32.015180	-0.00000001
8	32.015180	-0.00000001	0.84790767	0.104810	32.015180	-0.00000000
	32.015180					

Let $\phi_1 = 30.0$ deg. Substituting $\phi_1 = 30.0$ deg into Eq. (3.25) gives $f(\phi_1) = -0.03979719$. Equations (3.40) to (3.42) give

$$u(30.0) = \tfrac{2}{5}[\tfrac{5}{3}\cos(40.0) + \tfrac{11}{6} - \cos(40.0 - 30.0)] = 0.850107 \qquad (3.43)$$

$$\phi_2 = g(\phi_1) = g(30.0) = \cos^{-1}[u(30.0)] = \cos^{-1}(0.850107) = 31.776742 \text{ deg}$$
$$(3.44)$$

$$g'(30.0) = \frac{2}{5}\frac{\sin(40.0 - 30.0)}{\sqrt{1 - (0.850107)^2}} = 0.131899 \qquad (3.45)$$

Substituting $\phi_2 = 31.776742$ into Eq. (3.25) gives $f(\phi_2) = -0.00491050$. The entire procedure is now repeated with $\phi_2 = 31.776742$ deg.

These results and the results of the subsequent iterations are presented in Table 3.6. The convergence criterion is $|\phi_{i+1} - \phi_i| \leq 0.000001$ deg, which requires eight iterations. This is a considerable improvement over the interval halving (bisection) method presented in Example 3.1, which requires 24 iterations. It is comparable to the false position (regula falsi) method presented in Example 3.2, which requires nine iterations.

Convergence of the fixed-point iteration method, or any iterative method of the form $x_{i+1} = g(x_i)$, is analyzed as follows. Consider the iteration formula:

$$x_{i+1} = g(x_i) \qquad (3.46)$$

Let $x = \alpha$ denote the solution and let $e = (x - \alpha)$ denote the error. Subtracting $\alpha = g(\alpha)$ from Eq. (3.46) gives

$$x_{i+1} - \alpha = e_{i+1} = g(x_i) - g(\alpha) \qquad (3.47)$$

Expressing $g(\alpha)$ in a Taylor series about x_i gives:

$$g(\alpha) = g(x_i) + g'(\xi)(\alpha - x_i) + \cdots \qquad (3.48)$$

where $x_i \leq \xi \leq \alpha$. Truncating Eq. (3.48) after the first-order term, solving for $[g(x_i) - g(\alpha)]$, and substituting the result into Eq. (3.47) yields

$$\boxed{e_{i+1} = g'(\xi)e_i} \qquad (3.49)$$

Equation (3.49) can be used to determine whether or not a method is convergent, and if it is convergent, its rate of convergence. For any iterative method to converge,

$$\left|\frac{e_{i+1}}{e_i}\right| = |g'(\xi)| < 1 \tag{3.50}$$

Consequently, the fixed-point iteration method is convergent only if $|g'(\xi)| < 1$. Convergence is linear since e_{i+1} is linearly dependent on e_i. If $|g'(\xi)| > 1$, the procedure diverges. If $|g'(\xi)| < 1$ but close to 1.0, convergence is quite slow. For the example presented in Table 3.5, $g'(\alpha) = 9.541086$, which explains why that form of $\phi = g(\phi)$ diverges. For the example presented in Table 3.6, $g'(\alpha) = 0.104810$, which means that the error decreases by a factor of approximately 10 at each iteration. Such rapid convergence does not occur when $|g'(\alpha)|$ is close to 1.0. For example, for $|g'(\alpha)| = 0.9$, approximately 22 times as many iterations would be required to reach the same convergence criterion.

If the nonlinear equation, $f(\phi) = 0$, is rearranged into the form $\phi = \phi + f(\phi) = g(\phi)$, the fixed-point iteration formula becomes

$$\phi_{i+1} = \phi_i + f(\phi_i) = g(\phi_i) \tag{3.51}$$

and $g'(\phi)$ is given by

$$g'(\phi) = 1 + R_2 \sin(\phi) - \sin(\alpha - \phi) \tag{3.52}$$

Substituting the final solution value, $\phi = 32.015180 \deg$, into Eq. (3.52) gives $g'(\phi) = 2.186449$, which is larger than 1.0. The iteration method would not converge to the desired solution for this rearrangement of $f(\phi) = 0$ into $\phi = g(\phi)$. In fact, the solution converges to $\phi = -9.747105 \deg$, for which $g'(\phi) = -0.186449$. This is also a solution to the four-bar linkage problem, but not the desired solution.

Methods which sometimes work and sometimes fail are undesirable. Consequently, the fixed-point iteration method for solving nonlinear equations is not recommended.

Methods for accelerating the convergence of iterative methods based on a knowledge of the convergence rate can be developed. *Aitkens Δ^2 acceleration method* applies to linearly converging methods, such as fixed-point iteration, in which $e_{i+1} = ke_i$. The method is based on starting with an initial approximation x_i for which

$$x_i = \alpha + e_i \tag{3.53a}$$

Two more iterations are made to give

$$x_{i+1} = \alpha + e_{i+1} = \alpha + ke_i \tag{3.53b}$$

$$x_{i+2} = \alpha + e_{i+2} = \alpha + ke_{i+1} = \alpha + k^2 e_i \tag{3.53c}$$

There are three unknowns in Eqs. (3.53a) to (3.53c): e_i, α, and k. These three equations can be solved for these three unknowns. The value of α obtained by the procedure is not the exact root, since higher-order terms have been neglected. However, it is an improved approximation of the root. The procedure is then repeated using α as the initial approximation. It can be shown that the successive approximations to the root, α, converge quadratically. When applied to the fixed-point iteration method for finding the roots of a nonlinear equation, this procedure is known as *Steffensen's method*. Steffensen's method is not developed in this book since Newton's method, which is presented in Section 3.4.2, is a more straightforward procedure for achieving quadratic convergence.

3.4.2. Newton's Method

Newton's method (sometimes called the *Newton-Rhapson method*) for solving nonlinear equations is one of the most well-known and powerful procedures in all of numerical analysis. It always converges if the initial approximation is sufficiently close to the root, and it converges quadratically. Its only disadvantage is that the derivative $f'(x)$ of the nonlinear function $f(x)$ must be evaluated.

Newton's method is illustrated graphically in Figure 3.10. The function $f(x)$ is nonlinear. Let's locally approximate $f(x)$ by the linear function $g(x)$, which is tangent to $f(x)$, and find the solution for $g(x) = 0$. Newton's method is sometimes called the tangent method. That solution is then taken as the next approximation to the solution of $f(x) = 0$. The procedure is applied iteratively to convergence. Thus,

$$f'(x_i) = \text{slope of } f(x) = \frac{f(x_{i+1}) - f(x_i)}{x_{i+1} - x_i} \tag{3.54}$$

Solving Eq. (3.54) for x_{i+1} with $f(x_{i+1}) = 0$ yields

$$\boxed{x_{i+1} = x_i - \frac{f(x_i)}{f'(x_i)}} \tag{3.55}$$

Equation (3.55) is applied repetitively until either one or both of the following convergence criteria are satisfied:

$$|x_{i+1} - x_i| \le \varepsilon_1 \qquad \text{and/or} \qquad |f(x_{i+1})| \le \varepsilon_2 \tag{3.56}$$

Newton's method also can be obtained from the Taylor series. Thus,

$$f(x_{i+1}) = f(x_i) + f'(x_i)(x_{i+1} - x_i) + \cdots \tag{3.57}$$

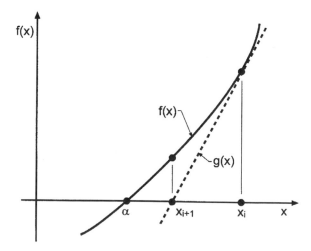

Figure 3.10   Newton's method.

Truncating Eq. (3.57) after the first derivative term, setting $f(x_{i+1}) = 0$, and solving for x_{i+1} yields

$$x_{i+1} = x_i - \frac{f(x_i)}{f'(x_i)} \qquad (3.58)$$

Equation (3.58) is the same as Eq. (3.55).

Example 3.4.   Newton's method.

To illustrate Newton's method, let's solve the four-bar linkage problem presented in Section 3.1. Recall Eq. (3.3):

$$f(\phi) = R_1 \cos(\alpha) - R_2 \cos(\phi) + R_3 - \cos(\alpha - \phi) = 0 \qquad (3.59)$$

The derivative of $f(\phi), f'(\phi)$ is

$$f'(\phi) = R_2 \sin(\phi) - \sin(\alpha - \phi) \qquad (3.60)$$

Thus, Eq. (3.55) becomes

$$\phi_{i+1} = \phi_i - \frac{f(\phi_i)}{f'(\phi_i)} \qquad (3.61)$$

For $R_1 = \frac{5}{3}, R_2 = \frac{5}{2}, R_3 = \frac{11}{6}$, and $\alpha = 40.0$ deg, Eqs. (3.59) and (3.60) yield

$$f(\phi) = \tfrac{5}{3}\cos(40.0) - \tfrac{5}{2}\cos(\phi) + \tfrac{11}{6} - \cos(40.0 - \phi) \qquad (3.62)$$

$$f'(\phi) = \tfrac{5}{2}\sin(\phi) - \sin(40.0 - \phi) \qquad (3.63)$$

For the first iteration let $\phi_1 = 30.0$ deg. Equations (3.62) and (3.63) give

$$f(\phi_1) = \tfrac{5}{3}\cos(40.0) - \tfrac{5}{2}\cos(30.0) + \tfrac{11}{6} - \cos(40.0 - 30.0) = -0.03979719 \quad (3.64)$$

$$f'(\phi_1) = \tfrac{5}{2}\sin(30.0) - \sin(40.0 - 30.0) = 1.07635182 \qquad (3.65)$$

Substituting these results into Eq. (3.61) yields

$$\phi_2 = 30.0 - \frac{(-0.03979719)(180/\pi)}{1.07635182} = 32.118463 \text{ deg} \qquad (3.66)$$

Substituting $\phi_2 = 32.118463$ deg into Eq. (3.62) gives $f(\phi_2) = 0.00214376$.

These results and the results of subsequent iterations are presented in Table 3.7. The convergence criterion, $|\phi_{i+1} - \phi_i| \leq 0.000001$ deg, is satisfied on the fourth iteration. This is a considerable improvement over the interval-halving method, the false position method, and the fixed-point iteration method.

Convergence of Newton's method is determined as follows. Recall Eq. (3.55):

$$x_{i+1} = x_i - \frac{f(x_i)}{f'(x_i)} \qquad (3.67)$$

Table 3.7. Newton's Method

i	ϕ_i, deg	$f(\phi_i)$	$f'(\phi_i)$	ϕ_{i+1}, deg	$f(\phi_{i+1})$
1	30.000000	− 0.03979719	1.07635164	32.118463	0.00214376
2	32.118463	0.00214376	1.19205359	32.015423	0.00000503
3	32.015423	0.00000503	1.18646209	32.015180	0.00000000
4	32.015180	0.00000000	1.18644892	32.015180	0.00000000
	32.015180	0.00000000			

Equation (3.67) is of the form

$$x_{i+1} = g(x_i) \tag{3.68}$$

where $g(x)$ is given by

$$g(x) = x - \frac{f(x)}{f'(x)} \tag{3.69}$$

As shown by Eq. (3.50), for convergence of any iterative method in the form of Eq. (3.68),

$$|g'(\xi)| \le 1 \tag{3.70}$$

where ξ lies between x_i and α. From Eq. (3.69),

$$g'(x) = 1 - \frac{f'(x)f'(x) - f(x)f''(x)}{[f'(x)]^2} = \frac{f(x)f''(x)}{[f'(x)]^2} \tag{3.71}$$

At the root, $x = \alpha$ and $f(\alpha) = 0$. Thus, $g'(\alpha) = 0$. Consequently, Eq.(3.70) is satisfied, and Newton's method is convergent.

The convergence rate of Newton's method is determined as follows. Subtract α from both sides of Eq. (3.67), and let $e = x - \alpha$ denote the error. Thus,

$$x_{i+1} - \alpha = e_{i+1} = x_i - \alpha - \frac{f(x_i)}{f'(x_i)} = e_i - \frac{f(x_i)}{f'(x_i)} \tag{3.72}$$

Expressing $f(x)$ in a Taylor series about x_i, truncating after the second-order term, and evaluating at $x = \alpha$ yields

$$f(\alpha) = f(x_i) + f'(x_i)(\alpha - x_i) + \tfrac{1}{2}f''(\xi)(\alpha - x_i)^2 = 0 \qquad x_i \le \xi \le \alpha \tag{3.73}$$

Letting $e_i = x_i - \alpha$ and solving Eq. (3.73) for $f(x_i)$ gives

$$f(x_i) = f'(x_i)e_i - \tfrac{1}{2}f''(\xi)e_i^2 \tag{3.74}$$

Substituting Eq. (3.74) into Eq. (3.72) gives

$$e_{i+1} = e_i - \frac{f'(x_i)e_i - \tfrac{1}{2}f''(\xi)e_i^2}{f'(x_i)} = \frac{1}{2}\frac{f''(\xi)}{f'(x_i)} e_i^2 \tag{3.75}$$

In the limit as $i \to \infty$, $x_i \to \alpha$, $f'(x_i) \to f'(\alpha)$, $f''(\xi) \to f''(\alpha)$, and Eq. (3.75) becomes

$$\boxed{e_{i+1} = \frac{1}{2}\frac{f''(\alpha)}{f'(\alpha)} e_i^2} \tag{3.76}$$

Equation (3.76) shows that convergence is second-order, or quadratic. The number of significant figures essentially doubles each iteration.

As the solution is approached, $f(x_i) \to 0$, and Eq. (3.70) is satisfied. For a poor initial estimate, however, Eq. (3.70) may not be satisfied. In that case, the procedure may converge to an alternate solution, or the solution may jump around wildly for a while and then converge to the desired solution or an alternate solution. The procedure will not diverge disastrously like the fixed-point iteration method when $|g'(x)| > 1$. Newton's method has excellent local convergence properties. However, its global convergence properties can be very poor, due to the neglect of the higher-order terms in the Taylor series presented in Eq. (3.57).

Newton's method requires the value of the derivative $f'(x)$ in addition to the value of the function $f(x)$. When the function $f(x)$ is an algebraic function or a transcendental function, $f'(x)$ can be determined analytically. However, when the function $f(x)$ is a general nonlinear relationship between an input x and an output $f(x)$, $f'(x)$ cannot be determined analytically. In that case, $f'(x)$ can be estimated numerically by evaluating $f(x)$ at x_i and $x_i + \varepsilon$, and approximating $f'(x_i)$ as

$$f'(x_i) = \frac{f(x_i + \varepsilon) - f(x_i)}{\varepsilon} \tag{3.77}$$

This procedure doubles the number of function evaluations at each iteration. However, it eliminates the evaluation of $f'(x)$ at each iteration. If ε is small, round-off errors are introduced, and if ε is too large, the convergence rate is decreased. This process is called the *approximate Newton method*.

In some cases, the efficiency of Newton's method can be increased by using the same value of $f'(x)$ for several iterations. As long as the sign of $f'(x)$ does not change, the iterates x_i move toward the root, $x = \alpha$. However, the second-order convergence is lost, so the overall procedure converges more slowly. However, in problems where evaluation of $f'(x)$ is more costly than evaluation of $f(x)$, this procedure may be less work. This is especially true in the solution of systems of nonlinear equations, which is discussed in Section 3.7. This procedure is called the *lagged Newton's method*.

A higher-order version of Newton's method can be obtained by retaining the second derivative term in the Taylor series presented in Eq. (3.57). This procedure requires the evaluation of $f''(x)$ and the solution of a quadratic equation for $\Delta x = x_{i+1} - x_i$. This procedure is not used very often.

Newton's method can be used to determine complex roots of real equations or complex roots of complex equations simply by using complex arithmetic. Newton's method also can be used to find multiple roots of nonlinear equations. Both of these applications of Newton's method, complex roots and multiple roots, are discussed in Section 3.5, which is concerned with polynomials, which can have both complex roots and multiple roots. Newton's method is also an excellent method for *polishing* roots obtained by other methods which yield results polluted by round-off errors, such as roots of deflated functions (see Section 3.5.2.2).

Newton's method has several disadvantages. Some functions are difficult to differentiate analytically, and some functions cannot be differentiated analytically at all. In such cases, the approximate Newton method defined by Eq. (3.77) or the secant method presented in Section 3.4.3 is recommended. When multiple roots occur, convergence drops to first order. This problem is discussed in Section 3.5.2 for polynomials. The presence of a local extremum (i.e., maximum or minimum) in $f(x)$ in the neighborhood of a root may cause oscillations in the solution. The presence of inflection points in $f(x)$ in the neighborhood of a root can cause problems. These last two situations are discussed in Section 3.6.1, which is concerned with pitfalls in root finding.

When Newton's method misbehaves, it may be necessary to bracket the solution within a closed interval and ensure that successive approximations remain within the interval. In extremely difficult cases, it may be necessary to make several iterations with the interval halving method to reduce the size of the interval before continuing with Newton's method.

3.4.3. The Secant Method

When the derivative function, $f'(x)$, is unavailable or prohibitively costly to evaluate, an alternative to Newton's method is required. The preferred alternative is the *secant method*.

The secant method is illustrated graphically in Figure 3.11. The nonlinear function $f(x)$ is approximated locally by the linear function $g(x)$, which is the secant to $f(x)$, and the root of $g(x)$ is taken as an improved approximation to the root of the nonlinear function $f(x)$. A *secant* to a curve is the straight line which passes through two points on the curve. The procedure is applied repetitively to convergence. Two initial approximations, x_0 and x_1, which are not required to bracket the root, are required to initiate the secant method. The slope of the secant passing through two points, x_{i-1} and x_i, is given by

$$g'(x_i) = \frac{f(x_i) - f(x_{i-1})}{x_i - x_{i-1}} \tag{3.78}$$

The equation of the secant line is given by

$$\frac{f(x_{i+1}) - f(x_i)}{x_{i+1} - x_i} = g'(x_i) \tag{3.79}$$

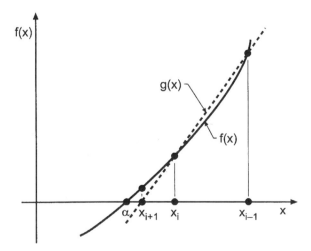

Figure 3.11   The secant method.

where $f(x_{i+1}) = 0$. Solving Eq. (3.79) for x_{i+1} yields

$$\boxed{x_{i+1} = x_i - \frac{f(x_i)}{g'(x_i)}} \tag{3.80}$$

Equation (3.80) is applied repetitively until either one or both of the following two convergence criteria are satisfied:

$$|x_{i+1} - x_i| \le \varepsilon_1 \qquad \text{and/or} \qquad |f(x_{i+1})| \le \varepsilon_2 \tag{3.81}$$

Example 3.5.   The secant method.

Let's solve the four-bar linkage problem presented in Section 3.1 by the secant method. Recall Eq. (3.3):

$$f(\phi) = R_1 \cos(\alpha) - R_2 \cos(\phi) + R_3 - \cos(\alpha - \phi) = 0 \tag{3.82}$$

Thus, Eq. (3.80) becomes

$$\boxed{\phi_{i+1} = \phi_i - \frac{f(\phi_i)}{g'(\phi_i)}} \tag{3.83}$$

where $g'(\phi_i)$ is given by

$$g'(\phi_i) = \frac{f(\phi_i) - f(\phi_{i-1})}{\phi_i - \phi_{i-1}} \tag{3.84}$$

For $R_1 = \frac{5}{3}, R_2 = \frac{5}{2}, R_3 = \frac{11}{6}$, and $\alpha = 40.0$ deg, Eq. (3.82) yields

$$\boxed{f(\phi) = \tfrac{5}{3}\cos(40.0) - \tfrac{5}{2}\cos(\phi) + \tfrac{11}{6} - \cos(40.0 - \phi)} \tag{3.85}$$

For the first iteration, let $\phi_0 = 30.0$ deg and $\phi_1 = 40.0$ deg. Equation (3.85) gives

$$f(\phi_0) = \tfrac{5}{3}\cos(40.0) - \tfrac{5}{2}\cos(30.0) + \tfrac{11}{6} - \cos(40.0 - 30.0) = -0.03979719 \tag{3.86a}$$

$$f(\phi_1) = \tfrac{5}{3}\cos(40.0) - \tfrac{5}{2}\cos(40.0) + \tfrac{11}{6} - \cos(40.0 - 40.0) = 0.19496296 \tag{3.86b}$$

Substituting these results into Eq. (3.84) gives

$$g'(\phi_1) = \frac{(0.19496296) - (-0.03979719)}{40.0 - 30.0} = 0.02347602 \tag{3.87}$$

Substituting $g'(\phi_1)$ into Eq. (3.83) yields

$$\phi_2 = 40.0 - \frac{0.19496296}{0.02347602} = 31.695228 \text{ deg} \tag{3.88}$$

Substituting $\phi_2 = 31.695228$ deg into Eq. (3.85) gives $f(\phi_2) = -0.00657688$.

These results and the results of subsequent iterations are presented in Table 3.8. The convergence criterion, $|\phi_{i+1} - \phi_i| \le 0.000001$ deg, is satisfied on the fifth iteration, which is one iteration more than Newton's method requires.

Table 3.8. The Secant Method

i	ϕ_i, deg	$f(\phi_i)$	$g'(\phi_i)$	ϕ_{i+1}, deg	$f(\phi_{i+1})$
0	30.000000	−0.03979719			
1	40.000000	0.19496296	0.02347602	31.695228	−0.00657688
2	31.695228	−0.00657688	0.02426795	31.966238	−0.00101233
3	31.966238	−0.00101233	0.02053257	32.015542	0.00000749
4	32.015542	0.00000749	0.02068443	32.015180	−0.00000001
5	32.015180	−0.00000001	0.02070761	32.015180	0.00000000
	32.015180	0.00000000			

The convergence rate of the secant method was analyzed by Jeeves (1958), who showed that

$$e_{i+1} = \left[\frac{1}{2} \frac{f''(\alpha)}{f'(\alpha)} \right]^{0.62\ldots} e_i^{1.62\ldots} \tag{3.89}$$

Convergence occurs at the rate 1.62..., which is considerably faster than the linear convergence rate of the fixed-point iteration method but somewhat slower than the quadratic convergence rate of Newton's method.

The question of which method is more efficient, Newton's method or the secant method, was also answered by Jeeves. He showed that if the effort required to evaluate $f'(x)$ is less than 43 percent of the effort required to evaluate $f(x)$, then Newton's method is more efficient. Otherwise, the secant method is more efficient.

The problems with Newton's method discussed at the end of Section 3.4.2 also apply to the secant method.

3.4.4. Muller's Method

Muller's method (1956) is based on locally approximating the nonlinear function $f(x)$ by a quadratic function $g(x)$, and the root of the quadratic function $g(x)$ is taken as an improved approximation to the root of the nonlinear function $f(x)$. The procedure is applied repetitively to convergence. Three initial approximations $x_1, x_2,$ and x_3, which are not required to bracket the root, are required to start the algorithm. The only difference between Muller's method and the secant method is that $g(x)$ is a quadratic function in Muller's method and a linear function in the secant method.

Muller's method is illustrated graphically in Figure 3.12. The quadratic function $g(x)$ is specified as follows:

$$g(x) = a(x - x_i)^2 + b(x - x_i) + c \tag{3.90}$$

The coefficients (i.e., $a, b,$ and c) are determined by requiring $g(x)$ to pass through the three known points $(x_i, f_i), (x_{i-1}, f_{i-1}),$ and (x_{i-2}, f_{i-2}). Thus

$$g(x_i) = f_i = a(x_i - x_i)^2 + b(x_i - x_i) + c = c \tag{3.91a}$$

$$g(x_{i-1}) = f_{i-1} = a(x_{i-1} - x_i)^2 + b(x_{i-1} - x_i) + c \tag{3.91b}$$

$$g(x_{i-2}) = f_{i-2} = a(x_{i-2} - x_i)^2 + b(x_{i-2} - x_i) + c \tag{3.91c}$$

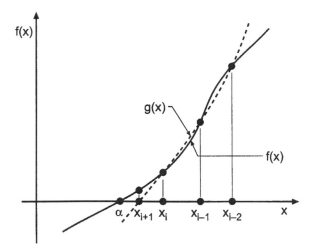

Figure 3.12 Muller's method.

Equation (3.91a) shows that $c = f_i$. Define the following parameters:

$$h_1 = (x_{i-1} - x_i) \quad \text{and} \quad h_2 = (x_{i-2} - x_i) \tag{3.92a}$$
$$\delta_1 = (f_{i-1} - f_i) \quad \text{and} \quad \delta_2 = (f_{i-2} - f_i) \tag{3.92b}$$

Then Eqs. (3.91b) and (3.91c) become

$$h_1^2 a + h_1 b = \delta_1 \tag{3.93a}$$
$$h_2^2 a + h_2 b = \delta_2 \tag{3.93b}$$

Solving Eq. (3.93) by Cramers rule yields

$$a = \frac{\delta_1 h_2 - \delta_2 h_1}{h_1 h_2 (h_1 - h_2)} \quad \text{and} \quad b = \frac{\delta_2 h_1^2 - \delta_1 h_2^2}{h_1 h_2 (h_1 - h_2)} \tag{3.94}$$

Now that the coefficients (i.e., a, b, and c) have been evaluated, Eq. (3.90) can be solved for the value of $(x_{i+1} - x_i)$ which gives $g(x_{i+1}) = 0$. Thus,

$$g(x_{i+1}) = a(x_{i+1} - x_i)^2 + b(x_{i+1} - x_i) + c = 0 \tag{3.95}$$

Solving Eq. (3.95) for $(x_{i+1} - x_i)$ by the rationalized quadratic formula, Eq. (3.112), gives

$$(x_{i+1} - x_i) = -\frac{2c}{b \pm \sqrt{b^2 - 4ac}} \tag{3.96}$$

Solving Eq. (3.96) for x_{i+1} yields

$$\boxed{x_{i+1} = x_i - \frac{2c}{b \pm \sqrt{b^2 - 4ac}}} \tag{3.97}$$

The sign, $+$ or $-$, of the square root term in the denominator of Eq. (3.97) is chosen to be the same as the sign of b to keep x_{i+1} close to x_i. Equation (3.97) is applied repetitively until either one or both of the following two convergence criteria are satisfied:

$$|x_{i+1} - x_i| \le \varepsilon_1 \quad \text{and/or} \quad |f(x_{i+1})| \le \varepsilon_2 \tag{3.98}$$

Example 3.6.   Muller's method.

Let's illustrate Muller's method by solving the four-bar linkage problem presented in Section 3.1. Recall Eq. (3.3):

$$f(\phi) = R_1 \cos(\alpha) - R_2 \cos(\phi) + R_3 - \cos(\alpha - \phi) = 0 \tag{3.99}$$

Thus, Eq. (3.97) becomes

$$\boxed{\phi_{i+1} = \phi_i - \frac{2c}{b \pm \sqrt{b^2 - 4ac}}} \tag{3.100}$$

where $c = f(\phi_i)$ and a and b are given by Eq. (3.94). For $R_1 = \frac{5}{3}, R_2 = \frac{5}{2}, R_3 = \frac{11}{6}$, and $\alpha = 40.0$ deg. Eq. (3.99) becomes

$$\boxed{f(\phi) = \tfrac{5}{3}\cos(40.0) - \tfrac{5}{2}\cos(\phi) + \tfrac{11}{6} - \cos(40.0 - \phi)} \tag{3.101}$$

For the first iteration, let $\phi_1 = 30.0$ deg, $\phi_2 = 30.5$ deg, and $\phi_3 = 31.0$ deg. Equation (3.101) gives $f(\phi_1) = f_1 = -0.03979719, f(\phi_2) = f_2 = -0.03028443$, and $f(\phi_3) = f_3 = -0.02053252$. Thus, $c = f_3 = -0.02053252$. Substituting these values of ϕ_1, ϕ_2, f_1, and f_2 into Eq. (3.92) gives

$$h_1 = (\phi_2 - \phi_3) = -0.50 \quad \text{and} \quad h_2 = (\phi_1 - \phi_3) = -1.00 \tag{3.102a}$$
$$\delta_1 = (f_2 - f_3) = -0.97519103 \quad \text{and} \quad \delta_2 = (f_1 - f_3) = -0.19264670 \tag{3.102b}$$

Substituting these results into Eq. (3.94) yields

$$a = \frac{\delta_1 h_2 - \delta_2 h_1}{h_1 h_2 (h_1 - h_2)} = -0.00047830092 \tag{3.103}$$

$$b = \frac{\delta_2 h_1^2 - \delta_1 h_2^2}{h_1 h_2 (h_1 - h_2)} = 0.019742971 \tag{3.104}$$

Substituting these results into Eq. (3.100) yields

$$\phi_{i+1} = 31.0$$

$$- \frac{2.0(-0.02053252)}{0.019742971 + \sqrt{(0.019742971)^2 - 4.0(0.00047830092)(-0.020532520)}} \tag{3.105}$$

which yields $\phi_{i+1} = 32.015031$ deg.

These results and the results of subsequent iterations are presented in Table 3.9. The convergence criterion, $|\phi_{i+1} - \phi_i| \le 0.000001$ deg, is satisfied on the third iteration.

Table 3.9. Muller's Method

i	ϕ_{i-2}, deg	ϕ_{i-1}, deg	ϕ_i, deg	ϕ_{i+1}, deg	$f(\phi_{i+1})$
1	30.000000				-0.03979717
2	30.500000				-0.03028443
3	31.000000				-0.02053252
4	30.000000	30.500000	31.000000	32.015031	-0.00000309
5	30.500000	31.000000	32.015031	32.015180	0.00000000
6	31.000000	32.015031	32.015180	32.015180	0.00000000
	32.015180				0.00000000

The convergence rate of Muller's method is 1.84, which is faster than the 1.62 rate of the secant method and slower than the 2.0 rate of Newton's method. Generally speaking, the secant method is preferred because of its simplicity, even though its convergence rate, 1.62, is slightly smaller than the convergence rate of Muller's method, 1.84.

3.4.5. Summary

Four open methods for finding the roots of a nonlinear equation are presented in this section: the fixed-point iteration method, Newton's method, the secant method, and Muller's; method. The fixed-point iteration method has a linear convergence rate and converges slowly, or not at all if $|g'(\alpha)| > 1.0$. Consequently, this method is not recommended.

Newton's method, the secant method, and Muller's method all have a higher-order convergence rate (2.0 for Newton's method, 1.62 for the secant method, and 1.84 for Muller's method). All three methods converge rapidly in the vicinity of a root. When the derivative $f'(x)$ is difficult to determine or time consuming to evaluate, the secant method is more efficient. In extremely sensitive problems, all three methods may misbehave and require some bracketing technique. All three of the methods can find complex roots simply by using complex arithmetic. The secant method and Newton's method are highly recommended for finding the roots of nonlinear equations.

3.5 POLYNOMIALS

The methods of solving for the roots of nonlinear equations presented in Sections 3.3 and 3.4 apply to any form of nonlinear equation. One very common form of nonlinear equation is a polynomial. Polynomials arise as the characteristic equation in eigenproblems, in curve-fitting tabular data, as the characteristic equation of higher-order ordinary differential equations, as the characteristic equation in systems of first-order-ordinary differential equations, etc. In all these cases, the roots of the polynomials must be determined. Several special features of solving for the roots of polynomials are discussed in this section.

3.5.1. Introduction

The basic properties of polynomials are presented in Section 4.2. The general form of an nth-degree polynomial is

$$P_n(x) = a_0 + a_1 x + a_2 x^2 + \cdots + a_n x^n \tag{3.106}$$

where n denotes the degree of the polynomial and a_0 to a_n are constant coefficients. The coefficients a_0 to a_n may be real or complex. The evaluation of a polynomial with real coefficients and its derivatives is straightforward, using nested multiplication and synthetic division as discussed in Section 4.2. The evaluation of a polynomial with complex coefficients requires complex arithmetic.

The *fundamental theorem of algebra* states that an nth-degree polynomial has exactly n zeros, or *roots*. The roots may be real or complex. If the coefficients are all real, complex roots always occur in conjugate pairs. The roots may be single (i.e., simple) or repeated (i.e., multiple). The single roots of a linear polynomial can be determined directly. Thus,

$$P_1(x) = ax + b \tag{3.107}$$

has the single root, $x = \alpha$, given by

$$\alpha = -\frac{b}{a} \tag{3.108}$$

The two roots of a second-degree polynomial can also be determined directly. Thus,

$$P_2(x) = ax^2 + bx + c = 0 \tag{3.109}$$

has two roots, α_1 and α_2, given by the *quadratic formula*:

$$\boxed{\alpha_1, \alpha_2 = \frac{-b \pm \sqrt{b^2 - 4ac}}{2a}} \tag{3.110}$$

Equation (3.110) yields two distinct real roots when $b^2 > 4ac$, two repeated real roots when $b^2 = 4ac$, and a pair of complex conjugate roots when $b^2 < 4ac$. When $b^2 \gg 4ac$, Eq. (3.110) yields two distinct real roots which are the sum and difference of two nearly identical numbers. In that case, a more accurate result can be obtained by rationalizing Eq. (3.110). Thus,

$$x = \frac{-b \pm \sqrt{b^2 - 4ac}}{2a} \left(\frac{-b \mp \sqrt{b^2 - 4ac}}{-b \mp \sqrt{b^2 - 4ac}} \right) \tag{3.111}$$

which yields the *rationalized quadratic formula*:

$$\boxed{x = -\frac{2c}{b \pm \sqrt{b^2 - 4ac}}} \tag{3.112}$$

Exact formulas also exist for the roots of third-degree and fourth-degree polynomials, but they are quite complicated and rarely used. Iterative methods are used to find the roots of higher-degree polynomials.

Descartes' rule of signs, which applies to polynomials having real coefficients, states that the number of positive roots of $P_n(x)$ is equal to the number of sign changes in the nonzero coefficients of $P_n(x)$, or is smaller by an even integer. The number of negative roots is found in a similar manner by considering $P_n(-x)$. For example, the fourth-degree polynomial

$$P_4(x) = -4 + 2x + 3x^2 - 2x^3 + x^4 \tag{3.113}$$

has three sign changes in the coefficients of $P_n(x)$ and one sign change in the coefficients of $P_n(-x) = -4 - 2x + 3x^2 + 2x^3 + x^2$. Thus, the polynomial must have either three positive real roots and one negative real root, or one positive real root, one negative real root, and two complex conjugate roots. The actual roots are $-1, 1, 1 + I1$, and $1 - I1$, where $I = \sqrt{-1.0}$.

The roots of high-degree polynomials can be quite sensitive to small changes in the values of the coefficients. In other words, high-degree polynomials can be *ill-conditioned*. Consider the factored fifth-degree polynomial:

$$P_5(x) = (x - 1)(x - 2)(x - 3)(x - 4)(x - 5) \tag{3.114}$$

which has five positive real roots, 1, 2, 3, 4, and 5. Expanding Eq. (3.114) yields the standard polynomial form:

$$P_5(x) = -120 + 274x - 225x^2 + 85x^3 - 15x^4 + x^5 \tag{3.115}$$

Descartes' rule of signs shows that there are either five positive real roots, or three positive real roots and two complex conjugate roots, or one positive real root and two pairs of complex conjugate roots. To illustrate the sensitivity of the roots to the values of the coefficients, let's change the coefficient of x^2, which is 225, to 226, which is a change of only 0.44 percent. The five roots are now $1.0514\ldots$, $1.6191\ldots$, $5.5075\ldots$, $3.4110\ldots + I1.0793\ldots$, and $3.4110\ldots - I1.0793\ldots$. Thus, a change of only 0.44 percent in one coefficient has made a major change in the roots, including the introduction of two complex conjugate roots. This simple example illustrates the difficulty associated with finding the roots of high-degree polynomials.

One procedure for finding the roots of high-degree polynomials is to find one root by any method, then deflate the polynomial one degree by factoring out the known root using synthetic division, as discussed in Section 4.2. The deflated $(n - 1)$st-degree polynomial is then solved for the next root. This procedure can be repeated until all the roots are determined. The last two roots should be determined by applying the quadratic formula to the $P_2(x)$ determined after all the deflations. This procedure reduces the work as the subsequent deflated polynomials are of lower and lower degree. It also avoids converging to an already converged root. The major limitation of this approach is that the coefficients of the deflated polynomials are not exact, so the roots of the deflated polynomials are not the precise roots of the original polynomial. Each deflation propagates the errors more and more, so the subsequent roots become less and less accurate. This problem is less serious if the roots are found in order from the smallest to the largest. In general, the roots of the deflated polynomials should be used as first approximations for those roots, which are then refined by solving the original polynomial using the roots of the deflated polynomials as the initial approximations for the refined roots. This process is known as *root polishing*.

The bracketing methods presented in Section 3.3, interval halving and false position, cannot be used to find repeated roots with an even multiplicity, since the nonlinear function $f(x)$ does not change sign at such roots. The first derivative $f'(x)$ does change sign at such roots, but using $f'(x)$ to keep the root bracketed increases the amount of work. Repeated roots with an odd multiplicity can be bracketed by monitoring the sign of $f(x)$, but even in this case the open methods presented in Section 3.4 are more efficient.

Three of the methods presented in Section 3.4 can be used to find the roots of polynomials: Newton's method, the secant method, and Muller's method. Newton's method for polynomials is presented in Section 3.5.2, where it is applied to find a simple root, a multiple root, and a pair of complex conjugate roots.

These three methods also can be used for finding the complex roots of polynomials, provided that complex arithmetic is used and reasonably good complex initial approximations are specified. Complex arithmetic is straightforward on digital computers. However, complex arithmetic is tedious when performed by hand calculation. Several methods exist for extracting complex roots of polynomials that have real coefficients which do not require complex arithmetic. Among these are Bairstow's method, the QD (quotient-difference) method [see Henrici (1964)], and Graeffe's method [see Hildebrand (1956)]. The QD method and Graeffe's method can find all the roots of a polynomial, whereas Bairstow's method extracts quadratic factors which can then be solved by the quadratic formula. Bairstow's method is presented in Section 3.5.3. These three methods use only real arithmetic.

When a polynomial has complex coefficients, Newton's method or the secant method using complex arithmetic and complex initial approximations are the methods of choice.

3.5.2. Newton's method

Newton's method for solving for the root, $x = \alpha$, of a nonlinear equation, $f(x) = 0$, is presented in Section 3.4. Recall Eq. (3.55):

$$\boxed{x_{i+1} = x_i - \frac{f(x_i)}{f'(x_i)}} \tag{3.116}$$

Equation (3.116) will be called Newton's basic method in this section to differentiate it from two variations of Newton's method which are presented in this section for finding multiple roots. Newton's basic method can be used to find simple roots of polynomials, multiple roots of polynomials (where the rate of convergence drops to first order), complex conjugate roots of polynomials with real coefficients, and complex roots of polynomials with complex coefficients. The first three of these applications are illustrated in this section.

3.5.2.1. Newton's Method for Simple Roots.

Newton's basic method can be applied directly to find simple roots of polynomials. Generally speaking, $f(x)$ and $f'(x)$ should be evaluated by the nested multiplication algorithm presented in Section 4.2 for maximum efficiency. No special problems arise. Accurate initial approximations are desirable, and in some cases they are necessary to achieve convergence.

Example 3.7.   Newton's method for simple roots.

Let's apply Newton's basic method to find the simple root of the following cubic polynomial in the neighborhood of $x = 1.5$:

$$f(x) = P_3(x) = x^3 - 3x^2 + 4x - 2 = 0 \tag{3.117}$$

Newton's basic method is given by Eq. (3.116). In this example, $f(x_i)$ and $f'(x_i)$ will be evaluated directly for the sake of clarity. The derivative of $f(x), f'(x)$, is given by the second-degree polynomial:

$$f'(x) = P_2(x) = 3x^2 - 6x + 4 \tag{3.118}$$

Table 3.10. Newton's Method for Simple Roots

i	x_i	$f(x_i)$	$f'(x_i)$	x_{i+1}	$f(x_{i+1})$
1	1.50	0.6250	1.750	1.142857	0.14577259
2	1.142857	0.14577259	1.06122449	1.005495	0.00549467
3	1.005495	0.00549476	1.00009057	1.000000	0.00000033
4	1.000000	0.00000033	1.00000000	1.000000	0.00000000
	1.000000	0.00000000			

Let $x_1 = 1.5$. Substituting this value into Eqs. (3.117) and (3.118) gives $f(1.5) = 0.6250$ and $f'(1.5) = 1.750$. Substituting these values into Eq. (3.116) yields

$$x_2 = x_1 - \frac{f(x_1)}{f'(x_1)} = 1.5 - \frac{0.6250}{1.750} = 1.142857 \tag{3.119}$$

These results and the results of subsequent iterations are presented in Table 3.10. Four iterations are required to satisfy the convergence tolerance, $|x_{i+1} - x_i| \le 0.000001$.

Newton's method is an extremely rapid procedure for finding the roots of a polynomial if a reasonable initial approximation is available.

3.5.2.2. Polynomial Deflation

The remaining roots of Eq. (3.117) can be found in a similar manner by choosing different initial approximations. An alternate approach for finding the remaining roots is to *deflate* the original polynomial by factoring out the linear factor corresponding to the known root and solving for the roots of the deflated polynomial.

Example 3.8. Polynomial deflation

Let's illustrate polynomial deflation by factoring out the linear factor, $(x - 1.0)$, from Eq. (3.117). Thus, Eq. (3.117) becomes

$$P_3(x) = (x - 1.0)Q_2(x) \tag{3.120}$$

The coefficients of the deflated polynomial $Q_2(x)$ can be determined by applying the synthetic division algorithm presented in Eq. (4.26). Recall Eq. (3.117):

$$P_3(x) = x^3 - 3x^2 + 4x - 2 \tag{3.121}$$

Applying Eq. (4.26) gives

$$b_3 = a_3 = 1.0 \tag{3.122.3}$$
$$b_2 = a_2 + xb_3 = -3.0 + (1.0)(1.0) = -2.0 \tag{3.122.2}$$
$$b_1 = a_1 + xb_2 = 4.0 + (1.0)(-2.0) = 2.0 \tag{3.122.1}$$

Thus, $Q_2(x)$ is given by

$$x^2 - 2.0x + 2.0 = 0 \tag{3.123}$$

Equation (3.123) is the desired deflated polynomial. Since Eq. (3.123) is a second-degree polynomial, its roots can be determined by the quadratic formula. Thus,

$$x = \frac{-b \pm \sqrt{b^2 - 4ac}}{2a} = \frac{-(-2.0) \pm \sqrt{(-2.0)^2 - 4.0(1.0)(2.0)}}{2(1.0)} \tag{3.124}$$

which yields the complex conjugate roots, $\alpha_{1,2} = 1 \pm I1$.

3.5.2.3. Newton's Method for Multiple Roots

Newton's method, in various forms, can be used to calculate multiple real roots. Ralston and Rabinowitz (1978) show that a nonlinear function $f(x)$ approaches zero faster than its derivative $f'(x)$ approaches zero. Thus, Newton's basic method can be used, but care must be exercised to discontinue the iterations as $f'(x)$ approaches zero. However, the rate of convergence drops to first-order for a multiple root. Two variations of Newton's method restore the second-order convergence of the basic method:

1. Including the multiplicity m in Eq. (3.116)
2. Solving for the root of the modified function, $u(x) = f(x)/f'(x)$

These two variations are presented in the following discussion.

First consider the variation which includes the multiplicity m in Eq. (3.116):

$$\boxed{x_{i+1} = x_i - m\frac{f(x_i)}{f'(x_{i+1})}} \tag{3.125}$$

Equation (3.125) is in the general iteration form, $x_{i+1} = g(x_i)$. Differentiating $g(x)$ and evaluating the result at $x = \alpha$ yields $g'(\alpha) = 0$. Substituting this result into Eq. (3.50) shows that Eq. (3.125) is convergent. Further analysis yields

$$e_{i+1} = \frac{g''(\xi)}{2} e_i^2 \tag{3.126}$$

where ξ is between x_i and α, which shows that Eq. (3.125) converges quadratically.

Next consider the variation where Newton's basic method is applied to the function $u(x)$:

$$u(x) = \frac{f(x)}{f'(x)} \tag{3.127}$$

If $f(x)$ has m repeated roots, $f(x)$ can be expressed as

$$f(x) = (x - \alpha)^m h(x) \tag{3.128}$$

where the deflated function $h(x)$ does not have a root at $x = \alpha$, that is, $h(\alpha) \neq 0$. Substituting Eq. (3.128) into Eq. (3.127) gives

$$u(x) = \frac{(x - r)^m h(x)}{m(x - \alpha)^{m-1} h(x) + (x - \alpha)^m g'(x)} \tag{3.129}$$

which yields

$$u(x) = \frac{(x - \alpha)h(x)}{mh(x) + (x - \alpha)g'(x)} \tag{3.130}$$

Equation (3.130) shows that $u(x)$ has a single root at $x = \alpha$. Thus, Newton's basic method, with second-order convergence, can be applied to $u(x)$ to give

$$\boxed{x_{i+1} = x_i - \frac{u(x_i)}{u'(x_i)}} \tag{3.131}$$

Differentiating Eq. (3.127) gives

$$u'(x) = \frac{f'(x)f'(x) - f(x)f''(x)}{[f'(x)]^2} \tag{3.132}$$

Substituting Eqs. (3.127) and (3.132) into Eq. (3.131) yields an alternate form of Eq. (3.131):

$$\boxed{x_{i+1} = x_i - \frac{f(x_i)f'(x_i)}{[f'(x_i)]^2 - f(x_i)f''(x_i)}} \tag{3.133}$$

The advantage of Eq. (3.133) over Newton's basic method for repeated roots is that Eq. (3.133) has second-order convergence. There are several disadvantages. There is an additional calculation for $f''(x_i)$. Equation (3.133) requires additional effort to evaluate. Round-off errors may be introduced due to the difference appearing in the denominator of Eq. (3.133). This method can also be used for simple roots, but it is less efficient than Newton's basic method in that case.

In summary, three methods are presented for evaluating repeated roots: Newton's basic method (which reduces to first-order convergence), Newton's basic method with the multiplicity m included, and Newton's basic method applied to the modified function, $u(x) = f(x)/f'(x)$. These three methods can be applied to any nonlinear equation. They are presented in this section devoted to polynomials simply because the problem of multiple roots generally occurs more frequently for polynomials than for other nonlinear functions. The three techniques presented here can also be applied with the secant method, although the evaluation of $f''(x)$ is more complicated in that case. These three methods are illustrated in Example 3.9.

Example 3.9. Newton's method for multiple roots.

Three versions of Newton's method for multiple roots are illustrated in this section:

1. Newton's basic method.
2. Newton's basic method including the multiplicity m.
3. Newton's basic method applied to the modified function, $u(x) = f(x)/f'(x)$.

These three methods are specified in Eqs. (3.116), (3.125), and (3.133), respectively, which are repeated below:

$$x_{i+1} = x_i - \frac{f(x_i)}{f'(x_i)} \tag{3.134}$$

$$x_{i+1} = x_i - m\frac{f(x_i)}{f'(x_i)} \tag{3.135}$$

where m is the multiplicity of the root, and

$$x_{i+1} = x_i - \frac{u(x_i)}{u'(x_i)} = x_i - \frac{f(x_i)f'(x_i)}{[f'(x_i)]^2 - f(x_i)f''(x_i)} \qquad (3.136)$$

where $u(x) = f(x)/f'(x)$ has the same roots as $f(x)$. Let's solve for the repeated root, $r = 1, 1$, of the following third-degree polynomial:

$$f(x) = P_3(x) = (x + 1)(x - 1)(x - 1) = 0 \qquad (3.137)$$

$$f(x) = x^3 - x^2 - x + 1 = 0 \qquad (3.138)$$

From Eq. (3.138),

$$f'(x) = 3x^2 - 2x - 1 \qquad (3.139)$$

$$f''(x) = 6x - 2 \qquad (3.140)$$

Let the initial approximation be $x_1 = 1.50$. From Eqs. (3.138) to (3.140), $f(1.50) = 0.6250, f'(1.50) = 2.750$, and $f''(1.5) = 7.0$. Substituting these values into Eqs. (3.134) to (3.136) gives

$$x_2 = 1.5 - \frac{0.6250}{2.750} = 2.272727 \qquad (3.141)$$

$$x_2 = 1.5 - 2.0\frac{0.6250}{2.750} = 1.045455 \qquad (3.142)$$

$$x_2 = 1.5 - \frac{(0.6250)(2.750)}{(2.750)^2 - (0.6250)(7.0)} = 0.960784 \qquad (3.143)$$

These results and the results of subsequent iterations required to achieve the convergence tolerance, $|\Delta x_{i+1}| \leq 0.000001$, are summarized in Table 3.11.

Newton's basic method required 20 iterations, while the two other methods required only four iterations each. The advantage of these two methods over the basic method for repeated roots is obvious.

3.5.2.4. Newton's Method for Complex Roots

Newton's method, the secant method, and Muller's method can be used to calculate complex roots simply by using complex arithmetic and choosing complex initial approximations.

Bracketing methods, such as interval halving and false position, cannot be used to find complex roots, since the sign of $f(x)$ generally does not change sign at a complex root. Newton's method is applied in this section to find the complex conjugate roots of a polynomial with real coefficients.

Example 3.10. Newton's method for complex roots.

The basic Newton method can find complex roots by using complex arithmetic and choosing a complex initial approximation. Consider the third-degree polynomial:

$$f(x) = P_3(x) = (x - 1)(x - 1 - I1)(x - 1 + I1) \qquad (3.144)$$

$$f(x) = x^3 - 3x^2 + 4x - 2 = 0 \qquad (3.145)$$

Table 3.11. Newton's Method for Multiple Real Roots

Newton's basic method, Eq. (3.134)

i	x_i	$f(x_i)$	x_{i+1}	$f(x_{i+1})$
1	1.50	0.6250	1.272727	0.16904583
2	1.272727	0.16904583	1.144082	0.04451055
3	1.144082	0.04451055	1.074383	0.01147723
...				
19	1.000002	0.00000000	1.000001	0.00000000
20	1.000000	0.00000000	1.000001	0.00000000
	1.000001	0.00000000		

Newton's multiplicity method, Eq. (3.135), with $m = 2$

i	x_i	$f(x_i)$	x_{i+1}	$f(x_{i+1})$
1	1.50	0.6250	1.045455	0.00422615
2	1.045455	0.00422615	1.00500	0.00000050
3	1.005000	0.00000050	1.000000	0.00000000
4	1.000000	0.00000000	1.000000	0.00000000
	1.000000	0.00000000		

Newton's modified method, Eq. (3.136)

i	x_i	$f(x_i)$	x_{i+1}	$f(x_{i+1})$
1	1.50	0.6250	0.960784	0.00301543
2	0.960784	0.00301543	0.999600	0.00000032
3	0.999600	0.00000032	1.000000	0.00000000
4	1.000000	0.00000000	1.000000	0.00000000
	1.000000	0.00000000		

Table 3.12. Newton's Method for Complex Roots

i	x_i	$f(x_i)$	$f'(x_i)$
1	$0.500000 + I0.500000$	$1.75000000 - I0.25000000$	$-1.00000000 + I0.50000000$
2	$2.000000 + I1.000000$	$1.00000000 + I7.00000000$	$5.00000000 + I10.00000000$
3	$1.400000 + I0.800000$	$0.73600000 + I1.95200000$	$1.16000000 + I5.12000000$
4	$1.006386 + I0.854572$	$0.53189072 + I0.25241794$	$-1.16521249 - I3.45103149$
5	$0.987442 + I1.015093$	$-0.03425358 - I0.08138309$	$-2.14100172 - I3.98388821$
6	$0.999707 + I0.999904$	$0.00097047 - I0.00097901$	$-2.00059447 - I3.99785801$
7	$1.000000 + I1.000000$	$-0.00000002 + I0.00000034$	$-1.99999953 - I4.00000030$
	$1.000000 + I1.000000$	$0.00000000 + I0.00000000$	

The roots of Eq. (3.144) are $r = 1, 1 + I1$, and $1 - I1$. Let's find the complex root $r = 1 + I1$ starting with $x_1 = 0.5 + I0.5$. The complex arithmetic was performed by a FORTRAN program for Newton's method. The results are presented in Table 3.12.

3.5.3. Bairstow's Method

A special problem associated with polynomials $P_n(x)$ is the possibility of complex roots. Newton's method, the secant method, and Muller's method all can find complex roots if complex arithmetic is used and complex initial approximations are specified. Fortunately, complex arithmetic is available in several programming languages, such as FORTRAN. However, hand calculation using complex arithmetic is tedious and time consuming. When polynomials with real coefficients have complex roots, they occur in conjugate pairs, which corresponds to a quadratic factor of the polynomial $P_n(x)$. Bairstow's method extracts quadratic factors from a polynomial using only real arithmetic. The quadratic formula can then be used to determine the corresponding pair of real roots or complex conjugate roots.

Consider the general nth-degree polynomial, $P_n(x)$:

$$P_n(x) = a_n x^n + a_{n-1} x^{n-1} + \cdots + a_0 \tag{3.146}$$

Let's factor out a quadratic factor from $P_n(x)$. Thus,

$$P_n(x) = (x^2 - rx - s)Q_{n-2}(x) + \text{remainder} \tag{3.147}$$

This form of the quadratic factor (i.e., $x^2 - rx - s$) is generally specified. Performing the division of $P_n(x)$ by the quadratic factor yields

$$P_n(x) = (x^2 - rx - s)(b_n x^{n-2} + b_{n-1} x^{n-3} + \cdots + b_3 x + b_2) + \text{remainder} \tag{3.148}$$

where the remainder is given by

$$\text{Remainder} = b_1(x - r) + b_0 \tag{3.149}$$

When the remainder is zero, $(x^2 - rx - s)$ is an exact factor of $P_n(x)$. The roots of the quadratic factor, real or complex, can be determined by the quadratic formula.

For the remainder to be zero, both b_1 and b_0 must be zero. Both b_1 and b_0 depend on both r and s. thus,

$$b_1 = b_1(r, s) \qquad \text{and} \qquad b_0 = b_0(r, s) \tag{3.150}$$

Thus, we have a two-variable root-finding problem. This problem can be solved by Newton's method for a system of nonlinear equations, which is presented in Section 3.7.

Expressing Eq. (3.150) in the form of two two-variable Taylor series in terms of $\Delta r = (r^* - r)$ as $\Delta s = (s^* - s)$, where r^* and s^* are the values of r and s which yield $b_1 = b_0 = 0$, gives

$$b_1(r^*, s^*) = b_1 + \frac{\partial b_1}{\partial r}\Delta r + \frac{\partial b_1}{\partial s}\Delta s + \cdots = 0 \tag{3.151a}$$

$$b_0(r^*, s^*) = b_0 \frac{\partial b_0}{\partial r}\Delta r + \frac{\partial b_0}{\partial s}\Delta s + \cdots = 0 \tag{3.151b}$$

where b_1, b_0, and the four partial derivatives are evaluated at point (r, s). Truncating Eq. (3.151) after the first-order terms and solving for Δr and Δs gives

$$\frac{\partial b_1}{\partial r}\Delta r + \frac{\partial b_1}{\partial s}\Delta s = -b_1 \tag{3.152}$$

$$\frac{\partial b_0}{\partial r}\Delta r + \frac{\partial b_0}{\partial s}\Delta s = -b_0 \tag{3.153}$$

Equations (3.152) and (3.153) can be solved for Δr and Δs by Cramer's rule or Gauss elimination. All that remains is to relate b_1, b_0, and the four partial derivatives to the coefficients of the polynomial $P_n(x)$, that is, a_i $(i = 0, 1, 2, \ldots, n)$.

Expanding the right-hand side of Eq. (3.148), including the remainder term, and comparing the two sides term by term, yields

$$b_n = a_n \tag{3.154.n}$$

$$b_{n-1} = a_{n-1} + rb_n \tag{3.154n-1}$$

$$b_{n-2} = a_{n-2} + rb_{n-1} + sb_n \tag{3.154n-2}$$

$$\cdots\cdots\cdots\cdots\cdots$$

$$b_1 = a_1 + rb_2 + sb_3 \tag{3.154.1}$$

$$b_0 = a_0 + rb_1 + sb_2 \tag{3.154.0}$$

Equation (3.154) is simply the synthetic division algorithm presented in Section 4.2 applied for a quadratic factor.

The four partial derivatives required in Eqs. (3.152) and (3.153) can be obtained by differentiating the coefficients b_i $(i = n, n-1, \ldots, b_1, b_0)$, with respect to r and s, respectively. Since each coefficient b_i contains b_{i+1} and b_{i+2}, we must start with the partial derivatives of b_n and work our way down to the partial derivatives of b_1 and b_0. Bairstow showed that the results are identical to dividing $Q_{n-2}(x)$ by the quadratic factor, $(x^2 - rx - s)$, using the synthetic division algorithm. The details are presented by Gerald and Wheatley (1999). The results are presented below.

$$c_n = b_n \tag{3.155.n}$$
$$c_{n-1} = b_{n-1} + rc_n \tag{3.155.n-1}$$
$$c_{n-2} = b_{n-2} + rc_{n-1} + sc_n \tag{3.155n-2}$$
$$\cdots\cdots\cdots\cdots\cdots\cdots$$
$$c_2 = b_2 + rc_3 + sc_4 \tag{3.155.2}$$
$$c_1 = b_1 + rc_2 + sc_3 \tag{3.155.1}$$

The required partial derivatives are given by

$$\frac{\partial b_1}{\partial r} = c_2 \quad \text{and} \quad \frac{\partial b_1}{\partial s} = c_3 \tag{3.156a}$$

$$\frac{\partial b_0}{\partial r} = c_1 \quad \text{and} \quad \frac{\partial b_0}{\partial s} c_2 \tag{3.156b}$$

Thus, Eqs. (3.152) and (3.153) become

$$\boxed{\begin{aligned} c_2 \, \Delta r + c_3 \, \Delta s &= -b_1 \\ c_1 \, \Delta r + c_2 \, \Delta s &= -b_0 \end{aligned}} \tag{3.157a}$$
$$\tag{3.157b}$$

where $\Delta r = (r^* - r)$ and $\Delta s = (s^* - s)$. Thus,

$$r_{i+1} = r_i + \Delta r_i \qquad\qquad\qquad (3.158a)$$

$$s_{i+1} = s_i + \Delta s_i \qquad\qquad\qquad (3.158b)$$

Equations (3.157) and (3.158) are applied repetitively until either one or both of the following convergence criteria are satisfied:

$$|\Delta r_i| \le \varepsilon_1 \qquad \text{and} \qquad |\Delta s_i| \le \varepsilon_1 \qquad\qquad (3.159a)$$

$$|(b_1)_{i+1} - (b_1)_i| \le \varepsilon_2 \qquad \text{and} \qquad |(b_0)_{i+1} - (b_0)_i| \le \varepsilon_2 \qquad\qquad (3.159b)$$

Example 3.11. Bairstow's method for quadratic factors.

Let's illustrate Bairstow's method by solving for a quadratic factor of Eq. (3.145):

$$f(x) = x^3 - 3x^2 + 4x - 2 = 0 \qquad\qquad (3.160)$$

The roots of Eq. (3.160) are $r = 1, 1 + I1$, and $1 - I1$.

To initiate the calculations, let $r_1 = 1.5$ and $s_1 = -2.5$. Substituting these values into Eq. (3.154) gives

$$b_3 = a_3 = 1.0 \qquad\qquad (3.161.3)$$
$$b_2 = a_2 + rb_3 = (-3.0) + (1.5)(1) = -1.50 \qquad\qquad (3.161.2)$$
$$b_1 = a_1 + rb_2 + sb_3 = 4.0 + (1.5)(-1.5) + (-2.5)(1.0) = -0.750 \qquad\qquad (3.161.1)$$
$$b_0 = a_0 + rb_1 + sb_2 = -2.0 + (1.5)(-0.75) + (-2.5)(-1.5) = 0.6250 \qquad (3.161.0)$$

Substituting these results into Eq. (3.155) gives

$$c_3 = b_3 = 1.0 \qquad\qquad (3.162.8)$$
$$c_2 = b_2 + rc_3 = -(1.5) + (1.5)(1.0) = 0.0 \qquad\qquad (3.162.2)$$
$$c_1 = b_1 + rc_2 + sc_3 = (-0.750) + (1.5)(0.0) + (-2.5)(1.0) = -3.250 \qquad (3.162.1)$$

Substituting the values of b_1, b_0, c_3, c_2, and c_1 into Eq. (3.157) gives

$$(0.0)\Delta r + (1.0)\Delta s = -(-0.75) = 0.750 \qquad\qquad (3.163a)$$
$$-3.250\Delta r + (0.0)\Delta s = -0.6250 \qquad\qquad (3.163b)$$

Table 3.13. Bairstow's Method for Quadratic Factors

i	r	s	Δr	Δs
1	1.50	-2.50	0.192308	0.750000
2	1.692308	-1.750	0.278352	-0.144041
3	1.970660	-1.894041	0.034644	-0.110091
4	2.005304	-2.004132	-0.005317	0.004173
5	1.999988	-1.999959	0.000012	-0.000041
6	2.000000	-2.000000	0.000000	0.000000

Solving Eq. (3.163) gives

$$\Delta r = 0.192308 \qquad \text{and} \qquad \Delta s = 0.750 \tag{3.164}$$

Substituting Δr and Δs into Eq. (3.158) gives

$$r_2 = r_1 + \Delta r = 1.50 + 0.192308 = 1.692308 \tag{3.165a}$$
$$s_2 = s_1 + \Delta s = -2.50 + 0.750 = -1.750 \tag{3.165b}$$

These results and the results of subsequent iterations are presented in Table 3.13. The convergence criteria, $|\Delta r_i| \le 0.000001$ and $|\Delta s_i| \le 0.000001$, are satisfied on the sixth iteration, where $r = 2.0$ and $s = -2.0$. Thus, the desired quadratic factor is

$$x^2 - rx - s = x^2 - 2.0x + 2.0 = 0 \tag{3.166}$$

Solving for the roots of Eq. (3.166) by the quadratic formula yields the pair of complex conjugate roots:

$$x = \frac{-b \pm \sqrt{b^2 - 4ac}}{2a} = \frac{-(-2.0) \pm \sqrt{(-2.0)^2 - 4.0(1.0)(2.0)}}{2(1.0)} = 1 + I1, 1 - I1$$
$$\tag{3.167}$$

3.5.4. Summary

Polynomials are a special case of nonlinear equation. Any of the methods presented in Sections 3.3 and 3.4 can be used to find the roots of polynomials. Newton's method is especially well suited for this purpose. It can find simple roots and multiple roots directly. However, it drops to first-order for multiple roots. Two variations of Newton's method for multiple roots restore the second-order convergence. Newton's method, like the secant method and Muller's method, can be used to find complex roots simply by using complex arithmetic with complex initial approximations. Bairstow's method can find quadratic factors using real arithmetic, and the quadratic formula can be used to find the two roots of the quadratic factor. Good initial guesses are desirable and may be necessary to find the roots of high-degree polynomials.

3.6 PITFALLS OF ROOT FINDING METHODS AND OTHER METHODS OF ROOT FINDING

The root-finding methods presented in Sections 3.3 to 3.5 generally perform as described. However, there are several pitfalls, or problems, which can arise in their application. Most of these pitfalls are discussed in Sections 3.3 to 3.5. They are summarized and discussed in Section 3.6.1.

The collection of root-finding methods presented in Sections 3.3 to 3.5 includes the more popular methods and the most well-known methods. Several less well-known root-finding methods are listed in Section 3.6.2.

3.6.1. Pitfalls of Root Finding Methods

Numerous pitfalls, or problems, associated with root finding are noted in Sections 3.3 to 3.5. These include:

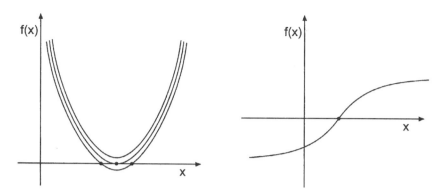

Figure 3.13   Pitfalls of root finding. (a) Closely spaced roots. (b) Inflection point.

1. Lack of a good initial approximation
2. Convergence to the wrong root
3. Closely spaced roots
4. Multiple roots
5. Inflection points
6. Complex roots
7. Ill-conditioning of the nonlinear equation
8. Slow convergence

These problems, and some strategies to avoid the problems, are discussed in this section.

Probably the most serious pitfall associated with root finding is the lack of a good initial approximation. Lack of a good initial approximation can lead to convergence to the wrong root, slow convergence, or divergence. The obvious way to avoid this problem is to obtain a better initial approximation. This can be accomplished by either graphing the function or a fine incremental search.

Closely spaced roots can be difficult to evaluate. Consider the situation illustrated in Figure 3.13. It can be difficult to determine where there are no roots, a double root, or two closely spaced distinct roots. This dilemma can be resolved by an enlargement of a graph of the function or the use of a smaller increment near the root in an incremental search.

Multiple roots, when known to exist, can be evaluated as described for Newton's method in Section 3.5.2. The major problem concerning multiple roots is not knowing they exist. Graphing the function or an incremental search can help identify the possibility of multiple roots.

Roots at an inflection point can send the root-finding procedure far away from the root. A better initial approximation can eliminate this problem.

Complex roots do not present a problem if they are expected. Newton's method or the secant method using complex arithmetic and complex initial approximations can find complex roots in a straightforward manner. However, if complex roots are not expected, and the root-finding method is using real arithmetic, complex roots cannot be evaluated. One solution to this problem is to use Bairstow's method for quadratic factors.

Ill-conditioning of the nonlinear function can cause serious difficulties in root finding. The problems are similar to those discussed in section 1.6.2 for solving ill-conditioned systems of linear algebraic equations. In root-finding problems, the best approach for finding the roots of ill-conditioned nonlinear equations is to use a computing device with more precision (i.e., a larger number of significant digits).

The problem of slow convergence can be addressed by obtaining a better initial approximation or by a different root-finding method.

Most root-finding problems in engineering and science are well behaved and can be solved in a straightforward manner by one or more of the methods presented in this chapter. Consequently, each problem should be approached with the expectation of success. However, one must always be open to the possibility of unexpected difficulties and be ready and willing to pursue other approaches.

3.6.2. Other Methods of Root Finding

Most of the straightforward popular methods for root finding are presented in Sections 3.3 to 3.5. Several additional methods are identified, but not developed in this section.

Brent's (1978) method uses a superlinear method (i.e., inverse quadratic interpolation) and monitors its behavior to ensure that it is behaving properly. If not, some interval halving steps are used to ensure at least linear behavior until the root is approached more closely, at which time the procedure reverts to the superlinear method. Brent's method does not require evaluation of the derivative. This approach combines the efficiency of open methods with the robustness of closed methods.

Muller's method (1956), mentioned in Section 3.4.4, is an extension of the secant method which approximates the nonlinear function $f(x)$ with a quadratic function $g(x)$, and uses the root of $g(x)$ as the next approximation to the root of $f(x)$. A higher-order version of Newton's method, mentioned in Section 3.4.2, retains the second-order term in the Taylor series for $f(x)$. This method is not used very often because the increased complexity, compared to the secant method and Newton's method, respectively, is not justified by the slightly increased efficiency.

Several additional methods have been proposed for finding the roots of polynomials. Graeff's root squaring method (see Hildebrand, 1956), the Lehmer-Schur method (see Acton, 1970), and the QD (quotient-difference) method (see Henrici, 1964) are three such methods. Two of the more important additional methods for polynomials are Laguerre's method (Householder, 1970) and the Jenkins-Traub method. Ralston and Rabinowitz (1979) present a discussion of these methods. An algorithm for Laguerre's method is presented by Press et al. (1989). The Jenkins-Traub method is implemented in the IMSL library.

3.7 SYSTEMS OF NONLINEAR EQUATIONS

Many problems in engineering and science require the solution of a system of nonlinear equations. Consider a system of two nonlinear equations:

$$f(x, y) = 0 \tag{3.168a}$$

$$g(x, y) = 0 \tag{3.168b}$$

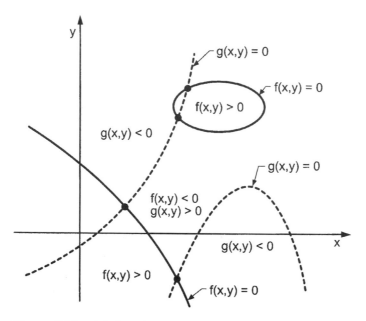

Figure 3.14   Solution of two nonlinear equations.

The problem can be stated as follows:

> Given the continuous functions $f(x, y)$ and $g(x, y)$, find the
> values $x = x^*$ and $y = y^*$ such that $f(x^*, y^*) = 0$ and $g(x^*, y^*) = 0$.

The problem is illustrated graphically in Figure 3.14. The functions $f(x, y)$ and $g(x, y)$ may be algebraic equations, transcendental equations, the solution of differential equations, or any nonlinear relationships between the inputs x and y and the outputs $f(x, y)$ and $g(x, y)$. The $f(x, y) = 0$ and $g(x, y) = 0$ contours divide the xy plane into regions where $f(x, y)$ and $g(x, y)$ are positive or negative. The solutions to Eq. (3.168) are the intersections of the $f(x, y) = g(x, y) = 0$ contours, if any. The number of solutions is not known a priori. Four such intersections are illustrated in Figure 3.14. This problem is considerably more complicated than the solution of a single nonlinear equation.

Interval halving and fixed-point iteration are not readily extendable to systems of nonlinear equations. Newton's method, however, can be extended to solve systems of nonlinear equations. In this section, Newton's method is extended to solve the system of two nonlinear equations specified by Eq. (3.168).

Assume that an approximate solution to Eq. (3.168) is known: (x_i, y_i). Express $f(x, y)$ and $g(x, y)$ in two-variable Taylor series about (x_i, y_i), and evaluate the Taylor series at (x^*, y^*). Thus,

$$f(x^*, y^*) = f_i + f_x|_i(x^* - x_i) + f_y|_i(y^* - y_i) + \cdots = 0 \tag{3.169a}$$

$$g(x^*, y^*) = g_i + g_x|_i(x^* - x_i) + g_y|_i(y^* - y_i) + \cdots = 0 \tag{3.169b}$$

Truncating Eq. (3.169) after the first derivative terms and rearranging yields

$$f_x|_i \, \Delta x_i + f_y|_i \, \Delta y_i = -f_i \qquad\qquad (3.170a)$$

$$g_x|_i \, \Delta x_i + g_y|_i \, \Delta y_i = -g_i \qquad\qquad (3.170b)$$

where Δx_i and Δy_i denote $(x^* - x_i)$ and $(y^* - y_i)$, respectively. Thus,

$$x_{i+1} = x_i + \Delta x_i \qquad\qquad (3.171a)$$

$$y_{i+1} = y_i + \Delta y_i \qquad\qquad (3.171b)$$

Equations (3.170) and (3.171) are applied repetitively until either one or both of the following convergence criteria are satisfied:

$$|\Delta x_i| \le \varepsilon_x \qquad \text{and} \qquad |\Delta y_i| \le \varepsilon_y \qquad\qquad (3.172a)$$

$$|f(x_{i+1}, y_{i+1})| \le \varepsilon_f \qquad \text{and} \qquad |g(x_{i+1}, y_{i+1})| \le \varepsilon_g \qquad\qquad (3.172b)$$

Example 3.12. Newton's method for two coupled nonlinear equations.

As an example of Newton's method for solving two nonlinear equations, let's solve the four-bar linkage problem presented in Section 3.1. Recall the two scalar components of the vector loop equation, Eq. (3.2):

$$f(\theta_2, \theta_3) = r_2 \cos(\theta_2) + r_3 \cos(\theta_3) + r_4 \cos(\theta_4) - r_1 = 0 \qquad\qquad (3.173a)$$

$$g(\theta_2, \theta_3) = r_2 \sin(\theta_2) + r_3 \sin(\theta_3) + r_4 \sin(\theta_4) = 0 \qquad\qquad (3.173b)$$

where r_1 to r_4 are specified, θ_4 is the input angle, and θ_2 and θ_3 are the two output angles.

Let θ_2^* and θ_3^* be the solution to Eq. (3.173), and θ_2 and θ_3 be an approximation to the solution. Writing Taylor series for $f(\theta_2, \theta_3)$ and $g(\theta_2, \theta_3)$ about (θ_2, θ_3) and evaluating at (θ_2^*, θ_3^*) gives

$$f(\theta_2^*, \theta_3^*) = f|_{\theta_2, \theta_3} + f_{\theta_2}|_{\theta_2, \theta_3} \, \Delta\theta_2 + f_{\theta_3}|_{\theta_2, \theta_3} \, \Delta\theta_3 + \cdots = 0 \qquad\qquad (3.174a)$$

$$g(\theta_2^*, \theta_3^*) = g|_{\theta_2, \theta_3} + g_{\theta_2}|_{\theta_2, \theta_3} \, \Delta\theta_2 + g_{\theta_3}|_{\theta_2, \theta_3} \, \Delta\theta_3 + \cdots = 0 \qquad\qquad (3.174b)$$

where $\Delta\theta_2 = (\theta_2^* - \theta_2)$ and $\Delta\theta_3 = (\theta_3^* - \theta_3)$. From Eq. (3.173),

$$f_{\theta_2} = -r_2 \sin(\theta_2) \qquad \text{and} \qquad f_{\theta_3} = -r_3 \sin(\theta_3) \qquad\qquad (3.175a)$$

$$g_{\theta_2} = r_2 \cos(\theta_2) \qquad \text{and} \qquad g_{\theta_3} = r_3 \cos(\theta_3) \qquad\qquad (3.174b)$$

Solving Eqs. (3.174) for $\Delta\theta_2$ and $\Delta\theta_3$ yields the following equations:

$$(f_{\theta_2}|_{\theta_2, \theta_3}) \, \Delta\theta_2 + (f_{\theta_3}|_{\theta_2, \theta_3}) \, \Delta\theta_3 = -f(\theta_2, \theta_3) \qquad\qquad (3.176a)$$

$$(g_{\theta_2}|_{\theta_2, \theta_3}) \, \Delta\theta_2 + (g_{\theta_3}|_{\theta_2, \theta_3}) \, \Delta\theta_3 = -g(\theta_2, \theta_3) \qquad\qquad (3.176b)$$

Equations (3.176a) and (3.176b) can be solved by Cramer's rule or Gauss elimination.

Table 3.14. Newton's Method for Two Coupled Nonlinear Equations

i	θ_2, deg	θ_3, deg	$f(\theta_2, \theta_3)$	$g(\theta_2, \theta_3)$	$\Delta\theta_2$, deg	$\Delta\theta_3$, deg
1	30.000000	0.000000	0.131975E + 00	0.428850E + 00	2.520530	−4.708541
2	32.520530	−4.708541	−0.319833E − 01	−0.223639E − 02	−0.500219	0.333480
3	32.020311	−4.375061	−0.328234E − 03	−0.111507E − 03	−0.005130	0.004073
4	32.015181	−4.370988	−0.405454E − 07	−0.112109E − 07	−0.000001	0.000000
	32.015180	−4.370987				

For the problem presented in Section 3.1, $r_1 = 10, r_2 = 6, r_3 = 8$, and $r_4 = 4$. Consider the case where $\theta_4 = 220.0$ deg. Let $\theta_2^{(1)} = 30.0$ deg and $\theta_3^{(1)} = 0.0$ deg. From Eq. (3.173):

$$f(30.0, 0.0) = 6.0\cos(30.0) + 8.0\cos(0.0) + 4.0\cos(220.0) - 10.0 = 0.131975$$

(3.177a)

$$g(30.0, 0.0) = 6.0\sin(30.0) + 8.0\sin(0.0) + 4.0\sin(220.0) = 0.428850 \quad (3.177b)$$

Equations (3.175a) and (3.175b) give

$$f_{\theta_2} = -6.0\sin(30.0) = -3.000000 \quad \text{and} \quad f_{\theta_3} = -8.0\sin(0.0) = 0.0 \quad (3.178a)$$

$$g_{\theta_2} = 6.0\cos(30.0) = 5.196152 \quad \text{and} \quad g_{\theta_3} = 8.0\cos(0.0) = 8.0 \quad (3.178b)$$

Substituting these results into Eq. (3.176) gives

$$-3.000000\,\Delta\theta_2 + 0.0\,\Delta\theta_3 = -0.131975$$

(3.179a)

$$5.196152\,\Delta\theta_2 + 8.0\,\Delta\theta_3 = -0.428850$$

(3.179b)

Solving Eq. (3.179) gives

$$\Delta\theta_2 = 0.043992(180/\pi) = 2.520530 \text{ deg}$$

(3.180a)

$$\Delta\theta_3 = -0.082180(180/\pi) = -4.708541 \text{ deg}$$

(3.180b)

where the factor $(180/\pi)$ is needed to convert radians to degrees. Thus,

$$\theta_2 = 32.520530 \text{ deg} \quad \text{and} \quad \theta_3 = -4.708541 \text{ deg}$$

(3.181)

These results and the results of subsequent iterations are presented in Table 3.14. Four iterations are required to satisfy the convergence criteria $|\Delta\theta_2| \leq 0.000001$ and $|\theta_3| \leq 0.000001$. These results were obtained on a 13-digit precision computer. As illustrated in Figure 3.1, $\theta_2 = \phi$. From Table 3.1, $\phi = 32.015180$ deg, which is the same as θ_2.

In the general case,

$$\boxed{\mathbf{f}(\mathbf{x}) = 0}$$

(3.182)

where $\mathbf{f}(\mathbf{x})^T = [f_1(\mathbf{x}) \quad f_2(\mathbf{x}) \quad \cdots \quad f_n(\mathbf{x})]$ and $\mathbf{x}^T = [x_1 \quad x_2 \quad \cdots \quad x_n]$. In this case, Eqs. (3.170) and (3.171) become

$$\boxed{\mathbf{A}\Delta = \mathbf{f}}$$

(3.183)

where $\mathbf{A}$ is the $n \times n$ matrix of partial derivatives,

$$\mathbf{A} = \begin{bmatrix} (f_1)_{x_1} & (f_1)_{x_2} & \cdots & (f_1)_{x_n} \\ (f_2)_{x_1} & (f_2)_{x_2} & \cdots & (f_2)_{x_n} \\ \cdots\cdots & \cdots\cdots & \cdots\cdots & \cdots\cdots \\ (f_n)_{x_1} & (f_n)_{x_2} & \cdots & (f_n)_{x_n} \end{bmatrix} \tag{3.184}$$

Δ is the column vector of corrections,

$$\Delta^T = [\Delta x_1 \quad \Delta x_2 \quad \cdots \quad \Delta x_n] \tag{3.185}$$

and $\mathbf{f}$ is the column vector of function values

$$\mathbf{f}^T = [f_1 \quad f_2 \quad \cdots \quad f_n] \tag{3.186}$$

The most costly part of solving systems of nonlinear equations is the evaluation of the matrix of partial derivatives, $\mathbf{A}$. Letting $\mathbf{A}$ be constant may yield a much less costly solution. However, $\mathbf{A}$ must be reasonably accurate for this procedure to work. A strategy based on making several corrections using constant $\mathbf{A}$, then reevaluating $\mathbf{A}$, may yield the most economical solution.

In situations where the partial derivatives of $\mathbf{f}(\mathbf{x})$ cannot be evaluated analytically, the above procedure cannot be applied. One alternate approach is to estimate the partial derivatives in Eq. (3.184) numerically. Thus,

$$\frac{\partial f_i}{\partial x_j} = \frac{f_i(\mathbf{x} + \Delta x_j) - f_i(\mathbf{x})}{\Delta x_j} \qquad (i, j = 1, 2, \ldots, n) \tag{3.187}$$

This procedure has two disadvantages. First, the number of calculations is increased. Second, if Δx_j is too small, round-off errors pollute the solution, and if Δx_j is too large, the convergence rate can decrease to first order. Nevertheless, this is one procedure for solving systems of nonlinear equations where the partial derivatives of $\mathbf{f}(\mathbf{x})$ cannot be determined analytically.

An alternate approach involves constructing a single nonlinear function $F(\mathbf{x})$ by adding together the sums of the squares of the individual functions $f_i(\mathbf{x})$. The nonlinear function $F(\mathbf{x})$ has a global minimum of zero when all of the individual functions are zero. Multidimensional minimization techniques can be applied to minimize $F(\mathbf{x})$, which yields the solution to the system of nonlinear equations, $\mathbf{f}(\mathbf{x}) = 0$. Dennis et al. (1983) discuss such procedures.

3.8 PROGRAMS

Three FORTRAN subroutines for solving nonlinear equations are presented in this section:

1. Newton's method
2. The secant method
3. Newton's method for two simultaneous equations

The basic computational algorithms are presented as completely self-contained subroutines suitable for use in other programs. Input data and output statements are contained in a main (or driver) program written specifically to illustrate the use of each subroutine.

3.8.1. Newton's Method

The general algorithm for Newton's method is given by Eq. (3.55):

$$x_{i+1} = x_i - \frac{f(x_i)}{f'(x_i)} \tag{3.188}$$

A FORTRAN subroutine, *subroutine newton*, for implementing Newton's method is presented in Program 3.1. *Subroutine newton* requires a function subprogram, *function funct*, which evaluates the nonlinear equation of interest. *Function funct* must be completely self-contained, including all numerical values and conversion factors. The value of x is passed to *function funct*, and the values of $f(x)$ and $f'(x)$ are returned as f and fp, respectively. Every nonlinear equation requires its own individual *function funct*. *Subroutine newton* calls *function funct* to evaluate $f(x)$ and $f'(x)$ for a specified value of x, applies Eq. (3.188), checks for convergence, and either continues or returns. After *iter* iterations, an error message is printed and the solution is terminated. Program 3.1 defines the data set and prints it, calls *subroutine newton* to implement the solution, and prints the solution.

Program 3.1. Newton's method program.

```
      program main
c     main program to illustrate nonlinear equation solvers
c     x1     first guess for the root
c     iter   number of iterations allowed
c     tol    convergence tolerance
c     iw     intermediate results output flag:  0 no,  1 yes
      data x1,iter,tol,iw / 30.0, 10, 0.000001, 1 /
      write (6,1000)
      write (6,1010)
      call newton (x1,iter,tol,iw,i)
      call funct (x1,f1,fp1)
      write (6,1020) i,x1,f1
      stop
 1000 format (' Newtons method')
 1010 format (' '/'    i',6x,'xi',10x,'fi',12x,'fpi',11x,'xi+1'/' ')
 1020 format (i4,f12.4.g,f14.8)
      end


      subroutine newton (x1,iter,tol,iw,i)
c     Newton's method
      do i=1,iter
         call funct (x1,f1,fp1)
         dx=-f1/fp1
         x2=x1+dx
         if (iw.eq.1) write (6,1000) i,x1,f1,fp1,x2
         x1=x2
         if (abs(dx).le.tol) return
      end do
      write (6,1010)
      return
 1000 format (i4,f12.4.g,2f14.8,f12.6)
```

```
1010 format (' '/' Iterations failed to converge')
     end

     function funct (x,f,fp)
c    evaluates the nonlinear function
     data r1,r2,r3,alpha / 1.66666667, 2.5, 1.83333333, 40.0 /
     rad=acos(-1.0)/180.0
     f=r1*cos(alpha*rad)-r2*cos(x*rad)+r3-cos((alpha-x)*rad)
     fp=(r2*sin(x*rad)-sin((alpha-x)*rad))*rad
     return
     end
```

The data set used to illustrate *subroutine newton* is taken from Example 3.4. The output generated by the Newton method program is presented in Output 3.1.

Output 3.1.   Solution by Newton's method.

```
Newtons method
```

i	xi	fi	fpi	xi+1
1	30.000000	-0.03979719	0.01878588	32.118463
2	32.118463	0.00214376	0.02080526	32.015423
3	32.015423	0.00000503	0.02070767	32.015180
4	32.015180	0.00000000	0.02070744	32.015180
4	32.015180	0.00000000		

Subroutine newton can be used to solve most of the nonlinear equations presented in this chapter. The values in the *data* statement must be changed accordingly, and the function subprogram, *function funct*, must evaluate the desired nonlinear equation. Complex roots can be evaluated simply by declaring all variables to be complex variables. As an additional example, *function funct* presented below evaluates a real coefficient polynomial up to fourth degree. This *function funct* is illustrated by solving Example 3.7. Simple roots can be evaluated directly. Multiple roots can be evaluated in three ways: directly (which reduces the order to first order), by including the multiplicity m as a coefficient of $f(x)$ in *function funct*, or by defining $u(x) = f(x)/f'(x)$ in *function funct*.

Polynomial *function funct.*

```
     function funct (x,f,fp)
c    evaluates a polynomial of up to fourth degree
     data a0,a1,a2,a3,a4 / -2.0, 4.0, -3.0, 1.0, 0.0 /
     f=a0+a1*x+a2*x**2+a3*x**3+a4*x**4
     fp=a1+2.0*a2*x+3.0*a3*x**2+4.0*a4*x**3
     return
     end
```

The data set used to illustrate the polynomial *function funct* is taken from Example 3.7. The results are presented below.

Solution by Newton's method.

```
Newtons method
```

i	xi	fi	fpi	xi+1
1	1.500000	0.62500000	1.75000000	1.142857
2	1.142857	0.14577259	1.06122449	1.005495
3	1.005495	0.00549467	1.00009057	1.000000
4	1.000000	0.00000033	1.00000000	1.000000
4	1.000000	0.00000000		

3.8.2. The Secant Method

The general algorithm for the secant method is given by Eq. (3.80):

$$x_{i+1} = x_i - \frac{f(x_i)}{g'(x_i)} \tag{3.189}$$

A FORTRAN subroutine, *subroutine secant*, for implementing the secant method is presented below. *Subroutine secant* works essentially like *subroutine newton* discussed in Section 3.8.1, except two values of x, x_1 and x_2, are supplied instead of only one value. Program 3.2 defines the data set and prints it, calls *subroutine secant* to implement the secant method, and prints the solution. Program 3.2 shows only the statements which are different from the statements in Program 3.1.

Program 3.2. The secant method program.

```
      program main
c     main program to illustrate nonlinear equation solvers
c     x2     second guess for the root
      data x1,x2,iter,tol,iw / 30.0, 40.0, 10, 0.000001, 1 /
      call secant (x1,x2,iter,tol,iw,i)
      call funct (x1,f1)
 1000 format (' The secant method')
 1010 format (' '//   i',6x,'xi',10x,'fi',12x,'gpi',11x,'xi+1'/' ')
      end

      subroutine secant (x1,x2,iter,tol,iw,i)
c     the secant method
      call funct (x1,f1)
      if (iw.eq.1) write (6,1000) i,x1,f1
      do i=1,iter
         call funct (x2,f2)
         gp2=(f2-f1)/(x2-x1)
         dx=-f2/gp2
         x3=x2+dx
```

```
      if (iw.eq.1) write (6,1000) i,x2,f2,gp2,x3
      x1=x3
      if (abs(dx).le.tol) return
      x1=x2
      f1=f2
      x2=x3
   end do
   write (6,1010)
   return
1000 format (i4,f12.4.g,2f14.8,f12.6)
1010 format (' '/' Iterations failed to converge')
   end

      function funct (x,f)
c     evaluates the nonlinear function
      end
```

The data set used to illustrate *subroutine secant* is taken from Example 3.5. The output generated by the secant method program is presented in Output 3.2.

Output 3.2.   Solution by the secant method.

The secant method

i	xi	fi	gpi	xi+1
0	30.000000	-0.03979719		
1	40.000000	0.19496296	0.02347602	31.695228
2	31.695228	-0.00657688	0.02426795	31.966238
3	31.966238	-0.00101233	0.02053257	32.015542
4	32.015542	0.00000749	0.02068443	32.015180
5	32.015180	-0.00000001	0.02070761	32.015180
5	32.015180	0.00000000		

3.8.3.   Newton's Method for Two Coupled Nonlinear Equations

The general algorithm for Newton's method for two simultaneous nonlinear equations is given by Eqs. (3.170) and (3.171).

$$f_x|_i \, \Delta x_i + f_y|_i \, \Delta y_i = -f_i \tag{3.190a}$$

$$g_x|_i \, \Delta x_i + g_y|_i \, \Delta y_i = -g_i \tag{3.190b}$$

$$x_{i+1} = x_i + \Delta x_i \tag{3.191a}$$

$$y_{i+1} = y_i + \Delta y_i \tag{3.191b}$$

The general approach to this problem is the same as the approach presented in Section 3.8.1 for a single nonlinear equation. A FORTRAN subroutine, *subroutine simul*, for implementing the procedure is presented below. Program 3.3 defines the data set and prints it, calls *subroutine simul* to implement the solution, and prints the solution.

Program 3.3.   Newton's method for simultaneous equations program.

```
      program main
c     main program to illustrate nonlinear equation solvers
c     x1,y1 first guess for the root
c     iter  number of iterations allowed
c     tol   convergence tolerance
c     iw    intermediate results output flag:  0 no,  1 yes
      data x1,y1,iter,tol,iw / 30.0, 0.0, 10, 0.000001, 1 /
      write (6,1000)
      write (6,1010)
      call simul (x1,y1,iter,tol,iw,i)
      call funct (x1,y1,f1,g1,fx,fy,gx,gy)
      write (6,1020) i,x1,y1,f1,g1
      stop
 1000 format (' Newtons method for two coupled nonlinear equations')
 1010 format (' '/'  i',6x,'xi',10x,'yi',9x,'fi',10x,'gi',9x,'dx',
     1 6x,'dy'/' ')
 1020 format (i3,2f12.6,2e12.4)
      end


      subroutine simul (x1,y1,iter,tol,iw,i)
c     Newton's method for two coupled nonlinear equations
      do i=1,iter
         call funct (x1,y1,f1,g1,fx,fy,gx,gy)
         del=fx*gy-fy*gx
         dx=(fy*g1-f1*gy)/del
         dy=(f1*gx-fx*g1)/del
         x2=x1+dx
         y2=y1+dy
         if (iw.eq.1) write (6,1000) i,x1,y1,f1,g1,dx,dy
         x1=x2
         y1=y2
         if ((abs(dx).le.tol).and.(abs(dy).le.tol)) return
      end do
      write (6,1010)
      return
 1000 format (i3,2f12.6,2e12.4,2f8.4)
 1010 format (' '/' Iteration failed to converge')
      end


      function funct (x,y,f,g,fx,fy,gx,gy)
c     evaluates the two nonlinear functions
      data r1,r2,r3,r4,theta4 / 10.0, 6.0, 8.0, 4.0, 220.0 /
      rad=acos(-1.0)/180.0
      f=r2*cos(x*rad)+r3*cos(y*rad)+r4*cos(theta4*rad)-r1
      g=r2*sin(x*rad)+r3*sin(y*rad)+r4*sin(theta4*rad)
      fx=(-r2*sin(x*rad))*rad
      fy=(-r3*sin(y*rad))*rad
      gx=(r2*cos(x*rad))*rad
      gy=(r3*cos(y*rad))*rad
      return
      end
```

The data set used to illustrate *subroutine simul* is taken from Example 3.12. The output is presented in Output 3.3.

Output 3.3. Solution by Newton's method for simultaneous equations.

```
Newtons method for two coupled nonlinear equations

  i      xi         yi          fi           gi         dx       dy

  1   30.000000    0.000000   0.1320E+00   0.4288E+00   2.5205  -4.7085
  2   32.520530   -4.708541  -0.3198E-01  -0.2236E-02  -0.5002   0.3335
  3   32.020311   -4.375061  -0.3282E-03  -0.1115E-03  -0.0051   0.0041
  4   32.015181   -4.370988  -0.4055E-07  -0.1121E-07   0.0000   0.0000
  4   32.015180   -4.370987   0.0000E+00   0.0000E+00
```

3.8.4. Packages for Nonlinear Equations

Numerous libraries and software packages are available for solving nonlinear equations. Many workstations and mainframe computers have such libraries attached to their operating systems.

Many commercial software packages contain nonlinear equation solvers. Some of the more prominent packages are Matlab and Mathcad. More sophisticated packages, such as IMSL, Mathematica, Macsyma, and Maple, also contain nonlinear equation solvers. Finally, the book *Numerical Recipes* (Press et al., 1989) contains numerous subroutines for solving nonlinear equations.

3.9 SUMMARY

Several methods for solving nonlinear equations are presented in this chapter. The nonlinear equation may be an algebraic equation, a transcendental equation, the solution of a differential equation, or any nonlinear relationship between an input x and a response $f(x)$.

Interval halving (bisection) and false position (regula falsi) converge very slowly, but are certain to converge because the root lies in a closed interval. These methods are not recommended unless the nonlinear equation is so poorly behaved that all other methods fail.

Fixed-point iteration converges only if the derivative of the nonlinear function is less than unity in magnitude. Consequently, it is not recommended.

Newton's method and the secant method are both effective methods for solving nonlinear equations. Both methods generally require reasonable initial approximations. Newton's method converges faster than the secant method (i.e., second order compared to 1.62 order), but Newton's method requires the evaluation of the derivative of the nonlinear function. If the effort required to evaluate the derivative is less than 43 percent of the effort required to evaluate the function itself, Newton's method requires less total effort than the secant method. Otherwise, the secant method requires less total effort. For functions whose derivative cannot be evaluated, the secant method is recommended. Both methods can find complex roots if complex arithmetic is used. The secant method is recommended as the best general purpose method.

The higher-order variations of Newton's method and the secant method, that is, the second-order Taylor series method and Muller's method, respectively, while quite effective, are not used frequently. This is probably because Newton's method and the secant method are so efficient that the slightly more complicated logic of the higher-order methods is not justified.

Polynomials can be solved by any of the methods for solving nonlinear equations. However, the special features of polynomials should be taken into account.

Multiple roots can be evaluated using Newton's basic method or its variations.

Complex roots can be evaluated by Newton's method or the secant method by using complex arithmetic and complex initial approximations. Complex roots can also be evaluated by Bairstow's method for quadratic factors.

Solving systems of nonlinear equations is a difficult task. For systems of nonlinear equations which have analytical partial derivatives, Newton's method can be used. Otherwise, multidimensional minimization techniques may be preferred. No single approach has proven to be the most effective. Solving systems of nonlinear equations remains a difficult problem.

After studying Chapter 3, you should be able to:

1. Discuss the general features of root finding for nonlinear equations
2. Explain the concept of bounding a root
3. Discuss the benefits of graphing a function
4. Explain how to conduct an incremental search
5. Explain the concept of refining a root
6. Explain the difference between closed domain (bracketing methods) and open domain methods
7. List several closed domain (bracketing) methods
8. List several open domain methods
9. Discuss the types of behavior of nonlinear equations in the neighborhood of a root
10. Discuss the general philosophy of root finding
11. List two closed domain (bracketing) methods
12. Explain how the internal halving (bisection) method works
13. Apply the interval halving (bisection) method
14. List the advantages and disadvantages of the interval halving (bisection) method
15. Explain how the false position (regula falsi) method works
16. Apply the false position (regula falsi) method
17. List the advantages and disadvantages of the false position (regula falsi) method
18. List several open domain methods
19. Explain how the fixed-point iteration method works
20. Apply the fixed-point iteration method
21. List the advantages and disadvantages of the fixed-point iteration method
22. Explain how Newton's method works
23. Apply Newton's method
24. List the advantages and disadvantages of Newton's method
25. Explain and apply the approximate Newton's method
26. Explain and apply the lagged Newton's method
27. Explain how the secant method works
28. Apply the secant method
29. List the advantages and disadvantages of the secant method
30. Explain the lagged secant method

31. Explain how Muller's method works
32. Apply Muller's method
33. List the advantages and disadvantages of Muller's method
34. Discuss the special features of polynomials
35. Apply the quadratic formula and the rationalized quadratic formula
36. Discuss the applicability (or nonapplicability) of closed domain methods and open domain methods for finding the roots of polynomials
37. Discuss the problems associated with finding multiple roots and complex roots
38. Apply Newton's basic method and its variations to find all the roots of a polynomial
39. Apply deflation to a polynomial
40. Explain the concepts underlying Bairstow's method for finding quadratic factors
41. Apply Bairstow's method to find real or complex roots
42. Discuss and give examples of the pitfalls of root finding
43. Suggest ways to get around the pitfalls
44. Explain the concepts underlying Newton's method for a system of nonlinear equations
45. Apply Newton's method to a system of nonlinear equations

EXERCISE PROBLEMS

In all of the problems in this chapter, carry at least six significant figures in all calculations, unless otherwise noted. Continue all iterative procedures until four significant figures have converged, unless otherwise noted. Consider the following four nonlinear equations:

$$f(x) = x - \cos(x) = 0 \quad \text{(A)} \qquad f(x) = e^x - \sin(\pi x/3) = 0 \quad \text{(B)}$$

$$f(x) = e^x - 2x - 2 = 0 \quad \text{(C)} \qquad f(x) = x^3 - 2x^2 - 2x + 1 = 0 \quad \text{(D)}$$

3.2   Closed Domain Methods

Interval Halving

1. Solve Eq. (A) by interval-halving starting with $x = 0.5$ and 1.0.
2. Solve Eq. (B) by interval-halving starting with $x = -3.5$ and -2.5.
3. Solve Eq. (C) by interval-halving starting with $x = 1.0$ and 2.0.
4. Solve Eq. (D) by interval-halving starting with $x = 0.0$ and 1.0.
5. Find the two points of intersection of the two curves $y = e^x$ and $y = 3x + 2$ using interval halving. Use $(-1.0$ and $0.0)$ and $(2.0$ and $3.0)$ as starting values.
6. Find the two points of intersection of the two curves $y = e^x$ and $y = x^4$ using interval halving. Use $(-1.0$ and $0.0)$ and $(1.0$ and $2.0)$ as starting values.
7. Problems 1 to 6 can be solved using any two initial values of x that bracket the root. Choose other sets of initial values of x to gain additional experience with interval halving.

False Position

8. Solve Eq. (A) by false position starting with $x = 0.5$ and 1.0.
9. Solve Eq. (B) by false position starting with $x = -3.5$ and -2.5.
10. Solve Eq. (C) by false position starting with $x = 1.0$ and 2.0.

11. Solve Eq. (D) by false position starting with $x = 0.0$ and 1.0.
12. Find the two points of intersection of the two curves $y = e^x$ and $y = 3x + 2$ using false position. Use $(-1.0$ and $0.0)$ and $(2.0$ and $3.0)$ as starting values.
13. Find the two points of intersection of the two curves $y = e^x$ and $y = x^4$ using false position. Use $(-1.0$ and $0.0)$ and $(1.0$ and $2.0)$ as starting values.
14. Problems 8 to 13 can be solved using any two initial values of x that bracket the root. Choose other sets of initial values of x to gain additional experience with false position.

3.4 Open Domain Methods

Fixed-Point Iteration

15. Solve Eq. (A) by fixed-point iteration with $x_0 = 0.5$.
16. Solve Eq. (A) by fixed-point iteration with $x_0 = 1.0$.
17. Solve Eq. (B) by fixed-point iteration with $x_0 = -3.5$.
18. Solve Eq. (B) by fixed-point iteration with $x_0 = -2.5$.
19. Solve Eq. (C) by fixed-point iteration with $x_0 = 1.0$.
20. Solve Eq. (C) by fixed-point iteration with $x_0 = 2.0$.
21. Solve Eq. (D) by fixed-point iteration with $x_0 = 0.0$.
22. Solve Eq. (D) by fixed-point iteration with $x_0 = 1.0$.
23. Problem 5 considers the function $f(x) = e^x - (3x + 2) = 0$, which can be rearranged into the following three forms: (a) $x = e^x - (2x + 2)$, (b) $x = (e^x - 2)/3$, and (c) $x = \ln(3x + 2)$. Solve for the positive root by fixed-point iteration for all three forms, with $x_0 = 1.0$.
24. Solve Problem 6 by fixed-point iteration with $x_0 = -1.0$ and $x_0 = 1.0$.
25. The function $f(x) = (x + 2)(x - 4) = x^2 - 2x - 8 = 0$ has the two roots $x = -2$ and 4. Rearrange $f(x)$ into the form $x = g(x)$ to obtain the root (a) $x = -2$ and (b) $x = 4$, starting with $x_0 = -1$ and 3, respectively. The function $f(x)$ can be rearranged in several ways, for example, (a) $x = 8/(x - 2)$, (b) $x = (2x + 8)^{1/2}$, and (c) $x = (x^2 - 8)/2$. One form always converges to $x = -2$, one form always converges to $x = 4$, and one form always diverges. Determine the behavior of the three forms.
26. For what starting values of x might the expression $x = 1/(x + 1)$ not converge?
27. The function $f(x) = e^x - 3x^2 = 0$ has three roots. The function can be rearranged into the form $x = \pm[e^x/3]^{1/2}$. Starting with $x_0 = 0.0$, find the roots corresponding to (a) the $+$ sign (near $x = 1.0$) and (b) the $-$ sign (near -0.5). (c) The third root is near $x = 4.0$. Show that the above form will not converge to this root, even with an initial guess close to the exact root. Develop a form of $x = g(x)$ that will converge to this root, and solve for the third root.
28. The cubic polynomial $f(x) = x^3 + 3x^2 - 2x - 4 = 0$ has a root near $x = 1.0$. Find two forms of $x = g(x)$ that will converge to this root. Solve these two forms for the root, starting with $x_0 = 1.0$.

Newton's Method

29. Solve Eq. (A) by Newton's method. Use $x_0 = 1.0$ as the starting value.
30. Solve Eq. (B) by Newton's method. Use $x_0 = -3.0$ as the starting value.

31. Solve Eq. (C) by Newton's method. Use $x_0 = 1.0$ as the starting value.
32. Solve Eq. (D) by Newton's method. Use $x_0 = 1.0$ as the starting value.
33. Find the positive root of $f(x) = x^{15} - 1 = 0$ by Newton's method, starting with $x_0 = 1.1$.
34. Solve Problem 33 using $x = 0.5$ as the initial guess. You may want to solve this problem on a computer, since a large number of iterations may be required.
35. The nth root of the number N can be found by solving the equation $x^n - N = 0$. (a) For this equation, show that Newton's method gives

$$x_{i+1} = \frac{1}{n}\left[(n-1)x_i + \frac{N}{x_1^{n+1}}\right] \tag{E}$$

Use the above result to solve the following problems: (a) $(161)^{1/3}$, (b) $(21.75)^{1/4}$, (c) $(238.56)^{1/5}$. Use $x = 6.0, 2.0$, and 3.0, respectively, as starting values.

36. Consider the function $f(x) = e^x - 2x^2 = 0$. (a) Find the two positive roots using Newton's method. (b) Find the negative root using Newton's method.

The Secant Method

37. Solve Eq. (A) by the secant method. Use $x = 0.5$ and 1.0 as starting values.
38. Solve Eq. (B) by the secant method. Use $x_0 = -3.0$ and -2.5 as starting values.
39. Solve Eq. (C) by the secant method. Use $x_0 = 1.0$ and 2.0 as starting values.
40. Solve Eq. (D) by the secant method. Use $x_0 = 0.0$ and 1.0 as starting values.
41. Find the positive root of $f(x) = x^{15} - 1 = 0$ by the secant method using $x = 1.2$ and 1.1 as starting values.
42. Solve Problem 41 by the secant method with $x = 0.5$ and 0.6 as starting values. You may want to solve this problem on a computer, since a large number of iterations may be required.
43. Solve Problems 35(a) to (c) by the secant method. Use the starting values given there for x_0 and let $x_1 = 1.1x_0$.
44. Solve Problem 36 using the secant method.

3.5. Polynomials

45. Use Newton's method to find the real roots of the following polynomials:

   (a) $x^3 - 5x^2 + 7x - 3 = 0$ (b) $x^4 - 9x^3 + 24x^2 - 36x + 80 = 0$
   (c) $x^3 - 2x^2 - 2x + 1 = 0$ (d) $3x^3 + 4x^2 - 8x - 2 = 0$

46. Use Newton's method to find the complex roots of the following polynomials:

   (a) $x^4 - 9x^3 + 24x^2 - 36x + 80 = 0$ (b) $x^3 + 2x^2 + x + 2 = 0$
   (c) $x^5 - 15x^4 + 85x^3 - 226x^2 + 274x - 120 = 0$

3.7 Systems of Nonlinear Equations

Solve the following systems of nonlinear equations using Newton's method.

47. $(x-1)^2 + (y-2)^2 = 3$ and $x^2/4 + y^2/3 = 1$. Solve for all roots.
48. $y = \cosh(x)$ and $x^2 + y^2 = 2$. Solve for both roots.
49. $x^2 + y^2 = 2x + y$ and $x^2/4 + y^2 = 1$. Solve for all four roots.
50. $y^2(1-x) = x^3$ and $x^2 + y^2 = 1$.
51. $x^3 + y^3 - 3xy = 0$ and $x^2 + y^2 = 1$.
52. $(x^2 + y^2)^2 = 2xy$ and $y = x^3$.
53. $(2x)^{2/3} + y^{2/3} = (9)^{1/3}$ and $x^2/4 + y^2 = 1$.

3.8. Programs

54. Implement the Newton method program presented in Section 3.8.1. Check out the program using the given data set.
55. Work any of Problems 29 to 36 using the Newton method program.
56. Implement the secant method program presented in Section 3.8.2. Check out the program using the given data set.
57. Work any of Problems 37 to 44 using the secant method program.
58. Implement the Newton method program presented in Section 3.8.3 for solving simultaneous equations. Check out the program using the given data set.
59. Work any of Problems 5, 6, or 47 to 53 using the Newton method program.
60. Write a computer program to solve Freudenstein's equation, Eq. (3.3), by the secant method. Calculate ϕ for $\alpha = 40$ deg to 90 deg in increments $\Delta\alpha = 10$ deg. For $\alpha = 40$ deg, let $\phi_0 = 25$ deg and $\phi_1 = 30$ deg. For subsequent values of α, let ϕ_0 be the solution value for the previous value of α, and $\phi_1 = \phi_0 + 1.0$. Continue the calculations until ϕ changes by less than 0.00001 deg. Design the program output as illustrated in Example 3.5.
61. Write a computer program to solve the van der Waal equation of state, Eq. (G) in Problem 69, by the secant method. Follow the procedure described in Problem 69 to initiate the calculations. Design the program output as illustrated in Example 3.5. For $P = 10,000$ kPa, calculate v corresponding to $T = 700$, 800, 900, 1000, 1100, 1200, 1300, 1400, 1500, and 1600 K. Write the program so that all the cases can be calculated in one run by stacking input data decks.
62. Write a computer program to solve the Colebrook equation, Eq. (I) in Problem 70, by the secant method for the friction coefficient f for specified value of the roughness ratio ε/D and the Reynolds number, Re. Use the approximation proposed by Genereaux (1939), Eq. (J), and 90 percent of that value as the initial approximations. Solve Problem 70 using the program.
63. Write a computer program to solve the $M - \varepsilon$ equation, Eq. (K) in Problem 71, by Newton's method for specified values of γ and ε. (a) Solve Problem 71, using the program. (b) Construct a table of M versus ε for $1.0 \leq \varepsilon \leq 10$, for subsonic flow, in increments $\Delta\varepsilon = 0.1$. For $\varepsilon = 1.1$, let $M_0 = 0.8$. For subsequent values of ε, let M_0 be the previous solution value.
64. Write a computer program to solve the $M - \varepsilon$ equation, Eq. (K) in Problem 71, by the secant method for specified values of γ and ε. (a) Solve Problem 71 using the program. (b) Construct a table of M versus ε for $1.0 \leq \varepsilon \leq 10$, for

supersonic flow, in increments $\Delta\varepsilon = 0.1$. For $\varepsilon = 1.1$ let $M_0 = 1.2$ and $M_1 = 1.3$. For subsequent values of ε, let M_0 be the previous solution value and $M_1 = 1.1M_0$.

65. Write a computer program to solve Eq. (L) in Problem 73 by the secant method for specified values of γ, δ, and M_1. (a) Solve Problem 73 using the program. (b) Construct a table of M versus δ for $\gamma = 1.4$ and $M_1 = 1.0$, for $0 < \delta \leq 40\,\text{deg}$, in increments $\Delta\delta = 1.0\,\text{deg}$. For $\delta = 1.0\,\text{deg}$, let $M_0 = 1.06$ and $M_1 = 1.08$. For subsequent values of δ, let M_0 be the previous solution value and $M_1 = 1.01M_0$.

APPLIED PROBLEMS

Several applied problems from various disciplines are presented in this section. All of these problems can be solved by any of the methods presented in this chapter. An infinite variety of exercises can be constructed by changing the numerical values of the parameters of the problem, by changing the starting values, or both.

66. Consider the four-bar linkage problem presented in Section 3.1. Solve for any (or all) of the results presented in Table 3.1.

67. Consider the four-bar linkage problem presented in Section 3.1. Rearrange this problem to solve for the value of r_1 such that $\phi = 60\,\text{deg}$ when $\alpha = 75\,\text{deg}$. Numerous variations of this problem can be obtained by specifying combinations of ϕ and α.

68. Solve the four-bar linkage problem for $\theta_4 = 210\,\text{deg}$ by solving the two scalar components of the vector loop equation, Eq. (3.2), by Newton's method. Let the initial guesses be $\theta_2 = 20\,\text{deg}$ and $\theta_3 = 0\,\text{deg}$. Continue the calculations until θ_2 and θ_3 change by less than $0.00001\,\text{deg}$. Show all calculations for the first iteration. Summarize the first iteration and subsequent iterations in a table, as illustrated in Example 3.12.

69. The van der Waal equation of state for a vapor is

$$\left(P + \frac{a}{v^2}\right)(v - b) = RT \qquad\qquad (F)$$

where P is the pressure ($\text{Pa} = \text{N/m}^2$), v is the specific volume (m^3/kg), T is the temperature (K), R is the gas constant (J/kg-K), and a and b are empirical constants. Consider water vapor, for which $R = 461.495\,\text{J/kg-K}$, $a = 1703.28\,\text{Pa-(m}^3/\text{kg})^3$, and $b = 0.00169099\,(\text{m}^3/\text{kg})$. Equation (F) can be rearranged into the form

$$Pv^3 - (Pb + RT)v^2 + av - ab = 0 \qquad\qquad (G)$$

Calculate the specific volume v for $P = 10{,}000\,\text{kPa}$ and $T = 800\,\text{K}$. Use the ideal gas law, $Pv = RT$, to obtain the initial guess (or guesses). Present the results in the format illustrated in the examples.

70. When an incompressible fluid flows steadily through a round pipe, the pressure drop due to the effects of wall friction is given by the empirical formula:

$$\Delta P = -0.5 f \rho V^2 \left(\frac{L}{D}\right) \qquad\qquad (H)$$

where ΔP is the pressure drop, ρ is the density, V is the velocity, L is the pipe length, D is the pipe diameter, and f is the D'Arcy friction coefficient. Several empirical formulas exist for the friction coefficient f as a function of the dimensionless Reynolds number, $\mathrm{Re} = DV\rho/\mu$, where μ is the viscosity. For flow in the turbulent regime between completely smooth pipe surfaces and wholly rough pipe surfaces, Colebrook (1939) developed the following empirical equation for the friction coefficient f:

$$\frac{1}{f^{1/2}} = -2\log_{10}\left(\frac{\varepsilon/D}{3.7} + \frac{2.51}{\mathrm{Re}\,f^{1/2}}\right) \tag{I}$$

where ε is the pipe surface roughness. Develop a procedure to determine f for specified values of ε/D and Re. Use the approximation proposed by Genereaux (1939) to determine the initial approximation(s):

$$f = 0.16\,\mathrm{Re}^{-0.16} \tag{J}$$

Solve for f for a pipe having $\varepsilon/D = 0.001$ for $\mathrm{Re} = 10^n$, for $n = 4, 5, 6$, and 7.

71. Consider quasi-one-dimensional isentropic flow of a perfect gas through a variable-area channel. The relationship between the Mach number M and the flow area A, derived by Zucrow and Hoffman [1976, Eq. (4.29)], is given by

$$\varepsilon = \frac{A}{A^*} = \frac{1}{M}\left[\frac{2}{\gamma+1}\left(1 + \frac{\gamma-1}{2}M^2\right)\right]^{(\gamma+1)/2(\gamma-1)} \tag{K}$$

where A^* is the choking area (i.e., the area where $M = 1$) and γ is the specific heat ratio of the flowing gas. For each value of ε, two values of M exist, one less than unity (i.e., subsonic flow) and one greater than unity (i.e., supersonic flow). Calculate both values of M for $\varepsilon = 10.0$ and $\gamma = 1.4$ by Newton's method. For the subsonic root, let $M_0 = 0.2$. For the supersonic root, let $M_0 = 5.0$.

72. Solve Problem 71 by the secant method. For the subsonic root, let $M_0 = 0.4$ and $M_1 = 0.6$. For the supersonic root, let $M_0 = 3.0$ and $M_1 = 4.0$.

73. Consider isentropic supersonic flow around a sharp expansion corner. The relationship between the Mach number before the corner (i.e., M_1) and after the corner (i.e., M_2), derived by Zucrow and Hoffman [1976, Eq. (8.11)], is given by

$$\delta = b^{1/2}\left(\tan^{-1}\left(\frac{M_2^2-1}{b}\right)^{1/2} - \tan^{-1}\left(\frac{M_1^2-1}{b}\right)^{1/2}\right)$$
$$- \left((\tan^{-1}((M_2^2-1)^{1/2}) - \tan^{-1}((M_1^2-1)^{1/2})\right) \tag{L}$$

where $b = (\gamma+1)/(\gamma-1)$ and γ is the specific heat ratio of the gas. Develop a procedure to solve for M_2 for specified values of γ, δ, and M_1. For $\gamma = 1.4$, solve for M_2 for the following combinations of M_1 and δ: (a) 1.0 and 10.0 deg, (b) 1.0 and 20.0 deg, (c) 1.5 and 10.0 deg, and (d) 1.5 and 20.0 deg. Use $M_2^{(0)} = 2.0$ and $M_2^{(1)} = 1.5$.

4

Polynomial Approximation and Interpolation

4.1 INTRODUCTION

Figure 4.1 illustrates a set of tabular data in the form of a set of $[x, f(x)]$ pairs. The function $f(x)$ is known at a finite set (actually eight) of discrete values of x. The value of the function can be determined at any of the eight values of x simply by a table lookup. However, a problem arises when the value of the function is needed at any value of x between the discrete values in the table. The actual function is not known and cannot be determined from the tabular values. However, the actual function can be approximated by some known function, and the value of the approximating function can be determined at any desired value of x. This process, which is called *interpolation*, is the subject of Chapter 4. The discrete data in Figure 4.1 are actually values of the function $f(x) = 1/x$, which is used as the example problem in this chapter.

In many problems in engineering and science, the data being considered are known only at a set of discrete points, not as a continuous function. For example, the continuous function

$$\boxed{y = f(x)} \tag{4.1}$$

may be known only at n discrete values of x:

$$\boxed{y_i = y(x_i) \qquad (i = 1, 2, \ldots, n)} \tag{4.2}$$

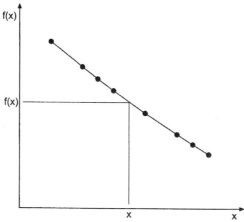

x	f(x)
3.20	0.312500
3.30	0.303030
3.35	0.298507
3.40	0.294118
3.50	0.285714
3.60	0.277778
3.65	0.273973
3.70	0.270270

Figure 4.1 Approximation of tabular data.

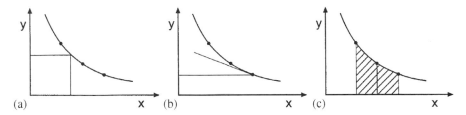

Figure 4.2 Applications of approximating functions. (a) Interpolation. (b) Differentiation. (c) Integration.

Discrete data, or tabular data, may consist of small sets of smooth data, large sets of smooth data, small sets of rough data, or large sets of rough data.

In many applications, the values of the discrete data at the specific points are not all that is needed. Values of the function at points other than the known discrete points may be needed (i.e., interpolation). The derivative of the function may be required (i.e., differentiation). The integral of the function may be of interest (i.e., integration). Thus, the processes of *interpolation, differentiation*, and *integration* of a set of discrete data are of interest. These processes are illustrated in Figure 4.2. These processes are performed by fitting an *approximating function* to the set of discrete data and performing the desired process on the approximating function.

Many types of approximating functions exist. In fact, any analytical function can be used as an approximating function. Three of the more common approximating functions are:

1. Polynomials
2. Trigonometric functions
3. Exponential functions

Approximating functions should have the following properties:

1. The approximating function should be easy to determine.
2. It should be easy to evaluate.
3. It should be easy to differentiate.
4. It should be easy to integrate.

Polynomials satisfy all four of these properties. Consequently, polynomial approximating functions are used in this book to fit sets of discrete data for interpolation, differentiation, and integration.

There are two fundamentally different ways to fit a polynomial to a set of discrete data:

1. Exact fits
2. Approximate fits

An *exact fit* yields a polynomial that passes exactly through all of the discrete points, as illustrated in Figure 4.3a. This type of fit is useful for small sets of smooth data. Exact polynomial fits are discussed in Sections 4.3 to 4.9. An *approximate fit* yields a polynomial that passes through the set of data in the best manner possible, without being required to pass exactly through any of the data points, as illustrated in Figure 4.3b. Several definitions of best manner possible exist. Approximate fits are useful for large sets of smooth data and

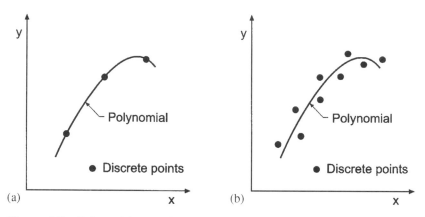

Figure 4.3   Polynomial approximation. (a) Exact fit. (b) Approximate fit.

small or large sets of rough data. In this book, the *least squares method* is used for approximate fits.

A set of discrete data may be *equally spaced* or *unequally spaced* in the independent variable x. In the general case where the data are unequally spaced, several procedures can be used to fit approximating polynomials, for example, (a) direct fit polynomials, (b) Lagrange polynomials, and (c) divided difference polynomials. Methods such as these require a considerable amount of effort. When the data are equally spaced, procedures based on differences can be used, for example, (a) the Newton forward-difference polynomial, (b) the Newton backward-difference polynomial, and (c) several other difference polynomials. These methods are quite easy to apply. Both types of methods are considered in this chapter.

Several procedures for polynomial approximation are developed in this chapter. Application of these procedures for interpolation is illustrated by examples. Numerical differentiation and numerical integration are discussed in Chapters 5 and 6, respectively.

Figure 4.4 illustrates the organization of Chapter 4. After the brief introduction in this section, the properties of polynomials which make them useful as approximating functions are presented. The presentation then splits into methods for fitting unequally spaced data and methods for fitting equally spaced data. A discussion of inverse interpolation follows next. Multivariate interpolation is then discussed. That is followed by an introduction to cubic splines. The final topic is a presentation of least squares approximation. Several programs for polynomial fitting are then presented. The chapter closes with a Summary which summarizes the main points of the chapter and presents a list of what you should be able to do after studying Chapter 4.

4.2   PROPERTIES OF POLYNOMIALS

The general form of an nth-degree polynomial is

$$P_n(x) = a_0 + a_1 x + a_2 x^2 + \cdots + a_n x^n \tag{4.3}$$

where n denotes the degree of the polynomial and a_0 to a_n are constant coefficients. There are $n + 1$ coefficients, so $n + 1$ discrete data points are required to obtain unique values for the coefficients.

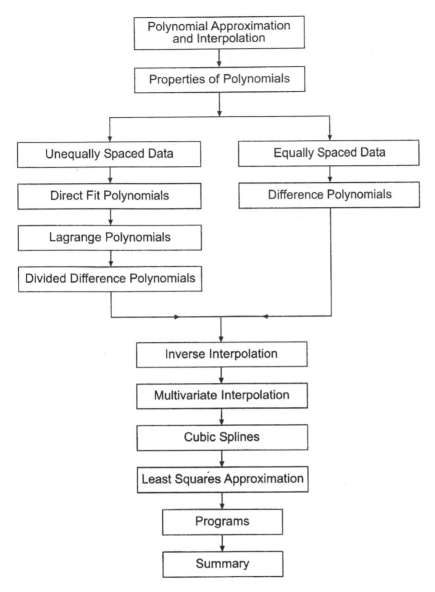

Figure 4.4 Organization of Chapter 4.

The property of polynomials that makes them suitable as approximating functions is stated by the *Weierstrass approximation theorem*:

> If $f(x)$ is a continuous function in the closed interval $a \le x \le b$, then for every $\varepsilon > 0$ there exists a polynomial $P_n(x)$, where the the value of n depends on the value of ε, such that for all x in the closed interval $a \le x \le b$,
>
> $$|P_n(x) - f(x)| < \varepsilon$$

Consequently, any continuous function can be approximated to any accuracy by a polynomial of high enough degree. In practice, low-degree polynomials are employed, so care must be taken to achieve the desired accuracy.

Polynomials satisfy a *uniqueness theorem*:

> A polynomial of degree n passing exactly through $n + 1$ discrete points is *unique*

The polynomial through a specific set of points may take many different forms, but all forms are equivalent. Any form can be manipulated into any other form by simple algebraic rearrangement.

The Taylor series is a polynomial of infinite order. Thus,

$$f(x) = f(x_0) + f'(x_0)(x - x_0) + \frac{1}{2!} f''(x_0)(x - x_0)^2 + \cdots \tag{4.4}$$

It is, of course, impossible to evaluate an infinite number of terms. The Taylor polynomial of degree n is defined by

$$f(x) = P_n(x) + R_{n+1}(x) \tag{4.5}$$

where the Taylor polynomial $P_n(x)$, and the remainder term $R_{n+1}(x)$ are given by

$$P_n(x) = f(x_0) + f'(x_0)(x - x_0) + \cdots + \frac{1}{n!} f^{(n)}(x_0)(x - x_0)^n \tag{4.6}$$

$$R_{n+1}(x) = \frac{1}{(n+1)!} f^{(n+1)}(\xi)(x - x_0)^{n+1} \qquad x_0 \leq \xi \leq x \tag{4.7}$$

The Taylor polynomial is a truncated Taylor series, with an explicit remainder, or error, term. The Taylor polynomial cannot be used as an approximating function for discrete data because the derivatives required in the coefficients cannot be determined. It does have great significance, however, for polynomial approximation, because it has an explicit error term.

When a polynomial of degree $n, P_n(x)$, is fit exactly to a set of $n + 1$ discrete data points, $(x_0, f_0), (x_1, f_1), \ldots, (x_n, f_n)$, as illustrated in Figure 4.5, the polynomial has no error at the data points themselves. However, at the locations between the data points, there is an error which is defined by

$$\text{Error}(x) = P_n(x) - f(x) \tag{4.8}$$

It can be shown that the *error term*, Error(x), has the form

$$\boxed{\text{Error}(x) = \frac{1}{(n+1)!} (x - x_0)(x - x_1) \cdots (x - x_n) f^{(n+1)}(\xi)} \tag{4.9}$$

where $x_0 \leq \xi \leq x_n$. This form of the error term is used extensively in the error analysis of procedures based on approximating polynomials. Equation (4.9) shows that the error in any polynomial approximation of discrete data (e.g., interpolation, differentiation, or integration) will be the smallest possible when the approximation is centered in the discrete data because that makes the $(x - x_i)$ terms as small as possible, which makes the product of those terms the smallest possible.

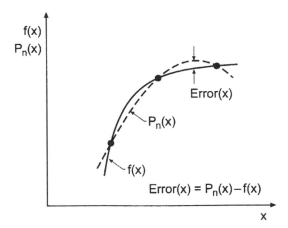

Figure 4.5 Error in polynomial approximation.

Differentiation of polynomials is straightforward. For the general term $a_i x^i$,

$$\frac{d}{dx}(a_i x^i) = i a_i x^{i-1} \tag{4.10}$$

The derivatives of the nth-degree polynomial $P_n(x)$ are

$$\frac{dP_n(x)}{dx} = P'_n(x) = a_1 + 2a_2 x + \cdots + n a_n x^{n-1} = P_{n-1}(x) \tag{4.11a}$$

$$\frac{d^2 P_n(x)}{dx^2} = \frac{d}{dx}\left[\frac{dP_n(x)}{dx}\right] = P''_n(x) = 2a_2 + 6a_3 x + \cdots + n(n-1)a_n x^{n-2} = P_{n-2}(x) \tag{4.11b}$$

. . . .

$$P_n^{(n)}(x) = n! a_n \tag{4.11n}$$

$$P_n^{(n+1)}(x) = 0 \tag{4.12}$$

Integration of polynomials is equally straightforward. For the general term $a_i x^i$,

$$\int a_i x^i \, dx = \frac{a_i}{i+1} x^{i+1} + \text{constant} \tag{4.13}$$

The integral of the nth-degree polynomial $P_n(x)$ is

$$I = \int P_n(x)\, dx = \int (a_0 + a_1 x + \cdots + a_n x^n)\, dx \tag{4.14}$$

$$I = a_0 x + \frac{a_1}{2} x^2 + \cdots + \frac{a_n}{n+1} x^{n+1} + \text{constant} = P_{n+1}(x) \tag{4.15}$$

The *evaluation of a polynomial*, $P_n(x)$, its derivative, $P'_n(x)$, or its integral, $\int P_n(x)\,dx$, for a particular value of x, is straightforward. For example, consider the fourth-degree polynomial, $P_4(x)$, its derivative, $P'_4(x)$, and its integral, $\int P_4(x)\,dx$:

$$P_4(x) = a_0 + a_1 x + a_2 x^2 + a_3 x^3 + a_4 x^4 \tag{4.16a}$$

$$P'_4(x) = a_1 + 2a_2 x + 3a_3 x^2 + 4a_4 x^3 = P_3(x) \tag{4.16b}$$

$$\int P_4(x)\,dx = a_0 x + \frac{a_1}{2} x^2 + \frac{a_2}{3} x^3 + \frac{a_3}{4} x^4 + \frac{a_4}{5} x^5 + \text{constant} = P_5(x) \tag{4.16c}$$

The evaluation of Eq. (4.16a) requires $(0 + 1 + 2 + 3 + 4) = 10$ multiplications and four additions; the evaluation of Eq. (4.16b) requires $(2 + 3 + 4) = 9$ multiplications and three additions; and the evaluation of Eq. (4.16c) requires $(1 + 2 + 3 + 4 + 5) = 15$ multiplications, four divisions, and five additions. This is a modest amount of work, even for polynomials of degree as high as 10. However, if a polynomial must be evaluated many times, or many polynomials must be evaluated, or very high degree polynomials must be evaluated, a more efficient procedure is desirable. The nested multiplication algorithm is such a procedure.

The *nested multiplication algorithm* is based on the following rearrangement of Eq. (4.16a):

$$P_4(x) = a_0 + x\{a_1 + x[a_2 + x(a_3 + a_4 x)]\} \tag{4.17}$$

which requires four multiplications and four additions. For a polynomial of degree n, $P_n(x)$, *nested multiplication* is given by

$$P_n(x) = a_0 + x(a_1 + x\{a_2 + x[a_3 + \cdots + x(a_{n-1} + a_n x)]\}) \tag{4.18}$$

which requires n multiplications and n additions. Equation (4.18) can be evaluated by constructing the *nested multiplication* sequence:

$$\begin{aligned} b_n &= a_n \\ b_i &= a_i + x b_{i+1} \qquad (i = n-1, n-2, \dots, 0) \end{aligned} \tag{4.19}$$

where $P_n(x) = b_0$. Equations (4.16b) and (4.16c) can be evaluated in a similar manner with minor modifications to account for the proper coefficients. Nested multiplication is sometimes called *Horner's algorithm*.

Several other properties of polynomials are quite useful. The *division algorithm* states that

$$P_n(x) = (x - N)Q_{n-1}(x) + R \tag{4.20}$$

where N is any number, $Q_{n-1}(x)$ is a polynomial of degree $n-1$, and R is a constant remainder. The *remainder theorem* states that

$$P_n(N) = R \tag{4.21}$$

The *factor theorem* states that if $P_n(N) = 0$, then $(x - N)$ is a factor of $P_n(x)$, which means that N is a root, α, or zero, of $P_n(x)$. That is, $(x - N) = 0$, and $\alpha = N$.

The *derivative of a polynomial* $P'_n(x)$ can be obtained from Eq. (4.20). Thus,

$$P'_n(x) = Q_{n-1}(x) + (x - N)Q'_{n-1}(x) \tag{4.22}$$

At $x = N$,

$$P'_n(N) = Q_{n-1}(N) \qquad (4.23)$$

Consequently, first derivatives of an nth-degree polynomial can be evaluated from the $(n-1)$st-degree polynomial $Q_{n-1}(x)$. Higher-order derivatives can be determined by applying the synthetic division algorithm to $Q_{n-1}(x)$, etc.

The $(n-1)$st-degree polynomial $Q_{n-1}(x)$, which can be used to evaluate the derivative $P'_n(x)$ and the remainder R, which yields $P_n(N) = R$, can be evaluated by the *synthetic division algorithm*. Consider $P_n(x)$ in the form given by Eq. (4.3):

$$P_n(x) = a_0 + a_1 x + a_2 x^2 + \cdots + a_n x^n \qquad (4.24a)$$

and $Q_{n-1}(x)$ in the form

$$Q_{n-1}(x) = b_1 + b_2 x + b_3 x^2 + \cdots + b_{n-1} x^{n-2} + b_n x^{n-1} \qquad (4.24b)$$

Substituting Eqs. (4.24a) and (4.24b) into Eq. (4.20) and equating coefficients of like powers of x yields:

$$b_n = a_n \qquad (4.25.n)$$
$$b_{n-1} = a_{n-1} + x b_n \qquad (4.25.n\text{-}1)$$
$$\cdots\cdots\cdots\cdots\cdots$$
$$b_1 = a_1 + x b_2 \qquad (4.25.1)$$
$$b_0 = a_0 + x b_1 = R \qquad (4.25.0)$$

Equation (4.25) can be written as

$$\boxed{\begin{aligned} b_n &= a_n \\ b_i &= a_i + x b_{i+1} \qquad (i = n-1, n-2, \ldots, 0) \end{aligned}} \qquad (4.26)$$

Equation (4.26) is identical to the nested multiplication algorithm presented in Eq. (4.19). Substituting $x = N$ into Eq. (4.24b) yields the value of $P'_n(N)$.

If a root, α, or zero, of $P_n(x)$ is known, $P_n(x)$ can be *deflated* by removing the factor $(x - \alpha)$ to yield the $(n-1)$st-degree polynomial, $Q_{n-1}(x)$. From Eq. (4.20), if α is a root of $P_n(x)$, then $P_n(\alpha) = 0$ and $R = 0$, and Eq. (4.20) yields

$$Q_{n-1}(x) = 0 \qquad (4.27)$$

The deflated polynomial $Q_{n-1}(x)$ has $n - 1$ roots, or zeros, which are the remaining roots, or zeros, of the original polynomial, $P_n(x)$.

The properties of polynomials presented in this section make them extremely useful as approximating functions.

Example 4.1. Polynomial evaluation.

Let's illustrate polynomial evaluation using nested multiplication, polynomial derivative evaluation using synthetic division, and polynomial deflation using synthetic division. Consider the fifth-degree polynomial considered in Section 3.5.1, Eq. (3.115):

$$P_5(x) = -120 + 274x - 225x^2 + 85x^3 - 15x^4 + x^5 \qquad (4.28)$$

Recall that the roots of Eq. (4.28) are $x = 1, 2, 3, 4$, and 5. Evaluate $P_5(2.5)$ and $P'_5(2.5)$, and determine the deflated polynomial $P_4(x)$ obtained by removing the factor $(x - 2)$.

Evaluating $P_5(2.5)$ by nested multiplication using Eq. (4.19) yields

$$b_5 = 1.0 \tag{4.29.5}$$
$$b_4 = -15.0 + 2.5(1.0) = -12.50 \tag{4.29.4}$$
$$b_3 = 85.0 + 2.5(-12.5) = 53.750 \tag{4.29.3}$$
$$b_2 = -225.0 + 2.5(53.750) = -90.6250 \tag{4.29.2}$$
$$b_1 = 274.0 + 2.5(-90.6250) = 47.43750 \tag{4.29.1}$$
$$b_0 = -120.0 + 2.5(47.43750) = -1.406250 \tag{4.29.0}$$

Thus, $P_5(2.50) = b_0 = -1.406250$. This result can be verified by direct evaluation of Eq. (4.28) with $x = 2.5$.

From Eq. (4.23), $P'_5(2.5) = Q_4(2.5)$, where $Q_4(x)$ is given by

$$Q_4(x) = 47.43750 - 90.6250x + 53.750x^2 - 12.50x^3 + x^4 \tag{4.30}$$

Evaluating $Q_4(2.5)$ by nested multiplication using Eq. (4.19), with the b_i replaced by c_i, gives

$$c_4 = 1.0 \tag{4.31.4}$$
$$c_3 = -12.5 + 2.5(1.0) = -10.0 \tag{4.31.3}$$
$$c_2 = 53.75 + 2.5(-10.0) = 28.750 \tag{4.31.2}$$
$$c_1 = -90.625 + 2.5(28.750) = -18.750 \tag{4.31.1}$$
$$c_0 = 47.4375 + 2.5(-18.750) = 0.56250 \tag{4.31.0}$$

Thus, $P'_5(2.5) = Q_4(2.5) = c_0 = 0.56250$. This result can be verified by direct evaluation of Eq. (4.30) with $x = 2.5$.

To illustrate polynomial deflation, let's deflate $P_5(x)$ by removing the factor $(x - 2)$. Applying the synthetic division algorithm, Eq. (4.26), with $x = 2.0$ yields

$$b_5 = 1.0 \tag{4.32.5}$$
$$b_4 = -15.0 + 2.0(1.0) = -13.0 \tag{4.32.4}$$
$$b_3 = 85.0 + 2.0(-13.0) = 59.0 \tag{4.32.3}$$
$$b_2 = -225.0 + 2.0(59.0) = -107.0 \tag{4.32.2}$$
$$b_1 = 274.0 + 2.0(-107.0) = 60.0 \tag{4.32.1}$$
$$b_0 = -120.0 + 2.0(60.0) = 0.0 \tag{4.32.0}$$

Thus, the deflated fourth-degree polynomial is

$$Q_4(x) = 60.0 - 107.0x + 59.0x^2 - 13.0x^3 + x^4 \tag{4.33}$$

This result can be verified directly by expanding the product of the four remaining linear factors, $Q_4(x) = (x - 1)(x - 3)(x - 4)(x - 5)$.

4.3   DIRECT FIT POLYNOMIALS

First let's consider a completely general procedure for fitting a polynomial to a set of equally spaced or unequally spaced data. Given $n+1$ sets of data $[x_0, f(x_0)]$, $[x_1, f(x_1)], \ldots, [x_n, f(x_n)]$, which will be written as (x_0, f_0), $(x_1, f_1), \ldots, (x_n, f_n)$, determine the unique nth-degree polynomial $P_n(x)$ that passes exactly through the $n+1$ points:

$$P_n(x) = a_0 + a_1 x + a_2 x^2 + \cdots + a_n x^n \qquad (4.34)$$

For simplicity of notation, let $f(x_i) = f_i$. Substituting each data point into Eq. (4.34) yields $n+1$ equations:

$$f_0 = a_0 + a_1 x_0 + a_2 x_0^2 + \cdots + a_n x_0^n \qquad (4.35.0)$$
$$f_1 = a_0 + a_1 x_1 + a_2 x_1^2 + \cdots + a_n x_1^n \qquad (4.35.1)$$
$$\cdots\cdots\cdots\cdots\cdots\cdots\cdots\cdots\cdots\cdots\cdots\cdots\cdots$$
$$f_n = a_0 + a_1 x_n + a_2 x_n^2 + \cdots + a_n x_n^n \qquad (4.35.n)$$

There are $n+1$ linear equations containing the $n+1$ coefficients a_0 to a_n. Equation (4.35) can be solved for a_0 to a_n by Gauss elimination. The resulting polynomial is the unique nth-degree polynomial that passes exactly through the $n+1$ data points. The direct fit polynomial procedure works for both equally spaced data and unequally spaced data.

Example 4.2.   Direct fit polynomials.

To illustrate interpolation by a direct fit polynomial, consider the simple function $y = f(x) = 1/x$, and construct the following set of six significant figure data:

x	$f(x)$
3.35	0.298507
3.40	0.294118
3.50	0.285714
3.60	0.277778

Let's interpolate for y at $x = 3.44$ using linear, quadratic, and cubic interpolation. The exact value is

$$y(3.44) = f(3.44) = \frac{1}{3.44} = 0.290698\ldots \qquad (4.36)$$

Let's illustrate the procedure in detail for a quadratic polynomial:

$$P_2(x) = a + bx + cx^2 \qquad (4.37)$$

To center the data around $x = 3.44$, the first three points are used. Applying $P_2(x)$ at each of these data points gives the following three equations:

$$0.298507 = a + b(3.35) + c(3.35)^2 \qquad (4.38.1)$$

$$0.294118 = a + b(3.40) + c(3.40)^2 \qquad (4.38.2)$$

$$0.285714 = a + b(3.50) + c(3.50)^2 \qquad (4.38.3)$$

Solving Eqs. (4.38) for a, b, and c by Gauss elimination without scaling or pivoting yields

$$P_2(x) = 0.876561 - 0.256080x + 0.0249333x^2 \qquad (4.39)$$

Substituting $x = 3.44$ into Eq. (4.39) gives

$$P_2(3.44) = 0.876561 - 0.256080(3.44) + 0.0249333(3.44)^2 = 0.290697 \qquad (4.40)$$

The error is Error $(3.44) = P_2(3.44) - f(3.44) = 0.290697 - 0.290698 = -0.000001$.

For a linear polynomial, use $x = 3.40$ and 3.50 to center that data around $x = 3.44$. The resulting linear polynomial is

$$P_1(x) = 0.579854 - 0.0840400x \qquad (4.41)$$

Substituting $x = 3.44$ into Eq. (4.41) gives $P_1(3.44) = 0.290756$. For a cubic polynomial, all four points must be used. The resulting cubic polynomial is

$$P_3(x) = 1.121066 - 0.470839x + 0.0878000x^2 - 0.00613333x^3 \qquad (4.42)$$

Substituting $x = 3.44$ into Eq. (4.42) gives $P_3(3.44) = 0.290698$.

The results are summarized below, where the results of linear, quadratic, and cubic interpolation, and the errors, Error$(3.44) = P(3.44) - 0.290698$, are tabulated. The advantages of higher-degree interpolation are obvious.

$P(3.44) = 0.290756$	linear interpolation	Error $=$	0.000058
$= 0.290697$	quadratic interpolation	$=$	-0.000001
$= 0.290698$	cubic interpolation	$=$	0.000000

The main advantage of direct fit polynomials is that the explicit form of the approximating function is obtained, and interpolation at several values of x can be accomplished simply by evaluating the polynomial at each value of x. The work required to obtain the polynomial does not have to be redone for each value of x. A second advantage is that the data can be unequally spaced.

The main disadvantage of direct fit polynomials is that each time the degree of the polynomial is changed, all of the work required to fit the new polynomial must be redone. The results obtained from fitting other degree polynomials is of no help in fitting the next polynomial. One approach for deciding when polynomial interpolation is accurate enough is to interpolate with successively higher-degree polynomials until the change in the result is within an acceptable range. This procedure is quite laborious using direct fit polynomials.

4.4   LAGRANGE POLYNOMIALS

The direct fit polynomial presented in Section 4.3, while quite straightforward in principle, has several disadvantages. It requires a considerable amount of effort to solve the system of equations for the coefficients. For a high-degree polynomial (n greater than about 4), the system of equations can be ill-conditioned, which causes large errors in the values of the coefficients. A simpler, more direct procedure is desired. One such procedure is the *Lagrange polynomial*, which can be fit to unequally spaced data or equally spaced data. The Lagrange polynomial is presented in Section 4.4.1. A variation of the Lagrange polynomial, called *Neville's algorithm*, which has some computational advantages over the Lagrange polynomial, is presented in Section 4.4.2.

4.4.1. Lagrange Polynomials

Consider two points, $[a, f(a)]$ and $[b, f(b)]$. The linear Lagrange polynomial $P_1(x)$ which passes through these two points is given by

$$\boxed{P_1(x) = \frac{(x-b)}{(a-b)}f(a) + \frac{(x-a)}{(b-a)}f(b)} \tag{4.43}$$

Substituting $x = a$ and $x = b$ into Eq. (4.43) yields

$$P_1(a) = \frac{(a-b)}{(a-b)}f(a) + \frac{(a-a)}{(b-a)}f(b) = f(a) \tag{4.44a}$$

$$P_1(b) = \frac{(b-b)}{(a-b)}f(a) + \frac{(b-a)}{(b-a)}f(b) = f(b) \tag{4.44b}$$

which demonstrates that Eq. (4.43) passes through the two points. Given three points, $[a, f(a)]$, $[b, f(b)]$, and $[c, f(c)]$, the quadratic Lagrange polynomial $P_2(x)$ which passes through the three points is given by:

$$\boxed{P_2(x) = \frac{(x-b)(x-c)}{(a-b)(a-c)}f(a) + \frac{(x-a)(x-c)}{(b-a)(b-c)}f(b) + \frac{(x-a)(x-b)}{(c-a)(c-b)}f(c)} \tag{4.45}$$

Substitution of the values of x substantiates that Eq. (4.45) passes through the three points.

This procedure can be applied to any set of $n + 1$ points to determine an nth-degree polynomial. Given $n + 1$ points, $[a, f(a)], [b, f(b)], \dots, [k, f(k)]$. the nth degree Lagrange polynomial $P_n(x)$ which passes through the $n + 1$ points is given by:

$$\boxed{\begin{aligned} P_n(x) &= \frac{(x-b)(x-c)\cdots(x-k)}{(a-b)(a-c)\cdots(a-k)}f(a) + \frac{(x-a)(x-c)\cdots(x-k)}{(b-a)(b-c)\cdots(b-k)}f(b) \\ &\quad + \cdots + \frac{(x-a)(x-b)\cdots(x-j)}{(k-a)(k-b)\cdots(k-j)}f(k) \end{aligned}} \tag{4.46}$$

The Lagrange polynomial can be used for both unequally spaced data and equally spaced data. No system of equations must be solved to evaluate the polynomial. However, a considerable amount of computational effort is involved, especially for higher-degree polynomials.

The form of the Lagrange polynomial is quite different in appearance from the form of the direct fit polynomial, Eq. (4.34). However, by the uniqueness theorem, the two forms both represent the unique polynomial that passes exactly through a set of points.

Example. 4.3.   Lagrange polynomials.

Consider the four points given in Example 4.2, which satisfy the simple function $y = f(x) = 1/x$:

x	$f(x)$
3.35	0.298507
3.40	0.294118
3.50	0.285714
3.60	0.277778

Let's interpolate for $y = f(3.44)$ using linear, quadratic, and cubic Lagrange interpolating polynomials. The exact value is $y = 1/3.44 = 0.290698\cdots$.

Linear interpolation using the two closest points, $x = 3.40$ and 3.50, yields

$$P_1(3.44) = \frac{(3.44 - 3.50)}{(3.40 - 3.50)}(0.294118) + \frac{(3.44 - 3.40)}{(3.50 - 3.40)}(0.285714) = 0.290756$$

(4.47)

Quadratic interpolation using the three closest points, $x = 3.35$, 3.40, and 3.50, gives

$$P_2(3.44) = \frac{(3.44 - 3.40)(3.44 - 3.50)}{(3.35 - 3.40)(3.35 - 3.50)}(0.298507)$$
$$+ \frac{(3.44 - 3.35)(3.44 - 3.50)}{(3.40 - 3.35)(3.40 - 3.50)}(0.294118)$$
$$+ \frac{(3.44 - 3.35)(3.44 - 3.40)}{(3.50 - 3.35)(3.50 - 3.40)}(0.285714) = 0.290697 \qquad (4.48)$$

Cubic interpolation using all four points yields

$$P_3(3.44) = \frac{(3.44 - 3.40)(3.44 - 3.50)(3.44 - 3.60)}{(3.35 - 3.40)(3.35 - 3.50)(3.35 - 3.60)}(0.298507)$$
$$+ \frac{(3.44 - 3.35)(3.44 - 3.50)(3.44 - 3.60)}{(3.40 - 3.35)(3.40 - 3.50)(3.40 - 3.60)}(0.294118)$$
$$+ \frac{(3.44 - 3.35)(3.44 - 3.40)(3.44 - 3.60)}{(3.50 - 3.35)(3.50 - 3.40)(3.50 - 3.60)}(0.285714)$$
$$+ \frac{(3.44 - 3.35)(3.44 - 3.40)(3.44 - 3.50)}{(3.60 - 3.35)(3.60 - 3.40)(3.60 - 3.50)}(0.277778) = 0.290698$$

(4.49)

The results are summarized below, where the results of linear, quadratic, and cubic interpolation, and the errors, Error(3.44) = $P(3.44) - 0.290698$, are tabulated. The advantages of higher-degree interpolation are obvious.

$$
\begin{aligned}
P(3.44) &= 0.290756 & \text{linear interpolation} & & \text{Error} &= 0.000058 \\
&= 0.290697 & \text{quadratic interpolation} & & &= -0.000001 \\
&= 0.290698 & \text{cubic interpolation} & & &= 0.000000
\end{aligned}
$$

These results are identical to the results obtained in Example 4.2 by direct fit polynomials, as they should be, since the same data points are used in both examples.

The main advantage of the Lagrange polynomial is that the data may be unequally spaced. There are several disadvantages. All of the work must be redone for each degree polynomial. All the work must be redone for each value of x. The first disadvantage is eliminated by Neville's algorithm, which is presented in the next subsection. Both disadvantages are eliminated by using divided differences, which are presented in Section 4.5.

4.4.2. Neville's Algorithm

Neville's algorithm is equivalent to a Lagrange polynomial. It is based on a series of linear interpolations. The data do not have to be in monotonic order, or in any structured order. However, the most accurate results are obtained if the data are arranged in order of closeness to the point to be interpolated.

Consider the following set of data:

x_i	f_i
x_1	f_1
x_2	f_2
x_3	f_3
x_4	f_4

Recall the linear Lagrange interpolating polynomial, Eq. (4.43):

$$f(x) = \frac{(x-b)}{(a-b)}f(a) + \frac{(x-a)}{(b-a)}f(b) \tag{4.50}$$

which can be written in the following form:

$$f(x) = \frac{(x-a)f(b) - (x-b)f(a)}{(b-a)} \tag{4.51}$$

In terms of general notation, Eq. (4.51) yields

$$f_i^{(n)} = \frac{(x-x_i)f_{i+1}^{(n-1)} - (x-x_{i+n})f_i^{(n-1)}}{x_{i+n} - x_i} \tag{4.52}$$

where the subscript i denotes the base point of the value (e.g., $i, i+1$, etc.) and the superscript (n) denotes the degree of the interpolation (e.g., zeroth, first, second, etc.).

A table of linearly interpolated values is constructed for the original data, which are denoted as $f_i^{(0)}$. For the first interpolation of the data,

$$f_i^{(1)} = \frac{(x-x_i)f_{i+1}^{(0)} - (x-x_{i+1})f_i^{(0)}}{x_{i+1} - x_i} \tag{4.53}$$

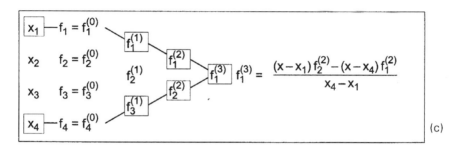

Figure 4.6 Neville's method. (a) First set of linear interpolations. (b) Second set of linear interpolation. (c) Third set of linear interpolations

as illustrated in Figure 4.6a. This creates a column of $n-1$ values of $f_i^{(1)}$. A second column of $n-2$ values of $f_i^{(2)}$ is obtained by linearly interpolating the column of $f_i^{(1)}$ values. Thus,

$$f_i^{(2)} = \frac{(x-x_i)f_{i+1}^{(1)} - (x - x_{i+2})f_i^{(1)}}{x_{i+2} - x_i} \qquad (4.54)$$

which is illustrated in Figure 4.6b. This process is repeated to create a third column of $f_i^{(3)}$ values, as illustrated in Figure 4.6c, and so on. The form of the resulting table is illustrated in Table 4.1.

It can be shown by direct substitution that each specific value in Table 4.1 is identical to a Lagrange polynomial based on the data points used to calculate the specific value. For example, $f_1^{(2)}$ is identical to a second-degree Lagrange polynomial based on points 1, 2, and 3.

The advantage of Neville's algorithm over direct Lagrange polynomial interpolation is now apparent. The third-degree Lagrange polynomial based on points 1 to 4 is obtained simply by applying the linear interpolation formula, Eq. (4.52), to $f_1^{(2)}$ and $f_2^{(2)}$ to obtain

Table 4.1. Table for Neville's
Algorithm

x_i	$f_i^{(0)}$	$f_i^{(1)}$	$f_i^{(2)}$	$f_i^{(3)}$
x_1	$f_1^{(0)}$			
		$f_1^{(1)}$		
x_2	$f_2^{(0)}$		$f_1^{(2)}$	
		$f_2^{(1)}$		$f_1^{(3)}$
x_3	$f_3^{(0)}$		$f_2^{(2)}$	
		$f_3^{(1)}$		
x_4	$f_4^{(0)}$			

$f_1^{(3)}$. None of the prior work must be redone, as it would have to be redone to evaluate a third-degree Lagrange polynomial. If the original data are arranged in order of closeness to the interpolation point, each value in the table, $f_i^{(n)}$, represents a centered interpolation.

Example 4.4.   Neville's algorithm.

Consider the four data points given in Example 4.3. Let's interpolate for $f(3.44)$ using linear, quadratic, and cubic interpolation using Neville's algorithm. Rearranging the data in order of closeness to $x = 3.44$ yields the following set of data:

x	$f(x)$
3.40	0.294118
3.50	0.285714
3.35	0.298507
3.60	0.277778

Applying Eq. (4.52) to the values of $f_i^{(0)}$ gives

$$f_1^{(1)} = \frac{(x - x_1)f_2^{(0)} - (x - x_2)f_1^{(0)}}{x_2 - x_1} = \frac{(3.44 - 3.40)0.285714 - (3.44 - 3.50)0.294118}{3.50 - 3.40}$$
$$= 0.290756 \qquad\qquad (4.55a)$$

Thus, the result of linear interpolation is $f(3.44) = f_1^{(1)} = 0.290756$. To evaluate $f_1^{(2)}, f_2^{(1)}$ must first be evaluated. Thus,

$$f_2^{(1)} = \frac{(x - x_2)f_3^{(0)} - (x - x_3)f_2^{(0)}}{x_3 - x_2} = \frac{(3.44 - 3.50)0.298507 - (3.44 - 3.35)0.285714}{3.35 - 3.50}$$
$$= 0.290831 \qquad\qquad (4.55b)$$

Evaluating $f_1^{(2)}$ gives

$$f_1^{(2)} = \frac{(x - x_1)f_2^{(1)} - (x - x_3)f_1^{(1)}}{x_3 - x_1} = \frac{(3.44 - 3.40)0.290831 - (3.44 - 3.35)0.290756}{3.35 - 3.40}$$
$$= 0.290696 \qquad\qquad (4.56)$$

Table 4.2. Neville's Algorithm

x_i	$f_i^{(0)}$	$f_i^{(1)}$	$f_i^{(2)}$	$f_i^{(3)}$
$x_1 = 3.40$	0.294118			
$x_2 = 3.50$	0.285714	0.290756	0.290697	
$x_3 = 3.35$	0.298507	0.290831	0.290703	0.290698
$x_4 = 3.60$	0.277778	0.291045		

Thus, the result of quadratic interpolation is $f(3.44) = f_1^{(2)} = 0.290696$. To evaluate $f_1^{(3)}, f_3^{(1)}$ and $f_2^{(2)}$ must first be evaluated. Then $f_1^{(3)}$ can be evaluated. These results, and the results calculated above, are presented in Table 4.2.

These results are the same as the results obtained by Lagrange polynomials in Example 4.3.

The advantage of Neville's algorithm over a Lagrange interpolating polynomial, if the data are arranged in order of closeness to the interpolated point, is that none of the work performed to obtain a specific degree result must be redone to evaluate the next higher degree result.

Neville's algorithm has a couple of minor disadvantages. All of the work must be redone for each new value of x. The amount of work is essentially the same as for a Lagrange polynomial. The divided difference polynomial presented in Section 4.5 minimizes these disadvantages.

4.5 DIVIDED DIFFERENCE TABLES AND DIVIDED DIFFERENCE POLYNOMIALS

A *divided difference* is defined as the ratio of the difference in the function values at two points divided by the difference in the values of the corresponding independent variable. Thus, the first divided difference at point i is defined as

$$f[x_i, x_{i+1}] = \frac{f_{i+1} - f_i}{x_{i+1} - x_i} \tag{4.57}$$

The second divided difference is defined as

$$f[x_i, x_{i+1}, x_{i+2}] = \frac{f[x_{i+1}, x_{i+2}] - f[x_i, x_{i+1}]}{x_{i+2} - x_i} \tag{4.58}$$

Similar expressions can be obtained for divided differences of any order. Approximating polynomials for nonequally spaced data can be constructed using divided differences.

4.5.1. Divided Difference Tables

Consider a table of data:

x_i	f_i
x_0	f_0
x_1	f_1
x_2	f_2
x_3	f_3

The first divided differences, in terms of standard notation, are

$$f[x_0, x_1] = \frac{(f_1 - f_0)}{(x_1 - x_0)} \tag{4.59a}$$

$$f[x_1, x_2] = \frac{(f_2 - f_1)}{(x_2 - x_1)} \tag{4.59b}$$

etc. Note that

$$f[x_i, x_{i+1}] = \frac{(f_{i+1} - f_i)}{(x_{i+1} - x_i)} = \frac{(f_i - f_{i+1})}{(x_i - x_{i+1})} = f[x_{i+1}, x_i] \tag{4.60}$$

The second divided difference is defined as follows:

$$f[x_0, x_1, x_2] = \frac{f[x_1, x_2] - f[x_0, x_1]}{(x_2 - x_0)} \tag{4.61}$$

In general,

$$f[x_0, x_1, \ldots, x_n] = \frac{f[x_1, x_2, \ldots, x_n] - f[x_0, x_1, \ldots, x_{n-1}]}{(x_n - x_0)} \tag{4.62}$$

By definition, $f[x_i] = f_i$.

The notation presented above is a bit clumsy. A more compact notation is defined in the same manner as the notation used in Neville's method, which is presented in Section 4.4.2. Thus,

$$f_i^{(1)} = f[x_i, x_{i+1}] \tag{4.63a}$$

$$f_i^{(2)} = f[x_i, x_{i+1}, x_{i+2}] \tag{4.63b}$$

In general,

$$f_i^{(n)} = f[x_i, x_{i+1}, \ldots, x_{i+n}] \tag{4.64}$$

Table 4.3 illustrates the formation of a divided difference table. The first column contains the values of x_i and the second column contains the values of $f(x_i) - f_i$, which are denoted by $f_i^{(0)}$. The remaining columns contain the values of the divided differences, $f_i^{(n)}$, where the subscript i denotes the base point of the value and the superscript (n) denotes the degree of the divided difference.

The data points do not have to be in any specific order to apply the divided difference concept. However, just as for the direct fit polynomial, the Lagrange polynomial, and Neville's method, more accurate results are obtained if the data are arranged in order of closeness to the interpolated point.

Table 4.3. Table of Divided differences

x_i	$f_i^{(0)}$	$f_i^{(1)}$	$f_i^{(2)}$	$f_i^{(3)}$
x_1	$f_1^{(0)}$			
		$f_1^{(1)}$		
x_2	$f_2^{(0)}$		$f_1^{(2)}$	
		$f_2^{(1)}$		$f_1^{(3)}$
x_3	$f_3^{(0)}$		$f_2^{(2)}$	
		$f_3^{(1)}$		
x_4	$f_4^{(0)}$			

Example 4.5. Divided difference Table.

Let's construct a six-place divided difference table for the data presented in Section 4.1. The results are presented in Table 4.4.

Table 4.4. Divided Difference Table

x_i	$f_i^{(0)}$	$f_i^{(1)}$	$f_i^{(2)}$	$f_i^{(3)}$	$f_i^{(4)}$
3.20	0.312500				
		-0.094700			
3.30	0.303030		0.028267		
		-0.090460		-0.007335	
3.35	0.298507		0.026800		-0.000667
		-0.087780		-0.009335	
3.40	0.294118		0.024933		0.010677
		-0.084040		-0.006132	
3.50	0.285714		0.023400		-0.001787
		-0.079360		-0.006668	
3.60	0.277778		0.021733		0.000010
		-0.076100		-0.006665	
3.65	0.273973		0.020400		
		-0.074060			
3.70	0.270270				

4.5.2. Divided Difference Polynomials

Let's define a power series for $P_n(x)$ such that the coefficients are identical to the divided differences, $f_i^{(n)}$. Thus,

$$
\begin{aligned}
P_n(x) = f_i^{(0)} &+ (x - x_0)f_i^{(1)} + (x - x_0)(x - x_1)f_i^{(2)} + \cdots \\
&+ (x - x_0)(x - x_1)\cdots(x - x_{n-1})f_i^{(n)}
\end{aligned}
\tag{4.65}
$$

$P_n(x)$ is clearly a polynomial of degree n. To demonstrate that $P_n(x)$ passes exactly through the data points, let's substitute the data points into Eq. (4.65). Thus,

$$
P_n(x_0) = f_i^{(0)} + (0)f_i^{(1)} + \cdots f_i^{(0)} = f_0
\tag{4.66}
$$

$$
P_n(x_1) = f_i^{(0)} + (x_1 - x_0)f_i^{(1)} + (x_1 - x_0)(0)f_i^{(2)} + \cdots
\tag{4.67a}
$$

$$
P_n(x_1) = f_0 + (x_1 - x_0)\frac{(f_1 - f_0)}{(x_1 - x_0)} = f_0 + (f_1 - f_0) = f_1
\tag{4.67b}
$$

$$
\begin{aligned}
P_n(x_2) = f_i^{(0)} &+ (x_2 - x_0)f_i^{(1)} + (x_2 - x_0)(x_2 - x_1)f_i^{(2)} \\
&+ (x_2 - x_0)(x_2 - x_1)(0)f_i^{(3)} + \cdots
\end{aligned}
\tag{4.68a}
$$

$$
\begin{aligned}
P_n(x_2) = f_0 &+ (x_2 - x_0)\frac{(f_1 - f_0)}{(x_1 - x_0)} \\
&+ (x_2 - x_0)(x_2 - x_1)\frac{(f_2 - f_1)/(x_2 - x_1) - (f_1 - f_0)/(x_1 - x_0)}{(x_2 - x_0)}
\end{aligned}
\tag{4.68b}
$$

$$
P_n(x_2) = f_0 + (f_2 - f_1) + (f_1 - f_0) = f_2 \quad \text{etc.}
\tag{4.68c}
$$

Since $P_n(x)$ is a polynomial of degree n and passes exactly through the $n + 1$ data points, it is obviously one form of the unique polynomial passing through the data points.

Example 4.6. Divided difference polynomials.

Consider the divided difference table presented in Example 4.5. Let's interpolate for $f(3.44)$ using the divided difference polynomial, Eq. (4.65), using $x_0 = 3.35$ as the base point. The exact solution is $f(3.44) = 1/3.44 = 0.290698$. From Eq. (4.65):

$$P_n(3.44) = f_0^{(0)} + (3.44 - x_0)f_0^{(1)} + (3.44 - x_0)(3.44 - x_1)f_0^{(2)}$$
$$+ (3.44 - x_0)(3.44 - x_1)(3.44 - x_2)f_0^{(3)} \tag{4.69}$$

Substituting the values of x_0 to x_2 and $f_0^{(0)}$ to $f_0^{(3)}$ into Eq. (4.69) gives

$$\begin{aligned}P_n(3.44) = &\ 0.298507 + (3.44 - 3.35)(-0.087780) \\ &+ (3.44 - 3.35)(3.44 - 3.4)(0.024933) \\ &+ (3.44 - 3.35)(3.44 - 3.4)(3.44 - 3.5)(-0.006132)\end{aligned} \tag{4.70}$$

Evaluating Eq. (4.70) term by term gives

$$P_n(3.44) = 0.298507 - 0.007900 + 0.000089 + 0.000001 \tag{4.71}$$

Summing the terms yields the following results and errors:

$$\begin{aligned}P(3.44) = 0.290607 &\quad \text{linear interpolation} &\quad \text{Error}(3.44) = -0.000091 \\ = 0.290696 &\quad \text{quadratic interpolation} &\quad = -0.000002 \\ = 0.290697 &\quad \text{cubic interpolation} &\quad = -0.000001\end{aligned}$$

The advantage of higher-degree interpolation is obvious.

The above results are not the most accurate possible since the data points in Table 4.4 are in monotonic order, which make the linear interpolation result actually linear extrapolation. Rearranging the data in order of closeness to $x = 3.44$ yields the results presented in Table 4.5. From Eq. (4.65):

$$\begin{aligned}P_n(3.44) = &\ 0.294118 + (3.44 - 3.40)(-0.084040) \\ &+ (3.44 - 3.40)(3.44 - 3.50)(0.024940) \\ &+ (3.44 - 3.40)(3.44 - 3.50)(3.44 - 3.35)(-0.006150)\end{aligned} \tag{4.72}$$

Evaluating Eq. (4.72) term by term gives

$$P_n(3.44) = 0.294118 - 0.003362 - 0.000060 + 0.000001 \tag{4.73}$$

Table 4.5. Rearranged Divided Difference Table

x_i	$f_i^{(0)}$	$f_i^{(1)}$	$f_i^{(2)}$	$f_i^{(3)}$
3.40	0.294118			
		-0.084040		
3.50	0.285714		0.024940	
		-0.085287		-0.006150
3.35	0.298507		0.023710	
		-0.082916		
3.60	0.277778			

Summing the terms yields the following results and errors:

$$
\begin{array}{llll}
P(3.44) = 0.290756 & \text{linear interpolation} & \text{Error} = & 0.000058 \\
 = 0.290697 & \text{quadratic interpolation} & = & -0.000001 \\
 = 0.290698 & \text{cubic interpolation} & = & 0.000000
\end{array}
$$

The linear interpolation value is much more accurate due to the centering of the data. The quadratic and cubic interpolation values are the same as before, except for round-off errors, because the same points are used in those two interpolations. These results are the same as the results obtained in the previous examples.

4.6 DIFFERENCE TABLES AND DIFFERENCE POLYNOMIALS

Fitting approximating polynomials to tabular data is considerably simpler when the values of the independent variable are equally spaced. Implementation of polynomial fitting for equally spaced data is best accomplished in terms of differences. Consequently, the concept of differences, difference tables, and difference polynomials are introduced in this section.

4.6.1. Difference Tables

A difference table is an arrangement of a set of data, $[x, f(x)]$, in a table with the x values in monotonic ascending order, with additional columns composed of the differences of the numbers in the preceding column. A triangular array is obtained, as illustrated in Table 4.6.

  The numbers appearing in a difference table are unique. However, three different interpretations can be assigned to these numbers, each with its unique notation. The forward difference relative to point i is $(f_{i+1} - f_i)$, the backward difference relative to point $i + 1$ is $(f_{i+1} - f_i)$, and the centered difference relative to point $i + 1/2$ is $(f_{i+1} - f_i)$. The forward difference operator Δ is defined as

$$
\Delta f(x_i) = \Delta f_i = (f_{i+1} - f_i) \tag{4.74}
$$

The backward difference operator ∇ is defined as

$$
\nabla f(x_{i+1}) = \nabla f_{i+1} = (f_{i+1} - f_i) \tag{4.75}
$$

The centered difference operator δ is defined as

$$
\delta f(x_{i+1/2}) = \delta f_{i+1/2} = (f_{i+1} - f_i) \tag{4.76}
$$

  A difference table, such as Table 4.6, can be interpreted as a forward-difference table, a backward-difference table, or a centered-difference table, as illustrated in Figure

Table 4.6.   Table of Differences

x	$f(x)$			
x_0	f_0			
		$(f_1 - f_0)$		
x_1	f_1		$(f_2 - 2f_1 + f_0)$	
		$(f_2 - f_1)$		$(f_3 - 3f_2 + 3f_1 - f_0)$
x_2	f_2		$(f_3 - 2f_2 + f_1)$	
		$(f_3 - f_2)$		
x_3	f_3			

x	f	Δf	$\Delta^2 f$	$\Delta^3 f$
x_0	f_0			
		Δf_0		
x_1	f_1		$\Delta^2 f_0$	
		Δf_1		$\Delta^3 f_0$
x_2	f_2		$\Delta^2 f_1$	
		Δf_2		
x_3	f_3			

x	f	∇f	$\nabla^2 f$	$\nabla^3 f$
x_{-3}	f_{-3}			
		∇f_{-2}		
x_{-2}	f_{-2}		$\nabla^2 f_{-1}$	
		∇f_{-1}		$\nabla^3 f_0$
x_{-1}	f_{-1}		$\nabla^2 f_0$	
		∇f_0		
x_0	f_0			

x	f	δf	$\delta^2 f$	$\delta^3 f$
x_{-1}	f_{-1}			
		$\delta f_{-1/2}$		
x_0	f_0		$\delta^2 f_0$	
		$\delta f_{1/2}$		$\delta^3 f_{1/2}$
x_1	f_1		$\delta^2 f_1$	
		$\delta f_{3/2}$		
x_2	f_2			

Figure 4.7 (a) Forward-difference table. (b) Backward-difference table. (c) Centered-difference table.

4.7. The numbers in the tables are identical. Only the notation is different. The three different types of interpretation and notation simplify the use of difference tables in the construction of approximating polynomials, which is discussed in Sections 4.6.2 to 4.6.4.

Example 4.7. Difference table.

Let's construct a six-place difference table for the function $f(x) = 1/x$ for $3.1 \le x \le 3.9$ with $\Delta x = 0.1$. The results are presented in Table 4.7, which uses the forward-difference notation to denote the columns of differences.

Table 4.7. Difference Table

x	$f(x)$	$\Delta f(x)$	$\Delta^2 f(x)$	$\Delta^3 f(x)$	$\Delta^4 f(x)$	$\Delta^5 f(x)$
3.1	0.322581					
3.2	0.312500	-0.010081				
			0.000611			
3.3	0.303030	-0.009470		-0.000053		
			0.000558		0.000003	
3.4	0.294118	-0.008912		-0.000050		0.000007
			0.000508		0.000010	
3.5	0.285714	-0.008404		-0.000040		-0.000010
			0.000468		0.000000	
3.6	0.277778	-0.007936		-0.000040		0.000008
			0.000428		0.000008	
3.7	0.270270	-0.007508		-0.000032		-0.000008
			0.000396		0.000000	
3.8	0.263158	-0.007112		-0.000032		
			0.000364			
3.9	0.256410	-0.006748				

Several observations can be made from Table 4.7. The first and second differences are quite smooth. The third differences, while monotonic, are not very smooth. The fourth differences are not monotonic, and the fifth differences are extremely ragged. The magnitudes of the higher-order differences decrease rapidly. If the differences are not smooth and decreasing, several possible explanations exist:

1. The original data set has errors.
2. The increment Δx may be too large.
3. There may be a singularity in $f(x)$ or its derivatives in the range of the table.

Difference tables are useful for evaluating the quality of a set of tabular data.

Tabular data have a finite number of digits. The last digit is typically rounded off. Round-off has an effect on the accuracy of the higher-order differences. To illustrate this effect, consider a difference table showing only the round-off error in the last significant digit. The worst possible round-off situation occurs when every other number is rounded off by one-half in opposing directions, as illustrated in Table 4.8. Table 4.8 shows that the errors due to round-off in the original data oscillate and double in magnitude for each higher-order difference. The maximum error in the differences is given by

$$\boxed{\text{Maximum round-off error in } \Delta^n f = \pm 2^{n-1}} \tag{4.77}$$

For the results presented in Table 4.7, $\Delta^5 f$ oscillates between -10 and $+8$. From Eq. (4.77), the maximum round-off error in $\Delta^5 f$ is $\pm 2^{5-1} = \pm 16$. Consequently, the $\Delta^5 f$ values are completely masked by the accumulated round-off error.

Polynomial fitting can be accomplished using the values in a difference table. The degree of polynomial needed to give a satisfactory fit to a set of tabular data can be estimated by considering the properties of polynomials. The nth-degree polynomial $P_n(x)$ is given by

$$P_n(x) = a_n x^n + (\text{Lower Degree Terms}) = a_n x^n + (\text{LDTs}) \tag{4.78}$$

In Section 4.2, it is shown that [see Eqs. (4.11n) and (4.12)]

$$P_n^{(n)}(x) = n! a_n = \text{constant} \tag{4.79}$$
$$P_n^{(n+1)}(x) = 0 \tag{4.80}$$

Table 4.8. Difference Table of Round-off Errors

x	f	Δf	$\Delta^2 f$	$\Delta^3 f$	$\Delta^4 f$
—	$+1/2$				
—	$-1/2$	-1			
—	$+1/2$	1	2	-4	8
—	$-1/2$	-1	-2	4	-8
—	$+1/2$	1	2	-4	
—	$-1/2$	-1	-2		

Let's evaluate the first forward difference of $P_n(x)$:

$$\Delta[P_n(x)] = \Delta(a_n x^n) + \Delta(\text{LDTs}) \tag{4.81}$$

$$\Delta[P_n(x)] = a_n(x + h)^n - a_n x^n + (\text{LDTs}) \tag{4.82}$$

Expanding $(x + h)^n$ in a binomial expansion gives

$$\Delta[P_n(x)] = [a_n x^n + a_n n x^{n-1} h + \cdots + a_n h^n] - a_n x^n + (\text{LDTs}) \tag{4.83}$$

which yields

$$\Delta[P_n(x)] = a_n n h x^{n-1} + (\text{LDTs}) = P_{n-1}(x) \tag{4.84}$$

Evaluating the second forward difference of $P_n(x)$ gives

$$\Delta^2[P_n(x)] = \Delta[(a_n n h) x^{n-1}] + \Delta(\text{LDTs}) \tag{4.85}$$

which yields

$$\Delta^2[P_n(x)] = a_n n(n - 1) h^2 x^{n-2} + (\text{LDTs}) \tag{4.86}$$

In a similar manner it can be shown that

$$\Delta^n P_n(x) = a_n n! h^n = \text{constant} \qquad \text{and} \qquad \Delta^{n+1} P_n(x) = 0 \tag{4.87}$$

Note the similarity between $P_n^{(n)}(x)$ and $\Delta^n P_n(x)$. In fact, $P_n^{(n)}(x) = \Delta^n P_n(x)/h^n$. Thus, if $f(x) = P_n(x)$, then $\Delta^n f(x) = \text{constant}$. Consequently, if $\Delta^n f(x) \approx \text{constant}$, $f(x)$ can be approximated by $P_n(x)$.

4.6.2. The Newton Forward-Difference Polynomial

Given $n + 1$ data points, $[x, f(x)]$, one form of the unique nth-degree polynomial that passes through the $n + 1$ points is given by

$$
\boxed{
\begin{aligned}
P_n(x) = f_0 + s\,\Delta f_0 &+ \frac{s(s - 1)}{2!} \Delta^2 f_0 + \frac{s(s - 1)(s - 2)}{3!} \Delta^3 f_0 \\
&+ \cdots + \frac{s(s - 1)(s - 2) \cdots [s - (n - 1)]}{n!} \Delta^n f_0
\end{aligned}
}
\tag{4.88}
$$

where s is the interpolating variable

$$s = \frac{x - x_0}{\Delta x} = \frac{x - x_0}{h} \qquad \text{and} \qquad x = x_0 + sh \tag{4.89}$$

Equation (4.88) does not look anything like the direct fit polynomial [see Eq. (4.34)], the Lagrange polynomial [see Eq. (4.46)], or the divided difference polynomial [see Eq. (4.65)]. However, if Eq. (4.88) is a polynomial of degree n and passes exactly through the $n + 1$ data points, it must be one form of the unique polynomial that passes through this set of data.

The interpolating variable, $s = (x - x_0)/h$, is linear in x. Consequently, the last term in Eq. (4.88) is order n, and Eq. (4.88) is an nth-degree polynomial. Let $s = 0$. Then $x = x_0, f = f_0$, and $P_n(x_0) = f_0$. Let $s = 1$. Then $x = x_0 + h = x_1, f = f_1$, and $P_n(x_1) = f_0 + \Delta f_0 = f_0 + (f_1 - f_0) = f_1$. In a similar manner, it can be shown that $P_n(x) = f(x)$ for the $n + 1$ discrete points. Therefore, $P_n(x)$ is the desired unique nth-degree polynomial. Equation (4.88) is called the *Newton forward-difference polynomial*.

The Newton forward-difference polynomial can be expressed in a more compact form by introducing the definition of the *binomial coefficient*. Thus,

$$\binom{s}{i} = \frac{s(s-1)(s-2)\cdots(s-[i-1])}{i!} \tag{4.90}$$

In terms of binomial coefficients, the Newton forward-difference polynomial is

$$P_n(x) = f_0 + \binom{s}{1}\Delta f_0 + \binom{s}{2}\Delta^2 f_0 + \binom{s}{3}\Delta^3 f_0 + \cdots \tag{4.91}$$

A major advantage of the Newton forward-difference polynomial, in addition to its simplicity, is that each higher-degree polynomial is obtained from the previous lower-degree polynomial simply by adding the next term. The work already performed for the lower-degree polynomial does not have to be repeated. This feature is in sharp contrast to the direct fit polynomial and the Lagrange polynomial, where all of the work must be repeated each time the degree of the polynomial is changed. This feature makes it simple to determine when the desired accuracy has been obtained. When the next term in the polynomial is less than some prespecified value, the desired accuracy has been obtained.

Example 4.8. Newton forward-difference polynomial.

From the six-place difference table for $f(x) = 1/x$, Table 4.7, calculate $P(3.44)$ by the Newton forward-difference polynomial. The exact solution is $f(3.44) = 1/3.44 = 0.290698\ldots$. In Table 4.7, $h = 0.1$. Choose $x_0 = 3.40$. Then,

$$s = \frac{x - x_0}{h} = \frac{3.44 - 3.40}{0.1} = 0.4 \tag{4.92}$$

Equation (4.88) gives

$$P(3.44) = f(3.4) + s\,\Delta f(3.4) + \frac{s(s-1)}{2!}\Delta^2 f(3.4) + \frac{s(s-1)(s-2)}{3!}\Delta^3 f(3.4) + \cdots \tag{4.93}$$

Substituting $s = 0.4$ and the values of the differences from Table 4.7 into Eq. (4.93) gives

$$P(3.44) = 0.294118 + (0.4)(-0.008404) + (0.4)\frac{(0.4-1)}{2}(0.000468)$$
$$+ \frac{(0.4)(0.4-1)(0.4-2)}{6}(-0.000040) + \cdots \tag{4.94}$$

Evaluating Eq. (4.94) term by term yields the following results and errors:

$P(3.44) = 0.290756$	linear interpolation	$\text{Error}(3.44) = 0.000058$
$= 0.290700$	quadratic interpolation	$= 0.000002$
$= 0.290698$	cubic interpolation	$= 0.000000$

The advantage of higher-degree interpolation is obvious.

In this example, the base point, $x_0 = 3.4$, was selected so that the point of interpolation, $x = 3.44$, falls within the range of data used to determine the polynomial, that is, interpolation occurs. If x_0 is chosen so that x does not fall within the range of fit, extrapolation occurs, and the results are less accurate. For example, let $x_0 = 3.2$, for which $s = 2.4$. The following results and errors are obtained:

$$P(3.44) = 0.289772 \qquad \text{linear extrapolation} \qquad \text{Error} = -0.000926$$
$$= 0.290709 \qquad \text{quadratic extrapolation} \qquad = 0.000011$$
$$= 0.290698 \qquad \text{cubic interpolation} \qquad = 0.000000$$

The increase in error is significant for linear and quadratic extrapolation. For $x_0 = 3.2$, the cubic yields an interpolating polynomial.

The error term for the Newton forward-difference polynomial can be obtained from the general error term [see Eq. (4.9)]:

$$\text{Error}(x) = \frac{1}{(n+1)!}(x - x_0)(x - x_1) \cdots (x - x_n) f^{(n+1)}(\xi) \tag{4.95}$$

From Eq. (4.89),

$$(x - x_0) = (x_0 + sh) - x_0 = sh \tag{4.96a}$$
$$(x - x_1) = (x_0 + sh) - x_1 = sh - (x_1 - x_0) = (s - 1)h \tag{4.96b}$$
$$\cdots \cdots \cdots \cdots \cdots \cdots \cdots \cdots \cdots \cdots \cdots \cdots \cdots \cdots \cdots \cdots$$
$$(x - x_n) = (x_0 + sh) - x_n = sh - (x_n - x_0) = (s - n)h \tag{4.96n}$$

Substituting Eq. (4.96) into Eq. (4.95) gives

$$\text{Error}(x) = \frac{1}{(n+1)!} s(s - 1)(s - 2) \cdots (s - n) h^{n+1} f^{(n+1)}(\xi) \tag{4.97}$$

which can be written as

$$\boxed{\text{Error}(x) = \binom{s}{n+1} h^{n+1} f^{(n+1)}(\xi)} \tag{4.98}$$

From Eq. (4.91), for $P_n(x)$, the term after the nth-term is

$$\binom{s}{n+1} \Delta^{n+1} f_0 \tag{4.99}$$

The error term, Eq. (4.98), can be obtained from Eq. (4.99) by the replacement

$$\Delta^{n+1} f_0 \to h^{n+1} f^{(n+1)}(\xi) \tag{4.100}$$

This procedure can be used to obtain the error term for all polynomials based on a set of discrete data points.

4.6.3 The Newton Backward-Difference Polynomial

The Newton forward-difference polynomial, Eq. (4.88), can be applied at the top or in the middle of a set of tabular data, where the downward-sloping forward differences illustrated in Figure 4.7a exist. However, at the bottom of a set of tabular data, the required forward differences do not exist, and the Newton forward-difference polynomial cannot be used. In that case, an approach that uses the upward-sloping backward differences illustrated in Figure 4.7b is required. Such a polynomial is developed in this section.

Given $n+1$ data points, $[x, f(x)]$, one form of the unique nth-degree polynomial that passes through the $n+1$ points is given by

$$
\begin{aligned}
P_n(x) = f_0 &+ s\,\nabla f_0 + \frac{s(s+1)}{2!}\nabla^2 f_0 + \frac{s(s+1)(s+2)}{3!}\nabla^3 f_0 \\
&+ \cdots + \frac{s(s+1)\cdots[s+(n-1)]}{n!}\nabla^n f_0
\end{aligned}
\tag{4.101}
$$

where s is the interpolating variable

$$
s = \frac{x - x_0}{\Delta x} = \frac{x - x_0}{h} \qquad \text{and} \qquad x = x_0 + sh
\tag{4.102}
$$

The interpolating variable, $s = (x - x_0)/h$, is linear in x. Consequently, the last term in Eq. (4.101) is order n, and Eq. (4.101) is an nth-degree polynomial. Let $s = 0$. Then $x = x_0, f = f_0$, and $P_n(x_0) = f_0$. Let $s = -1$. Then $x = x_0 - h = x_{-1}, f = f_{-1}$, and $P_n(x_{-1}) = f_0 - \nabla f_0 = f_0 - (f_0 - f_{-1}) = f_{-1}$ In a similar manner, it can be shown that $P_n(x) = f(x)$ for the $n+1$ discrete points. Therefore, $P_n(x)$ is the desired unique nth-degree polynomial. Equation (4.101) is called the *Newton backward-difference polynomial*.

The Newton backward-difference polynomial can be expressed in a more compact form using a nonstandard definition of the binomial coefficient. Thus,

$$
\binom{s^+}{i} = \frac{s(s+1)(s+2)\cdots(s+[i-1])}{i!}
\tag{4.103}
$$

In terms of the *nonstandard binomial coefficients*, the Newton backward-difference polynomial is

$$
P_n(x) = f_0 + \binom{s^+}{1}\nabla f_0 + \binom{s^+}{2}\nabla^2 f_0 + \binom{s^+}{3}\nabla^3 f_0 + \cdots
\tag{4.104}
$$

Example 4.9.   Newton backward-difference polynomial.

From the six-place difference table for $f(x) = 1/x$, Table 4.7, calculate $P(3.44)$ by the Newton backward-difference polynomial. The exact solution is $f(3.44) = 0.290698\ldots$. In Table 4.7, $h = 0.1$. Choose $x_0 = 3.50$. Then,

$$
s = \frac{x - x_0}{h} = \frac{3.44 - 3.50}{0.1} = -0.6
\tag{4.105}
$$

Equation (4.101) gives

$$
P(3.44) = f(3.5) + s\,\nabla f(3.5) + \frac{s(s+1)}{2!}\nabla^2 f(3.5) + \frac{s(s+1)(s+2)}{3!}\nabla^3 f(3.5) + \cdots
\tag{4.106}
$$

Substituting $s = -0.6$ and the values of the differences from Table 4.7 into Eq. (4.106) gives

$$P(3.44) = 0.285714 + (-0.6)(-0.008404) + \frac{(-0.6)(-0.6 + 1)}{2}(0.000508)$$
$$+ \frac{(-0.6)(-0.6 + 1)(-0.6 + 2)}{6}(-0.000050) + \cdots \tag{4.107}$$

Evaluating Eq. (4.107) term by term yields the following results and errors:

$$
\begin{array}{llll}
P(3.44) = 0.290756 & \text{linear interpolation} & \text{Error} = & 0.000058 \\
\quad\quad\quad = 0.290695 & \text{quadratic interpolation} & = & -0.000003 \\
\quad\quad\quad = 0.290698 & \text{cubic interpolation} & = & 0.000000
\end{array}
$$

The advantages of higher-degree interpolation are obvious.

The error term for the Newton backward-difference polynomial can be obtained from the general error term, Eq. (4.9), by making the following substitutions:

$$(x - x_0) = sh \tag{4.108a}$$
$$(x - x_1) = (x_0 + sh) - x_1 = sh + (x_0 - x_1) = (s + 1)h \tag{4.108b}$$
$$\cdots\cdots\cdots\cdots\cdots\cdots\cdots\cdots\cdots\cdots\cdots\cdots\cdots\cdots\cdots\cdots\cdots\cdots$$
$$(x - x_n) = (x_0 + sh) - x_n = sh + (x_0 - x_n) = (s + n)h \tag{4.108n}$$

Substituting Eq. (4.108) into Eq. (4.9) yields

$$\text{Error}(x) = \frac{1}{(n + 1)!} s(s + 1)(s + 2) \cdots (s + n - 1)(s + n)h^{n+1}f^{(n+1)}(\xi) \tag{4.109}$$

which can be written as

$$\boxed{\text{Error}(x) = \binom{s+}{n + 1} h^{n+1} f^{(n+1)}(\xi)} \tag{4.110}$$

Equation (4.110) can be obtained from Eq. (4.104) by the following replacement in the $(n + 1)$st term:

$$\nabla^{n+1} f_0 \to h^{n+1} f^{(n+1)}(\xi) \tag{4.111}$$

4.6.4. Other Difference Polynomials

The Newton forward-difference polynomial and the Newton-backward-difference polynomial presented in Sections 4.6.2 and 4.6.3, respectively, are examples of approximating polynomials based on differences. The Newton forward-difference polynomial uses forward differences, which follow a downward-sloping path in the forward-difference table illustrated in Figure 4.7a. The Newton backward-difference polynomial uses backward differences, which follow an upward-sloping path in the backward-difference table illustrated in Figure 4.7b. Numerous other difference polynomials based on other paths through a difference table can be constructed. Two of the more important ones are presented in this section.

The Newton forward-difference and backward-difference polynomials are essential for fitting an approximating polynomial at the beginning and end, respectively, of a set of tabular data. However, other forms of difference polynomials can be developed in the middle of a set of tabular data by using the centered-difference table presented in Figure 4.7c. The base point for the *Stirling centered-difference polynomial* is point x_0. This polynomial is based on the values of the centered differences with respect to x_0. That is, the polynomial follows a horizontal path through the difference table which is centered on point x_0. As illustrated in Figure 4.7c, even centered differences with respect to x_0 exist, but odd centered differences do not. Odd centered differences are based on the averages of the centered differences at the half points $x_{-1/2}$ and $x_{1/2}$. The Stirling centered-difference polynomial is

$$P_n(x) = f_0 + \binom{s}{1} \frac{1}{2}(\delta f_{1/2} + \delta f_{-1/2}) + \frac{1}{2}\left[\binom{s+1}{2} + \binom{s}{2}\right]\delta^2 f_0$$

$$+ \binom{s+1}{3} \frac{1}{2}(\delta^3 f_{1/2} + \delta^3 f_{-1/2}) + \frac{1}{2}\left[\binom{s+2}{4} + \binom{s+1}{4}\right]\delta^4 f_0$$

$$+ \cdots \tag{4.112}$$

It can be shown by direct substitution that the Stirling centered-difference polynomials of even degree are order n and pass exactly through the data points used to construct the differences appearing in the polynomial. The odd-degree polynomials use data from one additional point.

The base point for the *Bessel centered-difference polynomial* is point $x_{1/2}$. This polynomial is based on the values of the centered differences with respect to $x_{1/2}$. That is, the polynomial follows a horizontal path through the difference table which is centered on point $x_{1/2}$. As illustrated in Figure 4.7c, odd centered differences with respect to $x_{1/2}$ exist, but even centered differences do not. Even centered differences are based on the averages of the centered differences at points x_0 and x_1. The Bessel centered-difference polynomial is:

$$P_n(x) = \frac{1}{2}(f_0 + f_1) + \frac{1}{2}\left[\binom{s}{1} + \binom{s-1}{1}\right]\delta f_{1/2} + \binom{s}{2}\frac{1}{2}(\delta^2 f_0 + \delta^2 f_1)$$

$$+ \frac{1}{2}\left[\binom{s+1}{3} + \binom{s}{3}\right]\delta^3 f_{1/2} + \binom{s+1}{4}\frac{1}{2}(\delta^4 f_0 + \delta^4 f_1) + \cdots \tag{4.113}$$

It can be shown by direct substitution that the Bessel centered-difference polynomials of odd degree are order n and pass exactly through the data points used to construct the centered differences appearing in the polynomial. The even-degree polynomials use data from one additional point.

Centered-difference polynomials are useful in the middle of a set of tabular data where the centered differences exist to the full extent of the table. However, from the uniqueness theorem for polynomials (see Section 4.2), the polynomial of degree n that passes through a specific set of $n+1$ points is unique. Thus, the Newton polynomials and the centered-difference polynomials are all equivalent when fit to the same data points. The Newton polynomials are somewhat simpler to evaluate. Consequently, when the exact number of data points to be fit by a polynomial is prespecified, the Newton polynomials are recommended.

4.7 INVERSE INTERPOLATION

Interpolation is the process of determining the value of the dependent variable f corresponding to a particular value of the independent variable x when the function $f(x)$ is described by a set of tabular data. *Inverse interpolation* is the process of determining the value of the independent variable x corresponding to a particular value of the dependent variable f. In other words, inverse interpolation is evaluation of the inverse function $x(f)$. Inverse interpolation can be accomplished by:

    1.   Fitting a polynomial to the inverse function $x(f)$
    2.   Solving a direct polynomial $f(x)$ iteratively for $x(f)$

Fitting a polynomial to the inverse function $x(f)$ appears to be the obvious approach. However, some problems may occur. The inverse function $x(f)$ may not resemble a polynomial. The values of f most certainly are not equally spaced. In such cases, a direct fit polynomial $f(x)$ may be preferred, even though it must be solved iteratively for $x(f)$, for example, by Newton's method.

Example 4.10.   Inverse interpolation.

Consider the following set of tabular data, which corresponds to the function $f(x) = 1/x$:

x	$f(x)$	$\Delta f(x)$	$^2 f(x))$
3.3	0.303030	−0.008912	
3.4	0.294118		0.000508
3.5	0.285714	−0.008404	

Let's find the value of x for which $f(x) = 0.30$. The exact solution is $x = 1/f(x) = 1/0.3 = 3.333333\ldots$.
    Let's evaluate the quadratic Lagrange polynomial, $x = x(f)$, for $f = 0.30$. Thus,

$$x = \frac{(0.30 - 0.294118)(0.30 - 0.285714)}{(0.303030 - 0.294118)(0.303030 - 0.285714)} (3.3)$$

$$+ \frac{(0.30 - 0.303030)(0.30 - 0.285714)}{(0.294118 - 0.303030)(0.294118 - 0.285714)} (3.4)$$

$$+ \frac{(0.30 - 0.303030)(0.30 - 0.294118)}{(0.285714 - 0.303030)(0.285714 - 0.294118)} (3.5) \qquad (4.114)$$

which yields $x = 3.333301$. The error is Error $= 3.333301 - 3.333333 = -0.000032$.
    Alternatively, let's evaluate the quadratic Newton forward-difference polynomial for $f = f(x)$. Thus,

$$f(x) = f_0 + s\,\Delta f_0 + \frac{s(s-1)}{2}\Delta^2 f_0 \qquad (4.115)$$

Substituting the values from the table into Eq. (4.115) yields

$$0.30 = 0.303030 + s(-0.008912) + \frac{s(s-1)}{2}(0.000508) \qquad (4.116)$$

Simplifying Eq. (4.116) gives

$$0.000508s^2 - 0.018332s + 0.006060 = 0 \tag{4.117}$$

Solving Eq. (4.117) by the quadratic formula yields two solutions, $s = 0.333654$ and 35.752960. The second root is obviously extraneous. Solving for x from the first root yields

$$x = x_0 + sh = 3.3 + (0.333654)(0.1) = 3.333365 \tag{4.118}$$

The error is Error $= 3.333365 - 3.333333 = 0.000032$.

4.8 MULTIVARIATE APPROXIMATION

All of the approximating polynomials discussed so far are single-variable, or univariate, polynomials, that is, the dependent variable is a function of a single independent variable: $y = f(x)$. Many problems arise in engineering and science where the dependent variable is a function of two or more independent variables, for example, $z = f(x, y)$ is a two-variable, or bivariate, function. Such functions in general are called *multivariate functions*. When multivariate functions are given by tabular data, multivariate approximation is required for interpolation, differentiation, and integration. Two exact fit procedures for multivariate approximation are presented in this section:

1 Successive univariate polynomial approximation
2. Direct multivariate polynomial approximation

Approximate fit procedures for multivariate polynomial approximation are discussed in Section 4.10.4.

4.8.1. Successive Univariate Polynomial Approximation

Consider the bivariate function $z = f(x, y)$. A set of tabular data is illustrated in Table 4.9.

Simply stated, successive univariate approximation first fits a set of univariate approximating functions, for example, polynomials, at each value of one of the independent variables. For example, at each value of y_i, fit the univariate polynomials $z_i(x) = z(y_i, x)$. Interpolation, differentiation, or integration is then performed on each of these univariate approximating functions to yield values of the desired operation at the specified value of the other independent variable. For example, $z_i(x^*) = z_i(y_i, x^*)$. A univariate approximating function is then fit to those results as a function of the first independent variable. For example, fit the univariate polynomial $z = z(y)$ to the values

Table 4.9. Bivariate Tabular Data

	x_1	x_2	x_3	x_4
y_1	z_{11}	z_{12}	z_{13}	z_{14}
y_2	z_{21}	z_{22}	z_{23}	z_{24}
y_3	z_{31}	z_{32}	z_{33}	z_{34}
y_4	z_{41}	z_{42}	z_{43}	z_{44}

$z_i(x^*) = z_i(y_i, x^*)$. The final process of interpolation, differentiation, or integration is then performed on that univariate approximating function. The approximating functions employed for successive univariate approximation can be of any functional form. Successive univariate polynomial approximation is generally used.

Example 4.11.   Successive univariate quadratic interpolation.

Consider the following values of enthalpy, $h(P, T)$ Btu/lbm, from a table of properties of steam, Table 4.10.

Use successive univariate quadratic polynomial interpolation to calculate the enthalpy h at $P = 1225$ psia and $T = 1104$ F (the value from the steam tables is 1556.0 Btu/lbm).

First fit the quadratic polynomial

$$h(P_i, T) = a + bT + cT^2 \tag{4.119}$$

at each pressure level P_i, and interpolate for the values of h at $T = 1100$ F. At $P = 1150$ psia:

$$a + 800b + (800)^2 c = 1380.4 \tag{4.120a}$$
$$a + 1000b + (1000)^2 c = 1500.2 \tag{4.120b}$$
$$a + 1200b + (1200)^2 c = 1614.5 \tag{4.120c}$$

Solving for a, b, and c by Gauss elimination yields

$$h(1150, T) = 846.20 + 0.72275T - 0.00006875T^2 \tag{4.121}$$

Evaluating Eq. (4.121) at $T = 1100$ gives $h(1150, 1100) = 1558.04$ Btu/lbm. In a similar manner, at $P = 1200$ psia, $h(1200, T) = 825.50 + 0.75725T - 0.00008375T^2$, and $h(1200, 1100) = 1557.14$ Btu/lbm. At $P = 1250$ psia, $h(1250, T) = 823.60 + 0.75350T - 0.00008000T^2$, and $h(1250, 1100) = 1555.65$ Btu/lbm.

Next fit the quadratic polynomial

$$h(P, 1100) = a + bP + cP^2 \tag{4.122}$$

at $T = 1100$ F. and interpolate for the value of h at $P = 1225$ psia.

$$a + 1150b + (1150)^2 c = 1558.04 \tag{4.123a}$$
$$a + 1200b + (1200)^2 c = 1557.14 \tag{4.123b}$$
$$a + 1250b + (1250)^2 c = 1555.65 \tag{4.123c}$$

Table 4.10.   Enthalpy of Steam

P, psia	T, F		
	800	1000	1200
1150	1380.4	1500.2	1614.5
1200	1377.7	1499.0	1613.6
1250	1375.2	1497.1	1612.6

Solving for a, b, and c by Gauss elimination gives

$$h(P, 1100) = 1416.59 + 0.258125P - 0.00011750P^2 \qquad (4.124)$$

Evaluating Eq. (4.124) at $P = 1225$ yields $h(1225, 1100) = 1556.5$ Btu/lbm. The error in this result is Error $= 1556.5 - 1556.0 = 0.5$ Btu/lbm.

4.8.2. Direct Multivariate Polynomial Approximation

Consider the bivariate function, $z = f(x, y)$, and the set of tabular data presented in Table 4.10. The tabular data can be fit by a multivariate polynomial of the form

$$z = f(x, y) = a + bx + cy + dxy + ex^2 + fy^2 + gx^2y + hxy^2 + ix^3 + jy^3 + \cdots$$
$$(4.125)$$

The number of data points must equal the number of coefficients in the polynomial. A linear bivariate polynomial in x and y is obtained by including the first four terms in Eq. (4.125). The resulting polynomial is exactly equivalent to successive univariate linear polynomial approximation if the same four data points are used. A quadratic bivariate polynomial in x and y is obtained by including the first eight terms in Eq. (4.125). The number of terms in the approximating polynomial increases rapidly as the degree of approximation increases. This leads to ill-conditioned systems of linear equations for determining the coefficients. Consequently, multivariate high-degree approximation must be used with caution.

Example 4.12.   Direct multivariate linear interpolation.

Let's solve the interpolation problem presented in Example 4.11 by direct multivariate linear interpolation. The form of the approximating polynomial is

$$h = a + bT + cP + dTP \qquad (4.126)$$

Substituting the four data points that bracket $P = 1225$ psia and $T = 1100$ F into Eq. (4.126) gives

$$1499.0 = a + (1000)b + (1200)c + (1000)(1200)d \qquad (4.127a)$$
$$1497.1 = a + (1000)b + (1250)c + (1000)(1250)d \qquad (4.127b)$$
$$1613.6 = a + (1200)b + (1200)c + (1200)(1200)d \qquad (4.127c)$$
$$1612.6 = a + (1200)b + (1250)c + (1200)(1250)d \qquad (4.127d)$$

Solving for a, b, c, and d by Gauss elimination yields

$$h = 1079.60 + 0.4650T - 0.1280P + 0.0900 \times 10^{-3}TP \qquad (4.128)$$

Substituting $T = 1100$ and $P = 1225$ into Eq. (4.128) gives $h(1100, 1225) = 1555.5$ Btu/lbm. The error in this result is Error $= 1555.5 - 1556.0 = -0.5$ Btu/lbm. The advantage of this approach is that Eq. (4.128) can be evaluated for other values of (T, P), if required, without reevaluating the polynomial coefficients.

4.9 CUBIC SPLINES

Several procedures for fitting approximating polynomials to a set of tabular data are presented in Sections 4.3 to 4.6. Problems can arise when a single high-degree polynomial is fit to a large number of points. High-degree polynomials would obviously pass through all the data points themselves, but they can oscillate wildly between data points due to round-off errors and overshoot. In such cases, lower-degree polynomials can be fit to subsets of the data points. If the lower-degree polynomials are independent of each other, a *piecewise approximation* is obtained. An alternate approach is to fit a lower-degree polynomial to connect each pair of data points and to require the set of lower-degree polynomials to be consistent with each other in some sense. This type of polynomial is called a *spline function*, or simply a *spline*.

Splines can be of any degree. Linear splines are simply straight line segments connecting each pair of data points. Linear splines are independent of each other from interval to interval. Linear splines yield first-order approximating polynomials. The slopes (i.e., first derivatives) and curvature (i.e., second derivatives) are discontinuous at every data point. Quadratic splines yield second-order approximating polynomials. The slopes of the quadratic splines can be forced to be continuous at each data point, but the curvatures (i.e., the second derivatives) are still discontinuous.

A *cubic spline* yields a third-degree polynomial connecting each pair of data points. The slopes and curvatures of the cubic splines can be forced to be continuous at each data point. In fact, these requirements are necessary to obtain the additional conditions required to fit a cubic polynomial to two data points. Higher-degree splines can be defined in a similar manner. However, cubic splines have proven to be a good compromise between accuracy and complexity. Consequently, the concept of cubic splines is developed in this section.

The name *spline* comes from the thin flexible rod, called a spline, used by draftsmen to draw smooth curves through a series of discrete points. The spline is placed over the points and either weighted or pinned at each point. Due to the flexure properties of a flexible rod (typically of rectangular cross section), the slope and curvature of the rod are continuous at each point. A smooth curve is then traced along the rod, yielding a spline curve.

Figure 4.8 illustrates the discrete x space and defines the indexing convection. There are $n + 1$ total points, x_i ($i = 1, 2, \ldots, n + 1$), n intervals, and $n - 1$ interior grid points, x_i ($i = 2, 3, \ldots, n$). A cubic spline is to be fit to each interval. Thus,

$$\boxed{f_i(x) = a_i + b_i x + c_i x^2 + d_i x^3 \qquad (i = 1, 2, \ldots, n)}$$ (4.129)

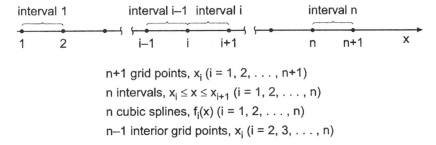

interval 1   interval i–1 interval i   interval n

1   2   i–1   i   i+1   n   n+1   x

n+1 grid points, x_i ($i = 1, 2, \ldots, n+1$)
n intervals, $x_i \le x \le x_{i+1}$ ($i = 1, 2, \ldots, n$)
n cubic splines, $f_i(x)$ ($i = 1, 2, \ldots, n$)
n–1 interior grid points, x_i ($i = 2, 3, \ldots, n$)

Figure 4.8   Cubic splines.

defines the cubic spline in interval $i, x_i \leq x \leq x_{i+1}$ $(i = 1, 2, \ldots, n)$. Since each cubic spline has four coefficients and there are n cubic splines, there are $4n$ coefficients to be determined. Thus, $4n$ boundary conditions, or constraints, must be available.

In the direct approach, the following constraints are applied.

1. The function values, $f(x_i) = f_i$ $(i = 2, 3, \ldots, n)$, must be the same in the two splines on either side of x_i at all of the $n - 1$ interior points. This constraint yields $2(n - 1)$ conditions.
2. The first derivative of the two splines on either side of point x_i must be equal at all of the $n - 1$ interior points. This constraint yields $(n - 1)$ conditions.
3. The second derivative of the two splines on either side of point x_i must be equal at all of the $n - 1$ interior points. This constraint yields $(n - 1)$ conditions.
4. The first and last spline must pass through the first (i.e., x_1) and last (i.e., x_{n+1}) points. That is, $f_1(x_1) = f_1$ and $f_n(x_{n+1}) = f_{n+1}$. This constraint yields 2 conditions.
5. The curvature [i.e., $f''(x)$] must be specified at the first (i.e., x_1) and last (i.e., x_{n+1}) points. That is, $f_1''(x_1) = f_1''$ and $f_n''(x_{n+1}) = f_{n+1}''$. This constraint yields 2 conditions.

When all of the conditions given above are assembled, $4n$ linear algebraic equations are obtained for the $4n$ spline coefficients a_i, b_i, c_i, and d_i $(i = 1, 2, \ldots, n)$. This set of equations can be solved by Gauss elimination. However, simpler approaches exist for determining cubic splines. Gerald and Wheatley (1999), Chapra and Canale (1998), and Press et al. (1989) present three such approaches. The approach presented by Chapra and Canale is followed below.

From the cubic equation, Eq. (4.129), it is obvious that the second derivative within each interval, $f_i''(x)$, is a linear function of x. The first-order Lagrange polynomial for the second derivative $f_i''(x)$ in interval $i, x_i \leq x \leq x_{i+1}$ $(i = 1, 2, \ldots, n)$, is given by

$$f_i''(x) = \frac{x - x_{i+1}}{x_i - x_{i+1}} f_i'' + \frac{x - x_i}{x_{i+1} - x_i} f_{i+1}'' \tag{4.130}$$

Integrating Eq. (4.130) twice yields expressions for $f_i'(x)$ and $f_i(x)$. Thus,

$$f_i'(x) = \frac{x^2/2 - xx_{i+1}}{x_i - x_{i+1}} f_i'' + \frac{x^2/2 - xx_i}{x_{i+1} - x_i} f_{i+1}'' + C \tag{4.131}$$

$$f_i(x) = \frac{x^3/6 - x^2 x_{i+1}/2}{x_i - x_{i+1}} f_i''(x) + \frac{x^3/6 - x^2 x_i/2}{x_{i+1} - x_i} f_{i+1}'' + Cx + D \tag{4.132}$$

Evaluating Eq. (4.132) at x_i and x_{i+1} and combining the results to eliminate the constants of integration C and D gives

$$\boxed{\begin{aligned} f_i(x) = {} & \frac{f_i''}{6(x_{i+1} - x_i)}(x_{i+1} - x)^3 + \frac{f_{i+1}''}{6(x_{i+1} - x_i)}(x - x_i)^3 \\ & + \left[\frac{f_i}{x_{i+1} - x_i} - \frac{f_i''(x_{i+1} - x_i)}{6} \right](x_{i+1} - x) \\ & + \left[\frac{f_{i+1}}{x_{i+1} - x_i} - \frac{f_{i+1}''(x_{i+1} - x_i)}{6} \right](x - x_i) \end{aligned}} \tag{4.133}$$

Equation (4.133) is the desired cubic spline for increment i expressed in terms of the two unknown second derivatives f_i'' and f_{i+1}''.

An expression for the second derivatives at the interior grid points, $f_i''(i = 2, 3, \ldots, n)$, can be obtained by setting $f_{i-1}'(x_i) = f_i'(x_i)$. An expression for $f_i'(x)$ can be obtained by differentiating Eq. (4.133). Applying that expression to intervals $i - 1$ and i and evaluating those results at $x = x_i''$ gives expressions for $f_{i-1}'(x_i)$ and $f_i'(x_i)$. Equating those expressions yields

$$(x_i - x_{i-1})f_{i-1}'' + 2(x_{i+1} - x_{i-1})f_i'' + (x_{i+1} - x_i)f_{i+1}'' = 6\frac{f_{i+1} - f_i}{x_{i+1} - x_i} - 6\frac{f_i - f_{i-1}}{x_i - x_{i-1}}$$

(4.134)

Applying Eq. (4.134) at the $n - 1$ interior points gives $n - 1$ coupled equations for the $n + 1$ second derivatives, $f_i''(i = 1, 2, \ldots, n + 1)$. Two more values of f_i'' are required to close the system of equations.

The two additional conditions are obtained by specifying the values of f_1'' and f_{n+1}''. Several approaches are available for specifying these two values:

1. Specify f_1'' and f_{n+1}'' if they are known. Letting $f_1'' = 0$ and/or $f_{n+1}'' = 0$ specifies a natural spline.
2. Specify f_1' and/or f_{n+1}' and use Eq. (4.131) to develop a relationship between f_1' and/or f_{n+1}' and f_1'', f_1'' and f_2'', etc. This requires the evaluation of the constant of integration C, which is not developed in this book.
3. Let $f_1'' = f_2''$ and $f_{n+1}'' = f_n''$.
4. Extrapolate f_1'' and f_{n+1}'' from interior values of f_i''.

The first approach, letting $f_1'' = f_{n+1}'' = 0$, is the most commonly employed approach.

Example 4.13.   Cubic splines.

Let's illustrate the cubic spline by applying it to the data in the following table, which is for the function $f(x) = e^x - x^3$.

i	x	$f(x)$	$f''(x)$
1	-0.50	0.731531	0.0
2	0.00	1.000000	
3	0.25	1.268400	
4	1.00	1.718282	0.0

There are $n + 1 = 4$ grid points, $n = 3$ intervals, $n = 3$ cubic splines to be fit, and $n - 1 = 2$ interior grid points at which f_i'' must be evaluated.

The first step is to determine the values of f_i'' at the four grid points, $i = 1$ to 4. Let's apply the natural spline boundary condition at $i = 1$ and $i = 4$. Thus, $f_1'' = f_4'' = 0.0$. Applying Eq. (4.134) at grid points 2 and 3 gives

$$i = 2: (x_2 - x_1)f_1'' + 2(x_3 - x_1)f_2'' + (x_3 - x_2)f_3'' = 6\frac{f_3 - f_2}{x_3 - x_2} - 6\frac{f_2 - f_1}{x_2 - x_1} \quad (4.135)$$

$$i = 3: (x_3 - x_2)f_2'' + 2(x_4 - x_2)f_3'' + (x_4 - x_3)f_4'' = 6\frac{f_4 - f_3}{x_4 - x_3} - 6\frac{f_3 - f_2}{x_3 - x_2} \quad (4.136)$$

Substituting values from the table into Eqs. (4.135) and (4.136) gives

$$(0.5)(0.0) + 2(0.75)f_2'' + (0.25)f_3'' = 6\left(\frac{0.268400}{0.25} - \frac{0.268469}{0.50}\right) \tag{4.137a}$$

$$(0.25)f_2 + 2(1.0)f_3'' + (0.75)(0.0) = 6\left(\frac{0.449882}{0.75} - \frac{0.268400}{0.25}\right) \tag{4.137b}$$

Evaluating Eqs. (4.137) gives

$$1.50f_2'' + 0.25f_3'' = 3.219972 \tag{4.138a}$$

$$0.25f_2'' + 2.0f_3'' = -2.842544 \tag{4.138b}$$

Solving Eq. (4.138) by Gauss elimination yields

$$f_2'' = 2.434240 \quad \text{and} \quad f_3'' = -1.725552 \tag{4.139}$$

The second step is to substitute the numerical values into Eq. (4.133) for the three intervals, $(x_1, x_2), (x_2, x_3)$, and (x_3, x_4), to determine the cubic splines for the three intervals. Thus, for interval 1, that is, (x_1, x_2), Eq. (4.133) becomes

$$f_1(x) = \frac{f_i''}{6(x_2 - x_1)}(x_2 - x)^3 + \frac{f_2''}{6(x_2 - x_1)}(x - x_1)^3$$
$$+ \left[\frac{f_1}{x_2 - x_1} - \frac{f_1''(x_2 - x_1)}{6}\right](x_2 - x) + \left[\frac{f_2}{x_2 - x_1} - \frac{f_2''(x_2 - x_1)}{6}\right](x - x_1) \tag{4.140}$$

Similar equations are obtained for intervals 2 and 3. Substituting values into these three equations gives

$$f_1(x) = (0.0) + \frac{2.434240}{6(0.5)}[x - (-0.5)]^3 + \left(\frac{0.731531}{0.5} - 0.0\right)(0.0 - x)$$
$$+ \left[\frac{1.0}{0.5} - \frac{2.434240(0.5)}{6}\right][x - (-0.5)] \tag{4.141a}$$

$$f_2(x) = \frac{2.434240}{6(0.25)}(0.25 - x)^3 + \frac{-1.725552}{6(0.25)}(x - 0.0)^3$$
$$+ \left[\frac{1.0}{0.25} - \frac{2.434240(0.25)}{6}\right](0.25 - x)$$
$$+ \left[\frac{1.268400}{0.25} - \frac{-1.725552(0.25)}{6}\right](x - 0.0) \tag{4.141b}$$

$$f_3(x) = \frac{-1.725552}{6(0.75)}(1.0 - x)^3 + (0.0) + \left[\frac{1.268400}{0.75} - \frac{-1.725552(0.75)}{6}\right](1.0 - x)$$
$$+ \left[\frac{1.718282}{0.75} - \frac{(0.0)(0.75)}{6}\right](x - 0.25) \tag{4.141c}$$

Evaluating Eq. (4.141) yields

$$f_1(x) = 0.811413(x + 0.5)^3 - 1.463062x + 1.797147(x + 0.5)^3 \qquad (4.142a)$$

$$f_2(x) = 1.622827(0.25 - x)^3 - 1.150368x^3 + 3.898573(0.25 - x) + 5.145498x$$
$$(4.142b)$$

$$f_3(x) = -0.383456(1.0 - x)^3 + 1.906894(1.0 - x) + 2.291043(x - 0.25) \quad (4.142c)$$

Equation (4.142) can be verified by substituting the values in the table into the equations.

4.10 LEAST SQUARES APPROXIMATION

Section 4.1 discusses the need for approximating functions for sets of discrete data (e.g., for interpolation, differentiation, and integration), the desirable properties of approximating functions, and the benefits of using polynomials for approximating functions. For small sets of smooth data, exact fits such as presented in Sections 4.3 to 4.9 are desirable. However, for large sets of data and sets of rough data, approximate fits are desirable. Approximate polynomial fits are the subject of this section.

4.10.1 Introduction

An approximate fit yields a polynomial that passes through the set of points in the *best possible manner* without being required to pass exactly through any of the points. Several definitions of best possible manner exist. Consider the set of discrete points, $[x_i, Y(x_i)] = (x_i, Y_i)$, and the approximate polynomial $y(x)$ chosen to represent the set of discrete points, as illustrated in Figure 4.9. The discrete points do not fall on the approximating polynomial. The deviations (i.e., distances) of the points from the approximating function must be minimized in some manner. Some ambiguity is possible in the definition of the deviation. For example, if the values of the independent variable x_i

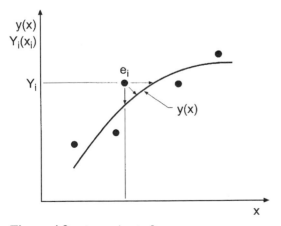

Figure 4.9 Approximate fit.

are considered exact, then all the deviation is assigned to the dependent variable Y_i, and the deviation e_i is the vertical distance between Y_i and $y_i = f(x_i)$. Thus,

$$e_i = Y_i - y_i \qquad\qquad (4.143)$$

It is certainly possible that the values of Y_i are quite accurate, but the corresponding values of x_i are in error. In that case, the deviation would be measured by the horizontal distance illustrated in Figure 4.9. If x_i and Y_i both have uncertainties in their values, then the perpendicular distance between a point and the approximating function would be the deviation. The usual approach in approximate fitting of tabular data is to assume that the deviation is the vertical distance between a point and the approximating function, as specified by Eq. (4.143).

Several best fit criteria are illustrated in Figure 4.10 for a straight line approximation. Figure 4.10a illustrates the situation where the sum of the deviations at two points is minimized. Any straight line that passes through the midpoint of the line segment connecting the two points yields the sum of the deviations equal to zero. Minimizing the sum of the absolute values of the deviations would yield the unique line that passes exactly through the two points. That procedure also has deficiencies, however, as illustrated in Figure 4.10b, where two points having the same value of the independent variable have different values of the dependent variable. The best straight line obviously passes midway between these two points, but any line passing between these two points yields the same value for the sum of the absolute values of the deviations. The minimax criterion is illustrated in Figure 4.10c, where the maximum deviation is minimized. This procedure gives poor results when one point is far removed from the other points. Figure

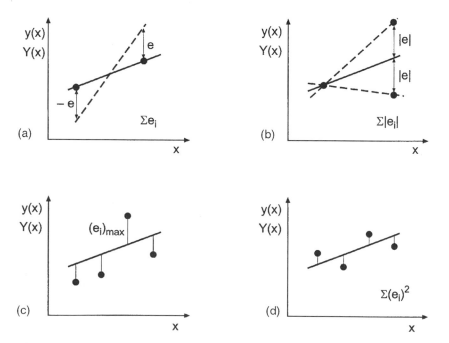

Figure 4.10 Best fit criteria. (a) Minimize $\sum e_i$. (b) Minimize $\sum |e_i|$. (c) Minimax. (d) Least squares.

4.10d illustrates the least squares criteria, in which the sum of the squares of the deviations is minimized. The least squares procedure yields a good compromise criterion for the best fit approximation.

The *least squares method* is defined as follows. Given N data points, $[x_i, Y(x_i)] = (x_i, Y_i)$, choose the functional form of the approximating function to be fit, $y = y(x)$, and minimize the sum of the squares of the deviations, $e_i = (Y_i - y_i)$.

4.10.2 The Straight Line Approximation

The simplest polynomial is a linear polynomial, the straight line. Least squares straight line approximations are an extremely useful and common approximate fit. The least squares straight line fit is determined as follows. Given N data points, (x_i, Y_i), fit the best straight line through the set of data. The approximating function is

$$\boxed{y = a + bx} \tag{4.144}$$

At each value of x_i, Eq. (4.144) gives

$$y_i = a + bx_i \qquad (i = 1, \dots, N) \tag{4.145}$$

The deviation e_i at each value of x_i is

$$e_i = Y_i - y_i \qquad (i = 1, \dots, N) \tag{4.146}$$

The sum of the squares of the deviations defines the function $S(a, b)$:

$$S(a, b) = \sum_{i=1}^{N} (e_i)^2 = \sum_{i=1}^{N} (Y_i - a - bx_i)^2 \tag{4.147}$$

The function $S(a, b)$ is a minimum when $\partial S/\partial a = \partial S/\partial b = 0$. Thus,

$$\frac{\partial S}{\partial a} = \sum_{i=1}^{N} 2(Y_i - a - bx_i)(-1) = 0 \tag{4.148a}$$

$$\frac{\partial S}{\partial b} = \sum_{i=1}^{N} 2(Y_i - a - bx_i)(-x_i) = 0 \tag{4.148b}$$

Dividing Eqs. (4.148) by 2 and rearranging yields

$$\boxed{\begin{array}{l} aN + b \sum_{i=1}^{N} x_i = \sum_{i=1}^{N} Y_i \\[2mm] a \sum_{i=1}^{N} x_i + b \sum_{i=1}^{N} x_i^2 = \sum_{i=1}^{N} x_i Y_i \end{array}} \tag{4.149a} \tag{4.149b}$$

Equations (4.149) are called the *normal equations* of the least squares fit. They can be solved for a and b by Gauss elimination.

Example 4.14. Least squares straight line approximation.

Consider the constant pressure specific heat for air at low temperatures presented in Table 4.11, where T is the temperature (K) and C_p is the specific heat (J/gm-K). The exact values, approximate values from the least squares straight line approximation, and the

Table 4.11. Specific Heat of Air at Low Temperatures

T, K	$C_{p,\text{exact}}$	$C_{p,\text{approx.}}$	Error, %
300	1.0045	0.9948	−0.97
400	1.0134	1.0153	0.19
500	1.0296	1.0358	0.61
600	1.0507	1.0564	0.54
700	1.0743	1.0769	0.24
800	1.0984	1.0974	−0.09
900	1.1212	1.1180	−0.29
1000	1.1410	1.1385	−0.22

percent error are also presented in the table. Determine a least squares straight line approximation for the set of data:

$$C_p = a + bT \tag{4.150}$$

For this problem, Eq. (4.149) becomes

$$8a + b\sum_{i=1}^{8} T_i = \sum_{i=1}^{8} C_{p,i} \tag{4.151}$$

$$a\sum_{i=1}^{8} T_i + b\sum_{i=1}^{8} T_i^2 = \sum_{i=1}^{8} T_i C_{p,i} \tag{4.152}$$

Evaluating the summations and substituting into Eqs. (4.151) and (4.152) gives

$$8a + 5200b = 8.5331 \tag{4.153a}$$

$$5200a + 3{,}800{,}000b = 5632.74 \tag{4.153b}$$

Solving for a and b by Gauss elimination without scaling or pivoting yields

$$\boxed{C_p = 0.933194 + 0.205298 \times 10^{-3}T} \tag{4.154}$$

Substituting the initial values of T into Eq. (4.154) gives the results presented in Table 4.11, which presents the exact data, the least squares straight line approximation, and the percent error. Figure 4.11 presents the exact data and the least squares straight line approximation. The straight line is not a very good approximation of the data.

4.10.3. Higher-Degree Polynomial Approximation

The least squares procedure developed in Section 4.10.2 can be applied to higher-degree polynomials. Given the N data points, (x_i, Y_i), fit the best nth-degree polynomial through the set of data. Consider the nth-degree polynomial:

$$\boxed{y = a_0 + a_1 x + a_2 x^2 + \cdots + a_n x^n} \tag{4.155}$$

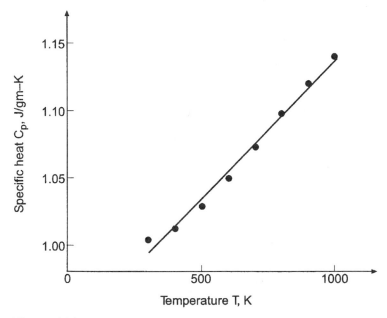

Figure 4.11   Least squares straight line approximation.

The sum of the squares of the deviations is given by

$$S(a_0, a_1, \ldots, a_n) = \sum_{i=1}^{N} (e_i)^2 = \sum_{i=1}^{N} (Y_i - a_0 - a_1 x_i - \cdots - a_n x_i^n)^2 \qquad (4.156)$$

The function $S(a_0, a_1, \ldots, a_n)$ is a minimum when

$$\frac{\partial S}{\partial a_0} = \sum_{i=1}^{N} 2(Y_i - a_0 - a_1 x_i - \cdots - a_n x_i^n)(-1) = 0 \qquad (4.157a)$$

$$\cdots\cdots\cdots\cdots\cdots\cdots\cdots\cdots\cdots\cdots\cdots\cdots\cdots\cdots$$

$$\frac{\partial S}{\partial a_n} = \sum_{i=1}^{N} 2(Y_i - a_0 - a_1 x_i - \cdots - a_n x_i^n)(-x_i^n) = 0 \qquad (4.157b)$$

Dividing Eqs. (4.157) by 2 and rearranging yields the normal equations:

$$a_0 N + a_1 \sum_{i=1}^{N} x_i + \cdots + a_n \sum_{i=1}^{N} x_i^n = \sum_{i=1}^{N} Y_i \qquad (4.158a)$$

$$\cdots\cdots\cdots\cdots\cdots\cdots\cdots\cdots\cdots\cdots\cdots\cdots\cdots\cdots$$

$$a_0 \sum_{i=1}^{N} x_i^n + a_1 \sum_{i=1}^{N} x_i^{n+1} + \cdots + a_n \sum_{i=1}^{N} x_i^{2n} = \sum_{i=1}^{N} x_i^n Y_i \qquad (4.158b)$$

Equation (4.158) can be solved for a_0 to a_n by Gauss elimination.

    A problem arises for high-degree polynomials. The coefficients in Eq. (4.158), N to $\sum x_i^{2n}$, can vary over a range of several orders of magnitude, which gives rise to ill-conditioned systems. Normalizing each equation helps the situation. Double precision calculations are frequently required. Values of n up to 5 or 6 generally yield good results, values of n between 6 and 10 may or may not yield good results, and values of n greater than approximately 10 generally yield poor results.

Example 4.15. Least squares quadratic polynomial approximation.

Consider the constant pressure specific heat of air at high temperatures presented in Table 4.12, where T is the temperature (K) and C_p is the specific heat (J/gm-K). The exact values, approximate values from a least squares quadratic polynomial approximation, and the percent error are also presented in the table. Determine a least squares quadratic polynomial approximation for this set of data:

$$C_p = a + bT + cT^2 \tag{4.159}$$

For this problem, Eq. (4.158) becomes

$$5a + b\sum T_i + c\sum T_i^2 = \sum C_{p,i} \tag{4.160a}$$

$$a\sum T_i + b\sum T_i^2 + c\sum T_i^3 = \sum T_i C_{p,i} \tag{4.160b}$$

$$a\sum T_i^2 + b\sum T_i^3 + c\sum T_i^4 = \sum T_i^2 C_{p,i} \tag{4.160c}$$

Evaluating the summations and substituting into Eq. (4.160) gives

$$5a + 10 \times 10^3 b + 22.5 \times 10^6 c = 6.1762 \tag{4.161a}$$

$$10 \times 10^3 a + 22.5 \times 10^6 b + 55 \times 10^9 c = 12.5413 \times 10^3 \tag{4.161b}$$

$$22.5 \times 10^6 a + 55 \times 10^9 b + 142.125 \times 10^{12} c = 288.5186 \times 10^6 \tag{4.161c}$$

Solving for a, b, and c by Gauss elimination yields

$$\boxed{C_p = 0.965460 + 0.211197 \times 10^{-3} T - 0.0339143 \times 10^{-6} T^2} \tag{4.162}$$

Substituting the initial values of T into Eq. (4.162) gives the results presented in Table 4.12. Figure 4.12 presents the exact data and the least squares quadratic polynomial approximation. The quadratic polynomial is a reasonable approximation of the discrete data.

Table 4.12. Specific Heat of Air at High Temperatures

T, K	$C_{p,\text{exact}}$	$C_{p,\text{approx.}}$	Error, %
1000	1.1410	1.1427	0.15
1500	1.2095	1.2059	−0.29
2000	1.2520	1.2522	0.02
2500	1.2782	1.2815	0.26
3000	1.2955	1.2938	−0.13

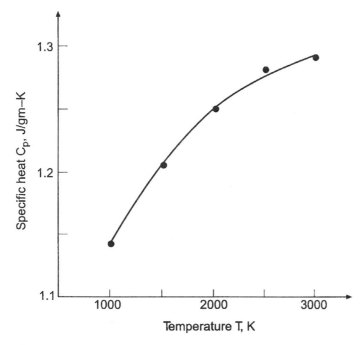

Figure 4.12   Least squares quadratic approximation.

4.10.4.   Multivariate Polynomial Approximation

Many problems arise in engineering and science where the dependent variable is a function of two or more independent variables, for example, $z = f(x, y)$ is a two-variable, or bivariate, function. Two exact fit procedures for multivariate approximation are presented in Section 4.8. Least squares multivariate approximation is considered in this section.

Given the N data points, (x_i, y_i, Z_i), fit the best linear bivariate polynomial through the set of data. Consider the linear polynomial:

$$\boxed{z = a + bx + cy} \tag{4.163}$$

The sum of the squares of the deviations is given by

$$S(a, b, c) = \sum (e_i)^2 = \sum (Z_i - a - bx_i - cy_i)^2 \tag{4.164}$$

The function $S(a, b, c)$ is a minimum when

$$\frac{\partial S}{\partial a} = \sum 2(Z_i - a - bx_i - cy_i)(-1) = 0 \tag{4.165a}$$

$$\frac{\partial S}{\partial b} = \sum 2(Z_i - a - bx_i - cy_i)(-x_i) = 0 \tag{4.165b}$$

$$\frac{\partial S}{\partial c} = \sum 2(Z_i - a - bx_i - cy_i)(-y_i) = 0 \tag{4.165c}$$

Dividing Eqs. (4.165) by 2 and rearranging yields the normal equations:

$$aN + b\sum x_i + c\sum y_i = \sum Z_i \tag{4.166a}$$

$$a\sum x_i + b\sum x_i^2 + c\sum x_i y_i = \sum x_i Z_i \tag{4.166b}$$

$$a\sum y_i + b\sum x_i y_i + c\sum y_i^2 = \sum y_i Z_i \tag{4.166c}$$

Equation (4.166) can be solved for a, b, and c by Gauss elimination.

A linear fit to a set of bivariate data may be inadequate. Consider the quadratic bivariate polynomial:

$$z = a + bx + cy + dx^2 + ey^2 + fxy \tag{4.167}$$

The sum of the squares of the deviations is given by

$$S(a, b, c, d, e, f) = \sum (Z_i - a - bx_i - cy_i - dx_i^2 - ey_i^2 - fx_i y_i)^2 \tag{4.168}$$

The function $S(a, b, \ldots, f)$ is a minimum when

$$\frac{\partial S}{\partial a} = \sum 2(Z_i - a - bx_i - cy_i - dx_i^2 - ey_i^2 - fx_i y_i)(-1) = 0 \tag{4.169a}$$

$$\cdots\cdots\cdots\cdots\cdots\cdots\cdots\cdots\cdots\cdots\cdots\cdots\cdots\cdots\cdots\cdots\cdots$$

$$\frac{\partial S}{\partial f} = \sum 2(Z_i - a - bx_i - cy_i - dx_i^2 - ey_i^2 - fx_i y_i)(-x_i y_i) = 0 \tag{4.169f}$$

Dividing Eqs. (4.169) by 2 and rearranging yields the normal equations:

$$aN + b\sum x_i + c\sum y_i + d\sum x_i^2 + e\sum y_i^2 + f\sum x_i y_i = \sum Z_i \tag{4.170a}$$

$$a\sum x_i + b\sum x_i^2 + c\sum x_i y_i + d\sum x_i^3 + e\sum x_i y_i^2 + f\sum x_i^2 y_i = \sum x_i Z_i \tag{4.170b}$$

$$a\sum y_i + b\sum x_i y_i + c\sum y_i^2 + d\sum x_i^2 y_i + e\sum y_i^3 + f\sum x_i y_i^2 = \sum y_i Z_i \tag{4.170c}$$

$$a\sum x_i^2 + b\sum x_i^3 + c\sum x_i^2 y_i + d\sum x_i^4 + e\sum x_i^2 y_i^2 + f\sum x_i^3 y_i = \sum x_i^2 Z_i \tag{4.170d}$$

$$a\sum y_i^2 + b\sum x_i y_i^2 + c\sum y_i^3 + d\sum x_i^2 y_i^2 + e\sum y_i^4 + f\sum x_i, y_i^3 = \sum y_i^2 Z_i \tag{4.170e}$$

$$a\sum x_i y_i + b\sum x_i^2 y_i + c\sum x_i y_i^2 + d\sum x_i^3 y_i + e\sum x_i y_i^3 + f\sum x_i^2 y_i^2 = \sum x_i y_i Z_i \tag{4.170f}$$

Equation (4.170) can be solved for a to f by Gauss elimination.

Example 4.16. Least squares quadratic bivariate polynomial approximation.

Let's rework Example 4.12 to calculate the enthalpy of steam at $P = 1225$ psia and $T = 1100$ F, based on the data in Table 4.10, using a least squares quadratic bivariate polynomial. Let the variables x, y, and Z in Eq. (4.170) correspond to P, T, and h, respectively. Evaluating the summations and substituting into Eq. (4.170) gives

$$
\begin{bmatrix}
9\ \text{E0} & 10.800\ \text{E3} & 9.000\ \text{E3} & 12.975\ \text{E6} & 9.240\ \text{E6} & 10.800\ \text{E6} \\
10.800\ \text{E3} & 12.975\ \text{E6} & 10.800\ \text{E6} & 15.606\ \text{E9} & 11.088\ \text{E9} & 12.975\ \text{E9} \\
9.000\ \text{E3} & 10.800\ \text{E6} & 9.240\ \text{E6} & 12.975\ \text{E9} & 9.720\ \text{E9} & 11.088\ \text{E9} \\
12.975\ \text{E6} & 15.606\ \text{E9} & 12.975\ \text{E9} & 18.792\ \text{E12} & 13.321\ \text{E12} & 15.606\ \text{E12} \\
9.240\ \text{E6} & 11.088\ \text{E9} & 9.720\ \text{E9} & 13.321\ \text{E12} & 10.450\ \text{E12} & 11.664\ \text{E12} \\
10.800\ \text{E6} & 12.975\ \text{E9} & 11.088\ \text{E9} & 15.606\ \text{E12} & 11.664\ \text{E12} & 13.321\ \text{E12}
\end{bmatrix}
\begin{bmatrix}
a \\ b \\ c \\ d \\ e \\ f
\end{bmatrix}
$$

$$
=
\begin{bmatrix}
13.4703\ \text{E3} \\
16.1638\ \text{E6} \\
13.6118\ \text{E6} \\
19.4185\ \text{E9} \\
14.1122\ \text{E9} \\
16.3337\ \text{E9}
\end{bmatrix}
\tag{4.171}
$$

Due to the large magnitudes of the coefficients, they are expressed in exponential format (i.e., x.xxx En = x.xxx $\times\ 10^n$). Each row in Eq. (4.171) should be normalized by the exponential term in the first coefficient of each row. Solving the normalized Eq. (4.171) by Gauss elimination yields

$$
h(T, P) = 914.033 - 0.0205000P + 0.645500T - 0.000040000P^2
$$
$$
- 0.000077500T^2 + 0.000082500PT
\tag{4.172}
$$

Evaluating Eq. (4.172) yields $h(1100.0, 1225.0) = 1556.3$ Btu/lbm. The error is Error $=$ $1556.3 - 1556.0 = 0.3$ Btu/lbm. which is smaller than the error incurred in Example 4.12 obtained by direct multivariate linear interpolation.

Equation (4.170) can be written as the matrix equation

$$
\mathbf{Ac} = \mathbf{b}
\tag{4.173}
$$

where $\mathbf{A}$ is the 6×6 coefficient matrix, $\mathbf{c}$ is the 6×1 column vector of polynomial coefficients (i.e., a to f), and $\mathbf{b}$ is the 6×1 column vector of nonhomogeneous terms. The solution to Eq. (4.173) is

$$
\mathbf{c} = \mathbf{A}^{-1}\mathbf{b}
\tag{4.174}
$$

where $\mathbf{A}^{-1}$ is the inverse of $\mathbf{A}$. In general, solving Eq. (4.173) for $\mathbf{c}$ by Gauss elimination is more efficient than calculating $\mathbf{A}^{-1}$. However, for equally spaced data (i.e. $\Delta x =$ constant and $\Delta y =$ constant) in a large table, a considerable simplification can be achieved. This is accomplished by locally transforming the independent variables so that $x = y = 0$ at the center point of each subset of data. In this case, $\mathbf{A}$ is constant for the entire table, so $\mathbf{A}^{-1}$ can be determined once for the entire table, and Eq. (4.174) can be used to calculate the coefficients a to f at any point in the table very efficiently. Only the nonhomogeneous vector $\mathbf{b}$ changes from point to point and must be recalculated.

4.10.5. Nonlinear Functions

One advantage of using polynomials for least squares approximation is that the normal equations are linear, so it is straightforward to solve for the coefficients of the polynomials. In some engineering and science problems, however, the underlying physics suggests forms of approximating functions other than polynomials. Examples of several such functions are presented in this section.

Many physical processes are governed by a nonlinear equation of the form

$$y = ax^b \tag{4.175}$$

Taking the natural logarithm of Eq. (4.175) gives

$$\ln(y) = \ln(a) + b \ln(x) \tag{4.176}$$

Let $y' = \ln(y)$, $a' = \ln(a)$, and $x' = \ln(x)$. Equation (4.176) becomes

$$y' = a' + bx' \tag{4.177}$$

which is a linear polynomial. Equation (4.177) can be used as a least squares approximation by applying the results developed in Section 4.10.2.

Another functional form that occurs frequently in physical processes is

$$y = ae^{bx} \tag{4.178}$$

Taking the natural logarithm of Eq. (4.178) gives

$$\ln(y) = \ln(a) + bx \tag{4.179}$$

which can be written as the linear polynomial

$$y' = a' + bx \tag{4.180}$$

Equations (4.175) and (4.178) are examples of nonlinear functions that can be manipulated into a linear form. Some nonlinear functions cannot be so manipulated. For example, consider the nonlinear function

$$y = \frac{a}{1 + bx} \tag{4.181}$$

For a given set of data, (x_i, Y_i) $(i = 1, 2, \ldots, N)$, the sum of the squares of the deviations is

$$S(a, b) = \sum \left(Y_i - \frac{a}{1 + bx_i} \right)^2 \tag{4.182}$$

The function $S(a, b)$ is a minimum when

$$\frac{\partial S}{\partial a} = \sum 2 \left(Y_i - \frac{a}{1 + bx_i} \right) \left(-\frac{1}{1 + bx_i} \right) = 0 \tag{4.183a}$$

$$\frac{\partial S}{\partial b} = \sum 2 \left(Y_i - \frac{a}{1 + bx_i} \right) \left(\frac{x_i}{(1 + bx_i)^2} \right) = 0 \tag{4.183b}$$

Equation (4.183) comprises a pair of nonlinear equations for determining the coefficients a and b. They can be solved by methods for solving systems of nonlinear equations, for example, by Newton's method, as discussed in Section 3.7.

4.11 PROGRAMS

Five FORTRAN subroutines for fitting approximating polynomials are presented in this section:

1. Direct fit polynomials
2. Lagrange polynomials
3. Divided difference polynomials
4. Newton forward-difference polynomials
5. Least squares polynomials

All of the subroutines are written for a *quadratic* polynomial, except the *linear* least squares polynomial. Higher-degree polynomials can be constructed by following the patterns of the quadratic polynomials. The variable *ndeg*, which specifies the degree of the polynomials, is thus not needed in the present programs. It is included, however, to facilitate the addition of other-dgree polynomials to the subroutines, in which case the degree of the polynomial to be evaluated must be specified.

The basic computational algorithms are presented as completely self-contained subroutines suitable for use in other programs. FORTRAN subroutines *direct*, *lagrange*, *divdiff* and *newtonfd* are presented in Sections 4.11.1 to 4.11.4 for implementing these procedures. A common *program main* defines the data sets and prints them, calls one of the subroutines to implement the solution, and prints the solution. The complete *program main* is presented with *subroutine direct*. For the other three subroutines, only the changed statements are included. *Program main* for *subroutine lesq* is completely self-contained.

4.11.1. Direct Fit Polynomials

The direct fit polynomial is specified by Eq. (4.34):

$$P_n(x) = a_0 + a_1 x + a_2 x^2 + \cdots + a_n x^n \tag{4.184}$$

A FORTRAN subroutine, *subroutine direct*, for implementing a *quadratic* direct fit polynomial is presented below. *Program main* (Program 4.1) defines the data set and prints it, calls *subroutine direct* to implement the solution, and prints the solution. *Subroutine gauss* is taken from Section 1.8.1.

Program 4.1. Direct fit polynomial program.

```
      program main
c     main program to illustrate polynomial fitting subroutines
c     ndim  array dimension, ndim = 6 in this example
c     ndeg  degree of the polynomial, ndeg = 2 in this example
c     n     number of data points and polynomial coefficients
c     xp    value of x at which the polynomial is evaluated
c     x     values of the independent variable, x(i)
c     f     values of the dependent variable, f(i)
c     fxp   interpolated value, f(xp)
c     c     coefficients of the direct fit polynomial, c(i)
      dimension x(6),f(6),a(6,6),b(6),c(6)
      data ndim,ndeg,n,xp / 6, 2, 3, 3.44 /
      data (x(i),i=1,3) / 3.35, 3.40, 3.50 /
      data (f(i),i=1,3) / 0.298507, 0.294118, 0.285714 /
```

```
      write (6,1000)
      do i=1,n
         write (6,1010) i,x(i),f(i)
      end do
      call direct (ndim,ndeg,n,x,f,xp,fxp,a,b,c)
      write (6,1020) xp,fxp
      stop
1000 format (' Quadratic direct fit polynomial'/' '/' x and f'/' ')
1010 format (i4,2f12.6)
1020 format (' '/' f(',f6.3,') = ',f12.6)
      end


      subroutine direct (ndim,ndeg,n,x,f,xp,fxp,a,b,c)
c     quadratic direct fit polynomial
      dimension x(ndim),f(ndim),a(ndim,ndim),b(ndim),c(ndim)
      do i=1,n
         a(i,1)=1.0
         a(i,2)=x(i)
         a(i,3)=x(i)**2
         b(i)=f(i)
      end do
      call gauss (ndim,n,a,b,c)
      write (6,1000) (c(i),i=1,n)
      fxp=c(1)+c(2)*xp+c(3)*xp**2
      return
1000 format (' '/' f(x) = ',e13.7,' + ',e13.7,' x + ',e13.7,' x**2')
      end


      subroutine gauss (ndim,n,a,b,x)
c     implements simple Gauss elimination
      end
```

The data set used to illustrate *subroutine direct* is taken from Example 4.2. The output generated by the direct fit polynomial program is presented below.

Output 4.1. Solution by a quadratic direct fit polynomial.

```
Quadratic direct fit polynomial

x and f

   1    3.350000    0.298507
   2    3.400000    0.294118
   3    3.500000    0.285714

f(x) = 0.8765607E+00 + -.2560800E+00 x + 0.2493333E-01 x**2

f( 3.440) =     0.290697
```

4.11.2. Lagrange Polynomials

The equation for the general Lagrange polynomial is given by Eq. (4.46). The quadratic Lagrange polynomial is given by Eq. (4.45):

$$P_2(x) = \frac{(x-b)(x-c)}{(a-b)(a-c)}f(a) + \frac{(x-a)(x-c)}{(b-a)(b-c)}f(b) + \frac{(x-a)(x-b)}{(c-a)(c-b)}f(c) \qquad (4.185)$$

A FORTRAN subroutine, *subroutine lagrange*, for implementing the *quadratic* Lagrange polynomial is presented below. *Program main* (Program 4.2) defines the data set and prints it, calls *subroutine lagrange* to implement the solution, and prints the solution. It shows only the statements which are different from the statements in Program 4.1.

Program 4.2.   Quadratic Lagrange polynomial program.

```
      program main
c     main program to illustrate polynomial fitting subroutines
      dimension x(6),f(6)
      call lagrange (ndim,n,x,f,xp,fxp)
 1000 format (' Quadratic Lagrange polynomial'/' '/' x and f'/' ')
      end

      subroutine lagrange (ndim,n,x,f,xp,fxp)
c     quadratic Lagrange polynomial
      dimension x(ndim),f(ndim)
      fp1=(xp-x(2))*(xp-x(3))/(x(1)-x(2))/(x(1)-x(3))*f(1)
      fp2=(xp-x(1))*(xp-x(3))/(x(2)-x(1))/(x(2)-x(3))*f(2)
      fp3=(xp-x(1))*(xp-x(2))/(x(3)-x(1))/(x(3)-x(2))*f(3)
      fxp=fp1+fp2+fp3
      return
      end
```

The data set used to illustrate *subroutine lagrange* is taken from Example 4.3. The output generated by the Lagrange polynomial program is presented in Output 4.2.

Output 4.2. Solution by a quadratic Lagrange polynomial.

```
Quadratic Lagrange polynomial

x and f

   1      3.350000      0.298507
   2      3.400000      0.294118
   3      3.500000      0.285714

f( 3.440) =      0.290697
```

4.11.3. Divided Difference Polynomial

The general formula for the divided difference polynomial is given by Eq. (4.65):

$$P_n(x) = f_i^{(0)} + (x - x_0)f_i^{(1)} + (x - x_0)(x - x_1)f_i^{(2)}$$
$$+ \cdots + (x - x_0)(x - x_1) \cdots (x - x_{n-1})f_i^{(n)} \tag{4.186}$$

A FORTRAN subroutine, *subroutine divdiff*, for implementing a *quadratic* divided difference polynomial is presented below. *Program main* (Program 4.3) defines the data set and prints it, calls *subroutine divdiff* to implement the solution, and prints the solution. It shows only the statements which are different from the statements in Program 4.1.

Program 4.3.   Quadratic divided difference polynomial program.

```
      program main
c     main program to illustrate polynomial fitting subroutines
      dimension x(6),f(6)
      call divdiff (ndim,n,x,f,xp,fxp)
 1000 format (' Quadratic divided diff. poly.'/' '/' x and f'/' ')
      end

      subroutine divdiff (ndim,n,x,f,xp,fxp)
c     quadratic divided difference polynomial
      dimension x(ndim),f(ndim)
      f11=(f(2)-f(1))/(x(2)-x(1))
      f21=(f(3)-f(2))/(x(3)-x(2))
      f12=(f21-f11)/(x(3)-x(1))
      fxp=f(1)+(xp-x(1))*f11+(xp-x(1))*(xp-x(2))*f12
      return
      end
```

The data set used to illustrate *subroutine divdiff* is taken from Example 4.6. The output generated by the divided difference polynomial program is presented in Output 4.3.

Output 4.3.   Solution by a quadratic divided difference polynomial.

```
Quadratic divided diff. polynomial

x and f

    1      3.350000      0.298507
    2      3.400000      0.294118
    3      3.500000      0.285714

f( 3.440) =      0.290697
```

4.11.4.   Newton Forward-Difference Polynomial

The general formula for the Newton forward-difference polynomial is given by Eq. (4.88):

$$P_n(x) = f_0 + s\,\Delta f_0 + \frac{s(s-1)}{2!}\,\Delta^2 f_0 + \frac{s(s-1)(s-2)}{3!}\,\Delta^3 f_0$$
$$+ \cdots + \frac{s(s-1)(s-2)\cdots[s-(n-1)]}{n!}\,\Delta^n f_0 \qquad (4.187)$$

A FORTRAN subroutine, *subroutine newtonfd*, for implementing a *quadratic* Newton forward difference polynomial is presented below. *Program main* (Program 4.4) defines the data set and prints it, calls *subroutine newtonfd* to implement the solution, and prints the solution. It shows only the statements which are different from the statements in Program 4.1.

Program 4.4.   Quadratic Newton forward-difference polynomial program.

```
      program main
c     main program to illustrate polynomial fitting subroutines
      dimension x(6),f(6)
      data (x(i),i=1,3) / 3.40, 3.50, 3.60 /
      data (f(i),i=1,3) / 0.294118, 0.285714, 0.27778 /
      call newtonfd (ndim,n,x,f,xp,fxp)
 1000 format (' Quadratic Newton FD polynomial'/' '/' ' x and f'/' ')
      end

      subroutine newtonfd (ndim,n,x,f,xp,fxp)
c     quadratic Newton forward-difference polynomial
      dimension x(ndim),f(ndim)
      d11=f(2)-f(1)
      d12=f(3)-f(2)
      d21=d12-d11
      s=(xp-x(1))/(x(2)-x(1))
      fxp=f(1)+s*d11+s*(s-1.0)/2.0*d21
      return
      end
```

The data set used to illustrate *subroutine newtonfd* is taken from Example 4.8. The output generated by the Newton forward-difference polynomial program is presented in Output 4.4.

Output 4.4.   Solution by quadratic Newton forward-difference polynomial.

```
Quadratic Newton FD polynomial

x and f

    1      3.400000      0.294118
    2      3.500000      0.285714
    3      3.600000      0.277780

f( 3.440) =      0.290700
```

4.11.5. Least Squares Polynomial

The normal equations for a general least squares polynomial are given by Eq. (4.158). The normal equations for a linear least squares polynomial are given by Eq. (4.149):

$$aN + b\sum_{i=1}^{N} x_i = \sum_{i=1}^{N} Y_i \tag{4.188a}$$

$$a\sum_{i=1}^{N} x_i + b\sum_{i=1}^{N} x_i^2 = \sum_{i=1}^{N} x_iY_i \tag{4.188b}$$

A FORTRAN subroutine, *subroutine lesq*, for implementing a *linear* least squares polynomial is presented below. *Program main* (Program 4.5) defines the data set and prints it, calls *subroutine lesq* to implement the solution, and prints the solution.

Program 4.5.   Linear least squares polynomial program.

```
      program main
c     main program to illustrate subroutine lesq
c     ndim  array dimension, ndim = 10 in this example
c     ndeg  degree of the polynomial
c     n     number of data points and polynomial coefficients
c     x     values of the independent variable, x(i)
c     f     values of the dependent variable, f(i)
c     c     coefficients of the least squares polynomials, c(i)
      dimension x(10),f(10),a(10,10),b(10),c(10)
      data ndim,ndeg,n / 10, 1, 8 /
      data (x(i), i=1,8) / 300.0, 400.0, 500.0, 600.0, 700.0,
     1 800.0, 900.0, 1000.0 /
      data (f(i), i=1,8) / 1.0045, 1.0134, 1.0296, 1.0507, 1.0743,
     1 1.0984, 1.1212, 1.1410 /
      write (6,1000)
      call lesq (ndim,ndeg,n,x,f,a,b,c)
      write (6,1010)
      do i=1,n
         fi=c(1)+c(2)*x(i)
         error=100.0*(f(i)-fi)
         write (6,1020) i,x(i),f(i),fi,error
      end do
      stop
 1000 format (' Linear least squares polynomial')
 1010 format (' '/'   i',5x,'x',9x,'f',9x,'fi',7x,'error'/' ')
 1020 format (i4,f10.3,2f10.4,f9.2)
      end

      subroutine lesq (ndim,ndeg,n,x,f,a,b,c)
c     linear least squares polynomial
      dimension x(ndim),f(ndim),a(ndim,ndim),b(ndim),c(ndim)
      sumx=0.0
      sumxx=0.0
      sumf=0.0
      sumxf=0.0
```

```
    do i=1,n
       sumx=sumx+x(i)
       sumxx=sumxx+x(i)**2
       sumf=sumf+f(i)
       sumxf=sumxf+x(i)*f(i)
    end do
    a(1,1)=float(n)
    a(1,2)=sumx
    a(2,1)=sumx
    a(2,2)=sumxx
    b(1)=sumf
    b(2)=sumxf
    call gauss (ndim,ndeg+1,a,b,c)
    write (6,1000) (c(i),i=1,ndeg+1)
    return
1000 format (' '/' f(x) = ',e13.7,' + ',e13.7,' x')
    end

    subroutine gauss (ndim,n,a,b,x)
c   implements simple gauss elimination
    end
```

The data set used to illustrate *subroutine lesq* is taken from Example 4.14. The output generated by the linear least squares polynomial program is presented in Output 4.5.

Output 4.5.   Solution by a linear least squares polynomial.

```
Linear least squares polynomial

f(x) = 0.9331940E+00 + 0.2052976E-03 x
```

i	x	f	fi	error
1	300.000	1.0045	0.9948	0.97
2	400.000	1.0134	1.0153	-0.19
3	500.000	1.0296	1.0358	-0.62
4	600.000	1.0507	1.0564	-0.57
5	700.000	1.0743	1.0769	-0.26
6	800.000	1.0984	1.0974	0.10
7	900.000	1.1212	1.1180	0.32
8	1000.000	1.1410	1.1385	0.25

4.11.6.   Packages for Polynomial Approximation

Numerous libraries and software packages are available for polynomial approximation and interpolation. Many workstations and mainframe computers have such libraries attached to their operating systems.

Many commercial software packages contain routines for fitting approximating polynomials and interpolation. Some of the more prominent packages are Matlab and Mathcad. More sophisticated packages, such as IMSL, Mathematica, Macsyma, and

Maple, also contain routines for fitting approximating polynomials and interpolation. Finally, the book *Numerical Recipes*, Pross et al. (1989) contains numerous subroutines for for fitting approximating polynomials and interpolation.

4.12 SUMMARY

Procedures for developing approximating polynomials for discrete data are presented in this chapter. For small sets of smooth data, exact fits are desirable. The direct fit polynomial, the Lagrange polynomial, and the divided difference polynomial work well for nonequally spaced data. For equally spaced data, polynomials based on differences are recommended. The Newton forward-difference and backward-difference polynomials are simple to fit and evaluate. These two polynomials are used extensively in Chapters 5 and 6 to develop procedures for numerical differentiation and numerical integration, respectively. Procedures are discussed for inverse interpolation and multivariate approximation.

Procedures for developing least squares approximations for discrete data also are presented in this chapter. Least squares approximations are useful for large sets of data and sets of rough data. Least squares polynomial approximation is straightforward, for both one independent variable and more than one independent variable. The least squares normal equations corresponding to polynomial approximating functions are linear, which leads to very efficient solution procedures. For nonlinear approximating functions, the least squares normal equations are nonlinear, which leads to complicated solution procedures. Least squares polynomial approximation is a straightforward, simple, and accurate procedure for obtaining approximating functions for large sets of data or sets of rough data.

After studying Chapter 4, you should be able to:

1. Discuss the general features of functional approximation
2. List the uses of functional approximation
3. List several types of approximating functions
4. List the required properties of an approximating function
5. Explain why polynomials are good approximating functions
6. List the types of discrete data which influence the type of fit required
7. Describe the difference between exact fits and approximate fits
8. Explain the significance of the Weirstrass approximation theorem
9. Explain the significance of the uniqueness theorem for polynomials
10. List the Taylor series and the Taylor polynomial
11. Discuss the error term of a polynomial
12. Evaluate, differentiate, and integrate a polynomial
13. Explain and apply the nested multiplication algorithm
14. State and apply the division algorithm, the remainder theorems, and the factor theorem
15. Explain and apply the synthetic division algorithm
16. Explain and implement polynomial deflation
17. Discuss the advantages and disadvantages of direct fit polynomials
18. Fit a direct fit polynomial of any degree to a set of tabular data
19. Explain the concept underlying Lagrange polynomials
20. Discuss the advantages and disadvantages of Lagrange polynomials
21. Fit a Lagrange polynomial of any degree to a set of tabular data

22. Discuss the advantages and disadvantages of Neville's algorithm
23. Apply Neville's algorithm
24. Define a divided difference and construct a divided difference table
25. Discuss the advantages and disadvantages of divided difference polynomials
26. Construct a divided difference polynomial
27. Define a difference and construct a difference table
28. Discuss the effects of round-off on a difference table
29. Discuss the advantages and disadvantages of difference polynomials.
30. Apply the Newton forward-difference polynomial
31. Apply the Newton backward-difference polynomial
32. Discuss the advantages of centered difference polynomials
33. Discuss and apply inverse interpolation
34. Describe approaches for multivariate approximation
35. Apply successive univariate polynomial approximation
36. Apply direct multivariate polynomial approximation
37. Describe the concepts underlying cubic splines
38. List the advantages and disadvantages of cubic splines
39. Apply cubic splines to a set of tabular data
40. Discuss the concepts underlying least squares approximation
41. List the advantages and disadvantages of least squares approximation
42. Derive the normal equations for a least squares polynomial of any degree
43. Apply least squares polynomials to fit a set of tabular data
44. Derive and solve the normal equations for a least squares approximation of a nonlinear function
45. Apply the computer programs presented in Section 4.11 to fit polynomials and to interpolate
46. Develop computer programs for applications not implemented in Section 4.11.

EXERCISE PROBLEMS

4.2 Properties of Polynomials

1. For the polynomial $P_3(x) = x^3 - 9x^2 + 26x - 24$, calculate (a) $P_3(1.5)$ by nested multiplication, (b) $P_3'(1.5)$ by constructing $Q_2(x)$ and evaluating $Q_2(1.5)$ by nested multiplication, and (c) the deflated polynomial $Q_2(x)$ obtained by removing the factor $(x - 2)$.
2. Work Problem 1 for $x = 2.5$ and remove the factor $(x - 3)$.
3. Work Problem 1 for $x = 3.5$ and remove the factor $(x - 4)$.
4. For the polynomial $P_4(x) = x^4 - 10x^3 + 35x^2 - 50x + 24$, calculate (a) $P_4(1.5)$ by nested multiplication, (b) $P_4'(1.5)$ by constructing $Q_3(x)$ and evaluating $Q_3(1.5)$ by nested multiplication, and (c) the deflated polynomial $Q_3(x)$ obtained by removing the factor $(x - 1)$.
5. Work Problem 4 for $x = 2.5$ and remove the factor $(x - 2)$.
6. Work Problem 4 for $x = 3.5$ and remove the factor $(x - 3)$.
7. Work Problem 4 for $x = 4.5$ and remove the factor $(x - 4)$.
8. For the polynomial $P_5(x) = x^5 - 20x^4 + 155x^3 - 580x^2 + 1044x - 720$, calculate (a) $P_5(1.5)$ by nested multiplication, (b) $P_5'(1.5)$ by constructing $Q_4(x)$

and evaluating $Q_4(1.5)$ by nested multiplication, and (c) the deflated polynomial $Q_4(x)$ obtained by removing the factor $(x - 2)$.

9. Work Problem 8 for $x = 2.5$ and remove the factor $(x - 3)$.
10. Work Problem 8 for $x = 3.5$ and remove the factor $(x - 4)$.
11. Work Problem 8 for $x = 4.5$ and remove the factor $(x - 5)$.
12. Work Problem 8 for $x = 5.5$ and remove the factor $(x - 6)$.

4.3 Direct Fit Polynomials

Table 1 is for the function $f(x) = 2/x + x^2$. This table is considered in several of the problems which follow.

Table 1. Tabular Data.

x	$f(x)$	x	$f(x)$	x	$f(x)$
0.4	5.1600	1.4	3.3886	2.2	5.7491
0.6	3.6933	1.6	3.8100	2.4	6.5933
0.8	3.1400	1.8	4.3511	2.6	7.5292
1.0	3.0000	2.0	5.0000	2.8	8.5543
1.2	3.1067				

13. The order n of an approximating polynomial specifies the rate at which the error of the polynomial approximation approaches zero as the increment in the tabular data approaches zero, that is, Error $= C \Delta x^n$. Estimate the order of a linear direct fit polynomial by calculating $f(2.0)$ for the data in Table 1 using $x = 1.6$ and 2.4 (i.e., $\Delta x = 0.8$), and $x = 1.8$ and 2.2 (i.e., $\Delta x = 0.4$), calculating the errors, and calculating the ratio of the errors.

14. Consider the tabular set of generic data shown in Table 2: Determine the direct fit quadratic polynomials for the tabular set of generic data for (a) $\Delta x_- = \Delta x_+ = \Delta x$ and (b) $\Delta x_- \neq \Delta x_+$.

Table 2. Generic Data

x	$f(x)$
$x_{i-1} = -\Delta x_-$	f_{i-1}
$x_i = 0$	f_i
$x_{i+1} = +\Delta x_+$	f_{i+1}

15. The formal order of a direct fit polynomial $P_n(x)$ can be determined by expressing all function values in the polynomial (i.e., f_{i-1}, f_{i+1}, etc.) in terms of a Taylor series at the base point and comparing that result to the Taylor series for $f(x)$ at the base point. For the direct fit polynomials developed in Problem 14, show that the order is $0(\Delta x^3)$ for part (a) and $0(\Delta x_-^2) + 0(\Delta x_+^2)$ for part (b).

16. Consider the data in the range $0.4 \leq x \leq 1.2$ in Table 1. Using direct fit polynomials, calculate (a) $P_2(0.9)$ using the first three points, (b) $P_2(0.9)$ using the last three points, (c) $P_3(0.9)$ using the first four points, (d) $P_3(0.9)$ using the last four points, and (e) $P_4(0.9)$ using all five data points.

17. Work Problem 16 for the range $1.2 \leq x \leq 2.0$ for $x = 1.5$.

18. Work Problem 16 for the range $2.0 \le x \le 2.8$ for $x = 2.5$.
19. The constant pressure specific heat C_p and enthalpy h of low pressure air are tabulated in Table 3. Using direct fit polynomials with the base point as close to the specified value of T as possible, calculate (a) $C_p(1120)$ using two points, (b) $C_p(1120)$ using three points, (c) $C_p(1480)$ using two points, and (d) $C_p(1480)$ using three points.

Table 3. Properties of Air

T, K	C_p, kJ/kg-K	h, kJ/kg	T, K	C_p, kJ/kg-K	h, kJ/kg
1000	1.1410	1047.248	1400	1.1982	1515.792
1100	1.1573	1162.174	1500	1.2095	1636.188
1200	1.1722	1278.663	1600	1.2197	1757.657
1300	1.1858	1396.578			

20. Work Problem 19 for $h(T)$ instead of $C_p(T)$.

4.4 Lagrange Polynomials

21. Work Problem 16 using Lagrange polynomials.
22. Work Problem 17 using Lagrange polynomials.
23. Work Problem 18 using Lagrange polynomials.
24. Work Problem 19 using Lagrange polynomials.
25. Work Problem 20 using Lagrange polynomials.
26. Work Problem 16 using Neville's algorithm.
27. Work Problem 17 using Neville's algorithm.
28. Work Problem 18 using Neville's algorithm.
29. Work Problem 19 using Neville's algorithm.
30. Work Problem 20 using Neville's algorithm.

4.5 Divided Difference Tables and Divided Difference Polynomials

Divided Difference Tables

31. Construct a six-place divided difference table for the function $f(x) = x^3 - 9x^2 + 26x - 24$ in the range $1.1 \le x \le 1.9$ for $x = 1.10$, 1.15, 1.20, 1.30. 1.35, 1.45, 1.6. 1.75, and 1.90.
32. Construct a divided difference table for the data in Table 1.
33. Construct a divided difference table for $C_p(T)$ for the data presented in Table 3.
34. Construct a divided difference table for $h(T)$ for the data presented in Table 3.

Divided Difference Polynomials

35. From the divided difference table constructed in Problem 31, interpolate for $f(1.44)$ using: (a) two points, (b) three points, and (c) four points. In each case use the closest points to $x = 1.44$.
36. Using the divided difference table constructed in Problem 32, evaluate $f(0.9)$ using: (a) two points, (b) three points, and (c) four points. In each case use the closest points to $x = 0.9$.

37. From the divided difference table constructed in Problem 33, calculate $C_p(1120)$ using: (a) two points, (b) three points, and (c) four points. In each case use the closest points to $T = 1120\,\text{K}$.

38. From the divided difference table constructed in Problem 34, calculate $h(1120)$ using: (a) two points, (b) three points, and (c) four points. In each case use the closest points to $T = 1120\,\text{K}$.

4.6 Difference Tables and Difference Polynomials

Difference Tables

39. Construct a six-place difference table for the function $f(x) = x^3 - 9x^2 + 26x - 24$ in the range $1.1 \le x \le 1.9$ for $\Delta x = 0.1$. Discuss the results. Analyze the effects of round off.

40. Construct a difference table for the data in Table 1. Discuss the results. Analyze the effects of round off. Comment on the degree of polynomial required to approximate these data at the beginning, middle, and end of the table.

41. Construct a difference table for $C_p(T)$ for the data presented in Table 3. Discuss the results. Analyze the effects of round-off. What degree of polynomial is required to approximate this set of data?

42. Work Problem 41 for $h(T)$.

The Newton Forward-Difference Polynomial

43. Work Problem 16 using Newton forward-difference polynomials.
44. Work Problem 17 using Newton forward-difference polynomials.
45. Work Problem 18 using Newton forward-difference polynomials.
46. Work Problem 19 using Newton forward-difference polynomials.
47. Work Problem 20 using Newton forward-difference polynomials.

The Newton Backward-Difference Polynomial

48. Work Problem 16 using Newton backward-difference polynomials.
49. Work Problem 17 using Newton backward-difference polynomials.
50. Work Problem 18 using Newton backward-difference polynomials.
51. Work Problem 19 using Newton backward-difference polynomials.
52. Work Problem 20 using Newton backward-difference polynomials.
53. For the data in Table 3 in the temperature range $1200 \le T \le 1400$, construct a quadratic direct fit polynomial, a quadratic Newton forward-difference polynomial, and a quadratic Newton backward-difference polynomial for $C_p(T)$. Rewrite both Newton polynomials in the form $C_p(T) = a + bT + CT^2$, and show that the three polynomials are identical.
54. Work Problem 53 including a quadratic Lagrange polynomial.

Other Difference Polynomials

55. Work Problem 16 using the Stirling centered-difference polynomial.
56. Work Problem 17 using the Stirling centered-difference polynomial.
57. Work Problem 18 using the Stirling centered-difference polynomial.
58. Work Problem 19 using the Stirling centered-difference polynomial.
59. Work Problem 20 using the Stirling centered-difference polynomial.

60. Work Problem 16 using the Bessel centered-difference polynomial.
61. Work Problem 17 using the Bessel centered-difference polynomial.
62. Work Problem 18 using the Bessel centered-difference polynomial.
63. Work Problem 19 using the Bessel centered-difference polynomial.
64. Work Problem 20 using the Bessel centered-difference polynomial.

4.8 Multivariate Interpolation

65. The specific volume v (m^3/kg) of steam, corresponding to the van der Waal equation of state (see Problem 3.69), as a function of pressure P (kN/m^2) and temperature T (K) in the neighborhood of $P = 10,000$ kN/m^2 and $T = 800$ K, is tabulated in Table 4.

Table 4. Specific Volume of Steam

	T, K		
P, kN/m^2	700	800	900
9,000	0.031980	0.037948	0.043675
10,000	0.028345	0.033827	0.039053
11,000	0.025360	0.030452	0.035270

Use successive quadratic univariate interpolation to calculate $v(9500,750)$. The exact value from the steam tables is $v = 0.032965$ m^3/kg.

66. Work Problem 65 for $v(9500,850)$. The exact value is 0.038534 m^3/kg.
67. Work Problem 65 for $v(10500,750)$. The exact value is 0.029466 m^3/kg.
68. Work Problem 65 for $v(10500,850)$. The exact value is 0.034590 m^3/kg.
69. Solve Problem 65 by direct linear bivariate interpolation for $v(9500.750)$:

$$v = a + bT + cP + dPT$$

70. Work Problem 69 for $v(9500,850)$.
71. Work Problem 69 for $v(10500,750)$.
72. Work Problem 69 for $v(10500,850)$.
73. Solve Problem 65 by direct quadratic bivariate interpolation for $v(9500,750)$:

$$v = a + bT + cP + dPT + eT^2 + fP^2$$

74. Work Problem 73 for $v(9500,850)$.
75. Work Problem 73 for $v(10500,750)$.
76. Work Problem 73 for $v(10500,850)$.

4.10 Least Squares Approximation

77. Consider the data for the specific heat C_p of air, presented in Table 3 for the range $1,000 \leq T \leq 1,400$. Find the best straight line approximation to this set of data. Compute the deviations at each data point.
78. Work Problem 77 using every other data point. Compare the results with the results of Problem 77.
79. Work Problem 77 for a quadratic polynomial. Compare the results with the results of Problem 77.

LIVERPOOL JOHN MOORES UNIVERSITY
LEARNING SERVICES

80. Consider the data for the specific volume of steam, $v = v(P, T)$, given in Table 4. Develop a least squares linear bivariate polynomial for the set of data in the form

$$v = a + bT + cP + dPT$$

Compute the derivation at each data point. Calculate $v(9500,750)$ and compare with the result from Problem 69.

81. Work Problem 80 for the least squares quadratic bivariate polynomial and compute the deviations.

$$v = a + bT + cP + dPT + eT^2 + f P^2$$

Compare the result with the result from Problem 73.

82. Fit the $C_p(T)$ data in Table 3 to a fourth-degree polynomial and compute the deviations.

$$C_p(T) = a + bT + cT^2 + dT^3 + eT^4$$

4.11 Programs

83. Implement the quadratic direct fit polynomial program presented in Section 4.11.1. Check out the program with the given data.
84. Solve any of Problems 16 to 20 with the program.
85. Modify the quadratic direct fit polynomial program to consider a linear direct fit polynomial. Solve any of Problems 16 to 20 with the modified program.
86. Modify the quadratic direct fit polynomial program to consider a cubic direct fit polynomial. Solve any of Problems 16 to 20 with the modified program.
87. Implement the quadratic Lagrange polynomial program presented in Section 4.11.2. Check out the program with the given data.
88. Solve any of Problems 16 to 20 with the program.
89. Modify the quadratic Lagrange polynomial program to consider a linear Lagrange polynomial. Solve any of Problems 16 to 20 with the modified program.
90. Modify the quadratic Lagrange polynomial program to consider a cubic Lagrange polynomial. Solve any of Problems 16 to 20 with the modified program.
91. Implement the quadratic divided difference polynomial program presented in Section 4.11.3. Check out the program with the given data.
92. Solve any of Problems 16 to 20 with the program.
93. Modify the quadratic divided difference polynomial program to consider a linear divided difference polynomial. Solve any of Problems 16 to 20 with the modified program.
94. Modify the quadratic divided difference polynomial program to consider a cubic divided difference polynomial. Solve any of Problems 16 to 20 with the modified program.
95. Implement the quadratic Newton forward-difference polynomial program presented in Section 4.11.4. Check out the program with the given data.
96. Solve any of Problems 16 to 20 with the program.
97. Modify the quadratic Newton forward-difference polynomial program to consider a linear Newton forward-difference polynomial. Solve any of Problems 16 to 20 with the modified program.

98. Modify the quadratic Newton forward-difference polynomial program to consider a cubic Newton forward-difference polynomial. Solve any of Problems 16 to 20 with the modified program A.

99. Implement the linear least squares polynomial program presented in Section 4.11.5. Check out the program with the given data.

100. Solve Problem 77 or 78 with the program.

101. Extend the linear least squares polynomial program to consider a quadratic least squares polynomial. Solve Problem 77 using the program.

102. Extend the linear least squares polynomial program to consider a fourth-degree least squares polynomial. Solve Problem 82 using the program.

103. Modify the linear least squares polynomial program to consider a linear bivariate least squares polynomial. Solve Problem 80 with the program.

104. Modify the linear least squares polynomial program to consider a quadratic bivariate least squares polynomial. Solve Problem 81 with the program.

APPLIED PROBLEMS

105. When an incompressible fluid flows steadily through a round pipe, the pressure drop ΔP due to friction is given by

$$\Delta P = -0.5 f \rho V^2 (L/D)$$

where ρ is the fluid density, V is the velocity, L/D is the pipe length-to-diameter ratio, and f is the D'Arcy friction coefficient. For laminar flow, the friction coefficient f can be related to the Reynolds number, Re, by a relationship of the form

$$f = a \, \mathrm{Re}^b$$

Use the measured data in Table 5 to determine a and b by a least squares fit.

Table 5. Friction Coefficient

Re	500	1000	1500	2000
f	0.0320	0.0160	0.0107	0.0080

106. Reaction rates for chemical reactions are usually expressed in the form

$$K = BT^\alpha \exp\left(\frac{-E}{RT}\right)$$

For a particular reaction, measured values of the forward and backward reaction rates K_f and K_b, respectively, are given by

Table 6. Reaction Rates

T, K	K_f	K_b
1000	7.5 E + 15	4.6 E + 07
2000	3.8 E + 15	5.9 E + 04
3000	2.5 E + 15	2.5 E + 08
4000	1.9 E + 15	1.4 E + 10
5000	1.5 E + 15	1.5 E + 11

(a) Determine B and α for the backward reaction rate K_b for which $E/R = 0$.

(b) Determine B, α and E/R for the forward reaction rate K_f.

107. The data in Table 1 can be fit by the expression

$$f = \frac{a}{x} + bx^2$$

Develop a least squares procedure to determine a and b. Solve for a and b and compute the deviations.

5

Numerical Differentiation and Difference Formulas

5.1 INTRODUCTION

Figure 5.1 presents a set of tabular data in the form of a set of $[x, f(x)]$ pairs. The function $f(x)$ is known only at discrete values of x. Interpolation within a set of discrete data is discussed in Chapter 4. *Differentiation* within a set of discrete data is presented in this chapter. The discrete data presented in Figure 5.1 are values of the function $f(x) = 1/x$, which are used as the example problem in this chapter.

The evaluation of a derivative is required in many problems in engineering and science:

$$\boxed{\frac{d}{dx}(f(x)) = f'(x) = f_x(x)} \tag{5.1}$$

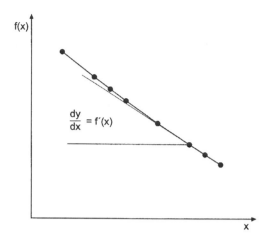

x	f(x)
3.20	0.312500
3.30	0.303030
3.35	0.298507
3.40	0.294118
3.50	0.285714
3.60	0.277778
3.65	0.273973
3.70	0.270270

Figure 5.1   Differentiation of tabular data.

where the alternate notations $f'(x)$ and $f_x(x)$ are used for the derivatives. The function $f(x)$, which is to be differentiated, may be a know function or a set of discrete data. In general, known functions can be differentiated exactly. Differentiation of discrete data, however, requires an approximate numerical procedure. The evaluation of derivatives by approximate numerical procedures is the subject of this chapter.

Numerical differentiation formulas can be developed by fitting approximating functions (e.g., polynomials) to a set of discrete data and differentiating the approximating function. Thus,

$$\frac{d}{dx}(f(x)) \cong \frac{d}{dx}(P_n(x)) \qquad (5.2)$$

This process is illustrated in Figure 5.2. As illustrated in Figure 5.2, even though the approximating polynomial $P_n(x)$ passes through the discrete data points exactly, the derivative of the polynomial $P_n'(x)$ may not be a very accurate approximation of the derivative of the exact function $f(x)$ even at the known data points themselves. In general, numerical differentiation is an inherently inaccurate process.

To perform numerical differentiation, an approximating polynomial is fit to the discrete data, or a subset of the discrete data, and the approximating polynomial is

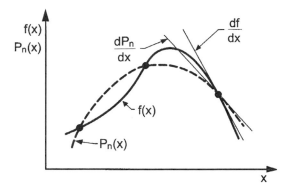

Figure 5.2 Numerical differentiation.

differentiated. The polynomial may be fit exactly to a set of discrete data by the methods presented in Sections 4.3 to 4.9, or approximately by a least squares fit as described in Section 4.10. In both cases, the degree of the approximating polynomial chosen to represent the discrete data is the only parameter under our control.

Several numerical differentiation procedures are presented in this chapter. Differentiation of direct fit polynomials, Lagrange polynomials, and divided difference polynomials can be applied to both unequally spaced data and equally spaced data. Differentiation formulas based on both Newton forward-difference polynomials and Newton backward-difference polynomials can be applied to equally spaced data. Numerical differentiation formulas can also be developed using Taylor series. This approach is quite useful for developing difference formulas for approximating exact derivatives in the numerical solution of differential equations.

The simple function

$$f(x) = \frac{1}{x} \tag{5.3}$$

which has the exact derivatives

$$\frac{d}{dx}\left(\frac{1}{x}\right) = f'(x) = -\frac{1}{x^2} \tag{5.4a}$$

$$\frac{d^2}{dx^2}\left(\frac{1}{x}\right) = f''(x) = \frac{2}{x^3} \tag{5.4b}$$

is considered in this chapter to illustrate numerical differentiation procedures. In particular, at $x = 3.5$:

$$f'(3.5) = -\frac{1}{(3.5)^2} = -0.081633\ldots \tag{5.5a}$$

$$f''(3.5) = \frac{2}{(3.5)^3} = 0.046647\ldots \tag{5.5b}$$

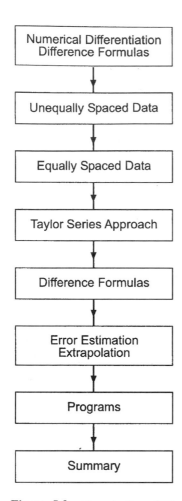

Figure 5.3   Organization of Chapter 5.

The organization of Chapter 5 is illustrated in Figure 5.3. Following the introductory discussion in this section, differentiation using direct fit polynomials, Lagrange polynomials, and divided difference polynomials as the approximating function is discussed. The development of differentiation formulas based on Newton difference polynomials is presented in Section 5.3. Section 5.4 develops difference formulas directly from the Taylor series. A table of difference formulas is presented in Section 5.5. A discussion of error estimation and extrapolation is presented in Section 5.6. Several programs for differentiating tabular data numerically are presented in Section 5.7. The chapter closes with a Summary which includes a list of what you should be able to do after studying Chapter 5.

5.2 UNEQUALLY SPACED DATA

Three straightforward numerical differentiation procedures that can be used for both unequally spaced data and equally spaced data are presented in this section:

1. Direct fit polynomials
2. Lagrange polynomials
3. Divided difference polynomials

5.2.1 Direct Fit Polynomials

A direct fit polynomial procedure is based on fitting the data directly by a polynomial and differentiating the polynomial. Recall the direct fit polynomial, Eq. (4.34):

$$P_n(x) = a_0 + a_1 x + a_2 x^2 + \cdots + a_n x^n \tag{5.6}$$

where $P_n(x)$ is determined by one of the following methods:

1. Given $N = n + 1$ points, $[x_i, f(x_i)]$, determine the exact nth-degree polynomial that passes through the data points, as discussed in Section 4.3.
2. Given $N > n + 1$ points, $[x_i, f(x_i)]$, determine the least squares nth-degree polynomial that best fits the data points, as discussed in Section 4.10.3.

After the approximating polynomial has been fit, the derivatives are determined by differentiating the approximating polynomial. Thus,

$$f'(x) \cong P_n'(x) = a_1 + 2a_2 x + 3a_3 x^2 + \cdots \tag{5.7a}$$

$$f''(x) \cong P_n''(x) = 2a_2 + 6a_3 x + \cdots \tag{5.7b}$$

Equations (5.7a) and (5.7b) are illustrated in Example 5.1.

5.2.2. Lagrange Polynomials

The second procedure that can be used for both unequally spaced data and equally spaced data is based on differentiating a Lagrange polynomial. For example, consider the second-degree Lagrange polynomial, Eq. (4.45):

$$P_2(x) = \frac{(x-b)(x-c)}{(a-b)(a-c)} f(a) + \frac{(x-a)(x-c)}{(b-a)(b-c)} f(b) + \frac{(x-a)(x-b)}{(c-a)(c-b)} f(c) \tag{5.8}$$

Differentiating Eq. (5.8) yields:

$$f'(x) \cong P_2'(x) = \frac{2x-(b+c)}{(a-b)(a-c)} f(a) + \frac{2x-(a+c)}{(b-a)(b-c)} f(b) + \frac{2x-(a+b)}{(c-a)(c-b)} f(c) \tag{5.9a}$$

Differentiating Eq. (5.9a) yields:

$$f''(x) \cong P_2''(x) = \frac{2f(a)}{(a-b)(a-c)} + \frac{2f(b)}{(b-a)(b-c)} + \frac{2f(c)}{(c-a)(c-b)} \tag{5.9b}$$

Equations (5.9a) and (5.9b) are illustrated in Example 5.1.

5.2.3. Divided Difference Polynomials

The third procedure that can be used for both unequally spaced data and equally spaced data is based on differentiating a divided difference polynomial, Eq. (4.65):

$$P_n(x) = f_i^{(0)} + (x-x_0)f_i^{(1)} + (x-x_0)(x-x_1)f_i^{(2)} + (x-x_0)(x-x_1)(x-x_2)f_i^{(3)} + \cdots \tag{5.10}$$

Differentiating Eq. (5.10) gives

$$f'(x) \cong P'_n(x) = f_i^{(1)} + [2x - (x_0 + x_1)]f_i^{(2)}$$
$$+ [3x^2 - 2(x_0 + x_1 + x_2)x + (x_0x_1 + x_0x_2 + x_1x_2)]f_i^{(3)} + \cdots \qquad (5.11a)$$

Differentiating Eq. (5.11a) gives

$$f''(x) \cong P''_n(x) = 2f_i^{(2)} + [6x - 2(x_0 + x_1 + x_2)]f_i^{(3)} + \cdots \qquad (5.11b)$$

Equations (5.10a) and (5.10b) are illustrated in Example 5.1.

Example 5.1.   Direct fit, Lagrange, and divided difference polynomials.

Let's solve the example problem presented in Section 5.1 by the three procedures presented above. Consider the following three data points:

x	$f(x)$
3.4	0.294118
3.5	0.285714
3.6	0.277778

First, fit the quadratic polynomial, $P_2(x) = a_0 + a_1x + a_2x^2$, to the three data points:

$$0.294118 = a_0 + a_1(3.4) + a_2(3.4)^2 \qquad (5.12a)$$

$$0.285714 = a_0 + a_1(3.5) + a_2(3.5)^2 \qquad (5.12b)$$

$$0.277778 = a_0 + a_1(3.6) + a_2(3.6)^2 \qquad (5.12c)$$

Solving for a_0, a_1, and a_2 by Gauss elimination gives $a_0 = 0.858314$, $a_1 = -0.245500$, and $a_2 = 0.023400$. Substituting these values into Eqs. (5.7a) and (5.7b) and evaluating at $x = 3.5$ yields the solution for the direct fit polynomial:

$$P'_2(3.5) = -0.245500 + (0.04680)(3.5) = -0.081700 \qquad (5.12d)$$

$$P''_2(x) = 0.046800 \qquad (5.12e)$$

Substituting the tabular values into Eqs. (5.9a) and (5.9b) and evaluating at $x = 3.5$ yields the solution for the Lagrange polynomial:

$$P'_2(3.5) = \frac{2(3.5) - (3.5 + 3.6)}{(3.4 - 3.5)(3.4 - 3.6)}(0.294118) + \frac{2(3.5) - (3.4 + 3.6)}{(3.5 - 3.4)(3.5 - 3.6)}(0.285714)$$

$$+ \frac{2(3.5) - (3.4 + 3.5)}{(3.6 - 3.4)(3.6 - 3.5)}(0.277778) = -0.081700 \qquad (5.13a)$$

$$P''_2(3.5) = \frac{2(0.294118)}{(3.4 - 3.5)(3.4 - 3.6)} + \frac{2(0.285714)}{(3.5 - 3.4)(3.5 - 3.6)} + \frac{2(0.277778)}{(3.6 - 3.4)(3.6 - 3.5)}$$

$$= 0.046800 \qquad (5.13b)$$

A divided difference table must be constructed for the tabular data to use the divided difference polynomial. Thus,

x_i	$f_i^{(0)}$	$f_i^{(1)}$	$f_i^{(2)}$
3.4	0.294118		
		-0.084040	
3.5	0.285714		0.023400
		-0.079360	
3.6	0.277778		

Substituting these values into Eqs. (5.11a) and (5.11b) yields the solution for the divided difference polynomial:

$$P_2'(3.5) = -0.084040 + [2(3.5) - (3.4 + 3.5)](0.023400) = -0.081700 \qquad (5.14a)$$

$$P_2''(3.5) = 2(0.023400) = 0.046800 \qquad (5.14b)$$

The results obtained by the three procedures are identical since the same three points are used in all three procedures. The error in $f'(3.5)$ is Error $= f'(3.5) - P_2'(3.5) = -0.081700 - (-0.081633) = -0.000067$, and the error in $f''(3.5)$ is Error $= f''(3.5) - P_2''(3.5) = 0.046800 - (0.046647) = 0.000153$.

5.3 EQUALLY SPACED DATA

When the tabular data to be differentiated are known at equally spaced points, the Newton forward-difference and backward-difference polynomials, presented in Section 4.6, can be fit to the discrete data with much less effort than a direct fit polynomial, a Lagrange polynomial, or a divided difference polynomial. This can significantly decrease the amount of effort required to evaluate derivatives. Thus,

$$\boxed{f'(x) \cong \frac{d}{dx}(P_n(x)) = P_n'(x)} \qquad (5.15)$$

where $P_n(x)$ is either the Newton forward-difference or backward-difference polynomial.

5.3.1. Newton Forward-Difference Polynomial

Recall the Newton forward-difference polynomial, Eq. (4.88):

$$P_n(x) = f_0 + s\,\Delta f_0 + \frac{s(s-1)}{2}\Delta^2 f_0 + \frac{s(s-1)(s-2)}{6}\Delta^3 f_0 + \cdots + \text{Error} \qquad (5.16)$$

$$\text{Error} = \binom{s}{n+1} h^{n+1} f^{(n+1)}(\xi) \qquad x_0 \le \xi \le x_n \qquad (5.17)$$

where the interpolating parameter s is given by

$$s = \frac{x - x_0}{h} \qquad \rightarrow \qquad x = x_0 + sh \qquad (5.18)$$

Equation (5.15) requires that the approximating polynomial be an explicit function of x, whereas Eq. (5.16) is implicit in x. Either Eq. (5.16) must be made explicit in x by introducing Eq. (5.18) into Eq. (5.16), or the differentiation operations in Eq. (5.15) must be transformed into explicit operations in terms of s, so that Eq. (5.16) can be used directly.

The first approach leads to a complicated procedure, so the second approach is taken. From Eq. (5.18), $x = x(s)$. Thus,

$$f'(x) \cong \frac{d}{dx}(P_n(x)) = P'_n(x) = \frac{d}{ds}(P_n(s))\frac{ds}{dx} \tag{5.19}$$

From Eq. (5.18),

$$\frac{ds}{dx} = \frac{1}{h} \tag{5.20}$$

Thus, Eq. (5.19) gives

$$f'(x) \cong P'_n(x) = \frac{1}{h}\frac{d}{ds}(P_n(s)) \tag{5.21}$$

Substituting Eq. (5.16) into Eq. (5.21) and differentiating gives

$$P'_n(x) = \frac{1}{h}\{\Delta f_0 + \frac{1}{2}[(s-1) + s]\Delta^2 f_0 + \frac{1}{6}[(s-1)(s-2) + s(s-2)$$
$$+ s(s-1)]\Delta^3 f_0 + \cdots\}$$

Simplifying Eq. (5.22) yields

$$P'_n(x) = \frac{1}{h}\left(\Delta f_0 + \frac{2s-1}{2}\Delta^2 f_0 + \frac{3s^2 - 6s + 2}{6}\Delta^3 f_0 + \cdots\right) \tag{5.23}$$

The second derivative is obtained as follows:

$$f''(x) \cong \frac{d}{dx}(P'_n(x)) = P''_n(x) = \frac{d}{ds}(P'_n(s))\frac{ds}{dx} = \frac{1}{h}\frac{d}{ds}(P'_n(s)) \tag{5.24}$$

Substituting Eq. (5.23) into Eq. (5.24), differentiating, and simplifying yields

$$P''_n(x) = \frac{1}{h}\frac{d}{ds}(P'_n(s)) = \frac{1}{h^2}(\Delta^2 f_0 + (s-1)\,\Delta^3 f_0 + \cdots) \tag{5.25}$$

Higher-order derivatives can be obtained in a similar manner. Recall that $\Delta^n f$ becomes less and less accurate as n increases. Consequently, higher-order derivatives become increasingly less accurate.

At $x = x_0$, $s = 0.0$, and Eqs. (5.23) and (5.25) becomes

$$P'_n(x_0) = \frac{1}{h}\left(\Delta f_0 - \frac{1}{2}\Delta^2 f_0 + \frac{1}{3}\Delta^3 f_0 - \frac{1}{4}\Delta^4 f_0 + \cdots\right) \tag{5.26}$$

$$P''_n(x_0) = \frac{1}{h^2}(\Delta^2 f_0 - \Delta^3 f_0 + \cdots) \tag{5.27}$$

Equations (5.26) and (5.27) are one-sided forward-difference formulas.

The error associated with numerical differentiation can be determined by differentiating the error term, Eq. (5.17). Thus,

$$\frac{d}{dx}(\text{Error}) = \frac{d}{ds}\left[\binom{s}{n+1}h^{n+1}f^{(n+1)}(\xi)\right]\frac{1}{h} \qquad (5.28)$$

From the definition of the binomial coefficient, Eq. (4.90):

$$\binom{s}{n+1} = \frac{s(s-1)(s-2)\cdots(s-n)}{(n+1)!} \qquad (5.29)$$

Substituting Eq. (5.29) into Eq. (5.28) and differentiating yields

$$\frac{d}{dx}(\text{Error}) = h^n f^{(n+1)}(\xi)\left[\frac{(s-1)(s-2)\cdots(s-n)+\cdots+s(s-1)\cdots(s-n+1)}{(n+1)!}\right] \qquad (5.30)$$

At $x = x_0, s = 0.0$, and Eq. (5.30) gives

$$\frac{d}{dx}[\text{Error}(x_0)] = \frac{(-1)^n}{(n+1)}h^n f^{(n+1)}(\xi) \neq 0 \qquad (5.31)$$

Even though there is no error in $P_n(x_0)$, there is error in $P_n'(x_0)$.

The order of an approximation is the rate at which the error of the approximation approaches zero as the interval h approaches zero. Equation (5.31) shows that the one-sided first derivative approximation $P_n'(x_0)$ is order n, which is written $0(h^n)$, when the polynomial $P_n(x)$ includes the nth forward difference. For example, $P_1'(x)$ is $0(h)$, $P_2'(x)$ is $0(h^2)$, etc. Each additional differentiation introduces another h into the denominator of the error term, so the order of the result drops by one for each additional differentiation. Thus, $P_n'(x_0)$ is $0(h^n)$, $P_n''(x_0)$ is $0(h^{n-1})$, etc.

A more direct way to determine the order of a derivative approximation is to recall that the error term of all difference polynomials is given by the first neglected term in the polynomial, with $\Delta^{(n+1)}f$ replaced by $h^{n+1}f^{(n+1)}(\xi)$. Each differentiation introduces an additional h into the denominator of the error term. For example, from Eqs. (5.26) and (5.27), if terms through $\Delta^2 f_0$ are accounted for, the error for $P_2'(x)$ is $0(h^3)/h = 0(h^2)$ and the error for $P_n''(x)$ is $0(h^3)/h^2 = 0(h)$. To achieve $0(h^2)$ for $P_n''(x_0)$, $P_3(x)$ must be used.

Example 5.2. Newton forward-difference polynomial, one-sided.

Let's solve the example problem presented in Section 5.1 using a Newton forward-difference polynomial with base point $x_0 = 3.5$, so $x_0 = 3.5$ in Eqs. (5.26) and (5.27). Selecting data points from Figure 5.1 and constructing the difference table gives

x	$f(x)$	$\Delta f(x)$	$\Delta^2 f(x)$	$\Delta^3 f(x)$
3.5	0.285714			
		-0.007936		
3.6	0.277778		0.000428	
		-0.007508		-0.000032
3.7	0.270270		0.000396	
		-0.007112		
3.8	0.263158			

Substituting values into Eq. (5.26) gives

$$P_n'(3.5) = \frac{1}{0.1}\left[(-0.007936) - \frac{1}{2}(0.000428) + \frac{1}{3}(-0.000032) + \cdots\right] \qquad (5.32)$$

The order of the approximation of $P_n'(x)$ is the same as the order of the highest-order difference included in the evaluation. The first term in Eq. (5.32) is Δf_0, so evaluating that term gives an $0(h)$ result. The second term in Eq. (5.32) is $\Delta^2 f_0$, so evaluating that term yields an $0(h^2)$ result, etc. Evaluating Eq. (5.32) term by term yields

$$
\begin{array}{llll}
P_n'(3.5) = -0.07936\_ & \text{first order} & \text{Error} = 0.00227\_ \\
\qquad\quad = -0.08150\_ & \text{second order} & \qquad\quad = 0.00013\_ \\
\qquad\quad = -0.08161\_ & \text{third order} & \qquad\quad = 0.00002\_
\end{array}
$$

The first-order result is quite inaccurate. The second- and third-order results are quite good. In all cases, only five significant digits after the decimal place are obtained.
   Equation (5.27) gives

$$P_n''(3.5) = \frac{1}{(0.1)^2}[0.000428 - (-0.000032) + \cdots] \qquad (5.33)$$

The order of the approximation of $P_n''(x)$ is one less than the order of the highest-order difference included in the evaluation. The first term in Eq. (5.33) is $\Delta^2 f_0$, so evaluating that term gives an $0(h)$ result. The second term in Eq. (5.33) is $\Delta^3 f_0$, so evaluating that term yields an $0(h^2)$ result, etc. Evaluating Eq. (5.33) term by term yields

$$
\begin{array}{llll}
P_n''(3.5) = 0.0428\_\_ & \text{first order} & \text{Error} = -0.0038\_\_ \\
\qquad\quad = 0.0460\_\_ & \text{second order} & \qquad\quad = -0.0006\_\_
\end{array}
$$

The first-order result is very poor. The second-order result, although much more accurate, has only four significant digits after the decimal place.
   The results presented in this section illustrate the inherent inaccuracy associated with numerical differentiation. Equations (5.26) and (5.27) are both one-sided formulas. More accurate results can be obtained with centered differentiation formulas.

Centred-difference formulas can be obtained by evaluating the Newton forward-difference polynomial at points within the range of fit. For example, at $x = x_1$, $s = 1.0$, and Eqs. (5.23) and (5.25) give

$$
\boxed{
\begin{aligned}
P_n'(x_1) &= \frac{1}{h}\left(\Delta f_0 + \frac{1}{2}\Delta^2 f_0 - \frac{1}{6}\Delta^3 f_0 + \cdots\right) \qquad (5.34) \\[2mm]
P_n''(x_1) &= \frac{1}{h^2}\left(\Delta^2 f_0 + \frac{1}{12}\Delta^4 f_0 + \cdots\right) \qquad (5.35)
\end{aligned}
}
$$

From Eq. (5.34), $P_1'(x_1)$ is $0(h)$ and $P_2'(x_1)$ is $0(h^2)$, which is the same as the one-sided approximation, Eq. (5.26). However, $P_2''(x_1)$ is $0(h^2)$ since the $\Delta^3 f_0$ term is missing, whereas the one-sided approximation, Eq. (5.27), is $0(h)$. The increased order of the approximation of $P_2''(x_1)$ is due to centering the polynomial fit at point x_1.

Example 5.3. **Newton forward-difference polynomial, centered.**

To illustrate a centered-difference formula, let's rework Example 5.2 using $x_0 = 3.4$ as the base point, so that $x_1 = 3.5$ is in the middle of the range of fit. Selecting data points from Figure 5.1 and constructing the difference table gives:

x	$f(x)$	$\Delta f(x)$	$\Delta^2 f(x)$	$\Delta^3 f(x)$
3.4	0.294118			
		-0.008404		
3.5	0.285714		0.000468	
		-0.007936		-0.000040
3.6	0.277778		0.000428	
		-0.007508		
3.7	0.270270			

Substituting values into Eq. (5.34) gives

$$P'_n(3.5) = \frac{1}{0.1}\left[-0.008404 + \frac{1}{2}(0.000468) - \frac{1}{6}(-0.000040) + \cdots\right] \qquad (5.36)$$

Evaluating Eq. (5.36) term by term yields

$$
\begin{aligned}
P'_n(3.5) &= -0.08404\_ & \text{first order} & & \text{Error} &= -0.00241\_ \\
&= -0.08170\_ & \text{second order} & & &= -0.00007\_ \\
&= -0.08163\_ & \text{third order} & & &= 0.00000\_
\end{aligned}
$$

Equation (5.35) gives

$$P''_n(3.5) = \frac{1}{(0.1)^2}(0.000468 + \cdots) \qquad (5.37)$$

which yields

$$P''_n(3.5) = 0.0468\_\_ \qquad \text{second order} \qquad \text{Error} = 0.0002\_\_$$

The error of the first-order result for $f'(3.5)$ is approximately the same magnitude as the error of the first-order result obtained in Example 5.2. The current result is a backward-difference approximation, whereas the result in Example 5.2 is a forward-difference approximation. The second-order centred-difference approximation of $f''(3.5)$ is more accurate than the second-order forward-difference approximation of $f''(3.5)$ in Example 5.2.

5.3.2. Newton Backward-Difference Polynomial

Recall the Newton backward-difference polynomial, Eq. (4.101):

$$P_n(x) = f_0 + s\,\nabla f_0 + \frac{s(s+1)}{2!}\nabla^2 f_0 + \frac{s(s+1)(s+2)}{3!}\nabla^3 f_0 + \cdots \qquad (5.38)$$

The first derivative $f'(x)$ is obtained from $P_n(x)$ as illustrated in Eq. (5.21):

$$f'(x) \cong P'_n(x) = \frac{1}{h}\frac{d}{ds}(P_n(s)) \tag{5.39}$$

Substituting Eq. (5.38) into Eq. (5.39), differentiating, and simplifying gives

$$P'_n(x) = \frac{1}{h}\left(\nabla f_0 + \frac{2s+1}{2}\nabla^2 f_0 + \frac{3s^2+6s+2}{6}\nabla^3 f_0 + \cdots\right) \tag{5.40}$$

The second derivative $P''_n(x)$ is given by

$$P''_n(x) = \frac{1}{h^2}(\nabla^2 f_0 + (s+1)\nabla^3 f_0 + \cdots) \tag{5.41}$$

Higher-order derivatives can be obtained in a similar manner. Recall that $\nabla^n f$ becomes less and less accurate as n increases. Consequently higher-order derivatives become increasingly less accurate.

At $x = x_0$, $s = 0.0$, and Eqs. (5.40) and (5.41) become

$$P'_n(x_0) = \frac{1}{h}\left(\nabla f_0 + \frac{1}{2}\nabla^2 f_0 + \frac{1}{3}\nabla^3 f_0 + \cdots\right) \tag{5.42}$$

$$P''_n(x_0) = \frac{1}{h^2}(\nabla^2 f_0 + \nabla^3 f_0 + \cdots) \tag{5.43}$$

Equations (5.42) and (5.43) are one-sided backward-difference formulas.

Centered-difference formulas are obtained by evaluating the Newton backward-difference polynomial at points within the range of fit. For example, at $x = x_{-1}$, $s = -1.0$, and Eqs. (5.40) and (5.41) give

$$P'_n(x_{-1}) = \frac{1}{h}\left(\nabla f_0 - \frac{1}{2}\nabla^2 f_0 - \frac{1}{6}\nabla^3 f_0 + \cdots\right) \tag{5.44}$$

$$P''_n(x_{-1}) = \frac{1}{h^2}\left(\nabla^2 f_0 - \frac{1}{12}\nabla^4 f_0 + \cdots\right) \tag{5.45}$$

The order of the derivative approximations are obtained by dividing the order of the first neglected term in each formula by the appropriate power of h, as discussed in Section 5.3.1 for derivative approximations based on Newton forward-difference polynomials.

5.3.3. Difference formulas

The formulas for derivatives developed in the previous subsections are expressed in terms of differences. Those formulas can be expressed directly in terms of function values if the order of the approximation is specified and the expressions for the differences in terms of function values are substituted into the formulas. The resulting formulas are called *difference formulas*. Several difference formulas are developed in this subsection to illustrate the procedure.

Consider the one-sided forward-difference formula for the first derivative, Eq. (5.26):

$$P'_n(x_0) = \frac{1}{h}\left(\Delta f_0 - \frac{1}{2}\Delta^2 f_0 + \frac{1}{3}\Delta^3 f_0 - \cdots \right) \tag{5.46}$$

Recall that the error term associated with truncating the Newton forward-difference polynomial is obtained from the leading truncated term by replacing $\Delta^n f_0$ by $f^{(n)}(\xi)h^n$. Truncating Eq. (5.46) after Δf_0 gives

$$P'_n(x_0) = \frac{1}{h}[\Delta f_0 + 0(h^2)] \tag{5.47}$$

where $0(h^2)$ denotes the error term, and indicates the dependence of the error on the step size, h. Substituting $\Delta f_0 = (f_1 - f_0)$ into Eq. (5.47) yields

$$P'_1(x_0) = \frac{f_1 - f_0}{h} + 0(h) \tag{5.48}$$

Equation (5.48) is a one-sided first-order forward-difference formula for $f'(x_0)$. Truncating Eq. (5.46) after the $\Delta^2 f_0$ term gives

$$P'_2(x_0) = \frac{1}{h}\left(\Delta f_0 - \frac{1}{2}\Delta^2 f_0 + 0(h^3) \right) \tag{5.49}$$

Substituting Δf_0 and $\Delta^2 f_0$ into Eq. (5.49) and simplifying yields

$$P'_2(x_0) = \frac{-3f_0 + 4f_1 - f_2}{2h} + 0(h^2) \tag{5.50}$$

Higher-order difference formulas can be obtained in a similar manner.

Consider the one-sided forward-difference formula for the second derivative, Eq. (5.27). The following difference formulas can be obtained in the same manner that Eqs. (5.48) and (5.50) are developed.

$$P''_2(x_0) = \frac{f_0 - 2f_1 + f_2}{h^2} + 0(h) \tag{5.51}$$

$$P''_3(x_0) = \frac{2f_0 - 5f_1 + 4f_2 - f_3}{h^2} + 0(h^2) \tag{5.52}$$

Centered-difference formulas for $P'_n(x_1)$ and $P''_n(x_1)$ can be derived from Eqs. (5.34) and (5.35). Thus,

$$P'_2(x_1) = \frac{f_2 - f_0}{2h} + 0(h^2) \tag{5.53}$$

$$P''_2(x_1) = \frac{f_0 - 2f_1 + f_2}{h^2} + 0(h^2) \tag{5.54}$$

Difference formulas of any order can be developed in a similar manner for derivatives of any order, based on one-sided, centred, or nonsymmetrical differences. A selection of difference formulas is presented in Table 5.1.

Example 5.4.   Newton polynomial difference formulas.

Let's illustrate the use of difference formulas obtained from Newton polynomials by solving the example problem presented in Section 5.1. Calculate the second-order centered-difference approximation of $f'(3.5)$ and $f''(3.5)$ using Eqs. (5.53) and (5.54). Thus,

$$P_2'(3.5) = \frac{f(3.6) - f(3.4)}{2(0.1)} = \frac{0.277778 - 0.294118}{2(0.1)} = -0.08170\_ \tag{5.55}$$

$$P_2''(3.5) = \frac{f(3.6) - 2f(3.5) + f(3.4)}{(0.1)^2} = \frac{0.277778 - 2(0.285714) + 0.294118}{(0.1)^2}$$

$$= 0.0468\_\_ \tag{5.56}$$

These results are identical to the second-order results obtained in Example 5.3.

5.4 TAYLOR SERIES APPROACH

Difference formulas can also be developed using Taylor series. This approach is especially useful for deriving finite difference approximations of exact derivatives (both total derivatives and partial derivatives) that appear in differential equations.

Difference formulas for functions of a single variable, for example, $f(x)$, can be developed from the Taylor series for a function of a single variable, Eq. (0.6):

$$f(x) = f_0 + f_0' \, \Delta x + \frac{1}{2} f_0'' \, \Delta x^2 + \cdots + \frac{1}{n!} f_0^{(n)} \, \Delta x^n + \cdots \tag{5.57}$$

where $f_0 = f(x_0), f_0' = f'(x_0)$, etc. The continuous spatial domain $D(x)$ must be dis-cretized into an equally space grid of discrete points, as illustrated in Figure 5.4. For the discretized x space,

$$f(x_i) = f_i \tag{5.58}$$

where the subscript i denotes a particular spatial location. The Taylor series for $f(x)$ at grid points surrounding point i can be combined to obtain difference formulas for $f'(x_i), f''(x_i)$, etc.

Difference formulas for functions of time, $f(t)$, can be developed from the Taylor series for $f(t)$:

$$f(t) = f_0 + f_0' \, \Delta t + \frac{1}{2} f_0'' \, \Delta t^2 + \cdots + \frac{1}{n!} f_0^{(n)} \, \Delta t^n + \cdots \tag{5.59}$$

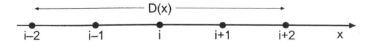

Figure 5.4   Continuous spatial domain $D(x)$ and discretized x space.

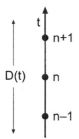

Figure 5.5 Continuous temporal domain $D(t)$ and discretized t space.

where $f_0 = f(t^0), f_0' = f'(t^0)$, etc. The continuous temporal domain $D(t)$ must be discretized into a grid of discrete points, as illustrated in Figure 5.5. For the discretized t space,

$$f(t^n) = f^n \tag{5.60}$$

where the superscript n denotes a particular temporal location. The Taylor series for $f(t)$ at grid points surrounding grid point n can be combined to obtain difference formulas for $f'(t^n), f''(t^n)$, etc.

Difference formulas for functions of several variables, for example, $f(x, t)$, can be developed from the Taylor series for a function of several variables, Eq. (0.12):

$$f(x, t) = f_0 + (f_x|_0 \, \Delta x + f_t|_0 \, \Delta t) + \frac{1}{2!}(f_{xx}|_0 \, \Delta x^2 + 2f_{xt}|_0 \, \Delta x \, \Delta t + f_{tt}|_0 \, \Delta t^2)$$
$$+ \cdots + \frac{1}{n!}\left(\Delta x \frac{\partial}{\partial x} + \Delta t \frac{\partial}{\partial t}\right)^n f_0 + \cdots \tag{5.61}$$

where $f_0 = f(x_0, t_0), f_x|_0 = f_x(x_0, t_0)$, etc. The expression $(\cdots)^n$ is expanded by the binomial expansion, the increments in Δx and Δt are raised to the indicated powers, and terms such as $(\partial/\partial x)^n$, etc., are interpreted as $\partial^n/\partial x^n$, etc. The continuous xt domain, $D(x, t)$, must be discretized into an orthogonal equally spaced grid of discrete points, as illustrated in Figure 5.6. For the discrete grid,

$$f(x_i, t^n) = f_i^n \tag{5.62}$$

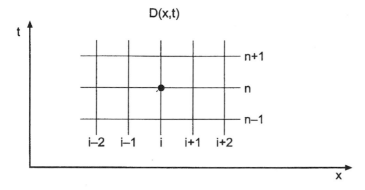

Figure 5.6 Continuous xt domain $D(x, t)$ and discretized xt space.

where the subscript i denotes a particular spatial location and the superscript n denotes a particular time level. The Taylor series for $f(x, t)$ at grid points surrounding point (i, n) can be combined to obtain difference formulas for f_x, f_t, f_{xt}, etc.

For partial derivatives of $f(x, t)$ with respect to x, $t = t_0 = $ constant, $\Delta t = 0$, and Eq. (5.61) becomes

$$f(x, t_0) = f_0 + f_x|_0 \, \Delta x + \frac{1}{2} f_{xx}|_0 \, \Delta x^2 + \cdots + \frac{1}{n!} f_{(n)x}|_0 \, \Delta x^n + \cdots \tag{5.63}$$

Equation (5.63) is identical in form to Eq. (5.57), where f_0' corresponds to $f_x|_0$, etc. The partial derivative $f_x|_0$ of the function $f(x, t)$ can be obtained from Eq. (5.63) in exactly the same manner as the total derivative, f_0', of the function $f(x)$ is obtained from Eq. (5.57). Since Eqs. (5.57) and (5.63) are identical in form, the difference formulas for f_0' and $f_x|_0$ are identical if the same discrete grid points are used to develop the difference formulas. Consequently, difference formulas for partial derivatives of a function of several variables can be derived from the Taylor series for a function of a single variable. To emphasize this concept, the following common notation for derivatives will be used in the development of difference formulas for total derivatives and partial derivatives:

$$\frac{d}{dx}(f(x)) = f_x \tag{5.64}$$

$$\frac{\partial}{\partial x}(f(x, t)) = f_x \tag{5.65}$$

In a similar manner, partial derivatives of $f(x, t)$ with respect to t with $x = x_0 = $ constant can be obtained from the expression

$$f(x_0, t) = f_0 + f_t|_0 \, \Delta t + \frac{1}{2} f_{tt}|_0 \, \Delta t^2 + \cdots + \frac{1}{n!} f_{(n)t}|_0 \, \Delta t^n + \cdots \tag{5.66}$$

Partial derivatives of $f(x, t)$ with respect to t are identical in form to total derivatives of $f(t)$ with respect to t.

This approach does not work for mixed partial derivatives, such as f_{xt}. Difference formulas for mixed partial derivatives must be determined directly from the Taylor series for several variables, Eq. (5.61).

The Taylor series for the function $f(x)$, Eq. (5.57), can be written as

$$f(x) = f_0 + f_x|_0 \, \Delta x + \frac{1}{2} f_{xx}|_0 \, \Delta x^2 + \cdots + \frac{1}{n!} f_{(n)x}|_0 \, \Delta x^n + \cdots \tag{5.67}$$

The Taylor formula with remainder is given by Eq. (0.10):

$$f(x) = f_0 + f_x|_0 \, \Delta x + \frac{1}{2} f_{xx}|_0 \, \Delta x^2 + \cdots + \frac{1}{n!} f_{(n)x}|_0 \, \Delta x^n + R^{n+1} \tag{5.68}$$

where the remainder term R^{n+1} is given by

$$R^{n+1} = \frac{1}{(n+1)!} f_{(n+1)x}(\xi) \, \Delta x^{n+1} \tag{5.69}$$

where $x_0 \leq \xi \leq x_0 + \Delta x$.

The infinite Taylor series, Eq. (5.67), and the Taylor formula with remainder, Eq. (5.68), are equivalent. The error incurred by truncating the infinite Taylor series after the nth derivative is exactly the remainder term of the nth-order Taylor formula. Truncating the Taylor series is equivalent to dropping the remainder term of the Taylor formula. Finite difference approximations of exact derivatives can be obtained by solving for the exact derivative from either the infinite Taylor series or the Taylor formula, and then either truncating the Taylor series or dropping the remainder term of the Taylor formula. These two procedures are identical. The terms which are truncated from the infinite Taylor series, which are identical to the remainder term of the Taylor formula, are called the *truncation error* of the finite difference approximation of the exact derivative. In most cases, our main concern is the *order* of the truncation error, which is the rate at which the truncation error approaches zero as $\Delta x \to 0$. The order of the truncation error, which is the order of the remainder term, is denoted by the notation $0(\Delta x^n)$.

Consider the equally spaced discrete finite difference grid illustrated in Figure 5.4. Choose point i as the base point, and write the Taylor series for f_{i+1} and f_{i-1}:

$$f_{i+1} = f_i + f_x|_i \, \Delta x + \tfrac{1}{2} f_{xx}|_i \, \Delta x^2 + \tfrac{1}{6} f_{xxx}|_i \, \Delta x^3 + \tfrac{1}{24} f_{xxxx}|_i \, \Delta x^4 + \cdots \qquad (5.70)$$

$$f_{i-1} = f_i - f_x|_i \, \Delta x + \tfrac{1}{2} f_{xx}|_i \, \Delta x^2 - \tfrac{1}{6} f_{xxx}|_i \, \Delta x^3 + \tfrac{1}{24} f_{xxxx}|_i \, \Delta x^4 - \cdots \qquad (5.71)$$

Subtracting Eq. (5.71) for f_{i-1} from Eq. (5.70) for f_{i+1} gives

$$f_{i+1} - f_{i-1} = 2 f_x|_i \, \Delta x + \tfrac{1}{3} f_{xxx}|_i \, \Delta x^3 + \cdots \qquad (5.72)$$

Letting the f_{xxx} term be the remainder term and solving for $f_x|_i$ yields

$$f_x|_i = \frac{f_{i+1} - f_{i-1}}{2 \, \Delta x} - \frac{1}{6} f_{xxx}(\bar{\xi}) \, \Delta x^2 \qquad (5.73)$$

where $x_{i-1} \leq \bar{\xi} \leq x_{i+1}$. Equation (5.73) is an exact expression for $f_x|_i$. If the remainder term is truncated, which is equivalent to truncating the infinite Taylor series, Eqs. (5.70) and (5.71), Eq. (5.73) yields an $0(\Delta x^2)$ finite difference approximation of $f_x|_i$. Thus,

$$\boxed{f_x|_i = \frac{f_{i+1} - f_{i-1}}{2 \, \Delta x}} \qquad (5.74)$$

The truncated result is identical to the result obtained from the Newton forward-difference polynomial, Eq. (5.53).

Adding Eq. (5.70) for f_{i+1} and Eq. (5.71) for f_{i-1} gives

$$f_{i+1} + f_{i-1} = 2 f_i + f_{xx}|_i \, \Delta x^2 + \frac{1}{12} f_{xxxx}|_i \, \Delta x^4 + \cdots \qquad (5.75)$$

Letting the f_{xxxx} term be the remainder term and solving for $f_{xx}|_i$ yields

$$f_{xx}|_i = \frac{f_{i+1} - 2 f_i + f_{i-1}}{2 \, \Delta x} - \frac{1}{12} f_{xxxx}(\bar{\xi}) \, \Delta x^2 \qquad (5.76)$$

where $x_{i-1} \leq \bar{\xi} \leq x_{i+1}$. Truncating the remainder term yields a finite difference approximation for $f_{xx}|_i$. Thus,

$$\boxed{f_{xx}|_i = \frac{f_{i+1} - 2 f_i + f_{i-1}}{\Delta x^2}} \qquad (5.77)$$

The truncated result is identical to the result obtained from the Newton forward-difference polynomial, Eq. (5.54).

Equations (5.74) and (5.77) are centered-difference formulas. They are inherently more accurate than one-sided difference formulas.

Example 5.5. Taylor series difference formulas.

Let's illustrate the use of difference formulas obtained from Taylor series by evaluating $f'(3.5)$ and $f''(3.5)$ for the data presented in Figure 5.1 using Eqs. (5.74) and (5.77), respectively. To obtain the most accurate results possible, use the closest data points to $x = 3.5$, that is, $x = 3.4$ and 3.6. Thus, from Eq. (5.74),

$$f'(3.5) = \frac{f(3.6) - f(3.4)}{2(0.1)} = \frac{0.279778 - 0.294118}{2(0.1)} = -0.08170_- \tag{5.78}$$

From Eq. (5.77),

$$f''(3.5) = \frac{f(3.6) - 2.0 f(3.5) + f(3.4)}{2.0(0.1)} = \frac{0.279778 - 2.0(0.285714) + 0.294118}{2.0(0.1)}$$

$$= 0.0468_{--} \tag{5.79}$$

These are the same results that are obtained with the difference formulas developed from the Newton forward-difference polynomials illustrated in Example 5.4, Eqs. (5.55) and (5.56).

Equations (5.74) and (5.77) are difference formulas for spatial derivatives. Difference formulas for time derivatives can be developed in a similar manner. The time dimension can be discretized into a discrete temporal grid, as illustrated in Figure 5.5, where the superscript n denotes a specific value of time. Thus, $f(t^n) = f^n$. Choose point n as the base point, and write the Taylor series for f^{n+1} and f^{n-1}:

$$f^{n+1} = f^n + f_t|^n \, \Delta t + \tfrac{1}{2} f_{tt}|^n \, \Delta t^2 + \cdots \tag{5.80}$$

$$f^{n-1} = f^n - f_t|^n \, \Delta t + \tfrac{1}{2} f_{tt}|^n \, \Delta t^2 - \cdots \tag{5.81}$$

Letting the $f_{tt}|^n$ term be the remainder term and solving Eq. (5.80) for $f_t|^n$ yields

$$\boxed{f_t|^n = \frac{f^{n+1} - f^n}{\Delta t} - \frac{1}{2} f_{tt}(\tau) \, \Delta t} \tag{5.82}$$

where $t^n \le \tau \le t^{n+1}$. Equation (5.82) is a first-order forward-difference formula for $f_t|^n$. Subtracting Eq. (5.81) for f^{n-1} from Eq. (5.80) for f^{n+1} gives

$$f^{n+1} - f^{n-1} = 2 f_t|^n \, \Delta t + \tfrac{1}{3} f_{ttt}|^n \, \Delta t^3 + \cdots \tag{5.83}$$

Letting the $f_{ttt}|^n$ term be the remainder term and solving for $f_t|^n$ yields

$$\boxed{f_t|^n = \frac{f^{n+1} - f^{n-1}}{2 \, \Delta t} - \frac{1}{6} f_{ttt}(\bar{\tau}) \, \Delta t^2} \tag{5.84}$$

where $t^{n-1} \leq \bar{\tau} \leq t^{n+1}$. Equation (5.84) is a second-order centred-difference formula for $f_t|^n$. Centred-difference formulas are inherently more accurate than one-sided difference formulas, such as Eq. (5.82).

Difference formulas of any order, based on one-sided forward differences, one-sided backward differences, centered differences, nonsymmetrical differences, etc., can be obtained by different combinations of the Taylor series for $f(x)$ or $f(t)$ at various grid points. Higher-order difference formulas require more grid points, as do formulas for higher-order derivatives.

Example 5.6. Third-order nonsymmetrical difference formula for f_x.

___

Let's develop a third-order, nonsymmetrical, backward-biased, difference formula for f_x. The Taylor series for $f(x)$ is:

$$f(x) = f_i + f_x|_i \, \Delta x + \tfrac{1}{2} f_{xx}|_i \, \Delta x^2 + \tfrac{1}{6} f_{xxx}|_i \, \Delta x^3 + \tfrac{1}{24} f_{xxxx}|_i \, \Delta x^4 + \cdots \tag{5.85}$$

Three questions must be answered before the difference formula can be developed: (a) What is to be the order of the remainder term, (b) how many grid points are required, and (c) which grid points are to be used? The coefficient of f_x is Δx. If the remainder term in the difference formula is to be $0(\Delta x^3)$, then the remainder term in the Taylor series must be $0(\Delta x^4)$, since the formulas must be divided by Δx when solving for f_x. Consequently, three grid points, in addition to the base point i, are required, so that f_{xx} and f_{xxx} can be eliminated from the Taylor series expansions, thus giving a third-order difference formula for f_x. For a backward-biased difference formula, choose grid points $i+1, i-1$, and $i-2$.

The fourth-order Taylor series for f_{i+1}, f_{i-1}, and f_{i-2} are:

$$f_{i+1} = f_i + f_x|_i \, \Delta x + \tfrac{1}{2} f_{xx}|_i \, \Delta x^2 + \tfrac{1}{6} f_{xxx}|_i \, \Delta x^3 + \tfrac{1}{24} f_{xxxx}(\xi_1) \, \Delta x^4 + \cdots \tag{5.86}$$

$$f_{i-1} = f_i - f_x|_i \, \Delta x + \tfrac{1}{2} f_{xx}|_i \, \Delta x^2 - \tfrac{1}{6} f_{xxx}|_i \, \Delta x^3 + \tfrac{1}{24} f_{xxxx}(\xi_{-1}) \, \Delta x^4 - \cdots \tag{5.87}$$

$$f_{i-2} = f_i - 2f_x|_i \, \Delta x + \tfrac{4}{2} f_{xx}|_i \, \Delta x^2 - \tfrac{8}{6} f_{xxx}|_i \, \Delta x^3 + \tfrac{16}{24} f_{xxxx}(\xi_{-2}) \, \Delta x^4 - \cdots \tag{5.88}$$

Forming the combination $(f_{i+1} - f_{i-1})$ gives

$$(f_{i+1} - f_{i-1}) = 2f_x|_i \, \Delta x + \tfrac{2}{6} f_{xxx}|_i \, \Delta x^3 + 0(\Delta x^5) \tag{5.89}$$

Forming the combination $(4f_{i+1} - f_{i-2})$ gives

$$(4f_{i+1} - f_{i-2}) = 3f_i + 6f_x|_i \, \Delta x + \tfrac{12}{6} f_{xxx}|_i \, \Delta x^3 - \tfrac{12}{24} f_{xxxx}(\bar{\xi}) \, \Delta x^4 + 0(\Delta x^5) \tag{5.90}$$

where $x_{i-2} \leq \bar{\xi} \leq x_{i+1}$. Multiplying Eq. (5.89) by 6 and subtracting Eq. (5.90) gives

$$6(f_{i+1} - f_{i-1}) - (4f_{i+1} - f_{i-2}) = -3f_i + 6f_x|_i \, \Delta x + \tfrac{12}{24} f_{xxxx}(\bar{\xi}) \, \Delta x^4 \tag{5.91}$$

Solving Eq. (5.91) for $f_x|_i$ yields

$$\boxed{f_x|_i = \frac{f_{i-2} - 6f_{i-1} + 3f_i + 2f_{i+1}}{6\Delta x} - \frac{1}{2} f_{xxxx}(\bar{\xi}) \, \Delta x^3} \tag{5.92}$$

Truncating Eq. (5.92) yields a third-order, nonsymmetrical, backward-biased, difference formula for $f_x|_i$.

In summary, the procedure for developing difference formulas by the Taylor series approach is as follows.

1. Specify the order n of the derivative $f_{(n)x}$ for which the difference formula is to be developed.
2. Choose the order m of the reminder term in the difference formula Δx^m.
3. Determine the order of the remainder term in the Taylor series, Δx^{m+n}.
4. Specify the type of difference formula desired: centered, forward, backward, or nonsymmetrical.
5. Determine the number of grid points required, which is at most $(m + n - 1)$.
6. Write the Taylor series of order $(m + n)$ at the $(m + n - 1)$ grid points.
7. Combine the Taylor series to eliminate the undesired derivatives, and solve for the desired derivative and the leading truncation error term.

For temporal derivatives, replace x by t in steps 1 to 7.

5.5 DIFFERENCE FORMULAS

Table 5.1 presents several difference formulas for both time derivatives and space derivatives. These difference formulas are used extensively in Chapters 7 and 8 in the numerical solution of ordinary differential equations, and in Chapters 9 to 11 in the numerical solution of partial differential equations.

5.6 ERROR ESTIMATION AND EXTRAPOLATION

When the functional form of the error of a numerical algorithm is known, the error can be estimated by evaluating the algorithm for two different increment sizes. The error estimate can be used both for error control and extrapolation.

Consider a numerical algorithm which approximates an exact calculation with an error that depends on an increment, h. Thus,

$$f_{\text{exact}} = f(h) + Ah^n + Bh^{n+m} + Ch^{n+2m} + \cdots \tag{5.113}$$

where n is the order of the leading error term and m is the increment in the order of the following error terms. Applying the algorithm at two increment sizes, $h_1 = h$ and $h_2 = h/R$, gives

$$f_{\text{exact}} = f(h) + Ah^n + 0(h^{n+m}) \tag{5.114a}$$

$$f_{\text{exact}} = f(h/R) + A(h/R)^n + 0(h^{n+m}) \tag{5.114b}$$

Subtracting Eq. (5.114b) from Eq. (5.114a) gives

$$0 = f(h) - f(h/R) + Ah^n + A(h/R)^n + 0(h^{n+m}) \tag{5.115}$$

Table 5.1. Difference Formulas

$$f_t|^n = \frac{f^{n+1} - f^n}{\Delta t} - \frac{1}{2}f_{tt}(\tau)\,\Delta t \tag{5.93}$$

$$f_t|^n = \frac{f^n - f^{n-1}}{\Delta t} + \frac{1}{2}f_{tt}(\tau)\,\Delta t \tag{5.94}$$

$$f_t|^{n+1/2} = \frac{f^{n+1} - f^n}{\Delta t} - \frac{1}{24}f_{ttt}(\tau)\,\Delta t^2 \tag{5.95}$$

$$f_t|^n = \frac{f^{n+1} - f^{n-1}}{2\,\Delta t} - \frac{1}{6}f_{ttt}(\tau)\,\Delta t^2 \tag{5.96}$$

$$f_x|_i = \frac{f_{i+1} - f_i}{\Delta x} - \frac{1}{2}f_{xx}(\xi)\,\Delta x \tag{5.97}$$

$$f_x|_i = \frac{f_i - f_{i-1}}{\Delta x} + \frac{1}{2}f_{xx}(\xi)\,\Delta x \tag{5.98}$$

$$f_x|_i = \frac{f_{i+1} - f_{i-1}}{2\,\Delta x} - \frac{1}{6}f_{xxx}(\xi)\,\Delta x^2 \tag{5.99}$$

$$f_x|_i = \frac{-3f_i + 4f_{i+1} - f_{i+2}}{2\,\Delta x} - \frac{1}{3}f_{xxx}(\xi)\,\Delta x^2 \tag{5.100}$$

$$f_x|_i = \frac{f_{i-2} - 4f_{i-1} + 3f_i}{2\,\Delta x} + \frac{1}{3}f_{xxx}(\xi)\,\Delta x^2 \tag{5.101}$$

$$f_x|_i = \frac{-11f_i + 18f_{i+1} - 9f_{i+2} + 2f_{i+3}}{6\,\Delta x} - \frac{1}{4}f_{xxxx}(\xi)\,\Delta x^3 \tag{5.102}$$

$$f_x|_i = \frac{-2f_{i-3} + 9f_{i-2} - 18f_{i-1} + 11f_i}{6\,\Delta x} + \frac{1}{4}f_{xxxx}(\xi)\,\Delta x^3 \tag{5.103}$$

$$f_x|_i = \frac{f_{i-2} - 6f_{i-1} + 3f_i + 2f_{i+1}}{6\,\Delta x} - \frac{1}{12}f_{xxxx}(\xi)\,\Delta x^3 \tag{5.104}$$

$$f_x|_i = \frac{-2f_{i-1} - 3f_i + 6f_{i+1} - f_{i+2}}{6\,\Delta x} + \frac{1}{12}f_{xxxx}(\xi)\,\Delta x^3 \tag{5.105}$$

$$f_x|_i = \frac{f_{i-2} - 8f_{i-1} + 8f_{i+1} - f_{i+2}}{12\,\Delta x} + \frac{1}{30}f_{xxxxx}(\xi)\,\Delta x^4 \tag{5.106}$$

$$f_{xx}|_i = \frac{f_i - 2f_{i+1} + f_{i+2}}{\Delta x^2} - f_{xxx}(\xi)\,\Delta x \tag{5.107}$$

$$f_{xx}|_i = \frac{f_{i-2} - 2f_{i-1} + f_i}{\Delta x^2} + f_{xxx}(\xi)\,\Delta x \tag{5.108}$$

$$f_{xx}|_i = \frac{f_{i+1} - 2f_i + f_{i-1}}{\Delta x^2} - \frac{1}{12}f_{xxxx}(\xi)\,\Delta x^2 \tag{5.109}$$

$$f_{xx}|_i = \frac{2f_i - 5f_{i+1} + 4f_{i+2} - f_{i+3}}{\Delta x^2} + \frac{11}{12}f_{xxxx}(\xi)\,\Delta x^2 \tag{5.110}$$

$$f_{xx}|_i = \frac{-f_{i-3} + 4f_{i-2} - 5f_{i-1} + 2f_i}{\Delta x^2} + \frac{11}{12}f_{xxxx}(\xi)\,\Delta x^2 \tag{5.111}$$

$$f_{xx}|_i = \frac{-f_{i-2} + 16f_{i-1} - 30f_i + 16f_{i+1} - f_{i+2}}{12\,\Delta x^2} + \frac{1}{90}f_{xxxxxx}(\xi)\,\Delta x^4 \tag{5.112}$$

Solving Eq. (5.115) for the leading error terms in Eqs. (5.114a) and (5.114b) yields

$$\text{Error}(h) = Ah^n = \frac{R^n}{R^n - 1}(f(h/R) - f(h)) \tag{5.116a}$$

$$\text{Error}(h/R) = A(h/R)^n = \frac{1}{R^n - 1}(f(h/R) - f(h)) \tag{5.116b}$$

Equation (5.116) can be used to estimate the leading error terms in Eq. (5.114).

The error estimates can be added to the approximate results to yield an improved approximation. This process is called *extrapolation*. Adding Eq. (5.116b) to Eq. (5.114b) gives

$$\text{Extrapolated value} = f(h/R) + \frac{1}{R^n - 1}(f(h/R) - f(h)) + 0(h^{n+m}) \tag{5.117}$$

The error of the extrapolated value is $0(h^{n+m})$. Two $0(h^{n+m})$ extrapolated results can be extrapolated to give an $0(h^{n+2m})$ result, where the exponent, n, in Eq. (5.117) is replaced with the exponent, $n + m$. Higher-order extrapolations can be obtained by successive applications of Eq. (5.117).

Example 5.7. Error estimation and extrapolation.

___

Example 5.5 evaluates $f'(3.5)$ using Eq. (5.73) with $\Delta x = 0.1$. A more accurate result could be obtained by evaluating Eq. (5.73) with $\Delta x = 0.05$, which requires data at $x = 3.45$ and 3.55. As seen in Figure 5.1, those points are not available. However, data are available at $x = 3.3$ and 3.7, for which $\Delta x = 0.2$. Applying Eq. (5.73) with $\Delta x = 0.2$ gives

$$f'(3.5) = \frac{f(3.7) - f(3.3)}{2(0.2)} = \frac{0.270270 - 0.303030}{2(0.2)} = -0.081900 \tag{5.118}$$

The exact error in this result is Error $= -0.081900 - (-0.081633) = -0.000267$, which is approximately four times larger than the exact error obtained in Example 5.5 where $f'(3.5) = -0.081700$, for which the exact error is Error $= -0.081700 - (-0.081633) = -0.000067$.

Now that two estimates of $f'(3.5)$ are available, the error estimate for the result with the smaller Δx can be calculated from Eq. (5.116b). Thus,

$$\text{Error}(\Delta x/2) = \frac{1}{3}[-0.081700 - (-0.081900)] = 0.000067 \tag{5.119}$$

Applying the extrapolation formula, Eq. (5.117), gives

$$\text{Extrapolated value} = -0.081700 + 0.000067 = -0.081633 \tag{5.120}$$

which is the exact value to six digits after the decimal place. The value of error estimation and extrapolation are obvious in this example.

___

5.7 PROGRAMS

Two procedures for numerical differentiation of tabular data are presented in this section:

1. Derivatives of unequally spaced data
2. Derivatives of equally space data

All of the subroutines are written for a *quadratic* polynomial. Higher-degree polynomials can be constructed by following the patterns of the quadratic polynomials. The variable *ndeg*, which specifies the degree of the polynomials, is thus not needed in the present programs. It is included, however, to facilitate the addition of other-degree polynomials to the subroutines, in which case the degree of the polynomial to be evaluated must be specified.

The basic computational algorithms are presented as completely self-contained subroutines suitable for use in other programs. Input data and output statements are contained in a main (or driver) program written specifically to illustrate the use of each subroutine.

5.7.1. Derivatives of Unequally Spaced Data

Three procedures are presented for differentiating unequally spaced data:

1. Differentiation of a direct fit polynomial
2. Differentiation of a Lagrange polynomial
3. Differentiation of a divided difference polynomial

FORTRAN *subroutines direct*, *lagrange*, and *divdiff* are presented in this subsection for implementing these procedures. A common *program main* defines the data sets and prints them, calls one of the subroutines to implement the solution, and prints the solution. The only change in *program main* for the three subroutines is the *call* statement to the particular subroutine and the output *format* statement identifying the particular subroutine.

5.7.1.1 Direct Fit Polynomial

The first- and second-order derivatives of a direct fit polynomial are given by Eq. (5.7):

$$P'_n(x) = a_1 + 2a_2x + 3a_3x^2 + \cdots + na_nx^{n-1} \tag{5.121a}$$

$$P''_n(x) = 2a_2 + 6a_3x + \cdots + n(n-1)a_nx^{n-2} \tag{5.121b}$$

A FORTRAN subroutine, *subroutine direct*, for implementing Eq. (5.121) for a quadratic direct fit polynomial is presented in Program 5.1.

Program 5.1. Differentiation of a direct fit polynomial program.

```
      program main
c     main program to illustrate numerical diff. subroutines
c     ndim  array dimension, n = 3 in this example
c     ndeg  degree of polynomial, ndeg = 2 in this example
c     n     number of data points
c     xp    value of x at which to evaluate the derivatives
c     x     independent variable, x(i)
```

```
c      f        dependent variable, f(i)
c      fx       numerical approximation of first derivative
c      fxx      numerical approximation of second derivative
       dimension x(3),f(3),a(3,3),b(3),c(3)
       data ndim,ndeg,n,xp / 3, 2, 3, 3.5 /
       data (x(i),i=1,3) / 3.4, 3.5, 3.6 /
       data (f(i),i=1,3) / 0.294118, 0.285714, 0.277778 /
       write (6,1000)
       do i=1,n
          write (6,1010) i,x(i),f(i)
       end do
       call direct (ndim,ndeg,n,x,f,xp,fx,fxx,a,b,c)
       write (6,1020) fx,fxx
       stop
 1000 format (' Direct fit polynomial'/' '/'  i',6x,'x',11x,'f'/' ')
 1010 format (i3,5f12.6)
 1020 format (' '/' fx =',f12.6,'  and  fxx =',f12.6)
       end

       subroutine direct (ndim,ndeg,n,x,f,xp,fx,fxx,a,b,c)
c      direct fit polynomial differentiation
       dimension x(ndim),f(ndim),a(ndim,ndim),b(ndim),c(ndim)
       do i=1,n
          a(i,1)=1.0
          a(i,2)=x(i)
          a(i,3)=x(i)**2
          b(i)=f(i)
       end do
       call gauss (ndim,n,a,b,c)
       fx=c(2)+2.0*c(3)*xp
       fxx=2.0*c(3)
       return
       end

       subroutine gauss (ndim,n,a,b,x)
c      implements simple gauss elimination
       end
```

The data set used to illustrate *subroutine direct* is taken from Example 5.1. The output generated by the program is presented in Output 5.1.

Output 5.1. Solution by differentiation of a direct fit polynomial.

```
Direct fit polynomial

  i      x            f

  1    3.400000     0.294118
  2    3.500000     0.285714
  3    3.600000     0.277778


fx =   -0.081700  and  fxx =    0.046800
```

5.7.1.2 Lagrange Polynomial

The first- and second-order derivatives of a quadratic Lagrange polynomial are given by Eq. (5.9):

$$f'(x) \cong P_2'(x) = \frac{2x - (b+c)}{(a-b)(a-c)}f(a) + \frac{2x - (a+c)}{(b-a)(b-c)}f(b) + \frac{2x - (a+b)}{(c-a)(c-b)}f(c)$$

(5.122a)

$$f''(x) \cong P_2''(x) = \frac{2f(a)}{(a-b)(a-c)} + \frac{2f(b)}{(b-a)(b-c)} + \frac{2f(c)}{(c-a)(c-b)}$$ (5.122b)

A FORTRAN subroutine, *subroutine lagrange*, for implementing Eq. (5.122) is presented in Program 5.2.

Program 5.2. Differentiation of a Lagrange polynomial program.

```
      program main
c     main program to illustrate numerical diff. subroutines
      dimension x(3),f(3)
      call lagrange (ndim,ndeg,n,x,f,xp,fx,fxx,a,b,c)
 1000 format (' Lagrange polynomial'/' '/'  i',6x,'x',11x,'f'/' ')
      end

      subroutine lagrange (ndim,ndeg,n,x,f,xp,fx,fxx)
c     Lagrange polynomial differentiation
      dimension x(ndim),f(ndim)
      a=x(1)
      b=x(2)
      c=x(3)
      fx=(2.0*xp-(b+c))*f(1)/(a-b)/(a-c)+(2.0*xp-(a+c))*f(2)/(b-a)
     1 /(b-c)+(2.0*xp-(a+b))*f(3)/(c-a)/(c-b)
      fxx=2.0*f(1)/(a-b)/(a-c)+2.0*f(2)/(b-a)/(b-c)+2.0*f(3)/(c-a)
     1 /(c-b)
      return
      end
```

The data set used to illustrate *subroutine Lagrange* is taken from Example 5.1. The output generated by the program is presented in Output 5.2.

Output 5.2. Solution by differentiation of a Lagrange polynomial.

```
Lagrange polynomial

  i      x            f

  1    3.400000    0.294118
  2    3.500000    0.285714
  3    3.600000    0.277778

fx =   -0.081700  and  fxx =    0.046800
```

5.7.1.3. Divided Difference Polynomial

The first- and second-order derivatives of a divided difference polynomial are given by Eq. (5.11):

$$f'(x) \cong P'_n(x) = f_i^{(1)} + [2x - (x_0 + x_1)] f_i^{(2)}$$

$$+ [3x^2 - 2(x_0 + x_1 + x_2)x + (x_0 x_1 + x_0 x_2 + x_1 x_2)] f_i^{(3)} + \cdots$$

(5.123a)

$$f''(x) \cong P''_n(x) = 2f_i^{(2)} + [6x - 2(x_0 + x_1 + x_2)] f_i^{(3)} + \cdots \qquad (5.123b)$$

A FORTRAN subroutine, *subroutine divdiff*, for implementing Eq. (5.123) for $P_2(x)$ is presented in Program 5.3.

Program 5.3.   Differentiation of a divided difference polynomial program.

```
      program main
c     main program to illustrate numerical diff. subroutines
      dimension x(3),f(3)
      call divdiff (ndim,ndeg,n,x,f,xp,fx,fxx,a,b,c)
 1000 format (' Divided diff. poly.'/' '/'  i',6x,'x',11x,'f'/' ')
      end

      subroutine divdiff (ndim,ndeg,n,x,f,xp,fx,fxx)
c     divided difference polynomial differentiation
      dimension x(ndim),f(ndim)
      f11=(f(2)-f(1))/(x(2)-x(1))
      f21=(f(3)-f(2))/(x(3)-x(2))
      f12=(f21-f11)/(x(3)-x(1))
      fx=f11+(2.0*xp-x(1)-x(2))*f12
      fxx=2.0*f12
      return
      end
```

The data set used to illustrate *subroutine divdiff* is taken from Example 5.1. The output generated by the program is presented in Output 5.3.

Output 5.3.   Solution by differentiation of a divided difference polynomial.

```
Divided diff. polynomial

  i      x            f

  1   3.400000    0.294118
  2   3.500000    0.285714
  3   3.600000    0.277778

fx =   -0.081700  and  fxx =    0.046800
```

5.7.2 Derivatives of Equally Spaced Data

The first and second derivatives of a Newton forward-difference polynomial are given by Eqs. (5.26) and (5.27), respectively:

$$P_n'(x_0) = \frac{1}{h}\left(\Delta f_0 - \frac{1}{2}\Delta^2 f_0 + \frac{1}{3}\Delta^3 f_0 + \cdots\right) \tag{5.124a}$$

$$P_n''(x_0) = \frac{1}{h^2}(\Delta^2 f_0 - \Delta^3 f_0 + \cdots) \tag{5.124b}$$

The extrapolation formulas are given by Eqs. (5.116b) and (5.117).

$$\text{Error}(h/2) = \tfrac{1}{3}[f'(h/2) - f'(h)] \tag{5.125a}$$

$$\text{Extrapolated value} = f'(h/2) + \text{Error}(h/2) \tag{5.125b}$$

A FORTRAN subroutine, *subroutine deriv,* for evaluating Eqs (5.124) and (5.125) for $P_2(x)$ is presented in Program 5.4. It defines the data set and prints it, calls *subroutine deriv* to implement the solution, and prints the solution.

Program 5.4.   Differentiation of a quadratic Newton forward-difference polynomial program.

```
      program main
c     main program to illustrate numerical diff. subroutines
c     ndim  array dimension, n = 5 in this example
c     ndeg  degree of polynomial, ndeg = 2 in this example
c     num   number of derivative evaluations for extrapolation
c     n     number of data points
c     x     independent variable, x(i)
c     f     dependent variable, f(i)
c     fx    numerical approximation of derivative, fx(i,j)
      dimension x(5),f(5),dx(2),fx(2,2),fxx(2,2)
      data ndim,ndeg,num,n / 5, 2, 2, 5 /
      data (x(i),i=1,5) / 3.3, 3.4, 3.5, 3.6, 3.7 /
      data (f(i),i=1,5) / 0.303030, 0.294118, 0.285714, 0.277778,
     1                    0.270270 /
      write (6,1000)
      do i=1,n
         write (6,1005) i,x(i),f(i)
      end do
      call deriv (ndim,ndeg,num,n,x,f,dx,fx,fxx)
      write (6,1010)
      write (6,1020) dx(1),fx(1,1),fx(1,2)
      write (6,1030) dx(2),fx(2,1)
      write (6,1050)
      write (6,1040) dx(1),fxx(1,1),fxx(1,2)
      write (6,1030) dx(2),fxx(2,1)
      stop
```

```
1000 format (' Equally spaced data'/' '/'    i',6x,'x',11x,'f'/' ')
1005 format (i4,2f12.6)
1010 format (' '/10x,'dx',8x,'0(h**2)',5x,'0(h**4)'/' ')
1020 format (' fx ',3f12.6)
1030 format (4x,3f12.6)
1040 format (' fxx',3f12.6)
1050 format (' ')
     end

     subroutine deriv (ndim,ndeg,num,n,x,f,dx,fx,fxx)
c    numerical differentiation and extrapolation for P2(x)
     dimension x(ndim),f(ndim),dx(num),fx(num,num),fxx(num,num)
     dx(1)=x(3)-x(1)
     dx(2)=x(2)-x(1)
     fx(1,1)=0.5*(f(5)-f(1))/dx(1)
     fx(2,1)=0.5*(f(4)-f(2))/dx(2)
     fx(1,2)=(4.0*fx(2,1)-fx(1,1))/3.0
     fxx(1,1)=(f(5)-2.0*f(3)+f(1))/dx(1)**2
     fxx(2,1)=(f(4)-2.0*f(3)+f(2))/dx(2)**2
     fxx(1,2)=(4.0*fxx(2,1)-fxx(1,1))/3.0
     return
     end
```

The data set used to illustrate *subroutine deriv* is taken from Examples 5.2 and 5.7. The output generated by the program is presented in Output 5.4.

Output 5.4.   Solution by differentiation of a Newton forward-difference polynomial.

```
Equally spaced data

  i      x            f

  1    3.300000    0.303030
  2    3.400000    0.294118
  3    3.500000    0.285714
  4    3.600000    0.277778
  5    3.700000    0.270270

        dx          0(h**2)      0(h**4)

fx   0.200000    -0.081900    -0.081633
     0.100000    -0.081700

fxx  0.200000     0.046800     0.046800
     0.100000     0.046800
```

5.7.3. Packages for Numerical Differentiation

Numerous libraries and software packages are available for numerical differentiation. Many workstations and mainframe computers have such libraries attached to their operating systems.

Many commercial software packages contain numerical differentiation algorithms. Some of the more prominent packages are Matlab and Mathcad. More sophisticated packages, such as IMSL, MATHEMATICA, MACSYMA, and MAPLE, also contain routines for numerical differentiation. Finally, the book *Numerical Recipes* (Press et al., 1989) contains a routine for numerical differentiation.

5.8 SUMMARY

Procedures for numerical differentiation of discrete data and procedures for developing difference formulas are presented in this chapter. The numerical differentiation formulas are based on approximating polynomials. The direct fit polynomial, the Lagrange polynomial, and the divided difference polynomial work well for both unequally spaced data and equally spaced data. The Newton polynomials yield simple differentiation formulas for equally spaced data. Least squares fit polynomials can be used for large sets of data or sets of rough data.

Difference formulas, which approximate derivatives in terms of function values in the neighborhood of a particular point, are derived by both the Newton polynomial approach and the Taylor series approach. Difference formulas are used extensively in the numerical solution of differential equations.

After studying Chapter 5, you should be able to:

1. Describe the general features of numerical differentiation
2. Explain the procedure for numerical differentiation using direct fit polynomials
3. Apply direct fit polynomials to evaluate a derivative in a set of tabular data
4. Apply Lagrange polynomials to evaluate a derivative in a set of tabular data
5. Apply divided difference polynomials to evaluate a derivative in a set of tabular data
6. Describe the procedure for numerical differentiation using Newton forward-difference polynomials
7. Describe the procedure for numerical differentiation using Newton backward-difference polynomials
8. Describe the procedure for developing difference formulas from Newton difference polynomials
9. Develop a difference formula of any order for any derivative from Newton polynomials
10. Describe the procedure for developing difference formulas from Taylor series
11. Develop a difference formula of any order for any derivative by the Taylor series approach
12. Be able to use the difference formulas presented in Table 5.1
13. Explain the concepts of error estimation and extrapolation
14. Apply error estimation
15. Apply extrapolation

EXERCISE PROBLEMS

Table 1 gives values of $f(x) = \exp(x)$. This table is used in several of the problems in this chapter.

Table 1. Values of $f(x)$

x	$f(x)$	x	$f(x)$	x	$f(x)$
0.94	2.55998142	0.99	2.69123447	1.03	2.80106584
0.95	2.58570966	1.00	2.71828183	1.04	2.82921701
0.96	2.61169647	1.01	2.74560102	1.05	2.85765112
0.97	2.63794446	1.02	2.77319476	1.06	2.88637099
0.98	2.66445624				

5.2 Unequally Spaced Data

Direct Fit Polynomials

1. For the data in Table 1, evaluate $f'(1.0)$ and $f''(1.0)$ using direct fit polynomials with the following data points: (a) 1.00 and 1.01, (b) 1.00, 1.01, and 1.02, and (c) 1.00, 1.01, 1.02, and 1.03. Compute and compare the errors.

2. For the data in Table 1, evaluate $f'(1.0)$ and $f''(1.0)$ using direct fit polynomials with the following data points: (a) 0.99 and 1.00, (b) 0.98, 0.99, and 1.00, and (c) 0.97, 0.98, 0.99, and 1.00. Compute and compare the errors.

3. For the data in Table 1, evaluate $f'(1.0)$ and $f''(1.0)$ using direct fit polynomials with the following data points: (a) 0.99 and 1.01, and (b) 0.99, 1.00, and 1.01. Compute and compare the errors.

4. Compare the errors in Problems 1 to 3 and discuss.

5. For the data in Table 1, evaluate $f'(1.0)$ and $f''(1.0)$ using direct fit polynomials with the following data points: (a) 0.98 and 1.02, (b) 0.98, 1.00, and 1.02, (c) 0.96 and 1.04, and (d) 0.96, 1.00, and 1.04. Compute the errors and compare the ratios of the errors for parts (a) and (c) and parts (b) and (d). Compare the results with the results of Problem 3.

Difference formulas can be derived from direct fit polynomials by fitting a polynomial to a set of symbolic data and differentiating the resulting polynomial. The truncation errors of such difference formulas can be obtained by substituting Taylor series into the difference formulas to recover the derivative being approximated accompanied by all of the neglected terms in the approximation. Use the symbolic Table 2, where Δx is considered constant, to work the following problems. Note that the algebra is simplified considerably by letting the base point value of x be zero and the other values of x be multiples of the constant increment size Δx.

Table 2. Symbolic Values of $f(x)$

x	$f(x)$	x	$f(x)$
x_{i-2}	f_{i-2}	x_{i+1}	f_{i+1}
x_{i-1}	f_{i-1}	x_{i+2}	f_{i+2}
x_i	f_i		

6. Derive difference formulas for $f'(x)$ by direct polynomial fit using the following data points: (a) i and $i+1$, (b) $i-1$ and i, (c) $i-1$ and $i+1$, (d) $i-1, i$, and $i+1$, (e) $i, i+1$, and $i+2$, and (f) $i-2, i-1, i, i+1$, and $i+2$.

For each result, determine the leading truncation error term. Compare with the results presented in Table 5.1.

7. Derive difference formulas for $f'''(x)$ by direct polynomial fit using the following data points: (a) $i-1, i$, and $i+1$, (b) $i, i+1$, and $i+2$, (c) $i-2, i-1, i, i+1$, and $i+2$, and (d) $i, i+1, i+2$, and $i+3$. For each result, determine the leading truncation error term. Compare with the results presented in Table 5.1.

Lagrange Polynomials

8. Work Problem 1 using Lagrange polynomials.
9. Work Problem 2 using Lagrange polynomials.
10. Work Problem 3 using Lagrange polynomials.
11. Work Problem 5 using Lagrange polynomials.
12. Derive differentiation formulas for the third-degree Lagrange polynomial.

Divided Difference Polynomials

13. Work Problem 1 using divided difference polynomials.
14. Work Problem 2 using divided difference polynomials.
15. Work Problem 3 using divided difference polynomials.
16. Work Problem 5 using divided difference polynomials.

5.3 Equally Spaced Data

The data presented in Table 1 are used in the following problems. Construct a difference table for that set of data through third differences for use in these problems.

17. For the data in Table 1, evaluate $f'(1.0)$ and $f''(1.0)$ using Newton forward-difference polynomials of orders 1, 2, and 3 with the following points: (a) 1.00 to 1.03, and (b) 1.00 to 1.06. Compare the errors and ratios of the errors for the two increment sizes.
18. For the data in Table 1, evaluate $f'(1.0)$ and $f''(1.0)$ using Newton backward-difference polynomials of orders 1, 2, and 3 with the following points: (a) 0.97 to 1.00, and (b) 0.94 to 1.00. Compare the errors and the ratios of the errors for the two increment sizes.
19. For the data in Table 1, evaluate $f'(1.0)$ and $f''(1.0)$ using Newton forward-difference polynomials of orders 1 and 2 with the following points: (a) 0.99 to 1.01, and (b) 0.98 to 1.02. Compare the errors and the ratios of the errors for these two increment sizes. Compare with the results of Problems 17 and 18 and discuss.
20. Derive Eq. (5.23).
21. Derive Eq. (5.25).
22. Derive Eq. (5.40).
23. Derive Eq. (5.41).

Difference formulas can be derived from Newton polynomials by fitting a polynomial to a set of symbolic data and differentiating the resulting polynomial. The truncation error can be determined from the error term of the Newton polynomial. The symbolic data in Table 2 are used in the following problems. Construct a difference table for that set of data.

24. Derive difference formulas for $f'(x)$ using Newton forward-difference polynomials using the following data points: (a) i and $i + 1$, (b) $i - 1$ and i, (c) $i - 1$ and $i + 1$, (d) $i - 1, i$, and $i + 1$, (e) $i, i + 1$, and $i + 2$, and (f) $i - 2, i - 1, i, i + 1$, and $i + 2$. For each result, determine the leading truncation error term. Compare with the results presented in Table 5.1.

25. Derive difference formulas for $f''(x)$ using Newton forward-difference polynomials using the following data points: (a) $i - 1, i$, and $i + 1$, (b) $i, i + 1$, and $i + 2$, (c) $i - 2, i - 1, i, i + 1$, and $i + 2$, and (d) $i, i + 1, i + 2$, and $i + 3$. For each result, determine the leading truncation error term. Compare with the results presented in Table 5.1

5.4 Taylor Series Approach

26. Derive Eqs. (5.93) to (5.96).
27. Derive Eqs. (5.97) to (5.106).
28. Derive Eqs. (5.107) to (5.112).

5.5 Difference Formulas

29. Verify Eq. (5.96) by substituting Taylor series for the function values to recover the first derivative and the leading truncation error term.
30. Verify Eq. (5.99) by substituting Taylor series for the function values to recover the first derivative and the leading truncation error term.
31. Verify Eq. (5.109) by substituting Taylor series for the function values to recover the second derivative and the leading truncation error term.
32. Verify Eq. (5.92) by substituting Taylor series for the function values to recover the first derivative and the leading truncation error term.

5.6 Error Estimation and Extrapolation

33. For the data in Table 1, evaluate $f'(1.0)$ using Eq. (5.99) for $\Delta x = 0.04, 0.02$, and 0.01. (a) Estimate the error for the $\Delta x = 0.02$ result. (b) Estimate the error for the $\Delta x = 0.01$ result. (c) Extrapolate the results to $0(\Delta x^4)$.
34. For the data in Table 1, evaluate $f''(1.0)$ using Eq. (5.109) for $\Delta x = 0.04, 0.02$, and 0.01. (a) Estimate the error for the $\Delta x = 0.02$ result. (b) Estimate the error for the $\Delta x = 0.01$ result. (c) Extrapolate the results to $0(\Delta x^4)$.

5.7 Programs

35. Implement the program presented in Section 5.7.1 for the differentiation of a quadratic direct fit polynomial. Check out the program using the given data.
36. Solve Problems 1b and 2b using the program.
37. Modify the program to consider a linear direct fit polynomial. Solve Problems 1a and 2a using the program.
38. Extend the program to consider a cubic direct fit polynomial. Solve Problems 1c and 2c using the program.
39. Implement the program presented in Section 5.7.2 for the differentiation of a quadratic Lagrange polynomial. Check out the program using the given data.
40. Solve Problems 1b and 2b using the program.

41. Modify the program to consider a linear Lagrange polynomial. Solve Problems 1a and 2a using the program.
42. Extend the program to consider a cubic Lagrange polynomial. Solve Problems 1c and 2c using the program.
43. Implement the program presented in Section 5.7.3 for the differentiation of a quadratic divided difference polynomial. Check out the program using the given data.
44. Solve Problems 1b and 2b using the program.
45. Modify the program to consider a linear divided difference polynomial. Solve Problems 1a and 2a using the program.
46. Extend the program to consider a cubic divided difference polynomial. Solve Problems 1c and 2c using the program.
47. Implement the program presented in Section 5.7.4 for the differentiation of a quadratic Newton forward-difference polynomial. Check out the program using the given data.
48. Solve Problems 1b and 2b using the program.
49. Modify the program to consider a linear Newton forward-difference polynomial. Solve Problems 1a and 2a using the program.
50. Extend the program to consider a cubic Newton forward-difference polynomial. Solve Problems 1c and 2c using the program.

APPLIED PROBLEMS

51. When a fluid flows over a surface, the shear stress τ (N/m^2) at the surface is given by the expression

$$\tau = \mu \frac{du}{dy}\bigg|_{\text{surface}} \tag{1}$$

where μ is the viscosity (N-s/m^2), u is the velocity parallel to the surface (m/s), and y is the distance normal to the surface (cm). Measurements of the velocity of an air stream flowing above a surface are made with an LDV (laser-Doppler-velocimeter). The values given in Table 3 were obtained.

Table 3. Velocity Measurements

y	u	y	u
0.0	0.00	2.0	88.89
1.0	55.56	3.0	100.00

At the local temperature, $\mu = 0.00024$ N-s/m^2. Calculate (a) the difference table for $u(y)$, (b) du/dy at the surface based on first-, second-, and third-order polynomials, (c) the corresponding values of the shear stress at the surface, and (d) the shear force acting on a flat plate 10 cm long and 5 cm wide.

52.   When a fluid flows over a surface, the heat transfer rate $\dot{q}$ (J/s) to the surface is given by the expression

$$\dot{q} = -kA \frac{dT}{dy}\bigg|_{\text{surface}} \qquad (2)$$

where k is the thermal conductivity (J/s-m-K), T is the temperature (K), and y is the distance normal to the surface (cm). Measurements of the temperature of an air stream flowing above a surface are made with a thermocouple. The values given in Table 4 were obtained.

Table 4.   Temperature Measurements

y	T	y	T
0.0	1000.00	2.0	355.56
1.0	533.33	3.0	300.00

At the average temperature, $k = 0.030$ J/s-m-K. Calculate (a) the difference table for $T(y)$, (b) dT/dy at the surface based on first-, second-, and third-order polynomials, (c) the corresponding values of the heat flux $\dot{q}/A$ at the surface, and (d) the heat transfer to a flat plate 10 cm long and 5 cm wide.

6

Numerical Integration

6.1 INTRODUCTION

A set of tabular data is illustrated in Figure 6.1 in the form of a set of $[x, f(x)]$ pairs. The function $f(x)$ is known at a finite set of discrete values of x. Interpolation within a set of discrete data is presented in Chapter 4. Differentiation within a set of tabular data is presented in Chapter 5. *Integration* of a set of tabular data is presented in this chapter. The discrete data in Figure 6.1 are actually values of the function $f(x) = 1/x$, which is used as the example problem in this chapter.

The evaluation of integrals, a process known as *integration* or *quadrature*, is required in many problems in engineering and science.

$$I = \int_a^b f(x)\,dx \tag{6.1}$$

285

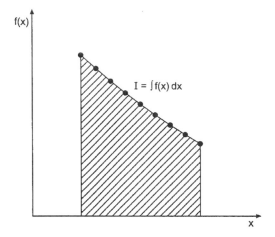

f(x)

$I = \int f(x)\,dx$

x	f(x)
3.1	0.32258065
3.2	0.31250000
3.3	0.30303030
3.4	0.29411765
3.5	0.28571429
3.6	0.27777778
3.7	0.27027027
3.8	0.26315789
3.9	0.25641026

Figure 6.1 Integral of tabular data.

The function $f(x)$, which is to be integrated, may be a known function or a set of discrete data. Some known functions have an exact integral, in which case Eq. (6.1) can be evaluated exactly in closed form. Many known functions, however, do not have an exact integral, and an approximate numerical procedure is required to evaluate Eq. (6.1). In many cases, the function $f(x)$ is known only at a set of discrete points, in which case an approximate numerical procedure is again required to evaluate Eq. (6.1). The evaluation of integrals by approximate numerical procedures is the subject of this chapter.

Numerical integration (quadrature) formulas can be developed by fitting approximating functions (e.g., polynomials) to discrete data and integrating the approximating function:

$$I = \int_a^b f(x)\,dx \cong \int_a^b P_n(x)\,dx \qquad (6.2)$$

This process is illustrated in Figure 6.2.

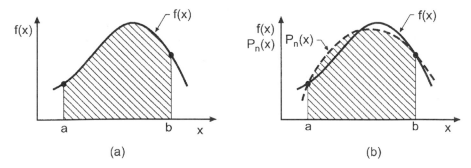

(a)

(b)

Figure 6.2 Numerical integration. (a) Exact integral. (b) Approximate integral.

Several types of problems arise. The function to be integrated may be known only at a finite set of discrete points. In that case, an approximating polynomial is fit to the discrete points, or several subsets of the discrete points, and the resulting polynomial, or polynomials, is integrated. The polynomial may be fit exactly to a set of points by the methods presented in Sections 4.3 to 4.6, or by a least squares fit as described in Section 4.10. In either case, the degree of the approximating polynomial chosen to represent the discrete data is the only parameter under our control.

When a known function is to be integrated, several parameters are under our control. The total number of discrete points can be chosen arbitrarily. The degree of the approximating polynomial chosen to represent the discrete data can be chosen. The locations of the points at which the known function is discretized can also be chosen to enhance the accuracy of the procedure.

Procedures are presented in this chapter for all of the situations discussed above. Direct fit polynomials are applied to prespecified unequally spaced data. Integration formulas based on Newton forward-difference polynomials, which are called *Newton-Cotes formulas*, are developed for equally spaced data. An important method, Romberg integration, based on extrapolation of solutions for successively halved increments, is presented. Adaptive integration, a procedure for minimizing the number of function evaluations required to integrate a known function, is discussed. Gaussian quadrature, which specifies the locations of the points at which known functions are discretized, is discussed. The numerical evaluation of multiple integrals is discussed briefly.

The simple function

$$f(x) = \frac{1}{x} \qquad\qquad (6.3)$$

is considered in this chapter to illustrate numerical integration methods. In particular,

$$I = \int_{3.1}^{3.9} \frac{1}{x}\, dx = \ln(x)\Big|_{3.1}^{3.9} = \ln\left(\frac{3.9}{3.1}\right) = 0.22957444\ldots \qquad (6.4)$$

The procedures presented in this chapter for evaluating integrals lead directly into integration techniques for ordinary differential equations, which is discussed in Chapters 7 and 8.

The organization of Chapter 6 is illustrated in Figure 6.3. After the background material presented in this section, integration using direct fit polynomials as the approximating function is discussed. This section is followed by a development of Newton-Cotes formulas for equally spaced data, among which the trapezoid rule and Simpson's 1/3 rule are the most useful. Romberg integration, an extremely accurate and efficient algorithm which is based on extrapolation of the trapezoid rule, is presented next. Adaptive integration is then discussed. The following section presents Gaussian quadrature, an extremely accurate procedure for evaluating the integral of known functions in which the points at which the integral is evaluated is chosen in a manner which doubles the order of the integration formula. A brief introduction to multiple integrals follows. A summary closes the chapter and presents a list of what you should be able to do after studying Chapter 6.

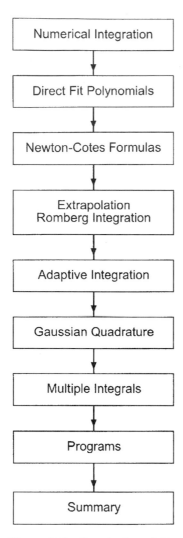

Figure 6.3   Organization of Chapter 6.

6.2 DIRECT FIT POLYNOMIALS

A straightforward numerical integration procedure that can be used for both unequally spaced data and equally spaced data is based on fitting the data by a direct fit polynomial and integrating that polynomial. Thus,

$$ f(x) \cong P_n(x) = a_0 + a_1 x + a_2 x^2 + \cdots \tag{6.5} $$

where $P_n(x)$ is determined by one of the following methods:

1. Given $N = n + 1$ sets of discrete data, $[x_i, f(x_i)]$, determine the exact nth-degree polynomial that passes through the data points, as discussed in Section 4.3.
2. Given $N > n + 1$ sets of discrete data, $[x_i, f(x_i)]$, determine the least squares nth-degree polynomial that best fits the data points, as discussed in Section 4.10.

3. Given a known function $f(x)$ evaluate $f(x)$ at N discrete points and fit a polynomial by an exact fit or a least squares fit.

After the approximating polynomial has been fit, the integral becomes

$$I = \int_a^b f(x)\,dx \cong \int_a^b P_n(x)\,dx \tag{6.6}$$

Substituting Eq. (6.5) into Eq. (6.6) and integrating yields

$$I = \left(a_0 x + a_1 \frac{x^2}{2} + a_2 \frac{x^3}{3} + \cdots \right)_a^b \tag{6.7}$$

Introducing the limits of integration and evaluating Eq. (6.7) gives the value of the integral.

Example 6.1. Direct fit polynomial

Let's solve the example problem presented in Section 6.1 by a direct fit polynomial. Recall:

$$I = \int_{3.1}^{3.9} \frac{1}{x}\,dx \cong \int_{3.1}^{3.9} P_n(x)\,dx \tag{6.8}$$

Consider the following three data points from Figure 6.1:

x	$f(x)$
3.1	0.32258065
3.5	0.28571429
3.9	0.25641026

Fit the quadratic polynomial, $P_2(x) = a_0 + a_1 x + a_2 x^2$, to the three data points by the direct fit method:

$$0.32258065 = a_0 + a_1(3.1) + a_2(3.1)^2 \tag{6.9a}$$
$$0.28571429 = a_0 + a_1(3.5) + a_2(3.5)^2 \tag{6.9b}$$
$$0.25641026 = a_0 + a_1(3.9) + a_2(3.9)^2 \tag{6.9c}$$

Solving for a_0, a_1, and a_2 by Gauss elimination gives

$$P_2(x) = 0.86470519 - 0.24813896x + 0.02363228x^2 \tag{6.10}$$

Substituting Eq. (6.10) into Eq. (6.8) and integrating gives

$$I = [(0.86470519)x + \tfrac{1}{2}(-0.24813896)x^2 + \tfrac{1}{3}(0.02363228)x^3]_{3.1}^{3.9} \tag{6.11}$$

Evaluating Eq. (6.11) yields

$$\boxed{I = 0.22957974} \tag{6.12}$$

The error is Error $= 0.22957974 - 0.22957444 = 0.00000530$.

6.3 NEWTON-COTES FORMULAS

The direct fit polynomial procedure presented in Section 6.2 requires a significant amount of effort in the evaluation of the polynomial coefficients. When the function to be integrated is known at equally spaced points, the Newton forward-difference polynomial presented in Section 4.6.2 can be fit to the discrete data with much less effort, thus significantly decreasing the amount of effort required. The resulting formulas are called *Newton-Cotes* formulas. Thus,

$$I = \int_a^b f(x)\, dx \cong \int_a^b P_n(x)\, dx \tag{6.13}$$

where $P_n(x)$ is the Newton forward-difference polynomial, Eq. (4.88):

$$P_n(x) = f_0 + s\, \Delta f_0 + \frac{s(s-1)}{2} \Delta^2 f_0 + \frac{s(s-1)(s-2)}{6} \Delta^3 f_0$$
$$+ \cdots + \frac{s(s-1)(s-2)\cdots[s-(n-1)]}{n!} \Delta^n f_0 + \text{Error} \tag{6.14}$$

where the interpolating parameter s is given by

$$s = \frac{x - x_0}{h} \rightarrow x = x_0 + sh \tag{6.15}$$

and the Error term is

$$\text{Error} = \binom{s}{n+1} h^{n+1} f^{(n+1)}(\xi) \qquad x_0 \le x \le x_n \tag{6.16}$$

Equation (6.13) requires that the approximating polynomial be an explicit function of x, whereas Eq. (6.14) is implicit in x. Either Eq. (6.14) must be made explicit in x by introducing Eq. (6.16) into Eq. (6.14), or the second integral in Eq. (6.13) must be transformed into an explicit function of s, so that Eq. (6.14) can be used directly. The first approach leads to a complicated result, so the second approach is taken. Thus,

$$I = \int_a^b f(x)\, dx \cong \int_a^b P_n(x)\, dx = h \int_{s(a)}^{s(b)} P_n(s)\, ds \tag{6.17}$$

where, from Eq. (6.15)

$$dx = h\, ds \tag{6.18}$$

The limits of integration, $x = a$ and $x = b$, are expressed in terms of the interpolating parameter s by choosing $x = a$ as the base point of the polynomial, so that $x = a$ corresponds to $s = 0$ and $x = b$ corresponds to $s = s$. Introducing these results into Eq. (6.17) yields

$$I = h \int_0^s P_n(x_0 + sh)\, ds \tag{6.19}$$

Each choice of the degree n of the interpolating polynomial yields a different Newton-Cotes formula. Table 6.1 lists the more common formulas. Higher-order formulas have been developed [see Abramowitz and Stegun (1964)], but those presented in Table 6.1 are sufficient for most problems in engineering and science. The rectangle rule has

Table 6.1 Newton-Cotes
Formulas

n	Formula
0	Rectangle rule
1	Trapezoid rule
2	Simpson's 1/3 rule
3	Simpson's 3/8 rule

poor accuracy, so it is not considered further. The other three rules are developed in this section.

Some terminology must be defined before proceeding with the development of the Newton-Cotes formulas. Figure 6.4 illustrates the region of integration. The distance between the lower and upper limits of integration is called the *range of integration*. The distance between any two data points is called an *increment*. A linear polynomial requires one *increment* and two data points to obtain a fit. A quadratic polynomial requires two increments and three data points to obtain a fit. And so on for higher-degree polynomials. The group of increments required to fit a polynomial is called an *interval*. A linear polynomial requires an interval consisting of only one increment. A quadratic polynomial requires an interval containing two increments. And so on. The total range of integration can consist of one or more intervals. Each interval consists of one or more increments, depending on the degree of the approximating polynomial.

6.3.1 The Trapezoid Rule

The trapezoid rule for a single interval is obtained by fitting a first-degree polynomial to two discrete points, as illustrated in Figure 6.5. The upper limit of integration x_1 corresponds to $s = 1$. Thus, Eq. (6.19) gives

$$\Delta I = h \int_0^1 (f_0 + s\,\Delta f_0)\,ds = h\left(sf_0 + \frac{s^2}{2}\Delta f_0\right)_0^1 \tag{6.20}$$

where $h = \Delta x$. Evaluating Eq. (6.20) and introducing $\Delta f_0 = (f_1 - f_0)$ yields

$$\Delta I = h(f_0 + \tfrac{1}{2}\Delta f_0) = h[f_0 + \tfrac{1}{2}(f_1 - f_0)] \tag{6.21}$$

where ΔI denotes the integral for a single interval. Simplifying yields the trapezoid rule for a single interval:

$$\boxed{\Delta I = \tfrac{1}{2}h(f_0 + f_1)} \tag{6.22}$$

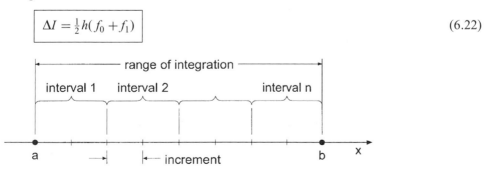

Figure 6.4 Range, intervals, and increments.

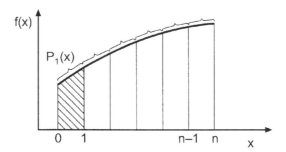

Figure 6.5 The trapezoid rule.

The composite trapezoid rule is obtained by applying Eq. (6.22) over all the intervals of interest. Thus,

$$I = \sum_{i=0}^{n-1} \Delta I_i = \sum_{i=0}^{n-1} \tfrac{1}{2} h_i (f_i + f_{i+1}) \tag{6.23}$$

where $h_i = (x_{i+1} - x_i)$. Equation (6.23) does not require equally spaced data. When the data are equally spaced, Eq. (6.23) simplifies to

$$I = \tfrac{1}{2} h (f_0 + 2f_1 + 2f_2 + \cdots + 2f_{n-1} + f_n) \tag{6.24}$$

where $\Delta x_i = \Delta x = h = \text{constant}$.

The error of the trapezoid rule for a single interval is obtained by integrating the error term given by Eq. (6.16). Thus,

$$\text{Error} = h \int_0^1 \frac{s(s-1)}{2} h^2 f''(\xi) \, ds = -\tfrac{1}{12} h^3 f''(\xi) = 0(h^3) \tag{6.25}$$

Thus, the local error is $0(h^3)$. The total error for equally spaced data is given by

$$\sum_{i=0}^{n-1} \text{Error} = \sum_{i=0}^{n-1} -\tfrac{1}{12} h^3 f''(\xi) = n[-\tfrac{1}{12} h^3 f''(\bar{\xi})] \tag{6.26}$$

where $x_0 \leq \bar{\xi} \leq x_n$. The number of steps $n = (x_n - x_0)/h$. Therefore,

$$\text{Total Error} = -\tfrac{1}{12}(x_n - x_0) h^2 f''(\bar{\xi}) = 0(h^2) \tag{6.27}$$

Thus, the global (i.e., total) error is $0(h^2)$.

Example 6.2. The trapezoid rule

Let's solve the example problem presented in Section 6.1 by the trapezoid rule. Recall that $f(x) = 1/x$. Solving the problem for the range of integration consisting of only one interval of $h = 0.8$ gives

$$I(h = 0.8) = \frac{0.8}{2}(0.32258065 + 0.25641026) = 0.23159636 \tag{6.28}$$

Let's break the total range of integration into two intervals of $h = 0.4$ and apply the composite rule. Thus,

$$I(h = 0.4) = \frac{0.4}{2}[0.32258065 + 2(0.28571429) + 0.25641026] = 0.23008389$$

(6.29)

For four intervals of $h = 0.2$, the composite rule yields

$$I(h = 0.2) = \frac{0.2}{2}[0.32258065 + 2(0.30303030 + 0.28571429$$
$$+ 0.27027027) + 0.25641026]$$
$$= 0.22970206$$

(6.30)

Finally, for eight intervals of $h = 0.1$,

$$I(h = 0.1) = \frac{0.1}{2}[0.32258065 + 2(0.31250000 + \cdots + 0.26315789)$$
$$+ 0.25641026]$$
$$= 0.22960636$$

(6.31)

Recall that the exact answer is $I = 0.22957444$.

The results are tabulated in Table 6.2, which also presents the errors and the ratios of the errors between successive interval sizes. The global error of the trapezoid rule is $0(h^2)$. Thus, for successive interval halvings,

$$\text{Ratio} = \frac{E(h)}{E(h/2)} = \frac{0(h^2)}{0(h/2)^2} = 2^2 = 4$$

(6.32)

The results presented in Table 6.2 illustrate the second-order behavior of the trapezoid rule.

6.3.2 Simpson's 1/3 Rule

Simpson's 1/3 rule is obtained by fitting a second-degree polynomial to three equally spaced discrete points, as illustrated in Figure 6.6. The upper limit of integration x_2 corresponds to $s = 2$. Thus, Eq. (6.19) gives

$$\Delta I = h \int_0^2 \left[f_0 + s\,\Delta f_0 + \frac{s(s-1)}{2}\Delta^2 f_0 \right] ds$$

(6.33)

Table 6.2 Results for the Trapezoid Rule

h	I	Error	Ratio
0.8	0.23159636	−0.00202192	
			3.97
0.4	0.23008389	−0.00050945	
			3.99
0.2	0.22970206	−0.00012762	
			4.00
0.1	0.22960636	−0.00003192	

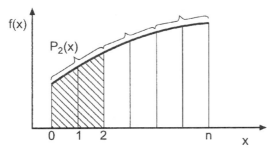

Figure 6.6   Simpson's 1/3 rule.

Performing the integration, evaluating the result, and introducing the expressions for Δf_0 and $\Delta^2 f_0$, yields Simpson's 1/3 rule for a single interval of two increments:

$$\Delta I = \tfrac{1}{3} h(f_0 + 4f_1 + f_2) \tag{6.34}$$

The composite Simpson's 1/3 rule for equally spaced points is obtained by applying Eq. (6.34) over the entire range of integration. Note that the total number of increments must be even. Thus,

$$I = \tfrac{1}{3} h(f_0 + 4f_1 + 2f_2 + 4f_3 + \cdots + 4f_{n-1} + f_n) \tag{6.35}$$

The error of Simpson's 1/3 rule for a single interval of two increments is obtained by evaluating the error term given by Eq. (6.16). Thus,

$$\text{Error} = h \int_0^2 \frac{s(s-1)(s-2)}{6} h^3 f'''(\xi) \, ds = 0 \tag{6.36}$$

This surprising result does not mean that the error is zero. It simply means that the cubic term is identically zero, and the error is obtained from the next term in the Newton forward-difference polynomial. Thus,

$$\text{Error} = h \int_0^2 \frac{s(s-1)(s-2)(s-3)}{24} h^4 f^{\text{iv}}(\xi) \, ds = -\tfrac{1}{90} h^5 f^{\text{iv}}(\xi) \tag{6.37}$$

Thus, the local error is $0(h^5)$. By an analysis similar to that performed for the trapezoid rule, it can be shown that the global error is $0(h^4)$.

Example 6.3.   Simpson's 1/3 Rule

Let's solve the example problem presented in Section 6.1 using Simpson's 1/3 rule. Recall that $f(x) = 1/x$. Solving the problem for two increments of $h = 0.4$, the minimum permissible number of increments for Simpson's 1/3 rule, and one interval yields

$$I(h = 0.4) = \frac{0.4}{3}[0.32258065 + 4(0.28571429) + 0.25641026] = 0.22957974$$

$$\tag{6.38}$$

Table 6.3   Results for Simpson's 1/3 Rule

h	I	Error	Ratio
0.4	0.22957974	−0.00000530	
			15.59
0.2	0.22967478	−0.00000034	
			15.45
0.1	0.22957446	−0.00000002	

Breaking the total range of integration into four increments of $h = 0.2$ and two intervals and applying the composite rule yields:

$$I(h = 0.2) = \frac{0.2}{3}[0.32258065 + 4(0.30303030) + 2(0.28571429)$$
$$+ 4(0.27027027) + 0.25641026]$$
$$= 0.22957478 \tag{6.39}$$

Finally, for eight increments of $h = 0.1$ and four intervals,

$$I(h = 0.1) = \frac{0.1}{3}[0.32258065 + 4(0.31250000) + 2(0.30303030)$$
$$+ 4(0.29411765) + 2(0.28571429) + 4(0.27777778)$$
$$+ 2(0.27027027) + 4(0.26315789) + 0.25641026]$$
$$= 0.22957446 \tag{6.40}$$

Recall that the exact answer is $I = 0.22957444$.

The results are tabulated in Table 6.3, which also presents the errors and the ratios of the errors between successive increment sizes. The global error of Simpson's 1/3 rule is $0(h^4)$. Thus, for successive increment halvings,

$$\text{Ratio} = \frac{E(h)}{E(h/2)} = \frac{0(h)^4}{0(h/2)^4} = 2^4 = 16 \tag{6.41}$$

The results presented in Table 6.3 illustrate the fourth-order behavior of Simpson's 1/3 rule.

6.3.3   Simpson's 3/8 Rule

Simpson's 3/8 rule is obtained by fitting a third-degree polynomial to four equally spaced discrete points, as illustrated in Figure 6.7. The upper limit of integration x_3 corresponds to $s = 3$. Thus, Eq. (6.19) gives

$$\Delta I = h \int_0^3 \left[f_0 + s\,\Delta f_0 + \frac{s(s-1)}{2}\Delta^2 f_0 + \frac{s(s-1)(s-2)}{6}\Delta^3 f_0 \right] ds \tag{6.42}$$

Performing the integration, evaluating the result, and introducing expressions for Δf_0, $\Delta^2 f_0$, and $\Delta^3 f_0$ yields Simpson's 3/8 rule for a single interval of three increments:

$$\boxed{\Delta I = \tfrac{3}{8}h(f_0 + 3f_1 + 3f_2 + f_3)} \tag{6.43}$$

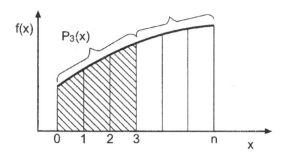

Figure 6.7 Simpson's 3/8 rule.

The composite Simpson's 3/8 rule for equally spaced points is obtained by applying Eq. (6.43) over the entire range of integration. Note that the total number of increments must be a multiple of three. Thus,

$$I = \tfrac{3}{8}h(f_0 + 3f_1 + 3f_2 + 2f_3 + 3f_4 + \cdots + 3f_{n-1} + f_n) \tag{6.44}$$

The error of Simpson's 3/8 rule for a single interval of three increments is obtained by evaluating the error term given by Eq. (6.16). Thus,

$$\text{Error} = h \int_0^3 \frac{s(s-1)(s-2)(s-3)}{24} h^4 f^{iv}(\xi)\, ds = -\tfrac{3}{80} h^5 f^{iv}(\xi) \tag{6.45}$$

Thus, the local error is $0(h^5)$ and the global error is $0(h^4)$.

Simpson's 1/3 rule and Simpson's 3/8 rule have the same order, $0(h^4)$, as shown by Eqs. (6.37) and (6.45). The coefficient in the local error of Simpson's 1/3 rule is $-1/90$, whereas the corresponding coefficient for Simpson's 3/8 rule is $-3/80$. Consequently, Simpson's 1/3 rule should be more accurate than Simpson's 3/8 rule. In view of this result, what use, if any, is Simpson's 3/8 rule? Simpson's 3/8 rule is useful when the total number of increments is odd. Three increments can be evaluated by the 3/8 rule, and the remaining even number of increments can be evaluated by the 1/3 rule.

6.3.4 Higher-Order Newton-Cotes Formulas

Numerical integration formulas based on equally spaced increments are called Newton-Cotes formulas. The trapezoid rule and Simpson's 1/3 and 3/8 rules are the first three Newton-Cotes formulas. The first 10 Newton-Cotes formulas are presented in Abramowitz and Stegun (1964). Newton-Cotes formulas can be expressed in the general form:

$$I = \int_a^b f(x)\, dx = n\beta h(\alpha_0 f_0 + \alpha_1 f_1 + \cdots) + \text{Error} \tag{6.46}$$

where n denotes both the number of increments and the degree of the polynomial, β and α_i are coefficients, and Error denotes the local error term. Table 6.4 presents n, β, α_i, and Error for the first seven Newton-Cotes formulas.

Table 6.4 Newton-Cotes Formulas

n	β	α_0	α_1	α_2	α_3	α_4	α_5	α_6	α_7	Local Error
1	1/2	1	1							$-1/12 f^{(2)} h^3$
2	1/6	1	4	1						$-1/90 f^{(4)} h^5$
3	1/8	1	3	3	1					$-3/80 f^{(4)} h^5$
4	1/90	7	32	12	32	7				$-8/945 f^{(6)} h^7$
5	1/288	19	75	50	50	75	19			$-275/12096 f^{(6)} h^7$
6	1/840	41	216	27	272	27	216	41		$-9/1400 f^{(8)} h^9$
7	1/17280	751	3577	1323	2989	2989	1323	3577	751	$-8183/518400 f^{(8)} h^9$

6.4 EXTRAPOLATION AND ROMBERG INTEGRATION

When the functional form of the error of a numerical algorithm is known, the error can be estimated by evaluating the algorithm for two different increment sizes, as discussed in Section 5.6. The error estimate can be used both for error control and extrapolation. Recall the error estimation formula, Eq. (5.116b), written for the process of numerical integration, that is, with $f(h) = I(h)$. Thus,

$$\text{Error}(h/R) = \frac{1}{R^n - 1}[I(h/R) - I(h)] \tag{6.47}$$

where R is the ratio of the increment sizes and n is the global order of the algorithm. The extrapolation formula is given by Eq. (5.117):

$$\text{Extrapolated value} = f(h/R) + \text{Error}(h/R) \tag{6.48}$$

When extrapolation is applied to numerical integration by the trapezoid rule, the result is called *Romberg integration*. Recall the composite trapezoid rule, Eq. (6.24):

$$I = \sum_{i=0}^{n-1} \Delta I_i = \tfrac{1}{2} h(f_0 + 2f_1 + 2f_2 + \cdots + 2f_{n-1} + f_n) \tag{6.49}$$

It can be shown that the error of the composite trapezoid rule has the functional form

$$\text{Error} = C_1 h^2 + C_2 h^4 + C_3 h^6 + \cdots \tag{6.50}$$

Thus, the basic algorithm is $0(h^2)$, so $n = 2$. The following error terms increase in order in increments of 2.

Let's apply the trapezoid rule for a succession of smaller and smaller increment sizes, where each successive increment size is one-half of the preceding increment size. Thus, $R = h/(h/2) = 2$. Applying the error estimation formula, Eq. (6.47), gives

$$\text{Error}(h/2) = \frac{1}{2^n - 1}[I(h/2) - I(h)] \tag{6.51}$$

For the trapezoid rule itself, $n = 2$, and Eq. (6.51) becomes

$$\text{Error}(h/2) = \tfrac{1}{3}[I(h/2) - I(h)] \tag{6.52}$$

Equation (6.52) can be used for error estimation and error control.

Applying the extrapolation formula, Eq. (6.48), for $R = 2$ gives

$$\text{Extrapolated value} = f(h/2) + \text{Error}(h/2) + 0(h^4) + \cdots \tag{6.53}$$

Equation (6.53) shows that the result obtained by extrapolating the $0(h^2)$ trapezoid rule is $0(h^4)$.

If two extrapolated $0(h^4)$ values are available, which requires three $0(h^2)$ trapezoid rule results, those two values can be extrapolated to obtain an $0(h^6)$ value by applying Eq. (6.47) with $n = 4$ to estimate the $0(h^4)$ error, and adding that error to the more accurate $0(h^4)$ value. Successively higher-order extrapolations can be performed until round-off error masks any further improvement. Each successive higher-order extrapolation begins with an additional application of the $0(h^2)$ trapezoid rule, which is then combined with the previously extrapolated values to obtain the next higher-order extrapolated result.

Example 6.4.   Romberg integration

Let's apply extrapolation to the results obtained in Example 6.2, in which the trapezoid rule is used to solve the example problem presented in Section 6.1. Recall that the exact answer is $I = 0.22957444$. Substituting $I(h = 0.8)$ and $I(h = 0.4)$ from Table 6.2 into Eq. (6.52) gives

$$\text{Error}(h/2) = \tfrac{1}{3}(0.23008389 - 0.23159636) = -0.00050416 \tag{6.54}$$

Substituting this result into Eq. (6.53) gives

$$\text{Extrapolated value} = 0.23008389 + (-0.00050416) = 0.22957973 \tag{6.55}$$

Repeating the procedure for the $h = 0.4$ and $h = 0.2$ results gives Error $= -0.00012728$ and Extrapolated value $= 0.22957478$. Both of the extrapolated values are $0(h^4)$. Substituting the two $0(h^4)$ extrapolated values into Eq. (6.51), with $n = 4$, gives

$$\text{Error}(h/2) = \frac{1}{2^4 - 1}(0.22957478 - 0.22957973) = -0.00000033 \tag{6.56}$$

Substituting this result into Eq. (6.53) gives

$$\text{Extrapolated value} = 0.22957478 + (-0.00000033) = 0.22957445 \tag{6.57}$$

These results, and the results of one more application of the trapezoid rule and its associated extrapolations, are presented in Table 6.5.

The $0(h^4)$ results are identical to the results for the $0(h^4)$ Simpson's 1/3 rule presented in Table 6.3. The second $0(h^6)$ result agrees with the exact value to eight significant digits.

Table 6.5   Romberg Integration

h	$I, 0(h^2)$	Error	$0(h^4)$	Error	$0(h^6)$
0.8	0.23159636				
		−0.00050416	0.22957973		
0.4	0.23008389			−0.00000033	0.22957445
		−0.00012728	0.22957478		
0.2	0.22970206			−0.00000002	0.22957444
		−0.00003190	0.22957446		
0.1	0.22960636				

The results presented in Table 6.5 illustrate the error estimation and extrapolation concepts. Error control is accomplished by comparing the estimated error to a specified error limit and terminating the process when the comparison is satisfied. For example, if an error limit of $|0.00000100|$ is specified, that limit is obtained for the first $0(h^4)$ error estimate presented in Table 6.5. One would continue and add the error estimate of -0.00000033 to the $0(h^4)$ value to obtain the first $0(h^6)$ value, 0.22957445, but the process would then be terminated and no further trapezoid rule values or extrapolations would be calculated. For the present case, the actual error for the more accurate $0(h^4)$ value is Error $= 0.22957478 - 0.22957444 = 0.00000034$, which is very close to the error estimate (except the signs are reversed as discussed below). The actual error in the corresponding $0(h^6)$ extrapolation is Error $= 0.22957445 - 0.22957444 = 0.00000001$.

The Error calculated by the extrapolation formula is based on the formula

$$f_{\text{exact}} = f(h) + \text{Error}(h)$$

which yields

$$\text{Error}(h) = f_{\text{exact}} - f(h)$$

The Error in the numerical results throughout this book is defined as

$$\text{Error}(h) = f(h) - f_{\text{exact}}$$

These two Error terms have the same magnitude, but opposite signs. Care must be exercised when both types of Error terms are discussed at the same time.

6.5 ADAPTIVE INTEGRATION

Any desired accuracy (within round-off limits) can be obtained by the numerical integration formulas presented in Section 6.3 by taking smaller and smaller increments. This approach is generally undesirable, since evaluation of the integrand function $f(x)$ is the most time-consuming portion of the calculation.

When the function to be integrated is known so that it can be evaluated at any location, the step size h can be chosen arbitrarily, so the increment can be reduced as far as desired. However, it is not obvious how to choose h to achieve a desired accuracy. Error estimation, as described in Section 6.4, can be used to choose h to satisfy a prespecified error criterion. Successive extrapolation, that is, Romberg integration, can be used to increase the accuracy further. This procedure requires the step size h to be a constant over the entire region of integration. However, the behavior of the integrand function $f(x)$ may not require a uniform step size to achieve the desired overall accuracy. In regions where the integrand function is varying slowly, only a few points should be required to achieve the desired accuracy. In regions where the integrand function is varying rapidly, a large number of points may be required to achieve the desired accuracy.

Consider the integrand function illustrated in Figure 6.8. In Region d–e, $f(x)$ is essentially constant, and the increment h may be very large. However, in region a–d, $f(x)$ varies rapidly, and the increment h must be very small. In fact, region a–d should be broken into three regions, as illustrated. Visual inspection of the integrand function can identify regions where h can be large or small. However, constructing a plot of the integrand function is time consuming and undesirable. A more straightforward automatic numerical procedure is required to break the overall region of integration into subregions in which the values of h may differ greatly.

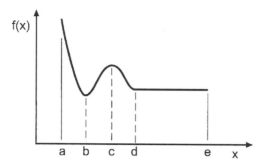

Figure 6.8 Integrand function.

Adaptive integration is a generic name denoting a strategy to achieve the desired accuracy with the minimum number of integrand function evaluations. A basic integration formula must be chosen, for example, the trapezoid rule or Simpson's 1/3 rule. The overall range of integration is broken into several subranges, and each subrange is evaluated to the desired accuracy by subdividing each individual subrange as required until the desired accuracy is obtained. Extrapolation may or may not be used in each subrange.

Example 6.5. Adaptive integration using the trapezoid rule

Let's illustrate adaptive integration by evaluating the following integral using the trapezoid rule:

$$I = \int_{0.1}^{0.9} \frac{1}{x} \, dx \tag{6.58}$$

The exact solution is

$$I = \ln(x)\big|_{0.1}^{0.9} = \ln(0.9/0.1) = 2.197225\ldots \tag{6.59}$$

First, let's evaluate Eq. (6.58) with a uniform increment h over the total range of integration, starting with $h = (0.9 - 0.1) = 0.8$, and successively having h until the error estimate given by Eq. (6.52) is less than 0.001 in absolute value. Recall Eq. (6.52):

$$\text{Error}(h/2) = \tfrac{1}{3}[I(h/2) - I(h)] \tag{6.60}$$

The results are presented in Table 6.6. To satisfy the error criterion, $|\text{Error}| \leq 0.001$, 129 integrand function evaluations with $h = 0.00625$ are required. Extrapolating the final result yields

$$I = 2.197546 + (-0.000321) = 2.197225 \tag{6.61}$$

which agrees with the exact value to six digits after the decimal place.

Next, let's break the total range of integration into two subranges, $0.1 \leq x \leq 0.5$ and $0.5 \leq x \leq 0.9$, and apply the trapezoid rule in each subrange. The error criterion for the two subranges is $0.001/2 = 0.0005$. The results for the two subranges are presented in Table 6.7. To satisfy the error criterion requires 65 and 17 function evaluations,

Table 6.6 Integration Using the Trapezoid Rule

n	h	I	Error
2	0.80000	4.444444	
3	0.40000	3.022222	−0.474074
5	0.20000	2.463492	−0.186243
9	0.10000	2.273413	−0.063360
17	0.05000	2.217330	−0.018694
33	0.02500	2.202337	−0.004998
65	0.01250	2.198509	−0.001276
129	0.00625	2.197546	−0.000321

respectively, in the two subranges, for a total of 82 function evaluations. This is 47 less than before, which is a reduction of 36 percent. Extrapolating the two final results yields

$$I_1 = 1.609750 + (-0.000312) = 1.609438 \qquad (6.62)$$
$$I_2 = 0.587931 + (-0.000144) = 0.587787 \qquad (6.63)$$

which yields

$$I = I_1 + I_2 = 2.197225 \qquad (6.64)$$

which agrees with the exact answer to six digits after the decimal place.

Example 6.5 is a simple example of adaptive integration. The extrapolation step increases the accuracy significantly. This suggests that using Romberg integration as the basic integration method within each subrange may yield a significant decrease in the number of function evaluations. Further increases in efficiency may be obtainable by subdividing the total range of integration into more than two subranges. More sophisti-

Table 6.7 Adaptive Integration Using the Trapezoid Rule

Subrange	n	h	I	Error
$0.1 \leq x \leq 0.5$	2	0.40000	2.400000	
	3	0.20000	1.866667	−0.177778
	5	0.10000	1.683333	−0.061111
	9	0.05000	1.628968	−0.018122
	17	0.02500	1.614406	−0.004854
	33	0.01250	1.610686	−0.001240
	65	0.00625	1.609750	−0.000312
$0.5 \leq x \leq 0.9$	2	0.40000	0.622222	
	3	0.20000	0.596825	−0.008466
	5	0.10000	0.590079	−0.002249
	9	0.05000	0.588362	−0.000572
	17	0.02500	0.587931	−0.000144

cated strategies can be employed to increase the efficiency of adaptive integration even further. The strategy employed in Example 6.5 is the simplest possible strategy.

6.6 GAUSSIAN QUADRATURE

The numerical integration methods presented in Section 6.3 are all based on equally spaced data. Consequently, if n points are considered, an $(n + 1)$st-degree polynomial can be fit to the n points and integrated. The resulting formulas have the form:

$$I = \int_a^b f(x)\, dx = \sum_{i=1}^n C_i f(x_i) \tag{6.65}$$

where x_i are the locations at which the integrand function $f(x)$ is known and C_i are weighting factors. When the locations x_i are prespecified, this approach yields the best possible result. However, when a known function is to be integrated, an additional degree of freedom exists: the locations x_i at which the integrand function $f(x)$ is evaluated. Thus, if n points are used, $2n$ parameters are available: x_i $(i = 1, 2, \ldots, n)$ and C_i $(i = 1, 2, \ldots, n)$. With $2n$ parameters it is possible to fit a polynomial of degree $2n - 1$. Consequently, it should be possible to obtain numerical integration methods of much greater accuracy by choosing the values of x_i appropriately. *Gaussian quadrature* is one such method.

Gaussian quadrature formulas are obtained by choosing the values of x_i and C_i in Eq. (6.65) so that the integral of a polynomial of degree $2n - 1$ is exact. To simplify the development of the formulas, consider the integral of the function $F(t)$ between the limits of $t = -1$ and $t = +1$:

$$I = \int_{-1}^1 F(t)\, dt = \sum_{i=1}^n C_i F(t_i) \tag{6.66}$$

First, consider two points (i.e., $n = 2$), as illustrated in Figure 6.9. Choose t_1, t_2, C_1, and C_2 so that I is exact for the following four polynomials: $F(t) = 1$, t, t^2, and t^3. Thus,

$$I[F(t) = 1] = \int_{-1}^1 (1)\, dt = t\big|_{-1}^1 = 2 = C_1(1) + C_2(1) = C_1 + C_2 \tag{6.67a}$$

$$I[F(t) = t] = \int_{-1}^1 t\, dt = \tfrac{1}{2}t^2\big|_{-1}^1 = 0 = C_1 t_1 + C_2 t_2 \tag{6.67b}$$

$$I[F(t) = t^2] = \int_{-1}^1 t^2\, dt = \tfrac{1}{3}t^3\big|_{-1}^1 = \tfrac{2}{3} = C_1 t_1^2 + C_2 t_2^2 \tag{6.67c}$$

$$I[F(t) = t^3] = \int_{-1}^1 t^3\, dt = \tfrac{1}{4}t^4\big|_{-1}^1 = 0 = C_1 t_1^3 + C_2 t_2^3 \tag{6.67d}$$

Solving Eqs. (6.67) yields

$$\boxed{C_1 = C_2 = 1 \qquad t_1 = -\frac{1}{\sqrt{3}} \qquad t_2 = \frac{1}{\sqrt{3}}} \tag{6.68}$$

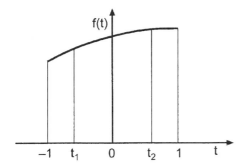

Figure 6.9 Gaussian quadrature.

Thus, Eq. (6.66) yields

$$I = \int_{-1}^{1} F(t)\, dt = F\left(-\frac{1}{\sqrt{3}}\right) + F\left(\frac{1}{\sqrt{3}}\right) \tag{6.69}$$

The actual problem of interest is

$$I = \int_{a}^{b} f(x)\, dx \tag{6.70}$$

The problem presented in Eq. (6.70) can be transformed from x space to t space by the transformation

$$x = mt + c \tag{6.71}$$

where $x = a \rightarrow t = -1$, $x = b \rightarrow t = 1$, and $dx = m\, dt$. Thus,

$$a = m(-1) + c \qquad \text{and} \qquad b = m(1) + c \tag{6.72}$$

which gives

$$m = \frac{b - a}{2} \qquad \text{and} \qquad c = \frac{b + a}{2} \tag{6.73}$$

Thus, Eq. (6.71) becomes

$$x = \left(\frac{b - a}{2}\right)t + \frac{b + a}{2} \tag{6.74}$$

and Eq. (6.70) becomes

$$I = \int_{a}^{b} f(x)\, dx = \int_{-1}^{1} f[x(t)]\, dt = \int_{-1}^{1} f(mt + c)m\, dt \tag{6.75}$$

Define the function $F(t)$:

$$F(t) = f[x(t)] = f(mt + c) \tag{6.76}$$

Table 6.8 Gaussian Quadrature Parameters

n	t_i	C_i	Order
2	$-1/\sqrt{3}$	1	3
	$1/\sqrt{3}$	1	
3	$-\sqrt{0.6}$	5/9	5
	0	8/9	
	$\sqrt{0.6}$	5/9	
4	-0.8611363116	0.3478548451	7
	-0.3399810436	0.6521451549	
	0.3399810436	0.6521451549	
	0.8611363116	0.3478548451	

Substituting Eqs. (6.73) and (6.76) into Eq. (6.75) yields

$$I = \frac{b-a}{2} \int_{-1}^{1} F(t)\, dt \tag{6.77}$$

Higher-order formulas can be developed in a similar manner. Thus,

$$\int_{a}^{b} f(x)\, dx = \frac{b-a}{2} \sum_{i=1}^{n} C_i F(t_i) \tag{6.78}$$

Table 6.8 presents t_i and C_i for $n = 2$, 3, and 4. Higher-order results are presented by Abramowitz and Stegun (1964).

Example 6.6. Gaussian quadrature

To illustrate Gaussian quadrature, let's solve the example problem presented in Section 6.1, where $f(x) = 1/x$, $a = 3.1$, and $b = 3.9$. Consider the two-point formula applied to the total range of integration as a single interval. From Eq. (6.73),

$$m = \frac{b-a}{2} = 0.4 \qquad \text{and} \qquad c = \frac{b+a}{2} = 3.5 \tag{6.79}$$

Equations (6.74) and (6.76) become

$$x = 0.4t + 3.5 \qquad \text{and} \qquad F(t) = \frac{1}{0.4t + 3.5} \tag{6.80}$$

Substituting these results into Eq. (6.78) with $n = 2$ gives

$$I = 0.4 \int_{-1}^{1} F(t)\, dt = 0.4\left[(1)F\left(\frac{-1}{\sqrt{3}}\right) + (1)F\left(\frac{1}{\sqrt{3}}\right) \right] \tag{6.81}$$

Evaluating $F(t)$ gives

$$F\left(-\frac{1}{\sqrt{3}}\right) = \frac{1}{0.4(-1/\sqrt{3}) + 3.5} = 0.30589834 \qquad (6.82a)$$

$$F\left(\frac{1}{\sqrt{3}}\right) = \frac{1}{0.4(1/\sqrt{3}) + 3.5} = 0.26802896 \qquad (6.82b)$$

Substituting Eq. (6.82) into Eq. (6.81) yields

$$I = 0.4[(1)(0.30589834) + (1)(0.26802896)] = 0.22957092 \qquad (6.83)$$

Recall that the exact value is $I = 0.22957444$. The error is Error $= 0.22957092 - 0.22957444 = -0.00000352$. This result is comparable to Simpson's 1/3 rule applied over the entire range of integration in a single step, that is, $h = 0.4$.

Next, let's apply the two-point formula over two intervals, each one-half of the total range of integration. Thus,

$$I = \int_{3.1}^{3.9} \frac{1}{x} \, dx = \int_{3.1}^{3.5} \frac{1}{x} \, dx + \int_{3.5}^{3.9} \frac{1}{x} \, dx = I_1 + I_2 \qquad (6.84)$$

For I_1, $a = 3.1$, $b = 3.5$, $(b - a)/2 = 0.2$, and $(b + a)/2 = 3.3$. Thus,

$$x = 0.2t + 3.3 \qquad F(t) = \frac{1}{0.2t + 3.3} \qquad (6.85)$$

$$I_1 = 0.2 \int_{-1}^{1} F(t) \, dt = 0.2\left[(1)F\left(-\frac{1}{\sqrt{3}}\right) + (1)F\left(\frac{1}{\sqrt{3}}\right)\right] \qquad (6.86)$$

$$I_1 = 0.2\left[\frac{1}{0.2(-1/\sqrt{3}) + 3.3} + \frac{1}{0.2(1/\sqrt{3}) + 3.3}\right] = 0.12136071 \qquad (6.87)$$

For I_2, $a = 3.5$, $b = 3.9$, $(b - a)/2 = 0.2$, and $(b + a)/2 = 3.7$. Thus,

$$x = 0.2t + 3.7 \qquad F(t) = \frac{1}{0.2t + 3.7} \qquad (6.88)$$

$$I_2 = 0.2 \int_{-1}^{1} F(t) \, dt = 0.2\left[(1)F\left(-\frac{1}{\sqrt{3}}\right) + (1)F\left(\frac{1}{\sqrt{3}}\right)\right] \qquad (6.89)$$

$$I_2 = 0.2\left[\frac{1}{0.2(-1/\sqrt{3}) + 3.7} + \frac{1}{0.2(1/\sqrt{3}) + 3.7}\right] = 0.10821350 \qquad (6.90)$$

Summing the results yields the value of the total integral:

$$I = I_1 + I_2 = 0.12136071 + 0.10821350 = 0.22957421 \qquad (6.91)$$

The error is Error $= 0.22957421 - 0.22957444 = -0.00000023$. This result is comparable to Simpson's 1/3 rule with $h = 0.2$.

Next let's apply the three-point formula over the total range of integration as a single interval. Thus,

$$a = 3.1 \qquad b = 3.9 \qquad \frac{b-a}{2} = 0.4 \qquad \text{and} \qquad \frac{b+a}{2} = 3. \tag{6.92}$$

$$x = 0.4t + 3.5 \qquad f(t) = \frac{1}{0.4t + 3.5}$$

$$I = 0.4[\tfrac{5}{9}F(-\sqrt{0.6}) + \tfrac{8}{9}F(0) + \tfrac{5}{9}F(\sqrt{0.6})] \tag{6.93}$$

$$I = 0.4\left[\frac{5}{9}\frac{1}{0.4(-\sqrt{0.6}) + 3.5} + \frac{8}{9}\frac{1}{0.4(0) + 3.5} + \frac{5}{9}\frac{1}{0.4(\sqrt{0.6}) + 3.5}\right]$$

$$= 0.22957443 \tag{6.94}$$

The error is Error $= 0.22957443 - 0.22957444 = -0.00000001$. This result is comparable to Simpson's 1/3 rule with $h = 0.1$.

As a final example, let's evaluate the integral using the sixth-order formula based on the fifth-degree Newton forward-difference polynomial. That formula is (see Table 6.4)

$$I = \frac{5h}{288}(19f_0 + 75f_1 + 50f_2 + 50f_3 + 75f_4 + 19f_5) + 0(h^7) \tag{6.95}$$

For five equally spaced increments, $h = (3.9 - 3.1)/5 = 0.16$. Thus,

$$I = \frac{5(0.16)}{288}[19(1/3.10) + 75(1/3.26) + 50(1/3.42) + 50(1/3.58)$$

$$+ 75(1/3.74) + 19(1/3.90)]$$

$$= 0.22957445 \tag{6.96}$$

The error is Error $= 0.22957445 - 0.22957444 = 0.00000001$. This result is comparable to Gaussian quadrature with three points.

6.7 MULTIPLE INTEGRALS

The numerical integration formulas developed in the preceding sections for evalulating single integrals can be used to evaluate multiple integrals. Consider the double integral:

$$\boxed{I = \int_c^d \int_a^b f(x, y)\, dx\, dy} \tag{6.97}$$

Equation (6.97) can be written in the form:

$$I = \int_c^d \left(\int_a^b f(x, y)\, dx \right) dy = \int_c^d F(y)\, dy \tag{6.98}$$

where

$$F(y) = \int_a^b f(x, y)\, dx \qquad y = \text{Constant} \tag{6.99}$$

The double integral is evaluated in two steps:

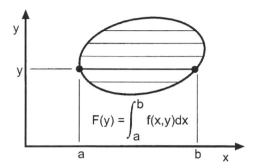

Figure 6.10   Double integration.

1. Evaluate $F(y)$ at selected values of y by any numerical integration formula.
2. Evaluate $I = \int F(y)\, dy$ by any numerical integration formula.

If the limits of integration are variable, as illustrated in Figure 6.10, that must be accounted for.

Example 6.7.   Double integral

To illustrate double integration with variable limits of integration, let's calculate the mass of water in a cylindrical container which is rotating at a constant angular velocity ω, as illustrated in Figure 6.11a. A meridional plane view through the axis of rotation is presented in Figure 6.11b. From a basic fluid mechanics analysis, it can be shown that the shape of the free surface is given by

$$z(r) = A + Br^2 \qquad\qquad (6.100)$$

From measured data, $z(0) = z_1$ and $z(R) = z_2$. Substituting these values into Eq. (6.100) gives

$$z(r) = z_1 + \frac{(z_2 - z_1)r^2}{R^2} \qquad\qquad (6.101)$$

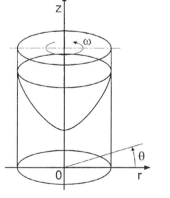

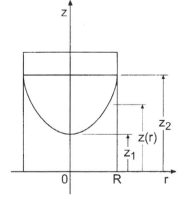

(a) Physical arrangement.   (b) Meridional plane view.

Figure 6.11   Spinning cylindrical container.

In a specific experiment, $R = 10.0\,\text{cm}$, $z_1 = 10.0\,\text{cm}$, and $z_2 = 20.0\,\text{cm}$. In this case, Eq. (6.101) gives

$$z(r) = 10.0 + 0.1r^2 \tag{6.102}$$

Let's calculate the mass of water in the container at this condition. The density of water is $\rho = 1.0\,\text{g/cm}^3$. Due to axial symmetry in the geometry of the container and the height distribution, the mass in the container can be expressed in cylindrical coordinates as

$$m = \int dm = \int_0^R \rho z(r)(2\pi r\, dr) = 2\pi\rho \int_0^R \left[z_1 + \frac{(z_2 - z_1)r^2}{R^2} \right] \tag{6.103}$$

which has the exact integral

$$m = \pi\rho \left[z_1 R^2 + \frac{(z_2 - z_1)R^2}{2} \right] \tag{6.104}$$

Substituting the specified values of ρ, R, z_1, and z_2 into Eq. (6.104) yields $m = 1500\pi\text{g} = 4712.388980\,\text{g}$.

To illustrate the process of numerical double integration, let's solve this problem in Cartesian coordinates. Figure 6.12(a) illustrates a discretized Cartesian grid on the bottom of the container. In Cartesian coordinates, $dA = dx\, dy$, and the differential mass in a differential column of height $z(r)$ is given by

$$dm = \rho z(r)\, dA \tag{6.105}$$

Substituting Eq. (6.101) into Eq. (6.105), where $r^2 = x^2 + y^2$, and integrating gives

$$m = \int dm = \rho \iint [z_1 + (z_2 - z_1)(x^2 + y^2)R^{-2}]\, dx\, dy \tag{6.106}$$

Substituting the specific values of ρ, R, z_1, and z_2 into Eq. (6.106) gives

$$m = 1.0 \iint [10 + 0.1(x^2 + y^2)]\, dx\, dy \tag{6.107}$$

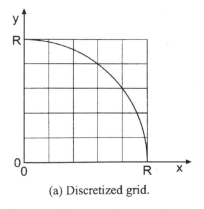

(a) Discretized grid.

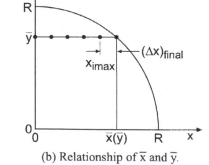

(b) Relationship of $\bar{x}$ and $\bar{y}$.

Figure 6.12 Cartesian coordinates.

Table 6.9 Geometrical Parameters

j	$\bar{y}$, cm	$\bar{x}(\bar{y})$, cm	$\mathrm{imax}(\bar{y})$	$x(\mathrm{imax}(\bar{y}))$	$(\Delta x)_{\mathrm{final}}$
1	0.0	10.000000	11	10.0	0.000000
2	1.0	9.949874	10	9.0	0.949874
3	2.0	9.797959	10	9.0	0.797959
4	3.0	9.539392	10	9.0	0.539392
5	4.0	9.165151	10	9.0	0.165151
6	5.0	8.660254	9	8.0	0.660254
7	6.0	8.000000	9	8.0	0.000000
8	7.0	7.141428	8	7.0	0.141428
9	8.0	6.000000	7	6.0	0.000000
10	9.0	4.358899	5	4.0	0.358899
11	10.0	0.000000	1	0.0	0.000000

Table 6.10 Integrand of Eq. (6.111) at $\bar{y} = 5.0\,\mathrm{cm}$

i	$F(x)$	i	$F(x)$
1	12.500	6	15.000
2	12.600	7	16.100
3	12.900	8	17.400
4	13.400	9	18.900
5	14.100	10	20.000

Due to symmetry about the x and y axes, Eq. (6.107) can be expressed as

$$m = 4(1.0) \int_0^R \left[\int_0^{\bar{x}(\bar{y})} [10 + 0.1(x^2 + \bar{y}^2)]\, dx \right] dy = 4 \int_0^R F(\bar{y})\, dy \tag{6.108}$$

where $F(\bar{y})$ is defined as

$$F(\bar{y}) = \int_0^{\bar{x}(\bar{y})} [10.0 + 0.1(x^2 + \bar{y}^2)]\, dx \tag{6.109}$$

and $\bar{y}$ and $\bar{x}(\bar{y})$ are illustrated in Figure 6.12(b). Thus,

$$\bar{x} = (R^2 - \bar{y}^2)^{1/2} \tag{6.110}$$

Table 6.11 Values of $F(\bar{y})$

j	$\bar{y}$, cm	$F(\bar{y})$	j	$\bar{y}$, cm	$F(\bar{y})$
1	0.0	133.500000	7	6.0	126.000000
2	1.0	133.492600	8	7.0	118.664426
3	2.0	133.410710	9	8.0	105.700000
4	3.0	133.068144	10	9.0	81.724144
5	4.0	132.128255	11	10.0	0.000000
6	5.0	130.041941			

Table 6.12. Results for $nx = 11, 21, 41, 81$ and 161

nx	m, g	Error, g	Error ratio
11	4643.920883	−68.468097	2.75
21	4687.527276	−24.861705	2.78
41	4703.442452	−8.946528	2.80
81	4709.188100	−3.200880	2.81
161	4711.248087	−1.140893	

Let's discretize the y axis into 10 equally spaced increments with $\Delta y = 1.0$ cm. For each value of $\bar{y} = (j-1)\,\Delta y\ (j = 1, 2, \ldots, 11)$ let's calculate $\bar{x}(\bar{y})$ from Eq. (6.110). At each value of $\bar{y}$, let's discretize the x axis into $(\text{imax}(\bar{y}) - 1)$ equally spaced increments with $\Delta x = 1.0$ cm, with a final increment $(\Delta x)_{\text{final}}$ between $x = (\text{imax}(\bar{y} - 1) - 1)\,\Delta x$ and $\bar{x}(\bar{y})$. The resulting geometrical parameters are presented in Table 6.9.

The values of $F(\bar{y})$, defined in Eq. (6.109), are evaluated by the trapezoid rule. As an example, consider $j = 6$ for which $\bar{y} = 5.0$ cm. Equation (6.109) becomes

$$F(5.0) = \int_{0.0}^{8.000000} [10.0 + 0.1(x^2 + 25.0)]\,dx = \int_{0.0}^{8.000000} F(x)\,dx \tag{6.111}$$

Table 6.10 presents the integrand $F(x)$ of Eq. (6.111).

Integrating Eq. (6.111) by the trapezoid rule gives

$$F(5.0) = \frac{1.0}{2}[12.500 + 2(12.600 + 12.900 + 13.400 + 14.000 + 15.000$$

$$+\ 16.100 + 17.400) + 18.900] + \frac{0.660254}{2}(18.900 + 20.000)$$

$$= 130.041941 \tag{6.112}$$

Repeating this procedure for every value of $\bar{y}$ in Table 6.9 yields the results presented in Table 6.11.

Integrating Eq. (6.108) by the trapezoid rule, using the values of $F(\bar{y})$ presented in Table 6.11, yields

$$m = 4.0\frac{1.0}{2}[133.500 + 2(133.492600 + 133.410710 + 133.068144 + 132.128255$$

$$+\ 130.041941 + 126.000000 + 118.664426 + 105.700000 + 81.724144)$$

$$+\ 0.000000]$$

$$= 4643.920883 \text{ g} \tag{6.113}$$

The error is Error $= 4643.920883 - 4712.388980 = -68.468047$ g. Repeating the calculations for $nx = 21, 41, 81,$ and 161 yields the results presented in Table 6.12. These results show the procedure is behaving better than first order, but not quite second order.

Example 6.7 illustrates the complexity of multiple integration by numerical methods, especially with variable limits of integration. The procedure is not difficult, just complicated. The procedure can be extended to three or more independent variables in a straightforward manner, with a corresponding increase in complexity.

Example 6.7 uses the trapezoid rule. Simpson's 1/3 rule could be used instead of the trapezoid rule. When the integrand is a known function, adaptive integration techniques can be employed, as can Gaussian quadrature.

6.8 PROGRAMS

Three FORTRAN subroutines for numerical integration are presented in this section:

1. The trapezoid rule
2. Simpson's 1/3 rule
3. Romberg integration

The basic computational algorithms are presented as completely self-contained subroutines suitable for use in other programs. Input data and output statements are contained in a main (or driver) program written specifically to illustrate the use of each subroutine.

6.8.1. The Trapezoid Rule

The general algorithm for the trapezoid rule is give by Eq. (6.24):

$$I = \tfrac{1}{2}h(f_0 + 2f_1 + 2f_2 + \cdots + 2f_{n-1} + f_n) \tag{6.114}$$

A FORTRAN subroutine, *subroutine trap*, for implementing the trapezoid rule is presented in Program 6.1. *Program main* defines the data set and prints it, calls *subroutine trap* to implement the solution, and prints the solution (Output 6.1).

Program 6.1   The trapezoid rule program.

```
      program main
c     main program to illustrate numerical integration subroutines
c     ndim  array dimension, ndim = 9 in this example
c     n     number of data points
c     x     independent variable array, x(i)
c     f     dependent variable array, f(i)
c     sum   value of the integral
      dimension x(9),f(9)
      data ndim,n / 9, 9 /
      data (x(i),i=1,9)/3.1, 3.2, 3.3, 3.4, 3.5, 3.6, 3.7, 3.8, 3.9/
      data (f(i),i=1,9) / 0.32258065, 0.31250000, 0.30303030,
     1 0.29411765, 0.28571429, 0.27777778, 0.27027027, 0.26315789,
     2 0.25641026 /
      write (6,1000)
      do i=1,n
         write (6,1010) i,x(i),f(i)
      end do
      call trap (ndim,n,x,f,sum)
      write (6,1020) sum
      stop
 1000 format (' Trapezoid rule'/' '/'    i',7x,'x',13x,'f'/' ')
 1010 format (i4,2f14.8)
 1020 format (' '/' I =',f14.8)
      end

      subroutine trap (ndim,n,x,f,sum)
c     trapezoid rule integration
      dimension x(ndim),f(ndim)
      sum=f(1)+f(n)
```

```
do i=2,n-1
   sum=sum+2.0*f(i)
end do
sum=sum*(x(n)-x(1))/float(n-1)/2.0
return
end
```

The data set used to illustrate *subroutine trap* is taken from Example 6.2. The output generated by the trapezoid rule program is presented in Output 6.1.

Output 6.1   Solution by the trapezoid rule

```
Trapezoid rule

   i       x              f

   1    3.10000000     0.32258065
   2    3.20000000     0.31250000
   3    3.30000000     0.30303030
   4    3.40000000     0.29411765
   5    3.50000000     0.28571429
   6    3.60000000     0.27777778
   7    3.70000000     0.27027027
   8    3.80000000     0.26315789
   9    3.90000000     0.25641026

I =    0.22960636
```

6.8.2. Simpson's 1/3 Rule

The general algorithm for Simpson's 1/3 rule is given by Eq. (6.35):

$$I = \tfrac{1}{3}h(f_0 + 4f_1 + 2f_2 + 4f_3 + \cdots + 4f_{n-1} + f_n) \tag{6.115}$$

A FORTRAN subroutine, *subroutine simpson*, for implementing Simpson's 1/3 rule is presented in Program 6.2. *Subroutine simpson* works essentially like *subroutine trap* discussed in Section 6.8.1, except Simpson's 1/3 rule is used instead of the trapezoid rule. *Program main* defines the data set and prints it, calls *subroutine simpson* to implement Simpson's 1/3 rule, and prints the solution. Only the statements in *program main* which are different from *program main* in Section 6.8.1 are presented.

Program 6.2   Simpson's 1/3 rule program

```
      program main
c     main program to illustrate numerical integration subroutines
      call simpson (ndim,n,x,f,sum)
 1000 format (' Simpsons 1/3 rule'/' '/'  i',7x,'x',13x,'f'/' ')
      end

      subroutine simpson (ndim,n,x,f,sum)
```

```
c       Simpson's 1/3 rule integration
        dimension x(ndim),f(ndim)
        sum2=0.0
        sum4=0.0
do i=3,n-1,2
   sum2=sum2+f(i)
end do
do i=2,n-1,2
   sum4=sum4+f(i)
end do
sum=(f(1)+2.0*sum2+4.0*sum4+f(n))*(x(2)-x(1))/3.0
return
end
```

The data set used to illustrate *subroutines simpson* is taken from Example 6.3. The output generated by Simpson's 1/3 rule program is presented in Output 6.2.

Output 6.2   Solution by Simpson's 1/3 rule

```
Simpsons 1/3 rule

   i       x                f

   1    3.10000000      0.32258065
   2    3.20000000      0.31250000
   3    3.30000000      0.30303030
   4    3.40000000      0.29411765
   5    3.50000000      0.28571429
   6    3.60000000      0.27777778
   7    3.70000000      0.27027027
   8    3.80000000      0.26315789
   9    3.90000000      0.25641026

I =     0.22957446
```

6.8.3.   Romberg Integration

Romberg integration is based on extrapolation of the trapezoid rule, Eq. (6.114). The general extrapolation formula for Romberg integration is given by Eq. (6.51):

$$\text{Error}(h/2) = \frac{1}{2^n - 1}[I(h/2) - I(h)] \tag{6.116}$$

A FORTRAN subroutine, *subroutine romberg*, for implementing the procedure is presented in Program 6.3. *Program main* defines the data set and prints it, calls *subroutine romberg* to implement the solution, and prints the solution (Output 6.3). Only the statements in *program main* which are different from *program main* in Section 6.8.1 are presented.

Program 6.3   Romberg integration program

```
      program main
c     main program to illustrate numerical integration subroutines
c     s       array of integrals and extrapolations, s(i,j)
      dimension x(9),f(9),s(9,9)
      data ndim,n,num / 9, 9, 4 /
      call romberg (ndim,n,num,x,f,s)
      write (6,1020)
      do i=1,num
         write (6,1010) i,(s(i,j),j=1,num+1-i)
      end do
      stop
 1000 format (' Romberg integration'/' '/'   i',7x,'x',13x,'f'/' ')
 1010 format (i3,4f14.8)
 1020 format (' '/'   i',6x,'s(i,1)',8x,'s(i,2)',8x,'s(i,3)',8x,
     1 's(i,4)'/' ')
      end


      subroutine romberg (ndim,n,num,x,f,s)
c     Romberg integration
      dimension x(ndim),f(ndim),s(ndim,ndim)
c     trapezoid rule integration
      k=n
      dx=(x(n)-x(1))
      s(1,1)=(f(1)+f(n))*dx/2.0
      do j=2,num
         dx=dx/2.0
         sum=0.0
         k=k/2
         do i=k+1,n-1,k
            sum=sum+f(i)
         end do
         s(j,1)=(f(1)+2.0*sum+f(n))*dx/2.0
      end do
c     Romberg extrapolation
      do j=2,num
         ex=float(2*(j-1))
         den=2.0**ex-1.0
         k=num+1-j
         do i=1,k
            s(i,j)=s(i+1,j-1)+(s(i+1,j-1)-s(i,j-1))/den
         end do
      end do
      return
      end
```

The data set used to illustrate *subroutine romberg* is taken from Example 6.5. The output is presented in Output 6.3.

Output 6.3   Solution by Romberg integration

Romberg integration

i	x	f
1	3.10000000	0.32258065
2	3.20000000	0.31250000
3	3.30000000	0.30303030
4	3.40000000	0.29411765
5	3.50000000	0.28571429
6	3.60000000	0.27777778
7	3.70000000	0.27027027
8	3.80000000	0.26315789
9	3.90000000	0.25641026

i	s(i,1)	s(i,2)	s(i,3)	s(i,4)
1	0.23159636	0.22957974	0.22957445	0.22957444
2	0.23008390	0.22957478	0.22957444	
3	0.22970206	0.22957446		
4	0.22960636			

6.8.4.   Packages for Numerical Integration

Numerous libraries and software packages are available for numerical integration. Many workstations and mainframe computers have such libraries attached to their operating systems.

Many commercial software packages contain numerical integration algorithms. Some of the more prominent are Matlab and Mathcad. More sophisticated packages, such as IMSL, MATHEMATICA, MACSYMA, and MAPLE, also contain numerical integration algorithms. Finally, the book *Numerical Recipes* [Press et al. (1989)] contains numerous routines for numerical integration.

6.9   SUMMARY

Procedures for the numerical integration of both discrete data and known functions are presented in this chapter. These procedures are based on fitting approximating polynomials to the data and integrating the approximating polynomials. The direct fit polynomial method works well for both equally spaced data and nonequally spaced data. Least squares fit polynomials can be used for large sets of data or sets of rough data. The Newton-Cotes formulas, which are based on Newton foward-difference polynomials, give simple integration formulas for equally spaced data.

Methods of error estimation and error control are presented. Extrapolation, or the deferred approach to the limit, is discussed. Romberg integration, which is extrapolation of the trapezoid rule, is developed. Adaptive integration procedures are suggested to reduce the effort required to integrate widely varying functions. Gaussian quadrature, which increases the accuracy of integrating known functions where the sampling locations can be chosen arbitrarily, is presented. An example of multiple integration is presented to illustrate the extension of the one-dimensional integration formulas to multiple dimensions.

Of all the methods considered, it is likely that Romberg integration is the most efficient. Simpson's rules are elegant and intellectually interesting, but the first extrapolation of Romberg integration gives comparable results. Subsequent extrapolations of Romberg integration increase the order at a fantastic rate. Simpson's 1/3 rule could be developed into an extrapolation procedure, but that yields no advantage over Romberg integration.

After studying Chapter 6, you should be able to:

1. Describe the general features of numerical integration
2. Explain the procedure for numerical integration using direct fit polynomials
3. Apply direct fit polynomials to integrate a set of tabular data
4. Describe the procedure for numerical integration using Newton forward-difference polynomials
5. Derive the trapezoid rule
6. Apply the trapezoid rule
7. Derive Simpson's 1/3 rule
8. Apply Simpson's 1/3 rule
9. Derive Simpson's 3/8 rule
10. Apply Simpson's 3/8 rule
11. Describe the relative advantage and disadvantages of Simpson's 1/3 rule and Simpson's 3/8 rule
12. Derive Newton-Cotes formulas of any order
13. Apply Newton-Cotes formulas of any order
14. Explain the concepts of error estimation, error control, and extrapolation
15. Describe Romberg integration
16. Apply Romberg integration
17. Describe the concept of adaptive integration
18. Develop and apply a simple adaptive integration algorithm
19. Explain the concepts underlying Gaussian quadrature
20. Derive the two-point Gaussian quadrature formula
21. Describe how to develop Gaussian quadrature formula for more than two points
22. Apply Gaussian quadrature
23. Describe the process for numerically evaluating multiple integrals
24. Develop a procedure for evaluating double integrals
25. Apply the double integral numerical integration procedure

EXERCISE PROBLEMS

6.1

The following integrals are used throughout this chapter to illustrate numerical integration methods. All of these integrals have exact solutions, which should be used for error analysis in the numerical problems.

(A) $\displaystyle\int_0^5 (3x^2 + 2)\, dx$

(B) $\displaystyle\int_{1.0}^{2.0} (x^3 + 3x^2 + 2x + 1)\, dx$

(C) $\displaystyle\int_0^{10} (5x^4 + 4x^3 + 2x + 3)\, dx$

(D) $\displaystyle\int_0^{\pi} (5 + \sin x)\, dx$

(E) $\displaystyle\int_{\pi}^{\pi/4} (e^{2x} + \cos x)\, dx$

(F) $\displaystyle\int_{1.5}^{2.5} \ln x\, dx$

(G) $\displaystyle\int_{0.1}^{1.0} e^x\, dx$

(H) $\displaystyle\int_{0.2}^{1.2} \tan x\, dx$

In the numerical integration problems, the term *range of integration* denotes the entire integration range, the word *interval* denotes subdivisions of the total range of integration, and the word *increment* denotes a single increment Δx.

6.2 Direct Fit Polynomials

1. Evaluate integrals (A), (B), and (C) by direct fit polynomials of order 2 and order 3 over the total range of integration. Repeat the calculations, breaking the total range of integration into two equal intervals. Compute the errors and the ratios of the errors for the two intervals.

2. Evaluate integrals (D), (E), and (F) by direct fit polynomials of order 2 and order 3 over the total range of integration. Repeat the calculations, breaking the total range of integration into two equal intervals. Compute the errors and the ratios of the errors for the two intervals.

3. Evaluate integrals (G) and (H) by direct fit polynomials of order 2 and order 3 over the total range of integration. Repeat the calculations, breaking the total range of integration into two equal intervals. Compute the errors and the ratios of the errors for the two intervals.

6.3 Newton-Cotes Formulas

The Trapezoid Rule

4. Evaluate integrals (A), (B), and (C) by the trapezoid rule for $n = 1, 2, 4$, and 8 intervals. Compute the errors and the ratios of the errors.

5*. Evaluate integrals (D), (E), and (F) by the trapezoid rule for $n = 1, 2, 4$, and 8 intervals. Compute the errors and the ratios of the errors.

6. Evaluate integrals (G) and (H) by the trapezoid rule for $n = 1, 2, 4$, and 8 intervals. Compute the errors and the ratios of the errors.

7*. Consider the function $f(x)$ tabulated in Table 1. Evaluate the integral $\int_{0.4}^{2.0} f(x)\, dx$ using the trapezoid rule with $n = 1, 2, 4$, and 8 intervals. The exact value is 5.86420916. Compute the errors and the ratios of the errors.

Table 1 Tabular Data

x	$f(x)$	x	$f(x)$	x	$f(x)$
0.4	5.1600	1.4	3.3886	2.2	5.7491
0.6	3.6933	1.6	3.8100	2.4	6.5933
0.8	3.1400	1.8	4.3511	2.6	7.5292
1.0	3.0000	2.0	5.0000	2.8	8.5543
1.2	3.1067				

8. Consider the function $f(x)$ tabulated in Table 1. Evaluate the integral $\int_{1.2}^{2.8} f(x)\, dx$ using the trapezoid rule with $n = 1$, 2, 4, and 8 intervals. The exact value is 8.43592905. Compute the errors and the ratios of the errors.

Table 2 Tabular Data

x	$f(x)$	x	$f(x)$	x	$f(x)$
0.4	6.0900	1.4	6.9686	2.2	4.4782
0.6	7.1400	1.6	6.5025	2.4	3.6150
0.8	7.4850	1.8	5.9267	2.6	2.6631
1.0	7.5000	2.0	5.2500	2.8	1.6243
1.2	7.3100				

9. Consider the function $f(x)$ tabulated in Table 2. Evaluate the integral $\int_{0.4}^{2.0} f(x)\, dx$ using the trapezoid rule with $n = 1$, 2, 4, and 8 intervals. The exact value is 10.92130980. Compute the errors and the ratios of the errors.

10. Consider the function $f(x)$ tabulated in Table 2. Evaluate the integral $\int_{1.2}^{2.8} f(x)\, dx$ using the trapezoid rule with $n = 1$, 2, 4, and 8 intervals.

Simpson's 1/3 Rule

11. Evaluate integrals (A), (B), and (C) using Simpson's 1/3 rule for $n = 1$, 2, and 4 intervals. Compute the errors and the ratio of the errors.

12*. Evaluate integrals (D), (E), and (F) using Simpson's 1/3 rule for $n = 1$, 2, and 4 intervals. Compute the errors and the ratio of the errors.

13. Evaluate integrals (G) and (H) using Simpson's 1/3 rule for $n = 1$, 2, and 4 intervals. Compute the errors and the ratio of the errors.

14. Consider the function $f(x)$ tabulated in Table 1. Evaluate the integral $\int_{0.4}^{2.0} f(x)\, dx$ using Simpson's 1/3 rule with $n = 1$, 2, and 4 intervals.

15. Consider the function $f(x)$ tabulated in Table 1. Evaluate the integral $\int_{1.2}^{2.8} f(x)\, dx$ using Simpson's 1/3 rule with $n = 1$, 2, and 4 intervals.

16. Consider the function $f(x)$ tabulated in Table 2. Evaluate the integral $\int_{0.4}^{2.0} f(x)\, dx$ using Simpson's 1/3 rule with $n = 1$, 2, and 4 intervals.

17*. Consider the function $f(x)$ tabulated in Table 2. Evaluate the integral $\int_{1.2}^{2.8} f(x)\, dx$ using Simpson's 1/3 rule with $n = 1$, 2, and 4 intervals.

Simpson's 3/8 Rule

18. Evaluate integrals (A), (B), and (C) using Simpson's 3/8 rule for $n = 1$, 2, and 4 intervals. Compute the errors and the ratio of the errors.

19*. Evaluate integrals (D), (E), and (F) using Simpson's 3/8 rule for $n = 1$, 2, and 4 intervals. Compute the errors and the ratio of the errors.

20. Evaluate integrals (G) and (H) using Simpson's 3/8 rule for $n = 1$, 2, and 4 intervals. Compute the errors and the ratio of the errors.

21. Consider the function $f(x)$ tabulated in Table 1. Evaluate the integral $\int_{0.4}^{1.6} f(x)\, dx$ using Simpson's 3/8 rule with $n = 1$ and 2 intervals.

22. Consider the function $f(x)$ tabulated in Table 1. Evaluate the integral $\int_{1.2}^{1.6} f(x)\, dx$ using Simpson's 3/8 rule with $n = 1$ and 2 intervals.

23. Consider the function $f(x)$ tabulated in Table 2. Evaluate the integral $\int_{0.4}^{1.6} f(x)\, dx$ using Simpson's 3/8 rule with $n = 1$ and 2 intervals.

24. Consider the function $f(x)$ tabulated in Table 2. Evaluate the integral $\int_{1.2}^{1.6} f(x)\, dx$ using Simpson's 3/8 rule with $n = 1$ and 2 intervals.

25. Evaluate integrals (A), (B), and (C) using Simpson's rules with $n = 5$ and 7 increments.

26. Evaluate integrals (D), (E), and (F) using Simpson's rules with $n = 5$ and 7 increments.

27. Evaluate integrals (G) and (H) using Simpson's rules with $n = 5$ and 7 increments.

Higher-Order Newton-Cotes Formulas

28. Derive the Newton-Cotes formulas for polynomials of degree $n = 4$, 5, 6, and 7.

29. Evaluate integrals (D), (E), and (F) by the Newton-Cotes fourth-order formula with one and two intervals. Compute the errors and the ratios of the errors.

30. Evaluate integrals (D), (E), and (F) by the Newton-Cotes fifth-order formula with one and two intervals. Compute the errors and the ratios of the errors.

31. Evaluate integrals (D), (E), and (F) by the Newton-Cotes sixth-order formula with one and two intervals. Compute the errors and the ratios of the errors.

32. Evaluate integrals (D), (E), and (F) by the Newton-Cotes seventh-order formula with one and two intervals. Compute the errors and the ratios of the errors.

6.4 Extrapolation and Romberg Integration

33. Evaluate the following integrals using Romberg integration with four intervals. Let the first interval be the total range of integration.
    (a) $\int_0^{\pi/4} \tan x\, dx$ (b) $\int_0^{0.5} e^{-x}\, dx$ (c) $\int_0^{3/4} e^{-x^2}\, dx$ (d) $\int_1^{2.5} (x^5 - x^2)\, dx$

34. Evaluate integrals (A), (B), and (C) using Romberg integration with four intervals. Let the first interval be the total range of integration.

35*. Evaluate integrals (D), (E), and (F) using Romberg integration with four intervals. Let the first interval be the total range of integration.

36. Evaluate integrals (G) and (H) using Romberg integration with four intervals. Let the first interval be the total range of integration.

37. Consider the function $f(x)$ tabulated in Table 1. Evaluate the integral $\int_{0.4}^{2.0} f(x)\, dx$ using Romberg integration with $n = 1$, 2, 4, and 8 intervals.

38. Consider the function $f(x)$ tabulated in Table 2. Evaluate the integral $\int_{0.4}^{2.0} f(x)\, dx$ using Romberg integration with $n = 1$, 2, 4, and 8 intervals.

39. Which row of a Romberg table yields $\int_a^b f(x)\, dx$ exactly if $f(x)$ is a polynomial of degree k, where (a) $k = 3$, (b) $k = 5$, and (c) $k = 7$? Verify your conclusion for $\int_0^2 x^k\, dx$. Start with one interval.

6.5 Adaptive Integration

40. Evaluate $I = \int_{-2.0}^{0.4} e^{-x}\, dx$ using Romberg integration. (a) Let the first interval be the total range of integration. (b) Divide the total range of integration into two equal intervals and use Romberg integration in each interval. Let the first subinterval in each interval be the total range of integration for that interval. (c) Divide the total range of integration into the two intervals, $-2.0 \le$

$x \le -0.4$ and $-0.4 \le x \le 0.4$, and repeat part (b). (d) Compare the errors incurred for the three cases.

41. Evaluate $I = \int_1^3 \ln(x)\, dx$ using Romberg integration. (a) Start with the total interval. Let the first subinterval be the total interval. (b) Divide the total range of integration into two equal intervals and use Romberg integration in each interval. Let the first subinterval in each interval be the total interval.

6.6 Gaussian Quadrature

42. Evaluate integrals (A), (B), and (C) by two-point Gaussian quadrature for $n = 1$, 2, and 4 intervals. Compute the errors and the ratios of the errors.

43*. Evaluate integrals (D), (E), and (F) by two-point Gaussian quadrature for $n = 1$, 2, and 4 intervals. Compute the errors and the ratios of the errors.

44. Evaluate integrals (G) and (H) by two-point Gaussian quadrature for $n = 1$, 2, and 4 intervals. Compute the errors and the ratios of the errors.

45. Evaluate integrals (A), (B), and (C) by three-point Gaussian quadrature for $n = 1$, 2, and 4 intervals. Compute the errors and the ratios of the errors.

46*. Evaluate integrals (D), (E), and (F) by three-point Gaussian quadrature for $n = 1$, 2, and 4 intervals. Compute the errors and the ratios of the errors.

47. Evaluate integrals (G) and (H) by three-point Gaussian quadrature for $n = 1$, 2, and 4 intervals. Compute the errors and the ratios of the errors.

48. Evaluate integrals (A), (B), and (C) by four-point Gaussian quadrature for $n = 1$, 2, and 4 intervals. Compute the errors and the ratios of the errors.

49*. Evaluate integrals (D), (E), and (F) by four-point Gaussian quadrature for $n = 1$, 2, and 4 intervals. Compute the errors and the ratios of the errors.

50. Evaluate integrals (G) and (H) by four-point Gaussian quadrature for $n = 1$, 2, and 4 intervals. Compute the errors and the ratios of the errors.

51. Evaluate the following integrals using k-point Gaussian quadrature for $k = 2$, 3, and 4, with $n = 1$, 2, and 4 intervals. Compare the results with the exact solutions.
    (a) $\int_1^{2.6} x e^{-x^2}\, dx$ (b) $1\int_1^3 \cosh x\, dx$ (c) $1\int_0^4 \sinh x\, dx$ (d) $\int_{-1}^2 e^{-x} \sin x\, dx$

52. (a) What is the smallest k for which k-point Gaussian quadrature is exact for a polynomial of degree 7? (b) Verify the answer to part (a) by evaluating $\int_0^2 x^7\, dx$.

6.7 Multiple Integrals

53. Evaluate the multiple integral $\int_{-1}^1 \int_0^2 (4x^3 - 2x^2 y + 3xy^2)\, dx\, dy$: (a) analytically, (b) using the trapezoid rule, and (c) using Simpson's 1/3 rule.

54. Evaluate the multiple integral $\int_0^1 \int_0^2 \sin(x^2 + y^2)\, dx\, dy$: (a) analytically, (b) using the trapezoid rule, and (c) using Simpson's 1/3 rule.

55. Evaluate the multiple integral $\int_0^1 \int_0^{e^x} (x^2 + 1/y)\, dy\, dx$: (a) analytically, (b) using the trapezoid rule for both integrals, (c) using three-point Gaussian quadrature for both integrals; (d) using the trapezoid rule for the y integral and three-point Gaussian quadrature for the x integral, and (e) using three-point Gaussian quadrature for the y integral and the trapezoid rule for the x integral.

56. Evaluate the multiple integral $\int_{-1}^1 \int_0^{x^2} xy\, dy\, dx$ by the procedures described in the previous problem.

6.8 Programs

57. Implement the trapezoid rule program presented in Section 6.8.1. Check out the program using the given data set.
58. Solve any of Problems 4 to 10 using the program.
59. Implement the Simpson's 1/3 rule program presented in Section 6.8.2. Check out the program using the given data set.
60. Solve any of Problems 11 to 17 using the program.
61. Implement the Simpson's 3/8 rule program presented in Section 6.8.3. Check out the program using the given data set.
62. Solve any of Problems 18 to 24 using the program.

APPLIED PROBLEMS

63. Evaluate the integral $\int\int_R \exp[(x + y^2)^{1/2}] \, dy \, dx$, where R is the area enclosed by the circle $x^2 + y^2 = 1$. Use Cartesian coordinates.
64. Find the volume of a circular pyramid with height and base radius equal to 1. Use Cartesian coordinates.

II

Ordinary Differential Equations

II.1 INTRODUCTION

Differential equations arise in all fields of engineering and science. Most real physical processes are governed by differential equations. In general, most real physical processes involve more than one independent variable, and the corresponding differential equations are *partial differential equations* (*PDEs*). In many cases, however, simplifying assumptions are made which reduce the PDEs to *ordinary differential equations* (*ODEs*). Part II is devoted to the solution of ordinary differential equations.

Some general features of ordinary differential equations are discussed in Part II. The two classes of ODEs (i.e., *initial-value ODEs* and *boundary-value ODEs*) are introduced. The two types of physical problems (i.e., *propagation problems* and *equilibrium problems*) are discussed.

The objectives of Part II are (a) to present the general features of ordinary differential equations; (b) to discuss the relationship between the type of physical problem being solved, the class of the corresponding governing ODE, and the type of numerical method required; and (c) to present examples to illustrate these concepts.

II.2 GENERAL FEATURES OF ORDINARY DIFFERENTIAL EQUATIONS

An *ordinary differential equation* (*ODE*) is an equation stating a relationship between a function of a single independent variable and the total derivatives of this function with respect to the independent variable. The variable y is used as a generic dependent variable throughout Part II. The dependent variable depends on the physical problem being modeled. In most problems in engineering and science, the independent variable is either time t or space x. Ordinary differential equations are the subject of Part II. If more than one independent variable exists, then partial derivatives occur, and partial differential equations (PDEs) are obtained. Partial differential equations are the subject of Part III of this book.

The *order* of an ODE is the order of the highest-order derivative in the differential equation. The general first-order ODE is

$$\frac{dy}{dt} = f(t, y) \tag{II.1}$$

where $f(t, y)$ is called the *derivative function*. For simplicity of notation, differentiation usually will be denoted by the superscript "prime" notation:

$$y' = \frac{dy}{dt} \tag{II.2}$$

Thus, Eq. (II.1) can be written as

$$y' = f(t, y) \tag{II.3}$$

The general nth-order ODE for $y(t)$ has the form

$$a_n y^{(n)} + a_{n-1} y^{(n-1)} + \cdots + a_2 y'' + a_1 y' + a_0 y = F(t) \tag{II.4}$$

where the superscripts (n), $(n-1)$, etc. denote nth-order differentiation, etc.

The solution of an ordinary differential equation is that particular function, $y(t)$ or $y(x)$, that identically satisfies the ODE in the domain of interest, $D(t)$ or $D(x)$, respectively, and satisfies the auxiliary conditions specified on the boundaries of the domain of interest. In a few special cases, the solution of an ODE can be expressed in closed form. In the majority of problems in engineering and science, the solution must be obtained by numerical methods. Such problems are the subject of Part II.

A *linear ODE* is one in which all of the derivatives appear in linear form and none of the coefficients depends on the dependent variable. The coefficients may be functions of the independent variable, in which case the ODE is a *variable-coefficient linear ODE*. For example,

$$y' + \alpha y = F(t) \tag{II.5}$$

is a linear, constant-coefficient, first-order ODE, whereas

$$y' + \alpha t y = F(t) \tag{II.6}$$

is a linear, variable-coefficient, first-order ODE. If the coefficients depend on the dependent variable, or the derivatives appear in a nonlinear form, then the ODE is *nonlinear*. For example,

$$yy' + \alpha y = 0 \tag{II.7}$$

$$(y')^2 + \alpha y = 0 \tag{II.8}$$

are *nonlinear* first-order ODEs.

A *homogeneous differential equation* is one in which each term involves the dependent variable or one of its derivatives. A *nonhomogeneous differential equation* contains additional terms, known as nonhomogeneous terms, source terms, or forcing functions, which do not involve the dependent variable. For example,

$$y' + \alpha y = 0 \tag{II.9}$$

is a linear, first-order, *homogeneous ODE*, and

$$y' + \alpha y = F(t) \tag{II.10}$$

is a linear, first-order, *nonhomogeneous ODE*, where $F(t)$ is the known nonhomogeneous term.

Many practical problems involve several dependent variables, each of which is a function of the same single independent variable and one or more of the dependent variables, and each of which is governed by an ordinary differential equation. Such coupled sets of ordinary differential equations are called *systems of ordinary differential equations*. Thus, the two coupled ODEs

$$y' = f(t, y, z) \tag{II.11a}$$
$$z' = g(t, y, z) \tag{II.11b}$$

comprise a system of two coupled first-order ordinary differential equations.

The general solution of a differential equation contains one or more constants of integration. Thus, a family of solutions is obtained. The number of constants of integration is equal to the order of the differential equation. The particular member of the family of solutions which is of interest is determined by auxiliary conditions. Obviously, the number of auxiliary conditions must equal the number of constants of integration, which is the same as the order of the differential equation.

As illustrated in the preceding discussion, a wide variety of ordinary differential equations exists. Each problem has its own special governing equation or equations and its own peculiarities which must be considered individually. However, useful insights into the general features of ODEs can be obtained by studying two special cases. The first special case is the *general nonlinear first-order ODE*:

$$\boxed{y' = f(t, y)} \tag{II.12}$$

where $f(t, y)$ is a nonlinear function of the dependent variable y. The second special case is the *general nonlinear second-order ODE*:

$$\boxed{y'' + P(x, y)y' + Q(x, y)y = F(x)} \tag{II.13}$$

These two special cases are studied in the following sections.

II.3 CLASSIFICATION OF ORDINARY DIFFERENTIAL EQUATIONS

Physical problems are governed by many different ordinary differential equations. There are two different types, or *classes*, of ordinary differential equations, depending on the type of auxiliary conditions specified. If all the auxiliary conditions are specified at the same value of the independent variable and the solution is to be marched forward from that initial point, the differential equation is an *initial-value ODE*. If the auxiliary conditions are specified at two different values of the independent variable, the end points or boundaries of the domain of interest, the differential equation is a *boundary-value ODE*.

Figure II.1 illustrates the solution of an initial-value ODE. The initial value of the dependent variable is specified at one value of the independent variable, and the solution domain $D(t)$ is open. Initial-value ODEs are solved by *marching numerical methods*.

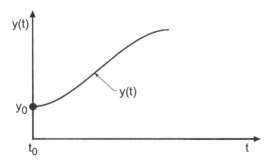

Figure II.1   Initial-value ODE.

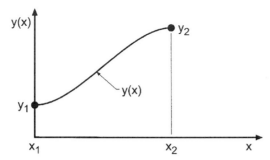

Figure II.2   Boundary-value ODE.

Figure II.2 illustrates the solution of a boundary-value ODE. The boundary values of the dependent variable are specified at two values of the independent variable, and the solution domain $D(x)$ is closed. Boundary-value ODEs can be solved by both *marching numerical methods* and *equilibrium numerical methods*.

II.4 CLASSIFICATION OF PHYSICAL PROBLEMS

Physical problems fall into one of the following three general classifications:

1. Propagation problems
2. Equilibrium problems
3. Eigenproblems

Each of these three types of physical problems has its own special features, its own particular type of ordinary differential equation, its own type of auxiliary conditions, and its own numerical solution methods. A clear understanding of these concepts is essential if meaningful numerical solutions are to be obtained.

Propagation problems are *initial-value problems* in *open domains* in which the known information (initial values) are *marched* forward in time or space from the initial state. The known information, that is, the initial values, are specified at one value of the independent variable. Propagation problems are governed by *initial-value ordinary differential equations*. The order of the governing ordinary differential equation may be one or greater. The number of initial values must be equal to the order of the differential equation.

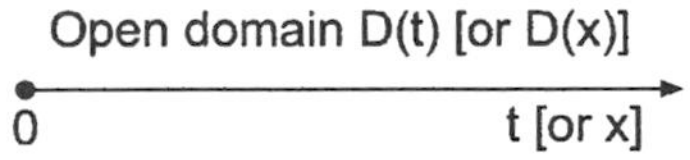

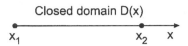

Open domain D(t) [or D(x)]

0   t [or x]

Figure II.3 Solution domain for propagation problems.

Closed domain D(x)

x_1 x_2 x

Figure II.4 Solution domain for equilibrium problems.

Propagation problems may be unsteady time (i.e., t) marching problems or steady space (i.e., x) marching problems. The marching direction in a steady space marching problem is sometimes called the *time-like direction*, and the corresponding coordinate is called the *time-like coordinate*. Figure II.3 illustrates the open solution domains $D(t)$ and $D(x)$ associated with time marching and space marching propagation problems, respectively.

Equilibrium problems are *boundary-value problems* in *closed domains* in which the known information (boundary values) are specified at two different values of the independent variable, the end points (boundaries) of the solution domain. Equilibrium problems are governed by *boundary-value ordinary differential equations*. The order of the governing differential equation must be at least 2, and may be greater. The number of boundary values must be equal to the order of the differential equation. Equilibrium problems are steady state problems in closed domains. Figure II.4 illustrates the closed solution domain $D(x)$ associated with equilibrium problems.

Eigenproblems are a special type of problem in which the solution exists only for special values (i.e., *eigenvalues*) of a parameter of the problem. The eigenvalues are to be determined in addition to the corresponding configuration of the system.

II.5 INITIAL-VALUE ORDINARY DIFFERENTIAL EQUATIONS

A classical example of an initial-value ODE is the general nonlinear first-order ODE:

$$y' = f(t, y) \qquad y(t_0) = y_0 \tag{II.14}$$

Equation (II.14) applies to many problems in engineering and science. In the following discussion, the general features of Eq. (II.14) are illustrated for the problem of transient heat transfer by radiation from a lumped mass to its surroundings and the transient motion of a mass-damper-spring system.

Consider the lumped mass m illustrated in Figure II.5. Heat transfer from the lumped mass m to its surroundings by radiation is governed by the *Stefan-Boltzmann law of radiation*:

$$\dot{q}_r = A\varepsilon\sigma(T^4 - T_a^4) \tag{II.15}$$

where $\dot{q}_r$ is the heat transfer rate (J/s), A is the surface area of the lumped mass (m^2), ε is the Stefan-Boltzmann constant (5.67×10^{-8} J/m^2-K^4-s), σ is the emissivity of the body (dimensionless), which is the ratio of the actual radiation to the radiation from a black body, T is the internal temperature of the lumped mass (K), and T_a is the ambient temperature (K) (i.e., the temperature of the surroundings). The energy E stored in the lumped mass is given by

$$E = mCT \qquad \qquad \text{(II.16)}$$

where m is the mass of the lumped mass (kg) and C is the specific heat of the material (J/kg-K). An energy balance states that the rate at which the energy stored in the lumped mass changes is equal to the rate at which heat is transferred to the surroundings. Thus,

$$\frac{d(mCT)}{dt} = -\dot{q}_r = -A\varepsilon\sigma(T^4 - T_a^4) \qquad \qquad \text{(II.17)}$$

The minus sign in Eq. (II.17) is required so that the rate of change of stored energy is negative when T is greater than T_a. For constant m and C, Eq. (II.17) can be written as

$$\frac{dT}{dt} = T' = -\alpha(T^4 - T_a^4) \qquad \qquad \text{(II.18)}$$

where

$$\alpha = \frac{A\varepsilon\sigma}{mC} \qquad \qquad \text{II.19)}$$

Consider the case where the temperature of the surroundings is constant and the initial temperature of the lumped mass is $T(0.0) = T_0$. The initial-value problem is stated as follows:

$$\boxed{T' = -\alpha(T^4 - T_a^4) = f(t, T) \qquad T(0) = T_0} \qquad \text{(II.20)}$$

Equation (II.20) is in the general form of Eq. (II.14). Equation (II.20) is a nonlinear first-order initial-value ODE. The solution of Eq. (II.20) is the function $T(t)$, which describes the temperature history of the lumped mass corresponding to the initial condition, $T(0) = T_0$. Equation (II.20) is an example of a nonlinear first-order initial-value ODE.

An example of a higher-order initial-value ODE is given by the nonlinear second-order ODE governing the vertical flight of a rocket. The physical system is illustrated in Figure II.6. Applying Newton's second law of motion, $\Sigma F = ma$, yields

$$\Sigma F = T - Mg - D = Ma = MV' = My'' \qquad \qquad \text{(II.21)}$$

where T is the thrust developed by the rocket motor (N), M is the instantaneous mass of the rocket (kg), g is the acceleration of gravity (m/s^2), which depends on the altitude y (m), D is the aerodynamic drag (N), a is the acceleration of the rocket (m/s^2), V is the velocity of the rocket (m/s), and y is the altitude of the rocket (m). The initial velocity, $V(0.0) = V_0$, is zero, and the initial elevation, $y(0.0) = y_0$, is zero. Thus, the initial conditions for Eq. (II.21) are

$$V(0.0) = y'(0.0) = 0.0 \qquad \text{and} \qquad y(0.0) = 0.0 \qquad \text{(II.22)}$$

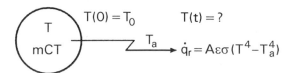

Figure II.5   Heat transfer by radiation from a lumped mass.

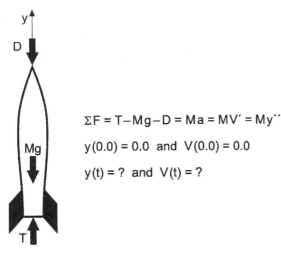

$$\Sigma F = T - Mg - D = Ma = MV' = My''$$

$$y(0.0) = 0.0 \quad \text{and} \quad V(0.0) = 0.0$$

$$y(t) = ? \quad \text{and} \quad V(t) = ?$$

Figure II.6   Vertical flight of a rocket.

In general, the thrust T is a variable, which depends on time and altitude. The instantaneous mass M is given by

$$M(t) = M_0 - \int_0^t \dot{m}(t)\, dt \tag{II.23}$$

where M_0 is the initial mass of the rocket (kg), and $\dot{m}(t)$ is the instantaneous mass flow rate being expelled by the rocket (kg/s). The instantaneous aerodynamic drag D is given by

$$D(\rho, V, y) = C_D(\rho, V, y)\tfrac{1}{2}\rho(y)AV^2 \tag{II.24}$$

where C_D is an empirical drag coefficient (dimensionless), which depends on the rocket geometry, the rocket velocity V and the properties of the atmosphere at altitude $y(m)$; ρ is the density of the atmosphere (kg/m^3), which depends on the altitude y (m); and A is the cross-sectional frontal area of the rocket (m^2).

Combining Eqs. (II.21) to (II.24) yields the following second-order nonlinear initial-value ODE:

$$y'' = \frac{F(t, y)}{M_0 - \int_0^t \dot{m}(t)\, dt} - g(y) - \frac{C_D(\rho, V, y)\tfrac{1}{2}\rho(y)AV^2}{M_0 - \int_0^t \dot{m}(t)\, dt} \tag{II.25}$$

Consider a simpler model where T, $\dot{m}$, and g are constant, and the aerodynamic drag D is neglected. In that case, Eq. (II.25) becomes

$$\boxed{y'' = \frac{F}{M_0 - \dot{m}t} - g \qquad y(0.0) = 0.0 \quad \text{and} \quad y'(0.0) = V(0.0) = 0.0} \tag{II.26}$$

The solution to Eq. (II.25) or (II.26) is the function $y(t)$, which describes the vertical motion of the rocket as a function of time t. Equations (II.25) and (II.26) are examples of second-order initial-value ODEs.

In summary, two classical examples of initial-value ODEs are represented by Eqs. (II.20) and (II.26). Many more complicated examples of initial-value ODEs arise in the fields of engineering and science. Higher-order ODEs and systems of ODEs occur frequently. Numerical procedures for solving initial-value ODEs are presented in Chapter 7.

II.6 BOUNDARY-VALUE ORDINARY DIFFERENTIAL EQUATIONS

A classical example of a boundary-value ODE is the general second-order ODE:

$$y'' + P(x, y)y' + Q(x, y)y = F(x) \qquad y(x_1) = y_1 \text{ and } y(x_2) = y_2 \qquad \text{(II.27)}$$

Equation (II.27) applies to many problems in engineering and science. In the following discussion, the general features of Eq. (II.27) are illustrated for the problem of steady one-dimensional heat diffusion (i.e., conduction) in a rod.

Consider the constant cross-sectional-area rod illustrated in Figure II.7. Heat diffusion transfers energy along the rod and energy is transferred from the rod to the surroundings by convection. An energy balance on the differential control volume yields

$$\dot{q}(x) = \dot{q}(x + dx) + \dot{q}_c(x) \tag{II.28}$$

which can be written as

$$\dot{q}(x) = \dot{q}(x) + \frac{d}{dx}[\dot{q}(x)] \, dx + \dot{q}_c(x) \tag{II.29}$$

which yields

$$\frac{d}{dx}[\dot{q}(x)] \, dx + \dot{q}_c(x) = 0 \tag{II.30}$$

Heat diffusion is governed by *Fourier's law of conduction*, which states that

$$\dot{q}(x) = -kA\frac{dT}{dx} \tag{II.31}$$

where $\dot{q}(x)$ is the energy transfer rate (J/s), k is the thermal conductivity of the solid (J/s-m-K), A is the cross-sectional area of the rod (m^2), and dT/dx is the temperature gradient (K/m). Heat transfer by convection is governed by *Newton's law of cooling*:

$$\dot{q}_c(x) = hA(T - T_a) \tag{II.32}$$

where h is an empirical heat transfer coefficient (J/s-m^2-K), A is the surface area of the rod ($A = P \, dx$, m^2), P is the perimeter of the rod (m), and T_a is the ambient temperature (K) (i.e., the temperature of the surroundings). Substituting Eqs. (II.31) and (II.32) into Eq. (II.30) gives

$$\frac{d}{dx}\left(-kA\frac{dT}{dx}\right) dx + h(P \, dx)(T - T_a) = 0 \tag{II.33}$$

For constant k, A, and P, Eq. (II.33) yields

$$\frac{d^2T}{dx^2} = \frac{hP}{kA}(T - T_a) = 0 \tag{II.34}$$

which can be written as

$$T'' - \alpha^2 T = -\alpha^2 T_a \tag{II.35}$$

where $\alpha^2 = hP/kA$.

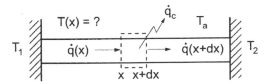

Figure II.7 Steady heat conduction in a rod.

Equation (II.35) is in the general form of Eq. (II.27). Equation (II.35) is a linear second-order boundary-value ODE. The solution of Eq. (II.35) is the function $T(x)$, which describes the temperature distribution in the rod corresponding to the boundary conditions

$$T(x_1) = T_1 \quad \text{and} \quad T(x_2) = T_2 \tag{II.36}$$

Equation (II.35) is an example of a second-order linear boundary-value problem.

An example of a higher-order boundary-value ODE is given by the fourth-order ODE governing the deflection of a laterally loaded symmetrical beam. The physical system is illustrated in Figure II.8. Bending takes place in the plane of symmetry, which causes deflections in the beam. The neutral axis of the beam is the axis along which the fibers do not undergo strain during bending. When no load is applied (i.e., neglecting the weight of the beam itself), the neutral axis is coincident with the x axis. When a distributed load $q(x)$ is applied, the beam deflects, and the neutral axis is displaced, as illustrated by the dashed line in Figure II.8. The shape of the neutral axis is called the deflection curve.

As shown in many strength of materials books (e.g., Timoshenko, 1955), the differential equation of the deflection curve is

$$EI(x)\frac{d^2y}{dx^2} = -M(x) \tag{II.37}$$

where E is the modulus of elasticity of the beam material, $I(x)$ is the moment of inertia of the beam cross section, which can vary along the length of the beam, and $M(x)$ is the bending moment due to transverse forces acting on the beam, which can vary along the

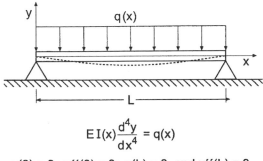

$$EI(x)\frac{d^4y}{dx^4} = q(x)$$

$y(0) = 0, \ y''(0) = 0; \ y(L) = 0, \text{ and } y''(L) = 0$

Figure II.8 Deflection of a beam.

length of the beam. The moment $M(x)$ is related to the shearing force $V(x)$ acting on each cross section of the beam as follows:

$$\frac{dM(x)}{dx} = V(x) \tag{II.38}$$

The shearing force $V(x)$ is related to the distributed load $q(x)$ as follows:

$$\frac{dV(x)}{dx} = -q(x) \tag{II.39}$$

Combining Eqs. (II.37) to (II.39) yields the differential equation for the beam deflection curve:

$$\boxed{EI(x)\frac{d^4 y}{dx^4} = q(x)} \tag{II.40}$$

Equation (II.40) requires four boundary conditions. For a horizontal beam of length L,

$$y(0.0) = y(L) = 0.0 \tag{II.41}$$

For a beam fixed (i.e., clamped) at both ends,

$$y'(0.0) = y'(L) = 0.0 \tag{II.42}$$

For a beam pinned (i.e., hinged) at both ends,

$$y''(0.0) = y''(L) = 0.0 \tag{II.43}$$

For a beam cantilevered (i.e., free) at either end,

$$y'''(0.0) = 0.0 \qquad \text{or} \qquad y'''(L) = 0.0 \tag{II.44}$$

Any two combinations of these four boundary conditions can be specified at each end.

Equation II.40 is a linear example of the general nonlinear fourth-order boundary-value ODE:

$$\boxed{y'''' = f(x, y, y', y'', y''')} \tag{II.45}$$

which requires four boundary conditions at the boundaries of the closed physical domain.

II.7 SUMMARY

The general features of ordinary differential equations are discussed in this section. Ordinary differential equations are classified as initial-value differential equations or boundary-value differential equations according to whether the auxiliary conditions are specified at a single initial time (or location) or at two locations, respectively. Examples of several ordinary differential equations that arise in engineering and science are presented in this section.

The classical example of a first-order nonlinear initial-value ODE is

$$\boxed{y' = f(t, y) \qquad y(t_0) = y_0} \tag{II.46}$$

Chapter 7 presents several numerical methods for solving Eq. (II.46) and illustrates those methods with numerical examples.

The classical example of a second-order nonlinear boundary-value ODE is

$$y'' + P(x, y)y' + Q(x, y)y = F(x) \qquad y(x_1) = y_1 \quad \text{and} \quad y(x_2) = y_2 \qquad \text{(II.47)}$$

Chapter 8 presents several numerical methods for solving Eq. (II.47) and illustrates those methods with numerical examples.

7

One-Dimensional Initial-Value Ordinary Differential Equations

Examples

7.1 INTRODUCTION

Most people have some physical understanding of heat transfer due to its presence in many aspects of our daily life. Consequently, the nonlinear first-order ODE governing unsteady radiation heat transfer from a lumped mass, illustrated at the top of Figure 7.1 and discussed in Section II.5, is considered in this chapter to illustrate finite difference methods for solving first-order initial-value ODEs. That equation, Eq. (II.20), is

$$\boxed{T' = -\alpha(T^4 - T_a^4) = f(t, T) \qquad T(0.0) = T_0 = 2500.0 \qquad T_a = 250.0} \tag{7.1}$$

where α is defined in Section II.5.

The exact solution to Eq. (7.1) can be obtained by separating variables, expressing $1/(T^4 - T_a^4)$ by a partial fraction expansion, integrating, and applying the initial condition. The result is

$$\tan^{-1}\left(\frac{T}{T_a}\right) - \tan^{-1}\left(\frac{T_0}{T_a}\right) + \frac{1}{2}\ln\left[\frac{(T_0 - T_a)(T + T_a)}{(T - T_a)(T_0 + T_a)}\right] = 2\alpha T_a^3 t \tag{7.2}$$

For $T_0 = 2500.0\,\text{K}$, $T_a = 250.0\,\text{K}$, and $\alpha = 4.0 \times 10^{-12}$ $(\text{K}^3\text{-s})^{-1}$, Eq. (7.2) becomes

$$\boxed{\begin{aligned} \tan^{-1}\left(\frac{T}{250}\right) &- \tan^{-1}\left(\frac{2500}{250}\right) + \frac{1}{2}\ln\left[\frac{(2500 - 250)(T + 250)}{(T - 250)(2500 + 250)}\right] \\ &= 2(4.0 \times 10^{-12})250^3 t \end{aligned}} \tag{7.3}$$

The exact solution at selected values of time, obtained by solving Eq. (7.3) by the secant method, is tabulated in Table 7.1 and illustrated in Figure 7.2.

The vertical flight of a rocket, illustrated at the bottom of Figure 7.1, is considered in this chapter to illustrate finite difference methods for solving higher-order initial-value ODEs. This problem is discussed in Section II.5 and solved in Sections 7.12 and 7.13.

The general features of *initial-value ordinary differential equations (ODEs)* are discussed in Section II.5. In that section it is shown that initial-value ODEs govern *propagation problems*, which are *initial-value problems* in *open domains*. Consequently, initial-value ODEs are solved numerically by *marching methods*. This chapter is devoted to presenting the basic properties of finite difference methods for solving initial-value (i.e., propagation) problems and to developing several specific finite difference methods.

The *objective* of a finite difference method for solving an ordinary differential equation (ODE) is to transform a calculus problem into an algebra problem by:

1. *Discretizing* the continuous physical domain into a discrete finite difference grid

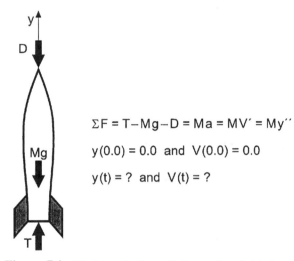

$$\frac{dT}{dt} = -\alpha(T^4 - T_a^4), \quad T(0) = T_0, \quad T(t) = ?$$

$$\Sigma F = T - Mg - D = Ma = MV' = My''$$

$$y(0.0) = 0.0 \quad \text{and} \quad V(0.0) = 0.0$$

$$y(t) = ? \quad \text{and} \quad V(t) = ?$$

Figure 7.1 Heat transfer by radiation and rocket trajectory.

2. *Approximating* the exact derivatives in the ODE by algebraic finite difference approximations (FDAs)
3. *Substituting* the FDAs into the ODE to obtain an algebraic finite difference equation (FDE)
4. *Solving* the resulting algebraic FDE

Table 7.1 Exact Solution of the Radiation Problem

t, s	$T(t)$, K	t, s	$T(t)$, K
0.0	2500.00000000	6.0	1944.61841314
1.0	2360.82998845	7.0	1890.58286508
2.0	2248.24731405	8.0	1842.09450785
3.0	2154.47079576	9.0	1798.22786679
4.0	2074.61189788	10.0	1758.26337470
5.0	2005.41636581		

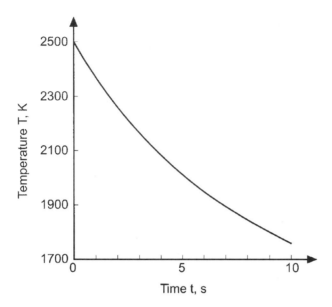

Figure 7.2   Exact solution of the radiation problem.

Numerous initial-value ordinary differential equations arise in engineering and science. Single ODEs governing a single dependent variable arise frequently, as do coupled systems of ODEs governing several dependent variables. Initial-value ODEs may be linear or nonlinear, first- or higher-order, and homogeneous or nonhomogeneous. In this chapter, the majority of attention is devoted to the general nonlinear first-order ordinary differential equation (ODE):

$$y' = \frac{dy}{dt} = f(t, y) \qquad y(t_0) = y_0 \tag{7.4}$$

where y' denotes the first derivative and $f(t, y)$ is the nonlinear *derivative function*. The solution to Eq. (7.4) is the function $y(t)$. This function must satisfy an initial condition at $t = t_0$, $y(t_0) = y_0$. In most physical problems, the initial time is $t = 0.0$ and $y(t_0) = y(0.0)$. The solution domain is open, that is, the independent variable t has an unspecified (i.e., open) final value. Several finite difference methods for solving Eq. (7.4) are developed in this chapter. Procedures for solving higher-order ODEs and systems of ODEs, based on the methods for solving Eq. (7.4), are discussed.

The objective of Chapter 7 is to develop *finite difference methods* for solving initial-value ODEs such as Eq. (7.4). Three major types of methods are considered:

1. Single-point methods
2. Extrapolation methods
3. Multipoint methods

Single-point methods advance the solution from one grid point to the next grid point using only the data at the single grid point. Several examples of single-point methods are presented. The most significant of these is the *fourth-order Runge-Kutta method*, which is presented in Section 7.7. Extrapolation methods evaluate the solution at a grid point for several values of grid size and extrapolate those results to obtain a more accurate solution. The *extrapolated modified midpoint method* is presented in Section 7.8 as an example of extrapolation methods. Multipoint methods advance the solution from one grid point to the next grid point using the data at several known points. The *fourth-order Adams-Bashforth-Moulton method* is presented in Section 7.9 as an example of multipoint methods. These three types of methods can be applied to solve linear and nonlinear single first-order ODEs. The *Gear method* for solving stiff ODEs is presented in Section 7.14.

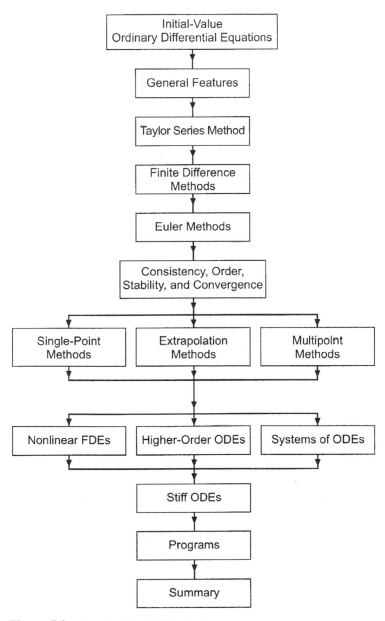

Figure 7.3   Organization of Chapter 7.

The organization of Chapter 7 is illustrated in Figure 7.3. The general features of initial-value ordinary differential equations (ODEs) are discussed first. The Taylor series method is then presented. This is followed by an introduction to finite difference methods. Two first-order methods are then developed: the explicit Euler method and the implicit Euler method. Definitions and discussion of the concepts of consistency, order, stability, and convergence are presented. The material then splits into a presentation of the three fundamentally different types of methods for solving initial-value ODEs: single-point methods, extrapolation methods, and multipoint methods. Then follows a discussion of nonlinear finite difference equations, higher-order ODEs, and systems of ODEs. A brief introduction to stiff ODEs follows. Programs are then presented for the fourth-order Runge-Kutta method, the extrapolated modified-midpoint method, and the fourth-order Adams-Bashforth-Moulton method. The chapter closes with a summary, which summarizes the contents of the chapter, discusses the advantages and disadvantages of the various methods, and lists the things you should be able to do after studying Chapter 7.

7.2 GENERAL FEATURES OF INITIAL-VALUE ODEs

Several general features of initial-value ordinary differential equations (ODEs) are presented in this section. The properties of the linear first-order ODE are discussed in some detail. The properties of nonlinear first-order ODEs, systems of first-order ODEs, and higher-order ODEs are discussed briefly.

7.2.1 The General Linear First-Order ODE

The general nonlinear first-order ODE is given by Eq. (7.4). The general linear first-order ODE is given by

$$\boxed{y' + \alpha y = F(t) \qquad y(t_0) = y_0} \tag{7.5}$$

where α is a real constant, which can be positive or negative.

The exact solution of Eq. (7.5) is the sum of the complementary solution $y_c(t)$ and the particular solution $y_p(t)$:

$$y(t) = y_c(t) + y_p(t) \tag{7.6}$$

The complementary solution $y_c(t)$ is the solution of the homogeneous ODE:

$$y_c' + \alpha y_c = 0 \tag{7.7}$$

The complementary solution, which depends only on the homogeneous ODE, describes the inherent properties of the ODE and the response of the ODE in the absence of external stimuli. The particular solution $y_p(t)$ is the function which satisfies the nonhomogeneous term $F(t)$:

$$y_p' + \alpha y_p = F(t) \tag{7.8}$$

The particular solution describes the response of the ODE to external stimuli specified by the nonhomogeneous term $F(t)$.

The complementary solution $y_c(t)$ is given by

$$y_c(t) = Ae^{-\alpha t} \tag{7.9}$$

which can be shown to satisfy Eq. (7.7) by direct substitution. The coefficient A can be determined by the initial condition, $y(t_0) = y_0$, after the complete solution $y(t)$ to the ODE has been obtained. The particular solution $y_p(t)$ is given by

$$y_p(t) = B_0 F(t) + B_1 F'(t) + B_2 F''(t) + \cdots \tag{7.10}$$

where the terms $F'(t)$, $F''(t)$, etc. are the derivatives of the nonhomogeneous term. These terms, $B_i F^{(i)}(t)$, continue until $F^{(i)}(t)$ repeats its functional form or becomes zero. The coefficients B_0, B_1, etc. can be determined by substituting Eq. (7.10) into Eq. (7.8), grouping similar functional forms, and requiring that the coefficient of each functional form be zero so that Eq. (7.8) is satisfied for all values of the independent variable t.

The total solution of Eq. (7.5) is obtained by substituting Eqs. (7.9) and (7.10) into Eq. (7.6). Thus,

$$\boxed{y(t) = Ae^{-\alpha t} + y_p(t)} \tag{7.11}$$

The constant of integration A can be determined by requiring Eq. (7.11) to satisfy the initial condition, $y(t_0) = y_0$.

The homogeneous ODE, $y' + \alpha y = 0$, has two completely different types of solutions, depending on whether α is positive or negative. Consider the pair of ODEs:

$$y' + \alpha y = 0 \tag{7.12}$$
$$y' - \alpha y = 0 \tag{7.13}$$

where α is a *positive* real constant. The solutions to Eqs. (7.12) and (7.13) are

$$y(t) = Ae^{-\alpha t} \tag{7.14}$$
$$y(t) = Ae^{\alpha t} \tag{7.15}$$

Equations (7.14) and (7.15) each specify a family of solutions, as illustrated in Figure 7.4. A particular member of either family is chosen by specifying the initial condition, $y(0) = y_0$, as illustrated by the dashed curves in Figure 7.4.

For the ODE $y' + \alpha y = 0$, the solution, $y(t) = Ae^{-\alpha t}$, decays exponentially with time. This is a *stable ODE*. For the ODE $y' - \alpha y = 0$, the solution, $y(t) = Ae^{\alpha t}$, grows exponentially with time without bound. This is an *unstable ODE*. Any numerical method for solving these ODEs must behave in a similar manner.

7.2.2 The General Nonlinear First-Order ODE

The general nonlinear first-order ODE is given by

$$\boxed{y' = f(t, y) \qquad y(t_0) = y_0} \tag{7.16}$$

where the derivative function $f(t, y)$ is a nonlinear function of y. The general features of the solution of a nonlinear first-order ODE in a small neighborhood of a point are similar to the general features of the solution of the linear first-order ODE at that point.

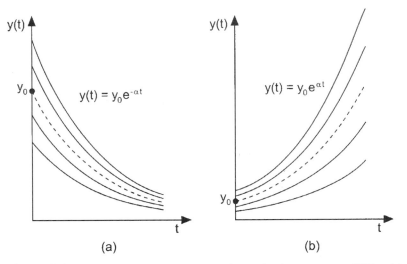

Figure 7.4 Exact solutions of the linear first-order homogeneous ODEs. (a) $y' + \alpha y = 0$. (b) $y' - \alpha y = 0$.

The nonlinear ODE can be linearized by expressing the derivative function $f(t, y)$ in a Taylor series at an initial point (t_0, y_0) and dropping all terms higher than first order. Thus,

$$f(t, y) = f_0 + f_t|_0(t - t_0) + f_y|_0(y - y_0) + \cdots \tag{7.17a}$$

$$f(t, y) = (f_0 - f_t|_0 t_0 - f_y|_0 y_0) + f_t|_0 t + f_y|_0 y + \cdots \tag{7.17b}$$

which can be written as

$$f(t, y) = -\alpha y + F(t) + \text{higher-order terms} \tag{7.18}$$

where α and $F(t)$ are defined as follows:

$$\boxed{\alpha = -f_y|_0} \tag{7.19}$$

$$F(t) = (f_0 - f_t|_0 t_0 - f_y|_0 y_0) + f_t|_0 t \tag{7.20}$$

Truncating the higher-order terms and substituting Eqs. (7.18) to (7.20) into the nonlinear ODE, Eq. (7.16), gives

$$\boxed{y' + \alpha y = F(t) \qquad y(t_0) = y_0} \tag{7.21}$$

Most of the general features of the numerical solution of a nonlinear ODE can be determined by analyzing the linearized form of the nonlinear ODE. Consequently, Eq. (7.21) will be used extensively as a model ODE to investigate the general behavior of numerical methods for solving initial-value ODEs, both linear and nonlinear.

7.2.3 Higher-Order ODEs

Higher-order ODEs generally can be replaced by a system of first-order ODEs. Each higher-order ODE in a system of higher-order ODEs can be replaced by a system of first-order ODEs, thus yielding coupled systems of first-order ODEs. Consequently, Chapter 7 is devoted mainly to solving first-order ODEs. A discussion of the solution of higher-order ODEs is presented in Section 7.12.

7.2.4 Systems of First-Order ODEs

Systems of coupled first-order ODEs can be solved by solving all the coupled first-order differential equations simultaneously using methods developed for solving single first-order ODEs. Consequently, Chapter 7 is devoted mainly to solving single first-order ODEs. A discussion of the solution of systems of coupled first-order ODEs is presented in Section 7.13.

7.2.5 Summary

Four types of initial-value ODEs have been considered:

1. The linear first-order ODE
2. The nonlinear first-order ODE
3. Higher-order ODEs
4. Systems of first-order ODEs

The general features of all four types of ODEs are similar. Consequently, Chapter 7 is devoted mainly to the numerical solution of nonlinear first-order ODEs.

The foregoing discussion considers initial-value (propagation) problems in time (i.e., t), which are unsteady time marching initial-value problems. Steady space marching (i.e., x) initial-value problems are also propagation problems. Both types of problems are initial-value (propagation) problems. Occasionally the space coordinate in a steady space marching problem is referred to as a timelike coordinate. Throughout Chapter 7, time marching initial-value problems are used to illustrate the general features, behavior, and solution of initial-value ODEs. All the results presented in Chapter 7 can be applied directly to steady space marching initial-value problems simply by changing t to x in all of the discussions, equations, and results.

7.3 THE TAYLOR SERIES METHOD

In theory, the infinite Taylor series can be used to evaluate a function, given its derivative function and its value at some point. Consider the nonlinear first-order ODE:

$$\boxed{y' = f(t, y) \qquad y(t_0) = y_0} \tag{7.22}$$

The Taylor series for $y(t)$ at $t = t_0$ is

$$y(t) = y(t_0) + y'(t_0)(t - t_0) + \tfrac{1}{2}y''(t_0)(t - t_0)^2 + \tfrac{1}{6}y'''(t_0)(t - t_0)^3$$
$$+ \cdots + \frac{1}{n!}y^{(n)}(t_0)(t - t_0)^n + \cdots \tag{7.23}$$

Equation (7.23) can be written in the simpler appearing form:

$$y(t) = y_0 + y'|_0 \, \Delta t + \tfrac{1}{2}y''|_0 \, \Delta t^2 + \tfrac{1}{6}y'''|_0 \, \Delta t^3 + \cdots \tag{7.24}$$

where $y'(t_0) = y'|_0$, etc., $\Delta t = (t - t_0)$, and, for convenience, Δt^n denotes $(\Delta t)^n$, not $\Delta(t^n)$.

Equation (7.24) can be employed to evaluate $y(t)$ if y_0 and the values of the derivatives at t_0 can be determined. The value of y_0 is the initial condition specified in Eq. (7.22). The first derivative $y'|_0$ can be determined by evaluating the derivative function $f(t, y)$ at t_0: $y'|_0 = f(t_0, y_0)$. The higher-order derivatives in Eq. (7.24) can be determined by successively differentiating the lower-order derivatives, starting with y'. Thus,

$$y'' = (y')' = \frac{d(y')}{dt} \tag{7.25a}$$

$$d(y') = d(y'(t, y)) = \frac{\partial y'}{\partial t} dt + \frac{\partial y'}{\partial y} dy = dt\left(\frac{\partial y'}{\partial t} + \frac{\partial y'}{\partial y}\frac{dy}{dt}\right) \tag{7.25b}$$

$$\frac{d(y')}{dt} = \frac{\partial y'}{\partial t} + \frac{\partial y'}{\partial y}\frac{dy}{dt} = y_t' + y_y'\frac{dy}{dt} \tag{7.25c}$$

Recall that $dy/dt = y'$. Substituting Eq. (7.25c) into Eq. (7.25a) yields

$$y'' = y_t' + y_y'y' \tag{7.25d}$$

In a similar manner,

$$y''' = (y'')' = \frac{d(y'')}{dt} = \frac{\partial}{\partial t}(y_t' + y_y'y') + \frac{\partial}{\partial y}(y_t' + y_y'y')\frac{dy}{dt} \tag{7.26a}$$

$$y''' = y_{tt}' + 2y_{ty}'y' + y_t'y_y' + (y_y')^2 y' + y_{yy}'(y')^2 \tag{7.26b}$$

Higher-order derivatives become progressively more complicated. It is not practical to evaluate a large number of the higher-order derivatives. Consequently, the Taylor series must be truncated. The remainder term in a finite Taylor series is:

$$\text{Remainder} = \frac{1}{(n+1)!}y^{(n+1)}(\tau)\,\Delta t^{n+1} \tag{7.27}$$

where $t_0 \leq \tau \leq t$. Truncating the remainder term yields a finite truncated Taylor series. Error estimation is difficult, since τ is unknown.

Example 7.1.   The Taylor series method

Let's solve the radiation problem presented in Section 7.1 by the Taylor series method. Recall Eq. (7.1):

$$T' = f(t, T) = -\alpha(T^4 - T_a^4) \qquad T(0.0) = 2500.0 \qquad T_a = 250.0 \tag{7.28}$$

where $\alpha = 4.0 \times 10^{-12}$ $(\text{K}^3\text{-s})^{-1}$. The Taylor series for $T(t)$ is given by

$$\boxed{T(t) = T_0 + T'|_0 t + \tfrac{1}{2}T''|_0 t^2 + \tfrac{1}{6}T'''|_0 t^3 + \tfrac{1}{24}T^{(4)}|_0 t^4 + \cdots} \tag{7.29}$$

where $\Delta t = t - t_0 = t$. From Eq. (7.28), $T_0 = 2500.0$, and

$$T'|_0 = -\alpha(T^4 - T_a^4)|_0 = -(4.0 \times 10^{-12})(2500.0^4 - 250.0^4) = -156.234375 \tag{7.30}$$

Solving for the higher-order derivatives yields:

$$T'' = (T')' = \frac{\partial T'}{\partial t} + \frac{\partial T'}{\partial T} T' = 0.0 - 4\alpha T^3 T' \tag{7.31a}$$

$$T_0'' = -4(4.0 \times 10^{-12})2500.0^3(-156.234375) = 39.058594 \tag{7.31b}$$

$$T'' = 4\alpha^2(T^7 - T^3 Ta^4) \tag{7.31c}$$

$$T''' = (T'')' = \frac{\partial T''}{\partial t} + \frac{\partial T''}{\partial T} T' = 0.0 + 4\alpha^2(7T^6 - 3T^2 T_a^4)T' \tag{7.32a}$$

$$T_0''' = 4(4.0 \times 10^{-12})^2$$
$$\times (7 \times 2500.0^6 - 3 \times 2500.0^2 \times 250.0^4)(-156.234375)$$
$$= -17.087402 \tag{7.32b}$$

$$T''' = -4\alpha^3(7T^{10} - 10T^6 T_a^4 + 3T^2 T_a^8) \tag{7.32c}$$

$$T^{(4)} = (T''')' = \frac{\partial T'''}{\partial t} + \frac{\partial T'''}{\partial T} T' = 0.0 - 4\alpha^3(70T^9 - 60T^5 T_a^4 + 6TT_a^8)T' \tag{7.33a}$$

$$T^{(4)} = -4(4.0 \times 10^{-12})^3(70 \times 2500.0^9 - 60 \times 2500.0^5 \times 250.0^4 + 6$$
$$\times 2500.0 \times 250.0^8)(-156.234375) = 10.679169 \tag{7.33b}$$

Substituting the above values into Eq. (7.29) yields

$$T(t) = 2500.0 - 156.284375t + 19.529297t^2 - 2.847900t^3 + 0.444965t^4 \tag{7.34}$$

The exact solution and the solution obtained from Eq. (7.34) are tabulated in Table 7.2, where $T(t)^{(1)}$, $T(t)^{(2)}$, etc. denote the Taylor series through the first, second, etc. derivative terms. These results are also illustrated in Figure 7.5.

From Figure 7.5, it is obvious that the accuracy of the solution improves as the number of terms in the Taylor series increases. However, even with four terms, the solution is not very accurate for $t > 2.0$ s. The Taylor series method is not an efficient method for solving initial-value ODEs.

Even though the Taylor series method is not an efficient method for solving initial-value ODEs, it is the basis of many excellent numerical methods. As illustrated in Figure 7.5, the solution by the Taylor series method is quite accurate for small values of t. Therein lies the basis for more accurate methods of solving ODEs. Simply put, use the Taylor series method for a small step in the neighborhood of the initial point. Then reevaluate the coefficients (i.e., the derivatives) at the new point. Successive reevaluation of the

Table 7.2 Solution by the Taylor Series Method

t, s	$T_{exact}(t)$, K	$T(t)^{(1)}$, K	$T(t)^{(2)}$, K	$T(t)^{(3)}$, K	$T(t)^{(4)}$, K
0.0	2500.00	2500.00	2500.00	2500.00	2500.00
1.0	2360.83	2343.77	2363.29	2360.45	2360.89
2.0	2248.25	2187.53	2265.65	2242.87	2249.98
3.0	2154.47	2031.30	2207.06	2130.17	2166.21
4.0	2074.61	1875.06	2187.53	2005.27	2119.18
5.0	2005.42	1718.83	2207.06	1851.07	2129.18

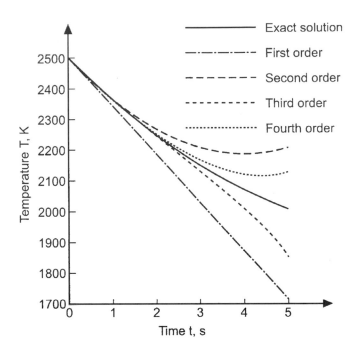

Figure 7.5 Solution by the Taylor series method.

coefficients as the solution progresses yields a much more accurate solution. This concept is the underlying basis of most numerical methods for solving initial-value ODEs.

7.4 THE FINITE DIFFERENCE METHOD

The objective of a finite difference method for solving an ordinary differential equation (ODE) is to transform a calculus problem into an algebra problem by:

1. *Discretizing* the continuous physical domain into a discrete finite difference grid
2. *Approximating* the exact derivatives in the initial-value ODE by algebraic finite difference approximations (FDAs)
3. *Substituting* the FDAs into the ODE to obtain an algebraic finite difference equation (FDE)
4. *Solving* the resulting algebraic FDE

Discretizing the continuous physical domain, approximating the exact derivatives by finite difference approximations, and developing finite difference equations are discussed in this section. Brief discussions of errors and smoothness close the section.

7.4.1 Finite Difference Grids

The solution domain $D(t)$ [or $D(x)$] and a discrete finite difference grid are illustrated in Figure 7.6. The solution domain is discretized by a one-dimensional set of discrete grid points, which yields the finite difference grid. The finite difference solution of the ODE is obtained at these grid points. For the present, let these grid points be equally spaced having uniform spacing Δt (or Δx). The resulting finite difference grid is illustrated in Figure 7.6.

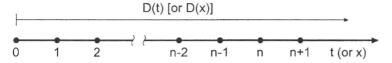

Figure 7.6   Solution domain, $D(t)$ [or $D(x)$], and discrete finite difference grid.

Nonuniform grids in which Δt (or Δx) is variable are considered later in the chapter. The subscript n is used to denote the physical grid points, that is, t_n (or x_n). Thus, grid point n corresponds to location t_n (or x_n) in the solution domain $D(t)$ [or $D(x)$]. The total number of grid points is denoted by *nmax*. For the remainder of this chapter, time t will be chosen as the independent variable. Similar results hold for steady space marching problems in which space x is the independent variable. All the results in terms of time t can be applied directly to space x simply by changing t to x everywhere.

    The dependent variable at grid point n is denoted by the same subscript notation that is used to denote the grid points themselves. Thus, the function $y(t)$ at grid point n is denoted by

$$y(t_n) = y_n \tag{7.35}$$

In a similar manner, derivates are denoted by

$$\frac{dy}{dt}(t_n) = y'(t_n) = y'|_n \tag{7.36}$$

7.4.2   Finite Difference Approximations

Now that the finite difference grid has been specified, finite difference approximations (FDAs) of the exact derivatives in the ODE must be developed. This is accomplished using the Taylor series approach developed in Section 5.4, where approximations of various types (i.e., forward, backward, and centered) of various orders (i.e., first order, second order, etc.) are developed for various derivatives (i.e., first derivative, second derivative, etc.). Those results are presented in Table 5.1.

    In the development of finite difference approximations of differential equations, a distinction must be made between the exact solution of the differential equation and the solution of the finite difference equation which is an approximation of the exact differential equation. For the remainder of this chapter, the exact solution of the ODE is denoted by an overbar on the symbol for the dependent variable [i.e., $\bar{y}(t)$], and the approximate solution is denoted by the symbol for the dependent variable without an overbar [i.e., $y(t)$]. Thus,

$$\bar{y}(t) = \text{exact solution}$$
$$y(t) = \text{approximate solution}$$

This very precise distinction between the exact solution of a differential equation and the approximate solution of a differential equation is required for studies of consistency, order, stability, and convergence, which are defined and discussed in Section 7.6.

Exact derivatives, such as $\bar{y}'$, can be approximated at a grid point in terms of the values of $\bar{y}$ at that grid point and adjacent grid points in several ways. Consider the derivative $\bar{y}'$. Writing the Taylor series for $\bar{y}_{n+1}$ using grid point n as the base point gives

$$\bar{y}_{n+1} = \bar{y}_n + \bar{y}'|_n \, \Delta t + \tfrac{1}{2}\bar{y}''|_n \, \Delta t^2 + \tfrac{1}{6}\bar{y}'''|_n \, \Delta t^3 + \cdots \tag{7.37}$$

where the convention $(\Delta t)^m \to \Delta t^m$ is employed for compactness. Equation (7.37) can be expressed as the Taylor polynomial with remainder:

$$\bar{y}_{n+1} = \bar{y}_n + \bar{y}'|_n \, \Delta t + \tfrac{1}{2}\bar{y}''|_n \, \Delta t^2 + \cdots + \frac{1}{m!}\bar{y}^{(m)}|_n \, \Delta t^m + R^{m+1} \tag{7.38a}$$

where the remainder term R^{m+1} is given by

$$R^{m+1} = \frac{1}{(m+1)!}\bar{y}^{(m+1)}(\tau) \, \Delta t^{m+1} \tag{7.38b}$$

where $t \leq \tau \leq t + \Delta t$. The remainder term is simply the next term in the Taylor series evaluated at $t = \tau$. If the infinite Taylor series is truncated after the mth derivative term to obtain an approximation of $\bar{y}^{n+1}$, the remainder term R_{m+1} is the error associated with the truncated Taylor series. In most cases, our main concern is the order of the error, which is the rate at which the error goes to zero as $\Delta t \to 0$.

Solving Eq. (7.37) for $\bar{y}'|_n$ yields

$$\bar{y}'|_n = \frac{\bar{y}_{n+1} - \bar{y}_n}{\Delta t} - \tfrac{1}{2}\bar{y}''|_n \, \Delta t - \tfrac{1}{6}\bar{y}'''|_n \, \Delta t^2 - \cdots \tag{7.39}$$

If Eq. (7.39) is terminated after the first term on the right-hand side, it becomes

$$\bar{y}'|_n = \frac{\bar{y}_{n+1} - \bar{y}_n}{\Delta t} - \tfrac{1}{2}\bar{y}''(\tau) \, \Delta t \tag{7.40}$$

A finite difference approximation of $\bar{y}'|_n$, which will be denoted by $y'|_n$, can be obtained from Eq. (7.40) by truncating the remainder term. Thus,

$$\boxed{y'|_n = \frac{y_{n+1} - y_n}{\Delta t} \qquad 0(\Delta t)} \tag{7.41}$$

where the $0(\Delta t)$ term is shown to remind us of the order of the remainder term which was truncated, which is the order of the approximation of $\bar{y}'|_n$. The remainder term which has been truncated to obtain Eq. (7.41) is called the *truncation error* of the finite difference approximation of $\bar{y}'|_n$. Equation (7.41) is a first-order forward-difference approximation of $\bar{y}'$ at grid point n.

A first-order backward-difference approximation of $\bar{y}'$ at grid point $n + 1$ can be obtained by writing the Taylor series for $\bar{y}_n$ using grid point $n + 1$ as the base point and solving for $\bar{y}'|_{n+1}$. Thus,

$$\bar{y}_n = \bar{y}_{n+i} + \bar{y}'_{n+1}(-\Delta t) + \tfrac{1}{2}\bar{y}''_{n+1}(-\Delta t)^2 + \cdots \tag{7.42}$$

$$\bar{y}'|_{n+1} = \frac{\bar{y}_{n+1} - \bar{y}_n}{\Delta t} + \tfrac{1}{2}\bar{y}''(\tau) \, \Delta t \tag{7.43}$$

Truncating the remainder term yields

$$\boxed{y'|_{n+1} = \frac{y_{n+1} - y_n}{\Delta t} \qquad 0(\Delta t)} \tag{7.44}$$

A second-order centered-difference approximation of $\bar{y}'$ at grid point $n + \frac{1}{2}$ can be obtained by writing the Taylor series for $\bar{y}_{n+1}$ and $\bar{y}_n$ using grid point $n + \frac{1}{2}$ as the base point, subtracting the two Taylor series, and solving for $\bar{y}'|_{n+1/2}$. Thus,

$$\bar{y}_{n+1} = \bar{y}_{n+1/2} + \bar{y}'_{n+1/2}(\Delta t/2) + \tfrac{1}{2}\bar{y}''_{n+1/2}(\Delta t/2)^2 + \tfrac{1}{6}\bar{y}'''_{n+1/2}(\Delta t/2)^3 + \cdots \qquad (7.45a)$$

$$\bar{y}_n = \bar{y}_{n+1/2} + \bar{y}'_{n+1/2}(-\Delta t/2) + \tfrac{1}{2}\bar{y}''_{n+1/2}(-\Delta t/2)^2 + \tfrac{1}{6}\bar{y}'''_{n+1/2}(-\Delta t/2)^3 + \cdots$$
$$(7.45b)$$

Subtracting Eq. (7.45b) from Eq. (7.45a) and solving for $\bar{y}'|_{n+1/2}$ yields

$$\bar{y}'|_{n+1/2} = \frac{\bar{y}_{n+1} - \bar{y}_n}{\Delta t} - \tfrac{1}{24}\bar{y}'''(\tau)\,\Delta t^2 \qquad (7.46)$$

Truncating the remainder term yields

$$\boxed{\, y'|_{n+1/2} = \frac{y_{n+1} - y_n}{\Delta t} \qquad 0(\Delta t^2) \,} \qquad (7.47)$$

Note that Eqs. (7.41), (7.44), and (7.47) are identical algebraic expressions. They all yield the same numerical value. The differences in the three finite difference approximations are in the values of the truncation errors, given by Eqs. (7.40), (7.43), and (7.46), respectively.

Equations (7.39) to (7.47) can be applied to steady space marching problems simply by changing t to x in all the equations.

Occasionally a finite difference approximation of an exact derivative is presented without its development. In such cases, the truncation error and order can be determined by a *consistency analysis* using Taylor series. For example, consider the following finite difference approximation (FDA):

$$\text{FDA} = \frac{y_{n+1} - y_n}{\Delta t} \qquad (7.48)$$

The Taylor series for the approximate solution $y(t)$ with base point n is

$$y_{n+1} = y_n + y'|_n\,\Delta t + \tfrac{1}{2}y''|_n\,\Delta t^2 + \cdots \qquad (7.49)$$

Substituting the Taylor series for y_{n+1} into the FDA, Eq. (7.48), gives

$$\text{FDA} = \frac{y_n + y'|_n\,\Delta t + \tfrac{1}{2}y'|_n\,\Delta t^2 + \cdots - y_n}{\Delta t} = y'|_n + \tfrac{1}{2}y''|_n\,\Delta t + \cdots \qquad (7.50)$$

As $\Delta t \to 0$, FDA $\to y'|_n$, which shows that FDA is an approximation of the exact derivative $\bar{y}$ at grid point n. The order of FDA is $0(\Delta t)$. The exact form of the truncation error relative to grid point n is determined. Choosing other base points for the Taylor series yields the truncation errors relative to those base points.

A finite difference approximation (FDA) of an exact derivative is *consistent* with the exact derivative if the FDA approaches the exact derivative as $\Delta t \to 0$, as illustrated in Eq. (7.50). Consistency is an important property of the finite difference approximation of derivatives.

7.4.3 Finite Difference Equations

Finite difference solutions of differential equations are obtained by discretizing the continuous solution domain and replacing the exact derivatives in the differential equation by finite difference approximations, such as Eqs. (7.41), (7.44), or (7.47), to obtain a finite difference approximation of the differential equation. Such approximations are called *finite difference equations* (FDEs).

Consider the general nonlinear initial-value ODE:

$$\bar{y}' = f(t, \bar{y}) \qquad \bar{y}(0) = \bar{y}_0 \tag{7.51}$$

Choose a finite difference approximation (FDA), y', for $\bar{y}'$. For example, from Eq. (7.41) or (7.44):

$$y_n' = \frac{y_{n+1} - y_n}{\Delta t} \qquad \text{or} \qquad y_{n+1}' = \frac{y_{n+1} - y_n}{\Delta t} \tag{7.52}$$

Substitute the FDA for $\bar{y}'$ into the exact ODE, $\bar{y}' = f(t, \bar{y})$, and solve for y_{n+1}:

$$y_n' = \frac{y_{n+1} - y_n}{\Delta t} = f(t_n, y_n) = f_n \tag{7.53a}$$

$$y_{n+1}' = \frac{y_{n+1} - y_n}{\Delta t} = f(t_{n+1}, y_{n+1}) = f_{n+1} \tag{7.53b}$$

Solving Eq. (7.53a) for y_{n+1} yields

$$y_{n+1} = y_n + \Delta t f(t_n, y_n) = y_n + \Delta t f_n \tag{7.54a}$$

Solving Eq. (7.53b) for y_{n+1} yields

$$y_{n+1} = y_n + \Delta t f(t_{n+1}, y_{n+1}) = y_n + \Delta t f_{n+1} \tag{7.54b}$$

Equation (7.54a) is an *explicit* finite difference equation, since f_n does not depend on y_{n+1}, and Eq. (7.54a) can be solved explicitly for y_{n+1}. Equation (7.54b) is an *implicit finite difference equation*, since f_{n+1} depends on y_{n+1}. If the ODE is linear, then f_{n+1} is linear in y_{n+1}, and Eq. (7.54b) can be solved directly for y_{n+1}. If the ODE is nonlinear, then f_{n+1} is nonlinear in y_{n+1}, and additional effort is required to solve Eq. (7.54b) for y_{n+1}.

7.4.4 Smoothness

Smoothness refers to the continuity of a function and its derivatives. The finite difference method of solving a differential equation employs Taylor series to develop finite difference approximations (FDAs) of the exact derivatives in the differential equation. If a problem has discontinuous derivatives of some order at some point in the solution domain, then FDAs based on the Taylor series may misbehave at that point.

For example, consider the vertical flight of a rocket illustrated in Figure 7.1 and Example 7.18. When the rocket engine is turned off, the thrust drops to zero instantly. This causes a discontinuity in the acceleration of the rocket, which causes a discontinuity in the second derivative of the altitude $y(t)$. The solution is not smooth in the neighborhood of the discontinuity in the second derivative of $y(t)$.

At a discontinuity, single-point methods such as presented in Section 7.7, or extrapolation methods such as presented in Section 7.8, should be employed, since the step size in the neighborhood of a discontinuity can be chosen so that the discontinuity occurs at a grid point. Multipoint methods such as presented in Section 7.9 should not be employed in the neighborhood of a discontinuity in the function or its derivatives.

Problems which do not have any discontinuities in the function or its derivatives are called *smoothly varying problems*. Problems which have discontinuities in the function or its derivatives are called *nonsmoothly varying problems*.

7.4.5 Errors

Five types of errors can occur in the numerical solution of differential equations:

1. Errors in the initial data (assumed nonexistent)
2. Algebraic errors (assumed nonexistent)
3. Truncation errors
4. Roundoff errors
5. Inherited error

These errors, their interactions, and their effects on the numerical solution are discussed in this section. This discussion is equally relevant to the numerical solution of ODEs, which is the subject of Part II, and the numerical solution of PDEs, which is the subject of Part III.

A differential equation has an infinite number of solutions, depending on the initial conditions. Thus, a family of solutions exists, as illustrated in Figure 7.7 for the linear first-order homogeneous ODEs given by Eqs. (7.12) and (7.13). Figure 7.7a illustrates a family of *converging solutions* for a *stable ODE*, and Figure 7.7b illustrates a family of *diverging solutions* for an *unstable ODE*. An error in the initial condition or an algebraic error simply moves the solution to a different member of the solution family. Such errors are assumed to be nonexistent.

Any error in the numerical solution essentially moves the numerical solution to a different member of the solution family. Consider the converging family of solutions illustrated in Figure 7.7a. Since the members of the solution family converge as t increases, errors in the numerical solution of any type tend to diminish as t increases. By contrast, for the diverging family of solutions illustrated in Figure 7.7b, errors in the numerical solution of any type tend to grow as t increases.

Truncation error is the error incurred in a single step caused by truncating the Taylor series approximations for the exact derivatives. Truncation error depends on the step size—$0(\Delta t^n)$. Truncation error decreases as the step size Δt decreases. Truncation errors propagate from step to step and accumulate as the number of steps increases.

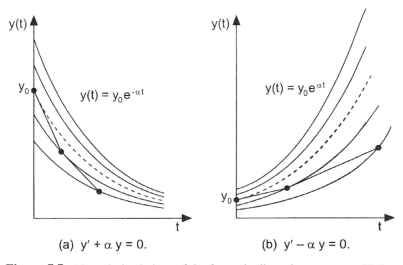

(a) $y' + \alpha y = 0$.   (b) $y' - \alpha y = 0$.

Figure 7.7 Numerical solutions of the first-order linear homogeneous ODEs.

Round-off error is the error caused by the finite word length employed in the calculations. Round-off error is more significant when small differences between large numbers are calculated. Consequently, round-off error increases as the step size Δt decreases, both because the changes in the solution are smaller and more steps are required. Most computers have either 32 bit or 64 bit word length, corresponding to approximately 7 or 13 significant decimal digits, respectively. Some computers have extended precision capability, which increases the number of bits to 128. Care must be exercised to ensure that enough significant digits are maintained in numerical calculations so that round-off is not significant. Round-off errors propagate from step to step and tend to accumulate as the number of calculations (i.e., steps) increases.

Inherited error is the sum of all accumulated errors from all previous steps. The presence of inherited error means that the initial condition for the next step is incorrect. Essentially, each step places the numerical solution on a different member of the solution family. Assuming that algebraic errors are nonexistent and that round-off errors are negligible, inherited error is the sum of all previous truncation errors. On the first step, the total error at the first solution point is the local truncation error. The initial point for the second step is on a different member of the solution family. Another truncation error is made on the second step. This truncation error is relative to the exact solution passing through the first solution point. The total error at the second solution point is due both to the truncation error at the first solution point, which is now called inherited error, and the local truncation error of the second step. This dual error source, inherited error and local truncation error, affects the solution at each step. For a converging solution family, inherited error remains bounded as the solution progresses. For a diverging solution family, inherited error tends to grow as the solution progresses. One practical consequence of these effects is that smaller step sizes may be required when solving unstable ODEs which govern diverging solution families than when solving stable ODEs which govern converging solution families.

7.5 THE FIRST-ORDER EULER METHODS

The *explicit Euler method* and the *implicit Euler method* are two first-order finite difference methods for solving initial-value ODEs. Although these methods are too inaccurate to be of much practical value, they are useful to illustrate many concepts relevant to the finite difference solution of initial-value ODEs.

7.5.1 The Explicit Euler Method

Consider the general nonlinear first-order ODE:

$$\bar{y}' = f(t, \bar{y}) \qquad \bar{y}(t_0) = \bar{y}_0 \tag{7.55}$$

Choose point n as the base point and develop a finite difference approximation of Eq. (7.55) at that point. The finite difference grid is illustrated in Figure 7.8, where the cross (i.e., $\times$) denotes the base point for the finite difference approximation of Eq. (7.55). The first-order forward-difference finite difference approximation of $\bar{y}'$ is given by Eq. (7.40):

$$\bar{y}'|_n = \frac{\bar{y}_{n+1} - \bar{y}_n}{\Delta t} - \tfrac{1}{2}\bar{y}''(\tau_n)\,\Delta t \tag{7.56}$$

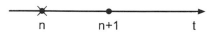

Figure 7.8 Finite difference grid for the explicit Euler method.

Substituting Eq. (7.56) into Eq. (7.55) and evaluating $f(t, \bar{y})$ at point n yields

$$\frac{\bar{y}_{n+1} - \bar{y}_n}{\Delta t} - \tfrac{1}{2}\bar{y}''(\tau_n)\,\Delta t = f(t_n, \bar{y}_n) = \bar{f}_n \qquad (7.57)$$

Solving Eq. (7.57) for $\bar{y}_{n+1}$ gives

$$\bar{y}_{n+1} = \bar{y}_n + \Delta t \bar{f}_n + \tfrac{1}{2}\bar{y}''(\tau_n)\,\Delta t^2 = \bar{y}_n + \Delta t \bar{f}_n + 0(\Delta t^2) \qquad (7.58)$$

Truncating the remainder term, which is $0(\Delta t^2)$, and solving for y_{n+1} yields the *explicit Euler finite difference equation (FDE)*:

$$\boxed{y_{n+1} = y_n + \Delta t f_n \qquad 0(\Delta t^2)} \qquad (7.59)$$

where the $0(\Delta t^2)$ term is included as a reminder of the order of the local truncation error. Several features of Eq. (7.59) are summarized below.

1. The FDE is explicit, since f_n does not depend on y_{n+1}.
2. The FDE requires only one known point. Hence, it is a single point method.
3. The FDE requires only one derivative function evaluation [i.e., $f(t, y)$] per step.
4. The error in calculating y_{n+1} for a single step, the local truncation error, is $0(\Delta t^2)$.
5. The global (i.e., total) error accumulated after N steps is $0(\Delta t)$. This result is derived in the following paragraph.

Equation (7.59) is applied repetitively to march from the initial point t_0 to the final point, t_N, as illustrated in Figure 7.9. The solution at point N is

$$y_N = y_0 + \sum_{n=0}^{N-1}(y_{n+1} - y_n) = y_0 + \sum_{n=0}^{N-1}\Delta y_{n+1} \qquad (7.60)$$

The total truncation error is given by

$$\text{Error} = \sum_{n=0}^{N-1}\tfrac{1}{2}y''(\tau_n)\,\Delta t^2 = N\tfrac{1}{2}y''(\tau)\,\Delta t^2 \qquad (7.61)$$

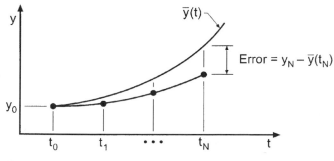

Figure 7.9 Repetitive application of the explicit Euler method.

where $t_0 \leq \tau \leq t_N$. The number of steps N is related to the step size Δt as follows:

$$N = \frac{t_N - t_0}{\Delta t} \tag{7.62}$$

Substituting Eq. (7.62) into Eq. (7.61) yields

$$\boxed{\text{Error} = \tfrac{1}{2}(t_N - t_0)y''(\tau)\,\Delta t = 0(\Delta t)} \tag{7.63}$$

Consequently, the global (i.e., total) error of the explicit Euler FDE is $0(\Delta t)$, which is the same as the order of the finite difference approximation of the exact derivative $\bar{y}'$, which is $0(\Delta t)$, as shown in Eq. (7.56).

The result developed in the preceding paragraph applies to all finite difference approximations of first-order ordinary differential equations. The order of the global error is always equal to the order of the finite difference approximation of the exact derivative $\bar{y}'$.

The algorithm based on the repetitive application of the explicit Euler FDE to solve initial-value ODEs is called the *explicit Euler method*.

Example 7.2.   The explicit Euler method

Let's solve the radiation problem presented in Section 7.1 using Eq. (7.59). The derivative function is $f(t, T) = -\alpha(T^4 - T_a^4)$. The explicit Euler FDE is given by Eq. (7.59). Thus,

$$T_{n+1} = T_n - \Delta t\,\alpha(T_n^4 - T_a^4) \tag{7.64}$$

Let $\Delta t = 2.0$ s. For the first time step,

$$f_0 = -(4.0 \times 10^{-12})(2500.0^4 - 250.0^4) = -156.234375 \tag{7.65a}$$
$$T_1 = 2500.0 + 2.0(-156.234375) = 2187.531250 \tag{7.65b}$$

These results and the results of subsequent time steps for t from 4.0 s to 10.0 s are summarized in Table 7.3. The results for $\Delta t = 1.0$ s are also presented in Table 7.3.

Table 7.3   Solution by the Explicit Euler Method

t_n t_{n+1}	T_n T_{n+1}	f_n	$\bar{T}_{n+1}$	Error
0.0	2500.000000	−156.234375		
2.0	2187.531250	−91.580490	2248.247314	−60.716064
4.0	2004.370270	−64.545606	2074.611898	−70.241628
6.0	1875.279058	−49.452290	1944.618413	−69.339355
8.0	1776.374478	−39.813255	1842.094508	−65.720030
10.0	1696.747960		1758.263375	−61.515406
0.0	2500.000000	−156.234375		
1.0	2343.765625	−120.686999	2360.829988	−17.064363
2.0	2223.078626	−97.680938	2248.247314	−25.168688
3.0	2125.397688	−81.608926	2154.470796	−29.073108
. . . . .	. . . .	. . . .	. . . .	. . . .
9.0	1768.780668	−39.136553	1798.227867	−29.447199
10.0	1729.644115		1758.263375	−28.619260

Several important features of the explicit Euler method are illustrated in Table 7.3. First, the solutions for both step sizes are following the general trend of the exact solution correctly. The solution for the smaller step size is more accurate than the solution for the larger step size. In fact, the order of the method can be estimated by comparing the errors at $t = 10.0$ s. From Eq. (7.63),

$$E(\Delta t = 2.0) = \tfrac{1}{2}(t_N - t_0)T''(\tau)(2.0) \tag{7.66a}$$

$$E(\Delta t = 1.0) = \tfrac{1}{2}(t_N - t_0)T''(\tau)(1.0) \tag{7.66b}$$

Assuming that the values of $T''(\tau)$ in Eqs. (7.66a) and (7.66b) are approximately equal, the ratio of the theoretical errors is

$$\text{Ratio} = \frac{E(\Delta t = 2.0)}{E(\Delta t = 1.0)} = \frac{2.0}{1.0} = 2.0 \tag{7.67}$$

From Table 7.3, at $t = 10.0$ s, the ratio of the numerical errors is

$$\text{Ratio} = \frac{E(\Delta t = 2.0)}{E(\Delta t = 1.0)} = \frac{-61.515406}{-28.619260} = 2.15 \tag{7.68}$$

Equation (7.68) shows that the method is first order. The value of 2.15 is not exactly equal to the theoretical value of 2.0 due to the finite step size. The theoretical value of 2.0 is achieved only in the limit as $\Delta t \to 0$. Order is discussed in more detail in Section 7.6.

Another feature illustrated in Table 7.3 is that the errors are relatively large. This is due to the large first-order, $0(\Delta t)$, truncation error. The errors are all negative, indicating that the numerical solution leads the exact solution. This occurs because the derivative function $f(t, T)$ decreases as t increases, as illustrated in Table 7.3. The derivative function in the FDE is evaluated at point n, the beginning of the interval of integration, where it has its largest value for the interval. Consequently, the numerical solution leads the exact solution.

The final feature of the explicit Euler method which is illustrated in Table 7.3 is that the numerical solution approaches the exact solution as the step size decreases. This property of a finite difference method is called *convergence*. Convergence is necessary for a finite difference method to be of any use in solving a differential equation. Convergence is discussed in more detail in Section 7.6.

When the base point for the finite difference approximation of an ODE is point n, the unknown value y_{n+1} appears in the finite difference approximation of $\bar{y}'$, but not in the derivative function $f(t, \bar{y})$. Such FDEs are called *explicit FDEs*. The explicit Euler method is the simplest example of an explicit FDE.

When the base point for the finite difference approximation of an ODE is point $n + 1$, the unknown value y_{n+1} appears in the finite difference approximation of $\bar{y}'$ and also in the derivative function $f(t, \bar{y})$. Such FDEs are called *implicit FDEs*. An example of an implicit FDE is presented in the next section.

7.5.2 The Implicit Euler Method

Consider the general nonlinear first-order ODE:

$$\boxed{\bar{y}' = f(t, \bar{y}) \qquad \bar{y}(t_0) = \bar{y}_0} \tag{7.69}$$

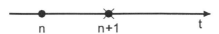

Figure 7.10 Finite difference grid for the implicit Euler method.

Choose point $n + 1$ as the base point and develop a finite difference approximation of Eq. (7.69) at that point. The finite difference grid is illustrated in Figure 7.10. The first-order backward-difference finite difference approximation of $\bar{y}'$ is given by Eq. (7.43):

$$\bar{y}'|_{n+1} = \frac{\bar{y}_{n+1} - \bar{y}_n}{\Delta t} + \tfrac{1}{2}\bar{y}''(\tau_{n+1})\,\Delta t \tag{7.70}$$

Substituting Eq. (7.70) into Eq. (7.69) and evaluating $f(t, \bar{y})$ at point $n + 1$ yields

$$\frac{\bar{y}_{n+1} - \bar{y}_n}{\Delta t} + \tfrac{1}{2}\bar{y}''(\tau_{n+1})\,\Delta t = f(t_{n+1}, \bar{y}_{n+1}) = \bar{f}_{n+1} \tag{7.71}$$

Solving Eq. (7.71) for $\bar{y}_{n+1}$ gives

$$\bar{y}_{n+1} = \bar{y}_n + \Delta t \bar{f}_{n+1} - \tfrac{1}{2}\bar{y}''(\tau_{n+1})\,\Delta t^2 = \bar{y}_n + \Delta t \bar{f}_{n+1} + 0(\Delta t^2) \tag{7.72}$$

Truncating the $0(\Delta t^2)$ remainder term yields the *implicit Euler FDE*:

$$\boxed{y_{n+1} = y_n + \Delta t f_{n+1} \qquad 0(\Delta t^2)} \tag{7.73}$$

Several features of Eq. (7.73) are summarized below.

1. The FDE is implicit, since f_{n+1} depends on y_{n+1}. If $f(t, y)$ is linear in y, then f_{n+1} is linear in y_{n+1}, and Eq. (7.73) is a linear FDE which can be solved directly for y_{n+1}. If $f(t, y)$ is nonlinear in y, then Eq. (7.73) is a nonlinear FDE, and additional effort is required to solve for y_{n+i}.
2. The FDE is a single-point FDE.
3. The FDE requires only one derivative function evaluation per step if $f(t, y)$ is linear in y. If $f(t, y)$ is nonlinear in y, Eq. (7.73) is nonlinear in y_{n+1}, and several evaluations of the derivative function may be required to solve the nonlinear FDE.
4. The single-step truncation error is $0(\Delta t^2)$, and the global error is $0(\Delta t)$.

The algorithm based on the repetitive application of the implicit Euler FDE to solve initial-value ODEs is called the *implicit Euler method*.

Example 7.3. The implicit Euler method

Let's solve the radiation problem presented in Section 7.1 using Eq. (7.73). The derivative function is $f(t, T) = -\alpha(T^4 - T_a^4)$. The implicit Euler FDE is given by Eq. (7.73). Thus,

$$T_{n+1} = T_n - \Delta t \alpha (T_{n+1}^4 - T_a^4) \tag{7.74}$$

Equation (7.74) is a nonlinear FDE. Procedures for solving nonlinear implicit FDEs are presented in Section 7.11. Let $\Delta t = 2.0$ s. For the first time step,

$$T_1 = 2500.0 - 2.0(4.0 \times 10^{-12})(T_1^4 - 250.0^4) \tag{7.75}$$

Equation (7.75) is a fourth-order polynomial. It is solved by Newton's method in Example 7.16. The result is $T_1 = 2373.145960$. This result and the results of subsequent time steps for t from 4.0 s to 10.0 s are presented in Table 7.4. The results for $\Delta t = 1.0$ s are also presented in Table 7.4.

The results presented in Table 7.4 behave generally the same as the results presented in Table 7.3 and discussed in Example 7.2. An error analysis at $t = 10.0$ s gives

$$\text{Ratio} = \frac{E(\Delta t = 2.0)}{E(\Delta t = 1.0)} = \frac{48.455617}{25.468684} = 1.90 \tag{7.76}$$

which shows that the method is first order. The errors are all positive, indicating that the numerical solution lags the exact solution. This result is in direct contrast to the error behavior of the explicit Euler method, where a leading error was observed. In the present case, the derivative function in the FDE is evaluated at point $n + 1$, the end of the interval of integration, where it has its smallest value. Consequently, the numerical solution lags the exact solution.

7.5.3   Comparison of the Explicit and Implicit Euler Methods

The explicit Euler method and the implicit Euler method are both first-order [i.e., $0(\Delta t)$] methods. As illustrated in Examples 7.2 and 7.3, the errors in these two methods are comparable (although of opposite sign) for the same step size. For nonlinear ODEs, the explicit Euler method is straightforward, but the implicit Euler method yields a nonlinear FDE, which is more difficult to solve. So what is the advantage, if any, of the implicit Euler method?

The implicit Euler method is *unconditionally stable*, whereas the explicit Euler method is *conditionally stable*. This difference can be illustrated by solving the linear first-order homogeneous ODE

$$\bar{y}' + \bar{y} = 0 \qquad \bar{y}(0) = 1 \tag{7.77}$$

for which $\bar{f}(t, \bar{y}) = -\bar{y}$, by both methods. The exact solution of Eq. (7.77) is

$$\bar{y}(t) = e^{-t} \tag{7.78}$$

Table 7.4.   Solution by the Implicit Euler Method

t_n t_{n+1}	T_n T_{n+1}	$\bar{T}_{n+1}$	Error
0.0	2500.000000		
2.0	2282.785819	2248.247314	34.538505
4.0	2120.934807	2074.611898	46.322909
6.0	1994.394933	1944.618413	49.776520
8.0	1891.929506	1842.094508	49.834998
10.0	1806.718992	1758.263375	48.455617
0.0	2500.000000		
1.0	2373.145960	2360.829988	12.315972
2.0	2267.431887	2248.247314	19.184573
. . . . . .	. . . .	. . . .	. . . .
9.0	1824.209295	1798.227867	25.981428
10.0	1783.732059	1758.263375	25.468684

Solving Eq. (7.77) by the explicit Euler method yields the following FDE:

$$y_{n+1} = y_n + \Delta t f_n = y_n + \Delta t (-y_n) \tag{7.79}$$

$$\boxed{y_{n+1} = (1 - \Delta t) y_n} \tag{7.80}$$

Solutions of Eq. (7.80) for several values of Δt are presented in Figure 7.11. The numerical solutions behave in a physically correct manner (i.e., decrease monotonically) for $\Delta t \leq 1.0$ as $t \to \infty$, and approach the exact asymptotic solution, $\bar{y}(\infty) = 0$. For $\Delta t = 1.0$, the numerical solution reaches the exact asymptotic solution, $\bar{y}(\infty) = 0$, in one step.

For $1.0 < \Delta t < 2.0$, the numerical solution overshoots and oscillates about the exact asymptotic solution, $\bar{y}(\infty) = 0$, in a damped manner and approaches the exact asymptotic solution as $t \to \infty$. For $\Delta t = 2.0$, the numerical solution oscillates about the exact asymptotic solution in a stable manner but never approaches the exact asymptotic solution. Thus, solutions are *stable* for $\Delta t \leq 2.0$.

For $\Delta t > 2.0$, the numerical solution oscillates about the exact asymptotic solution in an unstable manner that grows exponentially without bound. This is *numerical instability*. Consequently, the explicit Euler method is *conditionally stable* for this ODE, that is, it is stable only for $\Delta t \leq 2.0$.

The oscillatory behavior for $1.0 < \Delta t < 2.0$ is called *overshoot* and must be avoided. Overshoot is not instability. However, it does not model physical reality, thus it is unacceptable. The step size Δt generally must be 50 percent or less of the stable step size to avoid overshoot.

Solving Eq. (7.77) by the implicit Euler method gives the following FDE:

$$y_{n+1} = y_n + \Delta t f_{n+1} = y_n + \Delta t (-y_{n+1}) \tag{7.81}$$

Since Eq. (7.81) is linear in y_{n+1}, it can be solved directly for y_{n+1} to yield

$$\boxed{y_{n+1} = \frac{y_n}{1 + \Delta t}} \tag{7.82}$$

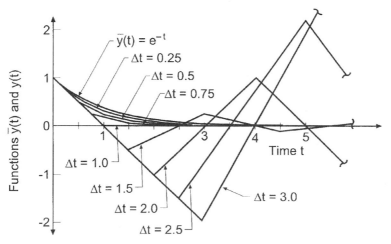

Figure 7.11   Behavior of the explicit Euler method.

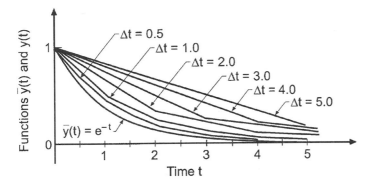

Figure 7.12   Behavior of the implicit Euler method.

Solutions of Eq. (7.82) for several values of Δt are presented in Figure 7.12. The numerical solutions behave in a physically correct manner (i.e., decrease monotonically) for all values of Δt. This is *unconditional stability*, which is the main advantage of implicit methods. The error increases as Δt increases, but this is an accuracy problem, not a stability problem.

Stability is discussed in more detail in Section 7.6.

7.5.4   Summary

The two first-order Euler methods presented in this section are rather simple *single-point* methods (i.e., the solution at point $n + 1$ is based only on values at point n). More accurate (i.e., higher-order) methods can be developed by sampling the derivative function, $f(t, y)$, at several locations between point n and point $n + 1$. One such method, the *Runge-Kutta method*, is developed in Section 7.7. More accurate results also can be obtained by *extrapolating* the results obtained by low-order single-point methods. One such method, the *extrapolated modified midpoint method*, is developed in Section 7.8. More accurate (i.e., higher-order) methods also can be developed by using more known points. Such methods are called *multipoint methods*. One such method, the *Adams-Bashforth-Moulton method* is developed in Section 7.9.

Before proceeding to the more accurate methods, however, several theoretical concepts need to be discussed in more detail. These concepts are *consistency, order, stability,* and *convergence*.

7.6   CONSISTENCY, ORDER, STABILITY, AND CONVERGENCE

There are several important concepts which must be considered when developing finite difference approximations of initial-value differential equations. They are (a) consistency, (b) order, (c) stability, and (d) convergence. These concepts are defined and discussed in this section.

A FDE is *consistent* with an ODE if the difference between them (i.e., the truncation error) vanishes as $\Delta t \rightarrow 0$. In other words, the FDE approaches the ODE.

The *order* of a FDE is the rate at which the global error decreases as the grid size approaches zero.

A FDE is *stable* if it produces a bounded solution for a stable ODE and is *unstable* if it produces an unbounded solution for a stable ODE.

A finite difference method is *convergent* if the numerical solution of the FDE (i.e., the numerical values) approaches the exact solution of the ODE as $\Delta t \to 0$.

7.6.1 Consistency and Order

All finite difference equations (FDEs) must be analyzed for consistency with the differential equation which they approximate. *Consistency analysis* involves changing the FDE back into a differential equation and determining if that differential equation approaches the exact differential equation of interest as $\Delta t \to 0$. This is accomplished by expressing all terms in the FDE by a Taylor series having the same base point as the FDE. This Taylor series is an infinite series. Thus, an infinite-order differential equation is obtained. This infinite-order differential equation is called the *modified differential equation (MDE)*. The MDE is the actual differential equation which is solved by the FDE.

Letting $\Delta t \to 0$ in the modified differential equation (MDE) yields a finite-order differential equation. If this finite order differential equation is identical to the exact differential equation whose solution is desired, then the FDE is a consistent approximation of that exact differential equation.

The *order* of a FDE is the order of the lowest-order terms in the modified differential equation (MDE).

Example 7.4. Consistency and order analysis of the explicit Euler FDE

As an example, consider the linear first-order ODE:

$$\bar{y}' + \alpha \bar{y} = F(t) \tag{7.83}$$

The explicit Euler FDE is:

$$y_{n+1} = y_n + \Delta t f_n \tag{7.84}$$

Substituting Eq. (7.83) into Eq. (7.84) gives

$$y_{n+1} = y_n - \alpha h y_n + h F_n \tag{7.85}$$

where $h = \Delta t$. Let grid point n be the base point, and write the Taylor series for y_{n+1}, the approximate solution. Thus,

$$y_{n+1} = y_n + h y'|_n + \tfrac{1}{2} h^2 y''|_n + \tfrac{1}{6} h^3 y'''|_n + \cdots \tag{7.86}$$

Substituting Eq. (7.86) into Eq. (7.85) gives

$$y_n + h y'|_n + \tfrac{1}{2} h^2 y''|_n + \tfrac{1}{6} h^3 y'''|_n + \cdots = y_n - \alpha h y_n + h F_n \tag{7.87}$$

Cancelling zero-order terms (i.e., the y_n terms), dividing through by h, and rearranging terms yields the modified differential equation (MDE):

$$\boxed{y'|_n + \alpha y_n = F_n - \tfrac{1}{2} h y''|_n - \tfrac{1}{6} h^2 y'''|_n - \cdots} \tag{7.88}$$

Equation (7.88) is the actual differential equation which is solved by the explicit Euler method.

Let $h = \Delta t \to 0$ in Eq. (7.88) to obtain the ODE with which Eq. (7.85) is consistent. Thus, Eq. (7.88) becomes

$$y'|_n + \alpha y_n = F_n - \tfrac{1}{2}(0)y''|_n - \tfrac{1}{6}(0)^2 y'''|_n - \cdots \tag{7.89}$$

$$\boxed{y'|_n + \alpha y_n = F_n} \tag{7.90}$$

Equation (7.90) is identical to the linear first-order ODE, $\bar{y}' + \alpha \bar{y} = F(t)$. Consequently, Eq. (7.85) is consistent with that equation.

The order of the FDE is the order of the lowest-order term in Eq. (7.88). From Eq. (7.88),

$$y'|_n + \alpha y_n = F_n + 0(h) + \cdots \tag{7.91}$$

Thus, Eq. (7.85) is an $0(\Delta t)$ approximation of the exact ODE, $\bar{y}' + \alpha \bar{y} = F(t)$.

7.6.2 Stability

The phenomenon of numerical instability was illustrated in Section 7.5 for the explicit Euler method. All finite difference equations must be analyzed for stability. A FDE is *stable* if it produces a bounded solution for a stable ODE and is *unstable* if it produces an unbounded solution for a stable ODE. When the ODE is unstable, the numerical solution must also be unstable. Stability is not relevant in that case.

Consider the exact ODE, $\bar{y}' = f(t, \bar{y})$, and a finite difference approximation to it, $y' = f(t, y)$. The exact solution of the FDE can be expressed as

$$\boxed{y_{n+1} = G y_n} \tag{7.92}$$

where G, which in general is a complex number, is called the *amplification factor* of the FDE.

The global solution of the FDE at $T = N \, \Delta t$ is

$$y_N = G^N y_0 \tag{7.93}$$

For y_N to remain bounded as $N \to \infty$,

$$\boxed{|G| \leq 1} \tag{7.94}$$

Stability analysis thus reduces to:

1. Determining the amplification factor G of the FDE
2. Determining the conditions to ensure that $|G| \leq 1$

Stability analyses can be performed only for linear differential equations. Nonlinear differential equations must be linearized locally, and the FDE that approximates the linearized differential equation is analyzed for stability. Experience has shown that applying the resulting stability criteria to the FDE which approximates the nonlinear differential equation yields stable numerical solutions. Recall the linearization of a nonlinear ODE, Eq. (7.21), presented in Section 7.2:

$$\boxed{\bar{y}' + \alpha \bar{y} = F(t) \qquad \text{where } \alpha = -\bar{f}_y|_0} \tag{7.95}$$

Example 7.5. Linearization of a nonlinear ODE

Consider the nonlinear first-order ODE governing the example radiation problem:

$$\bar{T}' = f(t, \bar{T}) = -\alpha(\bar{T}^4 - T_a^4) \qquad \bar{T}(0.0) = \bar{T}_0 \tag{7.96}$$

Express $f(t, \bar{T})$ in a Taylor series with base point t_0:

$$f(t, \bar{T}) = \bar{f}_0 + \bar{f}_t|_0(t - t_0) + \bar{f}_T|_0(\bar{T} - \bar{T}_0) + \cdots \tag{7.97}$$

$$\bar{f}_t = 0 \qquad \text{and} \qquad \bar{f}_T = -4\alpha\bar{T}^3 \tag{7.98}$$

$$f(t, \bar{T}) = \bar{f}_0 + (0)(t - t_0) + (-4\alpha\bar{T}_0^3)(\bar{T} - \bar{T}_0) + \cdots \tag{7.99}$$

$$f(t, \bar{T}) = -(4\alpha\bar{T}_0^3)\bar{T} + (\bar{f}_0 + 4\alpha\bar{T}_0^3\bar{T}_0) + \cdots \tag{7.100}$$

Substituting Eq. (7.100) into Eq. (7.96) and truncating the higher-order terms yields the linearized ODE:

$$\bar{T}' = f(t, \bar{T}) = -(4\alpha\bar{T}_0^3)\bar{T} + (\bar{f}_0 + 4\alpha\bar{T}_0^3\bar{T}_0) \tag{7.101}$$

Comparing this ODE with the model linear ODE, $\bar{y}' + \bar{\alpha}\bar{y} = 0$ (where $\bar{\alpha}$ is used instead of α to avoid confusion with α in Eq. (7.96)), gives

$$\bar{\alpha} = -\bar{f}_T|_0 = 4\alpha\bar{T}_0^3 = 4(4.0 \times 10^{-12})2500.0^3 = 0.2500 \tag{7.102}$$

Note that $\bar{\alpha}$ changes as T_0 changes. Thus, the stable step size changes as the solution progresses.

For a linear differential equation, the total solution is the sum of the complementary solution $y_c(t)$, which is the solution of the homogeneous differential equation, and the particular solution $y_p(t)$ which satisfies the nonhomogeneous term $F(t)$. The particular solution $y_p(t)$ can grow or decay, depending on $F(t)$. Stability is not relevant to the numerical approximation of the particular solution. The complementary solution $y_c(t)$ can grow or decay depending on the sign of α. Stability is relevant only when $\alpha > 0$.

Thus, the model differential equation for stability analysis is the linear first-order homogeneous differential equation:

$$\boxed{\bar{y}' + \alpha\bar{y} = 0} \tag{7.103}$$

Stability analysis is accomplished as follows:

1. Construct the FDE for the model ODE, $\bar{y}' + \alpha\bar{y} = 0$.
2. Determine the amplification factor, G, of the FDE.
3. Determine the conditions to ensure that $|G| \leq 1$.

In general, Δt must be 50 percent or less of the stable Δt to avoid overshoot and 10 percent or less of the stable Δt to obtain an accurate solution.

Stability analyses can be performed only for linear differential equations. Nonlinear differential equations must be linearized locally, and a stability analysis performed on the finite difference approximation of the linearized differential equation. Experience has shown that applying the stability criteria applicable to a linearized ODE to the corresponding nonlinear ODE yields stable numerical solutions.

Experience has also shown that, for most ODEs of practical interest, the step size required to obtain the desired accuracy is considerably smaller than the step size required for stability. Consequently, instability is generally not a problem for ordinary differential equations, except for stiff ODEs, which are discussed in Section 7.14. Instability is a serious problem in the solution of partial differential equations.

Example 7.6. Stability analysis of the Euler methods

Consider the *explicit Euler method*:

$$y_{n+1} = y_n + \Delta t \, f_n \tag{7.104}$$

Applying the explicit Euler method to the model ODE, $\bar{y} + \alpha \bar{y} = 0$, for which $f(t, \bar{y}) = -\alpha \bar{y}$, gives

$$y_{n+1} = y_n + \Delta t(-\alpha y_n) = (1 - \alpha \, \Delta t)y_n = Gy_n \tag{7.105}$$

$$\boxed{G = (1 - \alpha \, \Delta t)} \tag{7.106}$$

For stability, $|G| \leq 1$. Thus,

$$-1 \leq (1 - \alpha \, \Delta t) \leq 1 \tag{7.107}$$

The right-hand inequality is always satisfied for $\alpha \, \Delta t \geq 0$. The left-hand inequality is satisfied only if

$$\boxed{\alpha \, \Delta t \leq 2} \tag{7.108}$$

which requires that $\Delta t \leq 2/\alpha$. Consequently, the explicit Euler method is *conditionally stable*.

Consider the *implicit Euler method*:

$$y_{n+1} = y_n + \Delta t \, f_{n+1} \tag{7.109}$$

Applying the implicit Euler method to the model ODE, $\bar{y}' + \alpha \bar{y} = 0$, gives

$$y_{n+1} = y_n + \Delta t(-\alpha y_{n+1}) \tag{7.110}$$

$$y_{n+1} = \frac{1}{1 + \alpha \, \Delta t} y_n = Gy_n \tag{7.111}$$

$$\boxed{G = \frac{1}{1 + \alpha \, \Delta t}} \tag{7.112}$$

For stability, $|G| \leq 1$, which is true for all values of $\alpha \, \Delta t$. Consequently, the implicit Euler method is *unconditionally stable*.

7.6.3 Convergence

Convergence of a finite difference method is ensured by demonstrating that the finite difference equation is consistent and stable. For example, for the explicit Euler method, Example 7.4 demonstrates consistency and Example 7.6 demonstrates conditional stability. Consequently, the explicit Euler method is convergent.

7.6.4 Summary

In summary, the concepts of consistency, order, stability, and convergence must always be considered when solving a differential equation by finite difference methods. Consistency and order can be determined from the modified differential equation (MDE), as illustrated in Example 7.4. Stability can be determined by a stability analysis, as presented in Example 7.6. Convergence can be ensured by demonstrating consistency and stability.

In general, it is not necessary to actually develop the modified differential equation to ensure consistency and to determine order for a finite difference approximation of a *first-order* ODE. Simply by approximating the first derivative and the nonhomogeneous term at the same base point, the finite difference equation will always be consistent. The global order of the finite difference equation is always the same as the order of the finite difference approximation of the exact first derivative. Even so, it is important to understand the concept of consistency and to know that it is satisfied.

7.7 SINGLE-POINT METHODS

The explicit Euler method and the implicit Euler method are both *single-point* methods. Single-point methods are methods that use data at a single point, point n, to advance the solution to point $n + 1$. Single-point methods are sometimes called *single-step* methods or *single-value* methods. Both Euler methods are first-order single-point methods. Higher-order single-point methods can be obtained by using higher-order approximations of $\bar{y}'$.

Four second-order single-point methods are presented in the first subsection: (a) the midpoint method, (b) the modified midpoint method, (c) the trapezoid method, and (d) the modified trapezoid method (which is generally called the modified Euler method). The first, third, and fourth of these second-order methods are not very popular, since it is quite straightforward to develop fourth-order single-point methods. The second-order modified midpoint method, however, is very important, since it is the basis of the higher-order extrapolation method presented in Section 7.8.

Runge-Kutta methods are introduced in the second subsection. The fourth-order Runge-Kutta method is an extremely popular method for solving initial-value ODEs. Methods of error estimation and error control for single-point methods are also presented.

7.7.1 Second-Order Single-Point Methods

Consider the general nonlinear first-order ODE:

$$\boxed{\bar{y}' = f(t, \bar{y}) \qquad \bar{y}(t_0) = \bar{y}_0}$$

(7.113)

Choose point $n + 1/z$ as the base point. The finite difference grid is illustrated in Figure 7.13. Express $\bar{y}_{n+1}$ and $\bar{y}_n$ in Taylor series with base point $n + 1/2$:

$$\bar{y}_{n+1} = \bar{y}_{n+1/2} + \bar{y}'|_{n+1/2}\left(\frac{\Delta t}{2}\right) + \frac{1}{2}\bar{y}''|_{n+1/2}\left(\frac{\Delta t}{2}\right)^2 + \frac{1}{6}\bar{y}'''|_{n+1/2}\left(\frac{\Delta t}{2}\right)^3 + \cdots$$

(7.114)

$$\bar{y}_n = \bar{y}_{n+1/2} + \bar{y}'|_{n+1/2}\left(-\frac{\Delta t}{2}\right) + \frac{1}{2}\bar{y}''|_{n+1/2}\left(-\frac{\Delta t}{2}\right)^2 + \frac{1}{6}\bar{y}'''|_{n+1/2}\left(-\frac{\Delta t}{2}\right)^3 + \cdots$$

(7.115)

Figure 7.13 Finite difference grid for the midpoint method.

Subtracting Eq. (7.115) from Eq. (7.114) and solving for $\bar{y}'|_{n+1/2}$ gives

$$\bar{y}'|_{n+1/2} = \frac{\bar{y}_{n+1} - \bar{y}_n}{\Delta t} - \frac{1}{24} y'''(\tau)\, \Delta t^2 \tag{7.116}$$

where $t_n \leq \tau \leq t_{n+1}$. Substituting Eq. (7.116) into Eq. (7.113) gives

$$\frac{\bar{y}_{n+1} - \bar{y}_n}{\Delta t} + 0(\Delta t^2) = f(t_{n+1/2}, \bar{y}_{n+1/2}) = \bar{f}_{n+1/2} \tag{7.117}$$

Solving for $\bar{y}_{n+1}$ gives

$$\bar{y}_{n+1} = \bar{y}_n + \Delta t\, \bar{f}_{n+1/2} + 0(\Delta t^3) \tag{7.118}$$

Truncating the remainder term yields the *implicit midpoint FDE*:

$$\boxed{y_{n+1} = y_n + \Delta t\, f_{n+1/2} \qquad 0(\Delta t^3)} \tag{7.119}$$

where the $0(\Delta t^3)$ term is a reminder of the local order of the FDE. The implicit midpoint FDE itself is of very little use since $f_{n+1/2}$ depends on $y_{n+1/2}$, which is unknown.

However, if $y_{n+1/2}$ is first predicted by the first-order explicit Euler FDE, Eq. (7.59), and $f_{n+1/2}$ is then evaluated using that value of $y_{n+1/2}$, the *modified midpoint FDEs* are obtained:

$$\boxed{\begin{aligned} y^P_{n+1/2} &= y_n + \frac{\Delta t}{2} f_n \\ y^C_{n+1} &= y_n + \Delta t\, f^P_{n+1/2} \end{aligned}} \tag{7.120} \tag{7.121}$$

where the superscript P in Eq. (7.120) denotes that $y^P_{n+1/2}$ is a predictor value, the superscript P in Eq. (7.121) denotes that $f^P_{n+1/2}$ is evaluated using $y^P_{n+1/2}$, and the superscript C in Eq. (7.121) denotes that y^C_{n+1} is the corrected second-order result.

Consistency analysis and stability analysis of two-step predictor-corrector methods are performed by applying each step of the predictor-corrector method to the model ODE, $\bar{y}' + \alpha \bar{y} = 0$, for which $f(t, \bar{y}) = -\alpha \bar{y}$, and combining the resulting FDEs to obtain a single-step FDE. The single-step FDE is analyzed for consistency, order, and stability. From Eq. (7.58) for $\Delta t/2$,

$$\bar{y}_{n+1/2} = \bar{y}_n + \frac{\Delta t}{2}(-\alpha \bar{y}_n) + 0(\Delta t^2) = \left(1 - \frac{\alpha\, \Delta t}{2}\right)\bar{y}_n + 0(\Delta t^2) \tag{7.122}$$

Substituting Eq. (7.122) into Eq. (7.118) gives

$$\bar{y}_{n+1} = \bar{y}_n - \alpha\, \Delta t\left[\left(1 - \frac{\alpha\, \Delta t}{2}\right)\bar{y}_n + 0(\Delta t^2)\right] + 0(\Delta t^3) \tag{7.123}$$

$$\bar{y}_{n+1} = \left[1 - \alpha\, \Delta t + \frac{(\alpha\, \Delta t)^2}{2}\right]\bar{y}_n + 0(\Delta t^3) \tag{7.124}$$

Truncating the $0(\Delta t^3)$ remainder term yields the single-step FDE corresponding to the modified midpoint FDE:

$$y_{n+1} = \left[1 - \alpha\,\Delta t + \frac{(\alpha\,\Delta t)^2}{2} \right] y_n \qquad (7.125)$$

Substituting the Taylor series for y_{n+1} into Eq. (7.125) and letting $\Delta t \to 0$ gives $y'_n = -\alpha y_n$, which shows that Eq. (7.125) is consistent with the exact ODE, $\bar{y}' + \alpha\bar{y} = 0$. Equation (7.124) shows that the local truncation error is $0(\Delta t^3)$.

The amplification factor G for the modified midpoint FDE is determined by applying the FDE to solve the model ODE, $\bar{y}' + \alpha\bar{y} = 0$, for which $\bar{f}(t,\bar{y}) = -\alpha\bar{y}$. The single-step FDE corresponding to Eqs. (7.120) and (7.121) is given by Eq. (7.125). From Eq. (7.125), the amplification factor, $G = y_{n+1}/y_n$ is

$$G = 1 - \alpha\,\Delta t + \frac{(\alpha\,\Delta t)^2}{2} \qquad (7.126)$$

Substituting values of $(\alpha\,\Delta t)$ into Eq. (7.126) yields the following results:

$\alpha\,\Delta t$	G	$\alpha\,\Delta t$	G
0.0	1.000	1.5	0.625
0.5	0.625	2.0	1.000
1.0	0.500	2.1	1.105

These results show that $|G| \le 1$ if $\alpha\,\Delta t \le 2$.

The general features of the modified midpoint FDEs are presented below.

1. The FDEs are an explicit predictor-corrector set of FDEs which requires two derivative function evaluations per step.
2. The FDEs are consistent, $0(\Delta t^3)$ locally and $0(\Delta t^2)$ globally.
3. The FDEs are conditionally stable (i.e., $\alpha\,\Delta t \le 2$).
4. The FDEs are consistent and conditionally stable, and thus, convergent.

The algorithm based on the repetitive application of the modified midpoint FDEs is called the *modified midpoint method*.

Example 7.7. The modified midpoint method

To illustrate the modified midpoint method, let's solve the radiation problem presented in Section 7.1 using Eqs. (7.120) and (7.121). The derivative function is $f(t,T) = -\alpha(T^4 - T_a^4)$. Equations (7.120) and (7.121) yield

$$f_n = f(t_n, T_n) = -\alpha(T_n^4 - 250.0^4) \qquad (7.127)$$

$$T^P_{n+1/2} = T_n + \frac{\Delta t}{2} f_n \qquad (7.128)$$

$$f^P_{n+1/2} = f(t_{n+1/2}, T^P_{n+1/2}) = -\alpha[(T^P_{n+1/2})^4 - 250.0^4] \qquad (7.129)$$

$$T^C_{n+1} = T_n + \Delta t\, f^P_{n+1/2} \qquad (7.130)$$

Let $\Delta t = 2.0$ s. For the first time step, the predictor FDE gives

$$f_0 = -(4.0 \times 10^{-12})(2500.0^4 - 250.0^4) = -156.234375 \tag{7.131}$$

$$T_{1/2}^P = 2500.0 + (2.0/2)(-156.234375) = 2343.765625 \tag{7.132}$$

The corrector FDE yields

$$f_{1/2}^P = -(4.0 \times 10^{-12})(2343.765625^4 - 250.0^4) = -120.686999 \tag{7.133}$$

$$T_1^C = 2500.0 + 2.0(-120.686999) = 2258.626001 \tag{7.134}$$

These results, the results for subsequent time steps for $t = 4.0$ s to 10.0 s, and the solution for $\Delta t = 1.0$ s are presented in Table 7.5.

The errors presented in Table 7.5 for the second-order modified midpoint method for $\Delta t = 1.0$ s are approximately 15 times smaller than the errors presented in Table 7.3 for the first-order explicit Euler method. This illustrates the advantage of the second-order method. To achieve a factor of 15 decrease in the error for a first-order method requires a reduction of 15 in the step size, which increases the number of derivative function evaluations by a factor of 15. The same reduction in error was achieved with the second-order method at the expense of twice as many derivative function evaluations. An error analysis at $t = 10.0$ s gives

$$\text{Ratio} = \frac{E(\Delta t = 2.0)}{E(\Delta t = 1.0)} = \frac{8.855320}{1.908094} = 4.64 \tag{7.135}$$

Table 7.5 Solution by the Modified Midpoint Method

t_n t_{n+1}	T_n $T_{n+1/2}$ T_{n+1}	f_n $f_{n+1/2}$	$\bar{T}_{n+1}$	Error
0.0	2500.000000	−156.234375		
	2343.765625	−120.686999		
2.0	2258.626001	−104.081152	2248.247314	10.378687
	2154.544849	−86.179389		
4.0	2086.267223	−75.761780	2074.611898	11.655325
	2010.505442	−65.339704		
6.0	1955.587815	−58.486182	1944.618413	10.969402
	1897.101633	−51.795424		
8.0	1851.996968	−47.041033	1842.094508	9.902460
	1804.955935	−42.439137		
10.0	1767.118695		1758.263375	8.855320
0.0	2500.000000	−156.234375		
	2421.882812	−137.601504		
1.0	2362.398496	−124.571344	2360.829988	1.568508
	2300.112824	−111.942740		
2.0	2250.455756	−102.583087	2248.247314	2.208442
...	....	....	....	....
9.0	1800.242702	−41.997427	1798.227867	2.014835
	1779.243988	−40.071233		
10.0	1760.171468		1758.263375	1.908094

which demonstrates that the method is second order, since the theoretical error ratio for an $0(\Delta t^2)$ method is 4.0.

An alternate approach for solving the implicit midpoint FDE, Eq. (7.119), is obtained as follows. Recall Eq. (7.118):

$$\bar{y}_{n+1} = \bar{y}_n + \Delta t \, \bar{f}_{n+1/2} = 0(\Delta t^3) \tag{7.136}$$

Write Taylor series for $\bar{f}_{n+1}$ and $\bar{f}_n$ with base point $n + 1/2$:

$$\bar{f}_{n+1} = \bar{f}_{n+1/2} + \bar{f}'|_{n+1/2}\left(\frac{\Delta t}{2}\right) + 0(\Delta t^2) \tag{7.137}$$

$$\bar{f}_n = \bar{f}_{n+1/2} + \bar{f}'|_{n+1/2}\left(-\frac{\Delta t}{2}\right) + 0(\Delta t^2) \tag{7.138}$$

Adding Eqs. (7.137) and (7.138) and solving for $\bar{f}_{n+1/2}$ gives

$$\bar{f}_{n+1/2} = \tfrac{1}{2}(\bar{f}_n + \bar{f}_{n+1}) + 0(\Delta t^2) \tag{7.139}$$

Substituting Eq. (7.139) into Eq. (7.136) yields

$$\bar{y}_{n+1} = \bar{y}_n + \frac{\Delta t}{2}[\bar{f}_n + \bar{f}_{n+1} + 0(\Delta t^2)] + 0(\Delta t^3) \tag{7.140}$$

Truncating the third-order remainder terms yields the *implicit trapezoid FDE*:

$$\boxed{y_{n+1} = y_n + \frac{\Delta t}{2}(f_n + f_{n+1}) \qquad 0(\Delta t^3)} \tag{7.141}$$

Equation (7.141) can be solved directly for y_{n+1} for linear ODEs. For nonlinear ODEs, Eq. (7.141) must be solved iteratively for y_{n+1}.

However, if y_{n+1} is first predicted by the first-order explicit Euler FDE, Eq. (7.59), and f_{n+1} is then evaluated using that value of y_{n+1}, then the *modified trapezoid FDEs* are obtained:

$$\boxed{\begin{aligned} y_{n+1}^P &= y_n + \Delta t \, f_n \\ y_{n+1}^C &= y_n + \frac{\Delta t}{2}(f_n + f_{n+1}^P) \end{aligned}}$$

$$\tag{7.142}$$
$$\tag{7.143}$$

The superscript P in Eq. (7.142) denotes that y_{n+1}^P is a predictor value, the superscript P in Eq. (7.143) denotes that f_{n+1}^P is evaluated using y_{n+1}^P and the superscript C in Eq. (7.143) denotes that y_{n+1}^C is the corrected second-order result.

Equations (7.142) and (7.143) are usually called the *modified Euler FDEs*. In some instances, they have been called the *Heun FDEs*. We shall call them the modified Euler FDEs.

The corrector step of the modified Euler FDEs can be iterated, if desired, which may increase the absolute accuracy, but the method is still $0(\Delta t^2)$. Iteration is generally not as worthwhile as using a smaller step size.

Performing a consistency analysis of Eqs. (7.142) and Eq. (7.143) shows that they are consistent with the general nonlinear first-order ODE and that the global error is

$0(\Delta t^2)$. The amplification factor G for the modified Euler FDEs is determined by applying the algorithm to solve the model ODE $\bar{y}' + \alpha \bar{y} = 0$, for which $\bar{f}(t, \bar{y}) = -\alpha \bar{y}$. Thus,

$$y_{n+1}^P = y_n + \Delta t f_n = y_n + \Delta t(-\alpha y_n) = (1 - \alpha \, \Delta t)y_n \tag{7.144}$$

$$y_{n+1}^C = y_n + \frac{\Delta t}{2}(f_n + f_{n+1}) = y_n + \frac{\Delta t}{2}[(-\alpha y_n) - \alpha(1 - \alpha \, \Delta t)y_n] \tag{7.145}$$

$$G = \frac{y_{n+1}^C}{y_n} = 1 - \alpha \, \Delta t + \frac{(\alpha \, \Delta t)^2}{2} \tag{7.146}$$

This expression for G is identical to the amplification factor of the modified midpoint FDEs, Eq. (7.126), for which $|G| \le 1$ for $\alpha \, \Delta t \le 2$. Consequently, the same result applies to the modified Euler FDEs.

The general features of the modified Euler FDEs are presented below.

1. The FDEs are an explicit predictor-corrector set of FDEs which requires two derivative function evaluations per step.
2. The FDEs are consistent, $0(\Delta t^3)$ locally, and $0(\Delta t^2)$ globally.
3. The FDEs are conditionally stable (i.e., $\alpha \, \Delta t \le 2$).
4. The FDEs are consistent and conditionally stable, and thus, convergent.

The algorithm based on the repetitive application of the modified Euler FDEs is called the *modified Euler method*.

Example 7.8. The modified Euler method

To illustrate the modified Euler method, let's solve the radiation problem presented in Section 7.1 using Eqs. (7.142) and (7.143). The derivative function is $f(t, T) = -\alpha(T^4 - T_a^4)$. Equations (7.142) and (7.143) yield

$$f_n = f(t_n, T_n) = -\alpha(T_n^4 - 250.0^4) \tag{7.147}$$

$$T_{n+1}^P = T_n + \Delta t f_n \tag{7.148}$$

$$f_{n+1}^P = f(t_{n+1}, T_{n+1}^P) = -\alpha[(T_{n+1}^P)^4 - 250.0^4] \tag{7.149}$$

$$T_{n+1}^C = T_n + \frac{\Delta t}{2}(f_n + f_{n+1}^P) \tag{7.150}$$

Let $\Delta t = 2.0$ s. For the first time step, the predictor FDE gives

$$f_0 = -(4.0 \times 10^{-12})(2500.0^4 - 250.0^4) = -156.234375 \tag{7.151}$$

$$T_1^P = 2500.0 + 2.0(-156.234375) = 2187.531250 \tag{7.152}$$

The corrector FDE yields

$$f_1^P = -(4.0 \times 10^{-12})(2187.531250^4 - 250.0^4) = -91.580490 \tag{7.153}$$

$$T_1^C = 2500.0 + \tfrac{1}{2}(2.0)(-156.234375 - 91.580490) = 2252.185135 \tag{7.154}$$

These results, the results for subsequent time steps for $t = 4.0$ s to 10.0 s, and the solution for $\Delta t = 1.0$ s are presented in Table 7.6.

The errors presented in Table 7.6 for the second-order modified Euler method for $\Delta t = 1.0$ s are approximately 32 times smaller than the errors presented in Table 7.3 for the first-order explicit Euler method. This illustrates the advantage of the second-order method. To achieve a factor of 32 decrease in the error for a first-order method requires

Table 7.6   Solution by the Modified Euler Method

t_n t_{n+1}	T_n T^P_{n+1} T^C_{n+1}	f_n f^P_{n+1}	$\bar{T}_{n+1}$	Error
0.0	2500.000000	-156.234375		
	2187.531250	-91.580490		
2.0	2252.185135	-102.898821	2248.247314	3.937821
	2046.387492	-70.131759		
4.0	2079.154554	-74.733668	2074.611898	4.542656
	1929.687219	-55.447926		
6.0	1948.972960	-57.698650	1944.618413	4.354547
	1833.575660	-45.196543		
8.0	1846.077767	-46.442316	1842.094508	3.983259
	1753.193135	-37.774562		
10.0	1761.860889		1758.263375	3.597515
0.0	2500.000000	-156.234375		
	2343.765625	-120.686999		
1.0	2361.539313	-124.390198	2360.829988	0.709324
	2237.149114	-100.177915		
2.0	2249.255256	-102.364338	2248.247314	1.007942
...	....	....	....	....
9.0	1799.174556	-41.897804	1798.227867	0.946689
	1757.276752	-38.127884		
10.0	1759.161712		1758.263375	0.898337

a reduction of 32 in the step size, which increases the number of derivative function evaluations by a factor of 32. The same reduction in error was achieved with the second-order method at the expense of twice as many derivative function evaluations. An error analysis at $t = 10.0$ s gives

$$\text{Ratio} = \frac{E(\Delta t = 2.0)}{E(\Delta t = 1.0)} = \frac{3.597515}{0.898337} = 4.00 \tag{7.155}$$

which demonstrates that the method is second order, since the theoretical error ratio for an $0(\Delta t^2)$ method is 4.0.

The modified midpoint method and the modified Euler method are two of the simplest single-point methods. A more accurate class of single-point methods, called Runge-Kutta methods, is developed in the following subsection.

7.7.2 Runge-Kutta Methods

Runge-Kutta methods are a family of single-point methods which evaluate $\Delta y = (y_{n+1} - y_n)$ as the weighted sum of several Δy_i $(i = 1, 2, \ldots)$, where each Δy_i is evaluated as Δt multiplied by the derivative function $f(t, y)$, evaluated at some point in the range $t_n \leq t \leq t_{n+1}$, and the C_i $(i = 1, 2, \ldots)$ are the weighting factors. Thus,

$$y_{n+1} = y_n + \Delta y_n = y_n + (y_{n+1} - y_n) \tag{7.156}$$

where Δy is given by

$$\Delta y = C_1\,\Delta y_1 + C_2\,\Delta y_2 + C_3\,\Delta y_3 + \cdots \tag{7.157}$$

The second-order Runge-Kutta method is obtained by assuming that $\Delta y = (y_{n+1} - y_n)$ is a weighted sum of two Δy's:

$$y_{n+1} = y_n + C_1\,\Delta y_1 + C_2\,\Delta y_2 \tag{7.158}$$

where Δy_1 is given by the explicit Euler FDE:

$$\Delta y_1 = \Delta t\,f(t_n, y_n) = \Delta t\,f_n \tag{7.159}$$

and Δy_2 is based on $f(t,y)$ evaluated somewhere in the interval $t_n < t < t_{n+1}$:

$$\Delta y_2 = \Delta t\,f[t_n + (\alpha\,\Delta t), y_n + (\beta\,\Delta y_1)] \tag{7.160}$$

where α and β are to be determined. Let $\Delta t = h$. Substituting Δy_1 and Δy_2 into Eq. (7.158) gives

$$y_{n+1} = y_n + C_1(hf_n) + C_2 hf[t_n + (\alpha h), y_n + (\beta\,\Delta y_1)] \tag{7.161}$$

Expressing $f(t, \bar{y})$ in a Taylor series at grid point n gives

$$f(t, \bar{y}) = \bar{f}_n + \bar{f}_t|_n h + \bar{f}_y|_n\,\Delta y + \cdots \tag{7.162}$$

Evaluating $f(t, \bar{y})$ at $t = t_n + (\alpha h)$ (i.e., $\Delta t = \alpha h$) and $y = y_n + (\beta\,\Delta y_n)$ (i.e., $\Delta y = \beta hf_n$) gives

$$f(t_n + (\alpha h), y_n + (\beta\,\Delta y_n)) = f_n + (\alpha h) f_t|_n + (\beta hf_n) f_y|_n + 0(h^2) \tag{7.163}$$

Substituting this result into Eq. (7.161) and collecting terms yields

$$y_{n+1} = y_n + (C_1 + C_2)hf_n + h^2(\alpha C_2 f_t|_n + \beta C_2 f_n f_y|_n) + 0(h^3) \tag{7.164}$$

The four free parameters, C_1, C_2, α, and β, can be determined by requiring Eq. (7.164) to match the Taylor series for $\bar{y}(t)$ through second-order terms. That series is

$$\bar{y}_{n+1} = \bar{y}_n + \bar{y}'|_n h + \tfrac{1}{2}\bar{y}''|_n h^2 + \cdots \tag{7.165}$$

$$\bar{y}'|_n = \bar{f}(t_n, \bar{y}_n) = \bar{f}_n \tag{7.166}$$

$$\bar{y}''|_n = (\bar{y}')'|_n = \bar{f}'|_n = \left.\frac{d\bar{f}}{dt}\right|_n = \bar{f}_t|_n + \bar{f}_{\bar{y}}|_n \bar{y}'|_n + \cdots \tag{7.167}$$

Substituting Eqs. (7.166) and (7.167) into Eq. (7.165), where $\bar{y}'|_n = \bar{f}_n$, gives

$$\bar{y}_{n+1} = \bar{y}_n + h\bar{f}_n + \tfrac{1}{2}h^2(\bar{f}_t|_n + \bar{f}_n \bar{f}_y|_n) + 0(h^3) \tag{7.168}$$

Equating Eqs. (7.164) and (7.168) term by term gives

$$C_1 + C_2 = 1 \qquad \alpha C_1 = \tfrac{1}{2} \qquad \beta C_2 = \tfrac{1}{2} \tag{7.169}$$

There are an infinite number of possibilities. Letting $C_1 = \frac{1}{2}$ gives $C_2 = \frac{1}{2}$, $\alpha = 1$, and $\beta = 1$, which yields the modified Euler FDEs. Thus,

$$\Delta y_1 = hf(t_n, y_n) = hf_n \tag{7.170}$$

$$\Delta y_2 = hf(t_{n+1}, y_{n+1}) = hf_{n+1} \tag{7.171}$$

$$y_{n+1} = y_n + \tfrac{1}{2}\Delta y_1 + \tfrac{1}{2}\Delta y_2 = y_n + \frac{h}{2}(f_n + f_{n+1}) \tag{7.172}$$

Letting $C_1 = 0$ gives $C_2 = 1$, $\alpha = \frac{1}{2}$, and $\beta = \frac{1}{2}$, which yields the modified midpoint FDEs. Thus,

$$\Delta y_1 = hf(t_n, y_n) = hf_n \tag{7.173}$$

$$\Delta y_2 = hf\left(t_n + \frac{h}{2}, y_n + \frac{\Delta y_1}{2}\right) = hf_{n+1/2} \tag{7.174}$$

$$y_{n+1} = y_n + (0)\,\Delta y_1 + (1)\,\Delta y_2 = y_n + hf_{n+1/2} \tag{7.175}$$

Other methods result for other choices for C_1 and C_2.

In the general literature, Runge-Kutta formulas frequently denote the Δy_i's by k's $(i = 1, 2, \ldots)$. Thus, the second-order Runge-Kutta FDEs which are identical to the modified Euler FDEs, Eqs. (7.170) to (7.172), are given by

$$y_{n+1} = y_n + \tfrac{1}{2}(k_1 + k_2) \tag{7.176}$$

$$k_1 = hf(t_n, y_n) = hf_n \tag{7.178}$$

$$k_2 = hf(t_n + \Delta t, y_n + k_1) = hf_{n+1} \tag{7.178}$$

7.7.3 The Fourth-Order Runge-Kutta Method

Runge-Kutta methods of higher order have been devised. One of the most popular is the following fourth-order method:

$$\boxed{y_{n+1} = y_n + \tfrac{1}{6}(\Delta y_1 + 2\,\Delta y_2 + 2\,\Delta y_3 + \Delta y_4)} \tag{7.179}$$

$$\Delta y_1 = hf(t_n, y_n) \qquad \Delta y_2 = hf\left(t_n + \frac{h}{2}, y_n + \frac{\Delta y_1}{2}\right) \tag{7.180a}$$

$$\Delta y_3 = hf\left(t_n + \frac{h}{2}, y_n + \frac{\Delta y_2}{2}\right) \qquad \Delta y_4 = hf(t_n + h, y_n + \Delta y_3) \tag{7.180b}$$

To perform a consistency and order analysis and a stability analysis of the fourth-order Runge-Kutta method, Eqs. (7.179) and (7.180) must be applied to the model ODE,

$\bar{y}' + \alpha\bar{y} = 0$, for which $f(t, \bar{y}) = -\alpha\bar{y}$, and the results must be combined into a single-step FDE. Thus,

$$\Delta y_1 = hf(t_n, y_n) = h(-\alpha y_n) = -(\alpha h)y_n \tag{7.181}$$

$$\Delta y_2 = hf\left(t_n + \frac{\Delta t}{2}, y_n + \frac{\Delta y_1}{2}\right) = h(-\alpha\{y_n + \frac{1}{2}[-(\alpha h)y_n]\}) \tag{7.182a}$$

$$\Delta y_2 = -(\alpha h)y_n\left(1 - \frac{(\alpha h)}{2}\right) \tag{7.182b}$$

$$\Delta y_3 = hf\left(t_n + \frac{\Delta t}{2}, y_n + \frac{\Delta y_2}{2}\right) = h\left(-\alpha\left\{y_n + \frac{1}{2}\left[-(\alpha h)y_n\left(1 - \frac{(\alpha h)}{2}\right)\right]\right\}\right) \tag{7.183a}$$

$$\Delta y_3 = -(\alpha h)y_n\left[1 - \frac{(\alpha h)}{2} + \frac{(\alpha h)^2}{4}\right] \tag{7.183b}$$

$$\Delta y_4 = hf(t_n + \Delta t, y_n + \Delta y_3)$$
$$= h\left(-\alpha\left\{y_n + \left[-(\alpha h)y_n\left(1 - \frac{(\alpha h)}{2} + \frac{(\alpha h)^2}{4}\right)\right]\right\}\right) \tag{7.184a}$$

$$\Delta y_4 = -(\alpha h)y_n\left[1 - (\alpha h) + \frac{(\alpha h)^2}{2} - \frac{(\alpha h)^3}{4}\right] \tag{7.184b}$$

Substituting Eqs. (7.181), (7.182b), (7.183b), and (7.184b) into Eq. (7.179) yields the single-step FDE corresponding to Eqs. (7.179) and (7.180):

$$y_{n+1} = y_n - (\alpha h)y_n + \frac{1}{2}(\alpha h)^2 y_n - \frac{1}{6}(\alpha h)^3 y_n + \frac{1}{24}(\alpha h)^4 y_n \tag{7.185}$$

Example 7.9. Consistency and order analysis of the Runge-Kutta method

A consistency and order analysis of Eq. (7.185) is performed by substituting the Taylor series for y_{n+1} with base point n into Eq. (7.185). Thus

$$y_n + y'|_n h + \frac{1}{2}y''|_n h^2 + \frac{1}{6}y'''|_n h^3 + \frac{1}{24}y^{(iv)}|_n h^4 + \frac{1}{120}y^{(v)}|_n h^5 + \cdots$$
$$= y_n - (\alpha h)y_n + \frac{1}{2}(\alpha h)^2 y_n - \frac{1}{6}(\alpha h)^3 y_n + \frac{1}{24}(\alpha h)^4 y_n \tag{7.186}$$

For the model ODE, $\bar{y}' + \alpha\bar{y} = 0$, $y'|_n = -\alpha y_n$, $y''|_n = \alpha^2 y_n$, $y'''|_n = -\alpha^3 y_n$, $y^{(iv)}|_n = \alpha^4 y_n$, and $y^{(v)}|_n = -\alpha^5 y_n$. Substituting these results into the left-hand side of Eq. (7.186) gives

$$y_n + y'|_n h + \frac{1}{2}(\alpha h)^2 y_n - \frac{1}{6}(\alpha h)^3 y_n + \frac{1}{24}(\alpha h)^4 y_n + \frac{1}{120}(\alpha h)^5 y_n + \cdots$$
$$= y_n - (\alpha h)y_n + \frac{1}{24}(\alpha h)^2 y_n - \frac{1}{6}(\alpha h)^3 y_n + \frac{1}{24}(\alpha h)^4 y_n \tag{7.187}$$

Canceling like terms yields

$$y'|_n h + \frac{1}{120}y^{(v)}(\alpha h)^5 + \cdots = -(\alpha h)y_n \tag{7.188}$$

Dividing through by h yields the modified differential equation (MDE):

$$y'_n + \alpha y_n = -\frac{1}{120}y^{(v)}\alpha^5 h^4 + \cdots \tag{7.189}$$

Letting $h \to 0$ in Eq. (7.189) gives $y'_n + \alpha y_n = 0$, which is consistent with the model equation, $\bar{y}' + \alpha\bar{y} = 0$. From Eq. (7.189), Eq. (7.185) is $0(h^4)$.

Example 7.9 demonstrates that the fourth-order Runge-Kutta method is consistent and $O(h^4)$. A stability analysis is presented in Example 7.10.

Example 7.10. Stability analysis of the Runge-Kutta method

The single-step FDE corresponding to Eq. (7.179) is given by Eq. (7.185). Solving that equation for the amplification factor, $G = y_{n+1}|v_n$, yields

$$G = 1 - (\alpha h) + \tfrac{1}{2}(\alpha h)^2 - \tfrac{1}{6}(\alpha h)^3 + \tfrac{1}{24}(\alpha h)^4 \tag{7.190}$$

Recall that α is positive, so that (αh) is positive. Substituting values of (αh) into Eq. (7.190) gives the following results:

(αh)	G	(αh)	G
0.0	1.000000	2.5	0.648438
0.5	0.606771	2.6	0.754733
1.0	0.375000	2.7	0.878838
1.596...	0.270395	2.785...	1.000000
2.0	0.333333	2.8	1.022400

These results show that $|G| \leq 1$ if $(\alpha h) \leq 2.785\ldots$.

In summary, the fourth-order Runge-Kutta FDEs have the following characteristics:

1. The FDEs are explicit and require four derivative function evaluations per step.
2. The FDEs are consistent, $O(\Delta t^5)$ locally and $O(\Delta t^4)$ globally.
3. The FDEs are conditionally stable (i.e., $\alpha \Delta t \leq 2.785\ldots$).
4. The FDEs are consistent and conditionally stable, and thus, convergent.

Algorithms based on the repetitive application of Runge-Kutta FDEs are called Runge-Kutta methods.

Example 7.11. The fourth-order Runge-Kutta method

To illustrate the fourth-order Runge-Kutta method, let's solve the radiation problem presented in Section 7.1 using Eqs. (7.179) and (7.180). The derivative function is $f(t, T) = -\alpha(T^4 - T_a^4)$. Equations (7.179) and (7.180) yield

$$T_{n+1} = T_n + \tfrac{1}{6}(\Delta T_1 + 2\,\Delta T_2 + 2\,\Delta T_3 + \Delta T_4) \tag{7.191}$$

$$\Delta T_1 = \Delta t\, f(t_n, T_n) = \Delta t\, f_n \qquad \Delta T_2 = \Delta t\, f\left(t_n + \frac{\Delta t}{2}, T_n + \frac{\Delta T_1}{2}\right) \tag{7.192}$$

$$\Delta T_3 = \Delta t\, f\left(t_n + \frac{\Delta t}{2}, T_n + \frac{\Delta T_2}{2}\right) \qquad \Delta T_4 = \Delta t\, f(t_n + \Delta t, T_n + \Delta T_3) \tag{7.193}$$

Let $\Delta t = 2.0\,\text{s}$. For the first time step,

$$\Delta T_1 = (2.0)(-4.0 \times 10^{-12})(2500.0^4 - 250.0^4) = -312.46875000 \quad (7.194)$$

$$\Delta T_2 = (2.0)(-4.0 \times 10^{-12})\left[\left(2500.0 - \frac{312.46875000}{2}\right)^4 - 250.0^4\right]$$

$$= -241.37399871 \quad (7.195)$$

$$\Delta T_3 = (2.0)(-4.0 \times 10^{-12})\left[\left(2500.0 - \frac{241.37399871}{2}\right)^4 - 250.0^4\right]$$

$$= -256.35592518 \quad (7.196)$$

$$\Delta T_4 = (2.0)(-4.0 \times 10^{-12})[(2500.0 - 256.35592518)^4 - 250.0^4]$$

$$= -202.69306346 \quad (7.197)$$

$$T_1 = 2500.0 + \tfrac{1}{6}[-312.46875000 + 2(-241.37399871)$$
$$+ 2(-256.35592518) - 202.69306346]$$
$$= 2248.22972313 \quad (7.198)$$

These results, the results for subsequent time steps for $t = 4.0\,\text{s}$ to $10.0\,\text{s}$, and the solution for $\Delta t = 1.0\,\text{s}$ are presented in Table 7.7.

Table 7.7 Solution by the Fourth-Order Runge-Kutta Method

t_n	T_n	ΔT_1 ΔT_3	ΔT_2 ΔT_4	
t_{n+1}	T_{n+1}			Error
0.0	2500.000000000	−312.468750000 −256.355925175	−241.373998706 −202.693063461	
2.0	2248.229723129	−204.335495214 −175.210831240	−169.656671860 −147.710427814	−0.017590925
4.0	2074.596234925	−148.160603003 −130.689901700	−128.100880073 −114.201682488	−0.015662958
6.0	1944.605593419	−114.366138259 −102.884298188	−101.492235905 −92.010660768	−0.012819719
8.0	1842.083948884	−92.083178136 −84.038652301	−83.213391686 −76.389312398	−0.010558967
10.0	1758.254519132			−0.008855569
0.0	2500.000000000	−156.234375000 −139.731281344	−137.601503560 −124.122675002	
1.0	2360.829563365	−124.240707542 −112.896277161	−111.669705704 −102.123860057	−0.000425090
2.0	2248.246807810			−0.000506244
...	....	....	....	....
9.0	1798.227583359	−41.809631378 −39.984283934	−39.898370905 −38.211873099	−0.000283433
10.0	1758.263114333			−0.000260369

The error at $t = 10.0$ s for $\Delta t = 1.0$ s is approximately 110,000 times smaller than the error presented in Table 7.3 for the explicit Euler method and 3,500 times smaller than the error presented in Table 7.6 for the modified Euler method. Results such as these clearly demonstrate the advantage of higher-order methods. An error analysis at $t = 10.0$ s gives

$$\text{Ratio} = \frac{E(\Delta t = 2.0)}{E(\Delta t = 1.0)} = \frac{-0.008855569}{-0.000260369} = 34.01$$

which shows that the method is fourth order, since the theoretical error ratio for $0(\Delta t^4)$ method is 16.0. The value Ratio $= 34.01$ instead of 16.0 is obtained since the higher-order terms in the truncation error are still significant.

7.7.4 Error Estimation and Error Control for Single-Point Methods

Consider a FDE of $0(\Delta t^m)$. For a single step:

$$\bar{y}(t_{n+1}) = y(t_{n+1}, \Delta t) + A \, \Delta t^{m+1} \tag{7.199}$$

where $\bar{y}(t_{n+1})$ denotes the exact solution at t_{n+1}, $y(t_{n+1}, \Delta t)$ denotes the approximate solution at t_{n+1} with increment Δt, and $A \, \Delta t^{m+1}$ is the local truncation error. How can the magnitude of the local truncation error $A \, \Delta t^{m+1}$ be estimated? Repeat the calculation using step size $\Delta t/2$. Thus,

$$\bar{y}(t_{n+1}) = y\left(t_{n+1}, \frac{\Delta t}{2}\right) + 2\left[A\left(\frac{\Delta t}{2}\right)^{m+1}\right] \tag{7.200}$$

This process is illustrated in Figure 7.14. Subtract Eq. (7.200) from Eq. (7.199) and solve for $A \, \Delta t^{m+1}$, which is the local trucation error.

$$\text{Error} = A \, \Delta t^{m+1} = \left[y_{n+1}\left(t_{n+1}, \frac{\Delta t}{2}\right) - y_{n+1}(t_{n+1}, \Delta t)\right]\left(\frac{2^m}{2^m - 1}\right) \tag{7.201}$$

If $|\text{Error}| <$ (lower error limit), increase (double) the step size. If $|\text{Error}| >$ (upper error limit), decrease (halve) the step size. Care must be taken in the specification of the values of (lower error limit) and (upper error limit). This method of error estimation requires 200 percent more work. Consequently, it should be used only occasionally.

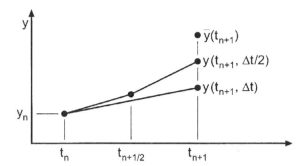

Figure 7.14 Step size halving for error estimation.

7.7.5 Runge-Kutta Methods with Error Estimation

Runge-Kutta methods with more function evaluations have been devised in which the additional results are used for error estimation. The *Runge-Kutta-Fehlberg method* [Fehlberg (1966)] uses six derivative function evaluations:

$$y_{n+1} = y_n + (\tfrac{16}{135}k_1 + \tfrac{6656}{12825}k_3 + \tfrac{28561}{56430}k_4 - \tfrac{9}{50}k_5 + \tfrac{2}{55}k_6) \qquad 0(h^6) \tag{7.202}$$

$$\tilde{y}_{n+1} = y_n + (\tfrac{25}{216}k_1 + \tfrac{1408}{2565}k_3 + \tfrac{2197}{4101}k_4 - \tfrac{1}{5}k_5) \qquad 0(h^5) \tag{7.203}$$

$$k_1 = \Delta t\, f(t_n, y_n) \tag{7.204a}$$

$$k_2 = \Delta t\, f(t_n + \tfrac{1}{4}h, y_n + \tfrac{1}{4}k_1) \tag{7.204b}$$

$$k_3 = \Delta t\, f(t_n + \tfrac{3}{8}h, y_n + \tfrac{3}{32}k_1 + \tfrac{9}{32}k_2) \tag{7.204c}$$

$$k_4 = \Delta t\, f(t_n + \tfrac{12}{13}h, y_n + \tfrac{1932}{2197}k_1 - \tfrac{7200}{2197}k_2 + \tfrac{7296}{2197}k_3) \tag{7.204d}$$

$$k_5 = \Delta t\, f(t_n + h, y_n + \tfrac{439}{216}k_1 - 8k_2 + \tfrac{3680}{513}k_3 - \tfrac{845}{4104}k_4) \tag{7.204e}$$

$$k_6 = \Delta t\, f(t_n + \tfrac{1}{2}h, y_n - \tfrac{8}{27}k_1 + 2k_2 - \tfrac{3544}{2565}k_3 + \tfrac{1859}{4104}k_4 - \tfrac{11}{40}k_5) \tag{7.204f}$$

The error is estimated as follows. Equations (7.202) and (7.203) can be expressed as follows:

$$\bar{y}_{n+1} = y_{n+1} + 0(h^6) \tag{7.205}$$

$$\bar{y}_{n+1} = \tilde{y}_{n+1} + 0(h^5) + 0(h^6) = \tilde{y}_{n+1} + \text{Error} + 0(h^6) \tag{7.206}$$

Substituting y_{n+1} and $\tilde{y}_{n+1}$, Eqs. (7.202) and (7.203), into Eqs. (7.205) and (7.206) and subtracting yields

$$\text{Error} = \tfrac{1}{360}k_1 - \tfrac{128}{4275}k_3 - \tfrac{2197}{75,240}k_4 + \tfrac{1}{50}k_5 + \tfrac{2}{55}k_6 + 0(h^6) \tag{7.207}$$

The error estimate, Error, is used for step size control. Use y_{n+1}, which is $0(h^6)$ locally, as the final value of y_{n+1}.

The general features of the Runge-Kutta-Fehlberg method are presented below.

1. The FDEs are explicit and require six derivative function evaluations per step.
2. The FDEs are consistent, $0(\Delta t^6)$ locally and $0(\Delta t^5)$ globally.

3. The FDEs are conditionally stable.
4. The FDEs are consistent and conditionally stable, and thus, convergent.
5. An estimate of the local error is obtained for step size control.

7.7.6 Summary

Several single-point methods have been presented. Single-point methods work well for both smoothly varying problems and nonsmoothly varying problems. The first-order Euler methods are useful for illustrating the basic features of finite difference methods for solving initial-value ODEs, but they are too inaccurate to be of any practical value. The second-order single-point methods are useful for illustrating techniques for achieving higher-order accuracy, but they also are too inaccurate to be of much practical value. Runge-Kutta methods can be developed for any order desired. The fourth-order Runge-Kutta method presented in this section is the method of choice when a single-point method is desired.

7.8 EXTRAPOLATION METHODS

The concept of extrapolation is applied in Section 5.6 to increase the accuracy of second-order finite difference approximations of derivatives, and in Section 6.4 to increase the accuracy of the second-order trapezoid rule for integrals (i.e., Romberg integration). Extrapolation can be applied to any approximation of an exact process $\bar{f}(t)$ if the functional form of the truncation error dependence on the increment in the numerical process $f(t, h)$ is known. Thus,

$$\bar{f}(t_i) = f(t_i, h) + 0(h^n) + 0(h^{n+r}) + \cdots \tag{7.208}$$

where $\bar{f}(t_i)$ denotes the exact value of $\bar{f}(t)$ at $t = t_i$, $f(t_i, h)$ denotes the approximate value of $f(t)$ at $t = t_i$ computed with increment h, n is the order of the leading truncation error term, and r is the increase in order of successive truncation error terms. In most numerical processes, $r = 1$, and the order of the error increases by one for successive error terms. In the two processes mentioned above, $r = 2$, and the order of the error increases by 2 for each successive error term. Thus, successive extrapolations increase the order of the result by 2 for each extrapolation. This effect accounts for the high accuracy obtainable by the two processes mentioned above. A similar procedure can be developed for solving initial-value ODEs. Extrapolation is applied to the modified midpoint method in this section to obtain the *extrapolated modified midpoint method*.

7.8.1 The Extrapolated Modified Midpoint Method

Gragg (1965) has shown that the functional form of the truncation error for the modified midpoint method is

$$\bar{y}(t_{n+1}) = y(t_{n+1}, \Delta t) + 0(\Delta t^2) + 0(\Delta t^4) + 0(\Delta t^6) + \cdots \tag{7.209}$$

Thus, $r = 2$. Consequently, extrapolation yields a rapidly decreasing error. The objective of the procedure is to march the solution from grid point n to grid point $n + 1$ with time step Δt. The step from n to $n + 1$ is taken for several substeps using the modified midpoint method. The substeps are taken for $h = \Delta t/M$, for $M = 2, 4, 8, 16$, etc. The results for

Figure 7.15   Grids for the extrapolated modified midpoint method.

$y(t_{n+1}, \Delta t/M)$ for the various substeps are then extrapolated to give a highly accurate result for y_{n+1}. This process is illustrated in Figure 7.15.

The version of the modified midpoint method to be used with extrapolation is presented below:

$$z_0 = y_n \tag{7.210a}$$

$$z_1 = z_0 + hf(t_n, z_0) \tag{7.210b}$$

$$z_i = z_{i-2} + 2hf[t_n + (i-1)h, z_{i-1}] \qquad (i = 2, \ldots, M) \tag{7.210c}$$

$$y_{n+1} = \tfrac{1}{2}[z_{M-1} + z_M + hf(t_n + \Delta t, z_M)] \tag{7.211}$$

A table of values for $y(t_{n+1}, \Delta t/M)$ is constructed for the selected number of values of M, and these results are successively extrapolated to higher and higher orders using the general extrapolation formula, Eq. (5.117).

$$IV = MAV + \frac{MAV - LAV}{2^n - 1} = \frac{2^n MAV - LAV}{2^n - 1} \tag{7.212}$$

where IV denotes the extrapolated (i.e., improved) value, MAV denotes the more accurate (i.e., smaller h result) value of the two results being extrapolated, and LAV denotes the less accurate (i.e., larger h result) value of the two values being extrapolated.

The algorithm based on the repetitive application of the extrapolated modified midpoint FDEs is called the *extrapolated modified midpoint method*.

Example 7.12.   The extrapolated modified midpoint method

Let's solve the radiation problem presented in Section 7.1 by the extrapolated modified midpoint method. Recall Eq. (7.1):

$$T' = -\alpha(T^4 - 250.0^4) = f(t, T) \qquad T(0.0) = 2500.0 \tag{7.213}$$

Apply the modified midpoint method with $\Delta t = 2.0\,\mathrm{s}$ to solve this problem. For each time step, let $M = 2, 4, 8,$ and 16. Thus, four values of T_{n+1} will be calculated at each time step Δt. These four values will be successively extrapolated from the $0(\Delta t^2)$ results obtained for the modified midpoint method itself to $0(\Delta t^4)$, $0(\Delta t^6)$, and $0(\Delta t^8)$. For the first time step for $M = 2$, $h = \Delta t/2 = 1.0$ s. The following results are obtained

$$z_0 = T_0 = 2500.0 \tag{7.214a}$$

$$z_1 = z_0 + hf_0 \tag{7.214b}$$

$$z_1 = 2500.0 + 1.0(-4.0 \times 10^{-12})(2500.0^4 - 250.0^4) = 2343.765625 \tag{7.214c}$$

$$z_2 = z_0 + 2hf_1 \tag{7.214d}$$

$$z_2 = 2500.0 + 2(1.0)(-4.0 \times 10^{-12})(2343.765625^4 - 250.0^4)$$
$$= 2258.626001 \tag{7.214e}$$

$$T_2 = \tfrac{1}{2}(z_1 + z_2 + hf_2) \tag{7.215}$$

$$T_2 = \tfrac{1}{2}[2343.765625 + 2258.626001$$
$$+ 1.0(-4.0 \times 10^{-12})(2258.626001^4 - 250.0^4)]$$
$$= 2249.15523693 \tag{7.216}$$

Repeating these steps for $M = 4$, 8, and 16 yields the second-order results for T_2 presented in the second column of Table 7.8.

Extrapolating the four $0(h^2)$ results yields the three $0(h^4)$ results presented in column 3. Extrapolating the three $0(h^4)$ results yields the two $0(h^6)$ results presented in column 4. Extrapolating the two $0(h^6)$ results yields the final $0(h^8)$ result presented in column 5. The $0(h^8)$ value of $T_2 = 2248.24731430$ K is accepted as the final result for T_2.

Repeating the procedure at $t = 4.0$, 6.0, 8.0, and 10.0 s yields the results presented in Table 7.9.

Let's compare these results with the results obtained by the fourth-order Runge-Kutta method in Example 7.11. The extrapolated modified midpoint method required 31 derivative function evaluations per overall time step, for a total of 151 derivative function evaluations with $\Delta t = 2.0$ s to march from $t = 0$ s to $t = 10.0$ s. The final error at $t = 10.0$ s is 0.00000010 K. The fourth-order Runge-Kutta method with $\Delta t = 2.0$ s required four derivative function evaluations per overall time step, for a total of 20 derivative function evaluations to march from $t = 0$ s to $t = 10.0$ s. The final error at $t = 10.0$ s for the Runge-Kutta solution is -0.00885557 K. To reduce this error to the size of the error of the extrapolated modified midpoint method would require a step size reduction of approximately $(0.00885557/0.00000010)^{1/4} = 17$, which would require a total of $5 \times 17 = 85$ time steps. Since each time step requires four derivative function evaluations, a total of 340 derivative function evaluations would be required by the Runge-

Table 7.8 First Step Solution for the Extrapolated Modified Midpoint Method

M	T_2, $0(h^2)$	$\dfrac{4\text{ MAV} - \text{LAV}}{3}$	$\dfrac{16\text{ MAV} - \text{LAV}}{15}$	$\dfrac{64\text{ MAV} - \text{LAV}}{63}$
2	2249.15523693	2248.26188883	2248.24740700	2248.24731430
4	2248.48522585	2248.24831212	2248.24731574	
8	2248.30754055	2248.24737802		
16	2248.26241865			

Table 7.9 Solution by the Extrapolated Modified Midpoint Method

t_n	T_n			
	$T(\Delta t/2)$	$0(\Delta t^4)$	$0(\Delta t^6)$	$0(\Delta t^8)$
	$T(\Delta t/4)$	$0(\Delta t^4)$	$0(\Delta t^6)$	
	$T(\Delta t/8)$	$0(\Delta t^4)$		
	$T(\Delta t/16)$			
t_{n+1}	T_{n+1}	$\bar{T}_{n+1}$	Error	
0.0	2500.00000000			
	2249.15523693	2248.26188883	2248.24740700	2248.24731430
	2248.48522585	2248.24831212	2248.24731574	
	2248.30754055	2248.24737802		
	2248.26241865			
2.0	2248.24731430	2248.24731405	0.00000024	
	2075.00025328	2074.61500750	2074.61190855	2074.61189807
	2074.71131895	2074.61210224	2074.61189824	
	2074.63690642	2074.61191099		
	2074.61815984			
4.0	2074.61189807	2074.61189788	0.00000019	
...	....	....	....	....
8.0	1842.09450797		1842.09450785	0.00000012
	1758.33199904	1758.26353381	1758.26337496	1758.26337480
	1758.28065012	1758.26338489	1758.26337480	
	1758.26770120	1758.26337543		
	1758.26445688			
10.0	1758.26337480	1758.26337470	0.00000010	

Kutta method to achieve the accuracy achieved by the extrapolated modified midpoint method with 151 derivative functions evaluations. This comparison is not intended to show that the extrapolated modified midpoint method is more efficient than the fourth-order Runge-Kutta method. Comparable accuracy and efficiency can be obtained with the two methods.

7.8.2 The Bulirsch-Stoer Method

Stoer and Bulirsch (1980) proposed a variation of the extrapolated modified midpoint method presented above in which the substeps are taken for $h = \Delta t/M$, for $M = 2, 4, 6, 8, 12, 16, \ldots$, and the extrapolation is performed using rational functions instead of the extrapolation formula used above. These modifications somewhat increase the efficiency of the extrapolated modified midpoint method presented above.

7.8.3 Summary

The extrapolated modified midpoint method is an excellent method for achieving high-order results with a rather simple second-order algorithm. This methods works well for both smoothly varying problems and nonsmoothly varying problems.

7.9 MULTIPOINT METHODS

The methods considered so far in this chapter are all *single-point methods*; that is, only one known point, point n, is required to advance the solution to point $n + 1$. Higher-order

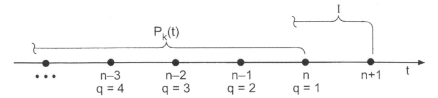

Figure 7.16   Finite difference grid for general explicit multipoint methods.

explicit and implicit methods can be derived by using more points to advance the solution (i.e., points n, $n-1$, $n-2$, etc.). Such methods are called *multipoint methods*. Multipoint methods are sometimes called *multistep methods* or *multivalue methods*.

There are several ways to derive multipoint methods, all of which are equivalent. We shall derive them by fitting Newton backward-difference polynomials to the selected points and integrating from some back point, $n+1-q$, to point $n+1$. Consider the general nonlinear first-order ODE:

$$\bar{y}' = \frac{d\bar{y}}{dt} = f(t, \bar{y}) \qquad \bar{y}(t_0) = \bar{y}_0 \tag{7.217}$$

which can be written in the form

$$d\bar{y} = f(t, \bar{y})\, dt = f[t, \bar{y}(t)]\, dt = F(t)\, dt \tag{7.218}$$

Consider the uniform finite difference grid illustrated in Figure 7.16. General explicit FDEs are obtained as follows:

$$I = \int_{\bar{y}_{n+1-q}}^{\bar{y}_{n+1}} d\bar{y} = \int_{t_{n+1-q}}^{t_{n+1}} [P_k(t)]_n\, dt \tag{7.219}$$

where the subscript q identifies the back point, the subscript k denotes the degree of the Newton backward-difference polynomial, and the subscript n denotes that the base point for the Newton backward-difference polynomial is point n. A two-parameter family of *explicit multipoint FDEs* results corresponding to selected combinations of q and k.

Consider the uniform finite difference grid illustrated in Figure 7.17. General implicit FDEs are obtained as follows:

$$I = \int_{\bar{y}_{n+1-q}}^{\bar{y}_{n+1}} d\bar{y} = \int_{t_{n+1-q}}^{t_{n+1}} [P_k(t)]_{n+1}\, dt \tag{7.220}$$

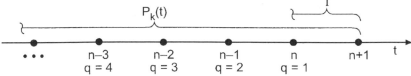

Figure 7.17   Finite difference grid for general implicit multipoint methods.

A two-parameter family of *implicit multipoint FDEs* results corresponding to selected combinations of q and k, where the subscript $n + 1$ denotes that the base point for the Newton backward-difference polynomial is point $n + 1$.

Predictor-corrector methods, such as the modified Euler method presented in Section 7.7, can be constructed using an explicit multipoint method for the predictor and an implicit multipoint method for the corrector.

When the lower limit of integration is point n (i.e., $q = 1$), the resulting FDEs are called *Adams FDEs*. Explicit Adams FDEs are called *Adams-Bashforth FDEs* (Bashforth and Adams, 1883), and implicit Adams FDEs are called *Adams-Moulton FDEs*. When used in a predictor-corrector combination, the set of equations is called an *Adams-Bashforth-Moulton FDE* set. The fourth-order Adams-Bashforth-Moulton FDEs are developed in the following subsection.

7.9.1   The Fourth-Order Adams-Bashforth-Moulton Method

The *fourth-order Adams-Bashforth FDE* is developed by letting $q = 1$ and $k = 3$ in the general explicit multipoint formula, Eq. (7.219). Consider the uniform finite difference grid illustrated in Figure 7.18. Thus,

$$I = \int_{\bar{y}_n}^{\bar{y}_{n+1}} d\bar{y} = \int_{t_n}^{t_{n+1}} [P_3(t)]_n \, dt \qquad (7.221)$$

Recall the third-degree Newton backward-difference polynomial with base point n, Eq. (4.101):

$$
\begin{aligned}
P_3(s) = \bar{f}_n + s\,\nabla\bar{f}_n + \frac{(s+1)s}{2}\nabla^2\bar{f}_n + \frac{(s+2)(s+1)s}{6}\nabla^3\bar{f}_n \\
+ \frac{(s+3)(s+2)(s+1)s}{24}h^4\bar{f}^{(4)}(\tau)
\end{aligned}
\qquad (7.222)
$$

The polynomial $P_3(s)$ is expressed in terms of the variable s, not the variable t. Two approaches can be taken to evaluate this integral: (a) substitute the expression for s in terms of t into the polynomial and integrate with respect to t, or (b) express the integral in terms of the variable s and integrate with respect to s. We shall take the second approach.

Recall the definition of the variable s, Eq. (4.102):

$$s = \frac{t - t_n}{h} \rightarrow t = t_n + sh \rightarrow dt = h\,ds \qquad (7.223)$$

The limits of integration in Eq. (7.221), in terms of s, are

$$t_n \rightarrow s = 0 \qquad \text{and} \qquad t_{n+1} \rightarrow s = 1 \qquad (7.224)$$

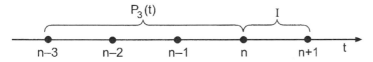

Figure 7.18   Finite difference grid for the fourth-order Adams-Bashforth method.

Thus, Eq. (7.221) becomes

$$\bar{y}_{n+1} - \bar{y}_n = h \int_0^1 P_3(s) \, ds + h \int_0^1 \text{Error}(s) \, ds \tag{7.225}$$

Substituting Eq. (7.222) into Eq. (7.225), where the second integral in Eq. (7.225) is the error term, gives.

$$\bar{y}_{n+1} - \bar{y}_n = h \int_0^1 \left(\bar{f}_n + s \, \nabla \bar{f}_n + \frac{s^2 + s}{2} \nabla^2 \bar{f}_n + \frac{s^3 + 3s^2 + 2s}{6} \nabla^3 \bar{f}_n \right) ds$$

$$+ h \int_0^1 \frac{s^4 + 6s^3 + 11s^2 + 6s}{24} h^4 \bar{f}^{(4)}(\tau) \, ds \tag{7.226}$$

Integrating Eq. (7.226) and evaluating the result at the limits of integration yields

$$\bar{y}_{n+1} = \bar{y}_n + h(\bar{f}_n + \tfrac{1}{2}\nabla \bar{f}_n + \tfrac{5}{12}\nabla^2 \bar{f}_n + \tfrac{3}{8}\nabla^3 \bar{f}_n) + \tfrac{251}{720} h^5 \bar{f}^{(4)}(\tau) \tag{7.227}$$

The backward differences in Eq. (7.227) can be expressed in terms of function values from the following backward-difference table:

t	$\bar{f}$	$\nabla \bar{f}$	$\nabla^2 \bar{f}$	$\nabla^3 \bar{f}$
t_{n-3}	$\bar{f}_{n-3}$			
		$(\bar{f}_{n-2} - \bar{f}_{n-3})$		
t_{n-2}	$\bar{f}_{n-2}$		$(\bar{f}_{n-1} - 2\bar{f}_{n-2} + \bar{f}_{n-3})$	
		$(\bar{f}_{n-1} - \bar{f}_{n-2})$		$(\bar{f}_n - 3\bar{f}_{n-1} + 3\bar{f}_{n-2} - \bar{f}_{n-3})$
t_{n-1}	$\bar{f}_{n-1}$		$(\bar{f}_n - 2\bar{f}_{n-1} + \bar{f}_{n-2})$	
		$(\bar{f}_n - \bar{f}_{n-1})$		$(\bar{f}_{n+1} - 3\bar{f}_n + 3\bar{f}_{n-1} - \bar{f}_{n-2})$
t_n	$\bar{f}_n$		$(\bar{f}_{n+1} - 2\bar{f}_n + \bar{f}_{n-1})$	
		$(\bar{f}_{n+1} - \bar{f}_n)$		
t_{n+1}	$\bar{f}_{n+1}$			

Substituting the expressions for the appropriate backward differences into Eq. (7.227) gives

$$\bar{y}_{n+1} = \bar{y}_n + h[\bar{f}_n + \tfrac{1}{2}(\bar{f}_n - \bar{f}_{n-1}) + \tfrac{5}{12}(\bar{f}_n - 2\bar{f}_{n-1} + \bar{f}_{n-2})$$

$$+ \tfrac{3}{8}(\bar{f}_n - 3\bar{f}_{n-1} + 3\bar{f}_{n-2} - \bar{f}_{n-3})] + \tfrac{251}{720} h^5 \bar{y}^{(5)}(\tau) \tag{7.228}$$

Collecting terms and truncating the remainder term yields the fourth-order Adams-Bashforth FDE:

$$\boxed{y_{n+1} = y_n + \frac{h}{24}(55f_n - 59f_{n-1} + 37f_{n-2} - 9f_{n-3})} \tag{7.229}$$

The general features of the fourth-order Adams-Bashforth FDE are summarized below.

1. The FDE is explicit and requires one derivative function evaluation per step.
2. The FDE is consistent, $0(\Delta t^5)$ locally and $0(\Delta t^4)$ globally.
3. The FDE is conditionally stable ($\alpha \, \Delta t \lesssim 0.3$).
4. The FDE is consistent and conditionally stable, and thus, convergent.

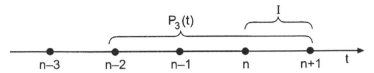

Figure 7.19 Finite difference grid for the fourth-order Adams-Moulton method.

The *fourth-order Adams-Moulton FDE* is developed by letting $q = 1$ and $k = 3$ in the general implicit multipoint equation, Eq. (7.220). Consider the uniform finite difference grid illustrated in Figure 7.19. Thus,

$$I = \int_{\bar{y}_n}^{\bar{y}_{n+1}} d\bar{y} = \int_{t_n}^{t_{n+1}} [P_3(t)]_{n+1} \, dt \qquad (7.230)$$

Recall the third-degree Newton backward-difference polynomial with base point $n + 1$, Eq. (4.101), with $n \to n + 1$:

$$P_3(s) = \bar{f}_{n+1} + s\,\nabla \bar{f}_{n+1} + \frac{(s+1)s}{2}\nabla^2 \bar{f}_{n+1} + \frac{(s+2)(s+1)s}{6}\nabla^3 \bar{f}_{n+1}$$
$$+ \frac{(s+3)(s+2)(s+1)s}{24} h^4 \bar{f}^{(4)}(\tau) \qquad (7.231)$$

As done for the Adams-Bashforth FDE, the integral will be expressed in terms of s. The limits of integration in Eq. (7.230), in terms of s, are

$$t_n \to s = -1 \qquad \text{and} \qquad t_{n+1} \to s = 0 \qquad (7.232)$$

Thus,

$$\bar{y}_{n+1} - \bar{y}_n = h \int_{-1}^{0} P_3(s) \, ds + h \int_{-1}^{0} \text{Error}(s) \, ds \qquad (7.233)$$

Substituting $P_3(s)$ into Eq. (7.233), integrating, substituting for the appropriate backward differences, collecting terms, and simplifying, gives

$$\bar{y}_{n+1} = \bar{y}_n + \frac{h}{24}(9\bar{f}_{n+1} + 19\bar{f}_n - 5\bar{f}_{n-1} + \bar{f}_{n-2}) - \tfrac{19}{720} h^5 \bar{y}^{(5)}(\tau) \qquad (7.234)$$

Truncating the remainder term yields the fourth-order Adams-Moulton FDE:

$$y_{n+1} = y_n + \frac{h}{24}(9f_{n+1} + 19f_n - 5f_{n-1} + f_{n-2}) \qquad (7.235)$$

The general features of the fourth-order Adams-Moulton FDE are summarized below.

1. The FDE is implicit and requires one derivative function evaluation per step.
2. The FDE is consistent, $0(\Delta t^5)$ locally and $0(\Delta t^4)$ globally.
3. The FDE is conditionally stable ($\alpha \, \Delta t \lesssim 3.0$).
4. The FDE is consistent and conditionally stable, and thus, convergent.

The Adams-Moulton FDE is implicit. It can be used as a corrector FDE with the Adams-Bashforth FDE as a predictor FDE. Thus, from Eqs. (7.229) and (7.235):

$$y_{n+1}^P = y_n + \frac{h}{24}(55f_n - 59f_{n-1} + 37f_{n-2} - 9f_{n-3}) \tag{7.236a}$$

$$y_{n+1}^C = y_n + \frac{h}{24}(9f_{n+1}^P + 19f_n - 5f_{n-1} + f_{n-2}) \tag{7.236b}$$

Stability analysis of multipoint methods is a little more complicated than stability analysis of single-point methods. To illustrate the procedure, let's perform a stability analysis of the fourth-order Adams-Bashforth FDE, Eq. (7.229):

$$y_{n+1} = y_n + \frac{h}{24}(55f_n - 59f_{n-1} + 37f_{n-2} - 9f_{n-3}) \tag{7.237}$$

Example 7.13.   Stability analysis of the fourth-order Adams-Bashforth FDE

The amplification factor G is determined by applying the FDE to solve the model ODE, $\bar{y}' + \alpha\bar{y} = 0$, for which $f(t, \bar{y}) = -\alpha\bar{y}$. Thus, Eq. (7.237) yields

$$y_{n+1} = y_n + \frac{h}{24}[55(-\alpha y_n) - 59(-\alpha y_{n-1}) + 37(-\alpha y_{n-2}) - 9(-\alpha y_{n-3})] \tag{7.238}$$

For a multipoint method applied to a linear ODE with constant Δt, the amplification factor G is the same for all time steps. Thus,

$$G = \frac{y_{n+1}}{y_n} = \frac{y_n}{y_{n-1}} = \frac{y_{n-1}}{y_{n-2}} = \frac{y_{n-2}}{y_{n-3}} \tag{7.239}$$

Solving Eq. (7.239) for y_{n-1}, y_{n-2}, and y_{n-3} gives

$$y_{n-1} = \frac{y_n}{G} \qquad y_{n-2} = \frac{y_n}{G^2} \qquad y_{n-3} = \frac{y_n}{G^3} \tag{7.240}$$

Substituting these values into Eq. (7.238) gives

$$y_{n+1} = y_n - \frac{(\alpha h)}{24}\left(55y_n - 59\frac{y_n}{G} + 37\frac{y_n}{G^2} - 9\frac{y_n}{G^3}\right) \tag{7.241}$$

Solving for $G = y_{n+1}/y_n$ gives

$$G = 1 - \frac{(\alpha h)}{24}\left(55 - \frac{59}{G} + \frac{37}{G^2} - \frac{9}{G^3}\right) \tag{7.242}$$

Multiplying Eq. (7.242) by G^3 and rearranging yields

$$G^4 + \left(\frac{55(\alpha h)}{24} - 1\right)G^3 - \frac{59(\alpha h)}{24}G^2 + \frac{37(\alpha h)}{24}G - \frac{9(\alpha h)}{24} = 0 \tag{7.243}$$

For each value of (αh), there are four values of G, G_i ($i = 1, \ldots, 4$). For stability, all four values of G_i must satisfy $|G_i| \le 1$. Solving Eq. (7.243) for the four roots by Newton's method gives the following results:

(αh)	G_1	G_2	G_3 and G_4	$|G_3|$ and $|G_4|$
0.00	1.000	0.000	$0.000 \pm I0.000$	0.000
0.10	0.905	-0.523	$0.195 \pm I0.203$	0.281
0.20	0.819	-0.773	$0.248 \pm I0.239$	0.344
0.30	0.742	-1.000	$0.285 \pm I0.265$	0.389
0.31	0.735	-1.022	$0.289 \pm I0.268$	0.394

These results show that $|G| \leq 1$ for $(\alpha \, \Delta t) \leq 0.3$.

The general features of the Adams-Bashforth-Moulton predictor-corrector FDEs are summarized below.

1. The FDEs are explicit and require two derivative function evaluations per step.
2. The FDEs are consistent, $0(\Delta t^5)$ locally and $0(\Delta t^4)$ globally.
3. The FDEs are conditionally stable ($\alpha \, \Delta t \lesssim 3.0$).
4. The FDEs are consistent and conditionally stable, and thus, convergent.
5. Four equally spaced starting points are needed. Use the fourth-order Runge-Kutta method.

The algorithm based on the repetitive application of the Adams-Bashforth-Moulton FDEs is called the *Adams-Bashforth-Moulton method*.

Example 7.14. The Fourth-Order Adams-Bashforth-Moulton method

---

To illustrate the Adams-Bashforth-Moulton method, let's solve the radiation problem presented in Section 7.1, for which the derivative function is $f(t, T) = -\alpha(T^4 - T_a^4)$. Recall Eq. (7.236):

$$f_n = f(t_n, T_n) = -\alpha(T_n^4 - 250.0^4) \tag{7.244a}$$

$$T_{n+1}^P = T_n + \frac{\Delta t}{24}(55f_n - 59f_{n-1} + 37f_{n-2} - 9f_{n-3}) \tag{7.244b}$$

$$f_{n+1}^P = f(t_{n+1}, T_{n+1}^P) = -\alpha[(T_{n+1}^P)^4 - 250.0^4] \tag{7.245a}$$

$$T_{n+1}^C = T_n + \frac{\Delta t}{24}(9f_{n+1}^P + 19f_n - 5f_{n-1} + f_{n-2}) \tag{7.245b}$$

Let $\Delta t = 1.0$ s. In general, starting values would be obtained by the fourth-order Runge-Kutta method. However, since the exact solution is known, let's use the exact solution at $t = 1.0, 2.0$, and 3.0 s for starting values. These values are given in Table 7.10, along with the corresponding values of f_n. For $t_4 = 4.0$ s,

$$T_4^P = 2154.47079576 + \frac{1.0}{24}[55(-86.16753966) - 59(-102.18094603)$$

$$+ 37(-124.24079704) - 9(-158.23437500)]$$

$$= 2075.24833822 \tag{7.246}$$

$$f_4^P = -(4.0 \times 10^{-12})(2075.24833822^4 - 250^4)$$

$$= -74.17350708 \tag{7.247}$$

$$T_4^C = 2154.47079576 + \frac{1.0}{24}[9(-74.17350708) + 19(-86.16753966)$$

$$- 5(-102.18094603) + (-124.24079704)]$$

$$= 2074.55075892 \tag{7.248}$$

These results and the results of the remaining time steps from $t = 5.0$ s to $t = 10.0$ s are presented in Table 7.10.

Let's compare these results with the results obtained in Table 7.7 by the fourth-order Runge-Kutta method. For $\Delta t = 1.0$ s, Table 7.7 gives an error at $t = 10.0$ s of -0.00026037, which is approximately 284 times smaller than the corresponding error

Table 7.10 Solution by the Fourth-Order Adams-Bashforth-Moulton
Method

t_n t_{n+1}	T_n T^P_{n+1} T^C_{n+1}	f_n f^P_{n+1}	$\bar{T}_{n+1}$	Error
0.0	2500.00000000	-156.23437500		
1.0	2360.82998845	-124.24079704	exact	
2.0	2248.24731405	-102.18094603	solution	
3.0	2154.47079576	-86.16753966		
	2075.24833822	-74.17350708		
4.0	2074.55075892	-74.07380486	2074.61189788	-0.06113896
	2005.68816488	-64.71557210		
5.0	2005.33468855	-64.66995205	2005.41636581	-0.08167726
	1944.70704983	-57.19500703		
6.0	1944.53124406	-57.17432197	1944.61841314	-0.08716907
...	....	....	....	....
9.0	1798.14913834	-41.80233360	1798.22786679	-0.07872845
	1758.21558333	-38.20946276		
10.0	1758.18932752		1758.26337470	-0.07404718

in Table 7.10. However, the Runge-Kutta method requires four derivative function
evaluations per step compared to two for the Adams-Bashforth-Moulton method. The
Runge-Kutta results for $\Delta t = 2.0$ s presented in Table 7.7 required the same number of
derivative function evaluations as the results in Table 7.10 for $\Delta t = 1.0$ s. The error at
$t = 10.0$ s for $\Delta t = 2.0$ s in Table 7.7 is -0.008855569. The corresponding error in Table
7.10 is -0.07404718, which is approximately 8.4 times larger. The fourth-order Runge-
Kutta method is more efficient in this problem.

7.9.2 General Adams Methods

Adams methods of any order can be derived by choosing different degree Newton
backward-difference polynomials to fit the solution at the data points. The finite difference
grid for the general explicit Adams-Bashforth FDEs is illustrated in Figure 7.20. The
general formula for the explicit Adams-Bashforth FDEs is:

$$\int_{\bar{y}_n}^{\bar{y}_{n+1}} d\bar{y} = h \int_0^1 [P_k(s)]_n \, ds \tag{7.249}$$

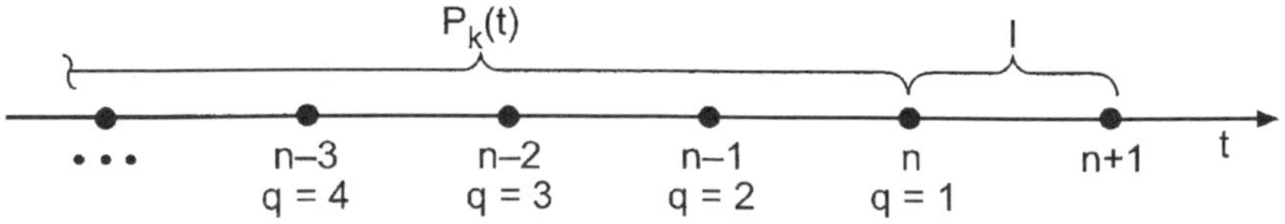

Figure 7.20 Finite difference grid for general Adams-Bashforth methods.

Table 7.11 Coefficients for the General Explicit Adams-Bashforth FDEs

k	β	α_0	α_{-1}	α_{-2}	α_{-3}	α_{-4}	α_{-5}	n	C
0	1	1						1	2.0
1	1/2	3	-1					2	1.0
2	1/12	23	-16	5				3	0.5
3	1/24	55	-59	37	-9			4	0.3
4	1/720	1901	-2774	2616	-1274	251		5	0.2
5	1/1440	4277	-7923	9982	-7298	2877	-475	6	

where k denotes the order of the Newton backward-difference polynomial fit at base point n. Integrating Eq. (7.249), evaluating the result for the limits of integration, introducing the expressions for the appropriate backward differences at point n, and simplifying the result yields the general explicit Adams-Bashforth FDE:

$$y_{n+1} = y_n + \beta h(\alpha_0 f_n + \alpha_{-1} f_{n-1} + \alpha_{-2} f_{n-2} + \cdots) \quad 0(h^n), \quad \alpha \, \Delta t \le C \qquad (7.250)$$

where the coefficients β and α_i $(i = 0, -1, -2, \ldots)$, the global order n and the stability limit, $\alpha \, \Delta t \le C$, are presented in Table 7.11.

The finite difference grid for the general implicit Adams-Moulton FDEs is illustrated in Figure 7.21. The general formula for the implicit Adams-Moulton FDEs is:

$$\int_{\bar{y}_n}^{\bar{y}_{n+1}} d\bar{y} = h \int_{-1}^{0} [P_k(s)]_{n+1} \, ds \qquad (7.251)$$

where k denotes the order of the Newton backward-difference polynomial fit at base point $n + 1$. Integrating Eq. (7.251), evaluating the results for the limits of integration, introducing the expressions for the appropriate backward differences at point $n + 1$, and simplifying the result yields the general implicit Adams-Moulton FDE:

$$y_{n+1} = y_n + \beta h(\alpha_1 f_{n+1} + \alpha_0 f_n + \alpha_{-1} f_{n-1} + \cdots) \qquad 0(h^n), \quad \alpha \, \Delta t \le C \qquad (7.252)$$

where the coefficients β and α_i $(i = 1, 0, -1, \ldots)$, the global order n, and the stability limit, $\alpha \, \Delta t \le C$, are presented in Table 7.12.

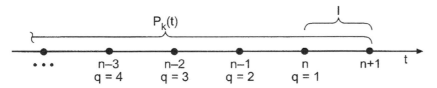

Figure 7.21 Finite difference grid for general Adams-Moulton methods.

Table 7.12   Coefficients for the General Implicit Adams-Moulton FDEs

k	β	α_1	α_0	α_{-1}	α_{-2}	α_{-3}	α_{-4}	n	C
0	1	1						1	∞
1	1/2	1	1					2	∞
2	1/12	5	8	-1				3	6.0
3	1/24	9	19	-5	1			4	3.0
4	1/720	251	646	-264	106	-19		5	1.9
5	1/1440	475	1427	-798	482	-173	27	6	

7.9.3   Error Estimation, Error Control, and Extrapolation

An efficient method of error estimation and error control can be developed for predictor-corrector methods. Consider the fourth-order Adams-Bashforth-Moulton FDEs given by Eq. (7.236). These equations can be expressed as

$$\bar{y}_{n+1} = y_{n+1}^P + \tfrac{251}{720}\Delta t^5\, y^{(v)}(\tau^P) \tag{7.253}$$

$$\bar{y}_{n+1} = y_{n+1}^C - \tfrac{19}{720}\Delta t^5\, y^{(v)}(\tau^C) \tag{7.254}$$

where $t_{n-3} \le \tau^P \le t_n$ and $t_{n-2} \le \tau^C \le t_{n+1}$. Assuming that $y^{(v)}(\tau^P) = y^{(v)}(\tau^C) = y^{(v)}(\tau)$, which is a reasonable approximation, Eqs. (7.253) and (7.254) can be combined to give

$$y_{n+1}^C - y_{n+1}^P = \Delta t^5\, y^{(v)}(\tau)(\tfrac{19}{720} + \tfrac{251}{720}) \tag{7.255}$$

from which

$$\Delta t^5\, y^{(v)}(\tau) = \frac{720}{19 + 251}(y_{n+1}^C - y_{n+1}^P) \tag{7.256}$$

Thus, the corrector error in Eq. (7.254) is given by

$$\text{Corrector Error} = -\frac{19}{720}\Delta t^5\, y^{(v)}(\tau) = -\frac{19}{19 + 251}(y_{n+1}^C - y_{n+1}^P) \tag{7.257}$$

If |Corrector Error| is less than a prescribed lower error limit, increase (double) the step size. If |Corrector Error| is greater than a prescribed upper error limit, decrease (halve) the step size. This method of error estimation requires no additional derivative function evaluations. When the step size is doubled, every other previous solution point is used to determine the four known points. When the step size is halved, the solution at the two additional points at $n + \tfrac{1}{2}$ and $n + \tfrac{3}{2}$ can be obtained by fourth-order interpolation, or the solution can be restarted by the fourth-order Runge-Kutta method at grid point n.

   Once an estimate of the error has been obtained, it can be used to extrapolate the solution. For the predictor, the truncation error is given by

$$\text{Predictor Error} = \frac{251}{720}\Delta t^5\, y^{(v)}(\tau) = \frac{251}{19 + 251}(y_{n+1}^C - y_{n+1}^P) \tag{7.258}$$

Unfortunately, y_{n+1}^C is not known until the corrector FDE is evaluated. An estimate of the predictor error can be obtained by lagging the term $(y_{n+1}^C - y_{n+1}^P)$. Thus, assume that

$(y_{n+1}^C - y_{n+1}^P) \cong (y_n^C - y_n^P)$. The mop up correction (i.e., extrapolation) for the fourth-order Adams-Bashforth predictor FDE is

$$\Delta y_{n+1}^{P,M} = \frac{251}{19+251}(y_n^C - y_n^P) \tag{7.259}$$

The mopped up (i.e., extrapolated) predictor solution is

$$y_{n+1}^{P,M} = y_{n+1}^P + \Delta y_{n+1}^{P,M} \tag{7.260}$$

On the first time step at t_0, Eq. (7.260) cannot be applied, since the values on the right-hand side are lagged one step.

The value of $y_{n+1}^{P,M}$ is used to evaluate $f_{n+1}^{P,M}$ for use in the fourth-order Adams-Moulton corrector FDE, Eq. (7.236b). Thus,

$$f_{n+1}^P = f_{n+1}^{P,M} = f(t_{n+1}, y_{n+1}^{P,M}) \tag{7.261}$$

$$y_{n+1}^C = y_n + \frac{h}{24}(9f_{n+1}^{P,M} + 19f_n - 5f_{n-1} + f_{n-2}) \tag{7.262}$$

The mop up correction (i.e., extrapolation) for the fourth-order Adams-Moulton corrector FDE is then

$$\Delta y_{n+1}^{C,M} = -\frac{19}{19+251}(y_{n+1}^C - y_{n+1}^P) \tag{7.263}$$

Note that y_{n+1}^P, not $y_{n+1}^{P,M}$ is used in Eq. (7.263). The mopped up (i.e., extrapolated) corrector solution is

$$y_{n+1}^{C,M} = y_{n+1}^C + \Delta y_{n+1}^{C,M} \tag{7.264}$$

7.9.4 Summary

Multipoint methods work well for smoothly varying problems. For nonsmoothly varying problems, single-point methods and extrapolation methods are preferred. The fourth-order Adams-Bashforth-Moulton method is an excellent example of a multipoint method. Multipoint methods other than Adams-Bashforth-Moulton methods can be derived by integrating from back points other than point n. The fourth-order Adams-Bashforth-Moulton method is one of the best, if not the best, example of this type of method. It has excellent stability limits, excellent accuracy, and a simple and inexpensive error estimation procedure. It is recommended as the method of choice when a multipoint method is desired.

7.10 SUMMARY OF METHODS AND RESULTS

Several finite difference methods for solving first-order initial-value ordinary differential equations are presented in Sections 7.5 to 7.9. Seven of the more prominent methods are summarized in Table 7.13.

The radiation problem presented in Section 7.1 was solved by these seven methods. The errors for a step size of $\Delta t = 1.0$ s are presented in Figure 7.22. The first-order explicit

Table 7.13 Summary of Selected Finite Difference Methods

The Explicit Euler Method

$$y_{n+1} = y_n + \Delta t\, f_n \tag{7.265}$$

The Implicit Euler Method

$$y_{n+1} = y_n + \Delta t\, f_{n+1} \tag{7.266}$$

The Modified Midpoint Method

$$y_{n+1/2}^P = y_n + \frac{\Delta t}{2} f_n, \qquad f_{n+1/2}^P = f(t_{n+1/2}, y_{n+1/2}^P),$$
$$y_{n+1}^C = y_n + \Delta t\, f_{n+1/2}^P \tag{7.267}$$

The Modified Euler Method

$$y_{n+1}^P = y_n + \Delta t\, f_n, \qquad f_{n+1}^P = f(t_{n+1}, y_{n+1}^P),$$
$$y_{n+1}^C = y_n + \tfrac{1}{2}\Delta t(f_n + f_{n+1}^P) \tag{7.268}$$

The Fourth-Order Runge-Kutta Method

$$y_{n+1} = y_n + \tfrac{1}{6}(\Delta y_1 + 2\,\Delta y_2 + 2\,\Delta y_3 + \Delta y_4) \tag{7.269}$$

$$\Delta y_1 = \Delta t\, f(t_n, y_n), \qquad \Delta y_2 = \Delta t\, f\left(t_n + \frac{\Delta t}{2}, y_n + \frac{\Delta y_1}{2}\right) \tag{7.270}$$

$$\Delta y_3 = \Delta t\, f\left(t_n + \frac{\Delta t}{2}, y_n + \frac{\Delta y_2}{2}\right),$$
$$\Delta y_4 = \Delta t\, f(t_n + \Delta t, y_n + \Delta y_3) \tag{7.271}$$

The Extrapolated Modified Midpoint Method

$$z_0 = y_n, \qquad z_1 = z_0 + hf(t_n, z_0) \tag{7.272}$$
$$z_i = z_{i-2} + 2hf[t_n + (i-1)h, z_{i-1}] \qquad (i = 2, \ldots, M) \tag{7.273}$$
$$y_{n+1} = \tfrac{1}{2}[z_{M-1} + z_M + hf(t_n + \Delta t, z_M)] \tag{7.274}$$
$$\text{IV} = \text{MAV} + \frac{\text{MAV} - \text{LAV}}{2^n - 1} = \frac{2^n \text{MAV} - \text{LAV}}{2^n - 1} \tag{7.275}$$

The Adams-Bashforth-Moulton Method

$$y_{n+1}^P = y_n + \frac{\Delta t}{24}(55f_n - 59f_{n-1} + 37f_{n-2} - 9f_{n-3}),$$
$$\Delta y_{n+1}^{P,M} = \tfrac{251}{270}(y_n^C - y_n^P) \tag{7.276}$$

$$y_{n+1}^C = y_n + \frac{\Delta t}{24}(9f_{n+1}^{P,M} + 19f_n - 5f_{n-1} + f_{n-2}),$$
$$\Delta y_{n+1}^{C,M} = -\tfrac{19}{270}(y_{n+1}^C - y_{n+1}^P) \tag{7.277}$$

Euler method, the second-order modified Euler method, and the second-order modified midpoint method are clearly inferior to the fourth-order methods. The fourth-order Runge-Kutta method is an excellent method for both smoothly varying and nonsmoothly varying problems. However, it lacks an efficient error control procedure.

The fourth-order Adams-Bashforth-Moulton method and the fourth-order extrapolated modified midpoint method yield comparable results. Both of these methods have excellent error control procedures. The higher-order (sixth- and eighth-order) extrapolated modified midpoint methods yield extremely accurate results for this smoothly varying problem.

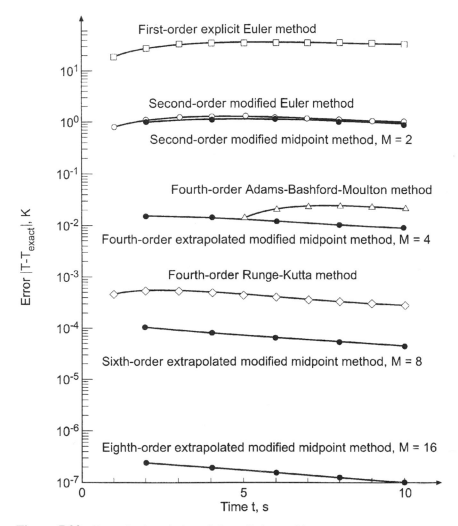

Figure 7.22   Errors in the solution of the radiation problem.

7.11 NONLINEAR IMPLICIT FINITE DIFFERENCE EQUATIONS

Several finite difference methods have been developed in this chapter for solving the general nonlinear first-order initial-value ordinary differential equation:

$$\bar{y}' = f(t, \bar{y}) \qquad \bar{y}(t_0) = \bar{y}_0 \tag{7.278}$$

The derivative function $f(t, \bar{y})$ may be linear or nonlinear in $\bar{y}$. When $f(t, \bar{y})$ is linear in $\bar{y}$, the corresponding FDE is linear in y_{n+1}, for both explicit FDEs and implicit FDEs. When $f(t, \bar{y})$ is nonlinear in $\bar{y}$, explicit FDEs are still linear in y_{n+1}. However, implicit FDEs are nonlinear in y_{n+1}, and special procedures are required to solve for y_{n+1}. Two procedures for solving nonlinear implicit FDEs are:

1. Time linearization
2. Newton's method

These two procedures are presented in this section.

7.11.1 Time Linearization

One approach for solving a nonlinear implicit FDE is time linearization, in which the nonlinear derivative function is expressed in a Taylor series about the known point n and truncated after the first derivative term. To illustrate this procedure, consider the implicit Euler method [see Eq. (7.73)]:

$$y_{n+1} = y_n + \Delta t \, f_{n+1} \tag{7.279}$$

Express $f(t, \bar{y})$ in a two-variable Taylor series. Thus,

$$\bar{f}_{n+1} = \bar{f}_n + \bar{f}_t|_n \, \Delta t + \bar{f}_y|_n (y_{n+1} - y_n) + \cdots \tag{7.280}$$

Truncating Eq. (7.280) and substituting into Eq. (7.279) yields

$$y_{n+1} = y_n + \Delta t (f_n + f_t|_n \, \Delta t + f_y|_n (y_{n+1} - y_n)] \tag{7.281}$$

Equation (7.281) is linear in y_{n+1}. Solving for y_{n+1} yields

$$y_{n+1} = \frac{y_n + \Delta t \, f_n + \Delta t^2 \, f_t|_n - \Delta t \, y_n f_y|_n}{1 - \Delta t \, f_y|_n} \tag{7.282}$$

Example 7.15.   Time linearization

In Example 7.3, the radiation problem presented in Section 7.1 is solved by the implicit Euler method. The FDE is

$$T_{n+1} = T_n + \Delta t \, f_{n+1} \tag{7.283}$$

Table 7.14   Solution by Time Linearization

t_n t_{n+1}	T_n T_{n+1}	f_n	$f_T|_n$	$\bar{T}_{n+1}$	Error
0.0	2500.000000	-156.234375	-0.250000		
2.0	2291.687500	-110.311316	-0.192569	2248.247314	43.440186
4.0	2132.409031	-82.691332	-0.155143	2074.611898	57.797133
6.0	2006.190233	-64.780411	-0.129192	1944.618413	61.571820
8.0	1903.232170	-52.468392	-0.110305	1842.094508	61.137662
10.0	1817.261400			1758.263375	58.998026
0.0	2500.000000	-156.234375	-0.250000		
1.0	2375.012500	-127.253656	-0.214347	2360.829988	14.182512
2.0	2270.220672	-106.235194	-0.187208	2248.247314	21.973358
$\cdots$	$\cdots\cdots\cdots$	$\cdots\cdots\cdots$	$\cdots\cdots\cdots$	$\cdots\cdots\cdots$	$\cdots\cdots$
9.0	1827.215365	-44.572473	-0.097609	1798.227867	28.987498
10.0	1786.606661			1758.263375	28.343286

The derivative function $f(t, T)$ is given by

$$f(t, T) = -\alpha(T^4 - T_a^4) \tag{7.284}$$

Thus,

$$f_t = 0 \quad \text{and} \quad f_T = -4\alpha T^3 \tag{7.285}$$

Substituting Eq. (7.285) into Eq. (7.282) yields

$$T_{n+1} = \frac{T_n + \Delta t\, f_n - \Delta t\, T_n f_T|_n}{1 - \Delta t\, f_T|_n} \tag{7.286}$$

Let $\Delta t = 2.0$ s. For the first time step,

$$f_0 = -(4.0 \times 10^{-12})(2500.0^4 - 250.0^4) = -156.234375 \tag{7.287}$$

$$f_T|_0 = -4(4.0 \times 10^{-12})2500.0^3 = -0.250000 \tag{7.288}$$

$$T_1 = \frac{2500.0 + 2.0(-156.234375) - 2.0(2500.0)(-0.250000)}{1 - 2.0(-0.250000)}$$

$$= 2291.687500 \tag{7.289}$$

These results and the results of subsequent time steps for t from 4.0 s to 10.0 s are presented in Table 7.14, which also presents the results for $\Delta t = 1.0$ s.

7.11.2 Newton's Method

Newton's method for solving nonlinear equations is presented in Section 3.4. A nonlinear implicit FDE can be expressed in the form

$$y_{n+1} = G(y_{n+1}) \tag{7.290}$$

Equation (7.290) can be rearranged into the form

$$F(y_{n+1}) = y_{n+1} - G(y_{n+1}) = 0 \tag{7.291}$$

Expanding $F(y_{n+1})$ in a Taylor series about the value y_{n+1} and evaluating at y_{n+1}^* yields

$$F(y_{n+1}^*) = F(y_{n+1}) + F'(y_{n+1})(y_{n+1}^* - y_{n+1}) + \cdots = 0 \tag{7.292}$$

where y_{n+1}^* is the solution of Eq. (7.290). Truncating Eq. (7.292) after the first-order term and solving for y_{n+1} yields

$$\boxed{y_{n+1}^{(k+1)} = y_{n+1}^{(k)} - \frac{F(y_{n+1}^{(k)})}{F'(y_{n+1}^{(k)})}} \tag{7.293}$$

Equation (7.293) must be solved iteratively. Newton's method works well for nonlinear implicit FDEs. A good initial guess may be required.

Example 7.16. Newton's method

Let's illustrate Newton's method by solving the radiation problem presented in Section 7.1 and solved in Example 7.3 by the implicit Euler method. Thus,

$$T_{n+1} = T_n + \Delta t\, f_{n+1} = T_n - \alpha\, \Delta t(T_{n+1}^4 - T_a^4) \tag{7.294}$$

Rearranging Eq. (7.294) into the form of Eq. (7.291) yields

$$F(T_{n+1}) = T_{n+1} - T_n + \Delta t \, \alpha(T_{n+1}^4 - T_a^4) = 0 \tag{7.295}$$

The derivative of $F(T_{n+1})$ is

$$F'(T_{n+1}) = 1 + 4 \, \Delta t \, \alpha T_{n+1}^3 \tag{7.296}$$

Equation (7.293) yields

$$T_{n+1}^{(k+1)} = T_{n+1}^{(k)} - \frac{F(T_{n+1}^{(k)})}{F'(T_{n+1}^{(k)})} \tag{7.297}$$

Let $\Delta t = 2.0$ s. For the first time step,

$$F(T_1) = T_1 - 2500.0 + (2.0)(4.0 \times 10^{-12})(T_1^4 - 250.0^4) \tag{7.298}$$
$$F'(T_1) = 1 + 4(2.0)(4.0 \times 10^{-12})T_1^3 \tag{7.299}$$

Let $T_1^{(0)} = 2500.0$ K. Then

$$F(T_1^{(0)}) = 2500.0 - 2500.0 + (2.0)(4.0 \times 10^{-12})(2500.0^4 - 250.0^4)$$
$$= 312.468250 \tag{7.300}$$
$$F'(T_1^{(0)}) = 1 + 4(2.0)(4.0 \times 10^{-12})2500.0^3 = 1.500000 \tag{7.301}$$

Substituting these values into Eq. (7.297) gives

$$T_1^{(1)} = 2500.0 - \frac{312.468750}{1.500000} = 2291.687500 \tag{7.302}$$

Repeating the procedure three more times yields the converged result $T_1^{(4)} = 2282.785819$. These results are presented in Table 7.15, along with the final results for the subsequent time steps from $t = 4.0$ s to 10.0 s.

The results presented in Tables 7.14 and 7.15 differ due to the additional truncation error associated with time linearization. However, the differences are quite small. Time linearization is quite popular for solving nonlinear implicit FDEs which approximate

Table 7.15 Solution by Newton's Method

t_n t_{n+1} t_{n+1}	k	T_n T_{n+1} T_{n+1}	F_n	F'_n	$\overline{T}_{n+1}$	Error
0.0	0	2500.000000				
	1	2500.000000	312.468750	1.500000		
	2	2291.687500	12.310131	1.385138		
	3	2282.800203	0.019859	1.380674		
	4	2282.785819	0.000000	1.380667		
2.0		2282.785819			2248.247314	34.538505
4.0		2120.934807			2074.611898	46.322909
6.0		1994.394933			1944.618413	49.776520
8.0		1891.929506			1842.094508	49.834998
10.0		1806.718992			1758.263375	48.455617

nonlinear PDEs. When the exact solution to an implicit nonlinear FDE is desired, Newton's method is recommended.

7.12 HIGHER-ORDER ORDINARY DIFFERENTIAL EQUATIONS

Sections 7.5 to 7.11 are devoted to the solution of first-order ordinary differential equations by finite difference methods. Many applications in engineering and science are governed by higher-order ODEs. In general, a higher-order ODE can be replaced by a system of first-order ODEs. When a system of higher-order ODEs is involved, each individual higher-order ODE can be replaced by a system of first-order ODEs, and the coupled system of higher-order ODEs can be replaced by coupled systems of first-order ODEs. The systems of first-order ODEs can be solved as described in Section 7.13.

Consider the second-order initial-value ODE developed in Section II.5 for the vertical flight of a rocket, Eq. (II.25), and the simpler model given by Eq. (II.26):

$$y'' = \frac{F(t, y)}{M_0 - \int_0^t \dot{m}(t)\, dt} - g(y) - \frac{C_D(\rho, V, y)\frac{1}{2}\rho(y)AV^2}{M_0 - \int_0^t \dot{m}(t)\, dt} \qquad (7.303)$$

$$y'' = \frac{F}{M_0 - \dot{m}t} - g \qquad y(0.0) = 0.0 \quad \text{and} \quad y'(0.0) = V(0.0) = 0.0 \qquad (7.304)$$

Equations (7.303) and (7.304) both can be reduced to a system of two coupled initial-value ODEs by the procedure described below.

Consider the general nth-order ODE:

$$y^{(n)} = f(t, y, y', y'', \ldots, y^{(n-1)}) \qquad (7.305)$$

$$\bar{y}(t_0) = \bar{y}_0 \quad \text{and} \quad \bar{y}^{(i)}(t_0) = \bar{y}_0^{(i)} \quad (i = 1, 2, \ldots, n-1) \qquad (7.306)$$

Equation (7.305) can be replaced by an equivalent system of n coupled first-order ODEs by defining n auxiliary variable. Thus,

$$y_1 = y \qquad (7.307.1)$$
$$y_2 = y' = y_1' \qquad (7.307.2)$$
$$y_3 = y'' = y_2' \qquad (7.307.3)$$

$$\ldots\ldots\ldots\ldots\ldots$$

$$y_n = y^{(n-1)} = y_{n-1}' \qquad (7.307.n)$$

Differentiating Eq. (7.307.n) gives

$$y_n' = y^{(n)} \qquad (7.308)$$

Rearranging Eqs. (7.307.2) to (7.307.n) and substituting these results and Eq. (7.308) into Eq. (7.305) yields the following system of n coupled first-order ODEs:

$$y_1' = y_2 \qquad\qquad y_1(0) = y_0 \qquad\qquad (7.309.1)$$
$$y_2' = y_3 \qquad\qquad y_2(0) = y_0' \qquad\qquad (7.309.2)$$

$$\ldots\ldots\ldots\ldots\ldots\ldots\ldots\ldots\ldots\ldots\qquad \ldots\ldots\ldots\ldots\ldots$$

$$y_{n-1}' = y_n \qquad\qquad y_{n-1}(0) = y_0^{(n-2)} \qquad (7.309.n-1)$$
$$y_n' = F(t, y_1, y_2, \ldots, y_n) \qquad y_n(0) = y_0^{(n-1)} \qquad (7.309.n)$$

where Eq. (7.309.n) is the original nth-order ODE, Eq. (7.305), expressed in terms of the auxiliary variables y_i $(i = 1, 2, \ldots, n)$.

The result is a system of n coupled first-order ODEs, which can be solved by the procedure discussed in Section 7.13. This reduction can nearly always be done. Thus, the general features of a higher-order ODE are similar to the general features of a first-order ODE.

Example 7.17.   Reduction of a second-order ODE to two coupled first-order ODEs

To illustrate the reduction of a higher-order ODE to a system of coupled first-order ODEs, let's reduce Eq. (7.304) to a system of two coupled first-order ODEs. Recall Eq. (7.304):

$$y'' = \frac{T}{M_0 - \dot{m}t} - g \qquad y(0.0) = 0.0 \quad \text{and} \quad y'(0.0) = V(0.0) = 0.0 \qquad (7.310)$$

Let $y' = V$. Then Eq. (7.310) reduces to the following pair of coupled first-order ODEs:

$$y' = V \qquad\qquad y(0.0) = 0.0 \qquad\qquad\qquad\qquad (7.311)$$

$$V' = \frac{T}{M_0 - \dot{m}t} - g \qquad V(0.0) = 0.0 \qquad\qquad\qquad (7.312)$$

Equations (7.311) and (7.312) comprise a system of two coupled first-order ODEs for $y(t)$ and $V(t)$. The solution to Eqs. (7.311) and (7.312) by the fourth-order Runge-Kutta method is presented in Example 7.18 in Section 7.13.

7.13 SYSTEMS OF FIRST-ORDER ORDINARY DIFFERENTIAL EQUATIONS

Sections 7.5 to 7.11 are devoted to the solution of a single first-order ordinary differential equation by finite difference methods. In many applications in engineering and science, systems of coupled first-order ODEs governing several dependent variables arise. The methods for solving a single first-order ODE can be used to solve systems of coupled first-order ODEs.

Consider the general system of n coupled first-order ODEs:

$$\boxed{\begin{aligned} \bar{y}_i' &= \bar{f}_i(t, \bar{y}_1, \bar{y}_2, \ldots, \bar{y}_n) \qquad (i = 1, 2, \ldots, n) \\ \bar{y}_i(0) &= Y_i \qquad (i = 1, 2, \ldots, n) \end{aligned}} \qquad \begin{aligned} (7.313) \\ (7.314) \end{aligned}$$

Each ODE in the system of ODEs can be solved by any of the methods developed for solving single ODEs. Care must be taken to ensure the proper coupling of the solutions. When predictor-corrector or multistep methods are used, each step must be applied to all the equations before proceeding to the next step. The step size must be the same for all the equations.

Example 7.18.   Solution of two coupled first-order ODEs

Consider the system of two coupled linear first-order initial-value ODEs developed in Example 7.17 in Section 7.12 for the vertical motion of a rocket, Eqs. (7.311) and (7.312):

$$y' = V \qquad\qquad y(0.0) = 0.0 \qquad\qquad (7.315)$$

$$V' = \frac{T}{M_0 - \dot{m}t} - g \qquad V(0.0) = 0.0 \qquad\qquad (7.316)$$

where M_0 is the initial mass, $\dot{m}$ is the mass expulsion rate, and g is the acceleration of gravity. These parameters are discussed in detail in Section II.6. The exact solution of Eq. (7.316) is

$$V(t) = -\frac{T}{\dot{m}} \ln\left(1 - \frac{\dot{m}t}{M_0}\right) - gt \qquad\qquad (7.317)$$

Substituting Eq. (7.317) into Eq. (7.315) and integrating yields the exact solution of Eq. (7.315):

$$y(t) = \frac{M_0}{\dot{m}}\left(\frac{T}{\dot{m}}\right)\left(1 - \frac{\dot{m}t}{M_0}\right)\ln\left(1 - \frac{\dot{m}t}{M_0}\right) + \frac{Tt}{\dot{m}} - \tfrac{1}{2}gt^2 \qquad\qquad (7.318)$$

As an example, let $T = 10,000\,\text{N}$, $M_0 = 100.0\,\text{kg}$, $\dot{m} = 5.0\,\text{kg/s}$, and $g = 9.8\,\text{m/s}^2$. Equations (7.315) and (7.316) become

$$y' = f(t, y, V) = V \qquad\qquad y(0.0) = 0.0 \qquad\qquad (7.319)$$

$$V' = g(t, y, V) = \frac{10,000.0}{100.0 - 5.0t} - 9.8 \qquad V(0.0) = 0.0 \qquad\qquad (7.320)$$

Equations (7.317) and (7.318) become

$$V(t) = -1,000 \ln(1 - 0.05t) - 9.8t \qquad\qquad (7.321)$$

$$y(t) = 10,000(1 - 0.05t)\ln(1 - 0.05t) + 2000t - 4.9t^2 \qquad\qquad (7.322)$$

Let's solve this problem by the fourth-order Runge-Kutta method, Eqs. (7.179) and (7.180), for $V(10.0)$ and $y(10.0)$ with $\Delta t = 1.0\,\text{s}$. Let Δy_i ($i = 1, 2, 3, 4$) denote the increments in $y(t)$ and ΔV_i ($i = 1, 2, 3, 4$) denote the increments in $V(t)$. Thus,

$$y_{n+1} = y_n + \tfrac{1}{6}(\Delta y_1 + 2\,\Delta y_2 + 2\,\Delta y_3 + \Delta y_4) \qquad\qquad (7.323)$$

$$V_{n+1} = V_n + \tfrac{1}{6}(\Delta V_1 + 2\,\Delta V_2 + 2\,\Delta V_3 + \Delta V_4) \qquad\qquad (7.324)$$

where Δy_i $(i = 1, 2, 3, 4)$ and ΔV_i $(i = 1, 2, 3, 4)$ are given by Eq. (7.180):

$$\Delta y_1 = \Delta t\, f(t_n, y_n, V_n) \qquad \Delta V_1 = \Delta t\, g(t_n, y_n, V_n) \tag{7.325a}$$

$$\Delta y_2 = \Delta t\, f\left(t_n + \frac{\Delta t}{2}, y_n + \frac{\Delta y_1}{2}, V_n + \frac{\Delta V_1}{2}\right)$$

$$\Delta V_2 = \Delta t\, g\left(t_n + \frac{\Delta t}{2}, y_n + \frac{\Delta y_1}{2}, V_n + \frac{\Delta V_1}{2}\right) \tag{7.325b}$$

$$\Delta y_3 = \Delta t\, f\left(t_n + \frac{\Delta t}{2}, y_n + \frac{\Delta y_2}{2}, V_n + \frac{\Delta V_2}{2}\right)$$

$$\Delta V_3 = \Delta t\, g\left(t_n + \frac{\Delta t}{2}, y_n + \frac{\Delta y_2}{2}, V_n + \frac{\Delta V_2}{2}\right) \tag{7.325c}$$

$$\Delta y_4 = \Delta t\, f(t_n + \Delta t, y_n + \Delta y_3, V_n + \Delta V_3)$$

$$\Delta V_4 = \Delta t\, g(t_n + \Delta t, y_n + \Delta y_3, V_n + \Delta V_3) \tag{7.325d}$$

Due to the coupling, Δy_1 and ΔV_1 both must be computed before Δy_2 and ΔV_2 can be computed, Δy_2 and ΔV_2 both must be computed before Δy_3 and ΔV_3 can be computed, etc.

The derivative functions, $f(t, y, V)$ and $g(t, y, V)$, are given by Eqs. (7.319) and (7.320), respectively. Thus, Eq. (7.325) reduces to

$$\Delta y_1 = \Delta t\, (V_n) \qquad \Delta V_1 = \Delta t\left(\frac{10{,}000.0}{100.0 - 5.0t_n} - 9.8\right) \tag{7.326a}$$

$$\Delta y_2 = \Delta t\left(V_n + \frac{\Delta V_1}{2}\right) \qquad \Delta V_2 = \Delta t\left[\frac{10{,}000.0}{100.0 - 5.0(t_n + \Delta t/2)} - 9.8\right] \tag{7.326b}$$

$$\Delta y_3 = \Delta t\left(V_n + \frac{\Delta V_2}{2}\right) \qquad \Delta V_3 = \Delta t\left[\frac{10{,}000.0}{100.0 - 5.0(t_n + \Delta t/2)} - 9.8\right] \tag{7.326c}$$

$$\Delta y_4 = \Delta t\left(V_n + \Delta V_3\right) \qquad \Delta V_3 = \Delta t\left[\frac{10{,}000.0}{100.0 - 5.0(t_n + \Delta t)} - 9.8\right] \tag{7.326d}$$

Let $\Delta t = 1.0$. For the first time step,

$$\Delta y_1 = 1.0(0.0) = 0.000000 \tag{7.327a}$$

$$\Delta V_1 = 1.0\left[\frac{10{,}000.0}{100.0 - 5.0(0.0)} - 9.8\right] = 90.200000 \tag{7.327b}$$

$$\Delta y_2 = 1.0\left(0.0 + \frac{90.200000}{2}\right) = 45.100000 \tag{7.327c}$$

$$\Delta V_2 = 1.0\left[\frac{10{,}000.0}{100.0 - 5.0(0.0 + 1.0/2)} - 9.8\right] = 95.463158 \tag{7.327d}$$

$$\Delta y_3 = 1.0\left(0.0 + \frac{95.463158}{2}\right) = 47.731579 \tag{7.327e}$$

$$\Delta V_3 = 1.0\left[\frac{10{,}000.0}{100.0 - 5.0(0.0 + 1.0/2)} - 9.8\right] = 95.463158 \tag{7.327f}$$

$$\Delta y_4 = 1.0(0.0 + 95.463158) = 95.463158 \tag{7.327g}$$

$$\Delta V_4 = 1.0\left[\frac{10{,}000.0}{100.0 - 5.0(0.0 + 1.0)} - 9.8\right] = 101.311111 \tag{7.327h}$$

Substituting these results into Eqs. (7.323) and (7.324) gives

$$y_1 = \tfrac{1}{6}[0.0 + 2(45.100000 + 47.731579) + 95.463158] = 46.854386 \text{ m} \qquad (7.328)$$

$$V_1 = \tfrac{1}{6}[90.200000 + 2(95.463158 + 95.463158) + 101.311111]$$
$$= 95.560624 \text{ m/s} \qquad (7.329)$$

These results and the results for the subsequent time steps for $t = 2.0$ to 10.0 s are presented in Table 7.16.

The results presented in Example 7.18 are deceptively simple since Eq. (7.316) is a simplified version of the actual problem specified by Eq. (7.303). Recall Eq. (7.303) for y'', from which $V' = y''$ is given by

$$V' = \frac{F(t, y)}{M_0 - \int_0^t \dot{m}(t)\, dt} - g(y) - \frac{C_D(\rho, V, y)\frac{1}{2}\rho(y)AV^2}{M_0 - \int_0^t \dot{m}(t)\, dt} \qquad (7.330)$$

The evaluation of ΔV_1 is done at (t_n, y_n, V_n). The evaluation of ΔV_2 is done at $(t_n + \Delta t/2, y_n + \Delta y_1/2, V_n + \Delta V_1/2)$. This requires that F, M, ρ, and C_D be evaluated at $t_n + \Delta t/2$, $y_n + \Delta y_1/2$, and $V_n + \Delta V_1/2$. This is a considerably more complicated calculation. However, the basic features of the Runge-Kutta method are unchanged.

7.14 STIFF ORDINARY DIFFERENTIAL EQUATIONS

A special problem arising in the numerical solution of ODEs is stiffness. This problem occurs in single linear and nonlinear ODEs, higher-order linear and nonlinear ODEs, and systems of linear and nonlinear ODEs. There are several definitions of stiffness:

1. An ODE is stiff if the step size required for stability is much smaller than the step size required for accuracy.

Table 7.16 Solution of Two Coupled First-Order ODEs

t_n	y_n V_n	Δy_1 ΔV_1	Δy_2 ΔV_2	Δy_3 ΔV_3	Δy_4 ΔV_4
t_{n+1}	y_{n+1} V_{n+1}				
0.00	0.00000000	0.00000000	45.10000000	46.38205128	92.76410256
	0.00000000	90.20000000	92.76410256	92.76410256	95.46315789
1.00	45.95470085	92.78659469	140.51817364	141.94064875	191.09470280
	92.78659469	95.46315789	98.30810811	98.30810811	101.31111111
2.00	187.41229123	191.12104493	241.77660049	243.36390207	295.60675922
	191.12104493	101.31111111	104.48571429	104.48571429	107.84705882
3.00	430.25599278	295.63788278	349.56141219	351.34394338	407.05000399
	295.63788278	107.84705882	111.41212121	111.41212121	115.20000000
...	....	....	....	....	....
9.00	4450.68315419	1107.47425871	1193.48334962	1197.81235395	1288.15044919
	1107.47425871	172.01818182	180.67619048	180.67619048	190.20000000
10.00	5647.05250670				
	1288.29474933				

2. An ODE is stiff if it contains some components of the solution that decay rapidly compared to other components of the solution.
3. A system of ODEs is stiff if at least one eigenvalue of the system is negative and large compared to the other eigenvalues of the system.
4. From a practical point of view, an ODE is stiff if the step size based on cost (i.e., computational time) is too large to obtain an accurate (i.e., stable) solution.

An example of a single stiff ODE is presented in the next subsection, and an example of a system of stiff ODEs is presented in the following subsection. The Gear (1971) method for solving stiff ODEs is presented in the final subsection.

7.14.1 A Single First-Order ODE

Although stiffness is usually associated with a system of ODEs, it can also occur in a single ODE that has more than one time scale of interest; one time scale associated with the complementary solution and one time scale associated with the particular solution. Gear (1971) considered the following ODE:

$$\bar{y}' = f(t, \bar{y}) = -\alpha(\bar{y} - F(t)) + F'(t), \qquad \bar{y}(t_0) = \bar{y}_0 \qquad (7.331)$$

which has the exact solution

$$\bar{y}(t) = (\bar{y}_0 - F(0))e^{-\alpha t} + F(t) \qquad (7.332)$$

When α is a large positive constant and $F(t)$ is a smooth slowly varying function, Eq. (7.332) exhibits two widely different time scales: a rapidly changing term associated with $\exp(-\alpha t)$ and a slowly varying term associated with $F(t)$.

As an example, let $\alpha = 1000$, $F(t) = t + 2$, and $\bar{y}(0) = 1$. Equation (7.331) becomes

$$\bar{y}' = \bar{f}(t, \bar{y}) = -1000(\bar{y} - (t + 2)) + 1, \qquad y(0) = 1 \qquad (7.333)$$

and Eq. (7.332) becomes

$$\bar{y}(t) = -e^{-1000t} + t + 2 \qquad (7.334)$$

The exact solution at small values of t, which is dominated by the $\exp(-1000t)$ term, is presented in Figure 7.23. The exact solution for large values of t, which is dominated by the $(t + 2)$ term, is presented in Figure 7.24. Note how rapidly the exponential term decays.

The error is controlled by Δt, but stability is controlled by $\alpha \Delta t$. For stability of many explicit methods, $\alpha \Delta t \leq 2$, which, for $\alpha = 1000$, gives $\Delta t < 2/\alpha = 2/1000 = 0.002$. To avoid overshoot, $\Delta t < 0.002/2 = 0.001$. For reasonable accuracy, $\Delta t \leq 0.002/10 = 0.0002$. To reach $t = 5$, $N = 5/0.0002 = 25{,}000$ time steps are required.

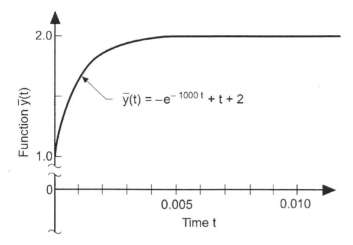

Figure 7.23 Exact solution of the stiff ODE at small time.

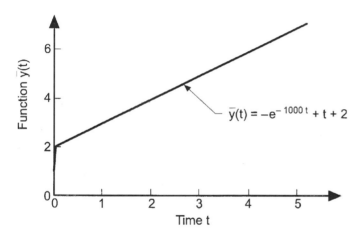

Figure 7.24 Exact solution of the stiff ODE at large time.

Example 7.19. Solution of the stiff ODE by the explicit Euler method

Let's solve Eq. (7.333) by the explicit Euler method, Eq. (7.59):

$$y_{n+1} = y_n + \Delta t\, f_n \tag{7.335}$$

Substituting the derivative function, $f(t, y)$, defined in Eq. (7.333), into Eq. (7.335) gives

$$y_{n+1} = y_n + \Delta t(-1000(y_n - (t_n + 2)) + 1) \tag{7.336}$$

Consider four different step sizes: $\Delta t = 0.0005$, 0.001, 0.002, and 0.0025. The results are tabulated in Table 7.17 and illustrated in Figure 7.25.

The stability limit is $\Delta t \leq 2/\alpha = 0.002$. For $\Delta t = 0.0005$, the solution is a reasonable approximation of the exact solution, although the errors are rather large. For $\Delta t = 0.001$, the solution reaches the asymptotic large time solution, $y(t) = t + 2$, in one step. For $\Delta t = 0.002$, the solution is stable but oscillates about the large time solution, $y(t) = t + 2$. For $\Delta t = 0.0025$, the solution is clearly unstable.

From this example, it is obvious that the stable step size is controlled by the rapid transient associated with the exponential term. At large times, that transient has completely died out and the solution is dominated totally by the asymptotic large time solution, $y(t) = t + 2$. If an accurate small time solution is required, then the small step size required for stability may be larger than the small step size required for accuracy. In that case, explicit methods can be used. However, if the early transient solution is of no interest, and an accurate large time solution is required, explicit methods are unsuitable because of the small stable step size.

Example 7.19 clearly illustrates the effect of stiffness of the ODE on the numerical solution by the explicit Euler method. In Example 7.20, the implicit Euler method is used to reduce the problems associated with stiffness.

Example 7.20. Solution of the stiff ODE by the implicit Euler method

Let's solve Eq. (7.333) by the implicit Euler method, Eq. (7.73):

$$y_{n+1} = y_n + \Delta t \, f_{n+1} \tag{7.337}$$

Substituting the derivative function, $f(t, y)$, defined in Eq. (7.333), into Eq. (7.337) yields

$$y_{n+1} = y_n + \Delta t (-1000(y_{n+1} - (t_{n+1} + 2)) + 1) \tag{7.338}$$

Table 7.17 Solution of the Stiff ODE by the Explicit Euler Method

	$\Delta t = 0.0005$			$\Delta t = 0.001$	
t_n	y_n	$\bar{y}_n$	t_n	y_n	$\bar{y}_n$
0.0000	1.000000		0.000	1.000000	
0.0005	1.500500	1.393969	0.001	2.001000	1.633121
0.0010	1.751000	1.633121	0.002	2.002000	1.866665
0.0015	1.876500	1.778370	0.003	2.003000	1.953213
0.0020	1.939500	1.866665	0.004	2.004000	1.985684
...	....	....	...	....	
0.0100	2.009999	2.009955	0.010	2.010000	2.009955

	$\Delta t = 0.002$			$\Delta t = 0.0025$	
t_n	y_n	$\bar{y}_n$	t_n	y_n	$\bar{y}_n$
0.000	1.000000		0.0000	1.000000	
0.002	3.002000	1.866665	0.0025	3.502500	1.920415
0.004	1.004000	1.985684	0.0050	−0.245000	1.998262
0.006	3.006000	2.003521	0.0075	5.382500	2.006947
0.008	1.008000	2.007665	0.0100	−3.052500	2.009955
0.010	3.010000	2.009955			

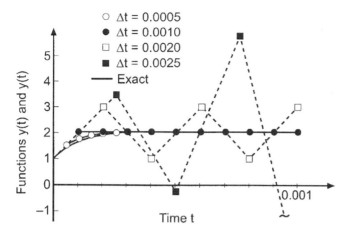

Figure 7.25   Solution of the stiff ODE by the explicit Euler method.

Equation (7.338) is implicit in y_{n+1}. However, since $f(t, y)$ is linear in y, Eq. (7.338) is linear in y_{n+1}, and it can be rearranged to give:

$$y_{n+1} = \left(\frac{1}{1 + 1000 \, \Delta t}\right)(y_n + 1000(t_{n+1} + 2) \, \Delta t + \Delta t) \tag{7.339}$$

Consider four different step sizes: 0.01, 0.05, 0.10, and 0.5. The results for the three small time steps are tabulated in Table 7.18, and the results for $\Delta t = 0.5$ are illustrated in Figure 7.26.

The implicit Euler method is unconditionally stable. Consequently, there is no limit on the stable step size. As illustrated in Table 7.18 and Figure 7.26, the solutions are all stable. However, as the step size is increased, the accuracy of the early time transient due to the exponential term suffers. In fact, the entire early time transient is completely lost for large values of Δt. However, even in those cases, the large time solution is predicted quite accurately. If the early time transient is of interest, then a small step size is required for

Table 7.18.   Solution of the Stiff ODE by the Implicit Euler Method

	$\Delta t = 0.01$			$\Delta t = 0.05$	
t_n	y_n	$\bar{y}_n$	t_n	y_n	$\bar{y}_n$
0.00	1.000000		0.00	1.000000	
0.01	1.919091	2.009955	0.05	2.030392	2.050000
0.02	2.011736	2.020000	0.10	2.099616	2.100000
0.03	2.029249	2.030000			
0.04	2.039932	2.040000		$\Delta t = 0.1$	
0.05	2.049994	2.050000			
0.06	2.059999	2.060000	t_n	y_n	$\bar{y}_n$
0.07	2.070000	2.070000			
0.08	2.080000	2.080000	0.0	1.000000	
0.09	2.090000	2.090000	0.1	2.090099	2.100000
0.10	2.100000	2.100000			

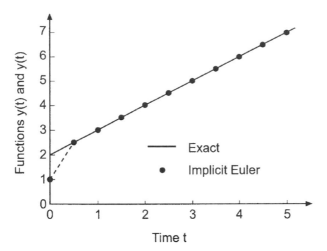

Figure 7.26 Solution of the stiff ODE by the implicit Euler method.

accuracy. If only the large time solution is of interest, then implicit methods can be employed to reduce the computational effort.

7.14.2 Systems of First-Order ODEs

Consider the general system of first-order ODEs discussed in Section 7.13, Eq. (7.313):

$$\bar{y}_i' = \bar{f}_i(t, \bar{y}_1, \bar{y}_2, \ldots, \bar{y}_n) \qquad (i = 1, 2, \ldots, n) \tag{7.340}$$

For a system of coupled linear ODEs, Eq. (7.340) can be expressed as

$$\bar{y}' = \mathbf{A}\bar{y} + \mathbf{F} \tag{7.341}$$

where $\bar{y}^T = [\bar{y}_1 \bar{y}_2 \ldots \bar{y}_n]$, $\mathbf{A}$ is an $n \times n$ matrix, and $\mathbf{F}^T = [F_1 F_2 \ldots F_n]$. Stability and stiffness are related to the eigenvalues, α_i $(i = 1, \ldots, n)$, of the matrix $\mathbf{A}$. For stability, $|\alpha_i| \leq 1$ $(i = 1, \ldots, n)$. A system of ODEs is stiff if at least one eigenvalue has a large negative real part which causes the corresponding component of the solution to vary rapidly compared to the typical scale of variation displayed by the rest of the solution. The stiffness ratio is defined as:

$$\text{Stiffness ratio} = \frac{\text{Max}|\text{Re}(\alpha_i)|}{\text{Min}|\text{Re}(\alpha_i)|} \tag{7.342}$$

A system of coupled linear ODEs can be uncoupled to yield a system of uncoupled linear ODEs. Consider the system of two coupled linear ODEs:

$$u' = F(t, u, v) \tag{7.343}$$
$$v' = G(t, u, v) \tag{7.344}$$

These two ODEs can be uncoupled to yield

$$y' = f(t, y) \tag{7.345}$$
$$z' = g(t, z) \tag{7.346}$$

where y and z are functions of u and v. Consider the uncoupled system of ODEs:

$$y' = -y, \qquad y(0) = 1 \tag{7.347}$$
$$z' = -1000z, \qquad z(0) = 1 \tag{7.348}$$

which corresponds to some coupled system of ODEs in terms of the variables u and v. Equations (7.347) and (7.348) can be analyzed separately for stability. Consider an explicit finite difference method for which the stability limit is $\alpha \, \Delta t \le C$, where C is a constant of order unity. For Eqs. (7.347) and (7.348), $\alpha_1 = 1$ and $\alpha_2 = 1000$, respectively. Thus, for $C = 1$,

$$\Delta t_1 \le \frac{C}{1} = C \quad \text{and} \quad \Delta t_2 \le \frac{C}{1000} = 0.001C \tag{7.349}$$

If a coupled system of equations such as Eqs. (7.343) and (7.344), which are equivalent to an uncoupled system of ODEs such as Eqs. (7.347) and (7.348), is solved by an explicit method, the common time step must be the smaller of the two values corresponding to Eqs. (7.347) and (7.348), that is, $\Delta t = \Delta t_2 \le 0.001C$. The exact solutions of Eqs. (7.347) and (7.348) are:

$$y(t) = e^{-t} \tag{7.350}$$
$$z(t) = e^{-1000t} \tag{7.351}$$

The function $z(t)$ decays to a negligible value after a few time steps, during which time the function $y(t)$ has changed only slightly. Small time steps must still be taken in the solution for $y(t)$ because of the stability limit associated with $z(t)$.

When a system of ODEs can be uncoupled, as in the previous paragraph, each ODE can be solved by a method appropriate to its own peculiarities. However, when the system of ODEs cannot be uncoupled, the problem of stiffness of the system becomes critical. In such cases, implicit finite difference methods are useful. However, when the derivative function is nonlinear, nonlinear implicit FDEs result. The system of nonlinear implicit FDEs can be solved by Newton's method, Section 3.7.

7.14.3 Higher-Order Implicit Methods

The problems of stiffness illustrated in the previous sections occur for both single ODEs and systems of ODEs. When even the most rapid transient component of the solution is of interest, small time steps are required for accuracy as well as stability, and explicit finite difference methods can be used to generate the solution. However, when the effects of the rapid transients have decayed to insignificant levels, small time steps must still be

Table 7.19 Coefficients for the Gear FDEs

k	γ	β	α_0	α_{-1}	α_{-2}	α_{-3}	α_{-4}	α_{-5}
1	1	1	1					
2	1/3	2	4	-1				
3	1/11	6	18	-9	2			
4	1/25	12	48	-36	16	-3		
5	1/137	60	300	-300	200	-75	12	
6	1/147	60	360	-450	400	-225	72	-10

employed due to the stability limit, not accuracy requirements. In that case, implicit finite difference methods can be used to take larger time steps.

The implicit Euler FDE, Eq. (7.73), is unconditionally stable. However, it is only first-order accurate. The implicit trapezoid FDE, Eq. (7.141), is also unconditionally stable, but it is only second-order accurate. Higher-order methods are desirable.

Any of the Adams-Moulton FDEs can be used to devise a higher-order implicit method. However, the stability limits of these FDEs are quite restrictive when applied to stiff ODEs.

Gear (1971) has devised a series of implicit FDEs that has much larger stability limits. The Gear formulas are presented below:

$$y_{n+1} = \gamma(\beta h f_{n+1} + (\alpha_0 y_n + \alpha_{-1} y_{n-1} + \alpha_{-2} y_{n-2} + \cdots)) \tag{7.352}$$

where k denotes the global order of the FDE and the coefficients γ, β, and α_i ($i = 0, 1, 2, \ldots$) are presented in Table 7.19.

The Gear FDEs have been incorporated into a FORTRAN package named LSODE, which was developed at the Lawrence Livermore National Laboratory. This package has an elaborate error control procedure based on using various-order Gear FDEs in conjunction with step size halving and doubling.

7.14.4 Summary

Stiff ODEs are especially challenging problems. Explicit methods are generally unsuitable for solving stiff ODEs. Implicit methods, especially the family of Gear methods, are recommended for solving stiff ODEs.

7.15 PROGRAMS

Three FORTRAN subroutines for integrating initial-value ordinary differential equations are presented in this section:

1. The fourth-order Runge-Kutta method
2. The extrapolated modified midpoint method
3. The fourth-order Adams-Bashforth-Moulton method

The basic computational algorithms are presented as completely self-contained subroutines suitable for use in other programs. Input data and output statements are contained in a main (or driver) program written specifically to illustrate the use of each subroutine.

7.15.1 The Fourth-Order Runge-Kutta Method

The general algorithm for the fourth-order Runge-Kutta method is given by Eqs. (7.179) and (7.180):

$$y_{n+1} = y_n + \tfrac{1}{6}(\Delta y_1 + 2\,\Delta y_2 + 2\,\Delta y_3 + \Delta y_4) \tag{7.353}$$

$$\Delta y_1 = hf(t_n, y_n) \qquad \Delta y_2 = hf\left(t_n + \frac{h}{2}, y_n + \frac{\Delta y_1}{2}\right) \tag{7.354a}$$

$$\Delta y_3 = hf\left(t_n + \frac{h}{2}, y_n + \frac{\Delta y_2}{2}\right) \qquad \Delta y_4 = hf(t_n + h, y_n + \Delta y_3) \tag{7.354b}$$

A FORTRAN subroutine, *subroutine rk*, for implementing the fourth-order Runge-Kutta method, Eqs. (7.353) and (7.354), is presented in Program 7.1. *Program main* defines the data set and prints it, calls *subroutine rk* to implement the solution, and prints the solution. A FORTRAN function, *function f*, specifies the derivative function, $y' = f(t, y)$.

Program 7.1   The fourth-order Runge-Kutta method program.

```
      program main
c     main program to illustrate ODE solvers
c     ndim  array dimension, ndim = 101 in this example
c     nmax  number of integration steps
c     t     independent variable array, t(n)
c     y     dependent variable array, y(n)
c     yp    derivative function array, yp(n)
c     iw    intermediate output flag: 0 no, 1 yes
c     t(1)  initial value of t, t0
c     y(1)  initial value of y, y0
c     dt    time step
      dimension t(101),y(101),yp(101)
      data ndim,nmax,n,iw,dt / 101, 11, 1, 1, 1.0 /
      data t(1) / 0.0 /
      data y(1) / 2500.0 /
      write (6,1000)
      if (iw.eq.0) write (6,1010)
      if (iw.eq.1) write (6,1020)
      write (6,1030) n,t(1),y(1)
      do n=1,nmax-1
         call rk (ndim,n,dt,t,y,iw)
         write (6,1030) n+1,t(n+1),y(n+1)
      end do
      stop
 1000 format (' Fourth-order Runge-Kutta'/' '/'  n',4x,'tn',9x,'yn',
 1010 format (' ')
 1020 format (18x,'dy1',11x,'dy2',11x,'dy3',11x,'dy4'/' ')
 1030 format (i3,f8.3,2f15.8)
      end


      subroutine rk (ndim,n,dt,t,y,iw)
c     implements the fourth-order Runge-Kutta method
      dimension t(ndim),y(ndim)
      dy1=dt*f(t(n),y(n))
      dy2=dt*f(t(n)+dt/2.0,y(n)+dy1/2.0)
      dy3=dt*f(t(n)+dt/2.0,y(n)+dy2/2.0)
      dy4=dt*f(t(n)+dt,y(n)+dy3)
      y(n+1)=y(n)+(dy1+2.0*(dy2+dy3)+dy4)/6.0
      t(n+1)=t(n)+dt
      if (iw.eq.1) write (6,1000) dy1,dy2,dy3,dy4
      return
 1000 format (11x,f15.8,3f14.8)
      end
```

```
      function f(t,y)
c     derivative function
      alpha=4.0e-12
      f=-alpha*(y**4-250.0**4)
      return
      end
```

The data set used to illustrate *subroutine rk* is taken from Example 7.11. The output generated by the fourth-order Runge-Kutta program is presented in Output 7.1.

Output 7.1   Solution by the fourth-order Runge-Kutta method

```
Fourth-order Runge-Kutta
```

n	tn	yn			
		dy1	dy2	dy3	dy4
1	0.000	2500.00000000			
		-156.23437500	-137.60150356	-139.73128134	-124.12267500
2	1.000	2360.82956337			
		-124.24070754	-111.66970570	-112.89627716	-102.12386006
3	2.000	2248.24680781			
		. . . .			
9	8.000	1842.09419793			
		-46.04261410	-43.78299210	-43.89190534	-41.80727846
10	9.000	1798.22758336			
		-41.80963138	-39.89837091	-39.98428393	-38.21187310
11	10.000	1758.26311433			

7.15.2   The Extrapolated Modified Midpoint Method

The general algorithm for the extrapolated modified midpoint method is given by Eqs. (7.210) to (7.212):

$$z_0 = y_n \tag{7.355a}$$

$$z_1 = z_0 + hf(t_n, z_0) \tag{7.355b}$$

$$z_i = z_{i-2} + 2hf(t_n + (i-1)h, z_{i-1}) \qquad (i = 2, \ldots, M) \tag{7.355c}$$

$$y_{n+1} = \tfrac{1}{2}[z_{M-1} + z_M + hf(t_n + \Delta t, z_M)] \tag{7.356}$$

$$\text{IV} = \text{MAV} + \frac{\text{MAV} - \text{LAV}}{2^n - 1} = \frac{2^n \text{MAV} - \text{LAV}}{2^n - 1} \tag{7.357}$$

A FORTRAN subroutine, *subroutine midpt*, for implementing the extrapolated modified midpoint method is presented in Program 7.2. *Subroutine midpt* works essentially like *subroutine rk* discussed in Section 7.15.1, except the extrapolated modified midpoint method is used instead of the fourth-order Runge-Kutta method. *Program main* defines the data set and prints it, calls *subroutine midpt* to implement the extrapolated modified midpoint method, and prints the solution. Only the statements in *program main* which are different from the statements in *program main* in Section 7.15.1 are presented.

Program 7.2   The extrapolated modified midpoint method program

```
      program main
c     main program to illustrate ODE solvers
      data ndim,nmax,n,iw,dt / 101, 6, 1, 1, 2.0 /
         call midpt (ndim,n,dt,t,y,iw)
 1000 format (' Extrapolated mod. midpoint method'/' '/'  n',4x,'tn',
     1 5x,'yn,  0(h**2)',6x,'0(h**4)',7x,'0(h**6)',7x,'0(h**8)'/' ')
 1030 format (i3,f8.3,f15.8,f14.8)
      end


      subroutine midpt (ndim,n,dt,t,y,iw)
c     implements the extrapolated modified midpoint method
c     kmax   number of segmentations, M, within each interval
      dimension t(ndim),y(ndim),w(4,4),z(33),g(33)
      kmax=4
      jmax=2
      dtk=dt
c     calculate w(1,k)=y(n+1) for k=kmax segments
      z(1)=y(n)
      g(1)=f(t(n),y(n))
      do k=1,kmax
         dtk=dtk/2.0
         jmax=2*jmax-1
         z(2)=z(1)+dtk*g(1)
         g(2)=f(t(n)+dtk,z(2))
         do j=3,jmax
           tj=t(n)+float(j-1)*dtk
           z(j)=z(j-2)+2.0*dtk*g(j-1)
           g(j)=f(tj,z(j))
         end do
         w(1,k)=0.5*(z(jmax-1)+z(jmax)+dtk*g(jmax))
      end do
c     extrapolation
      do k=2,kmax
         c=2.0**(2.0*float(k-1))
         do j=1,kmax+1-k
            w(k,j)=(c*w(k-1,j+1)-w(k-1,j))/(c-1)
         end do
      end do
      if (iw.eq.1) then
         do k=1,kmax
            write (6,1000) (w(j,k),j=1,kmax+1-k)
         end do
      end if
      t(n+1)=t(n)+dt
      y(n+1)=w(kmax,1)
      return
 1000 format (11x,f15.8,3f14.8)
      end
```

```
      function f(t,y)
c     derivative function
      end
```

The data set used to illustrate *subroutine midpt* is taken from Example 7.12. The output generated by the extrapolated modified midpoint method program is presented in Output 7.2.

Output 7.2   Solution by the extrapolated modified midpoint method

```
Extrapolated mod. midpoint method

  n    tn     yn, O(h**2)         O(h**4)         O(h**6)         O(h**8)

  1   0.000   2500.00000000
                2249.15523693  2248.26188883  2248.24740700  2248.24731430
                2248.48522585  2248.24831212  2248.24731574
                2248.30754055  2248.24737802
                2248.26241865
  2   2.000   2248.24731430
                2075.00025328  2074.61500750  2074.61190855  2074.61189807
                2074.71131895  2074.61210224  2074.61189824
                2074.63690642  2074.61191099
                2074.61815984
  3   4.000   2074.61189807
      . . . . . . . . . . . . . . . . . . . . . . . . . . . . . . . . . . . . . . . .
  5   8.000   1842.09450797
                1758.33199904  1758.26353381  1758.26337496  1758.26337480
                1758.28065012  1758.26338489  1758.26337480
                1758.26770120  1758.26337543
                1758.26445688
  6  10.000   1758.26337480
```

7.15.3   The Fourth-Order Adams-Bashforth-Moulton Method

The general algorithm for the fourth-order Adams-Bashforth-Moulton method is given by Eq. (7.236):

$$y_{n+1}^P = y_n + \frac{h}{24}(55f_n - 59f_{n-1} + 37f_{n-2} - 9f_{n-3}) \tag{7.358a}$$

$$y_{n+1}^C = y_n + \frac{h}{24}(9f_{n+1}^P + 19f_n - 5f_{n-1} + f_{n-2}) \tag{7.358b}$$

A FORTRAN subroutine, *subroutine abm*, for implementing the procedure is presented in Program 7.3. *Subroutine abm* works essentially like *subroutine rk* discussed in Section 7.15.1, except the fourth-order Adams-Bashforth-Moulton method is used instead of the fourth-order Runge-Kutta method. *Program main* defines the data set and prints it, calls *subroutine abm* to implement the solution, and prints the solution. Only the statements in *program main* which are different from the statements in *program main* in Section 7.15.1 are presented.

Program 7.3   The fourth-order Adams-Bashforth-Moulton method program

```
      program main
c     main program to illustrate ODE solvers
      data ndim,nmax,n,iw,dt / 101, 11, 1, 1, 1.0 /
      data (t(n),n=1,4) / 0.0, 1.0, 2.0, 3.0 /
      data (y(n),n=1,4) / 2500.0, 2360.82998845, 2248.24731405,
     1                    2154.47079576 /
      data (yp(n),n=1,4) / -156.23437500, -124.24079704,
     1                     -102.18094603, -86.16753966 /
      do n=1,4
         write (6,1030) n,t(n),y(n),yp(n)
      end do
      do n=4,nmax-1
         call abm (ndim,n,dt,t,y,yp,iw)
         write (6,1030) n+1,t(n+1),y(n+1),yp(n+1)
      end do
 1000 format (' Adams-B-M method'/' '/'n',5x,'tn',9x,'yn',13x,'fn')
 1020 format (16x,'yPred',10x,'fPred'/' ')
      end
      subroutine abm (ndim,n,dt,t,y,yp,iw)
c     the fourth-order Adams-Bashforth-Moulton method
      dimension t(ndim),y(ndim),yp(ndim)
      ypred=y(n)+dt*(55.0*yp(n)-59.0*yp(n-1)+37.0*yp(n-2)
     1 -9.0*yp(n-3))/24.0
      fpred=f(t(n)+dt, ypred)
      y(n+1)=y(n)+dt*(9.*fpred+19.*yp(n)-5.*yp(n-1)+yp(n-2))/24.0
      t(n+1)=t(n)+dt
      yp(n+1)=f(t(n+1),y(n+1))
      if (iw.eq.1) write (6,1000) ypred,fpred
      return
 1000 format (11x,2f15.8)
      end

      function f(t,y)
c     derivative function
      end
```

The data set used to illustrate *subroutine abm* is taken from Example 7.14. The output generated by the fourth-order Adams-Bashforth-Moulton program is presented in Output 7.3.

Output 7.3   Solution by the fourth-order Adams-Bashforth-Moulton method

```
Adams-B-M method

n      tn        yn              fn
                 yPred           fPred

  1    0.000   2500.00000000   -156.23437500
  2    1.000   2360.82998845   -124.24079704
  3    2.000   2248.24731405   -102.18094603
```

3	2.000	2248.24731405	-102.18094603
4	3.000	2154.47079576	-86.16753966
		2075.24833822	-74.17350708
5	4.000	2074.55075892	-74.07380486
		2005.68816487	-64.71557210
6	5.000	2005.33468855	-64.66995205
		. . . .	
10	9.000	1798.14913834	-41.80233360
		1758.21558333	-38.20946276
11	10.000	1758.18932752	-38.20717951

7.15.4   Packages for Integrating Initial-Value ODEs

Numerous libraries and software packages are available for integrating initial-value ordinary differential equations. Many work stations and main frame computers have such libraries attached to their operating systems.

Many commercial software packages contain algorithms for integrating initial-value ODEs. Some of the more prominent packages are Matlab and Mathcad. More sophisticated packages, such as IMSL, MATHEMATICA, MACSYMA, and MAPLE, also contain algorithms for integrating initial-value ODEs. Finally, the book *Numerical Recipes* [Press et al. (1989)] contains numerous subroutines for integrating initial-value ordinary differential equations.

7.16   SUMMARY

Methods for solving one-dimensional initial-value ordinary differential equations are presented in this chapter. Procedures for discretizing the continuous solution domain, representing exact derivatives by finite difference approximations, and developing finite difference equations are discussed. The concepts of consistency, order, stability, and convergence are defined and discussed. A procedure for investigating consistency and determining order by developing and analyzing a modified differential equation is presented. A procedure for determining stability criteria by analyzing the amplification factor G, the single-step exact solution of a finite difference equation, is presented.

Three types of methods for solving initial-value ODEs are presented:

1. Single-point methods
2. Extrapolation methods
3. Multipoint methods

These methods can be used to solve single first-order ODEs, higher-order ODEs, and systems of first-order ODEs.

Finite difference equations of any order can be developed for all three types of methods. Generally speaking, fourth-order methods are the best compromise between accuracy and simplicity for the single-point and multipoint methods. The extrapolated modified midpoint method can be extended easily to higher order.

The fourth-order Runge-Kutta method is an excellent general purpose single-point method, which works well for both smoothly-varying and nonsmoothly-varying problems. However, error estimation and error control can be expensive, so it may not be the most efficient method.

The extrapolated modified midpoint method is an excellent method for both smoothly varying problems and nonsmoothly varying problems. It can give extremely accurate results. Error estimation and error control are straightforward and efficient. However, for nonsmoothly varying problems, the Runge-Kutta method may be more straightforward.

Multipoint methods, such as the fourth-order Adams-Bashforth-Moulton method, work well for smoothly varying problems. Error estimation and error control are straightforward and efficient. However, for nonsmoothly varying problems, the Runge-Kutta method generally behaves better. Even for smoothly varying problems, the extrapolated modified midpoint method is generally more efficient than a multipoint method.

Stiff initial-value ODEs present an especially difficult problem. The Gear family of FDEs is recommended for stiff ODEs. For nonlinear stiff ODEs, the resulting nonlinear FDEs must be solved iteratively, usually by Newton's method. Although expensive, this procedure gives good results.

Any of the methods presented in this chapter for solving initial-value ODEs can be used to solve any initial-value ODE. The choice of a method for a particular problem depends on both the characteristics of the problem itself and the personal preference of the analyst.

After studying Chapter 7, you should be able to:

1. Describe the general features of initial-value ordinary differential equations
2. Discuss the general features of the linear first-order ODE, including the complimentary solution and the particular solution
3. Discuss the general features of a nonlinear first-order ODE
4. Linearize a nonlinear first-order ODE
5. Explain the difference between a stable ODE and an unstable ODE
6. Explain the concept of a family of solutions of an ODE and how a particular member is chosen
7. Describe how higher-order ODEs and systems of first-order ODEs can be solved using the procedures for solving a single first-order ODE
8. Explain the relationship between the solution of time-marching propagation problems and space marching propagation problems
9. Explain and implement the Taylor series method
10. Explain the objective of a finite difference method for solving an ODE
11. Describe the steps in the finite difference solution of an ODE
12. Discretize a continuous solution domain into a discrete finite difference grid
13. Develop a finite difference approximation of an exact derivative by the Taylor series approach
14. Explain how to develop an explicit FDE and an implicit FDE
15. Describe the effect of truncation error on the solution of an ODE
16. Derive and use the first-order explicit Euler method
17. Derive and use the first-order implicit Euler method
18. Explain the relative advantages and disadvantages of the explicit and implicit Euler methods
19. Define and discuss the concept of consistency
20. Define and discuss the concept of order
21. Define and discuss the concept of stability
22. Define and discuss the concept of convergence

23. Derive the modified differential equation (MDE) corresponding to a FDE
24. Analyze the MDE to determine consistency and order of a FDE
25. Develop the amplification factor, G, for a FDE
26. Analyze the amplification factor, G, to determine the stability criterion for a FDE
27. Determine whether or not a finite difference method is convergent
28. Explain the concept of a single-point method
29. Derive and use the midpoint and modified midpoint methods
30. Derive and use the trapezoid and modified Euler methods
31. Explain the concepts underlying Runge-Kutta methods
32. Apply the fourth-order Runge-Kutta method
33. Derive and apply error estimation and error control methods for single-point methods
34. Explain the concept underlying extrapolation methods
35. Derive and apply the extrapolated modified midpoint method
36. Explain the concepts underlying multipoint methods
37. Derive an explicit multipoint method
38. Derive an implicit multipoint method
39. Derive and apply the fourth-order Adams-Bashforth-Moulton method
40. Derive and apply error estimation, error control, and extrapolation methods for multipoint methods
41. Discuss the relative advantages and disadvantages of single-point methods, extrapolation methods, and multipoint methods
42. Apply time linearization to solve a nonlinear implicit FDE
43. Apply Newton's method to solve a nonlinear implicit FDE
44. Reduce a higher-order ODE to a system of first-order ODEs
45. Solve a system of first-order ODEs
46. Explain the concept of a stiff ODE
47. Discuss the problems arising in the solution of stiff ODEs
48. Describe and apply the Gear method for solving stiff ODEs
49. Choose and implement a finite difference method for solving initial-value ODEs

EXERCISE PROBLEMS

7.1 Introduction

1. Derive the exact solution of the radiation problem presented in Eq. (7.3).
2. Use the secant method to solve Eq. (7.3) for the times presented in Table 7.1.

7.2 General Features of Initial-Value ODEs

3. Derive the exact solution of Eq. (7.5).
4. Derive the exact solution of the following ODE:

$$\bar{y}' = a\bar{y} + b + ct + dt^2 \qquad \bar{y}(t_0) = \bar{y}_0$$

5. Derive the exact solution of Eq. (7.12). Let $\bar{y}(0,0) = 1.0$. Plot the solution from $t = 0.0$ to 5.0 for $\alpha = 1.0$ and 10.0.
6. Derive the exact solution of Eq. (7.13). Let $\bar{y}(0,0) = 1.0$. Plot the solution from $t = 0.0$ to 5.0 for $\alpha = 1.0$ and 10.0.
7. Express the following ODE in the linearized form of Eq. (7.21) and identify α.

$$\bar{y}' = \bar{y}^2 \sin t + 10 \qquad \bar{y}(0) = \bar{y}_0$$

8. Express the following ODE in the linearized form of Eq. (7.21) and identify α.

$$\bar{y}' = \bar{y}^3 t + t^2 \qquad \bar{y}(0) = \bar{y}_0$$

7.3 The Taylor Series Method

9. Solve the example radiation problem presented in Section 7.1 by the Taylor series method including (a) the fifth derivative term and (b) the sixth derivative term. Compare the results with the results presented in Table 7.2.
10. Solve the following ODE by the Taylor series method including the fourth derivative term for $t = 0.0$ to 10.0 at intervals of $\Delta t = 1.0$. Compare the results with the exact solution.

$$\bar{y}' = t + \bar{y} \qquad \bar{y}(0) = 1$$

11. Solve the following ODE by the Taylor series method including the fourth derivative term for $t = 0.0$ to 10.0 at intervals of $\Delta t = 1.0$. Compare the results with the exact solution.

$$\bar{y}' = t - \bar{y} \qquad \bar{y}(0) = 1$$

7.4 The Finite Difference Method

Finite Difference Approximations

12. Using the Taylor series approach, derive the following finite difference approximations (FDAs) of $\bar{y}' = d\bar{y}/dt$, including the leading truncation error term: (a) $y'|_n + 0(\Delta t)$, (b) $y'|_{n+1} + 0(\Delta t)$, (c) $y'|_n + 0(\Delta t^2)$, and (d) $y'|_{n+1/2} + 0(\Delta t^2)$.
13. Using Taylor series, determine what derivative is represented by the following finite difference approximation (FDA) and the leading truncation error term.

$$\text{FDA} = \frac{2y_{n+1} + 3y_n - 6y_{n-1} + y_{n-2}}{6 \, \Delta t}$$

The problems in Sections 7.5 and 7.7 to 7.9 are concerned with solving initial-value ODEs by finite difference methods. Those problems are concerned with the ODEs presented below, along with their exact solutions. Carry at least six digits after the decimal place in all calculations. For all problems, compare the errors for the two step sizes and calculate the ratio of the errors at $t = 1.0$.

$$\bar{y}' = 2 - 2t + 4t^2 - 4t^3 - 4t^4, \quad \bar{y}(0) = \bar{y}_0 = 1,$$
$$\bar{y}(t) = \bar{y}_0 + 2t - t^2 + \tfrac{4}{3}t^3 - t^4 - \tfrac{4}{5}t^5 \tag{A}$$

$$\bar{y}' = 1 - \bar{y}, \quad \bar{y}(0) = \bar{y}_0 = 0, \quad \bar{y}(t) = (\bar{y}_0 - 1)e^t + 1 \tag{B}$$

$$\bar{y}' = t + \bar{y}, \quad \bar{y}(0) = \bar{y}_0 = 1, \quad \bar{y}(t) = (\bar{y}_0 + 1)e^t - t - 1 \tag{C}$$

$$\bar{y}' = e^{-t} + \bar{y}, \quad \bar{y}(0) = \bar{y}_0 = 0, \quad \bar{y}(t) = (\bar{y}_0 + \tfrac{1}{2})e^t - \frac{e^{-t}}{2} \tag{D}$$

$$\bar{y}' = t^2\bar{y}, \quad \bar{y}(0) = \bar{y}_0 = 1, \quad \bar{y}(t) = \bar{y}_0 e^{t^3/3} \tag{E}$$

$$\bar{y}' = 2\sin t + \bar{y}, \quad \bar{y}(0) = \bar{y}_0 = 1, \quad \bar{y}(t) = e^t(\bar{y}_0 + 1) - \sin t - \cos t \tag{F}$$

$$\bar{y}' = 2\cos t + \bar{y}, \quad \bar{y}(0) = \bar{y}_0 = 1, \quad \bar{y}(t) = e^t(\bar{y}_0 - 1) + \sin t - \cos t \tag{G}$$

$$\bar{y}' = 1 + 0.5\bar{y}^2, \quad \bar{y}(0) = \bar{y}_0 = 0.5, \quad \bar{y}(t) = \frac{1}{\sqrt{0.5}}\tan[\sqrt{0.5}t + \tan^{-1}(\sqrt{0.5}\bar{y}_0)] \tag{H}$$

$$\bar{y}' = t\sqrt{\bar{y}}, \quad \bar{y}(0) = \bar{y}_0 = 1, \quad \bar{y}(t) = \tfrac{1}{4}\left(\frac{t}{2} + 2\sqrt{\bar{y}_0}\right)^2 \tag{I}$$

$$\bar{y}' = t\bar{y}^2, \quad \bar{y}(0) = \bar{y}_0 = 1, \quad \bar{y}(t) = -\frac{2}{t^2 - 2/\bar{y}_0} \tag{J}$$

$$\bar{y}' = \frac{1}{t + \bar{y}}, \quad \bar{y}(0) = \bar{y}_0 = 1, \quad e^{(y - \bar{y}_0)}(\bar{y}_0 + 1) - \bar{y} - 1 = t \tag{K}$$

An infinite variety of additional problems can be obtained from Eqs. (A) to (K) by (a) changing the coefficients in the ODEs, (b) changing the initial conditions, (c) changing the integration step size, (d) changing the range of integration, and (e) combinations of the above changes.

7.5 The First-Order Euler Methods

The Explicit Euler Method

14. Solve ODE (A) by the explicit Euler method from $t = 0.0$ to 1.0 with $\Delta t = 0.2$ and 0.1.

15.* Solve ODE (B) by the explicit Euler method from $t = 0.0$ to 1.0 with $\Delta t = 0.2$ and 0.1.

16. Solve ODE (C) by the explicit Euler method from $t = 0.0$ to 1.0 with $\Delta t = 0.2$ and 0.1.

17. Solve ODE (D) by the explicit Euler method from $t = 0.0$ to 1.0 with $\Delta t = 0.2$ and 0.1.

18. Solve ODE (E) by the explicit Euler method from $t = 0.0$ to 1.0 with $\Delta t = 0.2$ and 0.1.

19. Solve ODE (F) by the explicit Euler method from $t = 0.0$ to 1.0 with $\Delta t = 0.2$ and 0.1.

20. Solve ODE (G) by the explicit Euler method from $t = 0.0$ to 1.0 with $\Delta t = 0.2$ and 0.1.

21.* Solve ODE (H) by the explicit Euler method from $t = 0.0$ to 1.0 with $\Delta t = 0.2$ and 0.1.

22. Solve ODE (I) by the explicit Euler method from $t = 0.0$ to 1.0 with $\Delta t = 0.2$ and 0.1.
23. Solve ODE (J) by the explicit Euler method from $t = 0.0$ to 1.0 with $\Delta t = 0.2$ and 0.1.
24. Solve ODE (K) by the explicit Euler method from $t = 0.0$ to 1.0 with $\Delta t = 0.2$ and 0.1.

The Implicit Euler Method

25. Solve ODE (A) by the implicit Euler method from $t = 0.0$ to 1.0 with $\Delta t = 0.2$ and 0.1.
26.* Solve ODE (B) by the implicit Euler method from $t = 0.0$ to 1.0 with $\Delta t = 0.2$ and 0.1.
27. Solve ODE (C) by the implicit Euler method from $t = 0.0$ to 1.0 with $\Delta t = 0.2$ and 0.1.
28. Solve ODE (D) by the implicit Euler method from $t = 0.0$ to 1.0 with $\Delta t = 0.2$ and 0.1.
29. Solve ODE (E) by the implicit Euler method from $t = 0.0$ to 1.0 with $\Delta t = 0.2$ and 0.1.
30. Solve ODE (F) by the implicit Euler method from $t = 0.0$ to 1.0 with $\Delta t = 0.2$ and 0.1.
31. Solve ODE (G) the implicit Euler method from $t = 0.0$ to 1.0 with $\Delta t = 0.2$ and 0.1.

7.6 Consistency, Order, Stability, and Convergence

Consistency and Order

32. Develop the explicit Euler approximation of the linear first-order ODE, $\bar{y}' + \alpha\bar{y} = F(t)$. Derive the corresponding modified differential equation (MDE), including the leading truncation error term. Investigate consistency and order.
33. Solve Problem 32 for the implicit Euler method.
34. Solve Problem 32 for the implicit midpoint method.
35. Solve Problem 32 for the modified midpoint method.
36. Solve Problem 32 for the implicit trapezoid method.
37. Solve Problem 32 for the modified Euler method.
38. Solve Problem 32 for the fourth-order Runge-Kutta method.

The order of a finite difference equation is the order of the leading truncation error term. Order can be estimated numerically by solving an ODE that has an exact solution for two different steps sizes and comparing the ratio of the errors. For step size halving, the ratio of the errors, as $\Delta t \to 0$, is given by

$$\frac{\text{Error}(h)}{\text{Error}(h/2)} = 2^n$$

where n is the order of the FDE. For ODEs that do not have an exact solution, order can be estimated by solving the ODE numerically for three step sizes, each one being one-half the previous one, letting the most accurate solution (the solution for the smallest step size) be an approximation of the exact solution, and applying the procedure described above.

39.* Apply the above procedure to Eq. (H) solved by the explicit Euler method.

40.* Apply the above procedure to Eq. (H) solved by the modified Euler method.

41. Apply the above procedure to Eq. (H) solved by the fourth-order Runge-Kutta method.

42. Apply the above procedure to Eq. (H) solved by the Adams-Bashforth-Moulton method.

43. Apply the above procedure to Eq. (J) solved by the explicit Euler method.

44. Apply the above procedure to Eq. (J) solved by the modified Euler method.

45. Apply the above procedure to Eq. (J) solved by the fourth-order Runge-Kutta method.

46. Apply the above procedure to Eq. (J) solved by the Adams-Bashforth-Moulton method.

Stability

47. Perform a stability analysis of the explicit Euler FDE, Eq. (7.59).

48. Perform a stability analysis of the implicit Euler FDE, Eq. (7.73).

49. Perform a stability analysis of the implicit midpoint FDE, Eq. (7.119).

50. Perform a stability analysis for the modified midpoint FDEs, Eqs. (7.120) and (7.121). The predictor and corrector FDEs must be combined into a single-step FDE.

51. Perform a stability analysis of the implicit trapezoid FDE, Eq. (7.141).

52. Perform a stability analysis of the modified Euler FDEs, Eq. (7.142) and (7.143). The predictor and corrector FDEs must be combined into a single-step FDE.

53. Perform a stability analysis of the fourth-order Runge-Kutta FDEs, Eq. (7.179) and (7.180). The four-step FDEs must be combined into a single-step FDE.

54. Perform a stability analysis of the fourth-order Adams-Bashforth FDE, Eq. (7.229).

55. Perform a stability analysis of the nth-order Adams-Bashforth FDEs, Eq. (7.250), for (a) $n = 1$, (b) $n = 2$, and (c) $n = 3$ (see Table 7.11).

56. Perform a stability analysis of the fourth-order Adams-Moulton FDE, Eq. (7.235).

57. Perform a stability analysis of the nth-order Adams-Moulton FDEs, Eq. (7.252), for (a) $n = 1$, (b) $n = 2$, and (c) $n = 3$ (see Table 7.12).

7.7 Single-Point Methods

Second-Order Single-Point Methods

58. Solve ODE (A) by the modified midpoint method.

59.* Solve ODE (B) by the modified midpoint method.

60. Solve ODE (C) by the modified midpoint method.

61. Solve ODE (D) by the modified midpoint method.

62. Solve ODE (E) by the modified midpoint method.

63. Solve ODE (F) by the modified midpoint method.

64. Solve ODE (G) by the modified midpoint method.

65.* Solve ODE (H) by the modified midpoint method.

66. Solve ODE (I) by the modified midpoint method.

67. Solve ODE (J) by the modified midpoint method.

68. Solve ODE (K) by the modified midpoint method.
69. Solve ODE (A) by the modified Euler method.
70.* Solve ODE (B) by the modified Euler method.
71. Solve ODE (C) by the modified Euler method.
72. Solve ODE (D) by the modified Euler method.
73. Solve ODE (E) by the modified Euler method.
74. Solve ODE (F) by the modified Euler method.
75. Solve ODE (G) by the modified Euler method.
76.* Solve ODE (H) by the modified Euler method.
77. Solve ODE (I) by the modified Euler method.
78. Solve ODE (J) by the modified Euler method.
79. Solve ODE (K) by the modified Euler method.

Runge-Kutta Methods

80. Derive the general second-order Runge-Kutta method, Eqs. (7.158) and (7.169).
81. Derive the general third-order Runge-Kutta method:

$$y_{n+1} = y_n + C_1 k_1 + C_2 k_2 + C_3 k_3$$

Show that one such method is given by

$$y_{n+1} = y_n + \frac{(k_1 + 3k_2 + k_3)}{4} \qquad k_1 = \Delta t \, f(t_n, y_n)$$

$$k_2 = \Delta t \, f\left(t_n + \frac{\Delta t}{3}, y_n + \frac{k_1}{3}\right) \qquad k_3 = \Delta t \, f\left(t_n + \frac{2\,\Delta t}{3}, y_n + \frac{2k_2}{3}\right)$$

Fourth-Order Runge-Kutta Method

82. Derive the general fourth-order Runge-Kutta method. Show that Eqs. (7.179) and (7.180) comprise one such method.
83. Solve ODE (A) by the fourth-order Runge-Kutta method.
84.* Solve ODE (B) by the fourth-order Runge-Kutta method.
85. Solve ODE (C) by the fourth-order Runge-Kutta method.
86. Solve ODE (D) by the fourth-order Runge-Kutta method.
87. Solve ODE (E) by the fourth-order Runge-Kutta method.
88. Solve ODE (F) by the fourth-order Runge-Kutta method.
89. Solve ODE (G) by the fourth-order Runge-Kutta method.
90.* Solve ODE (H) by the fourth-order Runge-Kutta method.
91. Solve ODE (I) by the fourth-order Runge-Kutta method.
92. Solve ODE (J) by the fourth-order Runge-Kutta method.
93. Solve ODE (K) by the fourth-order Runge-Kutta method.

Runge-Kutta Methods with Error Estimation

94.* Solve ODE (J) by the Runge-Kutta-Fehlberg method, Eqs. (7.202) to (7.204). Compare the results with the results of Problem 92. Evaluate the error at each step.
95. Solve ODE (K) by the Runge-Kutta-Fehlberg method, Eqs. (7.202) to (7.204). Compare the results with the results of Problem 93. Evaluate the error at each step.

96. The Runge-Kutta-Merson method is as follows:

$$y_{n+1} = y_n + \tfrac{1}{6}(k_1 + 4k_4 + k_5), \quad \text{Error} = \tfrac{1}{30}(2k_1 - 9k_3 + 8k_4 - k_5),$$
$$k_1 = \Delta t\, f(t_n, y_n)$$
$$k_2 = \Delta t\, f(t_n + \tfrac{1}{3}\Delta t, y_n + \tfrac{1}{3}k_1),$$
$$k_3 = \Delta t\, f(t_n + \tfrac{1}{6}\Delta t, y_n + \tfrac{1}{6}k_1 + \tfrac{1}{6}k_2)$$
$$k_4 = \Delta t\, f(t_n + \tfrac{1}{2}\Delta t, y_n + \tfrac{1}{8}k_1 + \tfrac{3}{8}k_3),$$
$$k_5 = \Delta t\, f(t_n + \Delta t, y_n + \tfrac{1}{2}k_1 - \tfrac{3}{2}k_3 + 2k_4)$$

97. Solve ODE (J) by the Runge-Kutta-Merson method. Compare the results with the results of Problems 92 and 94.

98. Solve ODE (K) by the Runge-Kutta-Merson method. Compare the results with the results of Problems 93 and 95.

7.8 Extrapolation Methods

99. Solve Eq. (A) by the extrapolated modified midpoint method for $\Delta t = 0.2$ and $M = 2, 4, 8$, and 16.

100.* Solve Eq. (B) by the extrapolated modified midpoint method for $\Delta t = 0.2$ and $M = 2, 4, 8$, and 16.

101. Solve Eq. (C) by the extrapolated modified midpoint method for $\Delta t = 0.2$ and $M = 2, 4, 8$, and 16.

102. Solve Eq. (D) by the extrapolated modified midpoint method for $\Delta t = 0.2$ and $M = 2, 4, 8$, and 16.

103. Solve Eq. (E) by the extrapolated modified midpoint method for $\Delta t = 0.2$ and $M = 2, 4, 8$, and 16.

104. Solve Eq. (F) by the extrapolated modified midpoint method for $\Delta t = 0.2$ and $M = 2, 4, 8$, and 16.

105. Solve Eq. (G) by the extrapolated modified midpoint method for $\Delta t = 0.2$ and $M = 2, 4, 8$, and 16.

106.* Solve Eq. (H) by the extrapolated modified midpoint method for $\Delta t = 0.2$ and $M = 2, 4, 8$, and 16.

107. Solve Eq. (I) by the extrapolated modified midpoint method for $\Delta t = 0.2$ and $M = 2, 4, 8$, and 16.

108. Solve Eq. (J) by the extrapolated modified midpoint method for $\Delta t = 0.2$ and $M = 2, 4, 8$, and 16.

109. Solve Eq. (K) by the extrapolated modified midpoint method for $\Delta t = 0.2$ and $M = 2, 4, 8$, and 16.

7.9 Multipoint Methods

110. Derive the fourth-order Adams-Bashforth FDE, Eq. (7.229), with the leading truncation error term.

111. Derive the nth-order Adams-Bashforth FDEs, Eq. (7.250), for (a) $n = 1$, (b) $n = 2$, and (c) $n = 3$.

112. Derive the fourth-order Adams-Moulton FDE, Eq. (7.235), with the leading truncation error term.

113. Derive the nth-order Adams-Moulton FDEs, Eq. (7.252), for (a) $n = 1$, (b) $n = 2$, and (c) $n = 3$.
114. Develop the error estimation formulas for the fourth-order Adams-Bashforth-Moulton method, Eqs. (7.257) and (7.258).

In Problems 115 to 136, use the exact solution for starting values. Compare the results in Problems 126 to 136 with the results for the corresponding ODEs in Problems 115 to 125.

115. Solve ODE (A) by the fourth-order Adams-Bashforth-Moulton method.
116.* Solve ODE (B) by the fourth-order Adams-Bashforth-Moulton method.
117. Solve ODE (C) by the fourth-order Adams-Bashforth-Moulton method.
118. Solve ODE (D) by the fourth-order Adams-Bashforth-Moulton method.
119. Solve ODE (E) by the fourth-order Adams-Bashforth-Moulton method.
120. Solve ODE (F) by the fourth-order Adams-Bashforth-Moulton method.
121. Solve ODE (G) by the fourth-order Adams-Bashforth-Moulton method.
122.* Solve ODE (H) by the fourth-order Adams-Bashforth-Moulton method.
123. Solve ODE (I) by the fourth-order Adams-Bashforth-Moulton method.
124. Solve ODE (J) by the fourth-order Adams-Bashforth-Moulton method.
125. Solve ODE (K) by the fourth-order Adams-Bashforth-Moulton method.
126. Solve ODE (A) by the fourth-order Adams-Bashforth-Moulton method with mop up.
127. Solve ODE (B) by the fourth-order Adams-Bashforth-Moulton method with mop up.
128. Solve ODE (C) by the fourth-order Adams-Bashforth-Moulton method with mop up.
129. Solve ODE (D) by the fourth-order Adams-Bashforth-Moulton method with mop up.
130. Solve ODE (E) by the fourth-order Adams-Bashforth-Moulton method with mop up.
131. Solve ODE (F) by the fourth-order Adams-Bashforth-Moulton method with mop up.
132. Solve ODE (G) by the fourth-order Adams-Bashforth-Moulton method with mop up.
133.* Solve ODE (H) by the fourth-order Adams-Bashforth-Moulton method with mop up.
134. Solve ODE (I) by the fourth-order Adams-Bashforth-Moulton method with mop up.
135. Solve ODE (J) by the fourth-order Adams-Bashforth-Moulton method with mop up.
136. Solve ODE (K) by the fourth-order Adams-Bashforth-Moulton method with mop up.

In Problems 137 to 147, use the fourth-order Runge-Kutta method to obtain starting values. Compare the results with the results for the corresponding ODEs in Problems 115 to 125.

137. Solve ODE (A) by the fourth-order Adams-Bashforth-Moulton method.
138. Solve ODE (B) by the fourth-order Adams-Bashforth-Moulton method.
139. Solve ODE (C) by the fourth-order Adams-Bashforth-Moulton method.
140. Solve ODE (D) by the fourth-order Adams-Bashforth-Moulton method.

141. Solve ODE (E) by the fourth-order Adams-Bashforth-Moulton method.
142. Solve ODE (F) by the fourth-order Adams-Bashforth-Moulton method.
143. Solve ODE (G) by the fourth-order Adams-Bashforth-Moulton method.
144. Solve ODE (H) by the fourth-order Adams-Bashforth-Moulton method.
145. Solve ODE (I) by the fourth-order Adams-Bashforth-Moulton method.
146. Solve ODE (J) by the fourth-order Adams-Bashforth-Moulton method.
147. Solve ODE (K) by the fourth-order Adams-Bashforth-Moulton method.
148. Solve ODE (H) by the fourth-order Adams-Bashforth-Moulton method with mop up using the fourth-order Runge-Kutta method to obtain starting values. Compare the results with the results of Problems 122, 133, and 144.
149. Solve ODE (I) by the fourth-order Adams-Bashforth-Moulton method with mop up using the fourth-order Runge-Kutta method to obtain starting values. Compare the results with the results of Problems 123, 134, and 145.
150. Solve ODE (J) by the fourth-order Adams-Bashforth-Moulton method with mop up using the fourth-order Runge-Kutta method to obtain starting values. Compare the results with the results of Problems 124, 135, and 146.
151. Solve ODE (K) by the fourth-order Adams-Bashforth-Moulton method with mop up using the fourth-order Runge-Kutta method to obtain starting values. Compare the results with the results of Problems 125, 136, and 147.

7.11 Nonlinear Implicit Finite Difference Equations

Time Linearization

152. Solve ODE (A) by the implicit Euler method using time linearization.
153.* Solve ODE (B) by the implicit Euler method using time linearization.
154. Solve ODE (C) by the implicit Euler method using time linearization.
155. Solve ODE (D) by the implicit Euler method using time linearization.
156. Solve ODE (E) by the implicit Euler method using time linearization.
157. Solve ODE (F) by the implicit Euler method using time linearization.
158. Solve ODE (G) by the implicit Euler method using time linearization.
159. Solve ODE (H) by the implicit Euler method using time linearization.
160. Solve ODE (I) by the implicit Euler method using time linearization.
161. Solve ODE (J) by the implicit Euler method using time linearization.
162. Solve ODE (K) by the implicit Euler method using time linearization.

Newton's Method

163. Solve ODE (A) by the implicit Euler method using Newton's method.
164. Solve ODE (B) by the implicit Euler method using Newton's method.
165. Solve ODE (C) by the implicit Euler method using Newton's method.
166. Solve ODE (D) by the implicit Euler method using Newton's method.
167. Solve ODE (E) by the implicit Euler method using Newton's method.
168. Solve ODE (F) by the implicit Euler method using Newton's method.
169. Solve ODE (G) by the implicit Euler method using Newton's method.
170. Solve ODE (H) by the implicit Euler method using Newton's method.
171. Solve ODE (I) by the implicit Euler method using Newton's method.
172. Solve ODE (J) by the implicit Euler method using Newton's method.
173. Solve ODE (K) by the implicit Euler method using Newton's method.

7.12 Higher-Order Ordinary Differential Equations

174. Reduce the following ODE to a pair of first-order ODES:
$$ay'' + by' + cy = d \qquad y(0) = y_0 \text{ and } y'(0) = y_0'$$

175. Reduce the following ODE to a pair of first-order ODES:
$$ay'' + b|y'|y' + cy = F(t) \qquad y(0) = y_0 \text{ and } y'(0) = y_0'$$

176. Reduce the following ODE to a set of first-order ODES:
$$ay''' + by'' + cy' + dy = e \qquad y(0) = y_0, \quad y'(0) = y_0' \text{ and } y''(0) = y_0''$$

177. Reduce the following pair of ODEs to a set of four first-order ODES:
$$ay'' + by' + cz' + dy + ez = F(t) \qquad y(0) = y_0 \text{ and } y'(0) = y_0'$$
$$Az'' + By' + Cz' + Dy + Ez = G(t) \qquad z(0) = z_0 \text{ and } z'(0) = z_0'$$

178. The ODE governing the displacement $x(t)$ of a mass-damper-spring system is
$$mx'' + Cx' + Kx = F(t) \qquad x(0) = x_0 \text{ and } x'(0) = x_0'$$
Reduce this second-order ODE to a pair of coupled first-order ODEs.

179. The ODE governing the charge $q(t)$ in a series L (inductance), R (resistance), and C (capacitance) circuit is
$$Lq'' + Rq' + \frac{1}{C}q = \frac{dV(t)}{dt} \qquad q(0) = q_0 \text{ and } q'(0) = q_0'$$
Reduce this second-order ODE to two coupled first-order ODEs.

180. The angular displacement $\theta(t)$ of a frictionless pendulum is governed by the ODE
$$\theta'' + \frac{g}{L}\sin\theta = 0 \qquad \theta(0) = \theta_0 \text{ and } \theta'(0) = \theta_0'$$
Reduce this second-order ODE to a pair of first-order ODEs.

181. The governing equation for the displacement $y(t)$ of a projectile shot vertically upward is
$$my'' + C|V|V = -mg \qquad y(0) = 0, \quad y'(0) = V(0) = V_0$$
where $V = dy/dt$ is the projectile velocity, C is a drag parameter, and g is the acceleration of gravity. Reduce this second-order ODE to a pair of first-order ODEs.

182. The governing ODEs for the position, $x(t)$ and $y(t)$, of a projectile shot at an angle α with respect to the horizontal are
$$mx'' + C|V|V\cos\theta = 0, \qquad x(0) = 0, \quad x''(0) = u(0) = V_0\cos\alpha$$
$$my'' + C|V|V\sin\theta = -mg \qquad y(0) = 0, \quad y'(0) = v(0) = V_0\sin\alpha$$
where $V = (u^2 + v^2)^{1/2}$, $\theta = \tan^{-1}(v/u)$, $u = dx/dt$, $v = dy/dt$, C is a drag perameter, and g is the acceleration of gravity. Reduce this pair of coupled second-order ODEs to a set of four first-order ODEs.

183. The governing equation for the laminar boundary layer over a flat plate is

$$\frac{d^3f}{d\eta^3} + f\frac{d^2f}{d\eta^2} = 0 \qquad f(0) = 1, \quad f'(0) = 0, \text{ and } f'(\eta) \to 1 \text{ as } n \to \infty$$

Reduce this ODE to a set of three first-order ODEs.

184. The governing equation for a laminar mixing layer is

$$\frac{d^3f}{d\eta^3} + f\frac{df}{d\eta} + \left(\frac{df}{d\eta}\right)^2 = 0.0 \qquad f(0) = 0, f'(0) = 0, \text{ and } f'(\eta) \to 0$$
as $\eta \to \infty$

Reduce this ODE to a set of three first-order ODEs.

7.13 Systems of First-Order Ordinary Differential Equations

185. Solve the following pair of initial-value ODEs by the explicit Euler method from $t = 0.0$ to 1.0 with $\Delta t = 0.2$ and 0.1.

$$\bar{y}' = 2\bar{y} + \bar{z} + 1, \bar{y}(0) = 1 \text{ and } \bar{z}' = \bar{y} + \bar{z} + 1, \bar{z}(0) = 1 \qquad \text{(L)}$$

186. Solve ODE (L) by the modified Euler method.
187. Solve ODE (L) by the fourth-order Runge-Kutta method.
188. Solve the following pair of initial-value ODEs by the explicit Euler method from $t = 0.0$ to 1.0 with $\Delta t = 0.2$ and 0.1.

$$\bar{y}' = 2\bar{y} + \bar{z} + t, \bar{y}(0) = 1 \text{ and } \bar{z}' = \bar{y} + \bar{z} + t, \bar{z}(0) = 1 \qquad \text{(M)}$$

189. Solve ODE (M) by the modified Euler method.
190. Solve ODE (M) by the fourth-order Runge-Kutta method.
191. Solve the following pair of intial-value ODEs by the explicit Euler method from $t = 0.2$ to 1.0 with $\Delta t = 2.0$ and 0.1.

$$\bar{y}' = 2\bar{y} + \bar{z} + e^t, \bar{y}(0) = 1 \text{ and } \bar{z}' = \bar{y} + \bar{z} + 1, \bar{z}(0) = 1 \qquad \text{(N)}$$

192. Solve ODE (N) by the modified Euler method.
193. Solve ODE (N) by the fourth-order Runge-Kutta method.
194. Solve the following pair of initial-value ODEs by the explicit Euler method from $t = 0.0$ to 1.0 with $\Delta t = 0.2$ and 0.1.

$$\bar{y}' = 2\bar{y} + \bar{z} + e^t + 1 + t, \bar{y}(0) = 0 \text{ and } \bar{z}' = \bar{y} + \bar{z} + t, \bar{z}(0) = 1 \qquad \text{(O)}$$

195. Solve ODE (O) by the modified Euler method.
196. Solve ODE (O) by the fourth-order Runge-Kutta method.
197. Solve the following pair of initial-value ODEs by the explicit Euler method from $t = 0.0$ to 1.0 with $\Delta t = 0.2$ and 0.1.

$$\bar{y}' = \bar{z}, \bar{y}(0) = 1 \text{ and } \bar{z}' = -4\bar{y} - 5\bar{z} + 1 + t + e^t, \bar{z}(0) = 1 \qquad \text{(P)}$$

198. Solve ODE (P) by the modified Euler method.
199. Solve ODE (P) by the fourth-order Runge-Kutta method.
200. Solve the following pair of initial-value ODEs by the explicit Euler method from $t = 0.0$ to 1.0 with $\Delta t = 0.2$ and 0.1.

$$\bar{y}' = \bar{z}, \bar{y}(0) = 1 \text{ and } \bar{z}' = -6.25\bar{y} - 4\bar{z} + 1 + t + 2e^t, \bar{z}(0) = 1 \qquad \text{(Q)}$$

201. Solve ODE (Q) by the modified Euler method.
202. Solve ODE (Q) by the fourth-order Runge-Kutta method.
203. Solve the following pair of initial-value ODEs by the explicit Euler method from $t = 0.0$ to 1.0 with $\Delta t = 0.2$ and 0.1.

$$\bar{y}' = 0.1\bar{z}^2 + 1, \ \bar{y}(0) = 1 \text{ and } \bar{z}' = 0.1\bar{y}^2 + 0.1\bar{z}^2, \ \bar{z}(0) = 1 \qquad \text{(R)}$$

204. Solve ODE (R) by the modified Euler method.
205. Solve ODE (R) by the fourth-order Runge-Kutta method.

7.14 Stiff Ordinary Differential Equations

The following problems involving stiff ODEs require small step sizes and large numbers of steps. Consequently, programs should be written to solve these problems.

206. Consider the model stiff ODE:

$$\bar{y}' = -1000[\bar{y} - (t + 2)] + 1 \qquad \bar{y}(0) = 1 \qquad \text{(S)}$$

Solve ODE (S) by the explicit Euler method from $t = 0.0$ to 0.01 with $\Delta t = 0.0005, 0.001, 0.002,$ and 0.0025. Compare the solution with the exact solution.

207. Solve Eq. (S) by the implicit Euler method from $t = 0.0$ to 0.1 with $\Delta t = 0.01, 0.05,$ and 0.1.
208. Solve Eq. (S) by the implicit trapezoid method from $t = 0.0$ to 0.1 with $\Delta t = 0.01, 0.05,$ and 0.1.
209. Solve Eq. (S) by the modified Euler method from $t = 0.0$ to 0.01 with $\Delta t = 0.0005, 0.001, 0.002,$ and 0.0025.
210. Solve Eq. (S) by the second-order Gear method from $t = 0.0$ to 0.1 with $\Delta t = 0.01$ and 0.02. Use the exact solution for starting values.
211. Solve Eq. (S) by the fourth-order Gear method from $t = 0.0$ to 0.1 with $\Delta t = 0.01$ and 0.02. Use the exact solution for starting values.
212. Solve Eq. (S) by the fourth-order Gear method from $t = 0.0$ to 0.1 with $\Delta t = 0.01$ using the first-, second-, and third-order Gear methods to obtain starting values.
213. Solve the second-order ODE

$$\varepsilon \bar{y}'' + \bar{y}' + \bar{y} = 1 \qquad \bar{y}(0) = 0 \text{ and } \bar{y}'(0) = 1 \qquad \text{(T)}$$

by the explicit Euler method from $t = 0.0$ to 1.0 for $\varepsilon = 0.01$ with $\Delta t = 0.01, 0.02,$ and 0.025.

214. Solve Eq. (T) by the implicit Euler method from $t = 0.0$ to 1.0 with $\Delta t = 0.1$ $0.2,$ and 0.25.
215. Solve Eq. (T) by the implicit trapezoid method from $t = 0.0$ to 1.0 with $\Delta t = 0.1, 0.2$ and 0.25.
216. Solve Eq. (T) by the modified Euler method from $t = 0.0$ to 1.0 with $\Delta t = 0.01, 0.02,$ and 0.025.
217. Solve Eq. (T) by the first-order Gear method from $t = 0.0$ to 1.0 with $\Delta t = 0.1$ and 0.2.
218. Solve Eq. (T) by the second-order Gear method from $t = 0.0$ to 1.0 with $\Delta t = 0.1$. Use the first-order Gear method for starting values.

219. Solve Eq. (T) by the fourth-order Gear method from $t = 0.0$ to 1.0 with $\Delta t = 0.1$ using the first-, second-, and third-order Gear methods to obtain starting values.

220. Consider the pair of ODEs

$$\bar{y}' = -\bar{y}, \ \bar{y}(0) = 1 \text{ and } \bar{z}' = -100\bar{y}, \ \bar{z}(0) = 1 \qquad \text{(U)}$$

Assume that these two equations must be solved simultaneously with the same Δt as part of a larger problem. Solve these equations by the explicit Euler method from $t = 0.0$ to 1.0. Let $\Delta t = 0.1$, 0.2, and 0.25.

221. Solve Eq. (U) by the implicit Euler method from $t = 0.0$ to 1.0 with $\Delta t = 0.1$, 0.2, and 0.25.

222. Solve Eq. (U) by the implicit trapezoid method from $t = 0.0$ to 1.0 with $\Delta t = 0.1$ and 0.2, and 0.25.

223. Solve Eq. (U) by the modified Euler method from $t = 0.0$ to 1.0 with $\Delta t = 0.1$, 0.2, and 0.25.

224. Solve Eq. (U) by the first-order Gear method from $t = 0.0$ to 1.0 with $\Delta t = 0.1$ and 0.2.

225. Solve Eq. (U) by the second-order Gear method from $t = 0.0$ to 1.0 with $\Delta t = 0.1$ and 0.2. Use the first-order Gear method for starting values.

226. Solve Eq. (U) from $t = 0.0$ to 1.0 by the fourth-order Gear method with $\Delta t = 0.1$. Use the first-, second-, and third-order Gear methods to obtain starting values.

227. Consider the coupled pair of ODEs:

$$\bar{y}' = 998\bar{y} + 1998\bar{z}, \ \bar{y}(0) = 1 \text{ and } \bar{z}' = -999\bar{y} - 1999\bar{z}, \ \bar{z}(0) = 1 \quad \text{(V)}$$

Solve these two equations by the explicit Euler method from $t = 0.0$ to 0.1 with $\Delta t = 0.001$, 0.002, and 0.0025.

228. Solve Eq. (V) by the implicit Euler method from $t = 0.0$ to 0.1 with $\Delta t = 0.01$, 0.02, and 0.025.

229. Solve Eq. (V) by the implicit trapezoid method from $t = 0.0$ to 0.1 with $\Delta t = 0.01$, 0.02, and 0.025.

230. Solve Eq. (V) by the modified Euler method from $t = 0.0$ to 0.1 with $\Delta t = 0.001$, 0.002, and 0.0025.

231. Solve Eq. (V) by the first-order Gear method from $t = 0.0$ to 0.1 $\Delta t = 0.01$ and 0.02.

232. Solve Eq. (V) by the second-order Gear method from $t = 0.0$ to 0.1 with $\Delta t = 0.01$ and 0.02. Use the first-order Gear method for starting values.

233. Solve Eq. (V) by the fourth-order Gear method from $t = 0.0$ to 0.1 with $\Delta t = 0.01$ using the first-, second-, and third-order Gear methods to obtain starting values.

234. Solve the following coupled pair of ODEs by the explicit Euler method from $t = 0.0$ to 100.0:

$$\bar{y}' = -\bar{y} + 0.999\bar{z}, \ \bar{y}(0) = 1 \text{ and } \bar{z}' = 0.001\bar{z}, \ \bar{z}(0) = 1 \qquad \text{(W)}$$

Let $\Delta t = 1.0$, 2.0, and 2.5.

235. Solve Eq. (W) by the implicit Euler method from $t = 0.0$ to 100.0 with $\Delta t = 10.0$, 20.0, and 25.0.

236. Solve Eq. (W) by the implicit trapezoid method from $t = 0.0$ to 100.0 with $\Delta t = 10.0$, 20.0, and 25.0.

237. Solve Eq. (W) by the modified Euler method from $t = 0.0$ to 100.0 with $\Delta t = 1.0, 2.0,$ and 2.5.
238. Solve Eq. (W) by the first-order Gear method from $t = 0.0$ to 100.0 with $\Delta t = 10.0$ and 20.0.
239. Solve Eq. (W) by the second-order Gear method from $t = 0.0$ to 100.0 with $\Delta t = 10.0$ and 20.0. Use the first-order Gear method for starting values.
240. Solve Eq. (W) from $t = 0.0$ to 100.0 by the fourth-order Gear method with $\Delta t = 10.0$. Use the first-, second-, and third-order Gear methods to obtain starting values.
241. Solve the following coupled pair of ODEs by the explicit Euler method from $t = 0.0$ to 100.0:

$$\bar{y}' = -\bar{y} + 0.999\bar{z}, \ \bar{y}(0) = 2 \text{ and } \bar{z}' = -0.001\bar{z}, \ \bar{z}(0) = 1 \qquad \text{(X)}$$

Let $\Delta t = 1.0, 2.0,$ and 2.5.
242. Solve Eq. (X) by the implicit Euler method from $t = 0.0$ to 100.0 with $\Delta t = 10.0, 20.0,$ and 25.0.
243. Solve Eq. (X) by the implicit trapezoid method from $t = 0.0$ to 100.0 with $\Delta t = 10.0, 20.0,$ and 25.0.
244. Solve Eq. (X) by the modified Euler method from $t = 0.0$ to 100.0 with $\Delta t = 1.0, 2.0,$ and 2.5.
245. Solve Eq. (X) by the first-order Gear method from $t = 0.0$ to 100.0 with $\Delta t = 10.0$ and 20.0.
246. Solve Eq. (X) by the second-order Gear method from $t = 0.0$ to 100.0 with $\Delta t = 10.0$ and 20.0. Use the exact solution for starting values.
247. Solve Eq. (X) from $t = 0.0$ to 100.0 by the fourth-order Gear method with $\Delta t = 10.0$. Use the first-, second-, and third-order Gear methods to obtain starting values.

7.15 Programs

248. Implement the fourth-order Runge-Kutta method program presented in Section 7.15.1. Check out the program using the given data set.
249. Solve any of Problems 83 to 93 with the fourth-order Runge-Kutta method program.
250. Modify the fourth-order Runge-Kutta method program to implement the Runge-Kutta-Fehlberg method. Check out the modified program using the given data set.
251. Solve any of Problems 83 to 93 with the Runge-Kutta-Fehlberg method program.
252. Modify the fourth-order Runge-Kutta method program to implement the Runge-Kutta-Merson method. Check out the modified program using the given data set.
253. Solve any of Problems 83 to 93 with the Runge-Kutta-Merson method program.
254. Implement the extrapolated modified midpoint method program presented in Section 7.15.2. Check out the program using the given data set.
255. Solve any of Problems 99 to 109 with the extrapolated modified midpoint method program.

LIVERPOOL
JOHN MOORES UNIVERSITY
AVRIL ROBARTS LRC
TEL. 0151 231 4022

256. Implement the fourth-order Adams-Bashforth-Moulton method program presented in Section 7.15.3. Check out the program using the given data set.

257. Solve any of Problems 115 to 125 with the fourth-order Adams-Bashforth-Moulton method program.

APPLIED PROBLEMS

Several applied problems from various disciplines are presented in this section. These problems can be solved by any of the methods presented in this chapter. An infinite variety of exercises can be constructed by changing the numerical values of the parameters, the step size Δt, and so forth. Most of these problems require a large amount of computation to obtain accurate answers. Consequently, it is recommended that they be solve by computer programs.

258. Population growth of any species is frequently modeled by an ODE of the form

$$\frac{dN}{dt} = aN - bN^2 \qquad N(0) = N_0$$

where N is the population, aN represents the birthrate, and bN^2 represents the death rate due to all causes, such as disease, competition for food supplies, and so on. If $N_0 = 100{,}000$, $a = 0.1$, and $b = 0.0000008$, calculate $N(t)$ for $t = 0.0$ to 20.0 years.

259. A lumped mass m initially at the temperature T_0 is cooled by convection to its surroundings at the temperature T_a. From Newton's law of cooling, $\dot{q}_{\text{conv.}} = hA(T - T_a)$, where h is the convective cooling coefficient and A is the surface area of the mass. The energy E stored in the mass is $E = mCT$, where C is the specific heat. From an energy balance, the rate of change of E must equal the rate of cooling due to convection $\dot{q}_{\text{conv.}}$. Thus,

$$\frac{dT}{dt} = -\frac{hA}{mC}(T - T_a) \qquad T(0) = T_0$$

Consider a sphere of radius $r = 1.0$ cm made of an alloy for which $\rho = 3{,}000.0\,\text{kg/m}^3$ and $C = 1{,}000.0\,\text{J/(kg-K)}$. If $h = 500.0\,\text{J/(s-m}^2\text{-K)}$, $T(0) = 500.0\,\text{C}$, and $T_a = 50.0\,\text{C}$, calculate $T(t)$ for $t = 0.0$ to 10.0 s.

260. Consider the radiation problem presented in Part II.5, Eq. (II.20):

$$\frac{dT}{dt} = -\frac{A\varepsilon\sigma}{mC}(T^4 - T_a^4) \qquad T(0) = T_0$$

Consider a sphere of radius $r = 1.0$ cm made of an alloy for which $\rho = 8{,}000\,\text{kg/m}^3$ and $C = 500.0\,\text{J/(kg-K)}$. If $\varepsilon = 0.5$, $T(0) = 2{,}500.0\,\text{K}$, and $T_a = 100.0\,\text{K}$, calculate $T(t)$ for $t = 0.0$ to 10.0 s. The Stefan-Boltzmann constant $\sigma = 5.67 \times 10^{-8}\,\text{J/(s-m}^2\text{-K}^4)$.

261. Combine Problems 259 and 260 to consider simultaneous cooling by radiation and convection. Let the convective cooling coefficient $h = 600.0\,\text{J/(s-m}^2\text{-K)}$. Calculate $T(t)$ for $t = 0.0$ to 10.0 s and compare the results with the results of the previous problems.

262. When an ideal gas flows in a variable-area passage in the presence of friction and heat transfer, the Mach number M is governed by the following ODE [see Eq. (9.112) in Zucrow and Hoffman, Vol. 1 (1976)]:

$$\frac{dM}{dx} = \frac{M[1 + (\gamma - 1)M^2/2]}{1 - M^2}\left[-\frac{1}{A}\frac{dA}{dx} + \frac{1}{2}\gamma M^2\frac{4f}{D} + \frac{1}{2}(1 + \gamma M^2)\frac{1}{T}\frac{dT}{dx}\right]$$

where x is the distance along the passage (cm), γ is the ratio of specific heats (dimensionless), A is the cross-sectional flow area (cm^2), f is the friction coefficient (dimensionless), D is the diameter of the passage (cm), and T is the stagnation temperature (K). For a conical flow passage with a circular cross section, $A = \pi D^2/4$, where $D(x) = D_i + \alpha x$, where D_i is the inlet diameter. Thus,

$$\frac{dA}{dx} = \frac{d}{dx}\left(\frac{\pi}{4}D^2\right) = \frac{\pi}{4}\frac{d}{dx}(D_i + \alpha x)^2 = \frac{\pi}{2}(D_i + \alpha x)\alpha = \alpha\frac{\pi}{2}D$$

The stagnation temperature T is given by

$$T(x) = T_i + \frac{Q(x)}{C}$$

where $Q(x)$ is the heat transfer along the flow passage (J/cm) and C is the specific heat (kJ/kg-K). Thus,

$$\frac{dT}{dx} = \frac{1}{C}\frac{dQ}{dx}$$

For a linear heat transfer rate, $Q = Q_i + \beta x$, and

$$\frac{dT}{dx} = \frac{1}{C}\frac{d}{dx}(Q_i + \beta x) = \frac{\beta}{C}$$

The friction coefficient f is an empirical function of the Reynolds number and passage surface roughness. It is generally assumed to be constant for a specific set of flow conditions. Consider a problem where $f = \beta = 0.0$, $\alpha = 0.25$ cm/cm, $\gamma = 1.4$, and $D_i = 1.0$ cm. Calculate $M(x)$ for $x = 0.0$ to 5.0 cm for (a) $M_i = 0.7$ and (b) $M_i = 1.5$.

263. For Problem 262, let $\alpha = \beta = 0.0$, $f = 0.005$, $\gamma = 1.4$, and $D_i = 1.0$ cm, where $\alpha = 0.0$ specifies a constant-area tube. Calculate $M(x)$ for $x = 0.0$ to 5.0 cm for (a) $M_i = 0.7$ and (b) $M_i = 1.5$.

264. For Problem 262, let $\alpha = f = 0.0$, $T_i = 1,000.0$ K, $\beta = 50.0$ J/cm, $C = 1.0$ kJ/(kg-K), and $D = 1.0$ cm. Calculate $M(x)$ for $x = 0.0$ to 5.0 cm for (a) $M_i = 0.5$ and (b) $M_i = 2.0$.

265. Solve Problem 264 for $\beta = -50.0$ J/cm.

266. Consider combined area change and friction in the fluid mechanics problem described in Problem 262. Solve Problem 262 with the addition of $f = 0.005$.

267. Consider combined friction and heat transfer in the fluid mechanics problem described in Problem 262. Solve Problem 264 with the addition of $f = 0.005$.

268. The governing equation for a projectile shot vertically upward is

$$m\frac{d^2y}{dt^2} = -mg - C|V|V \qquad y(0) = 0 \text{ and } y'(0) = V_0$$

where m is the mass of the projectile (kg), $y(t)$ is the height (m), g is the acceleration of gravity ($9.80665 \, \text{m/s}^2$), C is an aerodynamic drag parameter, and $V = dy/dt$ is the velocity. For $m = 10.0 \, \text{kg}$, $C = 0.1 \, \text{N-s}^2/\text{m}^2$, and $V_0 = 500.0 \, \text{m/s}$, calculate (a) the maximum height attained by the projectile, (b) the time required to reach the maximum height, and (c) the time required to return to the original elevation.

269.   A machine of mass m (kg) rests on a support that exerts both a damping force and a spring force on the machine. The support is subjected to the displacement $y(t) = Y_0 \sin \omega t$. From Newton's second law of motion,

$$m\frac{d^2 y}{dt^2} = -C\left(\frac{dy}{dt} - \frac{dY}{dt}\right) - K(y - Y)$$

where $(y - Y)$ is the relative displacement between the machine and the support, C is the damping coefficient, and K is the spring constant. Determine the motion of the machine during the first cycle of oscillation of the support for $m = 1{,}000.0 \, \text{kg}$, $C = 5{,}000.0 \, \text{N-s/m}$, $K = 50{,}000.0 \, \text{N/m}$, $Y_0 = 1.0 \, \text{cm}$, $\omega = 100.0 \, \text{rad/s}$, and $y(0) = y'(0) = 0.0$.

270.   The current $i(t)$ in a series L-R-C circuit is governed by the ODE

$$L\frac{di}{dt} + Ri + \frac{1}{C}q = V(t) \qquad i(0) = i_0 \text{ and } q(0) = q_0$$

where i is the current (amps), q is the charge (coulombs), $dq/dt = i$, L is the inductance (henrys), C is the capacitance (farads), and V is the applied voltage (volts). Let $L = 100.0 \, \text{mH}$, $R = 10.0 \, \text{ohms}$, $C = 1 \, \text{mf}$, $V = 10.0 \, \text{volts}$, $i_0 = 0.0$, and $q_0 = 0.0$. Calculate $i(t)$ and $q(t)$ for $t = 0.0$ to 0.05 s. What is the maximum current, and at what time does it occur?

271.   Solve Problem 270 for $V = 10.0 \sin \omega t$, where ω is the frequency (1/s) of the applied voltage, which is given by $\omega = 2\pi f$, where $f = 60.0 \, \text{cycles/s}$.

272.   The angular displacement $\theta(t)$ (radians) of a frictionless pendulum is governed by the equation

$$\frac{d^2 \theta}{dt^2} + \frac{g}{L}\sin \theta = 0 \qquad \theta(0) = \theta_0 \text{ and } \theta'(0) = \theta'_0$$

where g is the acceleration of gravity ($9.80665 \, \text{m/s}^2$) and L is the length of the pendulum (m). For small θ, the governing equation simplifies to

$$\frac{d^2 \theta}{dt^2} + \frac{g}{L}\theta = 0$$

Solve for $\theta(t)$ for one period of oscillation for $\theta(0.0) = 0.1$ and 0.5 radians, $\theta'(0.0) = 0.0$, and $L = 0.1$, 1.0, and 10.0 m, using the simplified equation.

273.   Solve Problem 272 using the exact governing equation. Compare the results with the results of Problem 272.

274.   The population of two species competing for the same food supply can be modeled by the pair of ODEs:

$$\frac{dN_1}{dt} = N_1(A_1 - B_1 N_1 - C_1 N_2) \qquad N_1(0) = N_{1,0}$$

$$\frac{dN_2}{dt} = N_2(A_2 - B_2 N_2 - C_2 N_1) \qquad N_2(0) = N_{2,0}$$

where AN is the birthrate, BN^2 models the death rate due to disease, and CN_1N_2 models the death rate due to competition for the food supply. If $N_1(0.0) = N_2(0.0) = 100,000$, $A_1 = 0.1$, $B_1 = 0.0000008$, $C_1 = 0.0000010$, $A_2 = 0.1$, $B_2 = 0.0000008$, and $C_2 = 0.0000001$, calculate $N_1(t)$ and $N_2(t)$ for $t = 0.0$ to 10.0 years.

275. Consider a projectile of mass m (kg) shot upward at the angle α (radians) with respect to the horizontal at the initial velocity V_0 (m/s). The two ODEs that govern the displacement, $x(t)$ and $y(t)$ (m), of the projectile from the launch location are

$$m\frac{d^2x}{dt^2} = -C|V|V\cos\theta \qquad x(0) = 0 \text{ and } x'(0) = u(0) = V_0\cos\alpha$$

$$m\frac{d^2y}{dt^2} = -C|V|V\sin\theta - mg \qquad y(0) = 0 \text{ and } y'(0) = v(0) = V_0\sin\alpha$$

where the vector velocity $\mathbf{V} = \mathbf{i}u + \mathbf{j}v$, $u = dx/dt$ and $v = dy/dt$, C is a drag parameter, $\theta = \tan^{-1}(v/u)$, and g is the acceleration of gravity $(9.80665\,\text{m/s}^2)$. For $m = 10.0\,\text{kg}$, $C = 0.1\,\text{N-s}^2/\text{m}^2$, $V_0 = 500.0\,\text{m/s}$, $\alpha = 1.0\,\text{radian}$, and a level terrain, calculate (a) the maximum height attained by the projectile, (b) the corresponding time, (c) the maximum range of the projectile, (d) the corresponding time, and (e) the velocity $\mathbf{V}$ at impact.

276. The inherent features of finite rate chemical reactions can be modeled by the prototype rate equation

$$\frac{dC}{dt} = \frac{C_e - C}{\tau} \qquad C(0) = C_0$$

where C is the instantaneous nonequilibrium mass fraction of the species under consideration, C_e is its equilibrium mass fraction corresponding to the local conditions, and τ has the character of a chemical relaxation time. Assume that C_e varies quadratically with time. That is, $C_e = C_{e,0} + \alpha t^2$. Let $C(0.0) = 0.0$, $C_{e,0} = 0.1$, $\alpha = 0.5$, and $\tau = 0.0001$. Solve for $C(t)$ from $t = 0.0$ to 0.01.

8

One-Dimensional Boundary-Value Ordinary Differential Equations

Examples

8.1   INTRODUCTION

The steady one-dimensional heat transfer problem illustrated in Figure 8.1 consists of heat diffusion (i.e., heat conduction) along a constant-area rod with heat convection to the surroundings. The ends of the rod are maintained at the constant temperatures T_1 and T_2. An energy balance on a differential element of the rod, presented in Section II.6, yields the following second-order boundary-value ordinary differential equation (ODE):

$$T'' - \alpha^2 T = -\alpha^2 T_a \qquad T(x_1) = T(0.0) = T_1 \qquad T(x_2) = T(L) = T_2 \tag{8.1}$$

where T is the temperature of the rod (C), $\alpha^2 = hP/kA$ (cm^{-2}) (where h, P, k, and A are defined in Section II.6), and T_a is the ambient (i.e., surroundings) temperature (C).

The general solution of Eq. (8.1) is

$$T(x) = Ae^{\alpha x} + Be^{-\alpha x} + T_a \tag{8.2}$$

which can be demonstrated by direct substitution. Substituting the boundary conditions into Eq. (8.2) yields

$$A = \frac{(T_2 - T_a) - (T_1 - T_2)e^{-\alpha L}}{e^{\alpha L} - e^{-\alpha L}} \qquad \text{and} \qquad B = \frac{(T_1 - T_2)e^{\alpha L} - (T_2 - T_a)}{e^{\alpha L} - e^{-\alpha L}} \tag{8.3}$$

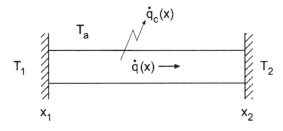

$$\frac{d^2T}{dx^2} - \alpha^2 T = -\alpha^2 T_a, \quad T(x_1) = T_1, \text{ and } T(x_2) = T_2$$

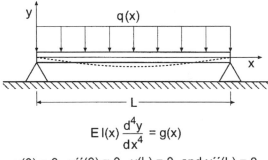

$$E\,I(x)\frac{d^4y}{dx^4} = g(x)$$

$$y(0) = 0, \ y''(0) = 0, \ y(L) = 0, \text{ and } y''(L) = 0$$

Figure 8.1   Steady one-dimensional heat transfer and deflection of a beam.

Consider a rod 1.0 cm long (i.e., $L = 1.0$ cm) with $T(0.0) = 0.0$ C and $T(1.0) = 100.0$ C. Let $\alpha^2 = 16.0$ cm^{-2} and $T_a = 0.0$ C. For these conditions, Eq. (8.2) yields

$$T(x) = 1.832179(e^{4x} - e^{-4x}) \qquad (8.4)$$

The exact solution at selected values of x is tabulated in Table 8.1 and illustrated in Figure 8.2.

The deflection of a laterally loaded symmetrical beam, illustrated at the bottom of Figure 8.1, is considered in this chapter to illustrate finite difference methods for solving fourth-order boundary value ODEs. This problem is discussed in Section II.6 and solved in Section 8.4.3.

The general features of *boundary-value ordinary differential equations* (*ODEs*) are discussed in Section II.6. In that section it is shown that boundary-value ODEs govern *equilibrium problems*, which are boundary-value problems in *closed domains*. This chapter is devoted to presenting the basic properties of boundary-value problems and to developing several methods for solving boundary-value ODEs.

Numerous boundary-value ordinary differential equations arise in engineering and science. Single ODEs governing a single dependent variable are quite common. Coupled systems of ODEs governing several dependent variables are also quite common. Boundary-value ODEs may be linear or nonlinear, second- or higher-order, and homogeneous or nonhomogeneous. In this chapter, the majority of attention is devoted to the

Table 8.1 Exact Solution of the Heat Transfer Problem

x, cm	$T(x)$, C	x, cm	$T(x)$, C
0.000	0.000000	0.625	22.170109
0.125	1.909479	0.750	36.709070
0.250	4.306357	0.875	60.618093
0.375	7.802440	1.000	100.000000
0.500	13.290111		

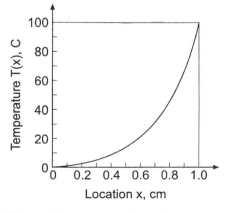

Figure 8.2 Exact solution of the heat transfer problem.

general second-order nonlinear boundary-value ODE and the general second-order linear boundary-value ODE:

$$y'' + P(x, y)y' + Q(x, y)y = F(x) \qquad y(x_1) = y_1 \text{ and } y(x_2) = y_2 \tag{8.5}$$

$$y'' + Py' + Qy = F(x) \qquad y(x_1) = y_1 \text{ and } y(x_2) = y_2 \tag{8.6}$$

The solution to these ODEs is the function $y(x)$. This function must satisfy two boundary conditions at the two boundaries of the solution domain. The solution domain $D(x)$ is closed, that is, $x_1 \leq x \leq x_2$. Several numerical methods for solving second-order boundary-value ODEs are presented in this chapter. Procedures for solving higher-order ODEs and systems of ODEs are discussed.

First, consider a class of methods called *finite difference methods*. There are two fundamentally different types of finite difference methods for solving boundary-value ODEs:

1. The shooting (initial-value) method
2. The equilibrium (boundary-value) method

Several variations of both of these methods are presented in this chapter.

The *shooting method* transforms the boundary-value ODE into a system of first-order ODEs, which can be solved by any of the initial-value (propagation) methods developed in Chapter 7. The boundary conditions on one side of the closed domain can be used as initial conditions. Unfortunately, however, the boundary conditions on the other side of the closed domain cannot be used as initial conditions. The additional initial conditions needed are assumed, the initial-value problem is solved, and the solution at the other boundary is compared to the known boundary conditions on that boundary. An iterative approach, called *shooting*, is employed to vary the assumed initial conditions on one boundary until the boundary conditions on the other boundary are satisfied.

The *equilibrium method* constructs a finite difference approximation of the exact ODE at every point on a discrete finite difference grid, including the boundaries. A system of coupled finite difference equations results, which must be solved simultaneously, thus relaxing the entire solution, including the boundary conditions, simultaneously.

A second class of methods for solving boundary-value ODEs is based on approximating the solution by a linear combination of trial functions, for example, polynomials, and determining the coefficients in the trial functions so as to satisfy the boundary-value ODE in some optimum manner. The most common examples of this type of method are

1. The Rayleigh-Ritz method
2. The collocation method
3. The Galerkin method
4. The finite element method

These four methods are discussed in Chapter 12.

The organization of Chapter 8 is illustrated in Figure 8.3. A discussion of the general features of boundary-value ordinary differential equations (ODEs) begins the chapter. The material then splits into a presentation of the two fundamentally different approaches for solving boundary-value ODEs: the shooting method and the equilibrium method. Procedures are then presented for implementing derivative (and other) boundary conditions. Then follows a discussion of higher-order equilibrium methods, nonlinear problems, and nonuniform grids. A brief introduction to eigenproblems arising from boundary-value

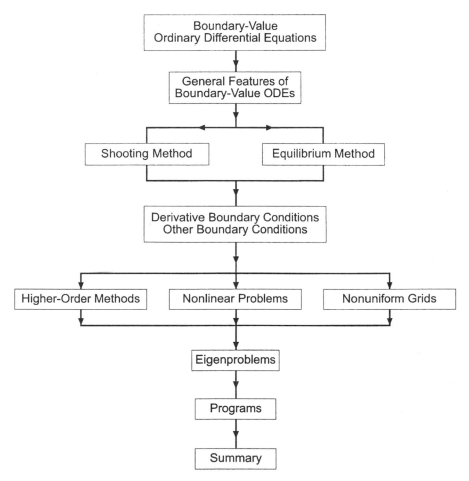

Figure 8.3 Organization of Chapter 8.

ODEs follows. The chapter closes with a Summary, which discusses the advantages and disadvantages of the shooting method and the equilibrium method, and lists the things you should be able to do after studying Chapter 8.

8.2 GENERAL FEATURES OF BOUNDARY-VALUE ODEs

Several general features of boundary-value ordinary differential equations (ODEs) are presented in this section. The general second-order linear ODE is discussed in some detail. Nonlinear second-order ODEs, systems of ODEs, and higher-order ODEs are discussed briefly.

8.2.1 The Linear Second-Order Boundary-Value ODE

The general linear second-order boundary-value ODE is given by Eq. (8.6):

$$y'' + Py' + Qy = F(x) \qquad y(x_1) = y_1 \text{ and } y(x_2) = y_2 \qquad (8.7)$$

where P and Q are constants.

As discussed in Section 7.2 for initial-value ODEs, the exact solution of an ordinary differential equation is the sum of the complementary solution $y_c(x)$ of the homogeneous ODE and the particular solution $y_p(x)$ of the nonhomogeneous ODE. The complementary solution $y_c(x)$ has the form

$$y_c(x) = Ae^{\lambda x} \tag{8.8}$$

Substituting Eq. (8.8) into Eq. (8.7) with $F(x) = 0$ yields the characteristic equation:

$$\lambda^2 + P\lambda + Q = 0 \tag{8.9}$$

Equation (8.9) has two solutions, λ_1 and λ_2. They can both be real, or they can be a complex conjugate pair. Thus, the complementary solution is given by

$$y_c(x) = Ae^{\lambda_1 x} + Be^{\lambda_2 x} \tag{8.10}$$

As discussed in Section 7.2, the particular solution has the form

$$y_p(x) = B_0\, F(x) + B_1\, F'(x) + B_2\, F''(x) + \cdots \tag{8.11}$$

where the terms $F'(x)$, $F''(x)$, etc., are the derivatives of the function $F(x)$. Thus, the total solution is

$$\boxed{y(x) = Ae^{\lambda_1 x} + Be^{\lambda_2 x} + y_p(x)} \tag{8.12}$$

The constants of integration A and B are determined by requiring Eq. (8.12) to satisfy the two boundary conditions.

8.2.2 The Nonlinear Second-Order Boundary-Value ODE

The general nonlinear second-order boundary-value ODE is given by Eq. (8.5):

$$\boxed{y'' + P(x, y)y' + Q(x, y)y = F(x) \qquad y(x_1) = y_1 \text{ and } y(x_2) = y_2} \tag{8.13}$$

where the coefficients $P(x, y)$ and $Q(x, y)$ may be linear or nonlinear functions of y. When solved by the shooting method, which is based on methods for solving initial-value ODEs, the nonlinear terms pose no special problems. However, when solved by the equilibrium method, in which the exact ODE is replaced by an algebraic finite difference equation which is applied at every point in a discrete finite difference grid, a system of nonlinear finite difference equations results. Solving such systems of nonlinear finite difference equations can be quite difficult.

8.2.3 Higher-Order Boundary-Value ODEs

Higher-order boundary-value ODEs can be solved by the shooting method by replacing each higher-order ODE by a system of first-order ODEs, which can then be solved by the shooting method. Some higher-order ODEs can be reduced to systems of second-order ODEs, which can be solved by the equilibrium method. Direct solution of higher-order ODEs by the equilibrium method can be quite difficult.

8.2.4 Systems of Second-Order Boundary-Value ODEs

Systems of coupled second-order boundary-value ODEs can be solved by the shooting method by replacing each second-order ODE by two first-order ODEs and solving the coupled systems of first-order ODEs. Alternatively, each second-order ODE can be solved by the equilibrium method, and the coupling between the individual second-order ODEs can be accomplished by relaxation. By either approach, solving coupled systems of second-order ODEs can be quite difficult.

8.2.5 Boundary Conditions

Boundary conditions (BCs) are required at the boundaries of the closed solution domain. Three types of boundary conditions are possible:

1. The function $y(x)$ may be specified (*Dirichlet* boundary condition)
2. The derivative $y'(x)$ may be specified (*Neumann* boundary condition)
3. A combination of $y(x)$ and $y'(x)$ may be specified (*mixed* boundary condition)

The terminology Dirichlet, Neumann, and mixed, when applied to boundary conditions, is borrowed from partial differential equation (PDE) terminology. The procedures for implementing these three types of boundary conditions are the same for ODEs and PDEs. Consequently, the same terminology is used to identify these three types of boundary conditions for both ODEs and PDEs.

8.2.6 Summary

Four types of boundary-value ordinary differential equations (ODEs) have been considered:

1. The general linear second-order boundary-value ODE
2. The general nonlinear second-order boundary-value ODE
3. Higher-order boundary-value ODEs
4. Systems of second-order boundary-value ODEs

Solution methods for ODE types 2, 3, and 4 are based on the solution methods for ODE type 1. Consequently, Chapter 8 is devoted mainly to the development of solution methods for the general linear second-order boundary-value ODE, Eq. (8.7), subject to boundary conditions at the boundaries of the closed solution domain. Keep in mind that equilibrium problems are steady state problems in closed solution domains. Equilibrium problems are not unsteady time dependent problems.

8.3 THE SHOOTING (INITIAL-VALUE) METHOD

The shooting method transforms a boundary-value ODE into a system of first-order ODEs, which can be solved by any of the initial-value methods presented in Chapter 7. The boundary conditions on one side of the closed solution domain $D(x)$ can be used as initial conditions for the system of initial-value ODEs. Unfortunately, however, the boundary conditions on the other side of the closed solution domain cannot be used as initial conditions on the first side of the closed solution domain. The additional initial conditions needed must be assumed. The initial-value problem can then be solved, and the solution at the other boundary can be compared to the known boundary conditions on that side of the closed solution domain. An iterative approach, called *shooting* is employed to vary the assumed initial conditions until the specified boundary conditions are satisfied. In this section, the shooting method is applied to the general nonlinear second-order boundary-value ODE with known function (i.e., Dirichlet) boundary conditions, Eq. (8.13). A brief discussion of the solution of higher-order boundary-value ODEs by the shooting method is presented. Derivative (i.e., Neumann) boundary conditions are discussed in Section 8.5.

As discussed in Section 7.4.2, when solving differential equations by an approximate method, a distinction must be made between the exact solution of the differential equation and the approximate solution of the differential equation. Analogously to the approach taken in Section 7.4, the exact solution of a boundary-value ODE is denoted by an overbar on the symbol for the dependent variable [i.e., $\bar{y}(x)$], and the approximate solution is denoted by the symbol for the dependent variable without an overbar [i.e., $y(x)$]. Thus,

$$\boxed{\begin{array}{l} \bar{y}(x) = \text{exact solution} \\ y(x) = \text{approximate solution} \end{array}}$$

This very precise distinction between the exact solution of a differential equation and the approximate solution of a differential equation is required for studies of consistency, order, stability, and convergence.

When solving boundary-value ODEs by the shooting method, consistency, order, stability, and convergence of the initial-value ODE solution method must be considered. These requirements are discussed thoroughly in Section 7.6 for marching methods for solving initial-value problems. Identical results are obtained when solving boundary-value problems by marching methods. Consequently, no further attention is given to these concepts in this section.

8.3.1 The Second-Order Boundary-Value ODE

Consider the general nonlinear second-order boundary-value ODE with Dirichlet boundary conditions, written in the following form:

$$\boxed{\bar{y}'' = f(x, \bar{y}, \bar{y}') \qquad \bar{y}(x_1) = \bar{y}_1 \text{ and } \bar{y}(x_2) = \bar{y}_2} \tag{8.14}$$

The exact solution of Eq. (8.14) is illustrated in Figure 8.4. The boundary conditions $\bar{y}(x_1) = \bar{y}_1$ and $\bar{y}(x_2) = \bar{y}_2$ are both specified. The first derivative $\bar{y}'(x_1) = \bar{y}'|_1$ is not specified as a boundary condition, but it does have a unique value which is obtained as part of the solution.

Let's rewrite the second-order ODE, Eq. (8.14), as two first-order ODEs. Define the auxiliary variable $\bar{z}(x) = \bar{y}'(x)$. Equation (8.14) can be expressed in terms of $\bar{y}$ and $\bar{z}$ as follows:

$$\bar{y}' = \bar{z} \qquad\qquad \bar{y}(x_1) = \bar{y}_1 \tag{8.15}$$

$$\bar{z}' = \bar{y}'' = f(x, \bar{y}, \bar{z}) \qquad \bar{z}(x_1) = \bar{y}'(x_1) = \bar{y}'|_1 = ? \tag{8.16}$$

An initial-value problem is created by assuming a value for $\bar{z}(x_1) = \bar{y}'|_1$. Choose an initial estimate for $\bar{y}'|_1$, denoted by $y'|_1^{(1)}$, and integrate the two coupled first-order ODEs, Eqs. (8.15) and (8.16), by any initial-value ODE integration method (e.g., the Runge-Kutta method). The solution is illustrated in Figure 8.5 by the curve labeled $y'|_1^{(1)}$. The solution at

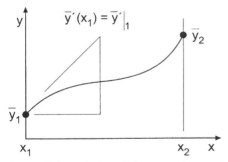

Figure 8.4 Solution of the general second-order boundary-value problem.

x_2 is $y(x_2) = y_2^{(1)}$, which is not equal to the specified boundary condition $\bar{y}(x_2) = \bar{y}_2$. Assume a second value for $\bar{z}(x_1) = y'|_1^{(2)}$, and repeat the process to obtain the solution labeled $y'|_1^{(2)}$. The solution at x_2 is $y(x_2) = y_2^{(2)}$, which again is not equal to the specified boundary condition $\bar{y}(x_2) = \bar{y}_2$. This procedure is continued until the value for $z(x_1) = y'|_1$ is determined for which $y(x_2) = \bar{y}_2$ within a specified tolerance.

For a nonlinear ODE, this is a zero finding problem for the function

$$\boxed{y(x_2) = f(z(x_1)) = \bar{y}_2} \tag{8.17}$$

which can be solved by the secant method (see Section 3.6). Thus,

$$\frac{\bar{y}_2 - y_2^{(n)}}{y'|_1^{(n+1)} - y'|_1^{(n)}} = \frac{y_2^{(n)} - y_2^{(n-1)}}{y'|_1^{(n)} - y'|_1^{(n-1)}} = \text{slope} \tag{8.18}$$

where the superscript (n) denotes the iteration number. Solving Eq. (8.18) for $y'|_1^{(n+1)}$ gives

$$\boxed{y'|_1^{(n+1)} = y'|_1^{(n)} + \frac{\bar{y}_2 - y_2^{(n)}}{\text{slope}}} \tag{8.19}$$

The entire initial-value problem is reworked with $z(x_1) = y'|_1^{(n+1)}$ until $y(x_2)$ approaches $\bar{y}_2$ within a specified tolerance.

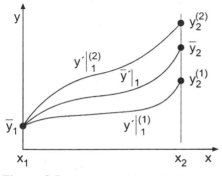

Figure 8.5 Iterative solution for the boundary condition.

Example 8.1   The second-order shooting method using iteration

As an example of the shooting method using iteration, let's solve the heat transfer problem presented in Section 8.1. The boundary-value ODE is [see Eq. (8.1)]:

$$T'' - \alpha^2 T = -\alpha^2 T_a \qquad T(0.0) = 0.0 \text{ C and } T(1.0) = 100.0 \text{ C} \qquad (8.20)$$

Rewrite the second-order ODE, Eq. (8.20), as two first-order ODEs:

$$T' = U \qquad T(0) = 0.0 \qquad (8.21)$$

$$U' = \alpha^2(T - T_a) \qquad U(0.0) = T'(0.0) \qquad (8.22)$$

Equations (8.21) and (8.22) can be solved by the implicit trapezoid method, Eq. (7.141):

$$T_{i+1} = T_i + \frac{\Delta x}{2}(U_i + U_{i+1}) \qquad (8.23a)$$

$$U_{i+1} = U_i + \frac{\Delta x}{2}[\alpha^2(T_i - T_a) + \alpha^2(T_{i+1} - T_a)] \qquad (8.23b)$$

For nonlinear ODEs, the implicit trapezoid method yields nonlinear implicit finite difference equations, which can be solved by the modified Euler (i.e., the modified trapezoid) method, Eqs. (7.141) and (7.142), or by Newton's method for systems of nonlinear equations, Section 3.7. However, for linear ODEs, such as Eqs. (8.21) and (8.22), the finite difference equations are linear and can be solved directly. Rearranging Eq. (8.23) yields

$$T_{i+1} - \frac{\Delta x}{2}U_{i+1} = T_i + \frac{\Delta x}{2}U_i \qquad (8.24a)$$

$$-\frac{\alpha^2 \Delta x}{2}T_{i+1} + U_{i+1} = \frac{\alpha^2 \Delta x}{2}(T_i - 2T_a) + U_i \qquad (8.24b)$$

Equation (8.24) can be solved directly for T_{i+1} and U_{i+1}.

Let $\alpha^2 = 16.0 \text{ cm}^{-2}$, $T_a = 0.0 \text{ C}$, and $\Delta x = 0.25 \text{ cm}$. To begin the solution, let $U(0.0)^{(1)} = T'(0.0)^{(1)} = 7.5 \text{ C/cm}$, and $U(0.0)^{(2)} = T'(0.0)^{(2)} = 12.5 \text{ C/cm}$. The solutions for these two values of $U(0.0)$ are presented in Table 8.2 and Figure 8.6. From Table 8.2, $T(1.0)^{(1)} = 75.925926 \text{ C}$, which does not equal the specified boundary condition $\bar{T}(1.0) = 100.0 \text{ C}$, and $T(1.0)^{(2)} = 126.543210 \text{ C}$, which also does not equal 100.0 C. Applying the secant method yields:

$$\text{Slope} = \frac{T(1.0)^{(2)} - T(1.0)^{(1)}}{U(0.0)^{(2)} - U(0.0)^{(1)}} = \frac{126.543210 - 75.925926}{12.5 - 7.5} = 10.123457 \qquad (8.25)$$

$$U(0.0)^{(3)} = U(0.0)^{(2)} + \frac{100.0 - T(1.0)^{(2)}}{\text{Slope}} = 9.878049 \qquad (8.26)$$

The solution for $U(0.0)^{(3)} = 9.878049 \text{ C/cm}$ is presented in Table 8.3 and illustrated in Figure 8.6. For this linear problem, $T(1.0)^{(3)} = 100.0 \text{ C}$, which is the desired value. The three solutions for $T'(0.0)^{(1)}$, $T'(0.0)^{(2)}$, and $T'(0.0)^{(3)}$ are presented in Figure 8.6. The final solution is presented in Table 8.4, along with the exact solution $\bar{T}(x)$ and the error, $\text{Error}(x) = [T(x) - \bar{T}(x)]$.

Repeating the solution for $\Delta x = 0.125 \text{ cm}$ yields the results presented in Table 8.5.

The accuracy of a numerical algorithm is generally assessed by the magnitude of the errors it produces. Individual errors, such as presented in Tables 8.4 and 8.5, present a

Table 8.2   Solution for $U(0.0)^{(1)} = 7.5\,\text{C/cm}$ and $U(0.0)^{(2)} = 12.5\,\text{C/cm}$

x	$T(x)^{(1)}$	$U(x)^{(1)}$	$T(x)^{(2)}$	$U(x)^{(2)}$
0.00	0.000000	7.500000	0.000000	12.500000
0.25	2.500000	12.500000	4.166667	20.833333
0.50	8.333333	34.166667	13.888889	56.944444
0.75	25.277778	101.388889	42.129630	168.981481
1.00	75.925926	303.796296	126.543210	506.327160

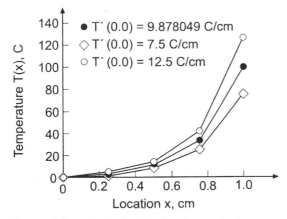

Figure 8.6   Solution by the shooting method.

Table 8.3   Solution for $U(0.0)^{(3)} = 9.878049\,\text{C/cm}$

x	$T(x)^{(3)}$	$U(x)^{(3)}$
0.00	0.000000	9.878049
0.25	3.292683	16.463415
0.50	10.975610	45.000000
0.75	33.292683	133.536585
1.00	100.000000	400.121951

Table 8.4   Solution by Shooting Method for $\Delta x = 0.25\,\text{cm}$

x, cm	$T(x)$, C	$\bar{T}(x)$, C	Error(x), C
0.00	0.000000	0.000000	
0.25	3.292683	4.306357	− 1.013674
0.50	10.975610	13.290111	− 2.314502
0.75	33.292683	36.709070	− 3.416388
1.00	100.000000	100.000000	

Table 8.5 Solution by the Shooting Method for $\Delta x = 0.125$ cm

x, cm	$T(x)$, C	$\bar{T}(x)$, C	Error(x), C
0.000	0.000000	0.000000	
0.125	1.792096	1.909479	−0.117383
0.250	4.062084	4.306357	−0.244273
0.375	7.415295	7.802440	−0.387145
0.500	12.745918	13.290111	−0.544194
0.625	21.475452	22.170109	−0.694658
0.750	35.931773	36.709070	−0.777298
0.875	59.969900	60.618093	−0.648193
1.000	100.000000	100.000000	

detailed picture of the error distribution over the entire solution domain. A measure of the overall, or global, accuracy of an algorithm is given by the *Euclidean norm of the errors*, which is the square root of the sum of the squares of the individual errors. The individual errors and the Euclidean norms of the errors are presented for all of the results obtained in Chapter 8.

The Euclidean norm of the errors in Table 8.4 is 4.249254 C. The Euclidean norm of the errors in Table 8.5 at the three common grid points (i.e., $x = 0.25$, 0.50, and 0.75 cm) is 0.979800 C. The ratio of the norms is 4.34. The ratios of the individual errors at the three common points in the two grids are 4.14, 4.25, and 4.39, respectively. Both of these results demonstrate that the method is second order (for step size halving the ratio of errors is 4.0 for a second-order method in the limit as $\Delta x \to 0$).

The errors in Table 8.5 are rather large, indicating that a smaller step size or a higher-order method is needed. The errors from Tables 8.4 and 8.5 are plotted in Figure 8.7, which also presents the errors from the extrapolation of the second-order shooting method results, which is presented in Example 8.3, as well as the errors for the solution by the

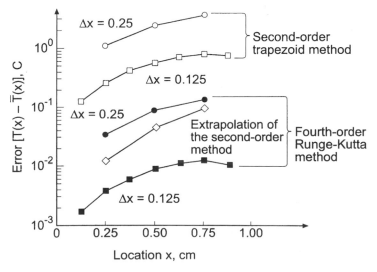

Figure 8.7 Errors in the solution by the shooting method.

fourth-order Runge-Kutta method (not presented here). For the Runge-Kutta method, the Euclidean norms of the errors at the three common grid points for the two step sizes are 0.159057 C and 0.015528 C, respectively. The ratio of the norms is 10.24. The ratios of the individual errors at the three common points in the two grids are 8.68, 10.05, and 10.46. Both of these results suggest that the method is fourth order (for step size halving the ratio of the errors is 16.0 for a fourth-order method in the limit as $\Delta x \to 0$).

8.3.2 Superposition

For a linear ODE, the principle of superposition applies. First, compute two solutions for $z(x_1) = y'|_1^{(1)}$ and $z(x_1) = y'|_1^{(2)}$, denoted by $y(x)^{(1)}$ and $y(x)^{(2)}$, respectively. Then form a linear combination of these two solutions:

$$y(x) = C_1 \, y(x)^{(1)} + C_2 \, y(x)^{(2)} \tag{8.27}$$

Apply Eq. (8.27) at $x = x_1$ and $x = x_2$. Thus,

$$\text{At } x = x_1: \qquad \bar{y}_1 = C_1 \bar{y}_1 + C_2 \bar{y}_1 \tag{8.28}$$

$$\text{At } x = x_2: \qquad \bar{y}_2 = C_1 y_2^{(1)} + C_2 y_2^{(2)} \tag{8.29}$$

Solving Eqs. (8.28) and (8.29) for C_1 and C_2 yields

$$C_1 = \frac{\bar{y}_2 - y_2^{(2)}}{y_2^{(1)} - y_2^{(2)}} \qquad \text{and} \qquad C_2 = \frac{y_2^{(1)} - \bar{y}_2}{y_2^{(1)} - y_2^{(2)}} \tag{8.30}$$

Substituting C_1 and C_2 into Eq. (8.27) yields the solution. No iteration is required for linear ODEs.

Example 8.2. The second-order shooting method using superposition

The heat transfer problem considered in Example 8.1 is governed by a linear boundary-value ODE. Consequently, the solution can be obtained by generating two solutions for two assumed values of $U(0.0) = T'(0.0)$ and superimposing those two solutions. The results of the first two solutions by the implicit trapezoid method with $\Delta x = 0.25$ cm are presented in Table 8.2 and repeated in Table 8.6. From Table 8.6, $T_2^{(1)} = 75.925926$ and $T_2^{(2)} = 126.543210$. The specified value of $\bar{T}_2$ is 100.0 C. Substituting these values into Eq. (8.30) gives

$$C_1 = \frac{100.000000 - 126.543210}{75.925926 - 126.543210} = 0.524389 \tag{8.31}$$

$$C_2 = \frac{75.925926 - 100.000000}{75.925926 - 126.543210} = 0.475611 \tag{8.32}$$

Substituting these results into Eq. (8.27) yields

$$T(x) = 0.524389 T(x)^{(1)} + 0.475611 T(x)^{(2)} \tag{8.33}$$

Table 8.6 Solution by Superposition for
$\Delta x = 0.25$ cm

x, cm	$T(x)^{(1)}$, C	$T(x)^{(2)}$, C	$T(x)$, C
0.00	0.000000	0.000000	0.000000
0.25	2.500000	4.166667	3.292683
0.50	8.333333	13.888889	10.975610
0.75	25.277778	42.129630	33.292683
1.00	75.925926	126.543210	100.000000

Solving Eq. (8.33) for $y(x)$ gives the values presented in the final column of Table 8.6, which are identical to the values presented in Table 8.4.

8.3.3 Extrapolation

The concept of *extrapolation* is presented in Section 5.6. As shown there, the error of any numerical algorithm that approximates an exact calculation by an approximate calculation having an error that depends on an increment h can be estimated if the functional dependence of the error on the increment h is known. The estimated error can be added to the approximate solution to obtain an improved solution. This process is known as *extrapolation*, or the *deferred approach to the limit*. The extrapolation formula is given by Eq. (5.117):

$$\boxed{\begin{aligned}\text{Improved value} &= \text{more accurate value} \\ &+ \frac{1}{R^n - 1} \text{ (more accurate value} - \text{less accurate value)}\end{aligned}}$$

(8.34)

where improved value is the extrapolated result, less accurate value and more accurate value are the results of applying the numerical algorithm for two increments h and h/R, respectively, where R is the ratio of the two step sizes (usually 2.0) and n is the order of the leading truncation error term of the numerical algorithm.

Example 8.3.   The second-order shooting method using extrapolation

Let's apply extrapolation to the heat transfer problem presented in Section 8.1. The boundary-value ODE is [see Eq. (8.1)]:

$$T'' - \alpha^2 T = -\alpha^2 T_a \qquad T(0.0) = 0.0 \text{ C and } T(1.0) = 100.0 \text{ C}$$

(8.35)

Let $\alpha^2 = 16.0$ cm^{-2} and $T_a = 0.0$ C.

This problem is solved in Example 8.1 by the second-order implicit trapezoid method for $\Delta x = 0.25$ and 0.125 cm. The results are presented in Tables 8.4 and 8.5, respectively. The results at the three common grid points in the two grids are summarized in Table 8.7. For these results, $R = 0.25/0.125 = 2.0$, and Eq. (8.34) gives

$$\text{IV} = \text{MAV} + \frac{1}{2^2 - 1}(\text{MAV} - \text{LAV}) = \frac{4\text{MAV} - \text{LAV}}{3}$$

(8.36)

Table 8.7 Solution by Extrapolation of the Second-Order Shooting Method Results

x, cm	T(LAV), C	T(MAV), C	T(IV), C	$\bar{T}(x)$, C	Error(x), C
0.00	0.000000	0.000000	0.000000	0.000000	
0.25	3.292683	4.062084	4.318551	4.306357	0.012195
0.50	10.975610	12.745918	13.336020	13.290111	0.045909
0.75	33.292683	35.931773	36.811469	36.709070	0.102399
1.00	100.000000	100.000000	100.000000	100.000000	

where IV denotes improved value, MAV denotes more accurate value corresponding to $h = 0.125$ cm, and LAV denotes less accurate value corresponding to $h = 0.25$ cm. Table 8.7 presents the results obtained from Eq. (8.36). The Euclidean norm of these errors is 0.112924 C, which is 8.68 times smaller than the Euclidean norm of the errors in Table 8.5. These results and the corresponding errors are presented in Figure 8.6.

8.3.4 Higher-Order Boundary-Value ODEs

Consider the third-order boundary-value problem:

$$\boxed{\bar{y}''' = f(x, \bar{y}, \bar{y}', \bar{y}'') \qquad \bar{y}(x_1) = \bar{y}_1, \ \bar{y}'(x_1) = \bar{y}'|_1, \ \text{and} \ \bar{y}(x_2) = \bar{y}_2} \qquad (8.37)$$

Rewriting Eq. (8.37) as three first-order ODEs gives:

$$\bar{y}' = \bar{z} \qquad\qquad\qquad \bar{y}(x_1) = \bar{y}_1 \qquad\qquad (8.38)$$
$$\bar{z}' = \bar{w} = \bar{y}'' \qquad\qquad \bar{z}(x_1) = \bar{y}'(x_1) = \bar{y}'|_1 \qquad (8.39)$$
$$\bar{w}' = \bar{z}'' = \bar{y}''' = f(x, \bar{y}, \bar{z}, \bar{w}) \qquad \bar{w}(x_1) = \bar{y}''(x_1) = \bar{y}''|_1 = ? \qquad (8.40)$$

An initial-value problem is created by assuming a value for $\bar{w}(x_1) = \bar{y}''(x_1) = \bar{y}''|_1$. Assume values for $y''|_1$ and proceed as discussed for the second-order boundary-value problem.

For fourth- and higher-order boundary-value problems, proceed in a similar manner. Start the initial-value problem from the boundary having the most specified boundary conditions. In many cases, more than one boundary condition may have to be iterated. In such cases, use Newton's method for systems of nonlinear equations (see Section 3.7) to conduct the iteration.

The solution of linear higher-order boundary-value problems can be determined by superposition. For an nth-order boundary-value problem, n solutions are combined linearly:

$$y(x) = C_1 \, y(x)^{(1)} + C_2 \, y(x)^{(2)} + \cdots + C_n \, y(x)^{(n)} \qquad (8.41)$$

The weighting factors C_i ($i = 1, 2, \ldots, n$) are determined by substituting the n boundary conditions into Eq. (8.41), as illustrated in Eqs. (8.28) to (8.30) for the second-order boundary-value problem.

8.4 THE EQUILIBRIUM (BOUNDARY-VALUE) METHOD

The solution of boundary-value problems by the equilibrium (boundary-value) method is accomplished by the following steps:

1. *Discretizing* the continuous solution domain into a discrete finite difference grid
2. Approximating the exact derivatives in the boundary-value ODE by algebraic finite difference approximations (FDAs)
3. *Substituting* the FDAs into the ODE to obtain an algebraic finite difference equation (FDE)
4. *Solving* the resulting system of algebraic FDEs

When the finite difference equation is applied at every point in the discrete finite difference grid, a system of coupled finite difference equations results, which must be solved simultaneously, thus relaxing the entire solution, including the boundary points, simultaneously. In this section, the equilibrium method is applied to the linear, variable coefficient, second-order boundary-value ODE with known function (i.e., Dirichlet) boundary conditions. Derivative (i.e., Neumann) boundary conditions are considered in Section 8.5, and nonlinear boundary-value problems are considered in Section 8.7.

When solving boundary-value problems by the equilibrium method, consistency, order, and convergence of the solution method must be considered. Stability is not an issue, since a relaxation procedure, not a marching procedure is employed. Consistency and order are determined by a Taylor series consistency analysis, which is discussed in Section 7.6 for marching methods. The same procedure is applicable to relaxation methods. Convergence is guaranteed for consistent finite difference approximations of a boundary-value ODE, as long as the system of FDEs can be solved. In principle, this can always be accomplished by direct solution methods, such as Gauss elimination.

8.4.1 The Second-Order Boundary-Value ODE

Consider the linear, variable coefficient, second-order boundary-value problem with Dirichlet boundary conditions:

$$\bar{y}'' + P(x)\,\bar{y}' + Q(x)\,\bar{y} = F(x) \qquad \bar{y}(x_1) = \bar{y}_1 \text{ and } \bar{y}(x_2) = \bar{y}_2 \tag{8.42}$$

The discrete finite difference grid for solving Eq. (8.42) by the equilibrium method is illustrated in Figure 8.8. Recall the second-order centered-difference approximations of $\bar{y}'|_i$ and $\bar{y}''|_i$ at grid point i developed in Section 5.4 [Eqs. (5.73) and (5.76), respectively]:

$$\bar{y}'|_i = \frac{\bar{y}_{i+1} - \bar{y}_{i-1}}{2\,\Delta x} + 0(\Delta x^2) \tag{8.43}$$

$$\bar{y}''|_i = \frac{\bar{y}_{i+1} - 2\bar{y}_i + \bar{y}_{i-1}}{\Delta x^2} + 0(\Delta x^2) \tag{8.44}$$

Substituting Eqs. (8.43) and (8.44) into Eq. (8.42) and evaluating the coefficients $P(x)$ and $Q(x)$ at grid point i yields

$$\left[\frac{\bar{y}_{i+1} - 2\bar{y}_i + \bar{y}_{i-1}}{\Delta x^2} + 0(\Delta x^2)\right] + P_i\left[\frac{\bar{y}_{i+1} - \bar{y}_{i-1}}{2\,\Delta x} + 0(\Delta x^2)\right] + Q_i\bar{y}_i = F_i \tag{8.45}$$

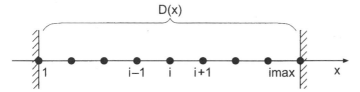

Figure 8.8 Solution domain $D(x)$ and finite difference grid.

All of the approximations in Eq. (8.45) are $0(\Delta x^2)$. Multiplying Eq. (8.45) through by Δx^2, gathering terms, and truncating the remainder terms yields:

$$\left(1 - \frac{\Delta x}{2}P_i\right)y_{i-1} + (-2 + \Delta x^2\,Q_i)y_i + \left(1 + \frac{\Delta x}{2}P_i\right)y_{i+1} = \Delta x^2\,F_i \qquad (8.46)$$

Applying Eq. (8.46) at each point in a discrete finite difference grid yields a tridiagonal system of FDEs, which can be solved by the Thomas algorithm (see Section 1.5).

Example 8.4. The second-order equilibrium method

Let's solve the heat transfer problem presented in Section 8.1 by the second-order equilibrium method. The boundary-value ODE is Eq. (8.1):

$$T'' - \alpha^2 T = -\alpha^2 T_a \qquad T(0.0) = 0.0 \text{ C and } T(1.0) = 100.0 \text{ C} \qquad (8.47)$$

Replacing T'' by the second-order centered-difference approximation, Eq. (8.44), and evaluating all the terms at grid point i gives

$$\frac{\bar{T}_{i+1} - 2\bar{T}_i + \bar{T}_{i-1}}{\Delta x^2} + 0(\Delta x^2) - \alpha^2 \bar{T}_i = -\alpha^2 T_a \qquad (8.48)$$

Multiplying through by Δx^2, gathering terms, and truncating the remainder term yields the FDE:

$$T_{i-1} - (2 + \alpha^2\,\Delta x^2)T_i + T_{i+1} = -\alpha^2\,\Delta x^2\,T_a \qquad (8.49)$$

Let $\alpha^2 = 16.0 \text{ cm}^{-2}$, $T_a = 0.0 \text{ C}$, and $\Delta x = 0.25 \text{ cm}$. Then Eq. (8.49) becomes

$$T_{i-1} - 3.0T_i + T_{i+1} = 0 \qquad (8.50)$$

Applying Eq. (8.50) at the three interior grid points, $x = 0.25$, 0.50, and 0.75 cm, gives

$$x = 0.25: \qquad T_1 - 3.0T_2 + T_3 = 0.0 \qquad T_1 = \bar{T}_1 = 0.0 \qquad (8.51a)$$
$$x = 0.50: \qquad T_2 - 3.0T_3 + T_4 = 0.0 \qquad (8.51b)$$
$$x = 0.75: \qquad T_3 - 3.0T_4 + T_5 = 0.0 \qquad T_5 = \bar{T}_5 = 100.0 \qquad (8.51c)$$

Transferring T_1 and T_5 to the right-hand sides of Eqs. (8.51a) and (8.51c), respectively, yields the following tridiagonal system of FDEs:

$$\begin{bmatrix} -3.0 & 1.0 & 0.0 \\ 1.0 & -3.0 & 1.0 \\ 0.0 & 1.0 & -3.0 \end{bmatrix} \begin{bmatrix} T_2 \\ T_3 \\ T_4 \end{bmatrix} = \begin{bmatrix} 0.0 \\ 0.0 \\ -100.0 \end{bmatrix} \qquad (8.52)$$

Table 8.8 Solution by the Equilibrium Method for
$\Delta x = 0.25$ cm

x, cm	$T(x)$, C	$\bar{T}(x)$, C	Error(x), C
0.00	0.000000	0.000000	
0.25	4.761905	4.306357	0.455548
0.50	14.285714	13.290111	0.995603
0.75	38.095238	36.709070	1.386168
1.00	100.000000	100.000000	

Solving Eq. (8.52) by the Thomas algorithm yields the results presented in Table 8.8. The exact solution and the errors are presented for comparison.

Let's repeat the solution for $\Delta x = 0.125$ cm. In this case Eq. (8.49) becomes

$$\boxed{T_{i-1} - 2.25T_i + T_{i+1} = 0} \tag{8.53}$$

Applying Eq. (8.53) at the seven interior grid points gives

$$x = 0.125: \quad T_1 - 2.25T_2 + T_3 = 0.0 \quad T_1 = \bar{T}_1 = 0.0 \tag{8.54a}$$
$$x = 0.250: \quad T_2 - 2.25T_3 + T_4 = 0.0 \tag{8.54b}$$
$$x = 0.375: \quad T_3 - 2.25T_4 + T_5 = 0.0 \tag{8.54c}$$
$$x = 0.500: \quad T_4 - 2.25T_5 + T_6 = 0.0 \tag{8.54d}$$
$$x = 0.625: \quad T_5 - 2.25T_6 + T_7 = 0.0 \tag{8.54e}$$
$$x = 0.750: \quad T_6 - 2.25T_7 + T_8 = 0.0 \tag{8.54f}$$
$$x = 0.875: \quad T_7 - 2.25T_8 + T_9 = 0.0 \quad T_9 = \bar{T}_9 = 100.0 \tag{8.54g}$$

Transferring T_1 and T_9 to the right-hand sides of Eqs. (8.54a) and (8.54g), respectively, yields a tridiagonal system of equations. That tridiagonal system of equations is solved by the Thomas algorithm in Example 1.17 in Section 1.5. The results are presented in Table 8.9.

Table 8.9 Solution by the Equilibrium Method for
$\Delta x = 0.125$ cm

x, cm	$T(x)$, C	$\bar{T}(x)$, C	Error(x), C
0.000	0.000000	0.000000	
0.125	1.966751	1.909479	0.057272
0.250	4.425190	4.306357	0.118833
0.375	7.989926	7.802440	0.187486
0.500	13.552144	13.290111	0.262033
0.625	22.502398	22.170109	0.332288
0.750	37.078251	36.709070	0.369181
0.875	60.923667	60.618093	0.305575
1.000	100.000000	100.000000	

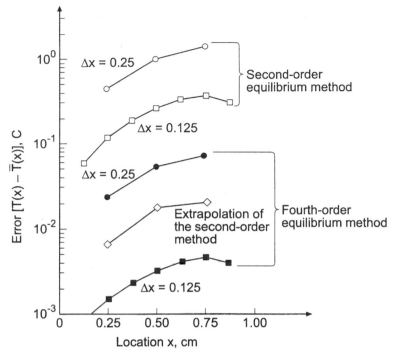

Figure 8.9   Errors in the solution by the equilibrium method.

The Euclidean norm of the errors in Table 8.8 is 1.766412 C. The Euclidean norm of the errors in Table 8.9 at the three common grid points is 0.468057 C. The ratio of the norms is 3.77. The ratios of the individual errors at the three common points in the two grids are 3.83, 3.80, and 3.75. Both of these results demonstrate that the method is second order.

The errors in Tables 8.8 and 8.9 are about 40 percent of the magnitude of the errors in Tables 8.4 and 8.5, respectively, which present the solution by the second-order shooting method. The errors in both cases can be decreased by using a smaller step size or a higher-order method. The errors are illustrated in Figure 8.9, which also presents the errors for the compact three-point fourth-order equilibrium method presented in Example 8.9 in Section 8.6, as well as the errors from extrapolation of the second-order method, which is presented in Example 8.5. For the fourth-order method, the Euclidean norms of the errors at the three common grid points in the two grids are 0.092448 C and 0.005919 C, respectively. The ratio of the norms is 15.62, which demonstrates the fourth-order behavior of the method.

8.4.2 Extrapolation

Extrapolation was applied in Example 8.3 to the results obtained by the second-order shooting method. The second-order results obtained in Example 8.4 by the equilibrium method can be extrapolated by the same procedure presented in Section 8.3, Eqs. (8.34) and (8.36).

Example 8.5   The second-order equilibrium method by extrapolation

Let's apply extrapolation to the results obtained in Example 8.4. Those results are presented in Tables 8.8 and 8.9. The results at the three common grid points in the two grids are summarized in Table 8.10. For these results, $R = 0.25/0.125 = 2.0$. The results obtained by applying Eq. (8.36) to the results presented in Table 8.10 are also presented in Table 8.10 and Figure 8.9. The Euclidean norm of these errors is 0.035514 C, which is 13.18 times smaller than the Euclidean norm of the errors presented in Table 8.9.

Table 8.10   Solution by Extrapolation of the Second-Order Equilibrium Method Results

x, cm	T(LAV), C	T(MAV), C	T(IV), C	$\bar{T}(x)$, C	Error(x), C
0.00	0.000000	0.000000	0.000000	0.000000	
0.25	4.761905	4.425190	4.312952	4.306357	0.006595
0.50	14.285714	13.552144	13.307621	13.290111	0.017510
0.75	38.095238	37.078251	36.739255	36.709070	0.030185
1.00	100.000000	100.000000	100.000000	100.000000	

8.4.3   Higher-Order Boundary-Value ODEs

Consider the general nonlinear fourth-order boundary-value problem presented in Section II.6, Eq. (II.45):

$$\boxed{y'''' = f(x, y, y', y'', y''')} \tag{8.55}$$

Since Eq. (8.55) is fourth-order, four boundary conditions are required. At least one boundary condition must be specified on each boundary of the closed solution domain. The two remaining boundary conditions can be specified on the same boundary, or one on each boundary. These boundary conditions can depend on y, y', y'', or y'''.

A finite difference approximation (FDA) must be developed for every derivative in Eq. (8.55). All the FDAs should be the same order. Second-order centered-difference FDAs can be developed for all four derivatives in Eq. (8.55). The first and second derivatives involve three grid points, points $i - 1$ to $i + 1$. The third and fourth derivatives involve five grid points, points $i - 2$ to $i + 2$. Consequently, the resulting finite difference equation (FDE) involves five grid points, points $i - 2$ to $i + 2$. Applying this FDE at each point in a finite difference grid yields a pentadiagonal system of FDEs, which can be solved by an algorithm similar to the Thomas algorithm for a tridiagonal system of FDEs.

Example 8.6.   A fourth-order ODE by the second-order equilibrium method

Let's solve the deflection problem presented in Section II.6 for a laterally loaded symmetrical beam, Eq. (II.40), expressed in the form of Eq. (8.55):

$$y'''' = \frac{q(x)}{EI(x)} \tag{8.56}$$

where E is the modulus of elasticity of the beam material, $I(x)$ is the moment of inertia of the beam cross section, and $q(x)$ is the distributed load on the beam. Let's consider a rectangular cross-section beam, for which $I = wh^3/12$, where w is the width of the beam and h is the height of the beam, and a uniform distributed load $q(x) = q = $ constant.

The ends of the beam are at the same elevation, which can be chosen as $y = 0.0$. Thus, $y(0.0) = y(L) = 0.0$, where L is the length of the beam. One additional boundary condition is required at each end of the beam. If the beam is fixed, then y' is specified. If the beam is pinned (i.e., clamped), then $y'' = 0.0$. If the beam is free (i.e., cantilevered), then $y''' = 0.0$. Let's assume that both ends of the beam are pinned. Thus, $y''(0.0) = y''(L) = 0.0$.

Thus, the problem to be solved, including the boundary conditions, is given by

$$y'''' = \frac{q}{EI} \qquad y(0.0) = y''(0.0) = y(L) = y''(L) = 0.0 \tag{8.57}$$

The exact solution of Eq. (8.57) is

$$y(x) = \frac{qx^4}{24EI} - \frac{qLx^3}{12EI} + \frac{qL^3x}{24EI} \tag{8.58}$$

As an example, let $q = -2000.0\,\text{N/m}$, $L = 5.0\,\text{m}$, $w = 5.0\,\text{cm}$, $h = 10.0\,\text{cm}$, and $E = 90 \times 10^9\,\text{N/m}^2$. Then Eqs. (8.57) and (8.58) become

$$y'''' = -0.0024 \qquad y(0.0) = y''(0.0) = y(5.0) = y''(5.0) = 0 \tag{8.59}$$

$$y(x) = 0.000100x^4 - 0.001000x^3 + 0.012500x \tag{8.60}$$

Let's solve this problem using a second-order centered-difference approximation for y''''. Write Taylor series for $\bar{y}_{i\pm1}$ and $\bar{y}_{i\pm2}$ with base point i:

$$\bar{y}_{i\pm1} = \bar{y}_i \pm \bar{y}'|_i\,\Delta x + \tfrac{1}{2}\bar{y}''|_i\,\Delta x^2 \pm \tfrac{1}{6}\bar{y}'''|_i\,\Delta x^3 + \tfrac{1}{24}\bar{y}''''|_i\,\Delta x^4$$
$$\pm \tfrac{1}{120}\bar{y}^{(v)}|_i\,\Delta x^5 + \tfrac{1}{720}\bar{y}^{(vi)}|_i\,\Delta x^6 \pm \cdots \tag{8.61}$$

$$\bar{y}_{i\pm2} = \bar{y}_i \pm 2\bar{y}'|_i\,\Delta x + \tfrac{4}{2}\bar{y}''|_i\,\Delta x^2 \pm \tfrac{8}{6}\bar{y}'''|_i\,\Delta x^3 + \tfrac{16}{24}\bar{y}''''|_i\,\Delta x^4$$
$$\pm \tfrac{32}{120}\bar{y}^{(v)}|_i\,\Delta x^5 + \tfrac{64}{720}\bar{y}^{(vi)}|_i\,\Delta x^6 \pm \cdots \tag{8.62}$$

Adding $\bar{y}_{i+1}$ and $\bar{y}_{i-1}$ gives

$$(\bar{y}_{i+1} + \bar{y}_{i-1}) = 2\bar{y}_i + \tfrac{2}{2}\bar{y}''|_i\,\Delta x^2 + \tfrac{2}{24}\bar{y}''''|_i\,\Delta x^4 + \tfrac{2}{720}\bar{y}^{(vi)}|_i\,\Delta x^6 + \cdots \tag{8.63}$$

Adding $\bar{y}_{i+2}$ and $\bar{y}_{i-2}$ gives

$$(\bar{y}_{i+2} + \bar{y}_{i-2}) = 2\bar{y}_i + \tfrac{8}{2}\bar{y}''|_i\,\Delta x^2 + \tfrac{32}{24}\bar{y}''''|_i\,\Delta x^4 + \tfrac{128}{720}\bar{y}^{(vi)}|_i\,\Delta x^6 + \cdots \tag{8.64}$$

Subtracting $4(\bar{y}_{i+1} + \bar{y}_{i-1})$ from $(\bar{y}_{i+2} + \bar{y}_{i-2})$ gives

$$(\bar{y}_{i+2} + \bar{y}_{i-2}) - 4(\bar{y}_{i+1} + \bar{y}_{i-1}) = 6\bar{y}_i + \bar{y}''''|_i\,\Delta x^4 + \tfrac{1}{6}\bar{y}^{(vi)}|_i\,\Delta x^6 + \cdots \tag{8.65}$$

Solving Eq. (8.64) for $\bar{y}''''|_i$ yields

$$\bar{y}''''|_i = \frac{\bar{y}_{i-2} - 4\bar{y}_{i-1} + 6\bar{y}_i - 4\bar{y}_{i+1} + \bar{y}_{i+2}}{\Delta x^4} - \tfrac{1}{6}\bar{y}^{(vi)}(\xi)\,\Delta x^2 \tag{8.66}$$

where $x_{i-2} \le \xi \le i+2$. Truncating the remainder term yields a second-order centered-difference approximation for $\bar{y}''''|_i$:

$$y''''|_i = \frac{y_{i-2} - 4y_{i-1} + 6y_t - 4y_{i+1} + y_{i+2}}{\Delta x^4} \tag{8.67}$$

Substituting Eq. (8.67) into Eq. (8.59) yields

$$y_{i-2} - 4y_{i-1} + 6y_i - 4y_{i+1} + y_{i+2} = -0.0024\, \Delta x^4 \tag{8.68}$$

Let $\Delta x = 1.0\,\text{m}$. Applying Eq. (8.68) at the four interior points illustrated in Figure 8.10 gives

$$x = 1.0: \quad y_A - 4y_1 + 6y_2 - 4y_3 + y_4 = -0.0024 \tag{8.69a}$$

$$x = 2.0: \quad y_1 - 4y_2 + 6y_3 - 4y_4 + y_5 = -0.0024 \tag{8.69b}$$

$$x = 3.0: \quad y_2 - 4y_3 + 6y_4 - 4y_5 + y_6 = -0.0024 \tag{8.69c}$$

$$x = 4.0: \quad y_3 - 4y_4 + 6y_5 - 4y_6 + y_B = -0.0024 \tag{8.69d}$$

Note that y_1 in Eq. (8.69a) and y_6 in Eq. (8.69b) are zero. Grid points A and B are outside the physical domain. Thus, y_A and y_B are unknown. These values are determined by applying the boundary conditions $\bar{y}''(0.0) = \bar{y}''(L) = 0.0$. Applying Eq. (8.44) at grid points 1 and 6 gives

$$y''|_1 = \frac{y_2 - 2y_1 + y_A}{\Delta x^2} = 0.0 \tag{8.70a}$$

$$y''|_6 = \frac{y_B - 2y_6 + y_5}{\Delta x^2} = 0.0 \tag{8.70b}$$

Solving Eq. (8.70) for y_A and y_B, with $y_1 = y_6 = 0.0$, gives

$$y_A = -y_2 \quad \text{and} \quad y_B = -y_5 \tag{8.71}$$

Substituting these values into Eq. (8.69a) and (8.69d), respectively, and rearranging those two equations yields the following system equation:

$$5y_2 - 4y_3 + \quad y_4 \qquad = -0.0024 \tag{8.72a}$$

$$-4y_2 + 6y_3 - 4y_4 + \quad y_5 = -0.0024 \tag{8.72b}$$

$$y_2 - 4y_3 + 6y_4 - 4y_5 = -0.0024 \tag{8.72c}$$

$$y_3 - 4y_4 + 5y_5 = -0.0024 \tag{8.72d}$$

Expressing Eq. (8.72) in matrix form yields

$$\begin{bmatrix} 5 & -4 & 1 & 0 \\ -4 & 6 & -4 & 1 \\ 1 & -4 & 6 & -4 \\ 0 & 1 & -4 & 5 \end{bmatrix} \begin{bmatrix} y_2 \\ y_3 \\ y_4 \\ y_5 \end{bmatrix} = \begin{bmatrix} -0.0024 \\ -0.0024 \\ -0.0024 \\ -0.0024 \end{bmatrix} \tag{8.73}$$

Although it is not readily apparent, Eq. (8.73) is a pentadiagonal matrix, which can be solved very efficiently by a modified Gauss elimination algorithm similar to the Thomas algorithm for tridiagonal matrices presented in Section 1.5. Solving Eq. (8.73) yields the

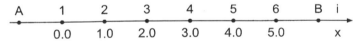

Figure 8.10 Finite difference grid for the beam.

Table 8.11 Solution by the Equilibrium Method with $\Delta x = 1.0$ m

x, m	$y(x)$, m	$\bar{y}(x)$, m	Error(x), m
0.00	0.000000	0.000000	
1.00	0.012000	0.011600	0.000400
2.00	0.019200	0.018600	0.000600
3.00	0.019200	0.018600	0.000600
4.00	0.012000	0.011600	0.000400
5.00	0.000000	0.000000	

results presented in Table 8.11. The exact solution and the errors are presented for comparison.

Let's repeat the solution for $\Delta x = 0.5$ m. In this case, Eq. (8.73) becomes

$$\begin{bmatrix} 5 & -4 & 1 & 0 & 0 & 0 & 0 & 0 & 0 \\ -4 & 6 & -4 & 1 & 0 & 0 & 0 & 0 & 0 \\ 1 & -4 & 6 & -4 & 1 & 0 & 0 & 0 & 0 \\ 0 & 1 & -4 & 6 & -4 & 1 & 0 & 0 & 0 \\ 0 & 0 & 1 & -4 & 6 & -4 & 1 & 0 & 0 \\ 0 & 0 & 0 & 1 & -4 & 6 & -4 & 1 & 0 \\ 0 & 0 & 0 & 0 & 1 & -4 & 6 & -4 & 1 \\ 0 & 0 & 0 & 0 & 0 & 1 & -4 & 6 & -4 \\ 0 & 0 & 0 & 0 & 0 & 0 & 1 & -4 & 5 \end{bmatrix} \begin{bmatrix} y_2 \\ y_3 \\ y_4 \\ y_5 \\ y_6 \\ y_7 \\ y_8 \\ y_9 \\ y_{10} \end{bmatrix} = \begin{bmatrix} 0.000150 \\ 0.000150 \\ 0.000150 \\ 0.000150 \\ 0.000150 \\ 0.000150 \\ 0.000150 \\ 0.000150 \\ 0.000150 \end{bmatrix}$$

(8.74)

The pentadiagonal structure of Eq. (8.74) is readily apparent. Solving Eq. (8.74) yields the results presented in Table 8.12.

The Euclidean norm of the errors in Table 8.11 is 0.123456 m. The Euclidean norm of the errors in Table 8.12 at the four common grid points is 0.012345 m. The ratio of the norms is 3.99. The ratios of the individual errors at the four common grid points in the two

Table 8.12 Solution by the Equilibrium Method with $\Delta x = 0.5$ m

x, m	$y(x)$, m	$\bar{y}(x)$, m	Error(x), m
0.0	0.000000	0.000000	
0.5	0.006188	0.006131	0.000056
1.0	0.011700	0.011600	0.000100
1.5	0.016013	0.015881	0.000131
2.0	0.018750	0.018600	0.000150
2.5	0.019688	0.019531	0.000156
3.0	0.018750	0.018600	0.000150
3.5	0.016013	0.015881	0.000131
4.0	0.011700	0.011600	0.000100
4.5	0.006188	0.006131	0.000056
5.0	0.000000	0.000000	

grids are 3.96, 3.97, 3.98, and 3.99. Both of these results demonstrate that the method is second order.

8.5 DERIVATIVE (AND OTHER) BOUNDARY CONDITIONS

The boundary-value problems considered so far in this chapter have all had Dirichlet (i.e., known function value) boundary conditions. Many problems in engineering and science have derivative (i.e., Neumann) boundary conditions. A procedure for implementing derivative boundary conditions for one-dimensional boundary-value problems is developed in this section. This procedure is directly applicable to derivative boundary conditions for elliptic and parabolic partial differential equations, which are discussed in Sections 9.6 and 10.7, respectively. Implementation of derivative boundary conditions by both the shooting method and the equilibrium method are discussed in this section.

The heat transfer problem presented in Section 8.1 is modified to create a derivative boundary condition by insulating the right end of the rod, so $T'(L) = 0.0$, as illustrated in Figure 8.11. The boundary-value problem is specified as follows:

$$T'' - \alpha^2 T = -\alpha^2 T_a \qquad T(0.0) = T_1 \text{ and } T'(L) = 0.0 \tag{8.75}$$

The general solution of Eq. (8.75) is [see Eq. (8.2)]

$$T(x) = A e^{\alpha x} + B e^{-\alpha x} + T_a \tag{8.76}$$

Substituting the boundary conditions into Eq. (8.76) yields

$$A = (T_1 - T_a)(1 + e^{2\alpha L})^{-1} \qquad \text{and} \qquad B = (T_1 - T_a)e^{2\alpha L}(1 + e^{2\alpha L})^{-1} \tag{8.77}$$

Let $\alpha^2 = 16.0$ cm^{-2}, $L = 1.0$ cm, $T(0.0) = 100.0$ C, $T'(1.0) = 0.0$ C/cm, and $T_a = 0.0$ C. Substituting these values into Eq. (8.77) and the results into Eq. (8.76) gives the exact solution:

$$T(x) = 0.03353501(e^{4x} + 99.96646499 e^{-4x}) \tag{8.78}$$

The solution at intervals of $\Delta x = 0.125$ cm is tabulated in Table 8.13 and illustrated in Figure 8.12.

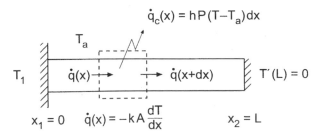

Figure 8.11   Heat transfer in a rod with an insulated end.

Table 8.13   Exact Solution for a Rod with an
Insulated End

x, cm	$\bar{T}(x)$, C	x, cm	$\bar{T}(x)$, C
0.000	100.000000	0.625	8.614287
0.125	60.688016	0.750	5.650606
0.250	36.866765	0.875	4.129253
0.375	22.455827	1.000	3.661899
0.500	13.776782		

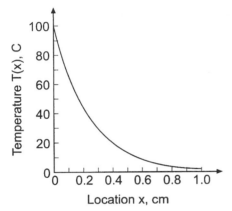

Figure 8.12   Exact solution for a rod with an insulated end.

8.5.1 The Shooting Method

The shooting method for derivative boundary conditions is analogous to the shooting
method for Dirichlet boundary conditions, except that we shoot for the value of the
derivative instead of the value of the function at the boundary. As an example, consider the
linear second-order boundary-value problem:

$$\boxed{\bar{y}'' + P\bar{y}' + Q\bar{y} = F(x) \qquad \bar{y}(x_1) = \bar{y}_1 \text{ and } \bar{y}'(x_2) = \bar{y}'|_2} \tag{8.79}$$

Rewrite Eq. (8.79) as a system of two first-order ODEs:

$$\bar{y}' = \bar{z} \qquad \bar{y}(x_1) = \bar{y}_1 \tag{8.80a}$$
$$\bar{z}' = F(x) - P\bar{z} - Q\bar{y} \qquad \bar{z}(x_1) = \bar{y}'(x_1) = \bar{y}'|_1 = ? \tag{8.80b}$$

The derivative boundary condition $\bar{y}'(x_2) = \bar{y}'|_2$ yields

$$\bar{z}(x_2) = \bar{y}'|_2 \tag{8.81}$$

Consequently, the shooting method described in Section 8.3 applies directly, with the
single change that as we vary $z(x_1) = y'|_1$, we are shooting for $\bar{z}(x_2) = \bar{y}'|_2$ rather than for
$y(x_2) = \bar{y}_2$.

Example 8.7   A derivative BC by the shooting method

As an example of the shooting method with a derivative BC, let's solve Eq. (8.75). The implicit trapezoid method for this problem is presented in Example 8.1, Eqs. (8.23) and (8.24). Let $\alpha^2 = 16.0$ cm^{-2}, $T_a = 0.0$ C, and $\Delta x = 0.25$ cm. Let $T(0.0) = 100.0$ C and $U(0.0) = T'(0.0)$ be the initial conditions at $x = 0.0$ cm, and shoot for the boundary condition at $x = 1.0$ cm, $\bar{T}'(1.0) = U(1.0) = 0.0$ C/cm. Let $U(0.0)^{(1)} = -405.0$ C/cm and $U(0.0)^{(2)} = 395.0$ C/cm. The solution for these two values of $U(0.0)$ are presented in Table 8.14 and Figure 8.13. For $U(0.0)^{(1)} = -405.0$ C/cm, $T'(1.0)^{(1)} = -207.469136$ C/cm, and for $U(0.0)^{(2)} = -395.0$ C/cm, $T'(1.0)^{(2)} = 197.592593$ C/cm. The final solution (obtained by superposition), the exact solution, and the errors are presented in Table 8.15. The final solution is also plotted in Figure 8.13.

Table 8.14   Solution for $U(0.0)^{(1)} = -405.0$ C/cm and $U(0.0)^{(2)} = -395.0$ C/cm

x, cm	$T(x)^{(1)}$, C	$U(x)^{(1)}$, C/cm	$T(x)^{(2)}$, C	$U(x)^{(2)}$, C/cm
0.00	100.000000	−405.000000	100.000000	−395.000000
0.25	31.666667	−141.666667	35.000000	−125.000000
0.50	5.555556	−67.222222	16.666667	−21.666667
0.75	−13.148148	−82.407407	20.555556	52.777778
1.00	−49.382716	−207.469136	51.851852	197.592593

Repeating the solution with $\Delta x = 0.125$ cm yields the results presented in Table 8.16.

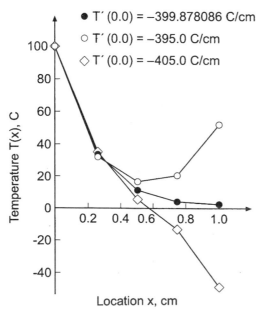

● T′ (0.0) = −399.878086 C/cm
○ T′ (0.0) = −395.0 C/cm
◇ T′ (0.0) = −405.0 C/cm

Figure 8.13   Solution by the shooting method.

Table 8.15 Solution with a Derivative BC by the Shooting Method for $\Delta x = 0.25$ cm

x, cm	$T(x)$, C	$\bar{T}(x)$, C	Error(x), C
0.00	100.000000	100.000000	
0.25	33.373971	36.866765	-3.492794
0.50	11.246571	13.776782	-2.530211
0.75	4.114599	5.650606	-1.536007
1.00	2.468760	3.661899	-1.193140

Table 8.16 Solution with a Derivative BC by the Shooting Method for $\Delta x = 0.125$ cm

x, cm	$T(x)^{(1)}$, C	$T(x)^{(2)}$, C	$T(x)$, C	$\bar{T}(x)$, C	Error(x), C
0.000	100.000000	100.000000	100.000000	100.000000	
0.125	60.000000	60.266667	60.030083	60.688016	-0.657932
0.250	36.000000	36.604444	36.068189	36.866765	-0.798576
0.375	21.600000	22.703407	21.724478	22.455827	-0.731349
0.500	12.960000	14.856612	13.173962	13.776782	-0.602820
0.625	7.776000	10.971581	8.136502	8.614287	-0.477786
0.750	4.665600	10.012304	5.268775	5.650606	-0.381831
0.875	2.799360	11.722974	3.806056	4.129253	-0.323197
1.000	1.679616	16.559771	3.358285	3.661899	-0.303615

The Euclidean norm of the errors in Table 8.15 for $\Delta x = 0.25$ cm is 4.731224. The Euclidean norm of the errors in Table 8.16 at the four common grid points is 1.113145 C. The ratio of the norms is 4.25. These results demonstrate that the method is second order. These Euclidean norms are comparable to the Euclidean norms for the errors in Tables 8.4 and 8.5, 4.249254 C and 0.979800 C, respectively, which were obtained for the heat transfer problem with Dirichlet boundary conditions.

8.5.2 The Equilibrium Method

When the equilibrium method is used to solve a boundary-value problem with a derivative boundary condition, a finite difference procedure must be developed to solve for the value of the function at the boundary where the derivative boundary condition is imposed. Consider the linear, variable coefficient, second-order boundary-value problem:

$$\bar{y}'' + P(x)\,\bar{y}' + Q(x)\,\bar{y} = F(x) \qquad \bar{y}(x_1) = \bar{y}_1 \text{ and } \bar{y}'(x_2) = \bar{y}'|_2 \tag{8.82}$$

A finite difference equation must be developed to evaluate $y(x_2)$.

Consider the finite difference grid in the neighborhood of point x_2, which is illustrated in Figure 8.14. The second-order centered-difference approximation of Eq. (8.82) at boundary point x_2 is given by Eq. (8.46) evaluated at $i = I$:

$$\left(1 - \frac{\Delta x}{2} P_I\right) y_{I-1} + (-2 + \Delta x^2\, Q_I) y_I + \left(1 + \frac{\Delta x}{2} P_I\right) y_{I+1} = \Delta x^2\, F_I \tag{8.83}$$

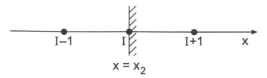

Figure 8.14   Finite difference grid at the right boundary.

where point $I + 1$ is outside of the solution domain. The value of y_{I+1} is unknown. It can be approximated by expressing the derivative boundary condition at point I in finite difference form as follows. From Eq. (8.43),

$$\bar{y}'|_I = \frac{\bar{y}_{I+1} - \bar{y}_{I-1}}{2\,\Delta x} + 0(\Delta x^2) \tag{8.84}$$

Solving Eq. (8.84) for $\bar{y}_{I+1}$ gives

$$\bar{y}_{I+1} = \bar{y}_{I-1} + 2\,\Delta x\,\bar{y}'|_I + \Delta x\,0(\Delta x^2) \tag{8.85}$$

Truncating the remainder term yields an expression for y_{I+1}:

$$\boxed{y_{I+1} = y_{I-1} + 2\,\Delta x\,\bar{y}'|_I} \tag{8.86}$$

Substituting Eq. (8.86) into Eq. (8.83) and simplifying yields the desired FDE:

$$\boxed{2y_{I-1} + (-2 + \Delta x^2 Q_I)y_I = \Delta x^2\,F_I - \Delta x(2 + \Delta x\,P_I)\bar{y}'|_I} \tag{8.87}$$

When Eq. (8.87) is used in conjunction with the centered-difference approximation at the interior points, Eq. (8.46), a tridiagonal system of equations results, which can be solved by the Thomas algorithm.

Example 8.8.   A derivative BC by the equilibrium method

Let's solve the heat transfer problem in a rod with an insulated end, Eq. (8.75), using the equilibrium method. Recall Eq. (8.75):

$$T'' - \alpha^2 T = -\alpha^2 T_a \qquad T(0.0) = 100.0 \text{ and } T'(1.0) = 0.0 \tag{8.88}$$

The interior point FDE is presented in Section 8.4, Eq. (8.46):

$$T_{i-1} - (2 + \alpha^2\,\Delta x^2)T_i + T_{i+1} = -\alpha^2\,\Delta x^2\,T_a \tag{8.89}$$

Applying Eq. (8.87) at the right end of the rod with $P = F = 0$ and $Q = -\alpha^2$ gives

$$2T_{I-1} - (2 + \alpha^2\,\Delta x^2)T_I = -2\,\Delta x\,\bar{T}'|_I \tag{8.90}$$

Let $\alpha^2 = 16.0 \text{ cm}^{-2}$, $T_a = 0.0\,\text{C}$, $\bar{T}'(1.0) = 0.0$, and $\Delta x = 0.25 \text{ cm}$. Equations (8.89) and (8.90) become

$$T_{i-1} - 3.0T_i + T_{i+1} = 0 \qquad (i = 1, 2, \dots, I - 1) \tag{8.91a}$$

$$2T_{I-1} - 3.0T_I = 0 \qquad (i = I) \tag{8.91b}$$

Applying Eqs. (8.91a) and (8.91b) at the four grid points yields the following tridiagonal system of equations, which can be solved by the Thomas algorithm:

$$x = 0.25: \quad T_1 - 3.0T_2 + T_3 = 0.0 \quad T_1 = \bar{T}_1 = 100.0 \tag{8.92a}$$

$$x = 0.50: \quad T_2 - 3.0T_3 + T_4 = 0.0 \tag{8.92b}$$

$$x = 0.75: \quad T_3 - 3.0T_4 + T_5 = 0.0 \tag{8.92c}$$

$$x = 1.00: \quad 2.0T_4 - 3.0T_5 = 0.0 \quad \bar{T}_x|_{1.0} = 0.0 \tag{8.92d}$$

The results are presented in Table 8.17. Repeating the solution with $\Delta x = 0.125$ cm yields the results presented in Table 8.18. The errors are presented in Figure 8.15, which also present the errors for extrapolation of these second-order results.

The Euclidean norm of the errors in Table 8.17 is 2.045460 C. The Euclidean norm of the errors in Table 8.18 at the four common grid points is 0.536997 C. The ratio of the errors is 3.81. These results demonstrate that the method is second order. These Euclidean norms are comparable to the Euclidean norms for the errors in Tables 8.7 and 8.8, 1.766412 C and 0.468057 C, respectively, which were obtained for the solution of the heat transfer problem with Dirichlet boundary conditions by the equilibrium method.

Table 8.17 Solution with a Derivative BC by the Equilibrium Method for $\Delta x = 0.25$ cm

x, cm	$T(x)$, C	$\bar{T}(x)$, C	Error(x), C
0.00	100.000000	100.000000	0.000000
0.25	38.297872	36.866765	1.431107
0.50	14.893617	13.776782	1.116835
0.75	6.382979	5.650606	0.732373
1.00	4.255319	3.661899	0.593420

Table 8.18 Solution with a Derivative BC by the Equilibrium Method for $\Delta x = 0.125$ cm

x, cm	$T(x)$, C	$\bar{T}(x)$, C	Error(x), C
0.000	100.000000	100.000000	
0.125	60.998665	60.688016	0.310649
0.250	37.246996	36.866765	0.380231
0.375	22.807076	22.455827	0.351249
0.500	14.068925	13.776782	0.292144
0.625	8.848006	8.614287	0.233719
0.750	5.839088	5.650606	0.188482
0.875	4.289942	4.129253	0.160690
1.000	3.813282	3.661899	0.151383

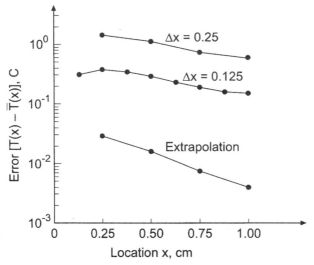

Figure 8.15 Errors in the solution by the equilibrium method.

8.5.3 Mixed Boundary Conditions

A mixed boundary condition at the right-hand boundary x_2 has the form

$$A\bar{y}(x_2) + B\bar{y}'(x_2) = C \qquad (8.93)$$

Mixed boundary conditions are implemented in the same manner as derivative boundary conditions, with minor modifications.

A mixed boundary condition is implemented in the shooting method simply by varying the assumed initial condition until the mixed boundary condition, Eq. (8.93), is satisfied at the other boundary.

A mixed boundary condition is implemented in the equilibrium method in the same manner as a derivative boundary condition is implemented. A finite difference approximation for y_{I+i} is obtained by approximating $\bar{y}'(x_2)$ by the second-order centered-difference approximation, Eq. (8.43). Thus, Eq. (8.93) becomes

$$A\bar{y}_I + B\frac{\bar{y}_{I+1} - \bar{y}_{I-1}}{2\,\Delta x} + 0(\Delta x^2) = C \qquad (8.94)$$

Solving Eq. (8.94) for $\bar{y}_{I+1}$ and truncating the remainder term yields

$$y_{I+1} = y_{I-1} + \frac{2\,\Delta x}{B}(C - Ay_I) \qquad (8.95)$$

Solving Eq. (8.95) in conjunction with the second-order centered-difference approximation of the boundary-value ODE at the interior points, Eq. (8.46), yields a tridiagonal system of FDEs, which can be solved by the Thomas algorithm.

8.5.4 Boundary Condition at Infinity

Occasionally one boundary condition is given at infinity, as illustrated in Figure 8.16. For example, for bodies moving through the atmosphere, infinity simply means very far away. In such a case, the boundary conditions might be

$$\bar{y}(0) = \bar{y}_0 \quad \text{and} \quad \bar{y}(\infty) = \bar{y}_\infty \tag{8.96}$$

Derivative boundary conditions can also be specified at infinity. Two procedures for implementing boundary conditions at infinity are:

1. Replace ∞ with a large value of x, say $x = X$.
2. An asymptotic solution at large values of x.

8.5.4.1 Finite Domain

In this approach, the boundary condition at $x = \infty$ is simply replaced by the same boundary condition applied at a finite location, $x = X$. Thus,

$$\bar{y}(\infty) = \bar{y}_\infty \rightarrow \bar{y}(X) = \bar{y}_\infty \quad \text{large } X \tag{8.97}$$

This procedure is illustrated in Figure 8.17. The boundary-value problem is then solved in the usual manner.

The major problem with this approach is determining what value of X, if any, yields a reasonable solution to the original problem. In most cases, our interest is in the near

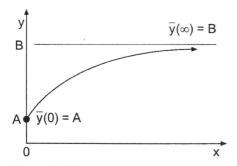

Figure 8.16 Boundary condition at infinity.

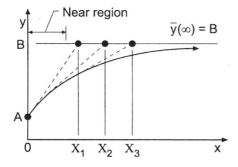

Figure 8.17 Finite domain approximation.

region far away from infinity. In that case, successively larger values of X, denoted by X_1, X_2, etc., can be chosen, and the boundary-value problem is solved for each value of X. The solution in the near region can be monitored as X increases, until successive solutions in the region of interest change by less than some prescribed tolerance.

8.5.4.2 Asymptotic Solution

A second approach for implementing boundary conditions at infinity is based on an asymptotic solution for large values of x. In many problems, the behavior of the solution near $x = \infty$ is much simpler than the behavior in the near region. The governing differential equation can be simplified, perhaps by linearization, and the simplified differential equation can be solved exactly, including the boundary condition at infinity, to yield the solution

$$\bar{y}_{\text{asymptotic}}(x) = F(x) \qquad X \le x \le \infty \tag{8.98}$$

The boundary condition for the solution of the original differential equation is determined by choosing a finite location, $x = X$, and substituting that value of x into Eq. (8.98) to obtain

$$\bar{y}_{\text{asymptotic}}(X) = F(X) = Y \tag{8.99}$$

The boundary condition $\bar{y}(\infty) = \bar{y}_\infty$ is replaced by the boundary condition $\bar{y}(X) = Y$, as illustrated in Figure 8.18. As discussed in the previous subsection, the value of X can be varied to determine its effect on the solution in the near region.

8.6 HIGHER-ORDER EQUILIBRIUM METHODS

Consider the general linear second-order boundary-value ODE:

$$\boxed{\bar{y}'' + P\bar{y}' + Q\bar{y} = F(x) \qquad \bar{y}(x_1) = \bar{y}_1 \text{ and } \bar{y}(x_2) = \bar{y}_2} \tag{8.100}$$

When solving ODEs such as Eq. (8.100) by the shooting method, it is quite easy to develop higher-order methods (e.g., the fourth-order Runge-Kutta method). However, it is more difficult to develop equilibrium methods higher than second order. Two procedures for obtaining fourth-order equilibrium methods are presented in this section:

  1. The five-point method
  2. The compact three-point method

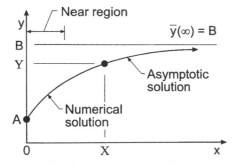

Figure 8.18 Asymptotic solution approximation.

8.6.1 The Five-Point Fourth-Order Equilibrium Method

Consider the five-point finite difference grid illustrated in Figure 8.19. Fourth-order approximations for $\bar{y}'$ and $\bar{y}''$ require five grid points: $i - 2$ to $i + 2$. The sixth-order Taylor series for $\bar{y}(x)$ at these points with base point i are given by

$$\bar{y}_{i+2} = \bar{y}_i + \bar{y}'|_i(2\,\Delta x) + \tfrac{1}{2}\bar{y}''|_i(2\,\Delta x)^2 + \tfrac{1}{6}\bar{y}'''|_i(2\,\Delta x)^3 + \tfrac{1}{24}\bar{y}^{iv}|_i(2\,\Delta x)^4$$
$$+ \tfrac{1}{120}\bar{y}^{v}|_i(2\,\Delta x)^5 + \tfrac{1}{720}\bar{y}^{vi}|_i(2\,\Delta x)^6 + \cdots \tag{8.101a}$$

$$\bar{y}_{i+1} = \bar{y}_i + \bar{y}'|_i\,\Delta x + \tfrac{1}{2}\bar{y}''|_i\,\Delta x^2 + \tfrac{1}{6}\bar{y}'''|_i\,\Delta x^3 + \tfrac{1}{24}\bar{y}^{iv}|_i\,\Delta x^4$$
$$+ \tfrac{1}{120}\bar{y}^{v}|_i\,\Delta x^5 + \tfrac{1}{720}\bar{y}^{vi}|_i\,\Delta x^6 + \cdots \tag{8.101b}$$

$$\bar{y}_{i-1} = \bar{y}_i - \bar{y}'|_i\,\Delta x + \tfrac{1}{2}\bar{y}''|_i\,\Delta x^2 - \tfrac{1}{6}\bar{y}'''|_i\,\Delta x^3 + \tfrac{1}{24}\bar{y}^{iv}|_i\,\Delta x^4$$
$$- \tfrac{1}{120}\bar{y}^{v}|_i\,\Delta x^5 + \tfrac{1}{720}\bar{y}^{vi}|_i\,\Delta x^6 + \cdots \tag{8.101c}$$

$$\bar{y}_{i-2} = \bar{y}_i - \bar{y}'|_i(2\,\Delta x) + \tfrac{1}{2}\bar{y}''|_i(2\,\Delta x)^2 - \tfrac{1}{6}\bar{y}'''|_i(2\,\Delta x)^3 + \tfrac{1}{24}\bar{y}^{iv}|_i(2\,\Delta x)^4$$
$$- \tfrac{1}{120}\bar{y}^{v}|_i(2\,\Delta x)^5 + \tfrac{1}{720}\bar{y}^{vi}|_i(2\,\Delta x)^6 + \cdots \tag{8.101d}$$

A finite difference approximation (FDA) for $\bar{y}'|_i$ can be obtained by forming the combination $-\bar{y}_{i+2} + 8\bar{y}_{i+1} - 8\bar{y}_{i-1} + \bar{y}_{i-2}$. Thus,

$$\bar{y}'|_i = \frac{-\bar{y}_{i+2} + 8\bar{y}_{i+1} - 8\bar{y}_{i-1} + \bar{y}_{i-2}}{12\,\Delta x} + \frac{48}{120}\bar{y}^{v}(\xi)\,\Delta x^4 \tag{8.102}$$

A FDA for $\bar{y}''|_i$ can be obtained by forming the combination $-\bar{y}_{i+2} + 16\bar{y}_{i+1} - 30\bar{y}_i + 16\bar{y}_{i-1} - \bar{y}_{i-2}$. The result is

$$\bar{y}''|_i = \frac{-\bar{y}_{i+2} + 16\bar{y}_{i+1} - 30\bar{y}_i + 16\bar{y}_{i-1} - \bar{y}_{i-2}}{12\,\Delta x^2} + \frac{96}{720}\bar{y}^{vi}(\xi)\,\Delta x^4 \tag{8.103}$$

These FDAs can be substituted into the boundary-value ODE to give an $0(\Delta x^4)$ finite difference equation (FDE).

Problems arise at the points adjacent to the boundaries, where the centered fourth-order FDA cannot be applied. Nonsymmetrical fourth-order FDAs can be used at these points, or second-order FDEs can be used with some loss of accuracy. A pentadiagonal matrix results. A method similar to the Thomas algorithm for tridiagonal matrices can be used to give an efficient solution of the pentadiagonal system of FDEs.

8.6.2 The Compact Three-Point Fourth-Order Equilibrium Method

Implicit three-point fourth-order FDAs can be devised. When the overall algorithm is implicit, as it is when solving boundary-value problems by the equilibrium method, implicitness is already present in the FDEs.

$i-2 \qquad i-1 \qquad i \qquad i+1 \qquad i+2 \qquad x$

Figure 8.19 Five-point finite difference grid.

First, let's develop the compact three-point fourth-order finite difference approximation for $\bar{y}'_i$. Subtract the Taylor series for $\bar{y}_{i-1}$, Eq. (8.101c), from the Taylor series for $\bar{y}_{i+1}$, Eq. (8.101b):

$$\bar{y}_{i+1} - \bar{y}_{i-1} = 2\,\Delta x\,\bar{y}'|_i + \tfrac{1}{3}\,\Delta x^3\,\bar{y}'''|_i + 0(\Delta x^5) \tag{8.104}$$

Dividing Eq. (8.104) by $2\,\Delta x$ gives

$$\frac{\bar{y}_{i+1} - \bar{y}_{i-1}}{2\,\Delta x} = \bar{y}'|_i + \tfrac{1}{6}\,\Delta x^2\,\bar{y}'''|_i + 0(\Delta x^4) \tag{8.105}$$

Write Taylor series for $\bar{y}'|_{i+1}$ and $\bar{y}'|_{i-1}$ with base point i. Thus,

$$\bar{y}'|_{i+1} = \bar{y}'|_i + \bar{y}''|_i\,\Delta x + \tfrac{1}{2}\bar{y}'''|_i\,\Delta x^2 + \tfrac{1}{6}\bar{y}^{\mathrm{iv}}|_i\,\Delta x^3 + \tfrac{1}{24}\bar{y}^{\mathrm{v}}|_i\,\Delta x^4 + \cdots \tag{8.106}$$

$$\bar{y}'|_{i-1} = \bar{y}'|_i - \bar{y}''|_i\,\Delta x + \tfrac{1}{2}\bar{y}'''|_i\,\Delta x^2 - \tfrac{1}{6}\bar{y}^{\mathrm{iv}}|_i\,\Delta x^3 + \tfrac{1}{24}\bar{y}^{\mathrm{vi}}|_i\,\Delta x^4 - \cdots \tag{8.107}$$

Adding Eqs. (8.106) and (8.107) yields

$$\bar{y}'|_{i+1} - 2\bar{y}'|_i + \bar{y}'|_{i-1} = \Delta x^2 \bar{y}'''|_i + 0(\Delta x^4) \tag{8.108}$$

Dividing Eq. (8.108) by 6, adding and subtracting $\bar{y}'|_i$, and rearranging gives

$$\bar{y}'|_i + \tfrac{1}{6}(\bar{y}'|_{i+1} - 2\bar{y}'|_i + \bar{y}'|_{i-1}) = \bar{y}'|_i + \tfrac{1}{6}\,\Delta x^2 \bar{y}'''|_i + 0(\Delta x^4) \tag{8.109}$$

The right-hand sides of Eqs. (8.105) and (8.109) are identical. Equating the left-hand sides of those two equations gives

$$\bar{y}'|_i + \tfrac{1}{6}(\bar{y}'|_{i+1} - 2\bar{y}'|_i + \bar{y}'|_{i-1}) = \frac{\bar{y}_{i+1} - \bar{y}_{i-1}}{2\,\Delta x} + 0(\Delta x^4) \tag{8.110}$$

Define the first-order and second-order centered differences $\delta \bar{y}_i$ and $\delta^2 \bar{y}'|_i$, respectively, as follows:

$$\delta \bar{y}_i = \bar{y}_{i+1} - \bar{y}_{i-1} \tag{8.111}$$

$$\delta^2 \bar{y}'|_i = \bar{y}'|_{i+1} - 2\bar{y}'|_i + \bar{y}'|_{i-1} \tag{8.112}$$

Substituting Eqs. (8.111) and (8.112) into Eq. (8.110) gives

$$\bar{y}'|_i + \tfrac{1}{6}\,\delta^2 \bar{y}'|_i = \frac{\delta \bar{y}_i}{2\,\Delta x} + 0(\Delta x^4) \tag{8.113}$$

Solving Eq. (8.113) for $\bar{y}'|_i$ yields

$$\bar{y}'|_i = \frac{\delta \bar{y}_i}{2\,\Delta x(1 + \delta^2/6)} + 0(\Delta x^4) \tag{8.114}$$

Truncating the remainder term yields an implicit three-point fourth-order centered-difference approximation for $\bar{y}'|_i$:

$$\boxed{y'|_i = \frac{\delta y_i}{2\,\Delta x(1 + \delta^2/6)}} \tag{8.115}$$

Now let's develop the compact three-point fourth-order finite difference approximation for $\bar{y}''_i$. Adding the Taylor series for $\bar{y}_{i+1}$, Eq. (8.101b), and $\bar{y}_{i-1}$, Eq. (8.101c), gives

$$\bar{y}_{i+1} - 2\bar{y}_i + \bar{y}_{i-1} = \Delta x^2\,\bar{y}''|_i + \tfrac{1}{12}\,\Delta x^4\,\bar{y}^{\mathrm{iv}}|_i + 0(\Delta x^6) \tag{8.116}$$

Dividing Eq. (8.116) by Δx^2 gives

$$\frac{\bar{y}_{i+1} - 2\bar{y}_i + \bar{y}_{i-1}}{\Delta x^2} = \bar{y}''|_i + \tfrac{1}{12} \Delta x^2 \, \bar{y}^{iv}|_i + 0(\Delta x^4) \tag{8.117}$$

Write Taylor series for $\bar{y}''|_{i+1}$ and $\bar{y}''|_{i-1}$ with base point i. Thus,

$$\bar{y}''|_{i+1} = \bar{y}''|_i + \bar{y}'''|_i \, \Delta x + \tfrac{1}{2}\bar{y}^{iv}|_i \, \Delta x^2 + \tfrac{1}{6}\bar{y}^v|_i \, \Delta x^3 + \tfrac{1}{24}\bar{y}^{vi}|_i \, \Delta x^4 + \cdots \tag{8.118}$$

$$\bar{y}''|_{i-1} = \bar{y}''|_i - \bar{y}'''|_i \, \Delta x + \tfrac{1}{2}\bar{y}^{iv}|_i \, \Delta x^2 - \tfrac{1}{6}\bar{y}^v|_i \, \Delta x^3 + \tfrac{1}{24}\bar{y}^{vi}|_i \, \Delta x^4 - \cdots \tag{8.119}$$

Adding Eqs. (8.118) and (8.119) yields

$$\bar{y}''|_{i+1} - 2\bar{y}''|_i + \bar{y}''|_{i-1} = \Delta x^2 \, \bar{y}^{iv}|_i + 0(\Delta x^4) \tag{8.120}$$

Dividing Eq. (8.120) by 12, adding and subtracting $\bar{y}''|_i$, and rearranging gives

$$\bar{y}''|_i + \tfrac{1}{12}(\bar{y}''|_{i+1} - 2\bar{y}''|_i + \bar{y}''|_{i-1}) = \bar{y}''|_i + \tfrac{1}{12} \Delta x^2 \, \bar{y}^{iv}|_i + 0(\Delta x^4) \tag{8.121}$$

The right-hand sides of Eqs. (8.117) and (8.121) are identical. Equating the left-hand sides of those equations gives

$$\bar{y}''|_i + \tfrac{1}{12}(\bar{y}''|_{i+1} - 2\bar{y}''|_i + \bar{y}''|_{i-1}) = \frac{1}{\Delta x^2}(\bar{y}_{i+1} - 2\bar{y}_i + \bar{y}_{i-1}) + 0(\Delta x^4) \tag{8.122}$$

Define the second-order centered-differences $\delta^2\bar{y}_i$ and $\delta^2\bar{y}''|_i$ as follows:

$$\delta^2\bar{y}_i = \bar{y}_{i+1} - 2\bar{y}_i + \bar{y}_{i-1} \tag{8.123}$$

$$\delta^2\bar{y}''|_i = \bar{y}''|_{i+1} - 2\bar{y}''|_i + \bar{y}''|_{i-1} \tag{8.124}$$

Substituting Eqs. (8.123) and (8.124) into Eq. (8.122) gives

$$\bar{y}''|_i + \tfrac{1}{12}\delta^2\bar{y}''|_i = \frac{\delta^2\bar{y}_i}{\Delta x^2} + 0(\Delta x^4) \tag{8.125}$$

Solving Eq. (8.125) for $\bar{y}''|_i$ yields

$$\bar{y}''|_i = \frac{\delta^2\bar{y}_i}{\Delta x^2(1 + \delta^2/12)} + 0(\Delta x^4) \tag{8.126}$$

Truncating the remainder term yields an implicit three-point fourth-order centered-difference approximation for $\bar{y}''|_i$:

$$\boxed{y''|_i = \frac{\delta^2 y_i}{\Delta x^2(1 + \delta^2/12)}} \tag{8.127}$$

Consider the second-order boundary-value ODE:

$$\bar{y}'' + P(x, \bar{y})\bar{y}' + Q(x, \bar{y})\bar{y} = F(x) \tag{8.128}$$

Substituting Eqs. (8.115) and (8.127) into Eq. (8.128) yields the implicit fourth-order finite difference equation:

$$\boxed{\frac{\delta^2 y_i}{\Delta x^2(1 + \delta^2/12)} + P_i\frac{\delta y_i}{2\,\Delta x(1 + \delta^2/6)} + Q_i y_i = F_i} \tag{8.129}$$

If P and/or Q depend on y, the system of FDEs is nonlinear. If not, the system is linear.

Example 8.9.   The compact three-point fourth-order equilibrium method

As an example of the compact three-point fourth-order method, let's solve the heat transfer problem presented in Section 8.1. Thus,

$$T'' - \alpha^2 T = -\alpha^2 T_a \qquad T(0.0) = 0.0 \text{ C and } T(1.0) = 100.0 \text{ C} \tag{8.130}$$

Substituting Eq. (8.127) into Eq. (8.130) gives

$$\frac{\delta^2 T_i}{\Delta x^2 (1 + \delta^2/12)} - \alpha^2 T_i = -\alpha^2 T_a \tag{8.131}$$

Simplifying Eq. (8.131) yields

$$\delta^2 T_i - \alpha^2 \, \Delta x^2 (1 + \delta^2/12) T_i = -\alpha^2 \, \Delta x^2 (1 + \delta^2/12) T_a \tag{8.132}$$

Expanding the second-order centered-difference operator δ^2 gives

$$(T_{i+1} - 2T_i + T_{i-1}) - \alpha^2 \, \Delta x^2 \, T_i - \frac{\alpha^2 \, \Delta x^2}{12}(T_{i+1} - 2T_i + T_{i-1}) = -\alpha^2 \, \Delta x^2 \, T_a \tag{8.133}$$

where $\delta^2 T_a = 0$ since T_a is constant. Defining $\beta = \alpha^2 \, \Delta x^2/12$ and gathering terms yields

$$(1 - \beta)T_{i-1} - (2 + 10\beta)T_i + (1 - \beta)T_{i+1} = -12\beta T_a \tag{8.134}$$

Let $\alpha^2 = 16.0 \text{ cm}^{-2}$, $T_a = 0.0 \text{ C}$, and $\Delta x = 0.25 \text{ cm}$, which gives $\beta = 16(0.25)^2/12 = 0.083333$. Equation (8.134) becomes

$$0.916667 T_{i-1} - 2.833333 T_i + 0.916667 T_{i+1} = 0 \tag{8.135}$$

Applying Eq. (8.15) at the three interior grid points, transferring the boundary conditions to the right-hand sides of the equations, and solving the tridiagonal system of FDEs by the Thomas algorithm yields the results presented in Table 8.19. Repeating the solution with $\Delta x = 0.125 \text{ cm}$ yields the results presented in Table 8.20.

The Euclidean norm of the errors in Table 8.19 is 0.092448 C, which is 19.11 times smaller than the Euclidean norm of 1.766412 C for the errors in Table 8.8. The Euclidean norm of the errors in Table 8.20 at the common grid points of the two grids is 0.005918 C,

Table 8.19   Solution by the Compact Fourth-Order Equilibrium Method for $\Delta x = 0.25$ cm

x, cm	$T(x)$, C	$\bar{T}(x)$, C	Error(x), C
0.00	0.000000	0.000000	
0.25	4.283048	4.306357	−0.023309
0.50	13.238512	13.290111	−0.051599
0.75	36.635989	36.709070	−0.073081
1.00	100.000000	100.000000	

Table 8.20 Solution by the Compact Fourth-Order Equilibrium Method for $\Delta x = 0.125$ cm

x, cm	$T(x)$, C	$\bar{T}(x)$, C	Error(x), C
0.000	0.000000	0.000000	
0.125	1.908760	1.909479	−0.000719
0.250	4.304863	4.306357	−0.001494
0.375	7.800080	7.802440	−0.002360
0.500	13.286807	13.290111	−0.003305
0.625	22.165910	22.170109	−0.004200
0.750	36.704394	36.709070	−0.004677
0.875	60.614212	60.618093	−0.003880
1.000	100.000000	100.000000	

which is 79.09 times smaller than the Euclidean norm of 0.468057 C for the errors in Table 8.9. The ratio of the norms is 15.62, which demonstrates that the method is fourth order.

8.7 THE EQUILIBRIUM METHOD FOR NONLINEAR BOUNDARY-VALUE PROBLEMS

Consider the general nonlinear second-order boundary-value ODE:

$$\bar{y}'' + P(x, \bar{y})\bar{y}' + Q(x, \bar{y})\bar{y} = F(x) \qquad \bar{y}(x_1) = \bar{y}_1 \text{ and } \bar{y}(x_2) = \bar{y}_2 \qquad (8.136)$$

The solution of Eq. (8.136) by the shooting method, as discussed in Section 8.3, is straightforward. The shooting method is based on finite difference methods for solving initial-value problems. Explicit methods, such as the Runge-Kutta method, solve nonlinear initial-values ODEs directly. Consequently, such methods can be applied directly to nonlinear boundary-value ODEs as described in Section 8.3 for linear boundary-value ODEs.

The solution of Eq. (8.136) by the equilibrium method is more complicated, since the corresponding finite difference equation (FDE) is nonlinear, which yields a system of nonlinear FDEs. Two methods for solving nonlinear boundary-value ODEs by the equilibrium method are presented in this section:

1. Iteration
2. Newton's method

8.7.1 Iteration

The iteration method for solving nonlinear boundary-value ODEs is similar to the fixed-point iteration method presented in Section 3.4 for solving single nonlinear equations. In that method, the problem $f(x) = 0$ is rearranged into the form $x = g(x)$, and an initial value of x is assumed and substituted into $g(x)$ to give the next approximation for x. The procedure is repeated to convergence.

The solution of a nonlinear boundary-value ODE by iteration proceeds in the following steps.

1. Develop a finite difference approximation of the ODE. Linearize the ODE by lagging all the nonlinear coefficients. Preserve the general character of the ODE by lagging the lowest-order terms in a group of terms. For example,

$$\bar{y}''\bar{y}'\bar{y}^{1/2} \rightarrow (\bar{y}'')^{(k+1)}(\bar{y}')^{(k)}(\bar{y}^{1/2})^{(k)} \qquad (8.137)$$

Choose finite difference approximations for the derivatives and construct the linearized FDE.

2. Assume an initial approximation for the solution: $y(x)^{(0)}$. A good initial approximation can reduce the number of iterations. A bad initial approximation may not converge. Choose the initial approximation similar in form to the expected form of the solution. For example, for the Dirichlet boundary conditions $\bar{y}(x_1) = \bar{y}_1$ and $\bar{y}(x_2) = \bar{y}_2$, the initial approximation $y(x)^{(0)}$ could be a step function at x_1, a step function at x_2, a linear variation from x_1 to x_2, or a quadratic variation from x_1 to x_2, as illustrated in Figure 8.20. The step functions are obviously less desirable than the linear or quadratic variations. The linear variation is quite acceptable in many cases. If additional insight into the problem suggests a quadratic variation, it should be used.

3. Calculate the lagged coefficients, $P(x, y) = P[x, y(x)^{(0)}] = P(x)^{(0)}$, etc.

4. Solve the system of linear FDEs to obtain $y(x)^{(1)}$.

5. Calculate the lagged coefficients based on the new solution, $y(x)^{(1)}$, $P(x)^{(1)}$, etc.

6. Repeat steps 4 and 5 to convergence.

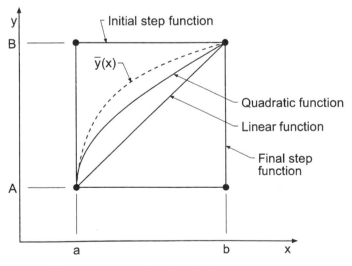

Figure 8.20   Initial approximations for iteration.

Example 8.10. A nonlinear implicit FDE using iteration

Consider the following nonlinear boundary-value ODE:

$$\bar{y}'' + 2\bar{y}\bar{y}' = 4 + 4x^3 \qquad \bar{y}(1.0) = 2.0 \text{ and } \bar{y}(2.0) = 4.5 \tag{8.138}$$

The exact solution to Eq. (8.138) is

$$\bar{y}(x) = x^2 + \frac{1}{x} \tag{8.139}$$

Let's solve Eq. (8.138) by iteration using the second-order equilibrium method.

Let's approximate $\bar{y}'$ and $\bar{y}''$ by second-order centered-difference approximations, Eqs. (8.43) and (8.44), and lag the nonlinear coefficient of $\bar{y}'$. This gives

$$\left(\frac{y_{i+1} - 2y_i + y_{i-1}}{\Delta x^2}\right)^{(k+1)} + 2y^{(k)}\left(\frac{y_{i+1} - y_{i-1}}{2\,\Delta x}\right)^{(k+1)} = 4 + 4x_i^3 \tag{8.140}$$

Multiplying Eq. (8.140) by Δx^2 and gathering terms yields the nonlinear FDE:

$$(1 - \Delta x\, y_i^{(k)})y_{i-1}^{(k+1)} - 2y_i^{(k+1)} + (1 + \Delta x\, y_i^{(k)})y_{i+1}^{(k+1)} = 4\,\Delta x^2(1 + x_i^3) \tag{8.141}$$

Let $\Delta x = 0.25$. Applying Eq. (8.141) at the three interior points of the uniform grid gives

$$x = 1.25: \qquad (1 - 0.25y_2^{(k)})y_1^{(k+1)} - 2y_2^{(k+1)} + (1 + 0.25y_2^{(k)})y_3^{(k+1)}$$
$$= 4(0.25)^2(1 + (1.25)^3) \tag{8.142a}$$

$$x = 1.50: \qquad (1 - 0.25y_3^{(k)})y_2^{(k+1)} - 2y_3^{(k+1)} + (1 + 0.25y_3^{(k)})y_4^{(k+1)}$$
$$= 4(0.25)^2(1 + (1.50)^3) \tag{8.142b}$$

$$x = 1.75: \qquad (1 - 0.25y_4^{(k)})y_3^{(k+1)} - 2y_4^{(k+1)} + (1 + 0.25y_4^{(k)})y_5^{(k+1)}$$
$$= 4(0.25)^2(1 + (1.75)^3) \tag{8.142c}$$

Transferring $\bar{y}(1.0) = 2.0$ and $\bar{y}(2.0) = 4.5$ to the right-hand sides of Eqs. (8.142a) and (8.142c), respectively, yields a tridiagonal system of FDEs, which can be solved by the Thomas algorithm.

For the initial approximation of $y(x)$, $y(x)^{(0)}$, assume a linear variation between the boundary values. Thus,

$$y(x)^{(0)} = -0.5 + 2.5x \tag{8.143}$$

The values of $y(x)^{(0)}$ are presented in the first line of Table 8.21, which corresponds to $k = 0$. Substituting these values into Eq. (8.142) yields the following system of linear algebraic FDEs, which can be solved by the Thomas algorithm:

$$\begin{bmatrix} -2.00000 & 1.65625 & 0.00000 \\ 0.18750 & -2.00000 & 1.81250 \\ 0.00000 & 0.03125 & -2.00000 \end{bmatrix} \begin{bmatrix} y_2 \\ y_3 \\ y_4 \end{bmatrix} = \begin{bmatrix} 0.050781 \\ 1.093750 \\ -7.269531 \end{bmatrix} \tag{8.144}$$

The solution $y(x)^{(1)}$ to Eq. (8.144) is presented in the second line of Table 8.21. Equation (8.142) is then reevaluated with $y(x)^{(1)}$, and a new tridiagonal system of linear FDEs is assembled and solved by the Thomas algorithm. This solution $y(x)^{(2)}$ is presented in the

Table 8.21 Solution of the Nonlinear Implicit FDE by Iteration

		x				
(k)	$y(x)$	1.00	1.25	1.50	1.75	2.00
0	$y(x)^{(0)}$	2.0	2.625000	3.250000	3.875000	4.5
1	$y(x)^{(1)}$	2.0	2.477350	3.022177	3.681987	4.5
2	$y(x)^{(2)}$	2.0	2.395004	2.943774	3.643216	4.5
...	...	...	...	...	...	...
11	$y(x)^{(11)}$	2.0	2.354750	2.911306	3.631998	4.5
	$\bar{y}(x)$	2.0	2.362500	2.916667	3.633929	4.5
	Error		-0.007750	-0.005361	-0.001931	

third line of Table 8.21. This procedure is repeated a total of 11 times until the solution satifies the convergence criterion $|y_i^{(k+1)} - y_i^{(k)}| \le 0.000001$.

8.7.2 Newton's Method

Newton's method for solving nonlinear boundary-value problems by the equilibrium method consists of choosing an approximate solution to the problem $Y(x)$ and assuming that the exact solution $\bar{y}(x)$ is the sum of the approximate solution $Y(x)$ and a small perturbation $\eta(x)$. Thus,

$$\bar{y}(x) = Y(x) + \eta(x) \tag{8.145}$$

Equation (8.145) is substituted into the nonlinear ODE, and terms involving products of $\eta(x)$ and its derivatives are assumed to be negligible compared to linear terms in $\eta(x)$ and its derivatives. This process yields a linear ODE for $\eta(x)$, which can be solved by the equilibrium method. The solution is not exact, since higher-order terms in $\eta(x)$ and its derivatives were neglected. The procedure is repeated with the approximate solution until $\eta(x)$ changes by less than some prescribed tolerance at each point in the finite difference grid.

The procedure is implemented as follows. Assume $Y(x)^{(0)}$. Solve the corresponding linear problem for $\eta(x)^{(0)}$. Substitute these results in Eq. (8.145) to obtain $y(x)^{(1)}$.

$$y(x)^{(1)} = Y(x)^{(0)} + \eta(x)^{(0)} = Y(x)^{(1)} \tag{8.146}$$

The next trial value for $Y(x)$ is $Y(x)^{(1)} = y(x)^{(1)}$. Solve for the corresponding $\eta(x)^{(1)}$. Substituting these results into Eq. (8.145) gives $y(x)^{(2)}$. This procedure is applied repetitively until $\eta(x)$ changes by less than some prescribed tolerance at each point in the finite difference grid. The general iteration algorithm is

$$Y(x)^{(k+1)} = y(x)^{(k+1)} = Y(x)^{(k)} + \eta(x)^{(k)} \tag{8.147}$$

Newton's method converges quadratically. Convergence is faster if a good initial approximation is available. The procedure can diverge for a poor (i.e., unrealistic) initial approximation. Newton's method, with its quadratic convergence, is especially useful if the same boundary-value problem must be worked many times with different boundary conditions, grid sizes, etc.

Example 8.11. A nonlinear implicit FDE using Newton's method

Let's solve the problem presented in Example 8.10 by Newton's method.

$$\bar{y}'' + 2\bar{y}\bar{y}' = 4 + 4x^3 \qquad \bar{y}(1.0) = 2.0 \text{ and } \bar{y}(2.0) = 4.5 \tag{8.148}$$

Let $Y(x)$ be an approximate solution. Then

$$\bar{y}(x) = Y(x) + \eta(x) \tag{8.149}$$
$$\bar{y}' = Y' + \eta' \tag{8.150}$$
$$\bar{y}'' = Y'' + \eta'' \tag{8.151}$$

Substituting Eqs. (8.149) to (8.151) into Eq. (8.148) gives

$$(Y'' + \eta'') + 2(Y + \eta)(Y' + \eta') = 4 + 4x^3 \tag{8.152}$$

Expanding the nonlinear term yields

$$(Y'' + \eta'') + 2(YY' + Y\eta' + \eta Y' + \eta\eta') = 4 + 4x^3 \tag{8.153}$$

Neglecting the nonlinear term $\eta\eta'$ in Eq. (8.153) yields the linear ODE:

$$\boxed{\eta'' + 2Y\eta' + 2Y'\eta = G(x)} \tag{8.154}$$

where $G(x)$ is given by

$$G(x) = 4 + 4x^3 - Y'' - 2YY' \tag{8.155}$$

The boundary conditions on $\bar{y}(x)$ must be transformed to boundary conditions on $\eta(x)$. At $x = 1.0$,

$$\eta(1.0) = \bar{y}(1.0) - Y(1.0) = 0.0 - 0.0 = 0.0 \tag{8.156}$$

since the approximate solution $Y(x)$ must also satisfy the boundary condition. In a similar manner, $\eta(2.0) = 0.0$. Equation (8.154), with the boundary conditions $\eta(1.0) = \eta(2.0) = 0.0$, can be solved by the equilibrium method to yield $\eta(x)$, which yields $y(x) = Y(x) + \eta(x)$. The procedure is applied repetitively to convergence.

Let's approximate η' and η'' by second-order centered-difference approximations. Equation (8.154) becomes

$$\frac{\eta_{i+1} - 2\eta_i + \eta_{i-1}}{\Delta x^2} + 2Y_i \frac{\eta_{i+1} - \eta_{i-1}}{2\Delta x} + 2Y'|_i \eta_i = G_i \tag{8.157}$$

Values of $Y'|_i$ and $Y''|_i$ are also approximated by second-order centered-difference approximations:

$$Y'|_i = \frac{Y_{i+1} - Y_{i-1}}{2\Delta x} \qquad \text{and} \qquad Y''|_i = \frac{Y_{i+1} - 2Y_i + Y_{i-1}}{\Delta x^2} \tag{8.158}$$

Multiplying Eq. (8.157) by Δx^2 and gathering terms yields the linear FDE:

$$(1 - \Delta x \, Y_i)\eta_{i-1} + (-2 + 2\Delta x^2 Y'|_i)\eta_i + (1 + \Delta x \, Y_i)\eta_{i+1} = \Delta x^2 \, G_i \tag{8.159}$$

LIVERPOOL JOHN MOORES UNIVERSITY
LEARNING SERVICES

Table 8.22 Values of x, $Y(x)^{(0)}$, $Y'(x)^{(0)}$, $Y''(x)^{(0)}$, and $G(x)$

x	$Y(x)^{(0)}$	$Y'(x)^{(0)}$	$Y''(x)^{(0)}$	$G(x)$
1.00	2.000			
1.25	2.625	2.50	0.0	-1.3125
1.50	3.250	2.50	0.0	1.2500
1.75	3.875	2.50	0.0	6.0625
2.00	4.500			

Let $\Delta x = 0.25$. For the initial approximation $y(x)^{(0)}$ assume a linear variation between the boundary values, as given by Eq. (8.153). These values of $y(x)^{(0)} = Y(x)^{(0)}$ are presented in column 2 of Table 8.22. The values of $Y'(x)^{(0)}$, $Y''(x)^{(0)}$, and $G(x)$ are also presented in Table 8.22.

Applying Eq. (8.159) at the three interior points of the uniform grid gives

$$x = 1.25: \quad 0.34375\eta_1 - 1.68750\eta_2 + 1.65625\eta_3 = -0.082031 \qquad (8.160a)$$

$$x = 1.50: \quad 0.18750\eta_2 - 1.68750\eta_3 + 1.81540\eta_4 = 0.078125 \qquad (8.160b)$$

$$x = 1.75: \quad 0.03125\eta_3 - 1.68750\eta_4 + 1.96875\eta_5 = 0.378906 \qquad (8.160c)$$

Substituting $\eta_1 = \eta_5 = 0$ into Eq. (8.160) and writing the result in matrix form gives

$$\begin{bmatrix} -1.68750 & 1.65625 & 0.00000 \\ 0.18750 & -1.68750 & 1.81250 \\ 0.00000 & 0.03125 & -1.68750 \end{bmatrix} \begin{bmatrix} \eta_2 \\ \eta_3 \\ \eta_4 \end{bmatrix} = \begin{bmatrix} -0.082031 \\ 0.078125 \\ 0.378906 \end{bmatrix} \qquad (8.161)$$

The results are presented in Table 8.23. The first line presents the values of x, and the second line presents the values of $y(x)^{(0)} = Y(x)^{(0)}$. Solving Eq. (8.161) by the Thomas algorithm yields the values of $\eta(x)^{(0)}$ shown in line 3 of Table 8.23. Adding lines 2 and 3 of Table 8.23 gives $y(x)^{(1)} = Y(x)^{(0)} + \eta(x)^{(0)}$ presented in line 4, which will be used as

Table 8.23 Solution of the Nonlinear Implicit FDE by Newton's Method

				x		
k	$y(x)^{(k)} = Y(x)^{(k)}$ $\eta(x)^{(k)}$	1.00	1.25	1.50	1.75	2.00
0	$y(x)^{(0)} = Y(x)^{(0)}$	2.0	2.625000	3.250000	3.875000	4.5
	$\eta(x)^{(0)}$	0.0	-0.269211	-0.323819	-0.230533	0.0
1	$y(x)^{(1)} = Y(x)^{(1)}$	2.0	2.355789	2.926181	3.644466	4.5
	$\eta(x)^{(1)}$	0.0	-0.001037	-0.014870	-0.012439	0.0
2	$y(x)^{(2)} = Y(x)^{(2)}$	2.0	2.354752	2.911311	3.632027	4.5
	$\eta(x)^{(2)}$	0.0	-0.000002	-0.000005	-0.000029	0.0
3	$y(x)^{(3)} = Y(x)^{(3)}$	2.0	2.354750	2.911306	3.631998	4.5
	$\eta(x)^{(3)}$	0.0	0.000000	0.000000	0.000000	0.0
4	$y(x)^{(4)}$	2.0	2.354750	2.911306	3.631998	4.5
	$\bar{y}(x)$	2.0	2.362500	2.916667	3.633929	4.5
	Error(x)		-0.007750	-0.005361	-0.001931	

$Y(x)^{(1)}$ for the next iteration. Repeating the solution with $Y(x)^{(1)}$ yields lines 5 and 6 in Table 8.23. Four iterations are required to reach convergence to $|y_i^{(k+1)} - y_i^{(k)}| \le 0.000001$. The final solution $y(x)^{(4)}$, the exact solution $\bar{y}(x)$, and the $\text{Error}(x) = y(x)^{(4)} - \bar{y}(x)$ are presented at the bottom of the table. Eleven iterations are required by the iteration method presented in Example 8.10.

8.8   THE EQUILIBRIUM METHOD ON NONUNIFORM GRIDS

All of the results presented in Sections 8.3 to 8.7 are based on a uniform grid. In problems where the solution is nonlinear, a nonuniform distribution of grid points generally yields a more accurate solution if the grid points are clustered closer together in regions of large gradients and spread out in regions of small gradients. Methods for implementing the solution by the finite difference approach on nonuniform grids are presented in this section. The use of nonuniform grids in the finite element method is discussed in Example 12.5 in Section 12.3.

Some general considerations apply to the use of nonuniform grids. Foremost of these considerations is that the nonuniform grid point distribution should reflect the nonuniform nature of the solution. For example, if the exact solution has a particular functional form, the nonuniform grid point distribution should attempt to match that functional form. Secondarily, the nonuniform grid point distribution should be relatively smooth. Abrupt changes in the grid point distribution can yield undesirable abrupt changes in the solution.

Once a nonuniform grid point distribution has been determined, there are two approaches for developing a finite difference solution on the nonuniform grid:

1.   Direct solution on the nonuniform grid using nonequally spaced finite difference approximations (FDAs)
2.   Solution on a transformed uniform grid using equally spaced FDAs

In the first approach, nonequally spaced finite difference approximations for all of the derivatives in the differential equation are developed directly on the nonuniform grid. In the second approach, the differential equation is transformed from the nonuniformly discretized physical space $D(x)$ to a uniformly discretized transformed space $\hat{D}(\xi)$, and equally spaced finite difference approximations are employed in the uniformly discretized transformed space. The first approach is illustrated in Example 8.12. The second approach is not developed in this book.

Let's develop centered-difference approximations for the first and second derivatives $\bar{f}'(x)$ and $\bar{f}''(x)$, respectively, on the nonuniform finite difference grid illustrated in Figure 8.21. Consider $\bar{f}''(x)$ first. Write Taylor series for $\bar{f}_{i+1}$ and $\bar{f}_{i-1}$:

$$\bar{f}_{i+1} = \bar{f}_i + \bar{f}_x|_i\, \Delta x_+ + \tfrac{1}{2}\bar{f}_{xx}|_i\, \Delta x_+^2 + \tfrac{1}{6}\bar{f}_{xxx}|_i\, \Delta x_+^3 + \cdots \tag{8.162}$$

$$\bar{f}_{i-1} = \bar{f}_i - \bar{f}_x|_i\, \Delta x_- + \tfrac{1}{2}\bar{f}_{xx}|_i\, \Delta x_-^2 - \tfrac{1}{6}\bar{f}_{xxx}|_i\, \Delta x_-^3 + \cdots \tag{8.163}$$

where $\Delta x_+ = (x_{i+1} - x_i)$ and $\Delta x_- = (x_i - x_{i-1})$. Multiply Eq. (8.162) by Δx_- and multiply Eq. (8.163) by Δx_+ and add the results to obtain

$$\Delta x_-\, \bar{f}_{i+1} + \Delta x_+\, \bar{f}_{i-1} = (\Delta x_- + \Delta x_+)\bar{f}_i + (0)\bar{f}_x|_i + \tfrac{1}{2}(\Delta x_-\, \Delta x_+^2 + \Delta x_+\, \Delta x_-^2)\bar{f}_{xx}|_i$$
$$+ \tfrac{1}{6}(\Delta x_-\, \Delta x_+^3 - \Delta x_+\, \Delta x_-^3)\bar{f}_{xxx}|_i + \cdots$$

$$\tag{8.164}$$

Figure 8.21   Nonuniform finite difference grid.

Dividing Eq. (8.164) by Δx_- and letting $\beta = \Delta x_+/\Delta x_-$ gives

$$\bar{f}_{i+1} + \beta \bar{f}_{i-1} = (1+\beta)\bar{f}_i + \tfrac{1}{2}(\Delta x_+^2 + \Delta x_+\,\Delta x_-)\bar{f}_{xx}|_i + \tfrac{1}{6}(\Delta x_+^3 - \Delta x_+\,\Delta x_-^2)\bar{f}_{xxx}|_i + \cdots \tag{8.165}$$

Rearranging Eq. (8.165) gives

$$\tfrac{1}{2}\bar{f}_{xx}|_i\,\Delta x_-\,\Delta x_+(\beta+1) = \bar{f}_{i+1} - (1+\beta)\bar{f}_i + \beta \bar{f}_{i-1} - \tfrac{1}{6}\Delta x_+\,\Delta x_-^2(\beta^2 - 1)\bar{f}_{xxx}|_i + \cdots \tag{8.166}$$

Solving Eq. (8.166) for $\bar{f}_{xx}|_i$ yields

$$\bar{f}_{xx}|_i = \frac{2[\bar{f}_{i+1} - (1+\beta)\bar{f}_i + \beta \bar{f}_{i-1}]}{\Delta x_+\,\Delta x_-(1+\beta)} - \tfrac{1}{3}\,\Delta x_+(\beta - 1)\bar{f}_{xxx}|_i + \cdots \tag{8.167}$$

Truncating the remainder term yields the desired result, which is first-order accurate:

$$\boxed{f_{xx}|_i = \frac{2[f_{i+1} - (1+\beta)f_i + \beta f_{i-1}]}{\Delta x_+\,\Delta x_-(1+\beta)}} \tag{8.168}$$

A finite difference approximation for the first derivative, $\bar{f}_x|_i$, can be developed in a similar manner. The result is

$$\bar{f}_x|_i = \frac{\bar{f}_{i+1} - (1-\beta^2)\bar{f}_i - \beta^2 \bar{f}_{i-1}}{\Delta x_+(1+\beta)} - \tfrac{1}{6}\Delta x_+\,\Delta x_-\,\bar{f}_{xxx}|_i + \cdots \tag{8.169}$$

Truncating the remainder term yields the desired result:

$$\boxed{f_x|_i = \frac{f_{i+1} - (1-\beta^2)f_i - \beta^2 f_{i-1}}{\Delta x_+(1+\beta)}} \tag{8.170}$$

Note that Eq. (8.170) is second-order accurate even on a nonuniform grid.

Example 8.12.   The equilibrium method on a nonuniform grid

Let's solve the heat transfer problem presented in Section 8.1 using the first-order nonuniform grid finite difference approximation given by Eq. (8.168). The boundary-value problem is

$$T'' - \alpha^2 T = -\alpha^2 T_a \qquad T(0.0) = 0.0 \text{ and } T(1.0) = 100.0 \tag{8.171}$$

Substituting Eq. (8.168) for T'' yields

$$\frac{2[T_{i+1} - (1+\beta)T_i + \beta T_{i-1}]}{\Delta x_-^2\,\beta(1+\beta)} - \alpha^2 T_i = -\alpha^2 T_a \tag{8.172}$$

Let $\alpha^2 = 16.0 \text{ cm}^{-2}$ and $T_a = 0.0 \text{ C}$. Rearranging Eq. (8.172) yields

$$\beta T_{i-1} - [(1 + \beta) + 8 \, \Delta x_-^2 \, \beta(1 + \beta)]T_i + T_{i+1} = 0 \tag{8.173}$$

From the results obtained in Sections 8.3 and 8.4, we see that the solution to the heat transfer problem changes slowly at the left end of the rod and rapidly at the right end of the rod. Thus, the grid points should be sparsely spaced near the left end of the rod and closely spaced near the right end of the rod. As a simple nonuniform grid example, let's choose a second-order function to relate the nonuniformly-spaced grid points, denoted by x, to the uniformly-spaced grid points, denoted by $\bar{x}$. Thus,

$$x = a + b\bar{x} + c\bar{x}^2 \tag{8.174}$$

Three sets of $(x, \bar{x})$ values are required to determine the coefficients a, b, and c. Two sets of values are obviously specified by the boundary points: $x = 0.0$ where $\bar{x} = 0.0$ and $x = 1.0$ where $\bar{x} = 1.0$. The third set is chosen to effect the desired grid point distribution. For example, let $x = 0.875$ where $\bar{x} = 0.75$. Substituting these three defining sets of $(x, \bar{x})$ values into Eq. (8.174) and solving the resulting system of equations for a, b, and c gives

$$\boxed{x = -0.458333 + 0.5\bar{x} - 0.041667\bar{x}^2} \tag{8.175}$$

Substituting $\bar{x} = 0.25$ and $\bar{x} = 0.5$ into Eq. (8.175) yields the complete nonuniform grid point distribution presented in Table 8.24. Figure 8.22 illustrates the nonuniform grid point distribution.

Table 8.25 presents the values of x, Δx_-, Δx_+, β, and the coefficient of T_i in Eq. (8.173) at grid points 2 to 4 specified in Table 8.24. Applying Eq. (8.173) at the three interior grid points gives:

$$x = 0.375000: \qquad 0.777778T_1 - 3.333333T_2 + T_3 = 0 \tag{8.176a}$$
$$x = 0.666667: \qquad 0.714286T_2 - 2.547619T_3 + T_4 = 0 \tag{8.176b}$$
$$x = 0.875000: \qquad 0.600000T_3 - 1.933333T_4 + T_5 = 0 \tag{8.176c}$$

Table 8.24 Nonuniform
Grid Geometry

$\bar{x}$	x, cm
0.00	0.000000
0.25	0.375000
0.50	0.666667
0.75	0.875000
1.00	1.000000

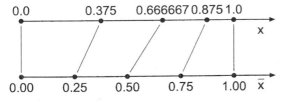

Figure 8.22 Nonuniform grid point distribution.

Table 8.25 Metric Data for the Nonuniform Grid

i	x, cm	Δx_-, cm	Δx_+, cm	β	$(\cdots)T_i$
1	0.0				
2	0.375000	0.375000	0.291667	0.777778	-3.333333
3	0.666667	0.291667	0.208333	0.714286	-2.547619
4	0.875000	0.208333	0.125000	0.600000	-1.933333
5	1.0				

Table 8.26 Solution by the Equilibrium Method on a Nonuniform Grid

ξ	x, cm	$T(x)$, C	$\bar{T}(x)$, C	Error(x), C
1	0.0	0.000000	0.000000	
2	0.375000	7.670455	7.802440	-0.131986
3	0.666667	25.568182	26.241253	-0.673072
4	0.875000	59.659091	60.618093	-0.959002
5	1.0	100.000000	100.000000	

Transferring $T_1 = 0.0$ and $T_5 = 100.0$ to the right-hand sides of Eqs. (8.176a) and (8.176c), respectively, yields the following tridiagonal system of FDEs:

$$\begin{bmatrix} -3.333333 & 1.000000 & 0.000000 \\ 0.714286 & -2.547610 & 1.000000 \\ 0.000000 & 0.600000 & -1.933333 \end{bmatrix} \begin{bmatrix} T_2 \\ T_3 \\ T_4 \end{bmatrix} = \begin{bmatrix} 0.0 \\ 0.0 \\ -100.0 \end{bmatrix} \tag{8.177}$$

Solving Eq. (8.177) by the Thomas algorithm yields the results presented in Table 8.26. The Euclidean norm of the errors in Table 8.26 is 1.179038 C, which is about 33 percent smaller than the Euclidean norm of 1.766412 C for the errors obtained for the uniform grid solution presented in Table 8.7.

8.9 EIGENPROBLEMS

Eigenproblems arise in equilibrium problems in which the solution exists only for special values (i.e., eigenvalues) of a parameter of the problem. Eigenproblems occur when homogeneous boundary-value ODEs also have homogeneous boundary conditions. The eigenvalues are to be determined in addition to the corresponding equilibrium configuration of the system. Shooting methods are not well suited for solving eigenproblems. Consequently, eigenproblems are generally solved by the equilibrium method.

8.9.1 Exact Eigenvalues

Consider the linear homogeneous boundary-value problem:

$$\boxed{\bar{y}'' + k^2 \bar{y} = 0 \qquad \bar{y}(0) = \bar{y}(1) = 0} \tag{8.178}$$

The exact solution to this problem is

$$\bar{y}(x) = A \sin(kx) + B \cos(kx) \tag{8.179}$$

where k is an unknown parameter to be determined. Substituting the boundary values into Eq. (8.179) gives

$$\bar{y}(0) = A \sin(k0) + B \cos(k0) = 0 \rightarrow B = 0 \tag{8.180}$$

$$\bar{y}(1) = A \sin(k1) = 0 \tag{8.181}$$

Either $A = 0$ (undesired) or $\sin(k) = 0$, for which

$$\boxed{k = \pm n\pi \qquad n = 1, 2, \ldots} \tag{8.182}$$

The values of k are the eigenvalues of the problem. There are an infinite number of eigenvalues. The solution of the differential equation is

$$\boxed{\bar{y}(x) = A \sin(n\pi x)} \tag{8.183}$$

The value of A is not uniquely determined. One of the major items of interest in eigenproblems are the eigenvalues of the system. For each eigenvalue of the system, there is an eigenfunction $\bar{y}(x)$ given by Eq. (8.183).

8.9.2 Approximate Eigenvalues

Eigenvalues of homogeneous boundary-value problems can also be obtained by numerical methods. In this approach, the boundary-value ODE is approximated by a system of finite difference equations, and the values of the unknown parameter (i.e., the eigenvalues) which satisfy the system of FDEs are determined. These values are approximations of the exact eigenvalues of the boundary-value problem.

Example 8.13. Approximation of eigenvalues by the equilibrium method

Let's solve Eq. (8.178) for its eigenvalues by the finite difference method. Choose an equally spaced grid with four interior points, as illustrated in Figure 8.23. Approximate $\bar{y}''$ with the second-order centered-difference approximation, Eq. (8.44). The corresponding finite difference equation is:

$$\frac{\bar{y}_{i+1} - 2\bar{y}_i + \bar{y}_{i-1}}{\Delta x^2} + 0(\Delta x^2) + k^2\bar{y}_i = 0 \tag{8.184}$$

Multiplying by Δx^2, truncating the remainder term, and rearranging gives the FDE:

$$\boxed{y_{i-1} - (2 - \Delta x^2 k^2)y_i + y_{i+1} = 0} \tag{8.185}$$

Apply the FDE, with $\Delta x = 0.2$, at the four interior points:

$x = 0.2$:	$y_1 - (2 - 0.04k^2)y_2 + y_3 = 0 \qquad y_1 = 0$	(8.186a)
$x = 0.4$:	$y_2 - (2 - 0.04k^2)y_3 + y_4 = 0$	(8.186b)
$x = 0.6$:	$y_3 - (2 - 0.04k^2)y_4 + y_5 = 0$	(8.186c)
$x = 0.8$:	$y_4 - (2 - 0.04k^2)y_5 + y_6 = 0 \qquad y_6 = 0$	(8.186d)

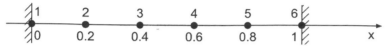

Figure 8.23 Finite difference grid for the eigenproblem.

Writing Eq. (8.186) in matrix form gives

$$
\begin{bmatrix}
(2 - 0.04k^2) & -1 & 0 & 0 \\
-1 & (2 - 0.04k^2) & -1 & 0 \\
0 & -1 & (2 - 0.04k^2) & -1 \\
0 & 0 & -1 & (2 - 0.04k^2)
\end{bmatrix}
[y_i] = 0
\qquad (8.187)
$$

which can be expressed as

$$
\boxed{(\mathbf{A} - \lambda \mathbf{I})\mathbf{y} = \mathbf{0}}
\qquad (8.188)
$$

where $\lambda = 0.04k^2$ and $\mathbf{A}$ is defined as

$$
\mathbf{A} =
\begin{bmatrix}
2 & -1 & 0 & 0 \\
-1 & 2 & -1 & 0 \\
0 & -1 & 2 & -1 \\
0 & 0 & -1 & 2
\end{bmatrix}
\qquad (8.189)
$$

This is a classical eigenproblem. The characteristic equation is given by

$$
\det(\mathbf{A} - \lambda \mathbf{I}) = 0
\qquad (8.190)
$$

Define $Z = (2 - 0.04k^2)$. The characteristic equation is determined by expanding the determinant $|\mathbf{A} - \lambda \mathbf{I}| = 0$, which gives

$$
Z^4 - 3Z^2 + 1 = 0
\qquad (8.191)
$$

which is quadratic in Z^2. Solving Eq. (8.191) by the quadratic formula yields

$$
Z = (2 - 0.04k^2) = \pm 1.618\ldots \pm 0.618\ldots
\qquad (8.192)
$$

The values of Z, k, k(exact), and percent error are presented in Table 8.27.

    The first eigenvalue is reasonably accurate. The higher-order eigenvalues become less and less accurate. To improve the accuracy of the eigenvalues and to obtain higher-order eigenvalues, more grid points are required. This is not without disadvantages, however. Expanding the determinant becomes more difficult, and finding the zeros of high-order polynomials is more difficult. Numerical methods for finding the eigenvalues, which are introduced in Chapter 2, may be used to determine the eigenvalues for large systems of FDEs.

Table 8.27 Solution of the Eigenproblem

Z	k	k(exact)	Error, %
1.618	± 3.090	$\pm \pi = \pm 3.142$	∓ 1.66
0.618	± 5.878	$\pm 2\pi = \pm 6.283$	∓ 6.45
-0.618	± 8.090	$\pm 3\pi = \pm 9.425$	∓ 14.16
-1.618	± 9.511	$\pm 4\pi = \pm 12.566$	∓ 24.31

8.10 PROGRAMS

Two FORTRAN subroutines for integrating boundary-value ordinary differential equations are presented in this section:

1. The fourth-order Runge-Kutta shooting method
2. The second-order equilibrium method

The basic computational algorithms are presented as completely self-contained subroutines suitable for use in other programs. Input data and output statements are contained in a main (or driver) program written specifically to illustrate the use of each subroutine.

8.10.1 The Fourth-Order Runge-Kutta Shooting Method

The general nonlinear second-order boundary-value ODE is given by Eq. (8.14):

$$\bar{y}'' = f(x, \bar{y}, \bar{y}') \qquad \bar{y}(x_1) = \bar{y}_1 \text{ and } \bar{y}(x_2) = \bar{y}_2 \tag{8.193}$$

Equation (8.193) can be written as a pair of coupled nonlinear first-order initial-value ODEs:

$$\bar{y}' = \bar{z} \qquad\qquad \bar{y}(x_1) = \bar{y}_1 \tag{8.194}$$

$$\bar{z}' = f(x, \bar{y}, \bar{z}) \qquad \bar{z}(x_1) = \bar{y}'(x_1) = \bar{y}'|_1 = ? \tag{8.195}$$

The general algorithm for the fourth-order Runge-Kutta method is given by Eqs. (7.179) and (7.180). These equations are implemented in *subroutine rk* in Section 7.15.1 for a single first-order initial-value ODE. That subroutine has been expanded in this section to solve the system of two coupled first-order ODEs, Eqs. (8.194) and (8.195), which arise from Eq. (8.193). The secant method for satisfying the right-hand side boundary condition is given by Eqs. (8.18) and (8.19):

$$\frac{\bar{y}_2 - y_2^{(n)}}{y'|_1^{(n+1)} - y'|_1^{(n)}} = \frac{y_2^{(n)} - y_2^{(n-1)}}{y'|_1^{(n)} - y'|_1^{(n-1)}} = \text{slope} \tag{8.196}$$

where the superscript (n) denotes the iteration number. Solving Eq. (8.196) for $y'|_1^{(n+1)}$ gives

$$y'|_1^{(n+1)} = y'|_1^{(n)} + \frac{\bar{y}_2 - y_2^{(n)}}{\text{slope}} \tag{8.197}$$

A FORTRAN subroutine, *subroutine shoot*, for implementing the fourth-order Runge-Kutta shooting method is presented in Program 8.1. *Program main* defines the data set and prints it, calls *subroutine shoot* to implement the solution, and prints the solution. Equations (8.194) and (8.195) are solved by the Runge-Kutta method in *subroutine rk2*. A FORTRAN function, *function f*, specifies the derivative function specified by Eq. (8.193).

Program 8.1.   The fourth-order Runge-Kutta shooting method program

```
      program main
c     main program to illustrate boundary-value ODE solvers
c     ndim  array dimension, ndim = 9 in this example
```

```
c       imax    number of grid points
c       x       independent variable array, x(i)
c       y       dependent variable array, y(i)
c       z       derivative dy/dx array, z(i)
c       y1,z1 left-hand side boundary condition values
c       y2,z2 right-hand side boundary condition values
c       by,bz right-hand-side boundary condition flags: 0.0 or 1.0
c       za,zb first and second guesses for z(1)
c       dx      grid increment
c       iter    maximum number of iterations
c       tol     convergence tolerance
c       iw      intermediate results output flag: 0 none, 1 some, 2 all
        dimension x(9),y(9),z(9)
        data ndim,imax,iter,tol,iw / 9, 5, 3, 1.0e-06, 1 /
        data (x(i),i=1,5),dx / 0.0, 0.25, 0.50, 0.75, 1.00, 0.25 /
        data y(1),za,zb / 0.0, 7.5, 12.5 /
        data by,bz,y1,z1,y2,z2 / 1.0, 0.0, 0.0, 0.0, 100.0, 0.0 /
        write (6,1000)
        call shoot (ndim,imax,x,y,z,by,bz,y1,z1,y2,z2,za,zb,dx,iter,
     1 tol,iw)
        if (iw.ne.1) write (6,1010) (i,x(i),y(i),z(i),i=1,imax)
        stop
 1000 format (' Shooting method'/' '/'   i',7x,'x',12x,'y',12x,'z'
     1 /' ')
 1010 format (i3,3f13.6)
 1020 format (' ')
        end


        subroutine shoot (ndim,imax,x,y,z,by,bz,y1,z1,y2,z2,za,zb,dx,
     1 iter,tol,iw)
c       the shooting method
        dimension x(ndim),y(ndim),z(ndim)
        rhs=by*y2+bz*z2
        z(1)=za
        do it=1,iter
           call rk2 (ndim,imax,x,y,z,dx,iw)
           if (iw.gt.1) write (6,1000)
           if (iw.gt.0) write (6,1010) (i,x(i),y(i),z(i),i=1,imax)
           if (iw.gt.0) write (6,1000)
           if (it.eq.1) then
              rhs1=by*y(imax)+bz*z(imax)
              z(1)=zb
           else
              rhs2=by*y(imax)+bz*z(imax)
              if (abs(rhs2-rhs).le.tol) return
              zb=z(1)
              slope=(rhs2-rhs1)/(zb-za)
              za=zb
              rhs1=rhs2
              z(1)=z(1)+(rhs-rhs2)/slope
           end if
        end do
```

```
      if (iter.gt.3) write (6,1020) iter
      return
 1000 format (' ')
 1010 format (i3,3f13.6)
 1020 format (' '/' The solution failed to converge, iter = ',i3)
      end

      subroutine rk2 (ndim,imax,x,y,z,dx,iw)
c     implements the fourth-order Runge-Kutta method for two odes
      dimension x(ndim),y(ndim),z(ndim)
      do i=2,imax
         dy1=dx*z(i-1)
         dz1=dx*f(x(i-1),y(i-1),z(i-1))
         dy2=dx*(z(i-1)+dz1/2.0)
         dz2=dx*f(x(i-1)+dx/2.0,y(i-1)+dy1/2.0,z(i-1)+dz1/2.0)
         dy3=dx*(z(i-1)+dz2/2.0)
         dz3=dx*f(x(i-1)+dx/2.0,y(i-1)+dy2/2.0,z(i-1)+dz2/2.0)
         dy4=dx*(z(i-1)+dz3)
         dz4=dx*f(x(i-1)+dx,y(i-1)+dy3,z(i-1)+dz3)
         y(i)=y(i-1)+(dy1+2.0*(dy2+dy3)+dy4)/6.0
         z(i)=z(i-1)+(dz1+2.0*(dz2+dz3)+dz4)/6.0
         if (iw.eq.2) write (6,1000) i,dy1,dy2,dy3,dy4,y(i)
         if (iw.eq.2) write (6,1000) i,dz1,dz2,dz3,dz4,z(i)
      end do
      return
 1000 format (i3,5f13.6)
      end

      function f(x,y,z)
c     derivative function
c     p       coefficient of yp in the ode
c     q       coefficient of y in the ode
c     fx      nonhomogeneous term
      data p,q,fx / 0.0, -16.0, 0.0 /
      f=fx-p*z-q*y
      return
      end
```

The data set used to illustrate *subroutine shoot* is taken from Example 8.1 with the fourth-order Runge-Kutta method replacing the second-order implicit trapezoid method. The output generated by the fourth-order Runge-Kutta program is presented in Output 8.1.

Output 8.1.   Solution by the fourth-order Runge-Kutta shooting method

Shooting method

i	x	y	z
1	0.000000	0.000000	7.500000
2	0.250000	2.187500	11.562500
3	0.500000	6.744792	28.033854
4	0.750000	18.574761	74.694553
5	1.000000	50.422002	201.836322

1	0.000000	0.000000	12.500000
2	0.250000	3.645833	19.270833
3	0.500000	11.241319	46.723090
4	0.750000	30.957935	124.490922
5	1.000000	84.036669	336.393870
1	0.000000	0.000000	14.874459
2	0.250000	4.338384	22.931458
3	0.500000	13.376684	55.598456
4	0.750000	36.838604	148.138810
5	1.000000	100.000000	400.294148

Example 8.7 is concerned with a derivative (i.e., Neumann) boundary condition. That problem can be solved by *subroutine shoot* by changing the following variable values: $y(1) = 100.0$, $by = 0.0$, $bz = 1.0$, $y2 = 0.0$, $za = -405.0$, and $zb = -395.0$.

8.10.2 The Second-Order Equilibrium Method

The general second-order nonlinear boundary-value ODE is given by Eq. (8.136):

$$\bar{y}'' = P(x, \bar{y})\bar{y}' + Q(x, \bar{y})\bar{y} = F(x) \qquad \bar{y}(x_1) = \bar{y}_1 \text{ and } \bar{y}(x_2) = \bar{y}_2 \tag{8.198}$$

The second-order centered-difference FDE which approximates Eq. (8.198) is given by Eq. (8.46):

$$\left(1 - \frac{\Delta x}{2}P_i\right)y_{i-1} + (-2 + \Delta x^2 Q_i)y_i + \left(1 + \frac{\Delta x}{2}P_i\right)y_{i+1} = \Delta x^2 F_i \tag{8.199}$$

Equation (8.199) is applied at every interior point in a finite difference grid. The resulting system of FDEs is solved by the Thomas algorithm. An initial approximation $y(x)^{(0)}$ must be specified. If the ODE is linear, the solution is obtained in one pass. If the ODE is nonlinear, the solution is obtained iteratively.

A FORTRAN subroutine, *subroutine equil*, for implementing the second-order equilibrium method is presented in Program 8.2. *Program main* defines the data set and prints it, calls *subroutine equil* to set up and solve the system of FDEs, and prints the solution. A first guess for the solution $y(i)$ must be supplied in a *data* statement. *Subroutine thomas*, Section 1.8.3, is used to solve the system equation.

Program 8.2. The second-order equilibrium method program

```
      program main
c     main program to illustrate boundary-value ODE solvers
c     insert comment statements from subroutine shoot main program
      dimension x(9),y(9),a(9,3),b(9),w(9)
      data ndim,imax,iter,tol,iw / 9, 5, 1, 1.0e-06, 1 /
      data (x(i), i=1,5),dx / 0.0, 0.25, 0.50, 0.75, 1.00, 0.25 /
      data (y(i), i=1,5) / 0.00, 25.0, 50.0, 75.0, 100.0 /
      data by,bz,y1,z1,y2,z2 / 1.0, 0.0, 0.0, 0.0, 100.0, 0.0 /
      write (6,1000)
      if (iw.gt.0) write (6,1010) (i,x(i),y(i),i=1,imax)
      call equil (ndim,imax,x,y,by,y1,z1,y2,z2,a,b,w,dx,iter,tol,iw)
      if (iw.gt.0) write (6,1020)
      if (iw.ne.1) write (6,1010) (i,x(i),y(i),i=1,imax)
      stop
```

```
 1000 format (' Equilibrium method'/' '/'   i',7x,'x',12x,'y'/' ')
 1010 format (i3,2f13.6)
 1020 format (' ')
      end

      subroutine equil (ndim,imax,x,y,by,y1,z1,y2,z2,a,b,w,dx,iter,
     1 tol,iw)
c     the equilibrium method for a nonlinear second-order ode
c     fx     nonhomogeneous term
c     p      coefficient of yp in the ode
c     q      coefficient of y in the ode
      dimension x(ndim),y(ndim),a(ndim,3),b(ndim),w(ndim)
      data fx,p,q / 0.0, 0.0, -16.0 /
      a(1,2)=1.0
      a(1,3)=0.0
      b(1)=y(1)
      if (by.eq.1) then
         a(imax,1)=0.0
         a(imax,2)=1.0
         b(imax)=y2
      else
         a(imax,1)=2.0
         a(imax,2)=-2.0+q*dx**2
         b(imax)=fx*dx**2-2.0*z2*dx**2
      end if
      do it=1,iter
         do i=2,imax-1
         a(i,1)=1.0-0.5*p*dx
         a(i,2)=-2.0+q*dx**2
         a(i,3)=1.0+0.5*p*dx
         b(i)=fx*dx**2
      end do
      call thomas (ndim,imax,a,b,w)
      dymax=0.0
      do i=1,imax
         dy=abs(y(i)-w(i))
         if (dy.gt.dymax) dymax=dy
         y(i)=w(i)
      end do
      if (iw.gt.0) write (6,1000)
      if (iw.gt.0) write (6,1010) (i,x(i),y(i),i=1,imax)
      if (dymax.le.tol) return
      end do
      if (iter.gt.1) write (6,1020) iter
      return
 1000 format (' ')
 1010 format (i3,2f13.6)
 1020 format (' '/' The solution failed to converge, it = ',i3)
      end

      subroutine thomas (ndim,n,a,b,x)
c     the Thomas algorithm for a tridiagonal system
      end
```

The data set used to illustrate *subroutine equil* is taken from Example 8.4. The output generated by the second-order equilibrium method program is presented in Output 8.2.

Output 8.2. Solution by the second-order equilibrium method

Equilibrium method

i	x	y
1	0.000000	0.000000
2	0.250000	25.000000
3	0.500000	50.000000
4	0.750000	75.000000
5	1.000000	100.000000
1	0.000000	0.000000
2	0.250000	4.761905
3	0.500000	14.285714
4	0.750000	38.095238
5	1.000000	100.000000

Example 8.8 is concerned with a derivative (i.e., Neumann) boundary condition. That problem can be solved by *subroutine equil* by changing the following variables in the data statements: $(y(i), i = 1,5)/100.0, 75.0, 50.0, 25.0, 0.0/, by = 0.0, bz = 1.0$, and $y1 = 100.0$.

8.10.3 Packages for Integrating Boundary-Value ODEs

Numerous libraries and software packages are available for integrating boundary-value ordinary differential equations. Many work stations and main frame computers have such libraries attached to their operating systems.

Many commercial software packages contain algorithms for integrating boundary-value ODEs. Some of the more prominent packages are Matlab and Mathcad. More sophisticated packages, such as IMSL, MATHEMATICA, MACSYMA, and MAPLE, also contain algorithms for integrating boundary-value ODEs. Finally, the book *Numerical Recipes* (Press et al., 1989) contains numerous routines for integrating boundary-value ordinary differential equations.

8.11 SUMMARY

Two finite difference approaches for solving boundary-value ordinary differential equations are presented in this chapter: (1) the shooting method and (2) the equilibrium method. The advantages and disadvantages of these methods are summarized in this section.

The shooting method is based on marching methods for solving initial-value ODEs. The advantages of the shooting method are:

1. Any initial-value ODE solution method can be used.
2. Nonlinear ODEs are solved directly.
3. It is easy to achieve fourth- or higher-order accuracy.
4. There is no system of FDEs to solve.

The disadvantages of the shooting method are:

1. One or more boundary conditions must be satisfied iteratively (by shooting).
2. Shooting for more than one boundary condition is time consuming.
3. Nonlinear problems require an iterative procedure (e.g., the secant method) to satisfy the boundary conditions.

The equilibrium method is based on relaxing a system of FDEs simultaneously, including the boundary conditions. The major advantage of the equilibrium method is that the boundary conditions are applied directly and automatically satisfied. The disadvantages of the equilibrium method are:

1. It is difficult to achieve higher than second-order accuracy.
2. A system of FDEs must be solved.
3. Nonlinear ODEs yield a system of nonlinear FDEs, which must be solved by iterative methods.

No rigid guidelines exist for choosing between the shooting method and the equilibrium method. Experience is the best guide. Shooting methods work well for nonsmoothly varying problems and oscillatory problems where their error control and variable grid size capacity are of great value. Shooting methods frequently require more computational effort, but they are generally more certain of producing a solution. Equilibrium methods work well for smoothly varying problems and for problems with complicated or delicate boundary conditions.

After studying Chapter 8, you should be able to:

1. Describe the general feature of boundary-value ordinary differential equations (ODEs)
2. Discuss the general features of the linear second-order ODE, including the complementary solution and the particular solution
3. Discuss the general features of the nonlinear second-order ODE
4. Describe how higher-order ODEs and systems of second-order ODEs can be solved using the procedures for solving a single second-order ODE
5. Discuss the number of boundary conditions required to solve a boundary-value ODE
6. Discuss the types of boundary conditions: Dirichlet, Neumann, and mixed
7. Explain the concept underlying the shooting (initial-value) method
8. Reformulate a boundary-value ODE as a system of initial-value ODEs
9. Apply any initial-value ODE finite difference method to solve a boundary-value ODE by the shooting method
10. Solve a nonlinear boundary-value ODE by the shooting method with iteration
11. Solve a linear boundary-value ODE by the shooting method with superposition
12. Explain the concept of extrapolation as it applies to boundary-value ODEs
13. Apply extrapolation to increase the accuracy of the solution of boundary-value problems by the shooting method
14. Explain the concepts underlying the equilibrium (boundary-value) method
15. Solve a linear second-order boundary-value ODE by the second-order equilibrium method
16. Apply extrapolation to increase the accuracy of the solution of boundary-value problems by the equilibrium method

17. Solve a boundary-value problem with derivative boundary conditions by the shooting method
18. Solve a boundary-value problem with derivative boundary conditions by the equilibrium method
19. Discuss mixed boundary conditions and boundary conditions at infinity
20. Derive the five-point fourth-order equilibrium method and discuss its limitations
21. Derive the compact three-point fourth-order finite difference approximations (FDAs) for $\bar{y}'$ and $\bar{y}''$
22. Apply the compact three-point fourth-order FDAs to solve a second-order boundary-value ODE
23. Explain the difficulties encountered when solving a nonlinear boundary-value problem by the equilibrium method
24. Explain and apply the iteration method for solving nonlinear boundary-value problems
25. Explain and apply Newton's method for solving nonlinear boundary-value problems
26. Derive finite difference approximations for $\bar{y}'$ and $\bar{y}''$ on a nonuniform grid
27. Solve a second-order boundary-value ODE on a nonuniform grid
28. Discuss the occurrence of eigenproblems in boundary-value ODEs
29. Solve simple eigenproblems arising in the solution of boundary-value ODEs
30. List the advantages and disadvantages of the shooting method for solving boundary-value ODEs
31. List the advantages and disadvantages of the equilibrium method for solving boundary-value ODEs
32. Choose and implement a finite difference method for solving boundary-value ODEs

EXERCISE PROBLEMS

8.1 Introduction

1. Derive the exact solution of the heat transfer problem presented in Section 8.1, Eq. (8.4). Calculate the solution presented in Table 8.1.

8.2 General Features of Boundary-Value ODEs

2. Derive the exact solution of the linear second-order boundary-value ODE, Eq. (8.7), where $y(x_1) = y_1$, $y(x_2) = y_2$, and $F(x) = a \exp(bx) + c + dx$.
3. Evaluate the exact solution of Problem 8.2 for $P = 5.0$, $Q = 4.0$, $F = 1.0$, $y(0.0) = 0.0$, and $y(1.0) = 1.0$. Tabulate the solution for $x = 0.0$ to 1.0 at intervals of $\Delta x = 0.125$. Plot the solution.

The following problems involve the numerical solution of boundary-value ODEs. These problems can be worked by hand calculation or computer programs. Carry at least six digits after the decimal place in all calculations. An infinite variety of additional problems can be obtained from these problems by (a) changing the coefficients in the ODEs, (b) changing the boundary conditions, (c) changing the step size, (d) changing the range of integration, and (e) combinations of the above changes.

For all problems solved by a shooting method, let $y'(0.0) = 0.0$ and 1.0 for the first two guesses, unless otherwise noted. For all problems solved by an equilibrium method, let the first approximation for $y(x)$ be a linear variation consistent with the boundary conditions. For all nonlinear ODEs, repeat the overall solution until $|\Delta y_{i,\max}| \leq 0.00|$.

8.3 The Shooting (Initial-Value) Method

4. *Solve the following ODE by the shooting method using the first-order explicit Euler method with (a) $\Delta x = 0.25$, (b) $\Delta x = 0.125$, and (c) $\Delta x = 0.0625$. Compare the errors and calculate the ratio of the errors at $x = 0.5$.

$$\bar{y}'' + 5\bar{y}' + 4\bar{y} = 1 \qquad \bar{y}(0) = 0 \text{ and } \bar{y}(1) = 1 \tag{A}$$

5. *Solve Problem 4 by the second-order modified Euler method.
6. *Solve Problem 4 by the fourth-order Runge-Kutta method.
7. Solve the following ODE by the shooting method using the first-order explicit Euler method with (a) $\Delta x = 0.25$, (b) $\Delta x = 0.125$, and (c) $\Delta x = 0.0625$. Compare the errors and calculate the ratio of the errors at $x = 0.5$.

$$\bar{y}'' + 4\bar{y}' + 6.25\bar{y} = 1 \qquad \bar{y}(0) = 0 \text{ and } \bar{y}(1) = 1 \tag{B}$$

8. Solve Problem 7 by the second-order modified Euler method.
9. Solve Problem 7 by the fourth-order Runge-Kutta method.
10. Solve the following ODE by the shooting method using the first-order explicit Euler method with (a) $\Delta x = 0.25$, (b) $\Delta x = 0.125$, and (c) $\Delta x = 0.0625$. Compare the errors and calculate the ratio of the errors at $x = 0.5$.

$$\bar{y}'' + 5\bar{y}' + 4\bar{y} = e^x \qquad \bar{y}(0) = 0 \text{ and } \bar{y}(1) = 1 \tag{C}$$

11. Solve Problem 10 by the second-order modified Euler method.
12. Solve Problem 10 by the fourth-order Runge-Kutta method.
13. Solve the following ODE by the shooting method using the first-order explicit Euler method with (a) $\Delta x = 0.25$, (b) $\Delta x = 0.125$, and (c) $\Delta x = 0.0625$. Compare the errors and calculate the ratio of the errors at $x = 0.5$.

$$\bar{y}'' + 4\bar{y}' + 6.25\bar{y} = e^x \qquad \bar{y}(0) = 0 \text{ and } \bar{y}(1) = 1 \tag{D}$$

14. Solve Problem 13 by the second-order modified Euler method.
15. Solve Problem 13 by the fourth-order Runge-Kutta method.
16. Solve the following ODE by the shooting method using the first-order explicit Euler method with (a) $\Delta x = 0.25$, (b) $\Delta x = 0.125$, and (c) $\Delta x = 0.0625$. Compare the errors and calculate the ratio of the errors at $x = 0.5$.

$$\bar{y}'' + 5\bar{y}' + 4\bar{y} = 2e^{x/2} + 1 + x \qquad \bar{y}(0) = 0 \text{ and } \bar{y}(1) = 1 \tag{E}$$

17. Solve Problem 16 by the second-order modified Euler method.
18. Solve Problem 16 by the fourth-order Runge-Kutta method.
19. Solve the following ODE by the shooting method using the first-order explicit Euler method with (a) $\Delta x = 0.25$, (b) $\Delta x = 0.125$, and (c) $\Delta x = 0.0625$. Compare the errors and calculate the ratio of the errors at $x = 0.5$.

$$\bar{y}'' + 4\bar{y}' + 6.25\bar{y} = 2e^{x/2} + 1 + x \qquad \bar{y}(0) = 0 \text{ and } \bar{y}(1) = 1 \tag{F}$$

20. Solve Problem 19 by the second-order modified Euler method.
21. Solve Problem 19 by the fourth-order Runge-Kutta method.

22. Solve the following ODE by the shooting method using the first-order explicit Euler method with (a) $\Delta x = 0.25$, (b) $\Delta x = 0.125$, and (c) $\Delta x = 0.0625$. Compare the errors and calculate the ratio of the errors at $x = 0.5$.

$$\bar{y}'' + (1+x)\bar{y}' + (1+x)\bar{y} = 1 \qquad \bar{y}(0) = 0 \text{ and } \bar{y}(1) = 1 \qquad \text{(G)}$$

23. Solve Problem 22 by the second-order modified Euler method.
24. Solve Problem 22 by the fourth-order Runge-Kutta method.
25. Solve the following ODE by the shooting method using the first-order explicit Euler method with (a) $\Delta x = 0.25$, (b) $\Delta x = 0.125$, and (c) $\Delta x = 0.0625$. Compare the errors and calculate the ratio of the errors at $x = 0.5$.

$$\bar{y}'' + (1+x)\bar{y}' + (1+x)\bar{y} = e^x \qquad \bar{y}(0) = 0 \text{ and } y(1) = 1 \qquad \text{(H)}$$

26. Solve Problem 25 by the second-order modified Euler method.
27. Solve Problem 25 by the fourth-order Runge-Kutta method.
28. Solve the following ODE by the shooting method using the first-order explicit Euler method with (a) $\Delta x = 0.25$, (b) $\Delta x = 0.125$, and (c) $\Delta x = 0.0625$. Compare the errors and calculate the ratio of the errors at $x = 0.5$.

$$\bar{y}'' + (1+x)\bar{y}' + (1+x)\bar{y} = 2e^{x/2} + 1 + x \qquad \bar{y}(0) = 0 \text{ and } \bar{y}(1) = 1$$
$$\text{(I)}$$

29. Solve Problem 28 by the second-order modified Euler method.
30. Solve Problem 28 by the fourth-order Runge-Kutta method.
31. Solve the following third-order boundary-value ODE by the shooting method using the first-order explicit Euler method with (a) $\Delta x = 0.25$ and (b) $\Delta x = 0.125$. Compare the solutions.

$$\bar{y}''' - 7\bar{y}'' + 14\bar{y}' - 8\bar{y} = 1 \qquad \bar{y}(0) = 0, \ \bar{y}'(0) = 1, \text{ and } \bar{y}(1) = 1 \qquad \text{(J)}$$

32. Solve Problem 31 by the second-order modified Euler method.
33. Solve Problem 31 by the fourth-order Runge-Kutta method.
34. Solve the following third-order boundary-value ODE by the shooting method using the first-order explicit Euler method with (a) $\Delta x = 0.25$ and (b) $\Delta x = 0.125$. Compare the solutions.

$$\bar{y}''' - 7\bar{y}'' + 14\bar{y}' - 8\bar{y} = 2e^{x/2} + 1 + x$$
$$\bar{y}(0) = 0, \ \bar{y}'(0) = 1, \text{ and } \bar{y}(1) = 1 \qquad \text{(K)}$$

35. Solve Problem 34 by the second-order modified Euler method.
36. Solve Problem 34 by the fourth-order Runge-Kutta method.
37. *Solve the following ODE by the shooting method using the first-order explicit Euler method with (a) $\Delta x = 0.25$, (b) $\Delta x = 0.125$, and (c) $\Delta x = 0.0625$. Compare the solutions at $x = 0.5$.

$$\bar{y}'' + (1+\bar{y})\bar{y}' + (1+\bar{y})\bar{y} = 1 \qquad \bar{y}(0) = 0 \text{ and } \bar{y}(1) = 1 \qquad \text{(L)}$$

38. Solve Problem 37 by the second-order modified Euler method.
39. Solve Problem 37 by the fourth-order Runge-Kutta method.
40. Solve the following ODE by the shooting method using the first-order explicit Euler method with (a) $\Delta x = 0.25$, (b) $\Delta x = 0.125$, and (c) $\Delta x = 0.0625$.

Compare the solutions at $x = 0.5$.

$$\bar{y}'' + (1 + x + \bar{y})\bar{y}' + (1 + x + \bar{y})\bar{y} = 2e^{x/2} + 1 + x$$
$$\bar{y}(0) = 0 \text{ and } \bar{y}(1) = 1 \quad \text{(M)}$$

41. Solve Problem 40 by the second-order modified Euler method.
42. Solve Problem 40 by the fourth-order Runge-Kutta method.

8.4 The Equilibrium (Boundary-Value) Method

43. *Solve Problem 4 by the second-order equilibrium method.
44. Solve Problem 7 by the second-order equilibrium method.
45. Solve Problem 10 by the second-order equilibrium method.
46. Solve Problem 13 by the second-order equilibrium method.
47. Solve Problem 16 by the second-order equilibrium method.
48. Solve Problem 19 by the second-order equilibrium method.
49. Solve Problem 22 by the second-order equilibrium method.
50. Solve Problem 25 by the second-order equilibrium method.
51. Solve Problem 28 by the second-order equilibrium method.
52. Solve Problem 31 by letting $\bar{z} = \bar{y}'$, thus reducing the third-order ODE to a second-order ODE for $\bar{z}(x)$. Solve this system of two coupled ODEs by solving the second-order ODE for $\bar{z}(x)$ by the second-order equilibrium method and the first-order ODE for $\bar{y}(x)$ by the second-order modified Euler method. Solve the problem for (a) $\Delta x = 0.25$ and (b) $\Delta x = 0.125$. Compare the solutions at $x = 0.5$.
53. Solve Problem 34 by the procedure described in Problem 52. Compare the solutions at $x = 0.5$.

8.5 Derivative (and Other) Boundary Conditions

Shooting Method

54. *Solve the following ODE by the shooting method using the second-order modified Euler method with $\Delta x = 0.125$.

$$\bar{y}'' + 5\bar{y}' + 4\bar{y} = 1 \qquad \bar{y}(0) = 1 \text{ and } \bar{y}'(1) = 0 \quad \text{(N)}$$

55. Solve the following ODE by the shooting method using the second-order modified Euler method with $\Delta x = 0.125$.

$$\bar{y}'' + 4\bar{y}' + 6.25\bar{y} = 1 \qquad \bar{y}(0) = 1 \text{ and } \bar{y}'(1) = 0 \quad \text{(O)}$$

56. Solve the following ODE by the shooting method using the second-order modified Euler method with $\Delta x = 0.125$.

$$\bar{y}'' + 5\bar{y}' + 4\bar{y} = e^x \qquad \bar{y}(0) = 1 \text{ and } \bar{y}'(1) = 0 \quad \text{(P)}$$

57. Solve the following ODE by the shooting method using the second-order modified Euler method with $\Delta x = 0.125$.

$$\bar{y}'' + 4\bar{y}' + 6.25\bar{y} = e^x \qquad \bar{y}(0) = 1 \text{ and } \bar{y}'(1) = 0 \quad \text{(Q)}$$

58. Solve the following ODE by the shooting method using the second-order modified Euler method with $\Delta x = 0.125$.

$$\bar{y}'' + (1+x)\bar{y}' + (1+x)\bar{y} = 1 \qquad \bar{y}(0) = 1 \text{ and } \bar{y}'(1) = 0 \qquad \text{(R)}$$

59. Solve the following ODE by the shooting method using the second-order modified Euler method with $\Delta x = 0.125$.

$$\bar{y}'' + (1+x)\bar{y}' + (1+x)\bar{y} = 2e^{x/2} + 1 + x \qquad \bar{y}(0) = 1 \text{ and } \bar{y}'(1) = 0 \qquad \text{(S)}$$

The Equilibrium Method

60. *Solve Problem 54 by the second-order equilibrium method.
61. Solve Problem 55 by the second-order equilibrium method.
62. Solve Problem 56 by the second-order equilibrium method.
63. Solve Problem 57 by the second-order equilibrium method.
64. Solve Problem 58 by the second-order equilibrium method.
65. Solve Problem 59 by the second-order equilibrium method.

Mixed Boundary Conditions

66. *Solve the following ODE by the shooting method using the first-order explicit Euler method with (a) $\Delta x = 0.125$ and (b) $\Delta x = 0.0625$.

$$\bar{y}'' + 5\bar{y}' + 4\bar{y} = 1 \qquad \bar{y}(0) = 0 \text{ and } \bar{y}(1) - 0.5\bar{y}'(1) = 0.5 \qquad \text{(T)}$$

67. Solve Problem 66 by the second-order modified Euler method.
68. Solve Problem 66 by the fourth-order Runge-Kutta method.
69. Solve the following ODE by the shooting method using the first-order explicit Euler method with (a) $\Delta x = 0.125$ and (b) $\Delta x = 0.0625$.

$$\bar{y}'' + 4\bar{y}' + 6.25\bar{y} = e^x \qquad \bar{y}(0) = 0 \text{ and } \bar{y}(1) - 0.5\bar{y}'(1) = 0.5 \qquad \text{(U)}$$

70. Solve Problem 69 by the second-order modified Euler method.
71. Solve Problem 69 by the fourth-order Runge-Kutta method.
72. Solve Problem 66 by the second-order equilibrium method. Implement the mixed boundary condition by applying the PDE at the boundary.
73. Solve Problem 69 by the second-order equilibrium method. Implement the mixed boundary condition by applying the PDE at the boundary.

Boundary Conditions at Infinity

74. *Solve the following ODE by the shooting method using the first-order explicit Euler method with (a) $\Delta x = 0.25$ and (b) $\Delta x = 0.125$.

$$\bar{y}'' - y = 0 \qquad \bar{y}(0) = 1 \text{ and } \bar{y}(\infty) = 0 \qquad \text{(V)}$$

Let $y'(0.0) = 0.0$ and -1.0 for the first two passes. Implement the boundary condition at infinity by applying that BC at $x = 2.0$, 5.0, 10.0, etc., until the solution at $x = 1.0$ changes by less than 0.001.

75. Work Problem 74 by the second-order modified Euler method.
76. Work Problem 74 by the fourth-order Runge-Kutta method.

77. Solve the following ODE by the shooting method using the first-order explicit Euler method with (a) $\Delta x = 0.25$ and (b) $\Delta x = 0.125$.

$$\bar{y}'' + \bar{y}' - 2\bar{y} = 1 \qquad \bar{y}(0) = 1 \text{ and } \bar{y}(\infty) = 0 \qquad \text{(W)}$$

78. Work Problem 77 by the second-order modified Euler method.
79. Work Problem 77 by the fourth-order Runge-Kutta method.
80. Solve Problem 74 by the second-order equilibrium method.
81. Solve Problem 77 by the second-order equilibrium method.

8.6 Higher-Order Equilibrium Methods

The Five-Point Fourth-Order Equilibrium Method

82. Solve ODE (A) using the five-point fourth-order equilibrium method for $\Delta x = 0.125$. Use the three-point second-order equilibrium method at points adjacent to the boundaries.
83. Solve ODE (B) by the procedure described in Problem 82.
84. Solve ODE (C) by the procedure described in Problem 82.
85. Solve ODE (D) by the procedure described in Problem 82.

The Compact Three-Point Fourth-Order Equilibrium Method

86. *Compact three-point fourth-order finite difference approximations are presented in Section 8.6.2 for $\bar{y}''(x)$ and $\bar{y}'(x)$. When used in a second-order boundary-value ODE, an implicit fourth-order FDE, Eq. (8.129), is obtained. This equation is straightforward in concept but difficult to implement numerically. When the first, derivative, $\bar{y}'(x)$ does not appear in the ODE, however, a simple FDE can be developed, such as Eq. (8.134) in Example 8.9. Apply this procedure to solve the following ODE with (a) $\Delta x = 0.25$ and (b) $\Delta x = 0.125$.

$$\bar{y}'' + \bar{y} = 1 \qquad \bar{y}(0) = 0 \text{ and } \bar{y}(1) = 1$$

87. Solve the following ODE by the procedure described in Problem 86:

$$\bar{y}'' + \bar{y} = 1 + x + e^x \qquad \bar{y}(0) = 0 \text{ and } \bar{y}(1) = 1$$

88. Solve the following ODE by the procedure described in Problem 86:

$$\bar{y}'' - \bar{y} = 1 \qquad \bar{y}(0) = 0 \text{ and } \bar{y}(1) = 1$$

89. Solve the following ODE by the procedure described in Problem 86:

$$\bar{y}'' - \bar{y} = 1 + x + e^x \qquad \bar{y}(0) = 0 \text{ and } \bar{y}(1) = 1$$

8.7 The Equilibrium Method for Nonlinear Boundary-Value Problems

Iteration

90. *Solve the following ODE by the second-order equilibrium method with $\Delta x = 0.25$ and 0.125 by iteration. Compare the solutions at $x = 0.5$.

$$\bar{y}'' + (1 + \bar{y})\bar{y}' + (1 + \bar{y})\bar{y} = 1 \qquad \bar{y}(0) = 0 \text{ and } \bar{y}(1) = 1 \qquad \text{(X)}$$

91. Solve the following ODE by the second-order equilibrium method with $\Delta x = 0.25$ and 0.125 by iteration. Compare the solutions at $x = 0.5$.

$$\bar{y}'' + (1 + x + \bar{y})\bar{y}' + (1 + x + \bar{y})\bar{y} = 2e^{x/2} + 1 + x$$

$$\bar{y}(0) = 0 \text{ and } \bar{y}(1) = 1 \quad (\text{Y})$$

Newton's Method

92. Solve the following ODE by the second-order equilibrium method with $\Delta x = 0.25$ and 0.125 by Newton's method. Compare the solutions at $x = 0.5$.

$$\bar{y}'' + (1 + \bar{y})\bar{y}' + (1 + \bar{y})\bar{y} = 1 \qquad \bar{y}(0) = 0 \text{ and } \bar{y}(1) = 1 \qquad (\text{Z})$$

93. Solve the following ODE by the second-order equilibrium method with $\Delta x = 0.25$ and 0.125 by Newton's method. Compare the solutions at $x = 0.5$.

$$\bar{y}'' + (1 + x + \bar{y})\bar{y}' + (1 + x + \bar{y})\bar{y} = 2e^{x/2} + 1 + x$$

$$\bar{y}(0) = 0 \text{ and } \bar{y}(1) = 1 \quad (\text{AA})$$

8.8 The Equilibrium Method on Nonuniform Grids

94. Solve the following ODE by the second-order equilibrium method on the nonuniform grid specified in Table 8.25.

$$\bar{y}'' + 5\bar{y}' + 4\bar{y} = 1 \qquad \bar{y}(0) = 0 \text{ and } \bar{y}(1) = 1 \qquad (\text{BB})$$

Compare the results with the results obtained in Problem 5.

95. Solve Problem by the second-order equilibrium method on the transformed grid specified in Table 8.25. Compare the results with the results obtained in Problem 5.

96. Solve Problem 94 by the second-order modified Euler method. Compare the results with the results obtained in Problem 5.

97. Solve the following ODE by the second-order equilibrium method on the nonuniform grid specified in Table 8.25.

$$\bar{y}'' + 4\bar{y}' + 6.25\bar{y} = 1 \qquad \bar{y}(0) = 0 \text{ and } \bar{y}(1) = 1 \qquad (\text{CC})$$

Compare the results with the results obtained in Problem 8.

98. Solve Problem 97 by the second-order equilibrium method on the transformed grid specified in Table 8.25. Compare the results with the results obtained in Problem 8.

99. Solve Problem 97 by the second-order modified Euler method. Compare the results with the results obtained in Problem 8.

100. Solve the following ODE by the second-order equilibrium method on the nonuniform grid specified in Table 8.25.

$$\bar{y}'' + 5\bar{y}' + 4\bar{y} = e^x \qquad \bar{y}(0) = 0 \text{ and } \bar{y}(1) = 1 \qquad (\text{DD})$$

Compare the results with the results obtained in Problem 11.

101. Solve Problem 100 by the second-order equilibrium method on the transformed grid specified in Table 8.25. Compare the results with the results obtained in Problem 11.

102. Solve Problem 100 by the second-order modified Euler method. Compare the results with the results obtained in Problem 11.

103. Solve the following ODE by the second-order equilibrium method on the nonuniform grid specified in Table 8.25.

$$\bar{y}'' + 4\bar{y}' + 6.25\bar{y} = e^x \qquad \bar{y}(0) = 0 \text{ and } \bar{y}(1) = 1 \qquad \text{(EE)}$$

Compare the results with the results obtained in Problem 14.

104. Solve Problem 103 by the second-order equilibrium method. Compare the results with the results obtained in Problem 14.

105. Solve Problem 103 by the second-order modified Euler method. Compare the results with the results obtained in Problem 14.

8.9 Eigenproblems

106. Consider the eigenproblem described by Eq. (8.178). The exact solution of this eigenproblem is $k = \pm n\pi$ $(n = 1, 2, \ldots)$. The finite difference equation corresponding to the eigenproblem is given by Eq. (8.185). The numerical solution of this eigenproblem for $\Delta x = \frac{1}{5}$ is presented in Table 8.27. Determine the solution to this eigenproblem for (a) $\Delta x = \frac{1}{6}$ and (b) $\Delta x = \frac{1}{8}$. Compare the three sets of results in normalized form, that is, k/π.

107. Consider the eigenproblem

$$\bar{y}'' - k^2\bar{y} = 0 \qquad \bar{y}(0) = \bar{y}(1) = 0$$

This problem has no solution except the trivial solution $\bar{y}(x) = 0$. (a) Demonstrate this result analytically. (b) Illustrate this result numerically by setting up a system of second-order finite difference equations with $\Delta x = 0.2$, and show that there are no real values of k.

108. Consider the eigenproblem

$$\bar{y}'' + \bar{y}' + k^2\bar{y} = 0 \qquad \bar{y}(0) = \bar{y}(1) = 0$$

Estimate the first three eigenvalues by setting up a system of second-order finite difference equations with $\Delta x = 0.25$.

109. Work Problem 108 for the eigenproblem

$$\bar{y}'' + (1 + x)\bar{y}' + k^2\bar{y} = 0 \qquad \bar{y}(0) = \bar{y}(1) = 0$$

110. Work Problem 108 for the eigenproblem

$$\bar{y}'' + \bar{y}' + k^2(1 + x)\bar{y} = 0 \qquad \bar{y}(0) = \bar{y}(1) = 0$$

8.10 Programs

111. Implement the fourth-order Runge-Kutta shooting method program presented in Section 8.10.1. Check out the program using the given data set.

112. Solve any of Problems 6, 9, 12, ..., 42 with the fourth-order Runge-Kutta method program.

113. Implement the second-order equilibrium method program presented in Section 8.10.2. Check out the program using the given data set.

114. Solve any of Problems 43 to 51 with the second-order equilibrium method program.

APPLIED PROBLEMS

Several applied problems from various disciplines are presented in this section. All these problems can be solved by any of the methods presented in this chapter. An infinite variety of exercises can be constructed by changing the numerical values of the parameters, the grid size Δx, and so on.

115. The temperature distribution in the wall of a pipe through which a hot liquid is flowing is given by the ODE

$$\frac{d^2T}{dr^2} + \frac{1}{r}\frac{dT}{dr} = 0 \qquad T(1) = 100 \text{ C and } T(2) = 0 \text{ C}$$

Determine the temperature distribution in the wall.

116. The pipe described in Problem 115 is cooled by convection on the outer surface. Thus, the heat conduction $\dot{q}_{cond}$ at the outer wall is equal to the heat convection $\dot{q}_{conv}$ to the surroundings:

$$\dot{q}_{cond} = -kA\frac{dT}{dr} = \dot{q}_{conv} = hA(T - T_a)$$

where the thermal conductivity $k = 100.0 \text{ J/(s-m-K)}$, the convective cooling coefficient $h = 500.0 \text{ J/(s-m}^2\text{-K)}$, and $T_a = 0.0 \text{ C}$ is the temperature of the surroundings. Determine the temperature distribution in the wall.

117. The temperature distribution in a cylindrical rod made of a radioactive isotope is governed by the ordinary differential equation

$$\frac{d^2T}{dr^2} + \frac{1}{r}\frac{dT}{dr} = A\left[1 + \left(\frac{r}{R}\right)^2\right] \qquad T'(0) = 0 \text{ and } T(R) = 0$$

Solve this problem for $T(r)$, where $R = 1.0$ and $A = -100.0$.

118. The velocity distribution in the laminar boundary layer formed when an incompressible fluid flows over a flat plate is related to the solution of the ordinary differential equation

$$\frac{d^3f}{d\eta^3} + f\frac{d^2f}{d\eta^2} = 0 \qquad f(0) = 0, \ f'(0) = 0, \text{ and } f'(\eta) \to 1 \text{ as } \eta \to \infty$$

where f is a dimensionless stream function, the velocity u is proportional to $f'(\eta)$, and η is proportional to distance normal to the plate. Solve this problem for $f(\eta)$.

119. The deflection of a simply supported and uniformly loaded beam is governed by the ordinary differential equation (for small deflections)

$$EI\frac{d^2y}{dx^2} = -\frac{qLx}{2} + \frac{qx^2}{2} \qquad y(0) = 0 \text{ and } y(L) = 0$$

where q is the uniform load per unit length, L is the length of the beam, I is the moment of inertia of the beam cross section, and E is the modulus of elasticity. For a rectangular beam, $I = wh^3/12$, where w is the width and h is the height. Consider a beam ($E = 200$ GN/m^2) 5.0 m long, 5.0 cm wide, and 10.0 cm high, which is subjected to the uniform load $q = -1,500.0$ N/m on the 5.0 cm face. Solve for the deflection $y(x)$.

120. When the load on the beam described in Problem 120 is applied on the 10.0 cm face, the deflection will be large. In that case, the governing differential equation is

$$\frac{EI(d^2y/dx^2)}{[1 + (dy/dx)^2]^{3/2}} = -\frac{qLx}{2} + \frac{qx^2}{2}$$

For the properties specified in Problem 120, determine $y(x)$.

III

Partial Differential Equations

III.1 INTRODUCTION

Partial differential equations (*PDEs*) arise in all fields of engineering and science. Most real physical processes are governed by partial differential equations. In many cases, simplifying approximations are made to reduce the governing PDEs to ordinary differential equations (ODEs) or even to algebraic equations. However, because of the ever increasing requirement for more accurate modeling of physical processes, engineers and scientists are more and more required to solve the actual PDEs that govern the physical problem being investigated. Part III is devoted to the solution of partial differential equations by *finite difference methods*.

For simplicity of notation, the phrase *partial differential equation* frequently will be replaced by the acronym *PDE* in Part III. This replacement generally makes the text flow more smoothly and more succinctly, without losing the meaning of the material.

Some general features of partial differential equations are discussed in this section. The three classes of PDEs (i.e., *elliptic*, *parabolic*, and *hyperbolic PDEs*) are introduced. The two types of physical problems (i.e., *equilibrium* and *propagation problems*) are discussed.

The objectives of Part III are:

1. To present the general features of partial differential equations
2. To discuss the relationship between the type of physical problem being solved, the classification of the corresponding governing partial differential equation, and the type of numerical method required
3. To present examples to illustrate these concepts.

III.2 GENERAL FEATURES OF PARTIAL DIFFERENTIAL EQUATIONS

A partial differential equation (PDE) is an equation stating a relationship between a function of two or more independent variables and the partial derivatives of this function with respect to these independent variables. The dependent variable f is used as a generic dependent variable throughout Part III. In most problems in engineering and science, the independent variables are either space (x, y, z) or space and time (x, y, z, t). The dependent variable depends on the physical problem being modeled. Examples of three simple partial differential equations having two independent variables are presented below:

$$\frac{\partial^2 f}{\partial x^2} + \frac{\partial^2 f}{\partial y^2} = 0 \tag{III.1}$$

$$\frac{\partial f}{\partial t} = \alpha \frac{\partial^2 f}{\partial x^2} \tag{III.2}$$

$$\frac{\partial^2 f}{\partial t^2} = c^2 \frac{\partial^2 f}{\partial x^2} \tag{III.3}$$

Equation (III.1) is the two-dimensional *Laplace equation*, Eq. (III.2) is the one-dimensional *diffusion equation*, and Eq. (III.3) is the one-dimensional *wave equation*. For simplicity of notation, Eqs. (III.1) to (III.3) usually will be written as

$$f_{xx} + f_{yy} = 0 \tag{III.4}$$
$$f_t = \alpha f_{xx} \tag{III.5}$$
$$f_{tt} = c^2 f_{xx} \tag{III.6}$$

where the subscripts denote partial differentiation.

The solution of a partial differential equation is that particular function, $f(x, y)$ or $f(x, t)$, which satisfies the PDE in the domain of interest, $D(x, y)$ or $D(x, t)$, respectively, and satisfies the initial and/or boundary conditions specified on the boundaries of the domain of interest. In a very few special cases, the solution of a PDE can be expressed in closed form. In the majority of problems in engineering and science, the solution must be obtained by numerical methods. Such problems are the subject of Part III.

Equations (III.4) to (III.6) are examples of partial differential equations in two independent variables, x and y, or x and t. Equation (III.4), which is the two-dimensional Laplace equation, in three independent variables is

$$\nabla^2 f = f_{xx} + f_{yy} + f_{zz} = 0 \tag{III.7}$$

where ∇^2 is the Laplacian operator, which in Cartesian coordinates is

$$\nabla^2 = \frac{\partial^2}{\partial x^2} + \frac{\partial^2}{\partial y^2} + \frac{\partial^2}{\partial z^2} \tag{III.8}$$

Equation (III.5), which is the one-dimensional diffusion equation, in four independent variables is

$$f_t = \alpha(f_{xx} + f_{yy} + f_{zz}) = \alpha \nabla^2 f \tag{III.9}$$

The parameter α is the diffusion coefficient. Equation (III.6), which is the one-dimensional wave equation, in four independent variables is

$$f_{tt} = c^2(f_{xx} + f_{yy} + f_{zz}) = c^2 \nabla^2 f \tag{III.10}$$

The parameter c is the wave propagation speed. Problems in two, three, and four independent variables occur throughout engineering and science.

Equations (III.4) to (III.10) are all *second-order* partial differential equations. The order of a PDE is determined by the highest-order derivative appearing in the equation. A large number of physical problems are governed by second-order PDEs. Some physical problems are governed by a first-order PDE of the form

$$af_t + bf_x = 0 \tag{III.11}$$

where a and b are constants. Other physical problems are governed by fourth-order PDEs such as

$$f_{xxxx} + f_{xxyy} + f_{yyyy} = 0 \tag{III.12}$$

Equations (III.4) to (III.12) are all *linear* partial differential equations. A linear PDE is one in which all of the partial derivatives appear in linear form and none of the coefficients depends on the dependent variable. The coefficients may be functions of the independent variables, in which case the PDE is a linear, variable coefficient, PDE. For example,

$$af_t + bxf_x = 0 \tag{III.13}$$

where a and b are constants, is a variable coefficient linear PDE, whereas Eqs. (III.4) to (III.12) are all linear PDEs. If the coefficients depend on the dependent variable, or the derivatives appear in a nonlinear form, then the PDE is *nonlinear*. For example,

$$ff_x + bf_y = 0 \tag{III.14}$$

$$af_x^2 + bf_y = 0 \tag{III.15}$$

are nonlinear PDEs.

Equations (III.4) to (III.15) are all *homogeneous* partial differential equations. An example of a nonhomogeneous PDE is given by

$$\nabla^2 f = f_{xx} + f_{yy} + f_{zz} = F(x, y, z) \tag{III.16}$$

Equation (III.16) is the *nonhomogeneous* Laplace equation, which is known as the *Poisson equation*. The nonhomogeneous term, $F(x, y, z)$, is a forcing function, a source term, or a dissipation function, depending on the application. The appearance of a nonhomogeneous term in a partial differential equation does not change the general features of the PDE, nor does it usually change or complicate the numerical method of solution.

Equations (III.4) to (III.16) are all examples of a single partial differential equation governing one dependent variable. Many physical problems are governed by a system of PDEs involving several dependent variables. For example, the two PDEs

$$af_t + bg_x = 0 \tag{III.17a}$$

$$Ag_t + Bf_x = 0 \tag{III.17b}$$

comprise a system of two coupled partial differential equations in two independent variables (x and t) for determining the two dependent variables, $f(x, t)$ and $g(x, t)$. Systems containing several PDEs occur frequently, and systems containing higher-order PDEs occur occasionally. Systems of PDEs are generally more difficult to solve numerically than a single PDE.

As illustrated in the preceding discussion, a wide variety of partial differential equations exists. Each problem has its own special governing equation or equations and its own peculiarities which must be considered individually. However, useful insights into the general features of PDEs can be obtained by studying three special cases. The first special case is the general quasilinear (i.e., linear in the highest-order derivative) second-order nonhomogeneous PDE in two independent variables, which is

$$\boxed{Af_{xx} + Bf_{xy} + Cf_{yy} + Df_x + Ef_y + Ff = G} \tag{III.18}$$

where the coefficients A to C may depend on x, y, f_x, and f_y, the coefficients D to F may depend on x, y, and f, and the nonhomogeneous term G may depend on x and y. The second special case is the general quasilinear first-order nonhomogeneous PDE in two independent variables, which is

$$\boxed{af_t + bf_x = c} \tag{III.19}$$

where a, b, and c may depend on x, t, and f. The third special case is the system of two general quasilinear first-order nonhomogeneous PDEs in two independent variables, which can be written as

$$\boxed{\begin{aligned} af_t + bf_x + cg_t + dg_x &= e \\ Af_t + Bf_x + Cg_t + Dg_x &= E \end{aligned}} \tag{III.20a} \tag{III.20b}$$

where the coefficients a to d and A to D and the nonhomogeneous terms e and E may depend on x, t, f, and g. The general features of these three special cases are similar to the general features of all the PDEs discussed in this book. Consequently, these three special cases are studied thoroughly in the following sections.

III.3 CLASSIFICATION OF PARTIAL DIFFERENTIAL EQUATIONS

Physical problems are governed by many different partial differential equations. A few problems are governed by a single first-order PDE. Numerous problems are governed by a system of first-order PDEs. Some problems are governed by a single second-order PDE, and numerous problems are governed by a system of second-order PDEs. A few problems are governed by fourth-order PDEs. The classification of PDEs is most easily explained for a single second-order PDE. Consequently, in the following discussion, the general quasilinear (i.e., linear in the highest-order derivative) second-order nonhomogeneous PDE in two independent variables [i.e., Eq. (III.18)] is classified first. The classification of the general first-order nonhomogeneous PDE in two independent variables [i.e., Eq. (III.19)] is studied next. Finally, the classification of the system of two quasilinear first-order nonhomogeneous PDEs [i.e., Eq. (III.20)] is studied. The classification of higher-order PDEs, larger systems of PDEs, and PDEs having more than two independent variables is considerably more complicated.

The general quasilinear second-order nonhomogeneous partial differential equation in two independent variables is [see Eq. (III.18)]

$$Af_{xx} + Bf_{xy} + Cf_{yy} + Df_x + Ef_y + Ff = G \tag{III.21}$$

The classification of Eq. (III.21) depends on the sign of the *discriminant*, $B^2 - 4AC$, as follows:

$B^2 - 4AC$	Classification
Negative	Elliptic
Zero	Parabolic
Positive	Hyperbolic

The terminology *elliptic*, *parabolic*, and *hyperbolic* chosen to classify PDEs reflects the analogy between the form of the discriminant, $B^2 - 4AC$, for PDEs and the form of the discriminant, $B^2 - 4AC$, which classifies conic sections. Conic sections are described by the general second-order algebraic equation

$$Ax^2 + Bxy + Cy^2 + Dx + Ey + F = 0 \tag{III.22}$$

The type of curve represented by Eq. (III.22) depends on the sign of the discriminant, $B^2 - 4AC$, as follows:

$B^2 - 4AC$	Type of curve
Negative	Ellipse
Zero	Parabola
Positive	Hyperbola

The analogy to the classification of PDEs is obvious. There is no other significance to the terminology.

What is the significance of the above classification? What impact, if any, does the classification of a PDE have on the allowable and/or required initial and boundary conditions? Does the classification of a PDE have any effect on the choice of numerical method employed to solve the equation? These questions are discussed in this section, and the results are applied to physical problems in the next section.

The classification of a PDE is intimately related to the *characteristics* of the PDE. Characteristics are $(n-1)$-dimensional hypersurfaces in n-dimensional hyperspace that have some very special features. The prefix *hyper* is used to denote spaces of more than three dimensions, that is, *xyzt* spaces, and curves and surfaces within those spaces. In two-dimensional space, which is the case considered here, characteristics are paths (curved, in general) in the solution domain along which information propagates. In other words, information propagates throughout the solution domain along the characteristics paths. Discontinuities in the derivatives of the dependent variable (if they exist) also propagate along the characteristics paths. If a PDE possesses real characteristics, then information propagates along these characteristics. If no real characteristics exist, then there are no preferred paths of information propagation. Consequently, the presence or absence of characteristics has a significant impact on the solution of a PDE (by both analytical and numerical methods).

A simple physical example can be used to illustrate the physical significance of characteristic paths. Convection is the process in which a physical property is propagated (i.e., convected) through space by the motion of the medium occupying the space. Fluid flow is a common example of convection. The convection of a property f of a fluid particle in one dimension is governed by the convection equation

$$f_t + u f_x = 0 \tag{III.23}$$

where u is the convection velocity. A moving fluid particle carries (convects) its mass, momentum, and energy with it as it moves through space. The location $x(t)$ of the fluid particle is related to its velocity $u(t)$ by the relationship

$$\frac{dx}{dt} = u \tag{III.24}$$

The path of the fluid particle, called its pathline, is given by

$$x = x_0 + \int_{t_0}^{t} u(t)\, dt \tag{III.25}$$

The pathline is illustrated in Figure III.1a.

Along the pathline, the convection equation [i.e., Eq. (III.23)] can be written as

$$f_t + u f_x = f_t + \frac{dx}{dt} f_x = \frac{df}{dt} = 0 \tag{III.26}$$

which can be integrated to yield $f = $ constant. Consequently, the fluid property f is convected along the pathline, which is the characteristic path associated with the convection equation. Equation (III.24), which is generally called the *characteristic equation*, is the differential equation of the characteristic path. The physical significance of the pathline (i.e., the characteristic path) as the path of propagation of the fluid property

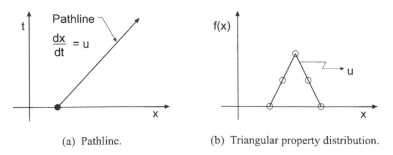

(a) Pathline.   (b) Triangular property distribution.

Figure III.1 Pathline as the characteristic for the convection equation.

f is quite apparent for fluid convection. Equation (III.26), which is generally called the *compatibility equation*, is the differential equation which applies along the characteristic path.

To illustrate further the property of a characteristic path as the path of propagation in a convection problem, consider the triangular property distribution illustrated in Figure III.1b. As the fluid particles move to the right at the constant convection velocity u, each particle carries with it its value of the property f. Consequently, the triangular property distribution simply moves (i.e., convects) to the right at the constant convection velocity u, unchanged in magnitude and shape. The apex of the triangle, which is a point of discontinuous slope in the property distribution, convects as a discontinuity in slope at the convection velocity u. This simple convection example illustrates the significance of characteristic paths.

Let's return to the classification of Eq. (III.21). Several procedures exist for determining the characteristics, and hence the classification, of PDEs. Because discontinuities in the derivatives of the solution, if they exist, must propagate along the characteristics, one approach is to answer the following question: Are there any paths in the solution domain $D(x, y)$ passing through a general point P along which the second derivatives of $f(x, y)$, that is, f_{xx}, f_{xy}, and f_{yy}, are multivalued or discontinuous? Such paths, if they exist, are the paths of information propagation, that is, the characteristics.

One relationship for determining the three second derivatives of $f(x, y)$ is given by the partial differential equation itself, Eq. (III.21). Two more relationships are obtained by applying the chain rule to determine the total derivatives of f_x and f_y, which are themselves functions of x and y. Thus,

$$d(f_x) = f_{xx}\, dx + f_{xy}\, dy \qquad\qquad\qquad\qquad\qquad (\text{III.27a})$$

$$d(f_y) = f_{yx}\, dx + f_{yy}\, dy \qquad\qquad\qquad\qquad\qquad (\text{III.27b})$$

Equations (III.21) and (III.27) can be written in matrix form as follows:

$$\begin{bmatrix} A & B & C \\ dx & dy & 0 \\ 0 & dx & dy \end{bmatrix} \begin{bmatrix} f_{xx} \\ f_{xy} \\ f_{yy} \end{bmatrix} = \begin{bmatrix} -Df_x - Ef_y - F + G \\ d(f_x) \\ d(f_y) \end{bmatrix} \qquad (\text{III.28})$$

Equation (III.28) can be solved by Cramer's rule to yield unique finite values of f_{xx}, f_{xy}, and f_{yy}, unless the determinant of the coefficient matrix vanishes. In that case, the second derivatives of $f(x, y)$ are either infinite, which is physically meaningless, or they are indeterminate, and thus multivalued or discontinuous.

Setting the determinant of the coefficient matrix of Eq. (III.28) equal to zero yields

$$A(dy)^2 - B(dx)(dy) + C(dx)^2 = 0 \qquad\qquad\qquad\qquad (\text{III.29})$$

Equation (III.29) is the *characteristic equation* corresponding to Eq. (III.21). Equation (III.29) can be solved by the quadratic formula to yield

$$\boxed{\frac{dy}{dx} = \frac{B \pm \sqrt{B^2 - 4AC}}{2A}} \qquad\qquad\qquad\qquad (\text{III.30})$$

Equation (III.30) is the differential equation for two families of curves in the xy plane, corresponding to the $\pm$ signs. Along these two families of curves, the second derivatives of

$f(x, y)$ may be multivalued or discontinuous. These two families of curves, if they exist, are the characteristic paths of the original PDE, Eq. (III.21).

The two families of characteristic curves may be complex, real and repeated, or real and distinct, according to whether the discriminant, $B^2 - 4AC$, is negative, zero, or positive, respectively. Accordingly, Eq. (III.21) is classified as follows:

$B^2 - 4AC$	Characteristic curves	Classification
Negative	Complex	Elliptic
Zero	Real and repeated	Parabolic
Positive	Real and distinct	Hyperbolic

Consequently, elliptic PDEs have no real characteristic paths, parabolic PDEs have one real repeated characteristic path, and hyperbolic PDEs have two real distinct characteristic paths.

The presence of characteristic paths in the solution domain leads to the concepts of *domain of dependence* and *range of influence*. Consider a point P in the solution domain, $D(x, y)$. The domain of dependence of point P is defined as the region of the solution domain upon which the solution at point P, $f(x_p, y_p)$, depends. In other words, $f(x_p, y_p)$ depends on everything that has happened in the domain of dependence. The range of influence of point P is defined as the region of the solution domain in which the solution $f(x, y)$ is influenced by the solution at point P. In other words, $f(x_p, y_p)$ influences the solution at all points in the range of influence.

Recall that parabolic and hyperbolic PDEs have real characteristic paths. Consequently, they have specific domains of dependence and ranges of influence. Elliptic PDEs, on the other hand, do not have real characteristic paths. Consequently, they have no specific domains of dependence or ranges of influence. In effect, the entire solution domain of an elliptic PDE is both the domain of dependence and the range of influence of every point in the solution domain. Figure III.2 illustrates the concepts of domain of dependence and range of influence for elliptic, parabolic, and hyperbolic PDEs.

Unlike the second-order PDE just discussed, a single first-order PDE is always hyperbolic. Consider the classification of the single general quasilinear first-order non-homogeneous PDE, Eq. (III.19):

$$af_t + bf_x = c \tag{III.31}$$

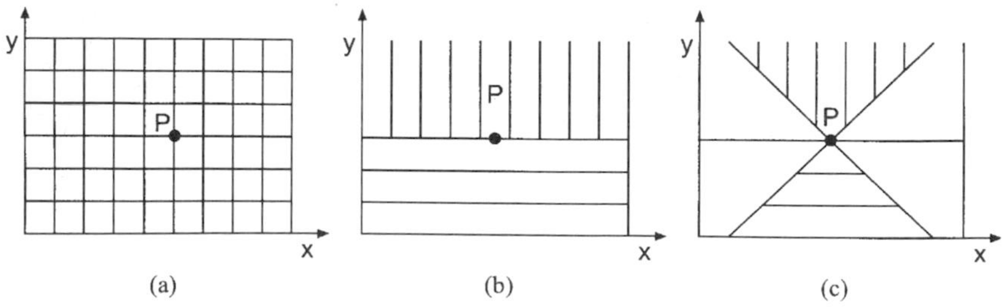

Figure III.2 Domain of dependence (horizontal hatching) and range of influence (vertical hatching) of PDEs: (a) eliptic PDE; (b) parabolic PDE; (c) hyperbolic PDE.

The characteristic paths, if they exist, are determined by answering the following question: Are there any paths in the solution domain $D(x, t)$ passing through a general point P along which the first derivatives of $f(x, t)$ may be discontinuous? Such paths, if they exist, are the characteristics of Eq. (III.31).

One relationship for determining f_t and f_x is given by Eq. (III.31). Another relationship is given by the total derivative of $f(t, x)$:

$$df = f_t \, dt + f_x \, dx \qquad \qquad \text{(III.32)}$$

Equations (III.31) and (III.32) can be written in matrix form as

$$\begin{bmatrix} a & b \\ dt & dx \end{bmatrix} \begin{bmatrix} f_t \\ f_x \end{bmatrix} = \begin{bmatrix} c \\ df \end{bmatrix} \qquad \qquad \text{(III.33)}$$

As before, the partial derivatives f_t and f_x are uniquely determined unless the determinant of the coefficient matrix of Eq. (III.33) is zero. Setting that determinant equal to zero gives the characteristic equation, which is

$$a \, dx - b \, dt = 0 \qquad \qquad \text{(III.34)}$$

Solving Eq. (III.34) for dx/dt gives

$$\boxed{\frac{dx}{dt} = \frac{b}{a}} \qquad \qquad \text{(III.35)}$$

Equation (III.35) is the differential equation for a family of paths in the solution domain along which f_t and f_x may be discontinuous, or multivalued. Since a and b are real functions, the characteristic paths always exist. Consequently, a single quasilinear first-order PDE is always hyperbolic. The convection equation, Eq. (III.23), is an example of such a PDE.

As a third example, consider the classification of the system of two general coupled quasilinear first-order nonhomogeneous partial differential equations, Eq. (III.20):

$$af_t + bf_x + cg_t + dg_x = e \qquad \qquad \text{(III.36a)}$$
$$Af_t + Bf_x + Cg_t + Dg_x = E \qquad \qquad \text{(III.36b)}$$

The characteristic paths, if they exist, are determined by answering the following question: Are there any paths in the solution domain $D(x, t)$ passing through a general point P along which the first derivatives of $f(x, t)$ and $g(x, t)$ are not uniquely determined? Such paths, if they exist, are the characteristics of Eq. (III.36).

Two relationships for determining the four first derivatives of $f(x, t)$ and $g(x, t)$ are given by Eq. (III.36). Two more relationships are given by the total derivatives of f and g:

$$df = f_t \, dt + f_x \, dx \qquad \qquad \text{(III.37a)}$$
$$dg = g_t \, dt + g_x \, dx \qquad \qquad \text{(III.37b)}$$

Equations (III.36) and (III.37), which comprise a system of four equations for determining f_t, f_x, g_t, and g_x, can be written in matrix form as follows:

$$
\begin{bmatrix}
a & b & c & d \\
A & B & C & D \\
dt & dx & 0 & 0 \\
0 & 0 & dt & dx
\end{bmatrix}
\begin{bmatrix}
f_t \\
f_x \\
g_t \\
g_x
\end{bmatrix}
=
\begin{bmatrix}
e \\
E \\
df \\
dg
\end{bmatrix}
\tag{III.38}
$$

As before, the partial derivatives are uniquely determined unless the determinant of the coefficient matrix of Eq. (III.38) is zero. Setting that determinant equal to zero yields the characteristic equation, which is

$$
(aC - Ac)(dx)^2 - (aD - Ad + bC - Bc)(dx)(dt) + (bD - Bd)(dt)^2 = 0 \tag{III.39}
$$

Equation (III.39), which is a quadratic equation in dx/dt, may be written as

$$
\bar{A}(dx)^2 - \bar{B}(dx)(dt) + \bar{C}(dt)^2 = 0 \tag{III.40}
$$

where $\bar{A} = (aC - Ac)$, $\bar{B} = (aD - Ad + bC - Bc)$, and $\bar{C} = (bD - Bd)$. Equation (III.40) can be solved by the quadratic formula to yield

$$
\boxed{\frac{dx}{dt} = \frac{\bar{B} \pm \sqrt{\bar{B}^2 - 4\bar{A}\,\bar{C}}}{2\bar{A}}}
\tag{III.41}
$$

Equation (III.41) is the differential equation for two families of curves in the xt plane, corresponding to the $\pm$ signs. Along these two families of curves, the first derivatives of $f(x, t)$ and $g(x, t)$ may be multivalued. These two families of curves, if they exist, are the characteristic paths of the original system of PDEs, Eq. (III.36). The slopes of the two families of characteristic paths may be complex, real and repeated, or real and distinct, according to whether the discriminant, $\bar{B}^2 - 4\bar{A}\,\bar{C}$, is negative, zero, or positive, respectively. Accordingly, Eq. (III.36) is classified as follows:

$\bar{B}^2 - 4\bar{A}\,\bar{C}$	Classification
Negative	Elliptic
Zero	Parabolic
Positive	Hyperbolic

In summary, the physical interpretation of the classification of a partial differential equation can be explained in terms of its characteristics. If real characteristics exist, preferred paths of information propagation exist. The speed of propagation of information through the solution domain depends on the slopes of the characteristics. Specific domains of dependence and ranges of influence exist for every point in the solution domain. Physical problems governed by PDEs that have real characteristics are *propagation problems*. Thus, *parabolic* and *hyperbolic PDEs* govern propagation problems.

If the slopes of the characteristics are complex, then no real characteristics exist. There are no preferred paths of information propagation. The domain of dependence and range of influence of every point is the entire solution domain. The solution at every point depends on the solution at all the other points, and the solution at each point influences the solution at all the other points. Since there are no curves along which the derivatives may be discontinuous, the solution throughout the entire solution domain must be continuous. Physical problems governed by PDEs that have complex characteristics are *equilibrium problems*. Thus, *elliptic PDEs* govern equilibrium problems. These concepts are related to the classification of physical problems in the next section.

The significance of the classification of a PDE as it relates to the numerical approximation of the PDE is as follows. For an elliptic PDE which contains only second-order spatial derivatives, there are no preferred physical information propagation paths. Consequently, all points are dependent on all other points and all points influence all other points. This physical behavior should be accounted for in the numerical approximation of the PDE.

For a parabolic PDE which contains only second-order spatial derivatives, the preferred physical information propagation paths are lines (or surfaces) of constant time (or constant timelike variable). In other words, at each time (or timelike variable) level, all points are dependent on all other points and all points influence all other points. This physical behavior should be accounted for in the numerical approximation of the PDE.

For a hyperbolic PDE which contains only first-order spatial derivatives, distinct physical information propagation paths exist. Physical information propagates along these distinct physical propagation paths. This physical behavior should be accounted for in the numerical approximation of the PDE.

Elliptic and parabolic PDEs exist which contain first-order spatial derivatives in addition to second-order spatial derivatives. In such cases, the physical behavior associated with the second-order spatial derivatives is the same as before, but the physical behavior associated with the first-order spatial derivatives acts similarly to the behavior of the first-order spatial derivatives in a hyperbolic PDE. This physical behavior should be accounted for in the numerical approximation of such PDEs.

III.4 CLASSIFICATION OF PHYSICAL PROBLEMS

Physical problems fall into one of the following three general classifications:

1. Equilibrium problems
2. Propagation problems
3. Eigenproblems

Each of these three types of physical problems has its own special features, its own particular type of governing partial differential equation, and its own special numerical solution method. A clear understanding of these concepts is essential if meaningful numerical solutions are to be obtained.

III.4.1 Equilibrium Problems

Equilibrium problems are *steady-state problems* in *closed domains* $D(x, y)$ in which the solution $f(x, y)$ is governed by an elliptic PDE subject to boundary conditions specified at each point on the boundary B of the domain. Equilibrium problems are *jury problems* in which the entire solution is passed on by a jury requiring satisfaction of all internal requirements (i.e., the PDE) and all the boundary conditions simultaneously.

As illustrated in the previous section, elliptic PDEs have no real characteristics. Thus, the solution at every point in the solution domain is influenced by the solution at all the other points, and the solution at each point influences the solution at all the other

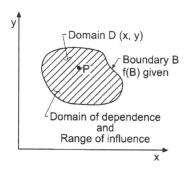

Figure III.3. Solution domain for an equilibrium problem.

points. Figure III.3 illustrates the closed solution domain $D(x, y)$ and its boundary B. Consequently, equilibrium problems are solved numerically by *relaxation methods*.

A classical example of an equilibrium problem governed by an elliptic PDE is steady heat diffusion (i.e., conduction) in a solid (see Section III.5). The governing PDE is the Laplace equation

$$\nabla^2 T = 0 \tag{III.42}$$

where T is the temperature of the solid. In two dimensions, Eq. (III.42) is

$$T_{xx} + T_{yy} = 0 \tag{III.43}$$

Along the boundary B, the temperature $T(x, y)$ is subject to the boundary condition

$$aT + bT_n = c \tag{III.44}$$

at each point on the boundary, where T_n denotes the derivative normal to the boundary.

Equilibrium problems arise in all fields of engineering and science. Equilibrium problems in partial differential equations are analogous to boundary-value problems in ordinary differential equations, which are considered in Chapter 8.

III.4.2 Propagation Problems

Propagation problems are *initial-value problems* in *open domains* (open with respect to one of the independent variables) in which the solution $f(x, t)$ in the domain of interest $D(x, t)$ is marched forward from the initial state, guided and modified by boundary conditions. Propagation problems are governed by parabolic or hyperbolic PDEs. Propagation problems in PDEs are analogous to initial-value problems in ODEs, which are considered in Chapter 7.

The majority of propagation problems are unsteady problems. The diffusion equation, Eq. (III.5), is an example of an unsteady propagation problem in which the initial property distribution at time t_0, $f(x, t_0) = F(x)$, is marched forward in time:

$$f_t = \alpha f_{xx} \tag{III.45a}$$

A few propagation problems are steady-state problems. An example of a steady-state propagation problem is

$$f_y = \beta f_{xx} \tag{III.45b}$$

in which the initial property distribution at location y_0, $f(x, y_0) = F(x)$, is marched forward in space in the y direction. The general features of these two PDEs, Eqs. (III.45a) and (III.45b), are identical, with the space coordinate y in Eq. (III.45b) taking on the character of the time coordinate t in the diffusion equation. Consequently, the marching direction in a steady-state space propagation problem is called the *timelike direction*, and the corresponding coordinate is called the *timelike coordinate*. The space direction in which diffusion occurs [i.e., the x direction in Eqs. (III.45a) and (III.45b)] is called the *spacelike direction*, and the corresponding coordinate is called the *spacelike coordinate*. In the present discussion, unsteady and steady propagation problems are considered simultaneously by considering the time coordinate t in the diffusion equation, Eq. (III.45a), to be a *timelike coordinate*, so that Eq. (III.45a) models both unsteady and steady propagation problems.

The solution of a propagation problem is subject to initial conditions specified at a particular value of the timelike coordinate and boundary conditions specified at each point on the spacelike boundary. The domain of interest $D(x, t)$ is open in the direction of the timelike coordinate. Figure III.4 illustrates the open solution domain $D(x, t)$ and its boundary B which is composed of the initial time boundary and the two physical boundaries. Propagation problems are initial-value problems, which are solved by marching methods.

A classical example of a propagation problem governed by a *parabolic PDE* is unsteady heat diffusion in a solid (see Section III.6). The governing PDE is the diffusion equation:

$$T_t = \alpha \, \nabla^2 T \tag{III.46}$$

where T is the temperature and α is the thermal diffusivity of the solid. In one space dimension, Eq. (III.46) is

$$T_t = \alpha T_{xx} \tag{III.47}$$

Since Eq. (III.47) is first order in time, values of T must be specified along the initial time boundary. Since Eq. (III.47) is second order in space, values of T must be specified along both space boundaries.

Parabolic PDEs have real repeated characteristics. As shown in Section III.6 for the diffusion equation, parabolic PDEs have specific domains of dependence and ranges of influence and infinite information propagation speed. Thus, the solution at each point in

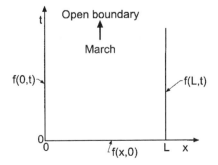

Figure III.4.   Solution domain for a propagation problem.

the solution domain depends on a specific domain of dependence and influences the solution in a specific range of influence.

In two variables (e.g., space x and time t), parabolic PDEs have two real repeated families of characteristics. As illustrated in Figure III.5, both families of characteristics have zero slope in the xt plane, which corresponds to an infinite information propagation speed. Consequently, parabolic PDEs behave like hyperbolic PDEs in the limit where the information propagation speed is infinite. Thus, the solution at point P depends on the entire solution domain upstream of and including the horizontal line through point P itself. The solution at point P influences the entire solution domain downstream of and including the horizontal line through point P itself. However, the solution at point P does not depend on the solution downstream of the horizontal line through point P, nor does the solution at point P influence the solution upstream of the horizontal line through point P. Numerical methods for solving propagation problems governed by parabolic PDEs must take the infinite information propagation speed into account.

A classical example of a propagation problem governed by a *hyperbolic PDE* is acoustic wave propagation (see Section III.7). The governing PDE is the wave equation

$$P'_{tt} = a^2 \, \nabla^2 P' \tag{III.48}$$

where P' is the acoustic pressure (i.e., the pressure disturbance) and a is the speed of propagation of small disturbances (i.e., the speed of sound). In one space dimension, Eq. (III.48) is

$$P'_{tt} = a^2 P'_{xx} \tag{III.49}$$

Since Eq. (III.49) is second order in time, initial values of both P' and P'_t must be specified along the initial time boundary. Since Eq. (III.49) is second order in space, values of P' must be specified along both space boundaries.

Hyperbolic PDEs have real distinct characteristics. As shown in Section III.7 for the wave equation, hyperbolic PDEs have finite domains of dependence and ranges of influence and finite information propagation speed. Thus, the solution at each point in the solution domain depends only on the solution in a finite domain of dependence and influences the solution only in a finite range of influence.

In two variables (e.g., space x and time t), hyperbolic PDEs have two real and distinct families of characteristics. As illustrated in Figure III.6, both families of

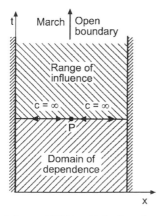

Figure III.5.   Solution domain for a parabolic propagation problem.

characteristics have finite slope in the x^t plane, which corresponds to a finite information speed. For acoustic fields, these two real families of characteristics are the right-running (i.e., in the positive x direction) and left-running (i.e., in the negative x direction) acoustic waves. These characteristics are illustrated in Figure III.6 at a particular point P. The characteristics have finite information propagation speed, thus giving rise to a finite domain of dependence and a finite range of influence for each point in the solution domain. The solution at point P depends only on the solution within the domain of dependence defined by the characteristics from the upstream portion of the solution domain. The solution at point P influences only the solution within the range of influence defined by the downstream propagating characteristics. The portion of the solution domain outside of the domain of dependence and the range of influence of point P neither influences the solution at point P nor depends on the solution at point P. Numerical methods for solving propagation problems governed by hyperbolic PDEs must take the finite information propagation speed into account.

From the above discussion, it is seen that propagation problems are governed by either a parabolic or a hyperbolic PDE. These two types of PDEs exhibit many similarities (e.g., an open boundary, initial data, boundary data, domains of dependence, and ranges of influence). Both types of problems are solved numerically by marching methods. However, there are significant differences in propagation problems governed by parabolic PDEs and hyperbolic PDEs, due to the infinite information propagation speed associated with parabolic PDEs and the finite information propagation speed associated with hyperbolic PDEs. These differences must be accounted for when applying marching methods to these two types of partial differential equations.

Propagation problems arise in all fields of engineering and science. Propagation problems governed by hyperbolic PDEs are somewhat analogous to initial-value problems in ODEs, while propagation problems governed by parabolic PDEs share some of the features of both initial-value and boundary-value problems in ODEs. Table III.1 summarizes the general features of PDEs as presented in this section.

III.4.3 Eigenproblems

Eigenproblems are special problems in which the solution exists only for special values (i.e., *eigenvalues*) of a parameter of the problem. The eigenvalues are to be determined in

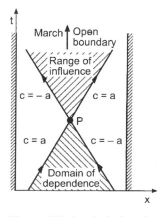

Figure III.6. Solution domain for a hyperbolic propagation problem.

Table III.1 General Features of Partial Differential Equations

	Type of physical problem		
	Equilibrium	Propagation	
	Elliptic	Parabolic	Hyperbolic
Mathematical classification of the PDE	Elliptic	Parabolic	Hyperbolic
Characteristics	Complex	Real and repeated	Real and distinct
Information propagation speed	Undefined	Infinite	Finite
Domain of dependence	Entire solution domain	Present and entire past solution domain	Past solution domain between characteristics
Range of influence	Entire solution domain	Present and entire future solution domain	Future solution domain between characteristics
Type of numerical method	Relaxation	Marching	Marching

addition to the corresponding configuration of the system. Eigenproblems for PDEs are analogous to eigenproblems for ODEs, which are considered in Section 8.9. Eigenproblems for PDEs are not considered in this book.

III.5 ELLIPTIC PARTIAL DIFFERENTIAL EQUATIONS

A classical example of an elliptic PDE is the *Laplace equation*:

$$\nabla^2 f = 0 \tag{III.50}$$

The Laplace equation applies to problems in ideal fluid flow, mass diffusion, heat diffusion, electrostatics, etc. In the following discussion, the general features of the Laplace equation are illustrated for the problem of steady two-dimensional heat diffusion in a solid.

Consider the differential cube of solid material illustrated in Figure III.7. Heat flow in a solid is governed by Fourier's law of conduction, which states that

$$\dot{q} = -kA\frac{dT}{dn} \tag{III.51}$$

where $\dot{q}$ is the energy transfer per unit time (J/s), T is the temperature (K), A is the area across which the energy flows (m^2), dT/dn is the temperature gradient normal to the area A (K/m), and k is the thermal conductivity of the solid (J/m-s-K), which is a physical property of the solid material. The net rate of flow of energy into the solid in the x direction is

$$\dot{q}_{\text{Net},x} = \dot{q}(x) - \dot{q}(x+dx) = \dot{q}(x) - \left[\dot{q}(x) + \frac{\partial\dot{q}(x)}{\partial x}dx\right] = -\frac{\partial\dot{q}(x)}{\partial x} \tag{III.52}$$

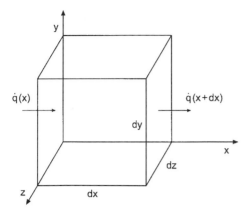

Figure III.7. Physical model of heat diffusion.

Introducing Eq. (III.51) into Eq. (III.52) yields

$$\dot{q}_{\text{Net},x} = -\frac{\partial}{\partial x}\left(-kA\frac{\partial T}{\partial x}\right)dx = \frac{\partial}{\partial x}\left(k\frac{\partial T}{\partial x}\right)dV \tag{III.53}$$

where $dV = A\,dx$ is the volume of the differential cube of solid material. Similarly,

$$\dot{q}_{\text{Net},y} = \frac{\partial}{\partial y}\left(k\frac{\partial T}{\partial y}\right)dV \tag{III.54}$$

$$\dot{q}_{\text{Net},z} = \frac{\partial}{\partial z}\left(k\frac{\partial T}{\partial z}\right)dV \tag{III.55}$$

For steady heat flow, there is no net change in the amount of energy stored in the solid, so the sum of the net rate of flow of energy in the three directions is zero. Thus,

$$\frac{\partial}{\partial x}\left(k\frac{\partial T}{\partial x}\right) + \frac{\partial}{\partial y}\left(k\frac{\partial T}{\partial y}\right) + \frac{\partial}{\partial z}\left(k\frac{\partial T}{\partial z}\right) = 0 \tag{III.56}$$

Equation (III.56) governs the steady diffusion of heat in a solid. When the thermal conductivity k is constant (i.e., neither a function of temperature or location), Eq. (III.56) simplifies to

$$\boxed{T_{xx} + T_{yy} + T_{zz} = \nabla^2 T = 0} \tag{III.57}$$

which is the *Laplace equation*.

For steady two-dimensional heat diffusion, Eq. (III.57) becomes

$$T_{xx} + T_{yy} = 0 \tag{III.58}$$

In terms of the general second-order PDE defined by Eq. (III.21), $A = 1$, $B = 0$, and $C = 1$. The discriminant, $B^2 - 4AC$, is

$$B^2 - 4AC = 0^2 - 4(1)(1) = -4 < 0 \tag{III.59}$$

Consequently, Eq. (III.58) is an elliptic PDE.

The characteristics associated with Eq. (III.58) are determined by performing a characteristic analysis. In this case, Eq. (III.28) becomes

$$
\begin{bmatrix} 1 & 0 & 1 \\ dx & dy & 0 \\ 0 & dx & dy \end{bmatrix} \begin{bmatrix} T_{xx} \\ T_{xy} \\ T_{yy} \end{bmatrix} = \begin{bmatrix} 0 \\ d(T_x) \\ d(T_y) \end{bmatrix} \tag{III.60}
$$

The characteristic equation corresponding to Eq. (III.58) is determined by setting the determinant of the coefficient matrix of Eq. (III.60) equal to zero and solving the resulting equation for the slopes of the characteristic paths. Thus,

$$
(1)(dy)^2 + (1)(dx)^2 = 0 \tag{III.61}
$$

$$
\frac{dy}{dx} = \pm\sqrt{-1} \tag{III.62}
$$

Equation (III.62) shows that there are no real characteristics associated with the steady two-dimensional heat conduction equation. Physically, this implies that there are no preferred paths of information propagation, and that the domain of dependence and range of influence of every point is the entire solution domain. The temperature at every point depends on the temperature at all the other points, including the boundaries of the solution domain, and the temperature at each point influences the temperature at all the other points. The temperature distribution is continuous throughout the solution domain because there are no paths along which the derivative of temperature may be discontinuous. The domain of dependence and the range of influence of point P are illustrated schematically in Figure III.3.

Another classical example of an elliptic PDE is the Poisson equation, which is the nonhomogeneous Laplace equation. Consider the problem of steady heat conduction in a solid with internal energy generation $\dot{E}$ (J/s) given by

$$
\dot{E} = \dot{Q}(x, y, z)\, dV \tag{III.63}
$$

where $\dot{Q}$ is the energy generation rate per unit volume (J/m^3-s). For steady heat flow, the sum of the energy transferred to the solid by conduction and the internal energy generation must equal zero. Thus, Eq. (III.56) becomes

$$
\frac{\partial}{\partial x}\left(k\frac{\partial T}{\partial x}\right) + \frac{\partial}{\partial y}\left(k\frac{\partial T}{\partial y}\right) + \frac{\partial}{\partial z}\left(k\frac{\partial T}{\partial z}\right) + \dot{Q} = 0 \tag{III.64}
$$

When the thermal conductivity k is constant (i.e., neither a function of temperature nor location), Eq. (III.64) becomes

$$
\boxed{T_{xx} + T_{yy} + T_{zz} = \nabla^2 T = -\frac{\dot{Q}}{k}} \tag{III.65}
$$

Equation (III.65) is the *Poisson equation*. The presence of the nonhomogeneous (i.e., source) term $\dot{Q}/k$ does not affect the classification of the nonhomogeneous Laplace equation. All the general features of the Laplace equation discussed above apply to the Poisson equation.

In summary, steady heat conduction is an equilibrium problem and must be solved by relaxation methods. The PDE governing steady heat conduction is a classical example of an elliptic PDE.

III.6 PARABOLIC PARTIAL DIFFERENTIAL EQUATIONS

A classical example of a parabolic PDE is the *diffusion equation*:

$$f_t = \alpha \nabla^2 f \tag{III.66}$$

The diffusion equation applies to problems in mass diffusion, momentum diffusion, heat diffusion, etc. The general features of the diffusion equation are illustrated for the problem of unsteady one-dimensional heat diffusion in a solid.

Consider the heat diffusion analysis presented in Section III.5. The net flow of heat in the x, y, and z directions is given by Eqs. (III.53) to (III.55), respectively. For steady-state heat flow, there is no net change in the amount of energy stored in the solid, so the sum of the net heat flow components is zero. In an unsteady situation, however, there can be a net change with time in the amount of energy stored in the solid. The energy E (J) stored in the solid mass dm (kg) is given by

$$E_{\text{stored}} = dm\ CT = (\rho\ dV)CT = (\rho CT)\ dV \tag{III.67}$$

where ρ is the density of the solid material (kg/m^3), dV is the differential volume (m^3), T is the temperature (K), and C is the specific heat (J/kg-K), which is a physical property of the solid material. The sum of the net heat flow components must equal the time rate of change of the stored energy. Thus,

$$\frac{\partial(\rho CT)}{\partial t} = \frac{\partial}{\partial x}\left(k\frac{\partial T}{\partial x}\right) + \frac{\partial}{\partial y}\left(k\frac{\partial T}{\partial y}\right) + \frac{\partial}{\partial z}\left(k\frac{\partial T}{\partial z}\right) \tag{III.68}$$

Equation (III.68) governs the unsteady diffusion of heat in a solid. When the thermal conductivity k, density ρ, and specific heat C are all constant (i.e., neither functions of temperature or position), Eq. (III.68) simplifies to

$$\boxed{T_t = \alpha(T_{xx} + T_{yy} + T_{zz}) = \alpha\nabla^2 T} \tag{III.69}$$

where $\alpha = k/\rho C$ is the thermal diffusivity (m^2/s). Equation (III.69) is the *diffusion equation*.

For unsteady one-dimensional heat diffusion, Eq. (III.69) becomes

$$T_t = \alpha T_{xx} \tag{III.70}$$

In terms of the general second-order PDE defined by Eq. (III.21), $A = \alpha$, $B = 0$, and $C = 0$. The discriminant, $B^2 - 4AC$, is

$$B^2 - 4AC = 0^2 - 4(\alpha)(0) = 0 \tag{III.71}$$

Consequently, Eq. (III.70) is a parabolic PDE.

The characteristics associated with Eq. (III.70) are determined by performing a characteristic analysis. In this case, Eq. (III.28) becomes

$$
\begin{bmatrix} \alpha & 0 & 0 \\ dx & dt & 0 \\ 0 & dx & dt \end{bmatrix} \begin{bmatrix} T_{xx} \\ T_{xt} \\ T_{tt} \end{bmatrix} = \begin{bmatrix} T_t \\ d(T_x) \\ d(T_t) \end{bmatrix}
\tag{III.72}
$$

The characteristic equation corresponding to Eq. (III.70) is determined by setting the determinant of the coefficient matrix of Eq. (III.72) equal to zero and solving for the slopes of the characteristic paths. In the present case, this yields

$$
\alpha \, dt^2 = 0
\tag{III.73}
$$

$$
dt = \pm 0
\tag{III.74}
$$

$$
t = \text{constant}
\tag{III.75}
$$

Equation (III.74) shows that there are two real repeated roots associated with the characteristic equation, and Eq. (III.75) shows that the characteristics are lines of constant time. The speed of propagation of information along these characteristic paths is

$$
c = \frac{dx}{dt} = \frac{dx}{\pm 0} = \pm \infty
\tag{III.76}
$$

Consequently, information propagates at an infinite speed along lines of constant time. This situation is illustrated schematically in Figure III.5. The information at point P propagates at an infinite speed in both directions. Consequently, the temperature at point P depends on the temperature at all other points in physical space at all times preceding and including the current time, and the temperature at point P influences the temperature at all other points in physical space at all times after and including the current time. In other words, the domain of dependence of point P is the finite region ahead of and including the current time line. The range of influence of point P is the semi-infinite region after and including the current time line. In this regard, the diffusion equation behaves somewhat like an elliptic PDE at each time level.

In summary, unsteady heat diffusion is a propagation problem which must be solved by marching methods. The PDE governing unsteady heat diffusion is a classical example of a parabolic PDE.

III.7 HYPERBOLIC PARTIAL DIFFERENTIAL EQUATIONS

A classical example of a hyperbolic PDE is the *wave equation*:

$$
f_{tt} = c^2 \, \nabla^2 f
\tag{III.77}
$$

The wave equation applies to problems in vibrations, electrostatics, gas dynamics, acoustics, etc. The general features of the wave equation are illustrated in this section for the problem of unsteady one-dimensional acoustic wave propagation.

Fluid flow is governed by the law of conservation of mass (the continuity equation), Newton's second law of motion (the momentum equation), and the first law of thermo-dynamics (the energy equation). As shown in any text on fluid dynamics [e.g., Fox and

McDonald (1985) or Zucrow and Hoffman (1976)], those basic physical laws yield the following system of quasi-linear first-order PDEs:

$$\rho_t + \nabla \cdot (\rho \mathbf{V}) = 0 \tag{III.78}$$

$$\rho \mathbf{V}_t + \rho (\mathbf{V} \cdot \nabla) \mathbf{V} + \nabla P = 0 \tag{III.79}$$

$$P_t + \mathbf{V} \cdot \nabla P - a^2 (\rho_t + \mathbf{V} \cdot \nabla \rho) = 0 \tag{III.80}$$

where ρ is the fluid density (kg/m^3), $\mathbf{V}$ is the fluid velocity vector (m/s), P is the static pressure (N/m^2), and a is the speed of propagation of small disturbances (m/s) (i.e., the speed of sound). Equations (III.78) to (III.80) are restricted to the flow of a pure substance with no body forces or transport phenomena (i.e., no mass, momentum, or energy diffusion). For unsteady one-dimensional flow, Eqs. (III.78) to (III.80) yield:

$$\rho_t + \rho u_x + u \rho_x = 0 \tag{III.81}$$

$$\rho u_t + \rho u u_x + P_x = 0 \tag{III.82}$$

$$P_t + u P_x - a^2 (\rho_t + u \rho_x) = 0 \tag{III.83}$$

Equations (III.81) to (III.83) are more general examples of the simple one-dimensional convection equation

$$f_t + u f_x = 0 \tag{III.84}$$

where the property f is being convected by the velocity u through the solution domain $D(x, t)$. Equation (III.84) in three independent variables is

$$f_t + u f_x + v f_y + w f_z = f_t + \mathbf{V} \cdot \nabla f = \frac{Df}{Dt} = 0 \tag{III.85}$$

where u, v, and w are the velocity components in the x, y, and z directions, respectively, and the vector operator D/Dt is called the *substantial derivative*:

$$\frac{D}{Dt} = \frac{\partial}{\partial t} + u \frac{\partial}{\partial x} + v \frac{\partial}{\partial y} + w \frac{\partial}{\partial z} = \frac{\partial}{\partial t} + \mathbf{V} \cdot \nabla \tag{III.86}$$

Equations (III.81) and (III.83) are frequently combined to eliminate the derivatives of density. Thus,

$$P_t + u P_x + \rho a^2 u_x = 0 \tag{III.87}$$

Equations (III.81) to (III.83), or Eqs. (III.82) and (III.87), are classical examples of a system of nonlinear first-order PDEs.

Acoustics is the science devoted to the study of the motion of small amplitude disturbances in a fluid medium. Consider the classical case of infinitesimally small perturbations in velocity, presure, and density in a stagnant fluid. In that case,

$$u = u_0 + u' = u' \qquad P = P_0 + P' \qquad \rho = \rho_0 + \rho' \qquad a = a_0 + a' \tag{III.88}$$

where u_0, P_0, ρ_0, and a_0 are the undisturbed properties of the fluid, and u', P', ρ', and a' are infinitesimal perturbations. For a stagnant fluid, $u_0 = 0$. Substituting Eq. (III.88) into Eqs. (III.82) and (III.87) and neglecting all products of perturbation quantities yields the following system of linear PDEs:

$$\rho_0 u'_t + P'_x = 0 \tag{III.89}$$

$$P'_t + \rho_0 a_0^2 u'_x = 0 \tag{III.90}$$

Equations (III.89) and (III.90) can be combined to solve explicitly for either the pressure perturbation P' or the velocity perturbation u'. Differentiating Eq. (III.89) with respect to x and Eq. (III.90) with respect to t and combining the results to eliminate u'_{xt} yields the wave equation for the pressure perturbation, P':

$$P'_{tt} = a_0^2 P'_{xx} \qquad\qquad\qquad\qquad\qquad \text{(III.91)}$$

Differentiating Eq. (III.89) with respect to t and Eq. (III.90) with respect to x and combining the results to eliminate P'_{xt} yields the wave equation for the velocity perturbation u':

$$u'_{tt} = a_0^2 u'_{xx} \qquad\qquad\qquad\qquad\qquad \text{(III.92)}$$

Equations (III.91) and (III.92) show that the properties of a linearized acoustic field are governed by the wave equation. In terms of the general second-order PDE defined by Eq. (III.21), $A = 1$, $B = 0$, and $C = -a_0^2$. The discriminant, $B^2 - 4AC$, is

$$B^2 - 4AC = 0 - 4(1)(-a_0^2) = 4a_0^2 > 0 \qquad\qquad\qquad \text{(III.93)}$$

Consequently, Eqs. (III.91) and (III.92) are hyperbolic PDEs.

Since Eqs. (III.91) and (III.92) both involve the same differential operators [i.e., $(\ )_{tt} = a_0^2 (\ )_{xx}$], they have the same characteristics. Consequently, it is necessary to study only one of them, so Eq. (III.91) is chosen. The characteristics associated with Eq. (III.91) are determined by performing a characteristics analysis. In this case, Eq. (III.28) becomes

$$\begin{bmatrix} 1 & 0 & -a_0^2 \\ dt & dx & 0 \\ 0 & dt & dx \end{bmatrix} \begin{bmatrix} P'_{tt} \\ P'_{xt} \\ P'_{xx} \end{bmatrix} = \begin{bmatrix} 0 \\ d(P'_x) \\ d(P'_t) \end{bmatrix} \qquad\qquad \text{(III.94)}$$

The characteristic equation corresponding to Eq. (III.91) is determined by setting the determinant of the coefficient matrix of Eq. (III.94) to zero and solving for the slopes of the characteristic paths. This yields

$$(dx)^2 - a_0^2 (dt)^2 = 0 \qquad\qquad\qquad\qquad \text{(III.95)}$$

Equation (III.95) is a quadratic equation for dx/dt. Solving for dx/dt gives

$$\frac{dx}{dt} = \pm a_0 \qquad\qquad\qquad\qquad\qquad \text{(III.96)}$$

$$x = x_0 \pm a_0 t \qquad\qquad\qquad\qquad\qquad \text{(III.97)}$$

Equation (III.96) shows that there are two real distinct roots associated with the characteristic equation, and Eq. (III.97) shows that the characteristic paths are straight lines having the slopes $\pm 1/a_0$ in the xt plane. The speed of propagation of information along these characteristic paths is

$$c = \frac{dx}{dt} = \pm a_0 \qquad\qquad\qquad\qquad\qquad \text{(III.98)}$$

Consequently, information propagates at the acoustic speed a_0 along the characteristic paths. This situation is illustrated schematically in Figure III.5. Information at point P propagates at a finite rate in physical space. Consequently, the perturbation pressure at point P depends only upon the solution within the finite domain of dependence illustrated in Figure III.6. Likewise, the perturbation pressure at point P influences the solution only within the finite range of influence illustrated in Figure III.6. The finite speed of propagation of information and the finite domain of dependence and range of influence must be accounted for when solving hyperbolic PDEs.

Equations (III.89) and (III.90) are examples of a system of two coupled first-order convection equations of the general form:

$$f_t + ag_x = 0 \tag{III.99a}$$
$$g_t + af_x = 0 \tag{III.99b}$$

Differentiating Eq. (III.99a) with respect to t, differentiating Eq. (III.99b) with respect to x and multiplying by a, and subtracting yields the wave equation:

$$f_{tt} = a^2 f_{xx} \tag{III.100}$$

Consequently, the second-order wave equation can be interpreted as a system of two coupled first-order convection equations.

In summary, unsteady wave motion is a propagation problem which must be solved by marching methods. The wave equation governing unsteady wave motion is a classical example of a hyperbolic PDE.

III.8 THE CONVECTION-DIFFUSION EQUATION

The Laplace equation and the Poisson equation govern steady diffusion processes. The diffusion equation governs unsteady diffusion processes. The convection equation and the wave equation govern unsteady convection processes. When convection and diffusion are both present in a physical process, the process is governed by the convection-diffusion equation. The unsteady convection-diffusion equation is given by

$$\boxed{f_t + \mathbf{V} \cdot \nabla f = \alpha \, \nabla^2 f} \tag{III.101}$$

and the steady convection-diffusion equation is given by

$$\boxed{\mathbf{V} \cdot \nabla f = \alpha \, \nabla^2 f} \tag{III.102}$$

Equations (III.101) and (III.102) are both second-order PDEs. Consider the unsteady one-dimensional convection-diffusion equation:

$$f_t + uf_x = \alpha f_{xx} \tag{III.103}$$

The discriminant of Eq. (III.103) is $B^2 - 4AC = 0$. Consequently, Eq. (III.103) is a parabolic PDE. Performing a characteristic analysis of Eq. (III.103) yields the characteristic paths $dt = \pm 0$, which shows that physical information propagates at an infinite rate.

However, as shown in Section III.3, the term uf_x models physical convection, which is a hyperbolic process with the distinct characteristic path

$$\frac{dx}{dt} = u \tag{III.104}$$

This characteristic path is not found in a classical characteristic analysis of the unsteady convection-diffusion equation.

Consider the steady two-dimensional convection-diffusion equation:

$$uf_x + vf_y = \alpha(f_{xx} + f_{yy}) \tag{III.105}$$

The discriminant of Eq. (III.105) is $B^2 - 4AC = -4$. Consequently, Eq. (III.105) is an elliptic PDE, with no real characteristic paths. However, the terms uf_x and vf_y model physical convection, which has distinct information propagation paths. These information propagation paths are not found in a classical characteristic analysis of the steady convection-diffusion equation.

The significance of the above discussion is as follows. When numerically approximating PDEs which contain both first-order and second-order spatial derivatives, the different physical behavior associated with the different spatial derivatives should be taken into account.

III.9 INITIAL VALUES AND BOUNDARY CONDITIONS

A differential equation governs a family of solutions. A particular member of the family of solutions is specified by the auxiliary conditions imposed on the differential equation.

For steady-state equilibrium problems, the auxiliary conditions consist of boundary conditions on the entire boundary of the closed solution domain. Three types of boundary conditions can be imposed:

1. *Dirichlet boundary condition*: The value of the function is specified.

    f is specified on the boundary. $\hspace{4cm}$ (III.106)

2. *Neumann boundary condition*: The value of the derivative normal to the boundary is specified.

    $\dfrac{\partial f}{\partial n}$ is specified on the boundary. $\hspace{3cm}$ (III.107)

3. *Mixed boundary condition*: A combination of the function and its normal derivative is specified on the boundary.

    $af + b\dfrac{\partial f}{\partial n}$ is specified on the boundary. $\hspace{2.5cm}$ (III.108)

One of the above types of boundary conditions must be specified at each point on the boundary of the closed solution domain. Different types of boundary conditions can be specified on different portions of the boundary.

For unsteady or steady propagation problems, the auxiliary conditions consist of an initial condition (or conditions) along the time (or timelike) boundary and boundary conditions on the physical boundaries of the solution domain. No auxiliary conditions can be applied on the open boundary in the time (or timelike) direction. For a PDE containing a first-order time (or timelike) derivative, one initial condition is required along the time (or timelike) boundary:

$$f(x, y, z, 0) = F(x, y, z) \text{ on the time boundary} \tag{III.109}$$

For a PDE containing a second-order time (or timelike) derivative, two initial conditions are required along the time (or timelike) boundary:

$$f(x, y, z, 0) = F(x, y, z) \text{ on the time boundary} \qquad \text{(III.110a)}$$

$$f_t(x, y, z, 0) = G(x, y, z) \text{ on the time boundary} \qquad \text{(III.110b)}$$

The required boundary conditions on the physical boundaries of the solution domain can be of the Dirichlet type, Eq. (III.106), the Neumann type, Eq. (III.107), or the mixed type, Eq. (III.108). Different types of boundary conditions can be specified on different portions of the boundary.

Proper specifications of the type and number of auxiliary conditions is a necessary condition to obtain a well-posed problem, as discussed in Section III.10.

III.10 WELL-POSED PROBLEMS

The general features of partial differential equations are discussed in the preceding sections. Elliptic PDEs govern equilibrium problems in closed domains. No real characteristics exist. Parabolic PDEs govern propagation problems in open domains. Real repeated characteristics exist. Hyperbolic PDEs govern propagation problems in open domains. Real distinct characteristics exist. In all three cases, auxiliary conditions (i.e., initial values and boundary conditions) are required to specify a particular solution of a PDE. The interrelationship between the type of PDE, the auxiliary data, and whether or not a solution exists and is unique gives rise to the concept of a *well-posed* problem.

Hadamard (1923) states that a physical problem is well posed if its solution exists, is unique, and depends continuously on the boundary and/or initial data.

For an elliptic PDE, the solution domain $D(x, y)$ must be closed, and continuous boundary conditions must be specified along the entire physical boundary B. The boundary conditions may be of three types: (a) Dirichlet boundary conditions, (b) Neumann boundary conditions, or (c) mixed boundary conditions.

For a parabolic PDE, the solution domain $D(x, t)$ must be open in the time (or timelike) direction, initial data must be specified along the time (or timelike) boundary, and continuous boundary conditions must be specified along the physical boundaries of the solution domain. The boundary conditions can be of the Dirichlet type, the Neumann type, or the mixed type.

For a hyperbolic PDE, the solution domain $D(x, t)$ must be open in the time (or timelike) direction, initial data must be specified along the time (or timelike) boundary, and continuous boundary conditions must be specified along the physical boundaries of the solution domain. The boundary conditions can be of the Dirichlet type, the Neumann type, or the mixed type. For a hyperbolic PDE, the initial data cannot be specified along only one characteristic curve (or surface). A pure initial-value problem (the Cauchy problem) can be defined for a hyperbolic PDE by specifying initial data along several characteristic curves (or surfaces). An initial-boundary-value problem is defined by specifying initial data along a noncharacteristic curve and boundary conditions along the physical boundaries of the solution domain.

Care must be exercised to ensure that a problem is well posed. Only well-posed problems are considered in this book.

III.11 SUMMARY

The general features of partial differential equations have been presented, and the concept of characteristics has been introduced. Characteristics are the physical paths along which information propagates. Partial differential equations are classified as elliptic, parabolic, or hyperbolic, according to whether there are no real characteristics, real repeated characteristics, or real distinct characteristics, respectively. Examples of several partial differential equations that arise in engineering and science have been presented.

The Laplace equation is a classical example of an elliptic PDE, which must be solved by relaxation methods:

$$\nabla^2 f = 0 \tag{III.111}$$

Chapter 9 is devoted to the solution of the Laplace equation. The diffusion equation is a classical example of a parabolic PDE, which must be solved by marching methods:

$$f_t = \alpha \, \nabla^2 f \tag{III.112}$$

Chapter 10 is devoted to the solution of the diffusion equation. The convection equation is a classical example of a hyperbolic PDE, which must be solved by marching methods:

$$f_t + \mathbf{V} \cdot \nabla f = 0 \tag{III.113}$$

Chapter 11 is devoted to the solution of the convection equation. When convection and diffusion are both present, the process is governed by the convection-diffusion equation:

$$f_t + \mathbf{V} \cdot \nabla f = \alpha \, \nabla^2 f \tag{III.114}$$

The convection-diffusion equation is a more complicated example of a parabolic PDE, which must be solved by marching methods. Section 10.9 is devoted to the solution of the convection-diffusion equation. Some physical problems are governed by a system of convection equations. In some cases, they can be recast as the wave equation:

$$f_{tt} = c^2 \, \nabla^2 f \tag{III.120}$$

The wave equation is a more complicated example of a hyperbolic PDE, which must be solved by marching methods. Section 11.8 is devoted to the solution of the wave equation.

9

Elliptic Partial Differential Equations

9.1 INTRODUCTION

The two thermal systems illustrated in Figure 9.1 are considered in this chapter to illustrate methods for solving elliptic partial differential equations (PDEs). The top figure illustrates a thin plate of width $w = 10$ cm, height $h = 15$ cm, and thickness $t = 1$ cm. The faces of the plate are insulated so that no heat flows in the direction of the thickness t. The top edge of the plate is maintained at temperature $T = 100\sin(\pi x/w)$ C, and the other three edges are maintained at $T = 0$ C. Heat flows into the plate through the top edge and out of the plate through the other three edges. There is no internal energy generation within the plate. The internal temperature distribution within the plate $T(x, y)$ is required, and the total heat transfer rate through the top of the plate $\dot{q}$ required. The temperature distribution within the plate is governed by the two-dimensional Laplace equation:

$$T_{xx} + T_{yy} = 0 \tag{9.1}$$

which is subject to the boundary conditions specified on the four edges of the plate.

The exact solution to this linear PDE is obtained by assuming a product solution of the form $T(x, y) = X(x)Y(y)$, substituting this functional form into the PDE and separating

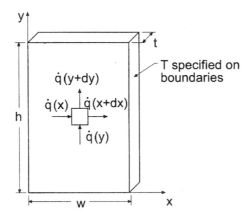

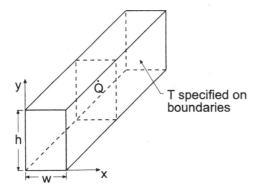

Figure 9.1   Steady heat diffusion problems.

variables, integrating the resulting two ordinary differential equations for $X(x)$ and $Y(y)$, and applying the boundary conditions at $x = 0$, $x = w$, $y = 0$, and $y = h$. The result is

$$T(x, y) = 100 \frac{\sinh(\pi y/w) \sin(\pi x/w)}{\sinh(\pi h/w)} \tag{9.2}$$

The exact solution at selected locations is tabulated in Table 9.1 for the left half of the plate. The solution is symmetrical about the vertical centerline. Selected isotherms (i.e., lines of constant temperature) are illustrated in Figure 9.2.

The total heat transfer rate through the top of the plate $\dot{q}$ is determined by integrating the local heat transfer rate across the top of the plate. From *Fourier's law of conduction*

$$d\dot{q} = -\frac{k\,dA}{\partial T/\partial y} = -\frac{k\,(t\,dx)}{\partial T/\partial y} - dA \frac{\partial T}{\partial y} = -k(t\,dx) \frac{\partial T}{\partial y} \tag{9.3}$$

where $k = 0.4\,\text{J/s-cm-C}$ is the thermal conductivity of the plate. Differentiating Eq. (9.2) gives

$$\frac{\partial T}{\partial y} = 100 \frac{\sin(\pi x/w)}{\sinh(\pi h/w)} \left(\frac{\pi}{w}\right) \cosh\left(\frac{\pi y}{w}\right) \tag{9.4}$$

Substituting Eq. (9.4) into Eq. (9.3), setting $y = h$, and integrating from $x = 0$ to $x = 10$ yields

$$\dot{q} = -\frac{200kt}{\tanh(\pi h/w)} \tag{9.5}$$

For the values of w, h, t, and k specified above, $\dot{q} = 80.012913\,\text{J/s}$.

The bottom figure in Figure 9.1 illustrates a long rectangular electrical conductor which has internal energy generation due to electrical resistance heating. The temperature

Table 9.1 Exact Solution of the Heat Diffusion Problem.

	Temperature $T(x, y)$, C				
y, cm	$x = 0.00$ cm	$x = 1.25$ cm	$x = 2.50$ cm	$x = 3.75$ cm	$x = 5.00$ cm
15.000000	0.000000	38.268343	70.710678	92.387953	100.000000
13.750000	0.000000	25.837518	47.741508	62.377286	67.516688
12.500000	0.000000	17.442631	32.229780	42.110236	45.579791
11.250000	0.000000	11.772363	21.752490	28.420998	30.762667
10.000000	0.000000	7.940992	14.673040	19.171250	20.750812
8.750000	0.000000	5.350040	9.885585	12.916139	13.980329
7.500000	0.000000	3.594789	6.642304	8.678589	9.393637
6.250000	0.000000	2.401061	4.436582	5.796673	6.274274
5.000000	0.000000	1.582389	2.923873	3.820225	4.134981
3.750000	0.000000	1.010893	1.867887	2.440512	2.641592
2.500000	0.000000	0.597304	1.103674	1.442020	1.560831
1.250000	0.000000	0.277016	0.511860	0.668777	0.723879
0.000000	0.000000	0.000000	0.000000	0.000000	0.000000

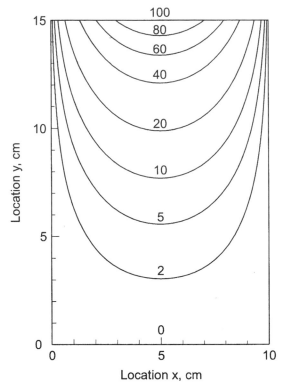

Figure 9.2 Exact solution of the heat diffusion problem.

distribution within the conductor $T(x, y)$ is required. The temperature distribution within the conductor is governed by the two-dimensional Poisson equation:

$$T_{xx} + T_{yy} = -\frac{\dot{Q}}{k} \tag{9.6}$$

where k is the thermal conductivity of the conductor and $\dot{Q}$ is the volumetric heating rate (J/cm³-s). This problem is considered in Section 9.8 to illustrate finite difference methods for solving the Poisson equation.

Numerous elliptic partial differential equations arise in engineering and science. Two of the more common ones are the *Laplace equation* and the *Poisson equation*, presented below for the generic dependent variable $f(x, y)$:

$$f_{xx} + f_{yy} = 0 \tag{9.7}$$
$$f_{xx} + f_{yy} = F(x, y) \tag{9.8}$$

where $F(x, y)$ is a known nonhomogeneous term. The Laplace equation applies to problems in mass diffusion, heat diffusion (i.e., conduction), neutron diffusion, electrostatics, inviscid incompressible fluid flow, etc. In fact, the Laplace equation governs the potential of many physical quantities where the rate of flow of a particular property is proportional to the gradient of a potential. The Poisson equation is simply the *nonhomogeneous Laplace equation*. The presence of the nonhomogeneous term $F(x, y)$ can greatly complicate the exact solution of the Poisson equation. However, the presence of this term does not complicate the numerical solution of the Poisson equation. The nonhomogeneous term is simply evaluated at each physical location and added to the finite difference approximation of the Laplace equation. Consequently, the present chapter is devoted mainly to the numerical solution of the Laplace equation. All the results apply directly to the numerical solution of the Poisson equation.

The solution of Eqs. (9.7) and (9.8) is the function $f(x, y)$. This function must satisfy a set of boundary conditions on the boundaries of the closed solution domain. These boundary conditions may be of the *Dirichlet type* (i.e., specified values of f), the *Neumann type* (i.e., specified values of the derivative of f), or the *mixed type* (i.e., a specified combination of f and its derivative). The basic properties of finite difference methods for solving equilibrium problems governed by elliptic PDEs are presented in this chapter.

Figure 9.3 presents the organization of Chapter 9. After the introductory discussion presented in this section, the general features of elliptic partial differential equations are reviewed. This is followed by a discussion of the finite difference method, which leads into the finite difference solution of the Laplace equation. The concepts of consistency, order, and convergence are introduced and discussed. Iterative methods of solution of the system of finite difference equations are presented. A procedure for implementing derivative boundary conditions is presented next. A brief introduction to finite difference methods for solving the Poisson equation follows. A discussion of several special topics then follows: higher-order methods, nonrectangular domains, nonlinear problems, and three-dimensional problems. All of the above discussions are based on the finite difference approach. A brief introduction to the control volume method is then presented. A presentation of a computer program for solving the Laplace equation and the Poisson equation follows. A summary wraps up the chapter. The numerical methods presented in this chapter are applied to solve the thermal systems presented in this section.

9.2 GENERAL FEATURES OF ELLIPTIC PDEs

The general features of *elliptic partial differential equations* (*PDEs*) are discussed in Section III.5. In that section it is shown that elliptic PDEs govern steady-state equilibrium problems, which are boundary-value problems in closed domains. Consequently, elliptic PDEs are solved numerically by relaxation methods. As shown in Section III.5, problems governed by elliptic PDEs have no real characteristic paths. Physically, this means that there are no preferred paths of information propagation and that the domain of dependence and the range of influence of every point is the entire solution domain. The solution at every point depends on the solution at all other points, including the boundaries of the solution domain, and the solution at every point influences the solution at all other points. The solution is continuous throughout the solution domain since there are no paths along which the derivatives of the solution may be discontinuous. These general features of elliptic PDEs are illustrated in Figure 9.4.

Every exact partial derivative in a PDE should be approximated in a manner consistent with the physical requirements of the problem. For an elliptic PDE, the solution at every point in the solution domain depends on the solution at all the other points, in particular, the immediate neighboring points. Thus, the exact partial derivatives in elliptic PDEs are approximated by centered-difference approximations, as discussed in Section 9.3.2.

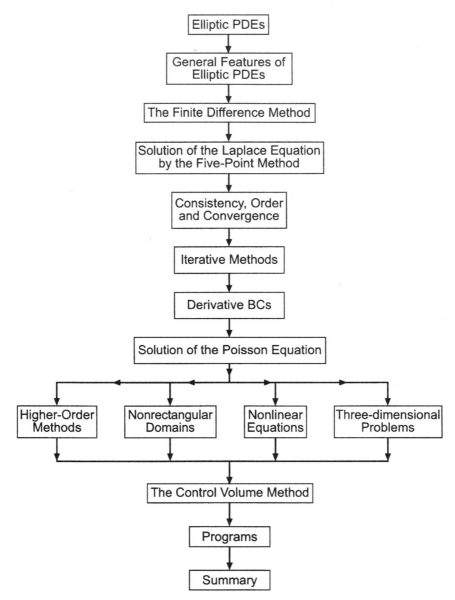

Figure 9.3 Organization of Chapter 9.

9.3 THE FINITE DIFFERENCE METHOD

The *finite difference method* is a numerical procedure which solves a partial differential equation (PDE) by discretizing the continuous physical domain into a discrete finite difference grid, approximating the individual exact partial derivatives in the PDE by algebraic finite difference approximations (FDAs), substituting the FDAs into the PDE to obtain an algebraic finite difference equation (FDE), and solving the resulting algebraic finite difference equations (FDEs) for the dependent variable. For simplicity of notation, the phrase "finite difference equation" frequently will be replaced by the acronym "FDE"

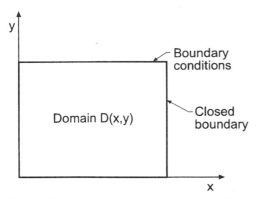

Figure 9.4   General features of elliptic PDEs.

in Part III of this book. Some general characteristics of finite difference grids and finite difference approximations for equilibrium problems governed by elliptic PDEs are discussed in this section.

9.3.1   Finite Difference Grids

The closed solution domain $D(x, y)$ in xy space for a two-dimensional equilibrium problem is illustrated in Figure 9.5. The solution domain must be covered by a two-dimensional grid of lines, called the *finite difference grid*. The intersections of these *grid lines* are the *grid points* at which the finite difference solution to the partial differential equation is to be obtained. For the present, let these grid lines be equally spaced lines perpendicular to the x and y axes having uniform spacings Δx and Δy, respectively, but with Δx and Δy not necessarily equal. The resulting finite difference grid is illustrated in Figure 9.5. The subscript i is used to denote the physical grid lines corresponding to constant values of x [i.e., $x_i = (i - 1)\,\Delta x$], and the subscript j is used to denote the physical grid lines corresponding to constant values of y [i.e., $y_j = (j - 1)\,\Delta y$]. Thus, grid point (i, j) corresponds to location (x_i, y_j) in the solution domain $D(x, y)$. The total number of x grid lines is denoted by imax, and the total number of y grid lines is denoted by jmax.

Three-dimensional physical spaces can be covered in a similar manner by a three-dimensional grid of planes perpendicular to the coordinate axes, where the subscripts i, j,

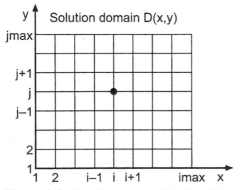

Figure 9.5   Solution domain $D(x, y)$ and discrete difference grid.

and k denote the physical grid planes perpendicular to the x, y, and z axes, respectively. Thus, grid point (i, j, k) corresponds to location (x_i, y_j, z_k) in the solution domain $D(x, y, z)$.

The dependent variable at a grid point is denoted by the same subscript notation that is used to denote the grid points themselves. Thus, the function $f(x, y)$ at grid point (i, j) is denoted by

$$f(x_i, y_i) = f_{i,j} \tag{9.9}$$

In a similar manner, derivatives are denoted by

$$\frac{\partial f(x_i, y_j)}{\partial x} = \frac{\partial f}{\partial x}\bigg|_{i,j} = f_x|_{i,j} \quad \text{and} \quad \frac{\partial^2 f(x_i, y_j)}{\partial x^2} = \frac{\partial^2 f}{\partial x^2}\bigg|_{i,j} = f_{xx}|_{i,j} \tag{9.10}$$

Analogous notation applies in three-dimensional spaces.

9.3.2 Finite Difference Approximations

Now that the finite difference grid has been specified, *finite difference approximations* (*FDAs*) of the individual exact partial derivatives appearing in the partial differential equation must be developed. This is accomplished by writing Taylor series for the dependent variable at several neighboring grid points using grid point (i, j) as the base point, and combining these Taylor series to solve for the desired partial derivatives. This is done in Chapter 5 for functions of one independent variable, where approximations of various types (i.e., forward, backward, and centered) having various orders of accuracy (i.e., first order, second order, etc.) are developed for various derivatives (i.e., first derivative, second derivative, etc.). Those results are presented in Table 5.1.

The forms of the finite difference approximations of the individual exact partial derivatives in a PDE should be governed by the physics being represented by the PDE. For elliptic PDEs containing only second derivatives, a characteristic analysis shows that there are no preferred physical informative propagation paths. Thus, the solution at all points depends on the solution at all other points, and the solution at all points influences the solution at all the other points. Consequently, centered-space finite difference approximations should be used for the second-order spatial derivatives in the Laplace equation and the Poisson equation.

In the development of finite difference approximations, a distinction must be made between the *exact solution* of a partial differential equation and the *approximate solution* of the partial differential equation. For the remainder of this chapter, exact solutions will be denoted by an overbar over the symbol for the dependent variable [i.e., $\bar{f}(x, y)$], and approximate solutions will be denoted by the symbol for the dependent variable without an overbar [i.e., $f(x, y)$]. Thus,

$$\boxed{\begin{aligned} \bar{f}(x, y) &= \text{exact solution} \\ f(x, y) &= \text{approximate solution} \end{aligned}}$$

Individual exact partial derivatives may be approximated at grid point (i, j) in terms of the values of $\bar{f}$ at grid point (i, j) itself and adjacent grid points in a number of ways. For example, consider the partial derivative $\bar{f}_{xx}$. Writing the Taylor series for $\bar{f}_{i+1,j}$ and $\bar{f}_{i-1,j}$ using grid point (i,j) as the base point gives

$$\bar{f}_{i+1,j} = \bar{f}_{i,j} + \bar{f}_x|_{i,j}\,\Delta x + \tfrac{1}{2}\bar{f}_{xx}|_{i,j}\,\Delta x^2 + \tfrac{1}{6}\bar{f}_{xxx}|_{i,j}\,\Delta x^3 + \tfrac{1}{24}\bar{f}_{xxxx}|_{i,j}\,\Delta x^4 + \cdots \qquad (9.11)$$

$$\bar{f}_{i-1,j} = \bar{f}_{i,j} - \bar{f}_x|_{i,j}\,\Delta x + \tfrac{1}{2}\bar{f}_{xx}|_{i,j}\,\Delta x^2 - \tfrac{1}{6}\bar{f}_{xxx}|_{i,j}\,\Delta x^3 + \tfrac{1}{24}\bar{f}_{xxxx}|_{i,j}\,\Delta x^4 + \cdots \qquad (9.12)$$

where the convention $(\Delta x)^n \to \Delta x^n$ is employed for compactness. Equations (9.11) and (9.12) can be expressed as Taylor formulas with remainders:

$$\bar{f}_{i+1,j} = \bar{f}_{i,j} + \bar{f}_x|_{i,j}\,\Delta x + \tfrac{1}{2}\bar{f}_{xx}|_{i,j}\,\Delta x^2 + \tfrac{1}{6}\bar{f}_{xxx}|_{i,j}\,\Delta x^3 + \tfrac{1}{24}\bar{f}_{xxxx}|_{i,j}\,\Delta x^4 + R_{n+1}(\xi_+) \quad (9.13)$$

$$\bar{f}_{i-1,j} = \bar{f}_{i,j} - \bar{f}_x|_{i,j}\,\Delta x + \tfrac{1}{2}\bar{f}_{xx}|_{i,j}\,\Delta x^2 - \tfrac{1}{6}\bar{f}_{xxx}|_{i,j}\,\Delta x^3 + \tfrac{1}{24}\bar{f}_{xxxx}|_{i,j}\,\Delta x^4 + R_{n+1}(\xi_-) \quad (9.13)$$

where the remainder term R_{n+1} is given by

$$R_{n+1} = \frac{1}{(n+1)!}\frac{\partial^{n+1}\bar{f}(\xi)}{\partial x^{n+1}}\,\Delta x^{n+1} \qquad (9.15)$$

where $x_i \le \xi_+ \le x_{i+1}$ and $x_{i-1} \le \xi_- \le x_i$. If the infinite Taylor series are truncated after the nth derivative to obtain approximations of $\bar{f}_{i+1,j}$ and $\bar{f}_{i-1,j}$, then the remainder term R_{n+1} is the error associated with the truncated Taylor series. In most cases, our main concern will be the *order* of the error, which is the rate at which the error goes to zero as $\Delta x \to 0$. The remainder term depends on Δx^{n+1}. Consequently, as $\Delta x \to 0$, the error goes to zero as Δx^{n+1}. Thus, the order of the truncated Taylor series approximations of $\bar{f}_{i\pm1,j}$ is $n+1$, which is denoted by the symbol $0(\Delta x^{n+1})$.

Adding Eqs. (9.13) and Eq. (9.14) and solving for $\bar{f}_{xx}|_{i,j}$ yields

$$\bar{f}_{xx}|_{i,j} = \frac{\bar{f}_{i+1,j} - 2\bar{f}_{i,j} + \bar{f}_{i-1,j}}{\Delta x^2} - \tfrac{1}{12}\bar{f}_{xxxx}(\xi)\,\Delta x^2 \qquad (9.16)$$

where $x_{i-1} \le \xi \le x_{i+1}$. Truncating the remainder term yields a second-order centered-space approximation of $\bar{f}_{xx}|_{i,j}$, which is denoted by $f_{xx}|_{i,j}$:

$$\boxed{f_{xx}|_{i,j} = \frac{f_{i+1,j} - 2f_{i,j} + f_{i-1,j}}{\Delta x^2}} \qquad (9.17)$$

The remainder term in Eq. (9.16) which was truncated to obtain Eq. (9.17) is called the *truncation error* of the finite difference approximation of $\bar{f}_{xx}|_{i,j}$. Equation (9.17) is a *second-order centered-space* approximation of $\bar{f}_{xx}$ at grid point (i,j).

Performing the analogous procedure in the y direction yields the following result:

$$\bar{f}_{yy}|_{i,j} = \frac{\bar{f}_{i,j+1} - 2\bar{f}_{i,j} + \bar{f}_{i,j-1}}{\Delta y^2} - \tfrac{1}{12}\bar{f}_{yyyy}(\eta)\,\Delta y^2 \qquad (9.18)$$

where $y_{i-1} \le \eta \le y_{i+1}$. Truncating the remainder term yields a second-order centered-space approximation of $\bar{f}_{yy}|_{i,j}$, which is denoted by $f_{yy}|_{i,j}$. Thus,

$$\boxed{f_{yy}|_{i,j} = \frac{f_{i,j+1} - 2f_{i,j} + f_{i,j-1}}{\Delta y^2}} \qquad (9.19)$$

9.3.3   Finite Difference Equations

Finite difference equations are obtained by replacing the individual exact partial derivatives in a partial differential equation by finite difference approximations, such as Eqs. (9.17) and (9.19), to obtain a finite difference approximation of the partial differential equation, which is called a *finite difference equation* (*FDE*). A finite difference approximation of the two-dimensional Laplace equation is developed in Section 9.4, and a finite difference approximation of the two-dimensional Poisson equation is developed in Section 9.8.

9.4   FINITE DIFFERENCE SOLUTION OF THE LAPLACE EQUATION

Consider the two-dimensional Laplace equation:

$$\bar{f}_{xx} + \bar{f}_{yy} = 0 \tag{9.20}$$

Replacing $\bar{f}_{xx}$ and $\bar{f}_{yy}$ by the second-order centered-difference approximations at grid point (i,j), Eqs. (9.17), and (9.19), respectively, yields

$$\frac{f_{i+1,j} - 2f_{i,j} + f_{i-1,j}}{\Delta x^2} + \frac{f_{i,j+1} - 2f_{i,j} + f_{i,j-1}}{\Delta y^2} = 0 \tag{9.21}$$

Equation (9.21) is a second-order centered-space approximation of Eq. (9.20). Equation (9.21) can be written as

$$f_{i+1,j} + \beta^2 f_{i,j+1} + f_{i-1,j} + \beta^2 f_{i,j-1} - 2(1 + \beta^2)f_{i,j} = 0 \tag{9.22}$$

where β is the grid aspect ratio:

$$\beta = \frac{\Delta x}{\Delta y} \tag{9.23}$$

Solving Eq. (9.22) for $f_{i,j}$ yields

$$f_{i,j} = \frac{f_{i+1,j} + \beta^2 f_{i,j+1} + f_{i-1,j} + \beta^2 f_{i,j-1}}{2(1 + \beta^2)} \tag{9.24}$$

The implicit nature of the finite difference equation is apparent in Eq. (9.24). The solution at every grid point depends on the solutions at the four neighboring grid points, which are not known until the entire solution is known. This implicit behavior of the finite difference equation is typical of the finite difference approximation of elliptic partial differential equations which govern equilibrium physical problems.

In the special case where $\Delta x = \Delta y$, the grid aspect ratio β is unity, and Eqs. (9.22) and (9.24) become

$$f_{i+1,j} + f_{i,j+1} + f_{i-1,j} + f_{i,j-1} - 4f_{i,j} = 0 \tag{9.25}$$

$$f_{i,j} = \tfrac{1}{4}(f_{i+1,j} + f_{i,j+1} + f_{i-1,j} + f_{i,j-1}) \tag{9.26}$$

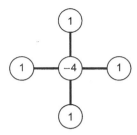

Figure 9.6   Finite difference stencil for the five-point method.

Although there is no formal mathematical advantage when β is unity, values of β greater than unity tend to produce less accurate solutions than values of β in the neighborhood of unity. Equation (9.26) has a very simple physical interpretation. It shows that, for a grid aspect ratio of unity, the solution at every point is the arithmetic average of the solutions at the four neighboring points. This result applies only to the Laplace equation (i.e., no nonhomogeneous term).

The finite difference solution of a partial differential equation is obtained by solving a finite difference equation, such as Eq. (9.22) or Eq. (9.25), at every point in the discretized solution domain.

A finite difference equation can be illustrated pictorially by a *finite difference stencil*, which is a picture of the finite difference grid used to develop the finite difference equation. The grid points employed in the finite difference approximations of the exact partial derivatives are denoted by open circles with the weighting factor associated with each grid point inside the circle. The finite difference stencil for Eq. (9.25) is illustrated in Figure 9.6. Equations (9.22) and (9.25) are called the *five-point approximation* of the Laplace equation.

The rate at which the truncation errors of the finite difference solution approach zero as Δx and Δy go to zero is called the *order* of the finite difference equation. The total error at any point in the solution domain is directly related to the local truncation errors throughout the entire solution domain. Consequently, the total error at a given point depends on Δx and Δy in exactly the same manner as the local truncation errors depend on Δx and Δy. Thus, the order of a finite difference solution of a partial differential equation is the order of the truncation errors of the finite difference approximations of the individual exact partial derivatives in the partial differential equation. For example, for Eqs. (9.22) and (9.25), the truncation error is $0(\Delta x^2) + 0(\Delta y^2)$. Consequently, the total error at a point decreases quadratically as Δx and $\Delta y \to 0$. The finite difference solution is second-order accurate in space.

Let's illustrate the five-point approximation of the Laplace equation by solving the heat diffusion problem presented in Section 9.1. The rectangular physical domain is 10 cm wide and 15 cm high. Discretizing the domain into a 5×7 finite difference grid yields the grid illustrated in Figure 9.7. For this grid, the grid aspect ratio β is unity, so Eq. (9.25) is the relevant FDE. The temperature across the top edge of the plate is $T(x, h) = 100 \sin(\pi x/w)$ C, and the temperatures on the other three edges of the plate are $T = 0$ C. The temperatures at the four corners of the plate are not required by the five-point FDE.

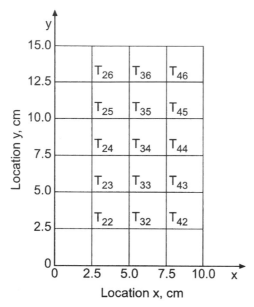

Figure 9.7   Finite difference grid for $\Delta x = \Delta y = 2.5\,\text{cm}$.

Applying Eq. (9.25) at every point in the finite difference grid yields the following system of FDEs:

$$0.0 + 0.0 - 4T_{22} + T_{23} + T_{32} = 0$$
$$0.0 + T_{22} - 4T_{23} + T_{24} + T_{33} = 0$$
$$0.0 + T_{23} - 4T_{24} + T_{25} + T_{34} = 0$$
$$0.0 + T_{24} - 4T_{25} + T_{26} + T_{35} = 0$$
$$0.0 + T_{25} - 4T_{26} + 70.710678 + T_{36} = 0$$
$$T_{22} + 0.0 - 4T_{32} + T_{33} + T_{42} = 0$$
$$T_{23} + T_{32} - 4T_{33} + T_{34} + T_{43} = 0$$
$$T_{24} + T_{33} - 4T_{34} + T_{35} + T_{44} = 0 \qquad (9.27)$$
$$T_{25} + T_{34} - 4T_{35} + T_{36} + T_{45} = 0$$
$$T_{26} + T_{35} - 4T_{36} + 100.0 + T_{46} = 0$$
$$T_{32} + 0.0 - 4T_{42} + T_{43} + 0.00 = 0$$
$$T_{33} + T_{42} - 4T_{43} + T_{44} + 0.0 = 0$$
$$T_{34} + T_{43} - 4T_{44} + T_{45} + 0.0 = 0$$
$$T_{35} + T_{44} - 4T_{45} + T_{46} + 0.0 = 0$$
$$T_{36} + T_{45} - 4T_{46} + 70.710678 + 0.0 = 0$$

where the subscripts i, j have been written as ij for simplicity, and the boundary point values of temperature are transferred to the right-hand sides of the equations before solving them. Equation (9.27) consists of 15 FDEs. Equation (9.27) can be written in matrix form as follows:

$$\begin{bmatrix}
-4 & 1 & 0 & 0 & 0 & 1 & & & & & & & & & \\
1 & -4 & 1 & 0 & 0 & 0 & 1 & & & & & & & & \\
0 & 1 & -4 & 1 & 0 & 0 & 0 & 1 & & & & & & & \\
0 & 0 & 1 & -4 & 1 & 0 & 0 & 0 & 1 & & & & & & \\
0 & 0 & 0 & 1 & -4 & 0 & 0 & 0 & 0 & 1 & & & & & \\
1 & 0 & 0 & 0 & 0 & -4 & 1 & 0 & 0 & 0 & 1 & & & & \\
& 1 & 0 & 0 & 0 & 1 & -4 & 1 & 0 & 0 & 0 & 1 & & & \\
& & 1 & 0 & 0 & 0 & 1 & -4 & 1 & 0 & 0 & 0 & 1 & & \\
& & & 1 & 0 & 0 & 0 & 1 & -4 & 1 & 0 & 0 & 0 & 1 & \\
& & & & 1 & 0 & 0 & 0 & 1 & -4 & 0 & 0 & 0 & 0 & 1 \\
& & & & & 1 & 0 & 0 & 0 & 0 & -4 & 1 & 0 & 0 & 0 \\
& & & & & & 1 & 0 & 0 & 0 & 1 & -4 & 1 & 0 & 0 \\
& & & & & & & 1 & 0 & 0 & 0 & 1 & -4 & 1 & 0 \\
& & & & & & & & 1 & 0 & 0 & 0 & 1 & -4 & 1 \\
& & & & & & & & & 1 & 0 & 0 & 0 & 1 & -4
\end{bmatrix}$$

$$\times \begin{bmatrix} T_{22} \\ T_{23} \\ T_{24} \\ T_{25} \\ T_{26} \\ T_{32} \\ T_{33} \\ T_{34} \\ T_{35} \\ T_{36} \\ T_{42} \\ T_{43} \\ T_{44} \\ T_{45} \\ T_{46} \end{bmatrix} = \begin{bmatrix} 0 \\ 0 \\ 0 \\ 0 \\ -70.710678 \\ 0 \\ 0 \\ 0 \\ 0 \\ -100 \\ 0 \\ 0 \\ 0 \\ 0 \\ -70.710678 \end{bmatrix} \qquad (9.28)$$

which is in the general form

$$\mathbf{AT} = \mathbf{b} \qquad (9.29)$$

where $\mathbf{A}$ is a (15×15) coefficient matrix, and $\mathbf{T}$ and $\mathbf{b}$ are (15×1) column vectors. All the terms not shown in the lower-left and upper-right portions of matrix $\mathbf{A}$ are zero. Equation (9.28) is a banded matrix.

Some discussion of terminology is called for at this point. The individual equation that approximates an exact PDE at a grid point, for example, Eqs. (9.22) and (9.25), is called a finite difference equation (FDE). When solving elliptic PDEs, all of the FDEs are coupled, and a system of FDEs must be solved, for example, Eq. (9.28). This system of FDEs is called the *system equation*. The system equation can be solved by the methods presented in Chapter 1.

The system equation obtained in this section, Eq. (9.28), can be solved by either a direct method (e.g., Gauss elimination) or an iterative method (e.g., successive-over-relaxation). The results presented in the remainder of this section were obtained by Gauss elimination. Iterative methods for solving the system equation are presented in Section 9.6.

Example 9.1.   Solution of the heat diffusion problem by the five-point method

Let's solve the heat diffusion problem on the 5×7 grid illustrated in Figure 9.7 by solving the system equation, Eq. (9.28), by Gauss elimination. The solution is tabulated in Table 9.2, which also presents the solution for a 9×13 grid at the common points of the two grids. Due to symmetry, the solution along the $x = 7.5$ cm line is identical to the solution along the $x = 2.5$ cm line. Consequently, the solution along the $x = 7.5$ cm line is not presented. Errors in the solution, Error$(x, y) = [T(x, y) - \bar{T}(x, y)]$, are presented directly below the solution values. The errors presented in Table 9.2 for the 5×7 grid, while not enormous, are rather large.

Let's rework the problem with $\Delta x = \Delta y = 1.25$ cm. In this case, imax = 9 and jmax = 13, and there are (imax $- 2) \times$ (jmax $- 2) = 7 \times 11 = 77$ unknown temperatures. The solution for the 9×13 grid is obtained as before. The amount of computational effort is increased considerably because of the larger number of equations to be solved. The solution at the common locations with the 5×7 grid is also tabulated in Table 9.2. The solution on the 9×13 grid is also presented in Figure 9.8, where the numerical solution is indicated by the dashed lines. The numerical solution is obviously a good approximation of the exact solution.

Table 9.2   Solution of the Heat Diffusion Problem by the Five-Point Method

	$T(x, y)$, C $\quad$ Error$(x, y) = [T(x, y) - \bar{T}(x, y)]$, C			
	$\Delta x = \Delta y = 2.5$ cm, 5×7 grid		$\Delta x = \Delta y = 1.25$ cm, 9×13 grid	
y, cm	$x = 2.5$ cm	$x = 5.0$ cm	$x = 2.5$ cm	$x = 5.0$ cm
15.0	70.710678	100.000000	70.710678	0.000000
12.5	33.459590	47.319006	32.549586	46.032067
	1.229810	1.739215	0.319806	0.452276
10.0	15.808676	22.356844	14.964426	21.162895
	1.135636	1.606032	0.291386	0.412083
7.5	7.418270	10.491019	6.838895	9.671658
	0.775966	1.097382	0.196591	0.278021
5.0	3.373387	4.770690	3.036510	4.294274
	0.449514	0.635709	0.112637	0.159293
2.5	1.304588	1.844967	1.153627	1.631475
	0.200914	0.284136	0.049953	0.070644
0.0	0.000000	0.000000	0.000000	0.000000

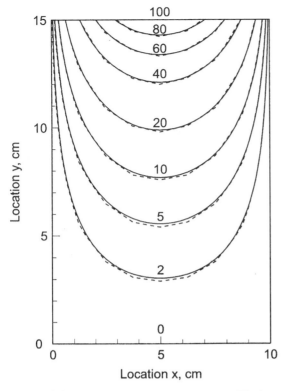

Figure 9.8   Numerical solution of the heat diffusion problem on a 9 × 13 grid.

Comparing the results in Table 9.2 for the two grids shows that the errors for the 9 × 13 grid are approximately one-fourth the size of the errors for the 5 × 7 grid. The second-order accuracy of the finite difference equation is quite apparent.

The total heat transfer rate $\dot{q}$ across the top of the plate is given by

$$\dot{q} = \int_0^w d\dot{q} = \int_0^w -k(t\,dx)\frac{\partial T(x,h)}{\partial y} \tag{9.30}$$

The temperature derivative $\partial T(x,h)/\partial y$ can be evaluated at each axial grid location by any of the one-sided difference formulas presented in Chapter 5. Let's apply the second-order one-sided-difference formula given by Eq. (5.101). Thus,

$$T_y|_{i,\text{jmax}} = \frac{T_{i,\text{jmax}-2} - 4T_{i,\text{jmax}-1} + 3T_{i,\text{jmax}}}{2\,\Delta y} \tag{9.31}$$

Evaluating Eq. (9.31) along the vertical centerline of the plate, $x = 5.0$ cm, using the data presented in Table 9.2 for the 5 × 7 grid gives

$$T_y(5.0, 15.0) = \frac{T(5.0, 10.0) - 4T(5.0, 12.5) + 3T(5.0, 15.0)}{2\,\Delta y} \tag{9.32}$$

$$T_y(5.0, 15.0) = \frac{22.356844 - 4(47.319006) + 3(100.0)}{2(2.5)} = -26.616164\ \text{C/cm}$$

$$\tag{9.33}$$

Table 9.3 Temperature Derivatives along the Top Edge of the Plate

$T_y(x, 15.0)$, C/cm					
$\Delta x = \Delta y = 2.5$ cm, 5×7 grid		$\Delta x = \Delta y = 1.25$ cm, 9×13 grid			
x, cm	T_y, C/cm	x, cm	T_y, C/cm	x, cm	T_y, C/cm
0.0	0.000000	0.00	0.000000	6.25	−27.578685
2.5	−18.820470	1.25	−11.423466	7.50	−21.107812
5.0	−26.616164	2.50	−21.107812	8.75	−11.423465
7.5	−18.820470	3.75	−27.578685	10.00	0.000000
10.0	0.000000	5.00	−29.850954		

Table 9.3 presents this result and the results at the remaining grid points along the top edge of the plate for the solutions presented in Table 9.2 for both grid sizes.

Equation (9.30) can be integrated by Simpson's 1/3 rule, Eq. (6.35). Applying Eq. (6.35) to integrate Eq. (9.30) for the 5×7 grid results gives

$$\dot{q} = -(0.4)(1.0)\frac{(2.5)}{3}[0.000000 + 4(-18.820470) + 2(-26.616164)$$

$$+ 4(-18.820470) + 0.000000] = 67.932030 \text{ J/s} \tag{9.34}$$

From Section 9.1, the exact solution is $\dot{q} = 80.012913$ J/s. Thus, the error is -12.080884 J/s. Repeating the calculation for the 9×13 grid results gives $\dot{q} = 76.025061$ J/s, for which the error is -3.987852 J/s. The ratio of the errors is

$$\text{Ratio} = \frac{E(\Delta x = \Delta y = 2.5)}{E(\Delta x = \Delta y = 1.25)} = \frac{-12.080884}{-3.987852} = 3.03 \tag{9.35}$$

which shows that the procedure is approximately second order.

The numerical solution obtained on the 9×13 grid is sufficiently accurate for most practical applications. If a more accurate solution is required, it can be obtained by decreasing the grid size even more and repeating the solution. In principle, this process can be repeated indefinitely. However, in practice, the amount of computational effort increases rapidly as the sizes of the grid spacings decrease (or conversely, as the number of grid points increases).

The results presented in this section were obtained by Gauss elimination. As mentioned in Section 1.3, the number of multiplicative operations, N, required for Gauss elimination to solve a system of linear algebraic equations is given by

$$N = \tfrac{1}{3}n^3 + n^2 - \tfrac{1}{3}n \tag{9.36}$$

where n is the number of equations to be solved. Thus, the amount of computational effort increases approximately as the cube of the number of grid points. For a two-dimensional problem, the number of grid points increases as the square of the reciprocal of the grid size (assuming that the grid aspect ratio β remains constant as the grid size is reduced). Consequently, the amount of computational effort increases approximately as the sixth power of the reciprocal of the grid size. For the two grid sizes chosen in this section (i.e., $\Delta x = \Delta y = 2.5$ cm and $\Delta x = \Delta y = 1.25$ cm), the number of operations $N = 1,345$ and

158,081, respectively. Clearly the amount of computational effort increases at an alarming rate.

Equation (9.36) is applicable to Gauss elimination for a full coefficient matrix. The coefficient matrices arising in the numerical solution of partial differential equations are banded matrices, as illustrated in Eq. (9.28). When such systems are solved by Gauss elimination, all of the zero coefficients outside of the outer bands remain zero and do not need to be computed. A significant amount of computational effort can be saved by modifying the Gauss elimination algorithm to omit the calculation of those terms. However, the bands of zeros between the central tridiagonal band of coefficients and the outer bands fill up with nonzero coefficients, which must be computed. Consequently, even when the banded nature of the coefficient matrix is accounted for, Gauss elimination still requires a large amount of computational effort for large systems of equations. In that case, iterative methods, which are discussed in Section 9.6, should be employed.

Another problem that arises in the direct solution of systems of linear algebraic equations is the large amount of computer memory required to store the coefficient matrix. From Eq. (9.28), it is apparent that many of the elements in the (15×15) coefficient matrix are zero. In fact only 59 of the 225 elements are nonzero, that is, 26 percent. For the (77×77) coefficient matrix associated with the 9×13 grid, only 169 of the 5,929 elements are nonzero, that is, 2.9 percent. The percent of nonzero elements decreases dramatically as the size of the coefficient matrix increases, since the total number of elements is n^2, whereas the number of nonzero elements is somewhat less than $5n$. To illustrate this point more dramatically, if $\Delta x = \Delta y = 0.1$ cm, then imax $= 101$ and jmax $= 151$, and there are $n = 99 \times 149 = 14{,}751$ interior grid points. The coefficient matrix for the corresponding system of finite difference equations would contain $14{,}751 \times 14{,}751 = 217{,}592{,}001$ elements. This exceeds the memory of all but the most advanced computers, and clearly cannot be considered. However, each finite difference equation contains at most five nonzero elements. Thus, the minimum amount of computer memory in which the coefficient matrix can be stored is $5n$, not n^2. However, if Gauss elimination is used to solve the system of FDEs, then computer memory must be reserved for all of the nonzero elements of the coefficient matrix within and including the outer bands.

Iterative methods, on the other hand, use only the nonzero elements of the coefficient matrix. As shown in Eq. (9.22), these nonzero elements are $1, \beta^2, 1, \beta^2$, and $-2(1 + \beta^2)$, corresponding to grid points $(i + 1, j)$, $(i, j + 1)$, $(i - 1, j)$, $(i, j - 1)$, and (i, j), respectively. These five coefficients are the same for all of the finite difference equations. Thus, there is no need to store even these $5n$ coefficients if iterative methods are used. Only the $n = $ imax $\times$ jmax values of $T_{i,j}$, the solution array itself, must be stored. Consequently, iterative methods, which are discussed in Section 9.6, should be employed when the number of grid points is large.

9.5 CONSISTENCY, ORDER, AND CONVERGENCE

There are three important properties of finite difference equations, for equilibrium problems governed by elliptic PDEs, that must be considered before choosing a specific numerical approach. They are

1. Consistency

2. Order
3. Convergence

These concepts are defined and discussed in this section.

9.5.1 Consistency

First consider the concept of *consistency*.

> *A finite difference equation is consistent with a partial differential equation if the difference between the FDE and the PDE (i.e., the truncation error) vanishes as the sizes of the grid spacings go to zero independently.*

When the truncation errors of the finite difference approximations of the individual exact partial derivatives are known, proof of consistency is straightforward. When the truncation errors of the finite difference approximations of the individual exact partial derivatives are not known, the complete finite difference equation must be analyzed for consistency. That is accomplished by expressing each term in the finite difference equation [i.e., $f(x, y)$, not $\bar{f}(x, y)$] in a Taylor series with base point (i, j). The resulting equation, which is called the *modified differential equation (MDE)*, can then be simplified to yield the exact form of the truncation error of the complete finite difference equation. Consistency can be investigated by letting the grid spacings go to zero.

Warming and Hyett (1974) developed a convenient technique for analyzing the consistency of finite difference equations which approximate propagation type PDEs. The technique can be applied to FDEs that approximate elliptic PDEs. The technique involves determining the actual partial differential equation that is solved by a finite difference equation. This actual partial differential equation is called the modified differential equation (MDE). Following Warming and Hyett, the MDE is determined by expressing each term in a finite difference equation in a Taylor series at some base point. Effectively, this changes the FDE back into a PDE.

Terms appearing in the MDE which do not appear in the original partial differential equation are truncation error terms. Analysis of the truncation error terms leads directly to the determination of *consistency* and *order*.

Example 9.2. Consistency and order analysis of the five-point method

As an example, consider the five-point approximation of the Laplace equation, $\bar{f}_{xx} + \bar{f}_{yy} = 0$, for $\Delta x = \Delta y$, Eq. (9.25):

$$f_{i+1,j} + f_{i,j+1} + f_{i-1,j} + f_{i,j-1} - 4f_{i,j} = 0 \tag{9.37}$$

Equation (9.37) can be rearranged as follows:

$$(f_{i+1,j} + f_{i-1,j}) + (f_{i,j+1} + f_{i,j-1}) - 4f_{i,j} = 0 \tag{9.38}$$

The Taylor series, with base point (i, j), for all values of $f(x, y)$ appearing in Eq. (9.38) are:

$$f_{i\pm1,j} = f_{i,j} \pm f_x|_{i,j}\,\Delta x + \tfrac{1}{2}f_{xx}|_{i,j}\,\Delta x^2 \pm \tfrac{1}{6}f_{xxx}|_{i,j}\,\Delta x^3 + \tfrac{1}{24}f_{xxxx}|_{i,j}\,\Delta x^4 \pm \cdots \tag{9.39}$$

$$f_{i,j\pm1} = f_{i,j} \pm f_y|_{i,j}\,\Delta y + \tfrac{1}{2}f_{yy}|_{i,j}\,\Delta y^2 \pm \tfrac{1}{6}f_{yyy}|_{i,j}\,\Delta y^3 + \tfrac{1}{24}f_{yyyy}|_{i,j}\,\Delta y^4 \pm \cdots \tag{9.40}$$

Dropping the notation $|_{i,j}$ for clarity and substituting Eqs. (9.39) and (9.40) into Eq. (9.38) gives

$$(2f + f_{xx} \Delta x^2 + \tfrac{1}{12} f_{xxxx} \Delta x^4 + \cdots) + (2f + f_{yy} \Delta y^2 + \tfrac{1}{12} f_{yyyy} \Delta y^4 + \cdots) - 4f = 0$$

$$(9.41)$$

Cancelling zero-order terms, dividing through by $\Delta x = \Delta y$, and rearranging terms yields the MDE:

$$\boxed{f_{xx} + f_{yy} = -\tfrac{1}{12} f_{xxxx} \Delta x^2 - \cdots - \tfrac{1}{12} f_{yyyy} \Delta y^2 - \cdots} \qquad (9.42)$$

As $\Delta x \to 0$ and $\Delta y \to 0$, Eq. (9.42) approaches $f_{xx} + f_{yy} = 0$, which is the Laplace equation. Consequently, Eq. (9.37) is a consistent approximation of the Laplace equation.

9.5.2 Order

Next consider the concept of *order*.

> *The order of a finite difference approximation of a partial dfferential equation is the rate at which the error of the finite difference solution approaches zero as the sizes of the grid spacings approach zero.*

The order of a finite difference equation is the order of the truncation error terms in the finite difference approximations of the individual exact partial derivatives in the PDE.

When the truncation errors of the finite difference approximations of the individual exact partial derivatives are known, as in Eqs. (9.16) and (9.18), the order of the FDE is obvious. When the truncation errors of the finite difference approximations of the individual exact partial derivatives are not known, the order can be determined from the modified differential equation. For example, the order of the five-point finite difference approximation of the Laplace equation, Eq. (9.37), can be obtained from the corresponding MDE, Eq. (9.42). From Eq. (9.42), Eq. (9.37) is an $0(\Delta x^2) + 0(\Delta y^2)$ approximation of the Laplace equation.

9.5.3 Convergence

Consider the concept of *convergence*.

> *A finite difference method is convergent if the solution of the finite difference equation approaches the exact solution of the partial differential equation as the sizes of the grid spacings go to zero.*

Let $\bar{f}_{i,j}$ denote the exact solution of the partial differential equation, $f_{i,j}$ denote the exact solution of the finite difference equation, and $E_{i,j}$ denote the difference between them. The statement of convergence is

$$f_{i,j} - \bar{f}_{i,j} = E_{i,j} \to 0 \qquad \text{as } \Delta x \to 0 \text{ and } \Delta y \to 0 \qquad (9.43)$$

The proof of convergence of a finite difference solution is in the domain of the mathematician. We shall not attempt to prove convergence directly. Numerous discussions of convergence appear in the literature, for example, Forsythe and Wasow (1960). We shall assume that if a finite difference approximation of an elliptic PDE is a consistent approximation of the PDE, then the finite difference method is convergent.

9.6 ITERATIVE METHODS OF SOLUTION

Direct solutions of the system equation approximating the Laplace equation for the two-dimensional heat diffusion problem, Eq. (9.28), are presented in Section 9.4. As discussed at the end of that section, direct solutions require excessive computational effort and computer memory as the number of grid points increases. In such cases, iterative methods should be employed.

Several iterative methods are discussed in Section 1.7, namely:

1. Jacobi iteration
2. Gauss-Seidel iteration
3. Successive-over-relaxation (SOR)

As discussed in Section 1.7, iterative methods require *diagonal dominance* to guarantee convergence. Diagonal dominance requires that

$$|a_{ii}| \geq \sum_{j=1, j\neq i}^{n} |a_{ij}| \qquad (i = 1, \ldots, n) \tag{9.44}$$

with the inequality satisfied for at least one equation. The system equation arising from the five-point second-order centered-space approximation of the Laplace equation is always diagonally dominant, as illustrated by the finite difference stencil presented in Figure 9.6 and the coefficient matrix presented in Eq. (9.28). Consequently, iterative methods can be employed to solve the system equation approximating the Laplace equation.

The Jacobi method converges slowly, so it will not be used here. The Gauss-Seidel method, which is the limiting case of the SOR method when the over-relaxation factor $\omega = 1.0$, will be used to demonstrate the general features of iterative methods. However, successive-over-relaxation (SOR) is the recommended method and should be used in general.

9.6.1 The Gauss-Seidel Method

The Gauss-Seidel method, applied to the finite difference approximation of the Laplace equation, is obtained by adding the term, $\pm f_{i,j}$, to Eq. (9.24) and rearranging as follows:

$$f_{i,j}^{k+1} = f_{i,j}^{k} + \Delta f_{i,j}^{k+1} \tag{9.45}$$

$$\Delta f_{i,j}^{k+1} = \frac{f_{i+1,j}^{k} + \beta^2 f_{i,j+1}^{k} + f_{i-1,j}^{k+1} + \beta^2 f_{i,j-1}^{k+1} - 2(1+\beta^2) f_{i,j}^{k}}{2(1+\beta^2)} \tag{9.46}$$

where the superscript k ($k = 0, 1, 2, \ldots$) denotes the iteration number. The term $\Delta f_{i,j}^{k+1}$ is called the *residual* in the relaxation method. Equation (9.46) is based on the sweep directions illustrated in Figure 9.9. The order of the sweeps is irrelevant, but once chosen, it should be maintained.

An initial approximation ($k = 0$) must be made for $f_{i,j}$ to start the process. Several choices are available. For example,

1. Let $f_{i,j} = 0.0$ at all of the interior points.
2. Approximate $f_{i,j}$ by some weighted average of the boundary values.
3. Construct a solution on a coarser grid, then interpolate for starting values on a finer grid. This procedure can be repeated on finer and finer grids.

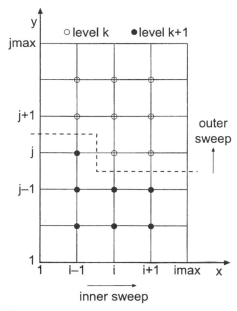

Figure 9.9   Sweep directions for the Gauss-Seidel method.

Iterative methods do not yield the exact solution of the system equation directly. They approach the exact solution asymptotically as the number of iterations increases. When the number of iterations increases without bound, the iterative solution yields the exact solution of the system equation, within the round-off limit of the computer. However, in most practical problems, such extreme precision is not warranted. Consequently, the iterative process is usually terminated when some form of convergence criterion has been achieved. Various convergence criteria are possible. For example:

$$|\Delta f_{i,j}^{k+1}| < \varepsilon \quad \text{(for all } i,j) \qquad |\Delta f_{i,j}^{k+1}/f_{i,j}^{k}| < \varepsilon \quad \text{(for all } i,j) \tag{9.47}$$

$$\sum_{i,j}^{n} |\Delta f_{i,j}^{k+1}| < \varepsilon \qquad \sum_{i,j}^{n} |\Delta f_{i,j}^{k+1}/f_{i,j}^{k}| < \varepsilon \tag{9.48}$$

$$\left[\sum_{i,j}^{n} (\Delta f_{i,j}^{k+1})^2 \right]^{1/2} < \varepsilon \qquad \left[\sum_{i,j}^{n} (\Delta f_{i,j}^{k+1}/f_{i,j}^{k})^2 \right]^{1/2} < \varepsilon \tag{9.49}$$

where ε is the convergence tolerance. The three criteria on the left are absolute criteria and are useful when the magnitude of the solution is known so that a meaningful value of ε can be specified. The three criteria on the right are relative criteria and are useful when the magnitude of the solution is not known, in which case a meaningful absolute convergence tolerance cannot be specified. Caution in the use of relative criteria is necessary if any of the $f_{i,j}$ are close to zero in magnitude.

Example 9.3.   Solution of the heat diffusion problem by the Gauss-Seidel method

Let's solve the two-dimensional heat diffusion problem presented in Section 9.1 by the Gauss-Seidel method using the 5×7 finite difference grid illustrated in Figure 9.7. Let $T_{i,j} = 0.0$ be the initial guess. The solution at selected iteration steps is presented in

Table 9.4 Temperatures at $y = 10$ cm for the 5×7 Grid as a Function of Iteration Number k for the Gauss-Seidel Method

Iteration number k	Temperature $T(x, 10.0)$, C			$|\Delta T_{i,j}|_{max}$
	$x = 2.5$ cm	$x = 5.0$ cm	$x = 7.5$ cm	
0	0.000000	0.000000		
1	4.419417	8.459709	8.373058	2.94E + 01
2	8.925485	14.899272	11.834527	1.05E + 01
3	11.920843	18.169362	13.551171	3.74E + 00
4	13.566276	19.921549	14.481619	1.75E + 00
5	14.484384	20.906785	15.012409	9.85E − 01
10	15.696510	22.232231	15.739435	7.74E − 02
20	15.807760	22.355825	15.808109	6.29E − 04
30	15.808669	22.356836	15.808672	5.16E − 06
40	15.808676	22.356844	15.808676	4.24E − 08
∞	15.808676	22.356844	15.808676	
Exact	14.673040	20.750812	14.673040	
Error	1.135636	1.606032	1.135636	

Table 9.4 for the three grid points along the horizontal grid line $y = 10$ cm, along with the maximum value of $|\Delta T_{i,j}| = |\Delta T_{i,j}|_{max}$ in the entire grid. The solution has converged to five significant figures on the 20th iteration. The maximum residual continues to decrease as the iterative process continues, as illustrated in Figure 9.10, but no changes occur in the solution in the first five significant digits. The converged (to 13 significant digits) solution is presented in the third line from the bottom of Table 9.4. These are the same values obtained by the Gauss-Seidel method in Example 9.1. The exact solution is presented in the second line from the bottom of the table. The difference between the converged solution of the system equation and the exact solution of the partial differential equation, which is the truncation error of the finite difference solution, is presented in the last line of Table 9.4.

The maximum residual, $|\Delta T_{i,j}|_{max}$, is presented in Figure 9.10 as a function of the iteration number k. The curve labeled "Gauss-Seidel" corresponds to the results presented in Table 9.4. The curves labeled "SOR" are discussed in the next subsection. The maximum residual decreases exponentially with increasing k.

9.6.2 The Successive-Over-Relaxation (SOR) Method

The convergence rate of the relaxation method can be greatly increased by using over-relaxation. The Gauss-Seidel method, Eq. (9.45), becomes over-relaxation simply by over-relaxing the residual, $\Delta f_{i,j}^{k+1}$, by the over-relaxation factor ω. Thus,

$$f_{i,j}^{k+1} = f_{i,j}^{k} + \omega \, \Delta f_{i,j}^{k+1} \tag{9.50}$$

where the residual, $\Delta f_{i,j}^{k+1}$, is given by Eq. (9.46), as before. When Eq. (9.50) is applied repetitively, it is called the successive-over-relaxation (SOR) method. When $\omega = 1.0$, the

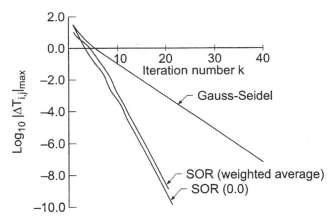

Figure 9.10   Maximum residual as a function of the iteration number, k.

SOR method reduces to the Gauss-Seidel method. The maximum rate of convergence is achieved for some optimum value of ω, denoted by ω_{opt}, which lies between 1.0 and 2.0.

In some special cases, the optimum over-relaxation factor ω_{opt} can be predicted theoretically. For a rectangular region with Dirichlet boundary conditions (i.e., specified values of the dependent variable), ω_{opt}, can be estimated from [Frankel (1950)]:

$$\omega_{opt} = 2\left(\frac{1 - \sqrt{1 - \xi}}{\xi}\right) \tag{9.51}$$

where

$$\xi = \left[\frac{\cos(\pi/I) + \beta^2 \cos(\pi/J)}{1 + \beta^2}\right]^2 \tag{9.52}$$

where $I = (imax - 1)$ is the number of spatial increments in the x direction, $J = (jmax - 1)$ is the number of spatial increments in the y direction, and $\beta = \Delta x/\Delta y$ is the grid aspect ratio. Values of ω_{opt} for the 10 cm by 15 cm physical space considered in the heat diffusion problem are presented in Table 9.5 for several grid sizes.

Example 9.4.   Solution of the heat diffusion problem by the SOR method

Table 9.6 presents the solution of the heat diffusion problem presented in Section 9.1 for the 5 × 7 grid using the SOR method with $\omega_{opt} = 1.23647138$. The convergence history is illustrated in Figure 9.10. The curve labeled "SOR (0.0)" uses $T_{i,j} = 0.0$ as the initial approximation. The curve labeled "SOR (weighted average)" uses the linearly weighted

Table 9.5   Values of ω_{opt} for Several Grid Sizes

Grid size imax × jmax	ω_{opt}	Grid size imax × jmax	ω_{opt}
3 × 4	1.01613323	17 × 25	1.71517254
5 × 7	1.23647138	33 × 49	1.84615581
9 × 13	1.50676003	65 × 97	1.91993173

Table 9.6 Temperatures at $y = 10$ cm for the 5×7 Grid as a Function of Iteration Number k for the SOR Method with ω_{opt}

Iteration number k	Temperature $T(x, 10.0)$, C			
	$x = 2.5$ cm	$x = 5.0$ cm	$x = 7.5$ cm	$|\Delta T_{i,j}|_{\mathrm{max}}$
0	0.000000	0.000000	0.000000	
1	6.756677	13.732603	14.601036	3.05E + 01
2	12.696767	20.924918	14.977353	6.93E + 00
3	15.197910	21.700940	15.539634	2.02E + 00
4	15.549183	22.137471	15.720817	3.53E − 01
5	15.714157	22.288290	15.784980	1.33E − 01
10	15.808558	22.356761	15.808648	2.78E − 04
15	15.808676	22.356844	15.808676	3.17E − 07
20	15.808676	22.356844	15.808676	3.09E − 10
∞	15.808676	22.356844	15.808676	
Exact	14.673040	20.750812	14.673040	
Error	1.135636	1.606032	1.135636	

average of the four corresponding boundary values as the initial approximation. As illustrated in Figure 9.10, the initial approximation has little effect on the convergence rate. Comparing Tables 9.4 and 9.6 shows that the solution converges much more rapidly using the SOR method. In fact, the solution has converged to five significant digits by the 10th iteration for the SOR method, as compared to 20 iterations for the Gauss-Seidel method. Figure 9.10 presents the maximum residual, $|\Delta T_{i,j}|_{\mathrm{max}}$, as a function of iteration number k for the SOR method with ω_{opt} for comparison with the results of the Gauss-Seidel method.

9.7 DERIVATIVE BOUNDARY CONDITIONS

All the finite difference solutions to the Laplace equation presented thus far have been for Dirichlet boundary conditions; that is, the values of $\bar{f}(x, y)$ are specified on the boundaries. In this section, a procedure for implementing a derivative, or Neumann, boundary condition is presented. We will solve the steady two-dimensional heat diffusion problem presented in Section 9.1 by recognizing that the vertical midplane of the rectangular plate considered in that problem is a plane of symmetry. Thus, no heat crosses the midplane, and the temperature gradient at that location is zero. This is true only when the boundary conditions on the edges of the plate are symmetrical, which is the case in this particular problem.

The formal statement of the problem is as follows. A rectangular plate of height $h = 15$ cm, width $w = 5$ cm, and thickness $t = 1$ cm is insulated on both faces so that no heat flows in the direction of the thickness t. The temperature on the top edge of the plate is held at $100 \sin(\pi x/w)$ C, and the temperatures on the left and bottom edges of the plate are held at 0 C. The right edge of the plate is insulated, so $\partial T/\partial x = 0$ along that edge. This boundary condition is

$$\bar{T}_x(5.0, y) = 0 \tag{9.53}$$

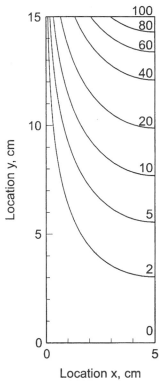

Figure 9.11   Exact solution of the heat diffusion problem with a derivative boundary condition.

The temperature distribution $\bar{T}(x, y)$ in the plate is required. The exact solution to this problem is the same as the exact solution to the problem presented in Section 9.1, which is given by Eq. (9.2). The plate, the boundary conditions, and the exact solution are presented in Figure 9.11, for selected isotherms.

In this section, we will solve this problem numerically using the finite difference method developed in Section 9.4, modified to account for the derivative boundary condition along the right edge of the plate, Eq. (9.53).

Let's apply the interior point finite difference equation, Eq. (9.22), at grid point (I, j) on the right edge of the plate, as illustrated in Figure 9.12.

$$f_{I+1,j} + \beta^2 f_{I,j+1} + f_{I-1,j} + \beta^2 f_{I,j-1} - 2(1 + \beta^2) f_{I,j} = 0 \qquad (9.54)$$

Grid point $(I + 1, j)$ is outside of the solution domain, so $f_{I+1,j}$ is not defined. However, a value for $f_{I+1,j}$ can be determined from the derivative boundary condition on the right edge of the plate.

The finite difference approximations employed in Eq. (9.54) for the spatial derivatives $\bar{f}_{xx}$ and $\bar{f}_{yy}$ are second order. It's desirable to match this truncation error by using a second-order centered-difference approximation for the derivative boundary condition, $\bar{f}_x|_{I,j} = \text{known}$. Apply Eq. (5.99) at grid point (I, j):

$$\bar{f}_x|_{I,j} = \frac{\bar{f}_{I+1,j} - \bar{f}_{I-1,j}}{2\,\Delta x} + 0(\Delta x^2) \qquad (9.55)$$

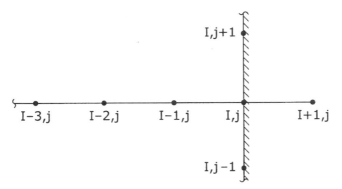

Figure 9.12 Finite difference grid at the boundary.

Solving Eq. (9.55) for $\bar{f}_{I+1,j}$ and truncating the remainder term gives

$$f_{I+1,j} = f_{I-1,j} + 2\bar{f}_x|_{I,j}\,\Delta x \tag{9.56}$$

Substituting Eq. (9.56) into Eq. (9.54) yields

$$\boxed{\beta^2 f_{I,j+1} + 2f_{I-1,j} + \beta^2 f_{I,j-1} - 2(1+\beta^2)f_{I,j} = -2\bar{f}_x|_{I,j}\,\Delta x} \tag{9.57}$$

In the present problem, $\bar{f}_x|_{I,j} = 0$. However, in general, $\bar{f}_x|_{I,j} \neq 0$.

Example 9.5. Solution of the heat diffusion problem with a derivative boundary condition

___

As an example, let's work the two-dimensional heat diffusion problem with a derivative boundary condition on a 5×13 grid using Eq. (9.57) along the right edge of the plate. The results are presented in Figure 9.13, where the numerical solution is indicated by the dashed lines. The numerical solution is obviously a good approximation of the exact solution. These results are identical to the results illustrated in Figure 9.8 and tabulated in Table 9.2 for the 9×13 grid for the full plate solution. This would not be the case, in general. The solution and the error, $\text{Error}(5.0, y) = [T(5.0, y) - \bar{T}(5.0, y)]$, along the right edge of the plate are presented in Table 9.7 for every other grid point.

___

The numerical results presented in this section were obtained by Gauss elimination for a 5×13 grid. The objective of the numerical studies presented in this section was to demonstrate procedures for implementing derivative boundary conditions. Accuracy and computational efficiency were not of primary interest. More accurate solutions can be obtained using finer grids. When a large number of grid points are considered, iterative methods should be employed.

9.8 FINITE DIFFERENCE SOLUTION OF THE POISSON EQUATION

The numerical solution of the Laplace equation is considered in Sections 9.4 to 9.7. In this section, the numerical solution of the Poisson equation is discussed. The Poisson equation is simply the nonhomogeneous Laplace equation. It applies to problems in mass diffusion, heat diffusion (i.e., conduction), incompressible fluid flow, etc., in which a nonhomogeneous term is present.

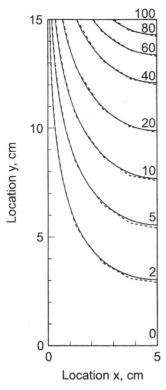

Figure 9.13 Solution of the heat diffusion problem with a derivative boundary condition.

Table 9.7 Solution Along the Right Edge of the Plate

Location	Temperature		
$(5.0, y)$, cm	$T(5.0, y)$, C	$\bar{T}(5.0, y)$, C	Error$(5.0, y)$, C
$(5.0, 15.0)$	100.000000	100.000000	
$(5.0, 12.5)$	46.032068	45.579791	0.452277
$(5.0, 10.0)$	21.162895	20.750812	0.412083
$(5.0, 7.5)$	9.671658	9.393637	0.278021
$(5.0, 5.0)$	4.294274	4.134981	0.159293
$(5.0, 2.5)$	1.631476	1.560831	0.070595
$(5.0, 0.0)$	0.000000	0.000000	

The thermal system illustrated in Section 9.1, consisting of a rectangular electrical conductor with internal heat generation due to electrical resistance heating, is governed by the following Poisson equation:

$$T_{xx} + T_{yy} = -\frac{\dot{Q}}{k} \tag{9.58}$$

which is subject to boundary conditions specified on the four sides of the conductor. The conductor is made of a copper alloy ($k = 0.4$ J/cm-s-C). The conductor has a width $w = 1.0$ cm and a height $h = 1.5$ cm. Negligible heat flows along the length of the conductor. Energy is generated within the conductor by electrical resistance heating at the rate $\dot{Q} = 400 \, \text{J/cm}^3$-s. The four sides of the conductor are held at $0\,\text{C}$. The conductor cross-section and boundary conditions are illustrated in Figure 9.14. The temperature distribution $\bar{T}(x, y)$ in the conductor cross section is required. The exact solution to this problem is

$$\bar{T}(x, y) = \frac{(\dot{Q}/k)(w^2/4 - x^2)}{2} - \frac{4w^2(\dot{Q}/k)}{\pi^3}$$
$$\times \sum_{n=0}^{\infty} \frac{(-1)^n \cos[(2n+1)\pi x/w]\cosh[(2n+1)\pi y/w]}{(2n+1)^3 \cosh[(2n+1)\pi h/2w]} \tag{9.59}$$

where x and y are measured from the center of the conductor. The exact solution for $w = 1.0 \, \text{cm}$ and $h = 1.5 \, \text{cm}$ is presented in Figure 9.14 for selected isotherms.

In this section the two-dimensional heat diffusion problem with internal energy generation is solved using the finite difference equations developed in Section 9.4 for solving the Laplace equation. Those equations must be modified to include the non-homogeneous (i.e., source) term.

Replacing $\bar{T}_{xx}$ and $\bar{T}_{yy}$ by the second-order centered-difference approximations at grid point (i, j) Eqs. (9.17) and (9.19), respectively, evaluating $\dot{Q}$ at grid point (i, j), multiplying by Δx^2, and introducing the grid aspect ratio β defined by Eq. (9.23) yields:

$$\boxed{T_{i+1,j} + \beta^2 T_{i,j+1} + T_{i-1,j} + \beta^2 T_{i,j-1} - 2(1 + \beta^2)T_{i,j} + \Delta x^2 \left(\frac{\dot{Q}_{i,j}}{k}\right) = 0} \tag{9.60}$$

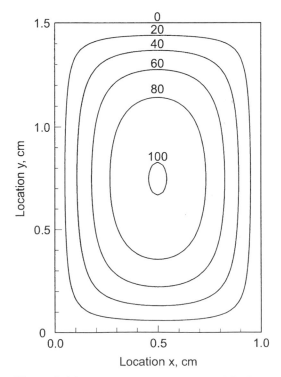

Figure 9.14 Exact solution of the heat diffusion problem with internal energy generation.

Solving Eq. (9.60) for $T_{i,j}$ gives

$$T_{i,j} = \frac{T_{i+1,j} + \beta^2 T_{i,j+1} + T_{i-1,j} + \beta^2 T_{i,j-1} + \Delta x^2(\dot{Q}_{i,j}/k)}{2(1 + \beta^2)} \qquad (9.61)$$

In the special case where $\Delta x = \Delta y$, the grid aspect ratio β is unity, and Eqs. (9.60) and (9.61) become

$$T_{i+1,j} + T_{i,j+1} + T_{i-1,j} + T_{i,j-1} - 4T_{i,j} + \Delta x^2\left(\frac{\dot{Q}_{i,j}}{k}\right) = 0 \qquad (9.62)$$

$$T_{i,j} = \frac{T_{i+1,j} + T_{i,j+1} + T_{i-1,j} + T_{i,j-1} + \Delta x^2(\dot{Q}_{i,j}/k)}{4} \qquad (9.63)$$

Equations (9.60) to (9.63) are analogous to Eqs. (9.22) to (9.26) presented in Section 9.4 for solving the Laplace equation numerically. All the general features of the numerical solution of the Laplace equation presented in Sections 9.4 to 9.7 apply directly to the numerical solution of the Poisson equation.

Example 9.6. **Solution of the heat diffusion problem with internal energy generation by the five-point method.**

As an example, let's solve the two-dimensional heat diffusion problem with internal energy generation described at the beginning of this section on a 5×7 grid, for $\Delta x = \Delta y = 0.25$ cm, for which $\beta = 1.0$. Equation (9.62) is the relevant FDE. For this grid, there are $3 \times 5 = 15$ interior points and 15 corresponding finite difference equations. This system of linear algebraic equations is solved by Gauss elimination. The results are presented in Table 9.8 for the upper left quadrant of the solution domain. Due to symmetry

Table 9.8 Solution of the Heat Diffusion Problem with Internal Energy Generation

	$T(x, y)$, C $\bar{T}(x, y)$, C $\text{Error}(x, y) = [T(x, y) - \bar{T}(x, y)]$, C			
	$\Delta x = \Delta y = 0.25$ cm, 5×7 grid		$\Delta x = \Delta y = 0.125$ cm, 9×13 grid	
y, cm	$x = 0.25$ cm	$x = 0.50$ cm	$x = 0.25$ cm	$x = 0.50$ cm
1.25	48.462813	62.407953	49.910484	64.045535
	50.442904	64.619671	50.442904	64.619671
	−1.980101	−2.211718	−0.532420	−0.574136
1.00	68.943299	90.206185	70.482926	92.228966
	71.018567	92.938713	71.018567	92.938713
	−2.075268	−2.732528	−0.535641	−0.709747
0.75	74.604197	98.030191	76.095208	100.064542
	76.606298	100.771385	76.606298	100.771383
	−2.002101	−2.741194	−0.511090	−0.706843

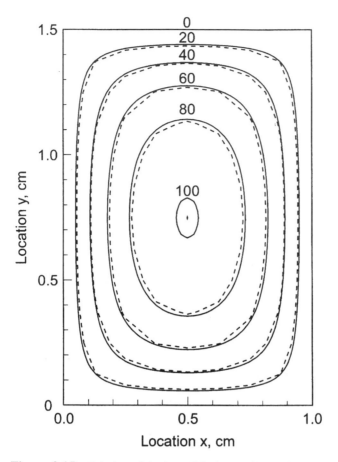

Figure 9.15 Solution of the heat diffusion problem with internal energy generation.

about the horizontal and vertical centerlines, the solutions in the other three quadrants are mirror images of the results presented in Table 9.8. The errors are rather large. Comparing these results with the results presented in Table 9.2 for the solution of a heat diffusion problem without internal energy generation shows that the errors in Table 9.8 are considerably larger in some regions. The larger errors are due to the presence of the source term.

Let's rework the problem on a 9×13 grid. The amount of computational effort is increased considerably because of the larger number of equations to be solved. The solution at the common locations with the 5×7 grid is presented in Table 9.8 and illustrated in Figure 9.15, where the numerical solution is indicated by the dashed lines. Comparing the results presented in Table 9.8 for the two grid sizes shows that the errors for the 9×13 grid are approximately one-fourth the size of the errors for the 5×7 grid, which demonstrates that the method is second order. For the 9×13 grid, the numerical solution is a good approximation of the exact solution. No special problems arise due to the presence of the source term.

9.9 HIGHER-ORDER METHODS

The five-point method developed in Section 9.4 and employed in Sections 9.4 to 9.8 is a second-order method. The objective of this section is to present some fourth-order methods. Three such methods are considered in this section:

1. The explicit fourth-order centered-difference FDA
2. The implicit compact fourth-order centered-difference FDA
3. Extrapolation of the five-point FDE developed in Section 9.4

Each of these methods has limitations. The explicit fourth-order FDA use five grid points along both the x and y axes to obtain fourth-order approximations of $\bar{f}_{xx}$ and $\bar{f}_{yy}$. The result for $\bar{f}_{xx}$ is given by Eq. (5.112) in Chapter 5. When applied at points adjacent to a boundary, these FDAs require a point outside of the boundary, and thus cannot be used. An unsymmetrical fourth-order FDA or the second-order five-point method must be used at these points, thus reducing the accuracy somewhat. Consequently, this approach is not developed in this book.

The implicit compact fourth-order approach is based on the implicit compact three-point FDAs developed in Chapter 8, Eqs. (8.115) and (8.127). This approach works well for Dirichlet boundary conditions. However, it is difficult to obtain fourth-order accuracy at the boundaries for Neumann (i.e., derivative) boundary conditions.

Extrapolation of the five-point FDE, Eqs. (9.22) and (9.25) developed in Section 9.4, is straightforward. However, successively reducing the grid size to obtain several solutions for extrapolation becomes computationally expensive, since each halving of the grid spacing increases the number of grid points by four for a two-dimensional problem and by eight for a three-dimensional problem.

9.9.1 Compact Fourth-Order Method

The compact fourth-order method is based on the compact three-point fourth-order approximations developed in Section 8.6.2. Consider the Laplace equations, $\bar{f}_{xx} + \bar{f}_{yy} = 0$. Recall Eq. (8.127) for $\bar{y}_{xx}$, written for $\bar{f}_{xx}$, and the corresponding expression for $\bar{f}_{yy}$:

$$f_{xx}|_{i,j} = \frac{\delta_x^2 f_{i,j}}{\Delta x^2 (1 + \delta_x^2/12)} \tag{9.64}$$

$$f_{yy}|_{i,j} = \frac{\delta_y^2 f_{i,j}}{\Delta y^2 (1 + \delta_y^2/12)} \tag{9.65}$$

where δ_x^2 and δ_y^2 are second-order centered differences in x and y, respectively. Substituting Eqs. (9.64) and (9.65) into the Laplace equation gives

$$\frac{\delta_x^2 f_{i,j}}{\Delta x^2 (1 + \delta_x^2/12)} + \frac{\delta_y^2 f_{i,j}}{\Delta y^2 (1 + \delta_y^2/12)} = 0 \tag{9.66}$$

Multiplying Eq. (9.66) by $(1 + \delta_x^2/12)(1 + \delta_y^2/12)$ gives

$$\left[\left(1 + \frac{\delta_y^2}{12} \right) \left(\frac{\delta_x^2}{\Delta x^2} \right) + \left(1 + \frac{\delta_x^2}{12} \right) \left(\frac{\delta_y^2}{\Delta y^2} \right) \right] f_{i,j} = 0 \tag{9.67}$$

Equation (9.67) can be rearranged as follows:

$$\left[\frac{\delta_x^2}{\Delta x^2} + \frac{\delta_y^2}{\Delta y^2} + \left(\frac{\Delta x^2 + \Delta y^2}{12} \right) \frac{\delta_x^2}{\Delta x^2} \frac{\delta_y^2}{\Delta y^2} \right] f_{i,j} = 0 \tag{9.68}$$

Applying the first two operators, $(\delta_x^2/\Delta x^2)$ and $(\delta_y^2/\Delta y^2)$, and the final operator, $(\delta_y^2/\Delta y^2)$, gives

$$\frac{f_{i+1,j} - 2f_{i,j} + f_{i-1,j}}{\Delta x^2} + \frac{f_{i,j+1} - 2f_{i,j} + f_{i,j-1}}{\Delta y^2}$$

$$+ \left(\frac{\Delta x^2 + \Delta y^2}{12} \right) \frac{\delta_x^2}{\Delta x^2} \left(\frac{f_{i,j+1} - 2f_{i,j} + f_{i,j-1}}{\Delta y^2} \right) = 0 \tag{9.69}$$

Applying the remaining operator, $(\delta_x^2/\Delta x^2)$, gives

$$\frac{f_{i+1,j} - 2f_{i,j} + f_{i-1,j}}{\Delta x^2} + \frac{f_{i,j+1} - 2f_{i,j} + f_{i,j-1}}{\Delta y^2}$$

$$+ \left(\frac{\Delta x^2 + \Delta y^2}{12 \Delta y^2} \right) \left(\frac{f_{i+1,j+1} - 2f_{i,j+1} + f_{i-1,j+1}}{\Delta x^2} \right.$$

$$\left. - 2 \frac{f_{i+1,j} - 2f_{i,j} + f_{i-1,j}}{\Delta x^2} + \frac{f_{i+1,j-1} - 2f_{i,j-1} + f_{i-1,j-1}}{\Delta x^2} \right) = 0 \tag{9.70}$$

Equation (9.70) can be simplified by multiplying by $12\Delta x^2 \, \Delta y^2/(\Delta x^2 + \Delta y^2)$, gathering terms, and introducing the grid aspect ratio, $\beta = \Delta x/\Delta y$, to yield

$$f_{i+1,j+1} + f_{i+1,j-1} + f_{i-1,j+1} + f_{i-1,j-1} + \frac{2(5 - \beta^2)}{\beta^2 + 1} (f_{i+1,j} + f_{i-1,j})$$

$$+ \frac{2(5\beta^2 - 1)}{\beta^2 + 1} (f_{i,j+1} + f_{i,j-1}) - 20f_{i,j} = 0 \tag{9.71}$$

For unity grid aspect ratio (i.e., $\beta = 1.0$), Eq. (9.71) becomes

$$f_{i+1,j+1} + f_{i+1,j-1} + f_{i-1,j+1} + f_{i-1,j-1} + 4(f_{i+1,j} + f_{i,j+1} + f_{i-1,j} + f_{i,j-1})$$

$$- 20f_{i,j} = 0 \tag{9.72}$$

Equations (9.71) and (9.72) are the compact fourth-order finite difference approximations of the Laplace equation. The finite difference stencil for Eq. (9.72) is illustrated Figure 9.16.

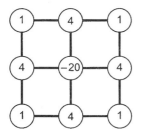

Figure 9.16 Finite difference stencil for the compact fourth-order method.

Example 9.7. Solution of the heat diffusion problem by the compact fourth-order method

---

Let's work the heat diffusion problem presented in Section 9.1 by the compact fourth-order method for the 5×7 grid illustrated in Figure 9.7, for which $\beta = 1.0$. Apply Eq. (9.72) at every point in the finite difference grid. For each unknown grid point, (i, j), apply Eq. (9.72) in grid point order $(i - 1, j - 1)$, $(i - 1, j)$, $(i - 1, j + 1)$, $(i, j - 1)$, etc. This yields the following system of FDEs:

$$0.0 + 4(0.0) + 0.0 + 0.0 - 20T_{22} + 4T_{23} + 0.0 + 4T_{32} + T_{33} = 0$$

$$0.0 + 4(0.0) + 0.0 + 4T_{22} - 20T_{23} + 4T_{24} + T_{32} + 4T_{33} + T_{34} = 0$$

$$0.0 + 4(0.0) + 0.0 + 4T_{23} - 20T_{24} + 4T_{25} + T_{33} + 4T_{34} + T_{35} = 0$$

$$0.0 + 4(0.0) + 0.0 + 4T_{24} - 20T_{25} + 4T_{26} + T_{34} + 4T_{35} + T_{36} = 0$$

$$0.0 + 4(0.0) + 0.0 + 4T_{25} - 20T_{26} + 4(70.710678) + T_{35} + 4T_{36} + 100.0 = 0$$

$$0.0 + 4T_{22} + T_{23} + 4(0.0) - 20T_{32} + 4T_{33} + 0.0 + 4T_{42} + T_{43} = 0$$

$$T_{22} + 4T_{23} + T_{24} + 4T_{32} - 20T_{33} + 4T_{34} + T_{42} + 4T_{43} + T_{44} = 0$$

$$T_{23} + 4T_{24} + T_{25} + 4T_{33} - 20T_{34} + 4T_{35} + T_{43} + 4T_{44} + T_{45} = 0$$

$$T_{24} + 4T_{25} + T_{26} + 4T_{34} - 20T_{35} + 4T_{36} + T_{44} + 4T_{45} + T_{46} = 0$$

$$T_{25} + 4T_{26} + 70.710678 + 4T_{35} - 20T_{36} + 4(100.0) + T_{45} + 4T_{46} + 4(70.710678) = 0$$

$$0.0 + 4T_{32} + T_{33} + 4(0.0) - 20T_{42} + 4T_{43} + 0.0 + 4(0.0) + 0.0 = 0$$

$$T_{32} + 4T_{33} + T_{34} + 4T_{42} - 20T_{43} + 4T_{44} + 0.0 + 4(0.0) + 0.0 = 0$$

$$T_{33} + 4T_{34} + T_{35} + 4T_{43} - 20T_{44} + 4T_{45} + 0.0 + 4(0.0) + 0.0 = 0$$

$$T_{34} + 4T_{35} + T_{36} + 4T_{44} - 20T_{45} + 4T_{46} + 0.0 + 4(0.0) + 0.0 = 0$$

$$T_{35} + 4T_{36} + 100.0 + 4T_{45} - 20T_{46} + 70.710678 + 0.0 + 4(0.0) + 0.0 = 0$$

$$(9.73)$$

where the subscripts i, j have been written as ij for simplicity. The boundary point values of temperature are transferred to the right-hand sides of the equations before solving them. Equation (9.73) consists of 15 FDEs. Writing Eq. (9.73) in matrix form yields

$$
\begin{bmatrix}
-20 & 4 & 0 & 0 & 0 & 4 & 1 & & & & & & & & \\
4 & -20 & 4 & 0 & 0 & 1 & 4 & 1 & & & & & & & \\
0 & 4 & -20 & 4 & 0 & 0 & 1 & 4 & 1 & & & & & & \\
0 & 0 & 4 & -20 & 4 & 0 & 0 & 1 & 4 & 1 & & & & & \\
0 & 0 & 0 & 4 & -20 & 0 & 0 & 0 & 1 & 4 & & & & & \\
4 & 1 & 0 & 0 & 0 & -20 & 4 & 0 & 0 & 0 & 4 & 1 & & & \\
1 & 4 & 1 & 0 & 0 & 4 & -20 & 4 & 0 & 0 & 1 & 4 & 1 & & \\
 & 1 & 4 & 1 & 0 & 0 & 4 & -20 & 4 & 0 & 0 & 1 & 4 & 1 & \\
 & & 1 & 4 & 1 & 0 & 0 & 4 & -20 & 4 & 0 & 0 & 1 & 4 & 1 \\
 & & & 1 & 4 & 0 & 0 & 0 & 4 & -20 & 0 & 0 & 0 & 1 & 4 \\
 & & & & & 4 & 1 & 0 & 0 & 0 & -20 & 4 & 0 & 0 & 0 \\
 & & & & & 1 & 4 & 1 & 0 & 0 & 4 & -20 & 4 & 0 & 0 \\
 & & & & & & 1 & 4 & 1 & 0 & 0 & 4 & -20 & 4 & 0 \\
 & & & & & & & 1 & 4 & 1 & 0 & 0 & 4 & -20 & 1 \\
 & & & & & & & & 1 & 4 & 0 & 0 & 0 & 4 & -20
\end{bmatrix}
$$

$$
\times
\begin{bmatrix}
T_{22} \\
T_{23} \\
T_{24} \\
T_{25} \\
T_{26} \\
T_{32} \\
T_{33} \\
T_{34} \\
T_{35} \\
T_{36} \\
T_{42} \\
T_{43} \\
T_{44} \\
T_{45} \\
T_{46}
\end{bmatrix}
=
\begin{bmatrix}
0 \\
0 \\
0 \\
0 \\
382.842712 \\
0 \\
0 \\
0 \\
0 \\
753.553390 \\
0 \\
0 \\
0 \\
0 \\
170.710678
\end{bmatrix}
\qquad (9.74)
$$

All the terms not shown in the lower-left and upper-right portions of matrix **A** are zero. Equation (9.74) is the system equation corresponding to Eq. (9.28) for the five-point method. Solving Eq. (9.74) by Gauss elimination yields the results tabulated in Table 9.9 which also presents the results for a 9×13 grid. The results are an excellent approximation of the exact solution. Comparing these results with the results of the second-order five-point method presented in Table 9.2 shows that the errors of the compact fourth-order method are approximately two orders of magnitude smaller than the errors of the second-

Table 9.9 Solution of the Heat Diffusion Problem by the Compact Fourth-Order Method

	$T(x, y)$, C			
	$\text{Error}(x, y) = [T(x, y) - \bar{T}(x, y)]$, C			
	$\Delta x = \Delta y = 2.5\,\text{cm}$, 5×7 grid		$\Delta x = \Delta y = 1.25\,\text{cm}$, 9×13 grid	
y, cm	$x = 2.5\,\text{cm}$	$x = 5.0\,\text{cm}$	$x = 2.5\,\text{cm}$	$x = 5.0\,\text{cm}$
12.5	32.230760	45.581178	32.229792	45.579809
	0.000980	−0.001387	0.000012	−0.000018
10.0	14.673929	20.752069	14.673049	20.750825
	0.000889	−0.001257	0.000009	−0.000013
7.5	6.642901	9.394481	6.642308	9.393644
	0.000597	0.000844	0.000004	0.000007
5.0	2.924214	4.135463	2.923875	4.134984
	0.000341	0.000482	0.000002	0.000003
2.5	1.103825	1.561044	1.103674	1.560831
	0.000151	0.000213	0.000000	0.000000

order five-point method. The solution obtained by the compact fourth-order method is sufficiently accurate for most engineering applications.

9.9.2 Extrapolation

The concept of extrapolation can be used to extrapolate the results of the second-order five-point method to fourth-order by calculating the solutions for a 5×7 grid and a 9×13 grid and applying the extrapolation formula at the common grid points. The extrapolation formula is:

$$\text{IV} = \text{MAV} + \frac{1}{2^n - 1}(\text{MAV} - \text{LAV}) \qquad (9.75)$$

where IV is the improved (i.e., extrapolated) value, MAV is the more accurate value (i.e., the 9×13 grid results), LAV is the less accurate value (i.e., the 5×7 grid results), and n is the order of the method. For the second-order five-point method, $n = 2$, and Eq. (9.75) becomes

$$\text{IV} = \text{MAV} + \frac{1}{2^2 - 1}(\text{MAV} - \text{LAV}) = \frac{4\,\text{MAV} - \text{LAV}}{3} \qquad (9.76)$$

Example 9.8. Solution of the heat diffusion problem by extrapolation

The solution of the heat diffusion problem presented in Section 9.1 by the second-order five-point method is presented in Section 9.4 for a 5×7 grid and a 9×7 grid in Table 9.2. Those results for $x = 2.5\,\text{cm}$ and $x = 5.0\,\text{cm}$, at $y = 12.5$, 10.0, 7.5, 5.0, and 2.5 cm, and the extrapolated values obtained using Eq. (9.76) are presented in Table 9.10. The results obtained by extrapolation are not as accurate as the results obtained by the compact fourth-

Table 9.10 Solution of the Heat Diffusion Problem by Extrapolation

	$T(x, y)$, C					
	Error$(x, y) = [T(x, y) - \bar{T}(x, y)]$, C					
	Second-order five-point method					
	5 × 7 grid		9 × 13 grid		Extrapolation	
y, cm	$x = 2.5$ cm	$x = 5.0$ cm	$x = 2.5$ cm	$x = 5.0$ cm	$x = 2.5$ cm	$x = 5.0$ cm
12.5	33.459590	47.319006	32.549586	46.032067	32.246251	45.603087
	1.229810	1.739215	0.319806	0.452276	0.016471	0.023296
10.0	15.808676	22.356844	14.964426	21.162895	14.683009	20.764912
	1.135636	1.606032	0.291386	0.412083	0.009969	0.014100
7.5	7.418270	10.491010	6.838895	9.671658	6.645770	9.398538
	0.775966	1.097382	0.196591	0.278021	0.003466	0.004901
5.0	3.373387	4.770690	3.036510	4.294274	2.924218	4.135469
	0.449514	0.635709	0.112637	0.159293	0.000345	0.000488
2.5	1.304588	1.844967	1.153627	1.631425	1.103307	1.560311
	0.200914	0.284136	0.049953	0.070644	−0.000367	−0.000520

order method on a 5 × 7 grid presented in Table 9.9. However, the results are in excellent agreement with the exact solution.

Extrapolation can be applied to successively higher-order results to obtain even higher-order results. For example, the heat diffusion problem can be solved a third time by the five-point method on a 17 × 33 grid. The results obtained on the 9 × 13 and 17 × 33 grids can be extrapolated to fourth order as done in Example 9.8 for the results obtained on the 5 × 7 and 9 × 13 grids. Both sets of extrapolated results are fourth order, that is, $n = 4$. Equation (9.75) can be applied with $n = 4$ to these two sets of fourth-order results to obtain sixth-order results. The extrapolation formula is given by

$$ \text{IV} = \text{MAV} + \frac{1}{2^4 - 1}(\text{MAV} - \text{LAV}) = \frac{16\,\text{MAV} - \text{LAV}}{15} \qquad (9.77) $$

However, sixth-order results are obtained only at the common points in the three grids, that is, at the points corresponding to the 5 × 7 grid.

9.10 NONRECTANGULAR DOMAINS

The methods presented so far have all been concerned with rectangular physical spaces and rectangular finite difference grids. Several significant simplifications result in this case:

1. Grid points of the finite difference grid fall on the boundary of the physical space, so boundary conditions can be specified.
2. The computational grid is uniform and orthogonal, so accurate finite difference approximations of exact partial derivatives can be derived.
3. The grid spacing adjacent to the boundaries is uniform and orthogonal.

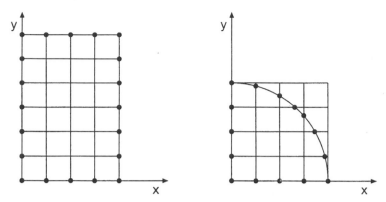

Figure 9.17 Rectangular and nonrectangular physical spaces. (a) Rectangular physical space. (b) Quarter-round physical space.

Figure 9.17a illustrates this situation. When the physical space is not rectangular, however, problems arise. Consider the quarter-round physical space illustrated in Figure 9.17b, which is discretized by a rectangular finite difference grid. Except for rare points, grid points do not fall on the curved boundary of the physical space, thus making it impossible to specify boundary conditions. The finite difference grid is not uniform at interior points adjacent to the curved boundary. Obviously, some new finite difference approach is required.

There are several approaches available for modeling nonrectangular physical spaces:

1. Approximate physical boundary
2. Other coordinate systems
3. Nonuniform finite difference approximations
4. Transformed spaces

Approximate physical boundaries can be developed by specifying a very dense finite difference grid (i.e., a large number of grid points) and letting the grid points closest to the physical boundary represent the physical boundary. This approach is not recommended. When the physical space has cylindrical or spherical symmetry, the governing partial differential equations can be expressed in *cylindrical* or *spherical coordinates*, respectively, and a uniform orthogonal cylindrical or spherical finite difference grid can be used. Alternately, a rectangular grid can be imposed on the nonrectangular physical space, as illustrated in Figure 9.17b, the boundary conditions can be imposed at the points where the grid lines intersect the physical boundary, and *nonuniform finite difference approximations* of the individual exact partial derivatives in the partial differential equation (PDE) can be applied at the interior points adjacent to the physical boundary. This approach is developed in this section. The final approach listed above, *transformed spaces*, is in general the preferred approach. That approach is described, but not developed, in Section 9.10.2.

9.10.1 Nonuniform Finite Difference Approximations

Figure 9.18 presents the local finite difference grid for a nonuniform finite difference grid. Let's develop nonuniform finite difference approximations of $\bar{f}_{xx}$ and $\bar{f}_{yy}$ for this nonuni-

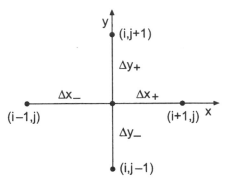

Figure 9.18 Finite difference grid for a nonuniform grid.

form grid. Let the grid increments be denoted by $\Delta x_\pm$ and $\Delta y_\pm$. First consider the finite difference approximation of $\bar{f}_{xx}|_{i,j}$. Writing Taylor series for $\bar{f}_{i+1,j}$ and $\bar{f}_{i-1,j}$ gives

$$\bar{f}_{i+1,j} = \bar{f}_{i,j} + \bar{f}_x|_{i,j}\,\Delta x_+ + \tfrac{1}{2}\bar{f}_{xx}|_{i,j}\,\Delta x_+^2 + \tfrac{1}{6}\bar{f}_{xxx}|_{i,j}\,\Delta x_+^3 + \cdots \qquad (9.78)$$

$$\bar{f}_{i-1,j} = \bar{f}_{i,j} - \bar{f}_x|_{i,j}\,\Delta x_- + \tfrac{1}{2}\bar{f}_{xx}|_{i,j}\,\Delta x_-^2 - \tfrac{1}{6}\bar{f}_{xxx}|_{i,j}\,\Delta x_-^3 + \cdots \qquad (9.79)$$

Multiplying Eq. (9.78) by Δx_- and Eq. (9.79) by Δx_+ and adding the results gives

$$\Delta x_-\,\bar{f}_{i+1,j} + \Delta x_+\,\bar{f}_{i-1,j} = (\Delta x_+ + \Delta x_-)\bar{f}_{i,j} + \tfrac{1}{2}(\Delta x_-\,\Delta x_+^2 + \Delta x_+\,\Delta x_-^2)\bar{f}_{xx}|_{i,j}$$
$$+ \tfrac{1}{6}(\Delta x_-\,\Delta x_+^3 - \Delta x_+\,\Delta x_-^3)\bar{f}_{xxx}|_{i,j} + \cdots \qquad (9.80)$$

Solving Eq. (9.80) for $\bar{f}_{xx}|_{i,j}$ gives

$$\bar{f}_{xx}|_{i,j} = \frac{2\,\Delta x_+}{(\Delta x_-\,\Delta x_+^2 + \Delta x_+\,\Delta x_-^2)}\bar{f}_{i-1,j} - \frac{2(\Delta x_- + \Delta x_+)}{(\Delta x_-\,\Delta x_+^2 + \Delta x_+\,\Delta x_-^2)}\bar{f}_{i,j}$$
$$+ \frac{2\,\Delta x_-}{(\Delta x_-\,\Delta x_+^2 + \Delta x_+\,\Delta x_-^2)}\bar{f}_{i+1,j} - \tfrac{1}{3}(\Delta x_+ - \Delta x_-)\bar{f}_{xxx}(\xi) \qquad (9.81)$$

where $x_{i-1} \le \xi \le x_{i+1}$. Truncating the remainder term yields the first-order centered-space approximation of $\bar{f}_{xx}|_{i,j}$, denoted by $f_{xx}|_{i,j}$:

$$\boxed{\begin{aligned} f_{xx}|_{i,j} &= \frac{2\,\Delta x_+}{(\Delta x_-\,\Delta x_+^2 + \Delta x_+\,\Delta x_-^2)}f_{i-1,j} - \frac{2(\Delta x_- + \Delta x_+)}{(\Delta x_-\,\Delta x_+^2 + \Delta x_+\,\Delta x_-^2)}f_{i,j} \\ &\quad + \frac{2\Delta x_-}{(\Delta x_-\,\Delta x_+^2 + \Delta x_+\,\Delta x_-^2)}f_{i+1,j} \end{aligned}} \qquad (9.82)$$

Repeating these steps for $\bar{f}_{yy}|_{i,j}$ yields

$$\boxed{\begin{aligned} f_{yy}|_{i,j} &= \frac{2\,\Delta y_+}{(\Delta y_-\,\Delta y_+^2 + \Delta y_+\,\Delta y_-^2)}f_{i,j-1} - \frac{2(\Delta y_- + \Delta y_+)}{(\Delta y_-\,\Delta y_+^2 + \Delta y_+\,\Delta y_-^2)}f_{i,j} \\ &\quad + \frac{2\Delta y_-}{(\Delta y_-\,\Delta y_+^2 + \Delta y_+\,\Delta y_-^2)}f_{i,j-1} \end{aligned}} \qquad (9.83)$$

Equations (9.82) and (9.83) are formally first-order approximations. However, as shown in Eq. (9.81), the leading truncation error term contains the difference of Δx_+ and Δx_-, which is significantly smaller than Δx_+ or Δx_- themselves. Consequently, while not formally second order, Eqs. (9.82) and (9.83) generally yields results which approach second order.

Example 9.9.   Solution of the heat diffusion problem with internal energy generation for a nonrectangular domain

---

As an example of a problem with a nonrectangular boundary, let's consider a circular electrical conductor with internal energy generation due to electrical resistance heating. As discussed in Section 9.8, this problem is governed by the Poisson equation:

$$T_{xx} + T_{yy} + \frac{\dot{Q}}{k} = 0 \tag{9.84}$$

For a circular conductor, Eq. (9.84), expressed in cylindrical coordinates, is

$$\frac{1}{r}\frac{\partial}{\partial r}(rT_r) + \frac{\dot{Q}}{k} = 0 \tag{9.85}$$

Let's assume that the internal energy generation is specified by the relationship

$$\dot{Q} = A\left(1.0 - 0.9\frac{r}{R}\right) \tag{9.86}$$

where $A = 400\,\text{J/cm}^3\text{-s}$ and the radius of the conductor is $R = 1.0\,\text{cm}$. The thermal conductivity of the conductor is $k = 0.4\,\text{J/cm-s-C}$. The exact solution of Eq. (9.85) is given by

$$T(r) = \frac{A}{k}\left[\frac{(r^2 - 1)}{4} - 0.9\frac{(r^3 - 1)}{9}\right] \tag{9.87}$$

For the specified values of $\dot{Q}$ and k, Eq. (9.87) gives

$$\boxed{T(r) = 250(1 - r^2) - 100(1 - r^3)} \tag{9.88}$$

Equation (9.84), expressed in Cartesian coordinates, is

$$T_{xx} + T_{yy} + \frac{\dot{Q}}{k} = 0 \tag{9.89}$$

Superimposing a rectangular finite difference grid on the quarter-round physical space leads to the problems discussed relative to Figure 9.17b. This problem obviously should be solved using Eq. (9.85). However, let's solve Eq. (9.89) to illustrate the application of a nonuniform finite difference approximation on a nonrectangular grid (i.e., a circular boundary in a rectangular space). Substituting Eqs. (9.82) and (9.83) into Eq. (9.89) yields the desired FDE:

$$A_{i,j}^+ T_{i+1,j} - 2A_{i,j}^0 T_{i,j} + A_{i,j}^- T_{i-1,j} + B_{i,j}^+ T_{i,j+1} - 2B_{i,j}^0 T_{i,j} + B_{i,j}^- T_{i,j-1} + \frac{\dot{Q}}{k} = 0 \tag{9.90}$$

where the coefficients, $A_{i,j}^+$, etc., are defined as the corresponding coefficients in Eqs. (9.82) and (9.83).

Let $\Delta x = \Delta y = 0.25$ cm. The resulting solution domain and finite difference grid are illustrated in Figure 9.19. The values of Δx and Δy between the grid points adjacent to the circular boundary are noted on the figure. For the present problem, $\dot{Q}/k$ is given by

$$\frac{\dot{Q}}{k} = \frac{400}{0.4}\left(1.0 - \frac{0.9r}{R}\right) = 1000\left(1.0 - \frac{0.9r}{R}\right) \tag{9.91}$$

Due to symmetry about the x and y axes, Eq. (9.90) can be applied there by setting $T_{i-1,j} = T_{i+1,j}$ along the y axis and $T_{i,j-1} = T_{i,j+1}$ along the x axis. Applying Eq. (9.90) at the 15 unknown points in Figure 9.19 yields the following system of FDEs:

$$
\begin{aligned}
32.0000T_{21} + 32.0000T_{12} - 64.0000T_{11} &= -1000.0000 \\
32.0000T_{22} + 16.0000T_{13} + 16.0000T_{11} - 64.0000T_{12} &= -775.0000 \\
32.0000T_{23} + 16.0000T_{14} + 16.0000T_{12} - 64.0000T_{13} &= -550.0000 \\
32.0000T_{24} + 16.0000(0) + 16.0000T_{13} - 64.0000T_{14} &= -325.0000 \\
16.0000T_{31} + 32.0000T_{22} + 16.0000T_{11} - 64.0000T_{21} &= -775.0000 \\
16.0000T_{32} + 16.0000T_{23} + 16.0000T_{12} + 16.0000T_{21} - 64.0000T_{22} &= -681.8019 \\
16.0000T_{33} + 16.0000T_{24} + 16.0000T_{13} + 16.0000T_{22} - 64.0000T_{23} &= -496.8847 \\
16.0000T_{34} + 19.5709(0) + 16.0000T_{14} + 17.0850T_{23} - 68.6559T_{24} &= -288.4875 \\
16.0000T_{41} + 32.0000T_{32} + 16.0000T_{21} - 64.0000T_{31} &= -550.0000 \\
16.0000T_{42} + 16.0000T_{33} + 16.0000T_{22} + 16.0000T_{31} - 64.0000T_{32} &= -496.8847 \\
16.0000T_{43} + 16.0000T_{34} + 16.0000T_{23} + 16.0000T_{23} - 64.0000T_{33} &= -363.6039 \\
30.1107(0) + 47.0940(0) + 19.4440T_{24} + 21.8564T_{33} - 118.5051T_{34} &= -188.7510 \\
16.0000(0) + 32.0000T_{42} + 16.0000T_{31} - 64.0000T_{41} &= -325.0000 \\
19.5709(0) + 16.0000T_{43} + 17.0850T_{32} + 16.0000T_{41} - 68.6559T_{42} &= -288.4875 \\
47.0940(0) + 30.1107(0) + 21.3564T_{33} + 19.4440T_{42} - 118.5051T_{43} &= -188.7510
\end{aligned}
$$

$$\tag{9.92}$$

Equation (9.92) is the system equation for this problem. Solving Eq. (9.92) by the SOR method with an absolute convergence tolerance of 0.000001 yields the results presented in Table 9.11.

Repeating the solution with $\Delta x = \Delta y = 0.125$ cm yields the results presented in Table 9.12, which presents the solution only at the common grid points of the two grids. Comparing Table 9.11 with Table 9.9 and Table 9.12 with Table 9.10 shows that the errors for the nonrectangular domain are comparable to the errors for the rectangular domain.

Comparing the errors in Tables 9.11 and 9.12 shows that the method is behaving second order.

9.10.2 Transformed Grids

The governing differential equations of engineering and science are generally derived and expressed in a Cartesian (i.e., rectangular) coordinate system. All the examples considered in Chapters 9 to 11 are expressed in Cartesian coordinates.

Finite difference methods for solving differential equations require that the continuous physical space be discretized into a uniform orthogonal computational space. The

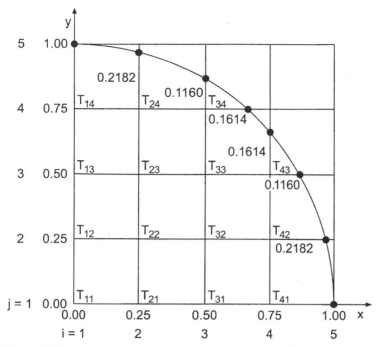

Figure 9.19 Finite difference grid for the quarter-round space.

Table 9.11 Solution of the Heat Diffusion Problem for a Nonrectangular Domain on a 5×5 Grid

	$T(x, y)$, C $\bar{T}(x, y)$, C $\text{Error}(x, y) = [T(x, y) - \bar{T}(x, y)]$, C			
y, cm	$x = 0.0$ cm	$x = 0.25$ cm	$x = 0.50$ cm	$x = 0.75$ cm
0.75	52.3761	43.6732	20.0274	
	51.5625	43.1606	20.1128	
	0.8136	0.5126	−0.0854	
0.50	101.8453	90.8093	61.0997	20.0274
	100.0000	89.3443	60.3553	20.1128
	1.8453	1.4650	0.7444	0.0854
0.25	139.0117	125.5637	90.8093	43.6732
	135.9375	123.1694	89.3443	43.1606
	3.0742	2.3943	1.4650	0.5126
0.00	154.6367	139.0117	101.8453	52.3761
	150.0000	135.9375	100.0000	51.5625
	4.6367	3.0742	1.8453	0.8136

Table 9.12 Solution of the Heat Diffusion Problem for a Nonrectangular Domain on a 9 × 9 Grid

y, cm	x = 0.0 cm	x = 0.25 cm	x = 0.50 cm	x = 0.75 cm
	$T(x,y)$, C
	$\text{Error}(x,y) = [T(x,y) - \bar{T}(x,y)]$, C
0.75	51.7754	43.3193	20.1659	
	0.2129	0.1587	0.0531	
0.50	100.4552	89.7267	60.5859	20.1659
	0.4552	0.3824	0.2306	0.0531
0.25	136.6561	123.7572	89.7267	43.3193
	0.7186	0.5878	0.3824	0.1587
0.00	151.0689	136.6561	100.4552	51.7754
	1.0689	0.7186	0.4552	0.2129

application of boundary conditions requires that the boundaries of the physical space fall on coordinate lines of the coordinate system. Accurate resolution of the solution requires that grid points be clustered in regions of large gradients. Economy requires that grid points be spread out in regions of small gradients. These requirements are generally incompatible with a Cartesian coordinate system.

As an example, consider the physical space illustrated in Figure 9.20, which is bounded by the four boundaries $x = X_1$, $x = X_2$, $y = 0$, and $y = Y(x)$, in which the function $f(x,y)$ is governed by a partial differential equation. Assume that the x gradient (i.e., f_x) is much larger near $x = 0$ than at any other location. Superimposing a uniform orthogonal Cartesian grid on the physical space, as illustrated in Figure 9.20, leads to the following problems:

1. The upper boundary of the physical space [i.e., the $y = Y(x)$ boundary] does not fall on a coordinate line (i.e., a line of constant y), so the application of boundary conditions is difficult.
2. The grid spacings adjacent to the upper boundary are not uniform, so centered-space finite difference approximations (FDAs) are not second-order.
3. In view of the large values of f_x near $x = 0$, the uniform grid spacing in the x direction (i.e., Δx) is either too large near $x = 0$ if Δx is chosen based on the smaller values of f_x near $x = X_1$ and X_2, or too small away from $x = 0$ if Δx is chosen based on the larger values of f_x near $x = 0$.

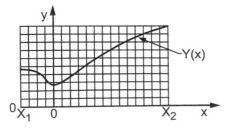

Figure 9.20 Physical space and a Cartesian grid.

The first two problems listed above can be eliminated by using the body-fitted clustered grid illustrated in Figure 9.21a. However, the grid spacings Δx and Δy are both nonuniform everywhere, and the grid is not orthogonal. This problem can be eliminated by transforming the nonuniform nonorthogonal physical space illustrated in Figure 9.21a into the uniform orthogonal computational space illustrated in Figure 9.21b.

The transformation relating the physical space xy and the computational space $\xi\eta$ is specified by the *direct transformation*:

$$\xi = \xi(x, y) \quad \text{and} \quad \eta = \eta(x, y) \tag{9.93}$$

The transformation from computational space $\xi\eta$ to physical space xy is specified by the *inverse transformation*:

$$x = x(\xi, \eta) \quad \text{and} \quad y = y(\xi, \eta) \tag{9.94}$$

The determination of the coordinate transformation is called *grid generation*.

Once the coordinate transformation has been determined, the differential equations must be transformed from physical space xy to computational space $\xi\eta$. For example, consider the first-order PDE:

$$a\frac{\partial f}{\partial x} + b\frac{\partial f}{\partial y} = c \tag{9.95}$$

Equation (9.95) is transformed from physical space xy to computational space $\xi\eta$ by applying the chain rule for partial derivatives. Thus,

$$\frac{\partial f}{\partial x} = \frac{\partial f}{\partial \xi}\frac{\partial \xi}{\partial x} + \frac{\partial f}{\partial \eta}\frac{\partial \eta}{\partial x} = \xi_x\frac{\partial f}{\partial \xi} + \eta_x\frac{\partial f}{\partial \eta} \tag{9.96}$$

$$\frac{\partial f}{\partial y} = \frac{\partial f}{\partial \xi}\frac{\partial \xi}{\partial y} + \frac{\partial f}{\partial \eta}\frac{\partial \eta}{\partial y} = \xi_y\frac{\partial f}{\partial \xi} + \eta_y\frac{\partial f}{\partial \eta} \tag{9.97}$$

where the derivatives ξ_x, ξ_y, η_x, and η_y are the *metrics* of the direct transformation. Substituting Eqs. (9.96) and (9.97) into Eq. (9.95) yields the transformed PDE. Thus,

$$(a\xi_x + b\xi_y)\frac{\partial f}{\partial \xi} + (a\eta_x + b\eta_y)\frac{\partial f}{\partial \eta} = c \tag{9.98}$$

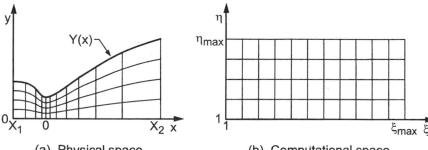

(a) Physical space.   (b) Computational space.

Figure 9.21 Body-fitted coordinate system.

Equation (9.98) is solved in the uniform orthogonal computational space, $\xi\eta$, using second-order centered-difference finite difference approximations (FDAs).

The advantages of transforming the differential equations to a body-fitted uniform orthogonal computational space are:

1. The boundaries of the physical space fall on coordinate lines of the computational space, so boundary conditions can be implemented accurately and easily.
2. The finite difference approximations of the exact partial derivatives are obtained on a uniform orthogonal grid.
3. Grid points can be clustered in regions of large gradients and spread out in regions of small gradients.

The most significant disadvantage is that the transformed PDEs are more complicated, so the resulting finite difference equations (FDEs) are also more complicated.

The book by Thompson et al. (1985) presents a thorough discussion of the transformed space concept.

9.11 NONLINEAR EQUATIONS AND THREE-DIMENSIONAL PROBLEMS

The partial differential equations considered so far in this chapter are linear (i.e., the Laplace equation and the Poisson equation). Consequently, the corresponding finite difference equations are linear. All the examples considered so far are for two-dimensional problems. Some of the problems that arise for nonlinear partial differential equations and for three-dimensional problems are discussed briefly in this section.

9.11.1 Nonlinear Equations

When a nonlinear elliptic partial differential equation is solved by finite difference methods, a system of nonlinear finite difference equations results. Several approaches can be used to solve the system of nonlinear equations.

Two procedures are presented in Section 8.7 for solving the systems of nonlinear finite difference equations that arise when one-dimensional boundary-value problems are solved by the equilibrium method:

1. The iteration method
2. Newton's method

In the *iteration method*, the finite difference equations are linearized, and the system of linearized finite difference equations is solved by an iterative technique, such as SOR. This approach can be extended to solve nonlinear elliptic partial differential equations in two or three space dimensions. This approach involves a two-step procedure. Step 1 involves the evaluation of the nonlinear coefficients in the system of finite difference equations, based on the current estimate of the solution. These coefficients are updated periodically, typically by under-relaxation. Step 2 involves the solution of the system of linearized FDEs by an iterative technique. Steps 1 and 2 are repeated until the system of nonlinear FDEs converges to the desired tolerance. It is not necessary to solve the linearized equations to a small tolerance during the early stages of the overall two-step procedure. As the two-step procedure approaches convergence, the tolerance of the linear equation solver should be decreased towards the desired final tolerance.

In *Newton's method*, the solution $f(x, y)$ is assumed to be the solution of a trial function $F(x, y)$ and a small perturbation $\eta(x, y)$. Thus,

$$f(x, y) = F(x, y) + \eta(x, y) \tag{9.99}$$

This expression is substituted into the nonlinear PDE, and nonlinear terms involving η and its derivatives are neglected. A linear elliptic PDE for $\eta(x, y)$ is obtained, which has $\eta(x, y) = 0.0$ on the boundaries. This linear elliptic PDE can be solved by the method presented in Section 9.4 or 9.6 for $\eta(x, y)$. Adding this solution for $\eta(x, y)$ to $F(x, y)$ yields an improved approximation for $f(x, y)$. This procedure is repeated to convergence.

The multigrid method discussed by Brandt (1977) can be applied directly to solve nonlinear elliptic partial differential equations. Unfortunately, this method is beyond the scope of this text. Any serious effort to solve nonlinear elliptic PDEs should consider the multigrid method very seriously. It works equally well for linear and nonlinear PDEs and for one-, two-, or three-space dimensions. The book by Hackbusch (1980) presents a comprehensive discussion of the multigrid method.

9.11.2 Three-Dimensional Problems

Three-dimensional problems can be solved by the same methods that are used to solve two-dimensional problems by including the finite difference approximations of the exact partial derivatives in the third direction. The major complication is that the size of the system of FDEs increases dramatically. The efficient solution of nonlinear three-dimensional elliptic PDEs is beyond the scope of this book. The multigrid method can be applied to three-dimensional problems. It is probably the most efficient procedure for solving three-dimensional elliptic PDEs, both linear and nonlinear.

9.12 THE CONTROL VOLUME METHOD

All the methods presented so far in Chapter 9 are based on the finite difference approach. The control volume approach is introduced in this section. The control volume approach is especially useful for problems where interfaces exist between regions having different physical properties. The control volume approach is based on flux balances on a finite control volume instead of the governing partial differential equations.

To illustrate the control volume approach, let's apply it to solve the heat diffusion problem presented in Section 9.1. The continuous physical domain $D(x, y)$ is first discretized into a discrete finite difference grid, as illustrated in Figure 9.5. Four cells surround grid point (i, j). A finite size control volume is drawn around each grid point, as illustrated in Figure 9.22 by the dashed lines. For simplicity, the grid points are denoted by the single subscripts 0 to 8. Two heat fluxes $\dot{q}$ cross the boundaries of the control volume in each cell, and a total of eight heat fluxes cross the boundary of the control volume. At steady state, the net heat flux into the control volume is zero. Thus,

$$\boxed{\dot{q}_{15} + \dot{q}_{18} + \dot{q}_{25} + \dot{q}_{26} - \dot{q}_{36} - \dot{q}_{37} - \dot{q}_{47} - \dot{q}_{48} = 0} \tag{9.100}$$

The heat flux $\dot{q}$ is specified by *Fourier's law of conduction*:

$$\dot{q} = -kA\frac{\partial T}{\partial n} \tag{9.101}$$

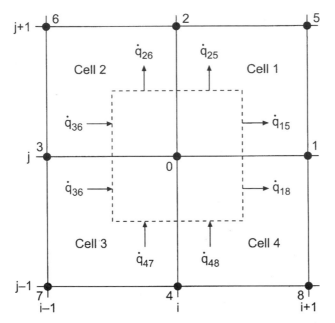

Figure 9.22 Control volume discretization.

where $\partial T/\partial n$ denotes the temperature gradient in the direction normal to a face of the control volume, k is the thermal conductivity of the substance, A is the area across which heat flows, and the negative sign arises because heat flows in the direction opposite to the sign of the temperature gradient. Note that the thermal conductivity k can vary from cell to cell in this approach. Expressions must be derived for the eight heat fluxes in terms of the temperatures at the nine grid points.

The eight heat fluxes depend on the temperature gradients at the midpoints of the eight segments of the control volume boundary illustrated in Figure 9.22. Thus, the temperature gradients at these eight points must be approximated. That is accomplished by developing an interpolating polynomial for each cell and differentiating the interpolating polynomials to determine $\partial T/\partial n$ at the midpoints of the eight segments.

Consider cell 1 in Figure 9.22. Assume an interpolating polynomial of the following form:

$$T(x, y) = a + bx + cy + dxy \tag{9.102}$$

Let the origin of the local coordinate system be at grid point 0. Thus, point 1 is at $(\Delta x, 0)$, point 2 is at $(0, \Delta x)$, and point 5 is at $(\Delta x, \Delta y)$. Substituting the four values of T at points 0, 1, 2, and 5 into Eq. (9.102) gives

$$T_0 = a \tag{9.103a}$$
$$T_1 = a + b\,\Delta x \tag{9.103b}$$
$$T_2 = a + c\,\Delta y \tag{9.103c}$$
$$T_5 = a + b\,\Delta x + c\,\Delta y + d\,\Delta x\,\Delta y \tag{9.103d}$$

Solving Eq. (9.103) by Gauss elimination gives

$$T(x, y) = T_0 + \left(\frac{T_1 - T_0}{\Delta x}\right)x + \left(\frac{T_2 - T_0}{\Delta y}\right)y + \left(\frac{T_0 + T_5 - T_1 - T_2}{\Delta x \, \Delta y}\right)xy \qquad (9.104)$$

Heat flux $\dot{q}_{15}$ is evaluated by differentiating Eq. (9.104) with respect to x and evaluating the result at $(\Delta x/2, \Delta y/4)$. Thus,

$$\frac{\partial T(x, y)}{\partial x} = \left(\frac{T_1 - T_0}{\Delta x}\right) + \left(\frac{T_0 + T_5 - T_1 - T_2}{\Delta x \, \Delta y}\right)y \qquad (9.105)$$

$$\frac{\partial T(\Delta x/2, \Delta y/4)}{\partial y} = \left(\frac{T_1 - T_0}{\Delta x}\right) + \left(\frac{T_0 + T_5 - T_1 - T_2}{\Delta x \, \Delta y}\right)\frac{\Delta y}{4} = \frac{3(T_1 - T_0) + T_5 - T_2}{4 \, \Delta x} \qquad (9.106)$$

Substituting Eq. (9.106) into Eq. (9.101), where $A = \Delta y/2$, gives

$$\dot{q}_{15} = -\left(\frac{k \, \Delta y}{8 \, \Delta x}\right)[3(T_1 - T_0) + T_5 - T_2] \qquad (9.107)$$

In a similar manner,

$$\frac{\partial T(x, y)}{\partial y} = \left(\frac{T_2 - T_0}{\Delta y}\right) + \left(\frac{T_0 + T_5 - T_1 - T_2}{\Delta x \, \Delta y}\right)x \qquad (9.108)$$

$$\frac{\partial T(\Delta x/4, \Delta y/2)}{\partial y} = \left(\frac{T_2 - T_0}{\Delta y}\right) + \left(\frac{T_0 + T_5 - T_1 - T_2}{\Delta x \, \Delta y}\right)\frac{\Delta x}{4} = \frac{3(T_2 - T_0) + T_5 - T_1}{4 \, \Delta y} \qquad (9.109)$$

$$\dot{q}_{25} = -\left(\frac{k \, \Delta x}{8 \, \Delta y}\right)[3(T_2 - T_0) + T_5 - T_1] \qquad (9.110)$$

Equations (9.107) and (9.110) specify the two heat fluxes in cell 1. Applying the same procedure to cells 2, 3, and 4 yields the other six heat fluxes at the surface of the control volume. Thus,

$$\dot{q}_{26} = -\left(\frac{k \, \Delta x}{8 \, \Delta y}\right)[3(T_2 - T_0) + T_6 - T_3] \qquad (9.111)$$

$$\dot{q}_{36} = -\left(\frac{k \, \Delta y}{8 \, \Delta x}\right)[3(T_0 - T_3) + T_2 - T_6] \qquad (9.112)$$

$$\dot{q}_{37} = -\left(\frac{k \, \Delta y}{8 \, \Delta x}\right)[3(T_0 - T_3) + T_4 - T_7] \qquad (9.113)$$

$$\dot{q}_{47} = -\left(\frac{k \, \Delta x}{8 \, \Delta y}\right)[3(T_0 - T_4) + T_3 - T_7] \qquad (9.114)$$

$$\dot{q}_{48} = -\left(\frac{k \, \Delta x}{8 \, \Delta y}\right)[3(T_0 - T_4) + T_1 - T_8] \qquad (9.115)$$

$$\dot{q}_{18} = -\left(\frac{k \, \Delta y}{8 \, \Delta x}\right)[3(T_1 - T_0) + T_8 - T_4] \qquad (9.116)$$

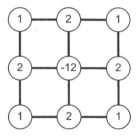

Figure 9.23 Computational stencil for the control volume method.

Substituting Eqs. (9.107), (9.110), and (9.111) to (9.116) into Eq. (9.100), collecting terms, and simplifying yields the control volume approximation of the heat diffusion equation:

$$2(3 - \beta^2)(T_1 + T_3) + 2(3\beta^2 - 1)(T_2 + T_4) + (\beta^2 + 1) \\ \times (T_5 + T_6 + T_7 + T_8) - 12(\beta^2 + 1)T_0 = 0 \tag{9.117}$$

where $\beta = \Delta x/\Delta y$ is the grid aspect ratio. For unity grid aspect ratio (i.e., $\beta = 1$), Eq. (9.117) becomes

$$2(T_1 + T_2 + T_3 + T_4) + (T_5 + T_6 + T_7 + T_8) - 12T_0 = 0 \tag{9.118}$$

The computational stencil corresponding to Eq. (9.118) is illustrated in Figure 9.23.

Example 9.10.   Solution of the heat diffusion problem by the control volume method

Table 9.13 Solution of the Heat Diffusion Problem by the Control Volume Method

| | $T(x, y)$, C $\quad\quad$ Error$(x, y) = [T(x, y) - \bar{T}(x, y)]$, C | | | |
| | $\Delta x = \Delta y = 2.5$ cm, 5×7 grid | | $\Delta x = \Delta y = 1.25$ cm, 9×13 grid | |
y, cm	$x = 2.5$ cm	$x = 5.0$ cm	$x = 2.5$ cm	$x = 5.0$ cm
12.5	31.561725	44.635020	3.062060	45.343460
	−0.668055	−0.944771	−0.167720	−0.236331
10.0	14.073316	19.902674	14.518283	20.533330
	−0.599724	−0.848138	−0.154757	−0.217482
7.5	6.243305	8.829366	6.535827	9.244523
	−0.398999	−0.564271	−0.106477	−0.149114
5.0	2.698021	3.815578	2.861504	4.047952
	−0.225852	−0.319403	−0.062369	−0.087029
2.5	1.004366	1.420389	1.075579	1.521725
	−0.099308	−0.140442	−0.028095	−0.039106

Let's solve the heat diffusion problem presented in Section 9.1 by the control volume method for the 5×7 grid illustrated in Figure 9.7, for which $\beta = 1.0$. Applying Eq. (9.118) at every point in the finite difference grid yields a system of equations similar to Eq. (9.27) and a system equation similar to Eq. (9.28). The solution and the errors, which were obtained by applying the SOR method to solve the system equation, are tabulated in Table 9.13. The results for a 9×13 grid are also presented. Comparing these results with the results obtained by the five-point method, which are presented in Table 9.2 in Example 9.1, shows that the errors are a little larger for the control volume method. Comparing the errors for the 5×7 and 9×13 grids in Table 9.13 shows that the method is second order.

9.13 PROGRAMS

Three FORTRAN subroutines for solving the Laplace equation and the Poisson equation are presented in this section:

1. The five-point method for the Laplace equation with Dirichlet BCs
2. The five-point method for the Laplace equation with Neumann BCs
3. The five-point method for the Poisson equation with Dirichlet BCs

The basic computational algorithms are presented as completely self-contained subroutines suitable for use in other programs. Input data and output statements are contained in a main (or driver) program written specifically to illustrate the use of each subroutine.

9.13.1 The Five-Point Method for the Laplace Equation with Dirichlet BCs

The Laplace equation is given by Eq. (9.7):

$$f_{xx} + f_{yy} = 0 \tag{9.119}$$

When Dirichlet (i.e., specified f) boundary conditions are imposed, those values must be specified at the boundary points. That type of boundary condition is considered in this section. The second-order centered-space approximation of Eq. (9.119) is given by Eq. (9.22):

$$f_{i+1,j} + \beta^2 f_{i,j+1} + f_{i-1,j} + \beta^2 f_{i,j-1} - 2(1 + \beta^2)f_{i,j} = 0 \tag{9.120}$$

where β is the grid aspect ratio, $\beta = \Delta x / \Delta y$.

A FORTRAN subroutine, *subroutine pde1*, for solving the system equation arising from the application of Eq. (9.120) at every point in a rectangular finite difference grid is presented in Program 9.1. The system equation is solved by successive-over-relaxation. A value for the over-relaxation factor ω must be specified. *Program main* defines the data set and prints it, calls *subroutine pde1* to implement the solution, and prints the solution.

Program 9.1. The five-point method Laplace equation solver with Dirichlet BCs program

```
        program main
c       main program to illustrate Laplace (Poisson) equation solvers
c       nxdim x-direction array dimension, nxdim = 9 in this program
```

```
c     nydim y-direction array dimension, nydim = 13 in this program
c     imax  number of grid points in the x direction
c     jmax  number of grid points in the y direction
c     iw    intermediate results output flag: 0 none, 1 all
c     ix    output increment: 1 every point, n every nth point
c     x     x direction array, x(i,j)
c     y     y direction array, y(i,j)
c     f     solution array, f(i,j)
c     fx    right-hand side derivative boundary condition
c     fxy   nonhomogeneous term in the Poisson equation
c     dx,dy x-direction and y-direction grid increments
c     iter  maximum number of iterations
c     tol   convergence tolerance
c     omega sor overrelaxation factor
      dimension x(9,13),y(9,13),f(9,13)
      data nxdim,nydim,imax,jmax,iw,ix / 9, 13, 5, 7, 0, 1 /
      data (f(i,1),i=1,5)/0.0,70.71067812,100.0,70.71067812,0.0/
      data (f(i,7),i=1,5) / 0.0, 0.0, 0.0, 0.0, 0.0 /
      data (f(1,j),j=2,6) / 0.0, 0.0, 0.0, 0.0, 0.0 /
      data (f(5,j),j=2,6) / 0.0, 0.0, 0.0, 0.0, 0.0 /
      data fx,fxy / 0.0, 0.0 /
      data dx,dy,iter,tol,omega/2.5, 2.5, 25, 1.0e-06, 1.23647138/
c     initialize interior points to 0.0 and print initial values
      do i=2,imax-1
         do j=2,jmax-1
            f(i,j)=0.0
         end do
      end do
      write (6,1000)
      if (iw.eq.1) then
         do j=1,jmax,ix
            write (6,1010) (f(i,j),i=1,imax,ix)
         end do
      end if
c     solve the pde and print the solution
      call pde1 (nxdim,nydim,imax,jmax,x,y,f,fx,fxy,dx,dy,iter,tol,
     1 omega,iw,ix)
      do j=1,jmax,ix
         write (6,1010) (f(i,j),i=1,imax,ix)
      end do
      stop
 1000 format (' Laplace equation with Dirichlet BCs'/' ')
 1010 format (5f12.6)
      end

      subroutine pde1 (nxdim,nydim,imax,jmax,x,y,f,fx,fxy,dx,dy,
     1 iter,tol,omega,iw,ix)
c     Laplace equation solver with Dirichlet BCs
      dimension x(nxdim,nydim),y(nxdim,nydim),f(nxdim,nydim)
      beta2=(dx/dy)**2
      d=2.0*(1.0+beta2)
      do it=1,iter
```

```
          dfmax=0.0
          do j=2,jmax-1
             do i=2,imax-1
                df=(f(i+1,j)+beta2*f(i,j+1)+f(i-1,j)+beta2*f(i,j-1)
    1              -d*f(i,j))/d
                if (abs(df).gt.dfmax) dfmax=df
                f(i,j)=f(i,j)+omega*df
             end do
          end do
          if (iw.eq.1) then
             do j=1,jmax,ix
                write (6,1000) (f(i,j),i=1,imax,ix)
             end do
          end if
          if (dfmax.le.tol) then
             write (6,1010) it
             return
          end if
       end do
       write (6,1000) iter
       return
1000 format (5f12.6)
1010 format (' The solution has converged, it = ',i3/' ')
1020 format (' The solution failed to converge, iter = ',i3/' ')
       end
```

The data set used to illustrate *subroutine pde1* is taken from Example 9.1. The output generated by the program is presented in Output 9.1.

Output 9.1. Solution of the Laplace equation with Dirichlet BCs by the five-point method

```
Laplace equation with Dirichlet BCs

The solution converged, it =   14

  0.000000   70.710678  100.000000   70.710678    0.000000
  0.000000   33.459590   47.319006   33.459590    0.000000
  0.000000   15.808676   22.356844   15.808676    0.000000
  0.000000    7.418270   10.491019    7.418271    0.000000
  0.000000    3.373387    4.770690    3.373387    0.000000
  0.000000    1.304588    1.844967    1.304589    0.000000
  0.000000    0.000000    0.000000    0.000000    0.000000
```

9.13.2 The Five-Point Method for the Laplace Equation with Neumann BCs

When Neumann boundary conditions are imposed, a finite difference equation based on that boundary condition must be solved at the boundary points. Equation (9.57) presents the relevant FDE for a derivative boundary condition on the right-hand side of a rectangular domain:

$$\beta^2 f_{I,j+1} + 2f_{I-1,j} + \beta^2 f_{I,j-1} - 2(1 + \beta^2)f_{I,j} = -2\bar{f}_x|_{I,j}\Delta x \qquad (9.121)$$

Subroutine pde1 presented in Section 9.13.1 can be modified to include the derivative BC on the right-hand side simply by solving Eq. (9.121) for $f(\text{imax}, j)$ ($j = 2, \text{jmax} - 1$). Similar procedures can be implemented on the other three sides of a rectangular domain.

A FORTRAN subroutine, *subroutine pde2*, for implementing the system equation arising from the application of Eq. (9.120) at every interior point in a rectangular finite difference grid and Eq. (9.121) at every right-hand side boundary point is presented in Program 9.2. Only the statements which are different from the statements in *program main* and *program pde1* in Section 9.13.1 are presented. *Program main* defines the data set and prints it, calls *subroutine pde2* to implement the solution, and prints the solution.

Program 9.2.   The five-point method Laplace equation solver with Neumann BCs program

```
      program main
c     main program to illustrate Laplace (Poisson) equation solver
      data nxdim,nydim,imax,jmax,iw,ix / 9, 13, 3, 7, 0, 1 /
      data (f(i,1),i=1,3) / 0.0, 70.71067812, 100.0 /
      data (f(i,13),i=1,5) / 0.0, 0.0, 0.0, 0.0, 0.0 /
      data (f(1,j),j=2,12) / 0.,0.,0.,0.,0.,0.,0.,0.,0.,0.,0. /
      data (f(5,j),j=2,12) / 0.,0.,0.,0.,0.,0.,0.,0.,0.,0.,0. /
      data dx,dy,iter,tol,omega /2.5, 2.5, 25, 1.0e-06, 1.23647138/
      call pde2 (nxdim,nydim,imax,jmax,x,y,f,fx,fxy,dx,dy,iter,tol,
     1 omega,iw,ix)
 1000 format (' Laplace equation with Neumann BCs'/' ')
      end

      subroutine pde2 (nxdim,nydim,imax,jmax,x,y,f,fx,fxy,dx,dy,
     1 iter,tol,omega,iw,ix)
c     Laplace equation solver with Neumann BCs
          do i=2,imax
              if(i.lt.imax) then
                  df=(f(i+1,j)+beta2*f(i,j+1)+f(i-1,j)+beta2
     1                *f(i,j-1)-d*f(i,j))/d
              else
                  df=(beta2*f(i,j+1)+2.0*f(i-1,j)+beta2*f(i,j-1)
     1                -d*f(i,j)+2.0*fx*dx)/d
              end if
          end do
```

The data set used to illustrate *subroutine pde2* is taken from Example 9.5. The output generated by the program is presented in Output 9.2.

Output 9.2.   Solution of the Laplace equation with Neumann BCs by the five-point method

```
Laplace equation with Neumann BCs

The solution has converged, it =  14

    0.000000    70.710678   100.000000
    0.000000    33.459590    47.319006
    0.000000    15.808676    22.356844
    0.000000     7.418270    10.491019
    0.000000     3.373387     4.770690
    0.000000     1.304589     1.844967
    0.000000     0.000000     0.000000
```

9.13.3   The Five-Point Method for the Poisson Equation with Dirichlet BCs

The Poisson equation is given by Eq. (9.8):

$$f_{xx} + f_{yy} = F(x, y) \tag{9.122}$$

where $F(x, y)$ is the nonhomogeneous term. The second-order centered-space approximation of Eq. (9.122) is given by Eq. (9.120) with the term $\Delta x^2 F_{i,j}$ included:

$$f_{i+1,j} + \beta^2 f_{i,j+1} + f_{i-1,j} + \beta^2 f_{i,j-1} - 2(1 + \beta^2)f_{i,j} = \Delta x^2 \, F_{i,j} \tag{9.123}$$

where β is the grid aspect ratio, $\beta = \Delta x / \Delta y$. *Subroutine pde1* presented in Section 9.13.1 for solving the Laplace equation includes the nonhomogeneous term $F(x, y)$. Consequently, that subroutine also solves the Poisson equation. The only difference is that the variable specifying the value of the nonhomogeneous term *fxy* must be specified.

A FORTRAN subroutine, *subroutine pde3*, for implementing the system equation arising from the application of Eq. (9.123) at every point in a rectangular finite difference grid is presented in Program 9.3. Only the statements which are different from the statements in *program main* and *program pde1* in Section 9.13.1 are presented. *Program main* defines the data set and prints it, calls *subroutine pde3* to implement the solution, and prints the solution.

Program 9.3.   The five-point method Poisson equation solver with Dirichlet BCs program

```
      program main
c     main program to illustrate Laplace (Poisson) equation solvers
      data fx,fxy / 0.0, -1000.0 /
      data dx,dy,iter,tol,omega /0.25,0.25,25,1.0e-06,1.23647138/
      call pde3 (nxdim,nydim,imax,jmax,x,y,f,fx,fxy,dx,dy,iter,tol,
     1 omega,iw,ix)
 1000 format ('  Poisson equation with Dirichlet BCs'/' ')
      end

      subroutine pde3 (nxdim,nydim,imax,jmax,x,y,f,fx,fxy,dx,dy,
     1 iter,tol,omega,iw,ix)
```

```
c      Poisson equation solver with Dirichlet BCs
             df=(f(i+1,j)+beta2*f(i,j+1)+f(i-1,j)+beta2*f(i,j-1)
    1            -d*f(i,j)-fxy*dx**2)/d
       end
```

The data set used to illustrate *subroutine pde3* is taken from Example 9.6. The output generated by the program is presented in Output 9.3.

Output 9.3. Solution of the Poisson equation with Dirichlet BCs by the five-point method

```
Poisson equation with Dirichlet BCs

The solution has converged, it =  17

    0.000000    0.000000    0.000000    0.000000    0.000000
    0.000000   48.462813   62.407953   48.462813    0.000000
    0.000000   68.943299   90.206185   68.943299    0.000000
    0.000000   74.604197   98.030191   74.604197    0.000000
    0.000000   68.943299   90.206186   68.943299    0.000000
    0.000000   48.462813   62.407953   48.462813    0.000000
    0.000000    0.000000    0.000000    0.000000    0.000000
```

The Poisson equation solver can be extended to account for Neumann (i.e., derivative) BCs in the manner described in Section 9.13.2 for the Laplace equation solver.

9.13.4 Packages for Integrating the Laplace and Poisson Equations

Numerous libraries and software packages are available for integrating the Laplace and Poisson equations. Many work stations and main frame computers have such libraries attached to their operating systems.

Many commercial software packages contain algorithms for integrating the Laplace and Poisson equations. Due to the wide variety of elliptic PDEs governing physical problems, many elliptic PDE solvers (i.e., programs) have been developed. For this reason, no specific programs are recommended in this section.

9.14 SUMMARY

The numerical solution of elliptic partial differential equations by finite difference methods is discussed in this chapter. Elliptic PDEs govern equilibrium problems, which have no preferred paths of information propagation. The domain of dependence and range of influence of every point is the entire closed solution domain. Such problems are solved numerically by relaxation methods. The two-dimensional Laplace equation is considered as the model elliptic PDE in this chapter.

Finite difference methods, as typified by the five-point method, yield a system of finite difference equations, called the system equation, which must be solved by relaxation methods. The successive-over-relaxation (SOR) method is generally the method of choice. The multigrid method (Brandt, 1977) shows the best potential for rapid convergence.

Nonlinear partial differential equations yield nonlinear finite difference equations. Systems of nonlinear FDEs can be extremely difficult to solve. The multigrid method can be applied directly to nonlinear PDEs. Three-dimensional PDEs are approximated simply by including the finite difference approximations of the spatial derivatives in the third direction. The relaxation techniques used to solve two-dimensional problems generally can be used to solve three-dimensional problems, at the expense of a considerable increase in computational effort.

After studying Chapter 9, you should be able to:

1. Discuss the general features of elliptic PDEs
2. Recognize the Laplace equation and the Poisson equation
3. Explain the general features of the Laplace equation and the Poisson equation
4. Describe the three types of boundary conditions applicable to PDEs
5. Discretize a continuous physical space into a discrete finite difference grid
6. Develop finite difference approximations (FDAs) of the individual exact partial derivatives appearing in PDEs
7. Develop a finite difference approximation of an elliptic PDE
8. Determine the order of a FDA of an elliptic PDE
9. Apply the five-point method to solve the Laplace equation and the Poisson equation
10. Develop the system equation for the finite difference approximation of an elliptic PDE
11. Express the system equation in matrix form
12. Discuss the advantages and disadvantages of both direct methods and iterative methods for solving the system equation
13. Solve the system equation by Gauss elimination
14. Solve the system equation by successive-over-relaxation (SOR)
15. Explain the importance of the optimum overrelaxation factor ω_{opt}
16. Apply the modified differential equation concept to develop the MDE corresponding to the finite difference approximation of an elliptic PDE
17. Determine the consistency of a FDE from the MDE
18. Determine the order of a FDE from the MDE
19. Determine if a finite difference method is convergent
20. Apply derivative boundary conditions for an elliptic PDE
21. Apply the compact fourth-order method to solve an elliptic PDE
22. Apply the extrapolation method to solve an elliptic PDE
23. Develop unequally spaced FDAs of partial derivatives
24. Solve an elliptic PDE on a nonrectangular domain
25. Explain the complications that arise in the finite difference solution of nonlinear PDEs
26. Suggest some approaches for solving nonlinear elliptic PDEs
27. Explain the complications that arise in the finite difference solution of three-dimensional problems
28. Suggest some approaches for solving three-dimensional problems
29. Explain the control volume concept
30. Apply the control volume concept to solve an elliptic PDE
31. Choose a method for solving a linear elliptic PDE and implement the method to obtain a numerical solution

EXERCISE PROBLEMS

Section 9.1   Introduction

1. Consider the two-dimensional Laplace equation, $\bar{f}_{xx} + \bar{f}_{yy} = 0$. Classify this PDE. Determine the characteristic curves. Discuss the significance of these results as regards domain of dependence, range of influence, signal propagation speed, auxiliary conditions, and numerical solution procedures.
2. Develop the exact solution of the heat diffusion problem presented in Section 9.1, Eq. (9.2).
3. By hand, calculate the exact solution for $T(5.0, 12.5)$.

Section 9.3   The Finite Difference Method

4. Develop the second-order centered-space approximations for $\bar{f}_{xx}$ and $\bar{f}_{yy}$, Eqs. (9.16) and (9.18), respectively, including the leading truncation error terms.
5. Develop a second-order centered-space approximation of the mixed partial derivative $\bar{f}_{xy}$ for the finite difference grid illustrated in Figure 9.5.

Section 9.4   Finite Difference Solution of the Laplace Equation

6. Develop the five-point finite difference approximation of the Laplace equation for (a) $\Delta x = \Delta y$, and (b) $\Delta x / \Delta y = \beta \neq 1$.
7. *Solve the heat diffusion problem presented in Section 9.1 by hand using the five-point method with $\Delta x = \Delta y = 5.0$ cm using Gauss elimination. Compare the results with the exact solution in Table 9.1.
8. Solve the heat diffusion problem presented in Section 9.1 by hand using the five-point method with $\Delta x = \Delta y = 2.5$ cm using Gauss elimination. Compare the results with the exact solution in Table 9.1. Compare the errors and the ratio of the errors with the results of Problem 7.
9. *Modify the heat diffusion problem presented in Section 9.1 by letting $T = 0.0$ C on the top boundary and $T = 100.0$ C on the right boundary. Solve this problem by hand using the five-point method with $\Delta x = \Delta y = 5.0$ cm using Gauss elimination.
10 Modify the heat diffusion problem presented in Section 9.1 by letting $T = 0.0$ C on the top boundary and $T = 100.0$ C on the right boundary. Solve this problem by hand using the five-point method with $\Delta x = \Delta y = 2.5$ cm using Gauss elimination.
11. *Consider steady heat diffusion in the unit square, $0.0 \leq x \leq 1.0$ and $0.0 \leq y \leq 1.0$. Let $T(0, y) = T(x, 0) = 100.0$ and $T(1, y) = T(x, 1) = 0.0$. Solve this problem by hand using the five-point method with $\Delta x = \Delta y = 0.25$ using Gauss elimination.

Section 9.5   Consistency, Order, and Convergence

12. Derive the modified differential equation (MDE) for the five-point approximation of the Laplace equation with $\Delta x = \Delta y$, Eq. (9.25). Discuss consistency and order of this FDE.

13. Consider the following finite difference approximations of the Laplace equation:

(a) $f_{i+1,j+1} + f_{i+1,j} + f_{i+1,j-1} + f_{i,j+1} + f_{i,j-1} + f_{i-1,j+1} + f_{i-1,j}$
$$+ f_{i-1,j-1} - 8f_{i,j} = 0 \qquad (A)$$

(b) $f_{i+1,j+1} + 2f_{i+1,j} + f_{i+1,j-1} + 2f_{i,j+1} + 2f_{i,j-1} + f_{i-1,j+1}$
$$+ 2f_{i-1,j} + f_{i-1,j-1} - 12f_{i,j} = 0 \qquad (B)$$

(c) $-f_{i+2,j} + 16f_{i+1,j} - f_{i,j+2} + 16f_{i,j+1} + 16f_{i,j-1} - f_{i,j-2}$
$$+ 16f_{i-1,j} - f_{i-2,j} - 60f_{i,j} = 0 \qquad (C)$$

(d) $f_{i+1,j+1} + 4f_{i+1,j} + f_{i+1,j-1} + 4f_{i,j+1} + 4f_{i,j-1} + f_{i-1,j+1}$
$$+ 4f_{i-1,j} + f_{i-1,j-1} - 20f_{i,j} = 0 \qquad (D)$$

Derive the MDE for each of these FDEs. Discuss consistency and order.

Section 9.6 Iterative Methods of Solution

The Gauss-Seidel Method

14. Solve the heat diffusion problem presented in Section 9.1 by hand using the five-point method with $\Delta x = \Delta y = 5.0$ cm using Gauss-Seidel iteration. Iterate until $|\Delta T_{max}| \le \varepsilon = 1.0$.

The Successive-Over-Relaxation (SOR) Method

15. Use the program presented in Section 9.13.1 to solve the Laplace equation in a rectangle with Dirichlet boundary conditions. Solve the heat diffusion problem presented in Section 9.1 with $\Delta x = \Delta y = 5.0$ cm and compare the results with Table 9.1 and Problem 7.
16. Use the program presented in Section 9.13.A to solve the Laplace equation in a rectangle with Dirichlet boundary conditions. Solve the heat diffusion problem presented in Section 9.1 with $\Delta x = \Delta y = 2.5$ cm and compare the results with Tables 9.1 and 9.2 and Problem 8.
17. Solve Problem 16 with $\omega = \omega_{opt}$ from Eq. (9.51). Compare the rates of convergence of Problems 16 and 17.
18. Solve Problem 11 with $\Delta x = \Delta y = 0.1$ using the program presented in Section 9.13.1 with (a) $\omega = 1.0$ and (b) $\omega = \omega_{opt}$ from Eq. (9.51), both for $\varepsilon = 0.000001$. Compare the rates of convergence.
19. Use the computer program presented in Section 9.13.1 to determine ω_{opt} numerically for the unit square, $0.0 \le x \le 1.0$ and $0.0 \le y \le 1.0$, Problem 18. Let $\Delta = \Delta x = \Delta y$ and $\varepsilon = 0.000001$. For $\Delta = 0.25,\ 0.125,$ and 0.0625, calculate ω_{opt} and compare with the values obtained from Eq. (9.51).

Section 9.7 Derivative Boundary Conditions

20. Work Example 9.5 by hand with $\Delta x = \Delta y = 2.5$ cm using Gauss elimination.
21. Consider steady heat diffusion in the unit square, $0.0 \le x \le 1.0$ and $0.0 \le y \le 1.0$. Let $T(0.0, y) = T(x, 0.0) = 100.0$ and $T_x(1.0, y) =$

$T_y(x, 1.0) = 0.0$. Solve this problem using the five-point method with $\Delta x = \Delta y = 0.25$ by hand using Gauss elimination.

22. Consider steady heat diffusion in the unit square, $0.0 \le x \le 1.0$ and $0.0 \le y \le 1.0$. Let $T(x, 0.0) = 0.0$, $T(x, 1.0) = 100.0$, and $T(1.0, 0.0) = 0.0$. The left side of the square is cooled by convection to the surroundings. For steady conditions, the rate of convection $\dot{q}_{conv}$ must equal the rate of conduction $\dot{q}_{cond}$ at the boundary. Thus,

$$\dot{q}_{conv} = hA(T - T_a) = \dot{q}_{cond} = \frac{kA}{\partial x}\frac{\partial T}{\partial x} \qquad (E)$$

(a) Develop a finite difference approximation of the convection boundary condition, Eq. (E). Let $h = 100.0\,J/(s\text{-}m^2\text{-}K)$, $k = 5.0\,J/(s\text{-}m\text{-}K)$, and $T_a = 10.0\,C$. Solve this problem by hand for $\Delta x = \Delta y = 0.25\,cm$ using Gauss elimination.

23. Implement the program presented in Section 9.13.2 to consider a derivative BC along the right side of a rectangle. Solve Example 9.5 using the program. Compare the results with the results presented in Table 9.7.

Section 9.8   Finite Difference Solution of the Poisson Equation

24. Develop the five-point finite difference approximation of the Poisson equation for (a) $\Delta x = \Delta y$, and (b) $\Delta x/\Delta y = \beta \ne 1$.
25. Derive the modified differential equation (MDE) for the five-point approximation of the Poisson equation with $\Delta x = \Delta y$. Discuss consistency and order of the FDE.
26. *Solve the heat diffusion problem presented in Section 9.8 by hand using the five-point method with $\Delta x = \Delta y = 0.5\,cm$ using Gauss elimination.
27. Solve the heat diffusion problem presented in Section 9.8 by hand using the five-point method with $\Delta x = \Delta y = 0.25\,cm$ using Gauss elimination.
28. Implement the program presented in Section 9.13.3 to solve the Poisson equation in a rectangle with Dirichlet boundary conditions using SOR. Solve the heat diffusion problem presented in Section 9.8 with $\omega = 1.0$ and $\varepsilon = 0.000001$.
29. Solve Problem 28 with $\omega = \omega_{opt}$ from Eq. (9.51).

Section 9.9   Higher-Order Methods

Compact Fourth-Order Method

30. Solve the heat diffusion problem presented in Section 9.1 using Eq. (D) from Problem 13 with $\Delta x = \Delta y = 5.0\,cm$. Compare the results with the exact solution in with Table 9.1 and the results obtained in Problem 7.
31. Solve the heat diffusion problem presented in Section 9.1 using Eq. (D) from Problem 13 with $\Delta x = \Delta y = 2.5\,cm$. Compare the results with the exact solution in Table 9.1 and the results obtained in Problem 8.

Extrapolation

32. Extrapolate the results obtained in Problems 15 and 16. Compare the results with the results of Problem 31.

Section 9.10   Nonrectangular Domains

33. Solve Example 9.9 for $\Delta x = \Delta y = 5.0$ cm by hand using Gauss elimination.
34. Write a program to solve the Poisson equation on a nonrectangular domain by successive-over-relaxation (SOR). Solve Example 9.9 using the program. Compare the results with the results presented in Example 9.9.

Section 9.11   Nonlinear Equations and Three-Dimensional Problems

35. Derive the seven-point finite difference approximation of the three-dimensional Laplace equation with $\Delta x = \Delta y = \Delta z$.
36. Consider steady heat diffusion in the unit cube, $0.0 \le x \le 1.0$, $0.0 \le y \le 1.0$, and $0.0 \le z \le 1.0$. Let $T = 100.0$ on the surface $z = 1.0$ and $T = 0.0$ on the other five surfaces. Solve this problem by hand by the seven-point method with $\Delta x = \Delta y = \Delta z = 0.5$.
37. Solve Problem 36 with $\Delta x = \Delta y = \Delta z = 1/3$ using Gauss elimination.
38. Solve Problem 36 with $\Delta x = \Delta y = \Delta z = 1/4$ using Gauss-Seidel iteration with $\varepsilon = 0.1$. Let $T_{i,j,k}^{(0)} = 0.0$.

Section 9.12   The Control Volume Method

39. Solve the heat diffusion problem presented in Section 9.1 by the control volume method using Eq. (9.118) with $\Delta x = \Delta y = 5.0$ cm. Compare the results with the exact solution presented in Table 9.1 and the results obtained in Problem 7.
40. Solve the heat diffusion problem presented in Section 9.1 by the control volume method using Eq. (9.118) with $\Delta x = \Delta y = 2.5$ cm. Compare the results with the exact solution presented in Table 9.1 and the results obtained in Problem 8.

Section 9.13   Programs

41. Implement the five-point method for the Laplace equation with Dirichlet BCs program presented in Section 9.13.1. Check out the program using the given data set.
42. Solve any of Problems 15 to 20 with the program.
43. Implement the five-point method for the Laplace equation with Neumann BCs program presented in Section 9.13.2. Check out the program using the given data set.
44. Solve any of Problems 20 to 23 with the program.
45. Implement the five-point method for the Poisson equation with Dirichlet BCs program presented in Section 9.13.3. Check out the program using the given data set.
46. Solve any of Problems 26 to 29 with the program.

10

Parabolic Partial Differential Equations

10.1 INTRODUCTION

Figure 10.1 illustrates two heat diffusion problems. The plate illustrated at the top of the figure has a thickness $L = 1.0$ cm and thermal diffusivity $\alpha = 0.01$ cm^2/s. The internal temperature distribution is governed by the unsteady one-dimensional heat diffusion equation:

$$T_t = \alpha T_{xx} \qquad\qquad (10.1)$$

587

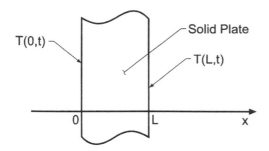

$$T_t = \alpha T_{xx}, \quad T(x,0) = F(x), \quad T(x,t) = ?$$

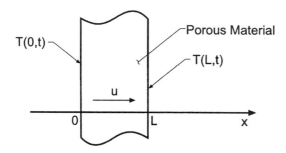

$$T_t + uT_x = \alpha T_{xx}, \quad T(x,0) = F(x), \quad T(x,t) = ?$$

Figure 10.1 Unsteady heat diffusion problems.

The plate is heated to an initial temperature distribution, $T(x, 0)$, at which time the heat source is turned off. The initial temperature distribution in the plate is specified by

$$T(x, 0.0) = 200.0x \qquad 0.0 \leq x \leq 0.5 \qquad (10.2a)$$
$$T(x, 0.0) = 200.0(1.0 - x) \qquad 0.5 \leq x \leq 1.0 \qquad (10.2b)$$

where T is measured in degrees Celcius (C). This initial temperature distribution is illustrated by the top curve in Figure 10.2. The temperatures on the two faces of the plate are held at $0.0\,C$ for all time. Thus,

$$T(0.0, t) = T(1.0, t) = 0.0 \qquad (10.2c)$$

The temperature distribution within the plate, $T(x, t)$, is required.

The exact solution to this problem is obtained by assuming a product solution of the form $T(x, t) = X(x)\hat{T}(t)$, substituting this functional form into the PDE and separating variables, integrating the two resulting ordinary differential equations for $X(x)$ and $\hat{T}(t)$, applying the boundary conditions at $x = 0$ and $x = L$, and superimposing an infinite

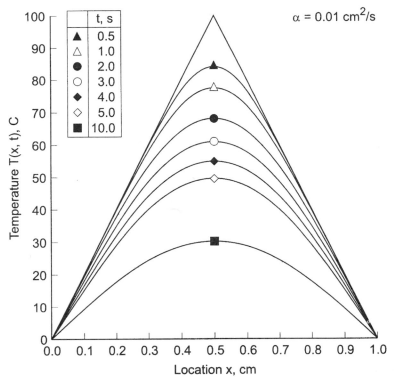

Figure 10.2   Exact solution of the heat diffusion problem.

number of harmonic functions (i.e., sines and cosines) in a Fourier series to satisfy the initial conditions $T(x, 0)$. The result is

$$T(x, t) = \frac{800}{\pi^2} \sum_{m=0}^{\infty} \frac{(-1)^m}{(2m+1)^2} \sin[(2m+1)\pi x] e^{-(2m+1)^2 \pi^2 \alpha t} \tag{10.3}$$

The exact solution at selected values of time is tabulated in Table 10.1 and illustrated in Figure 10.2. The solution is symmetrical about the midplane of the plate. The solution smoothly approaches the asymptotic steady state solution, $T(x, \infty) = 0.0$.

The second problem illustrated in Figure 10.1 is a combined convection-diffusion problem, which is governed by the convection-diffusion equation. This problem is similar to the first problem, with the added feature that the plate is porous and a cooling fluid flows through the plate. The exact solution and the numerical solution of this problem are presented in Section 10.10.

A wide variety of parabolic partial differential equations are encountered in engineering and science. Two of the more common ones are the *diffusion equation* and the *convection-diffusion equation*, presented below for the generic dependent variable $f(x, t)$:

$$f_t = \alpha f_{xx} \tag{10.4}$$
$$f_t + u f_x = \alpha f_{xx} \tag{10.5}$$

Table 10.1 Exact Solution of the Heat Diffusion Problem

	Temperature $T(x, t)$, C					
t, s	$x = 0.0$	$x = 0.1$	$x = 0.2$	$x = 0.3$	$x = 0.4$	$x = 0.5$
0.0	0.0	20.0000	40.0000	60.0000	80.0000	100.0000
0.5	0.0	19.9997	39.9847	59.6604	76.6674	84.0423
1.0	0.0	19.9610	39.6551	57.9898	72.0144	77.4324
2.0	0.0	19.3513	37.6601	53.3353	64.1763	68.0846
3.0	0.0	18.1235	34.8377	48.5749	57.7018	60.9128
4.0	0.0	16.6695	31.8585	44.1072	52.0966	54.8763
5.0	0.0	15.2059	28.9857	40.0015	47.1255	49.5912
10.0	0.0	9.3346	17.7561	24.4405	28.7327	30.2118
20.0	0.0	3.4794	6.6183	9.1093	10.7086	11.2597
50.0	0.0	0.1801	0.3427	0.4716	0.5544	0.5830
∞	0.0	0.0000	0.0000	0.0000	0.0000	0.0000

where α is the diffusivity and u is the convection velocity. The diffusion equation applies to problems in mass diffusion, heat diffusion (i.e., conduction), neutron diffusion, etc. The convection-diffusion equation applies to problems in which convection occurs in combination with diffusion, for example, fluid mechanics and heat transfer. The present chapter is devoted mainly to the numerical solution of the diffusion equation. All of the results also apply to the numerical solution of the convection-diffusion equation, which is considered briefly in Section 10.10.

The solution of Eqs. (10.4) and (10.5) is the function $f(x, t)$. This function must satisfy an initial condition at $t = 0, f(x, 0) = F(x)$. The time coordinate has an unspecified (i.e., open) final value. Since Eqs. (10.4) and (10.5) are second order in the spatial coordinate x, two boundary conditions are required. These may be of the Dirichlet type (i.e., specified values of f), the Neumann type (i.e., specified values of f_x) , or the mixed type (i.e., specified combinations of f and f_x). The basic properties of finite difference methods for solving propagation problems governed by parabolic PDEs are presented in this chapter.

The organization of Chapter 10 is presented in Figure 10.3. Following the Introduction, the general features of parabolic partial differential equations are discussed. This discussion is followed by a discussion of the finite difference method. The solution of the diffusion equation by the forward-time centered-space (FTCS) method is then presented. This presentation is followed by a discussion of the concepts of consistency, order, stability, and convergence. Two additional explicit methods, the Richardson (leapfrog) method and the DuFort-Frankel method are then presented to illustrate an unstable method and an inconsistent method. Two implicit methods are then presented: the backward-time centered-space (BTCS) method and the Crank-Nicholson method. A procedure for implementing derivative boundary conditions is presented next. A discussion of nonlinear equations and multidimensional problems follows. A brief introduction to the solution of the convection-diffusion equation is then presented. This is followed by a discussion of the asymptotic steady-state solution of propagation problems as a procedure for solving mixed elliptic-parabolic and mixed elliptic-hyperbolic problems. A brief presentation of a program for solving the diffusion equation follows. A summary wraps up the chapter.

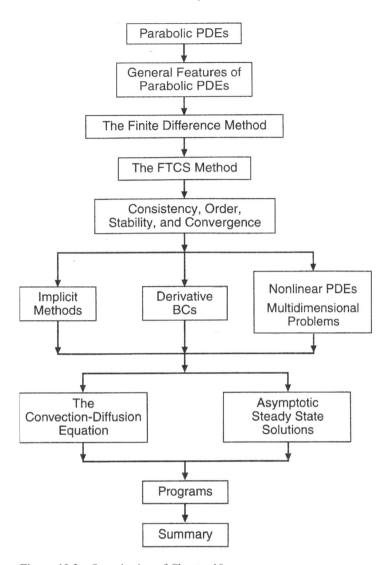

Figure 10.3 Organization of Chapter 10.

10.2 GENERAL FEATURES OF PARABOLIC PDEs

Several concepts must be considered before a propagation type PDE can be solved by a finite difference method. In this section, some fundamental considerations are discussed, the general features of diffusion are presented, and the concept of characteristics is introduced.

10.2.1 Fundamental Considerations

Propagation problems are *initial-boundary-value problems* in *open domains* (open with respect to time or a timelike variable) in which the solution in the domain of interest is marched forward from the initial state, guided and modified by the boundary conditions. Propagation problems are governed by parabolic or hyperbolic partial differential equations. The general features of parabolic and hyperbolic PDEs are discussed in Part III. Those features which are relevant to the finite difference solution of both parabolic and hyperbolic PDEs are presented in this section. Those features which are relevant only to the finite difference solution of hyperbolic PDEs are presented in Section 11.2.

The general features of *parabolic partial differential equations* (*PDEs*) are discussed in Section III.6. In that section it is shown that parabolic PDEs govern propagation problems, which are initial-boundary-value problems in open domains. Consequently, parabolic PDEs are solved numerically by marching methods. From the characteristic analysis presented in Section III.6, it is known that problems governed by parabolic PDEs have an *infinite physical information propagation speed*. As a result, the solution at a given point P at time level n depends on the solution at all other points in the solution domain at all times preceding and including time level n, and the solution at a given point P at time level n influences the solution at all other points in the solution domain at all times including and after time level n. Consequently, the physical information propagation speed $c = dx/dt$ is infinite. These general features of parabolic PDEs are illustrated in Figure 10.4.

10.2.2 General Features of Diffusion

Consider pure diffusion, which is governed by the diffusion equation:

$$\boxed{f_t = \alpha f_{xx}} \tag{10.6}$$

where α is the diffusion coefficient. Consider an initial property distribution, $f(x, 0) = \phi(x)$, given the general term of an exponential Fourier series:

$$\phi(x) = A_m e^{Ik_m x} \tag{10.7}$$

where $I = \sqrt{-1}$, $k_m = 2\pi m/2L$ is the wave number, and L is the width of the physical space. Assume that the exact solution of Eq. (10.6) is given by

$$f(x, t) = e^{-\alpha k_m^2 t}\phi(x) \tag{10.8}$$

Substituting Eq. (10.7) into Eq. (10.8) yields

$$f(x, t) = e^{-\alpha k_m^2 t}A_m e^{Ik_m x} \tag{10.9}$$

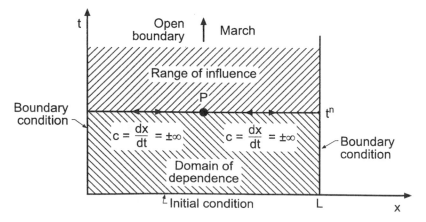

Figure 10.4 General features of parabolic PDEs.

Differentiating Eq. (10.9) with respect to t and x gives

$$f_t = -\alpha k_m^2 e^{-\alpha k_m^2 t} A_m e^{Ik_m x} = -\alpha k_m^2 e^{-\alpha k_m^2 t} \phi(x) \tag{10.10}$$

$$f_{xx} = e^{-\alpha k_m^2 t} A_m (Ik_m)^2 e^{Ik_m x} = -k_m^2 e^{-\alpha k_m^2 t} \phi(x) \tag{10.11}$$

Substituting Eqs. (10.10) and (10.11) into Eq. (10.6) demonstrates that Eq. (10.8) is the exact solution of the diffusion equation:

$$\boxed{f(x, t) = e^{-\alpha k_m^2 t} \phi(x)} \tag{10.12}$$

Equation (10.12) shows that the initial property distribution $\phi(x)$ simply decays with time at the exponential rate $\exp(-\alpha k_m^2 t)$. Thus, the rate of decay depends on the square of the wave number k_m. The initial property distribution does not propagate in space.

For an arbitrary initial property distribution represented by a Fourier series, Eq. (10.12) shows that each Fourier component simply decays exponentially with time, but that each component decays at a rate which depends on the square of its individual wave number k_m. Thus, the total property distribution changes shape. Consequently, pure diffusion causes the initial property distribution to decay and change shape, but the property distribution does not propagate in space.

10.2.3 Characteristic Concepts

The concept of characteristics of partial differential equations is introduced in Section III.3. In two-dimensional space, which is the case considered here (i.e., space x and time t), characteristics are paths (curved, in general) in the solution domain $D(x, t)$ along which physical information propagates. If a partial differential equation possesses real characteristics, then physical information propagates along the characteristic paths. The presence of characteristics has a significant effect on the solution of a partial differential equation (by both analytical and numerical methods).

Consider the unsteady one-dimensional diffusion equation $f_t = \alpha f_{xx}$. It is shown in Section III.6 that the characteristic paths for the unsteady one-dimensional diffusion equation are the lines of constant time. Thus, physical information propagates at an infinite rate throughout the entire physical solution domain. Every point influences all the other points, and every point depends on the solution at all the other points, including the boundary points. This behavior should be considered when solving parabolic PDEs by numerical methods.

10.3 THE FINITE DIFFERENCE METHOD

The objective of a finite difference method for solving a partial differential equation (PDE) is to transform a calculus problem into an algebra problem by

1. *Discretizing* the continuous physical domain into a discrete difference grid
2. *Approximating* the individual exact partial derivatives in the partial differential equation (PDE) by algebraic finite difference approximations (FDAs)
3. *Substituting* the FDAs into the PDE to obtain an algebraic finite difference equation (FDE)
4. *Solving* the resulting algebraic FDEs

There are several choices to be made when developing a finite difference solution to a partial differential equation. Foremost among these are the choice of the discrete finite

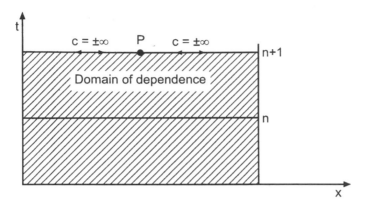

Figure 10.5 Physical domain of dependence of parabolic PDEs.

difference grid used to discretize the continuous physical domain and the choice of the finite difference approximations used to represent the individual exact partial derivatives in the partial differential equation. Some fundamental considerations relevant to the finite difference approach are discussed in the next subsection. The general features of finite difference grids and finite difference approximations, which apply to both parabolic and hyperbolic PDEs, are discussed in the following subsections.

10.3.1 Fundamental Considerations

The objective of the numerical solution of a PDE is to march the solution at time level n forward in time to time level $n + 1$, as illustrated in Figure 10.5, where the physical domain of dependence of a parabolic PDE is illustrated. In view of the infinite physical information propagation speed $c = dx/dt$ associated with parabolic PDEs, the solution at point P at time level $n + 1$ depends on the solution at all of the other points at time level $n + 1$.

Finite difference methods in which the solution at point P at time level $n + 1$ depends only on the solution at neighboring points at time level n have a finite numerical information propagation speed $c_n = \Delta x/\Delta t$. Such finite difference methods are called *explicit methods* because the solution at each point is specified explicitly in terms of the known solution at neighboring points at time level n. This situation is illustrated in Figure 10.6, which resembles the physical domain of dependence of a hyperbolic PDE. The numerical information propagation speed $c_n = \Delta x/\Delta t$ is finite.

Finite difference methods in which the solution at point P at time level $n + 1$ depends on the solution at neighboring points at time level $n + 1$ as well as the solution at time level n have an infinite numerical information propagation speed $c_n = \Delta x/\Delta t$. Such methods couple the finite difference equations at time level $n + 1$ and result in a system of finite difference equations which must be solved at each time level. Such finite difference methods are called *implicit methods* because the solution at each point is specified implicitly in terms of the unknown solution at neighboring points at time level $n + 1$. This situation is illustrated in Figure 10.7, which resembles the physical domain of dependence of a parabolic PDE. The numerical information propagation speed, $c_n = \Delta x/\Delta t$, is infinite.

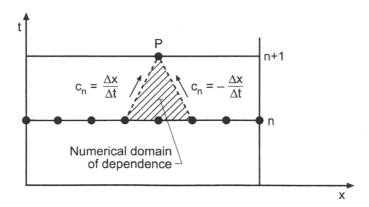

Figure 10.6 Numerical domain of dependence of explicit methods.

The similarities of and the differences between explicit and implicit numerical marching methods are illustrated in Figures 10.6 and 10.7. The major similarity is that both methods march the solution forward from one time level to the next time level. The major difference is that the numerical information propagation speed for explicit marching methods is finite, whereas the numerical information propagation speed for implicit marching methods is infinite.

Explicit methods are computationally faster than implicit methods because there is no system of finite difference equations to solve. Thus, explicit methods might appear to be superior to implicit methods. However, the finite numerical information propagation speed of explicit methods does not correctly model the infinite physical information propagation speed of parabolic PDEs, whereas the infinite numerical information propagation speed of implicit methods correctly models the infinite physical information propagation speed of parabolic PDEs. Thus, implicit methods appear to be well suited for solving parabolic PDEs, and explicit methods appear to be unsuitable for solving parabolic PDEs. In actuality, only an infinitesimal amount of physical information propagates at the infinite physical information propagation speed. The bulk of the physical information travels at a finite physical information propagation speed. Experience has shown that explicit methods as well as implicit methods can be employed to solve parabolic PDEs.

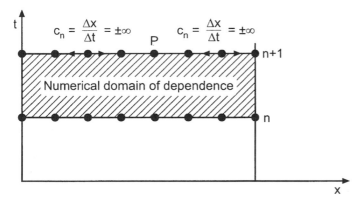

Figure 10.7 Numerical domain of dependence of implicit methods.

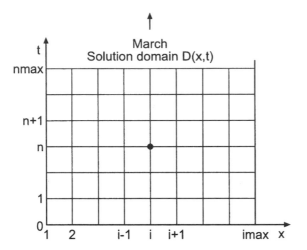

Figure 10.8 Solution domain, $D(x, t)$, and finite difference grid.

10.3.2 Finite Difference Grids

The solution domain $D(x, t)$ in xt space for an unsteady one-dimensional propagation problem is illustrated in Figure 10.8. The solution domain must be covered by a two-dimensional grid of lines, called the *finite difference grid*. The intersections of these *grid lines* are the *grid points* at which the finite difference solution of the partial differential equation is to be obtained. For the present, let the spatial grid lines be equally spaced lines perpendicular to the x axis having uniform spacing Δx. The temporal grid line spacing Δt may or may not be equally spaced. The resulting finite difference grid is also illustrated in Figure 10.8. The subscript i is used to denote the physical grid lines [i.e., $x_i = (i - 1)\Delta x$], and the superscript n is used to denote the time grid lines (i.e., $t^n = n\,\Delta t$ if Δt is constant). Thus, grid point (i, n) corresponds to location (x_i, t^n) in the solution domain $D(x, t)$. The total number of x grid lines is denoted by imax, and the total number of time steps is denoted by nmax.

Two-dimensional physical spaces can be covered in a similar manner by a three-dimensional grid of planes perpendicular to the coordinate axes, where the subscripts i and j denote the physical grid planes perpendicular to the x and y axes, respectively, and the superscript n denotes time planes. Thus, grid point (i, j, n) corresponds to location (x_i, y_j, t^n) in the solution domain $D(x, y, t)$. Similarly, in three-dimensioinal physical space, grid point (i, j, k, n) corresponds to location (x_i, y_j, z_k, t^n) in the solution domain $D(x, y, z, t)$.

The dependent variable at a grid point is denoted by the same subscript-superscript notation that is used to denote the grid points themselves. Thus, the function $f(x, t)$ at grid point (i, n) is denoted by

$$f(x_i, t^n) = f_i^n \tag{10.13}$$

In a similar manner, derivatives are denoted by

$$\frac{\partial f(x_i, t^n)}{\partial t} = \left.\frac{\partial f}{\partial t}\right|_i^n = f_t|_i^n \qquad \text{and} \qquad \frac{\partial^2 f(x_i, t^n)}{\partial x^2} = \left.\frac{\partial^2 f}{\partial x^2}\right|_i^n = f_{xx}|_i^n \tag{10.14}$$

Similar results apply in two- and three-dimensional spaces.

10.3.3 Finite Difference Approximations

Now that the finite difference grid has been specified, *finite difference approximations* of the individual exact partial derivatives in the partial differential equation must be obtained. This is accomplished by writing Taylor series for the dependent variable at one or more grid points using a particular grid point as the base point and combining these Taylor series to solve for the desired partial derivatives. This is done in Chapter 5 for functions of one independent variable, where approximations of various types (i.e., forward, backward, and centered) of various orders (i.e., first order, second order, etc.) are developed for various derivatives (i.e., first derivative, second derivative, etc.). Those results are presented in Table 5.1.

In the development of finite difference approximations, a distinction must be made between the *exact solution* of a partial differential equation and the solution of the finite difference equation which is an *approximate solution* of the partial differential equation. For the remainder of this chapter, the exact solution of a PDE is denoted by an overbar over the symbol for the dependent variable, that is, $\bar{f}(x, t)$, and the approximate solution is denoted by the symbol for the dependent variable without an ovarbar, that is, $f(x, t)$. Thus,

$$\begin{array}{l} \bar{f}(x, t) = \text{exact solution} \\ f(x, t) = \text{approximate solution} \end{array}$$

Exact partial derivatives, such as $\bar{f}_t$ and $\bar{f}_{xx}$, which appear in the parabolic diffusion equation can be approximated at a grid point in terms of the values of f at that grid point and adjacent grid points in several ways. The exact time derivative $\bar{f}_t$ can be approximated at time level n by a first-order forward-time approximation or a second-order centered-time approximation. It can also be approximated at time level $n + 1$ by a first-order backward-time approximation or at time level $n + 1/2$ by a second-order centered-time approximation. The spatial derivative $\bar{f}_{xx}$ must be approximated at the same time level at which the time derivative $\bar{f}_t$ is evaluated.

The second-order spatial derivative $\bar{f}_{xx}$ is a model of physical diffusion. From characteristic concepts, it is known that the physical information propagation speed associated with second-order spatial derivatives is infinite, and that the solution at a point at a specified time level depends on and influences all of the other points in the solution domain at that time level. Consequently, second-order spatial derivatives, such as $\bar{f}_{xx}$, should be approximated by centered-space approximations at spatial location i. The centered-space approximations can be second-order, fourth-order, etc. Simplicity of the resulting finite difference equation usually dictates the use of second-order centered-space approximations for second-order spatial derivatives.

10.3.3.1 Time Derivatives

Consider the partial derivative $\bar{f}_t$. Writing the Taylor series for $\bar{f}_i^{n+1}$ using grid point (i, n) as the base point gives

$$\bar{f}_i^{n+1} = \bar{f}_i^n + \bar{f}_t|_i^n \, \Delta t + \tfrac{1}{2}\bar{f}_{tt}|_i^n \, \Delta t^2 + \cdots \tag{10.15}$$

where the convention $(\Delta t)^m \rightarrow \Delta t^m$ is employed for compactness. Solving Eq. (10.15) for $\bar{f}_t|_i^n$ yields

$$\bar{f}_t|_i^n = \frac{\bar{f}_i^{n+1} - \bar{f}_i^n}{\Delta t} - \tfrac{1}{2}\bar{f}_{tt}(\tau)\,\Delta t \tag{10.16}$$

where $t \le \tau \le t + \Delta t$. Truncating the remainder term yields the *first-order forward-time* approximation of $\bar{f}_t|_i^n$, denoted by $f_t|_i^n$:

$$f_t|_i^n = \frac{f_i^{n+1} - f_i^n}{\Delta t} \tag{10.17}$$

The remainder term which has been truncated in Eq. (10.17) is called the *truncation error* of the finite difference approximation of $\bar{f}_t|_i^n$. A *first-order backward-time* approximation and a *second-order centered time* approximation can be developed in a similar manner by choosing base points $n+1$ and $n+1/2$, respectively.

10.3.3.2 Space Derivatives

Consider the partial derivatives $\bar{f}_x$ and $\bar{f}_{xx}$. Writing Taylor series for $\bar{f}_{i+1}^n$ and $\bar{f}_{i-1}^n$ using grid point (i, n) as the base point gives

$$\bar{f}_{i+1}^n = \bar{f}_i^n + \bar{f}_x|_i^n\,\Delta x + \tfrac{1}{2}\bar{f}_{xx}|_i^n\,\Delta x^2 + \tfrac{1}{6}\bar{f}_{xxx}|_i^n\,\Delta x^3 + \tfrac{1}{24}\bar{f}_{xxxx}|_i^n\,\Delta x^4 + \cdots \tag{10.18}$$

$$\bar{f}_{i-1}^n = \bar{f}_i^n - \bar{f}_x|_i^n\,\Delta x + \tfrac{1}{2}\bar{f}_{xx}|_i^n\,\Delta x^2 - \tfrac{1}{6}\bar{f}_{xxx}|_i^n\,\Delta x^3 + \tfrac{1}{24}\bar{f}_{xxxx}|_i^n\,\Delta x^4 + \cdots \tag{10.19}$$

Subtracting Eq. (10.19) from Eq. (10.18) and solving for $\bar{f}_x|_i^n$ gives

$$\bar{f}_x|_i^n = \frac{\bar{f}_{i+1}^n - \bar{f}_{i-1}^n}{2\,\Delta x} - \tfrac{1}{3}\bar{f}_{xxx}(\xi)\,\Delta x^2 \tag{10.20}$$

where $x_{i-1} \le \xi \le x_{i+1}$. Truncating the remainder term yields the *second-order centered-space* approximation of $\bar{f}_x|_i^n$, denoted by $f_x|_i^n$:

$$f_x|_i^n = \frac{f_{i+1}^n - f_{i-1}^n}{2\,\Delta t} \tag{10.21}$$

Adding Eqs. (10.18) and (10.19) and solving for $\bar{f}_{xx}|_i^n$ gives

$$\bar{f}_{xx}|_i^n = \frac{\bar{f}_{i+1}^n - 2\bar{f}_i^n + \bar{f}_{i-1}^n}{\Delta x^2} - \tfrac{1}{12}\bar{f}_{xxxx}(\xi)\,\Delta x^2 \tag{10.22}$$

where $x_{i-1} \le \xi \le x_{i+1}$. Truncating the remainder term yields the *second-order centered-space* approximation of $\bar{f}_{xx}|_i^n$, denoted by $f_{xx}|_i^n$:

$$f_{xx}|_i^n = \frac{f_{i+1}^n - 2f_i^n + f_{i-1}^n}{\Delta x^2} \tag{10.23}$$

Second-order centered-difference finite difference approximations (FDAs) of $\bar{f}_x$ and $\bar{f}_{xx}$ at time level $n+1$ are obtained simply by replacing n by $n+1$ in Eqs. (10.21) and (10.23).

10.3.4 Finite Difference Equations

Finite difference equations are obtained by substituting the finite difference approximations of the individual exact partial derivatives into the PDE. Two types of FDEs can be developed, depending on the base point chosen for the FDAs. If grid point (i, n) is chosen as the base point of the FDAs, then f_i^{n+1} appears only in the finite difference approximation of $\bar{f_t}$. In that case, the FDE can be solved directly for f_i^{n+1}. Such FDEs are called *explicit* FDEs. However, if grid point $(i, n+1)$ is chosen as the base point of the FDAs, then f_i^{n+1} appears in the finite difference approximations of both $\bar{f_t}$ and $\bar{f_{xx}}$, and f_{i+1}^{n+1} and f_{i-1}^{n+1} appear in the finite difference approximation of $\bar{f_{xx}}$. In that case, f_i^{n+1} cannot be solved for directly, since f_i^{n+1} depends on f_{i+1}^{n+1} and f_{i-1}^{n+1}, which are also unknown. Such FDEs are called *implicit* FDEs.

Explicit FDEs are obviously easier to evaluate numerically than implicit FDEs. However, there are advantages and disadvantages of both explicit and implicit FDEs. Examples of both types of FDEs for solving the unsteady one-dimensional diffusion equation are developed in this chapter.

10.4 THE FORWARD-TIME CENTERED-SPACE (FTCS) METHOD

In this section the unsteady one-dimensional parabolic diffusion equation $\bar{f_t} = \alpha \bar{f_{xx}}$ is solved numerically by the *forward-time centered-space (FTCS) method*. In the FTCS method, the base point for the finite difference approximation (FDA) of the partial differential equation (PDE) is grid point (i, n). The finite difference equation (FDE) approximates the partial derivative $\bar{f_t}$ by the first-order forward-time approximation, Eq. (10.17), and the partial derivative $\bar{f_{xx}}$ by the second-order centered-space approximation, Eq. (10.23). The finite difference stencil is illustrated in Figure 10.9, where the grid points used to approximate $\bar{f_t}$ are denoted by the symbol $\times$ and the grid points used to approximate $\bar{f_{xx}}$ are denoted by the symbol $\cdot$. Thus,

$$\frac{f_i^{n+1} - f_i^n}{\Delta t} = \alpha \frac{f_{i+1}^n - 2f_i^n + f_{i-1}^n}{\Delta x^2} \tag{10.24}$$

Solving for f_i^{n+1} yields the desired FDE:

$$\boxed{f_i^{n+1} = f_i^n + d(f_{i+1}^n - 2f_i^n + f_{i-1}^n)} \tag{10.25}$$

where $d = \alpha \Delta t / \Delta x^2$ is called the *diffusion number*. Equation (10.25) is the FTCS approximation of the unsteady one-dimensional diffusion equation.

The general features of the FTCS approximation of the diffusion equation can be illustrated by applying it to solve the heat diffusion problem described in Section 10.1. Several solutions are presented in Example 10.1.

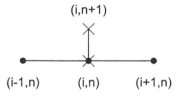

Figure 10.9 The FTCS method stencil.

Example 10.1. The FTCS method applied to the diffusion equation

Let's solve the heat diffusion problem presented in Section 10.1 by the FTCS method with $\Delta x = 0.1$ cm. Let $\Delta t = 0.1$ s, so $d = \alpha\, \Delta t/\Delta x^2 = (0.01)(0.1)/(0.1)^2 = 0.1$. The numerical solution $T(x, t)$ and errors, Error $= [T(x, t) - \bar{T}(x, t)]$, at selected times are presented in Table 10.2 and illustrated in Figure 10.10. Due to the symmetry of the solution, results are tabulated only for $x = 0.0$ to 0.5 cm. It is apparent that the numerical solution is a good approximation of the exact solution. The error at the midpoint (i.e., $x = 0.5$ cm) is the largest error at each time level. This is a direct result of the discontinuity in the slope of the initial temperature distribution at that point. However, the magnitude of this error decreases rapidly as the solution progresses, and the initial discontinuity in the slope is smoothed out. The errors at the remaining locations grow initially due to the accumulation of truncation errors, and reach a maximum value. As the solution progresses, however, the numerical solution approaches the exact asymptotic solution, $\bar{T}(x, \infty) = 0.0$, so the errors decrease and approach zero. The numerical results presented in Table 10.2 present a very favorable impression of the FTCS approximation of the diffusion equation.

The results obtained with $d = 0.1$ are quite good. However, a considerable amount of computational effort is required. The following question naturally arises: Can acceptable results be obtained with larger values of Δt, thus requiring less computational effort? To answer this question, let's rework the problem with $\Delta t = 0.5$ s (i.e., $d = 0.5$), which requires only one-fifth of the computational effort to reach a given time level. The results at selected times are illustrated in Figure 10.11. Although the solution is still

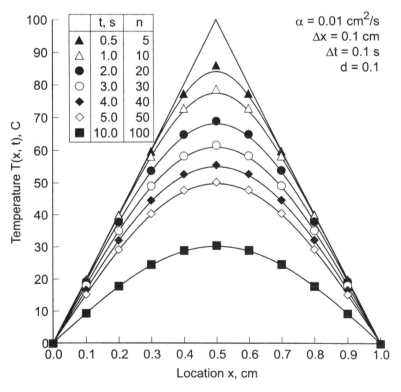

Figure 10.10 Solution by the FTCS method with $d = 0.1$.

Table 10.2   Solution by the FTCS Method for $d = 0.1$

| | | $T(x, t)$, C | | | | |
| | | Error$(x, t) = [T(x, t) - \bar{T}(x, t)]$, C | | | | |
t, s	$x = 0.0$	$x = 0.1$	$x = 0.2$	$x = 0.3$	$x = 0.4$	$x = 0.5$
0.0	0.0	20.0000	40.0000	60.0000	80.0000	100.0000
1.0	0.0	19.9577	39.6777	58.2210	72.8113	78.6741
		−0.0033	0.0226	0.2312	0.7969	1.2417
2.0	0.0	19.3852	37.8084	53.7271	64.8650	68.9146
		0.0339	0.1483	0.3918	0.6887	0.8300
3.0	0.0	18.2119	35.0634	48.9901	58.2956	61.5811
		0.0884	0.2257	0.4152	0.5938	0.6683
4.0	0.0	16.7889	32.1161	44.5141	52.6255	55.4529
		0.1994	0.2576	0.4069	0.5289	0.5766
5.0	0.0	15.3371	29.2500	40.3905	47.6070	50.1072
		0.1312	0.2643	0.3890	0.4815	0.5160
10.0	0.0	9.4421	17.9610	24.7230	29.0653	30.5618
		0.1075	0.2049	0.2825	0.3326	0.3500

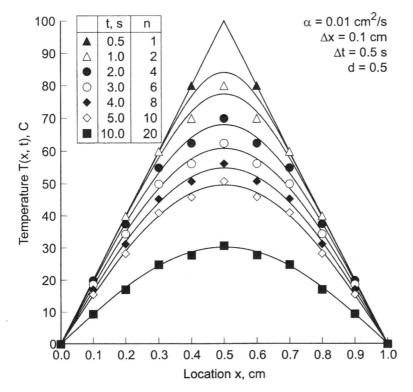

Figure 10.11   Solution by the FTCS method with $d = 0.5$.

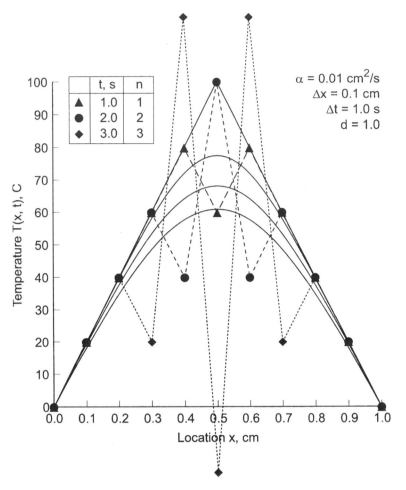

Figure 10.12 Solution by the FTCS method with $d = 1.0$.

reasonable, it is apparent that the solution is no longer smooth. A slight oscillation about the exact solution is apparent. Every numerically computed point is on the opposite side of the exact solution than its two neighbors.

Let's rework the problem again with $\Delta t = 1.0$ s (i.e., $d = 1.0$). The results after each of the first three time steps are illustrated in Figure 10.12. This solution is obviously physically incorrect. Severe oscillations have developed in the solution. These oscillations grow larger and larger as time increases. Values of $T(x, t)$ greater than the initial value of 100.0 and less than the boundary values of 0.0 are predicted. Both of these results are physically impossible. These results are *numerically unstable*.

The results presented in Figure 10.11, while qualitatively correct, appear on the verge of behaving like the results presented in Figure 10.12. The value $d = 0.5$ appears to be the boundary between physically meaningful results for d less than 0.5 and physically meaningless results for d greater than 0.5. To check out this supposition, let's rework the problem for two more values of d: $d = 04$ and $d = 0.6$. These results are illustrated in Figure 10.13 at $t = 6.0$ s. The numerical solution with $d = 0.4$ is obviously modeling physical reality, while the solution with $d = 0.6$ is not. These results support the

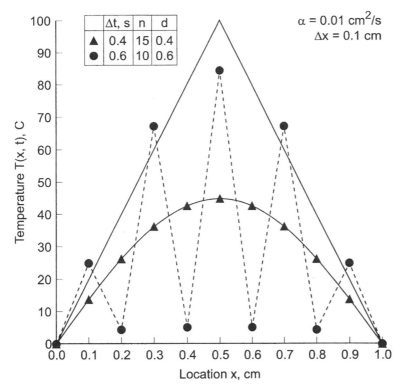

Figure 10.13   Solution by the FTCS method at $t = 6.0$ s.

supposition that the value $d = 0.5$ is the boundary between physically correct solutions and physically incorrect solutions.

The apparent stability restriction, $d \leq 0.5$, imposes a serious limitation on the usefulness of the FTCS method for solving the diffusion equation. One procedure for deciding whether or not a solution is accurate enough is to cut Δx in half and repeat the solution up to the same specified time level to see if the solution changes significantly. For the FTCS method, cutting Δx in half while holding d constant requires a factor of four decreases in Δt. Thus, four times as many time steps are required to reach the previously specified time level, and twice as much work is required for each time step since twice as many physical grid points are involved. Thus, the total computational effort increases by a factor of 8!

To further illustrate the FTCS approximation of the diffusion equation, consider a parametric study in which the temperature at $x = 0.4$ cm and $t = 5.0$ s, $T(0.4, 5.0)$, is calculated using values of $\Delta x = 0.1$, 0.05, 0.025 and 0.0125 cm for values of $d = 0.1$ and 0.5. The value of Δt for each solution is determined by the specified values of Δx and d. The exact solution is $\bar{T}(0.4, 5.0) = 47.1255$ C. The results are presented in Table 10.3. The truncation error of the FTCS method is $0(\Delta t) + 0(\Delta x^2)$. For a constant value of d, $\Delta t = d\,\Delta x^2 / \alpha$. Thus, as Δx is successively halved, Δt is quartered. Consequently, both the $0(\Delta t)$ error term and the $0(\Delta x^2)$ error term, and thus the total error, should decrease by a factor of approximately 4 as Δx is halved for a constant value of d. This result is clearly evident from the results presented in Table 10.3.

Table 10.3 Parametric Study of $T(0.4, 5.0)$ by the FTCS Method

Δx, cm	t, s	$d = 0.1$	t, s	$d = 0.5$
		$T(0.4, 5.0)$, C		
		Error $(0.4, 5.0) = [T(0.4, 5.0) - \bar{T}(0.4, 5.0)]$, C		
0.1	0.1	47.6070	0.5	45.8984
		0.4815		−1.2271
0.05	0.025	47.2449	0.125	47.4117
		0.1194		0.2862
0.025	0.00625	47.1553	0.03125	47.1970
		0.0298		0.0715
0.0125	0.0015625	47.1329	0.0078125	47.1434
		0.0074		0.0178

The forward-time centered-space (FTCS) method has a finite numerical information propagation speed $c_n = \Delta x/\Delta t$. Numerically, information propagates one physical grid increment in all directions during each time step. The diffusion equation has an infinite physical information propagation speed. Consequently, the FTCS method does not correctly model the physical information propagation speed of the diffusion equation. However, the bulk of the information propagates at a finite speed, and the FTCS method yields a reasonable approximation of the exact solution of the diffusion equation. For example, consider the results presented in this section. The solution at $t = 5.0$ s is presented in Table 10.4 for $d = 0.1$ and 0.5. The grid spacing, $\Delta x = 0.1$ cm, is the same for both solutions. The time step is determined from $\Delta t = d \, \Delta x^2/\alpha$. Thus, the numerical information propagation speed $c_n = \Delta x/\Delta t$ is given by

$$c_n = \frac{\Delta x}{\Delta t} = \frac{\Delta x}{d \, \Delta x^2/\alpha} = \frac{\alpha}{d \, \Delta x} = \frac{0.01}{d(0.1)} = \frac{0.1}{d} \text{ cm/s} \tag{10.26}$$

Thus, $c_n = 1.0$ cm/s for $d = 0.1$ and $c_n = 0.2$ cm/s for $d = 0.5$. Consequently, the numerical information propagation speed varies by a factor of five for the results presented in Table 10.4. Those results show very little influence of this large change in the numerical information propagation speed, thus supporting the observation that the bulk of the physical information travels at a finite speed.

The explicit FTCS method can be applied to nonlinear PDEs simply by evaluating the nonlinear coefficients at base point (i, n). Systems of PDEs can be solved simply by

Table 10.4 Solution by the FTCS Method at $t = 5.0$ s

d	$x = 0.0$	$x = 0.1$	$x = 0.2$	$x = 0.3$	$x = 0.4$	$x = 0.5$
			$T(x, 5.0)$, C			
			Error $= [T(x, 5.0) - \bar{T}(x, 5.0)]$, C			
0.1	0.0	15.3371	29.2500	40.3905	47.6070	50.1072
		0.1312	0.2643	0.3890	0.4815	0.5160
0.5	0.0	15.6250	28.3203	41.0156	45.8984	50.7812
		0.4191	−0.6654	1.0141	−1.2271	1.1900

solving the corresponding system of FDEs. Multdimensional problems can be solved simply by adding on the finite difference approximations of the y and z partial derivatives. Consequently, the FTCS method can be used to solve nonlinear PDEs, systems of PDEs, and multidimensional problems by a straightforward extension of the procedure presented in this section. The solution of nonlinear equations and multidimensional problems is discussed further in Section 10.9.

In summary, the forward-time centered-space (FTCS) approximation of the diffusion equation is explicit, single step, consistent, $0(\Delta t) + 0(\Delta x^2)$, conditionally stable, and convergent. It is somewhat inefficient because the time step varies as the square of the spatial grid size.

10.5 CONSISTENCY, ORDER, STABILITY, AND CONVERGENCE

There are four important properties of finite difference methods, for propagation problems governed by parabolic and hyperbolic PDEs, that must be considered before choosing a specific approach. They are:

1. Consistency
2. Order
3. Stability
4. Convergence

These concepts are defined and discussed in this section.

A finite difference equation is *consistent* with a partial differential equation if the difference between the FDE and the PDE (i.e., the truncation error) vanishes as the sizes of the grid spacings go to zero independently.

The *order* of a FDE is the rate at which the global error decreases as the grid sizes approach zero.

A finite difference equation is *stable* if it produces a bounded solution for a stable partial differential equation and is *unstable* if it produces an unbounded solution for a stable PDE.

A finite difference method is *convergent* if the solution of the finite difference equation (i.e., the numerical values) approaches the exact solution of the partial differential equation as the sizes of the grid spacings go to zero.

10.5.1 Consistency and Order

All finite difference equations must be analysed for consistency with the differential equation which they approximate. When the truncation errors of the finite difference approximations of the individual exact partial derivatives are known, proof of consistency is straightforward. When the truncation errors of the individual finite difference approximations are not known, the complete finite difference equation must be analyzed for consistency. That is accomplished by expressing each term in the finite difference equation [i.e., $f(x, t)$, not $\bar{f}(x, t)$] by a Taylor series with a particular base point. The resulting equation, which is called the *modified differential equation (MDE)*, can be simplified to yield the exact form of the truncation error of the complete finite difference equation. Consistency can be investigated by letting the grid spacings go to zero. The order of the FDE is given by the lowest order terms in the MDE.

Warming and Hyett (1974) developed a convenient technique for analyzing the consistency of finite difference equations. The technique involves determining the actual partial differential equation that is solved by a finite difference equation. This actual partial differential equation is called the *modified differential equation* (*MDE*). Following Warming and Hyett, the MDE is determined by expressing each term in a finite difference equation in a Taylor series at some base point. Effectively, this changes the FDE back into a PDE.

Terms expressing in the MDE which do not appear in the original partial differential equation are truncation error terms. Analysis of the truncation error terms leads directly to the determination of *consistency* and *order*. A study of these terms can also yield insight into the stability of the finite difference equation. However, that approach to stability analysis is not presented in this book.

Example 10.2.   Consistency and order analysis of the FTCS method.

As an example, consider the FTCS approximation of the diffusion equation $\bar{f}_t = \alpha \bar{f}_{xx}$ given by Eq. (10.25):

$$f_i^{n+1} = f_i^n + d(f_{i+1}^n - 2f_i^n + f_{i-1}^n) \tag{10.27}$$

where $d = \alpha \, \Delta t/\Delta x^2$ is the diffusion number. Let grid point (i, n) be the base point, and write Taylor series for all of the terms in Eq. (10.27). Thus,

$$f_i^{n+1} = f_i^n + f_t|_i^n \, \Delta t + \tfrac{1}{2} f_{tt}|_i^n \, \Delta t^2 + \tfrac{1}{6} f_{ttt}|_i^n \, \Delta t^3 + \cdots \tag{10.28}$$

$$f_{i\pm1}^n = f_i^n \pm f_x|_i^n \, \Delta x + \tfrac{1}{2} f_{xx}|_i^n \, \Delta x^2 \pm \tfrac{1}{6} f_{xxx}|_i^n \, \Delta x^3 + \tfrac{1}{24} f_{xxxx}|_i^n \, \Delta x^4$$
$$\pm \tfrac{1}{120} f_{xxxxx}|_i^n \, \Delta x^5 + \tfrac{1}{720} f_{xxxxxx}|_i^n \, \Delta x^6 \pm \cdots \tag{10.29}$$

Dropping the notation $|_i^n$ for clarity and substituting Eqs. (10.28) and (10.29) into Eq. (10.27) gives

$$f + f_t \, \Delta t + \tfrac{1}{2} f_{tt} \, \Delta t^2 + \tfrac{1}{6} f_{ttt} \, \Delta t^3 + \cdots$$
$$= f + \frac{\alpha \, \Delta t}{\Delta x^2}(2f + f_{xx} \, \Delta x^2 + \tfrac{1}{12} f_{xxxx} \, \Delta x^4 + \tfrac{1}{360} f_{xxxxxx} \, \Delta x^6 + \cdots - 2f) \tag{10.30}$$

Cancelling zero-order terms (i.e., f), dividing through by Δt, and rearranging terms yields the MDE:

$$\boxed{\begin{aligned} f_t &= \alpha f_{xx} - \tfrac{1}{2} f_{tt} \, \Delta t - \tfrac{1}{6} f_{ttt} \, \Delta t^2 - \cdots \\ &\quad + \tfrac{1}{12} \alpha f_{xxxx} \, \Delta x^2 + \tfrac{1}{360} \alpha f_{xxxxxx} \, \Delta x^4 + \cdots \end{aligned}} \tag{10.31}$$

As $\Delta t \to 0$ and $\Delta x \to 0$, Eq. (10.31) approaches $f_t = \alpha f_{xx}$, which is the diffusion equation. Consequently, Eq. (10.27) is a consistent approximation of the diffusion equation. Equation (10.31) shows that the FDE is $0(\Delta t) + 0(\Delta x^2)$.

10.5.2   Stability

First, the general behavior of the exact solution of the PDE must be considered. If the partial differential equation itself is unstable, then the numerical solution also must be unstable. The concept of stability does not apply in that case. However, if the PDE itself is stable, then the numerical solution must be bounded. The concept of stability applies in that case.

Several methods have been devised to analyze the stability of a finite difference approximation of a PDE. Three methods for analysing the stability of FDEs are

1. The discrete perturbation method
2. The von Neumann method
3. The matrix method

The von Neumann method will be used to analyze stability for all of the finite difference equations developed in this book.

Stability analyses can be peformed only for linear PDEs. Consequently, nonlinear PDEs must be linearized locally, and the FDE which approximates the linearized PDE is analyzed for stability. Experience has shown that the stability criteria obtained for the FDE approximating the linearized PDE also apply to the FDE approximating the nonlinear PDE. Instances of suspected *nonlinear instabilities* have been reported in the literature, but it is not clear whether those phenomena are due to actual instabilities, inconsistent finite difference equations, excessively large grid spacings, inadequate treatment of boundary conditions, or simply incorrect computations. Consequently, in this book, the stability analysis of the finite difference equation which approximates a linearized PDE will be considered sufficient to determine the stability criteria for the FDE, even for nonlinear partial differential equations.

The von Neumann method of stability analysis will be used exclusively in this book. In the von Neumann method, the exact solution of the finite difference equation is obtained for the general Fourier component of a complex Fourier series representation of the initial property distribution. If the solution for the general Fourier component is bounded (either conditionally or unconditionally), then the finite difference equation is stable. If the solution for the general Fourier component is unbounded, then the finite difference equation is unstable.

Consider the FTCS approximation of the unsteady diffusion equation, Eq. (10.25):

$$f_i^{n+1} = f_i^n + d(f_{i+1}^n - 2f_i^n + f_{i-1}^n) \tag{10.32}$$

The exact solution of Eq. (10.32) for a single step can be expressed as

$$\boxed{f_i^{n+1} = Gf_i^n} \tag{10.33}$$

where G, which is called the *amplification factor*, is in general a complex constant. The solution of the FDE at time $T = N\,\Delta t$ is then

$$f_i^N = G^N f_i^0 \tag{10.34}$$

where $f_i^N = f(x_i, T)$ and $f_i^0 = f(x_i, 0)$. For f_i^N to remain bounded,

$$\boxed{|G| \le 1} \tag{10.35}$$

Stability analysis thus reduces to the determination of the single step exact solution of the finite difference equation, that is, the amplification factor G, and an investigation of the conditions necessary to ensure that $|G| \le 1$.

From Eq. (10.32), it is seen that f_i^{n+1} depends not only on f_i^n, but also on f_{i-1}^n and f_{i+1}^n. Consequently, f_{i-1}^n and f_{i+1}^n must be related to f_i^n, so that Eq. (10.32) can be solved explicitly for G. That is accomplished by expressing $f(x, t^n) = F(x)$ in a complex Fourier series. Each component of the Fourier series is propagated forward in time independently of all of the other Fourier components. The complete solution at any subsequent time is simply the sum of the individual Fourier components at that time.

The complex Fourier series for $f(x, t^n) = F(x)$ is given by

$$f(x, t^n) = F(x) = \sum_{m=-\infty}^{\infty} A_m e^{Ik_m x} = \sum_{m=-\infty}^{\infty} F_m \tag{10.36}$$

where the wave number k_m is defined as

$$k_m = 2m\pi/2L$$

Let $f_i^n = f(x_i, t^n)$ consist of the general term F_m. Thus,

$$f_i^n = F_m = A_m e^{Ik_m x_i} = A_m e^{Ik_m(i\,\Delta x)} = A_m e^{Ii(k_m\,\Delta x)} \tag{10.37}$$

Then $f_{i\pm1}^n = f(x_{i\pm1}, t^n)$ is given by

$$f_{i\pm1}^n = A_m e^{Ik_m(x_i\pm\Delta x)} = A_m e^{Ik_m(i\pm1)(\Delta x)} = A_m e^{Ii(k_m\,\Delta x)} e^{\pm I(k_m\,\Delta x)} = f_i^n e^{\pm I(k_m\,\Delta x)} \tag{10.38}$$

Equation (10.38) relates $f_{i\pm1}^n$ to f_i^n. A similar analysis of $f(x_i, t^{n+1})$ gives

$$f_{i\pm1}^{n+1} = f_i^{n+1} e^{\pm I(k_m\,\Delta x)} \tag{10.39}$$

Substituting these results into a FDE expresses the FDE in terms of f_i^n and f_i^{n+1} only, which enables the exact solution, Eq. (10.33), to be determined.

Equations (10.38) and (10.39) apply to the mth component of the complex Fourier series, Eq. (10.36). To ensure stability for a general property distribution, all components of Eq. (10.36) must be considered, and all values of Δx must be considered. This is accomplished by letting m vary from $-\infty$ to $+\infty$ and letting Δx vary from 0 to L. Thus, the product $k_m\,\Delta x$ varies from $-\infty$ to $+\infty$.

The complex exponentials in Eqs. (10.38) and (10.39), that is, $\exp[\pm I(k_m\,\Delta x)]$, represent sine and cosine functions, which have a period of 2π. Consequently, the values of these exponentials repeat themselves with a period of 2π. Thus, it is only necessary to investigate the behavior of the amplification factor G over the range $0 \le (k_m\Delta x) \le 2\pi$. In view of this behavior, the term $(k_m\,\Delta x)$ will be denoted simply as θ, and Eqs. (10.38) and (10.39) can be written as

$$f_{i\pm1}^n = f_i^n e^{\pm I\theta} \quad \text{and} \quad f_{i\pm1}^{n+1} = f_i^{n+1} e^{\pm I\theta} \tag{10.40}$$

Equation (10.40) can be expressed in terms of $\sin\theta$ and $\cos\theta$ using the relationships

$$\cos\theta = \frac{e^{I\theta} + e^{-I\theta}}{2} \quad \text{and} \quad \sin\theta = \frac{e^{I\theta} - e^{-I\theta}}{2I} \tag{10.41}$$

The steps for performing a von Neumann stability analysis of a finite difference equation (FDE) are summarized below.

1. Substitute the complex Fourier components for $f_{i\pm1}^n$ and $f_{i\pm1}^{n+1}$ into the FDE.
2. Express $\exp(\pm I\theta)$ in terms of $\sin\theta$ and $\cos\theta$ and determine the amplification factor, G.
3. Analyze G (i.e., $|G| \le 1$) to determine the stability criteria for the FDE.

Example 10.3.   Stability analysis of the FTCS method

As an example of the von Neumann method of stability analysis, let's perform a stability analysis of the FTCS approximation of the diffusion equation, Eq. (10.25):

$$f_i^{n+1} = f_i^n + d(f_{i+1}^n - 2f_i^n + f_{i-1}^n) \tag{10.42}$$

The required Fourier components are given by Eq. (10.40). Substituting Eq. (10.40) into Eq. (10.42) gives

$$f_i^{n+1} = f_i^n + d(f_i^n e^{I\theta} - 2f_i^n + f_i^n e^{-I\theta}) \tag{10.43a}$$

which can be written as

$$f_i^{n+1} = f_i^n[1 + d(e^{I\theta} + e^{-I\theta} - 2)] = f_i^n\left[1 + 2d\left(\frac{e^{I\theta} + e^{-I\theta}}{2} - 1\right)\right] \tag{10.43b}$$

Introducing the relationship between the cosine and exponential functions, Eq. (10.41), yields

$$f_i^{n+1} = f_i^n[1 + 2d(\cos\theta - 1)] \tag{10.44}$$

Thus, the amplification factor G is defined as

$$\boxed{G = 1 + 2d(\cos\theta - 1)} \tag{10.45}$$

The amplification factor G is the single step exact solution of the finite difference equation for the general Fourier component, which must be less than unity in magnitude to ensure a bounded solution. For a specific wave number k_m and grid spacing Δx, Eq. (10.45) can be analysed to determine the range of values of the diffusion number d for which $|G| \leq 1$. In the infinite Fourier series representation of the property distribution, k_m ranges from $-\infty$ to $+\infty$. The grid spacing Δx can range from zero to any finite value up to L, where L is the length of the physical space. Consequently, the product $(k_m \Delta x) = \theta$ ranges continuously from $-\infty$ to $+\infty$. To ensure that the FDE is stable for an arbitrary property distribution and arbitrary Δx, Eq. (10.45) must be analysed to determine the range of values of d for which $|G| \leq 1$ as θ ranges continuously from $-\infty$ to $+\infty$.

Solving Eq. (10.45) for $|G| \leq 1$ yields

$$-1 \leq 1 + 2d(\cos\theta - 1) \leq 1 \tag{10.46}$$

Note that $d = \alpha \, \Delta t/\Delta x^2$ is always positive. The upper limit is always satisfied for $d \geq 0$ because $(\cos\theta - 1)$ varies between -2 and 0 as θ ranges from $-\infty$ to $+\infty$. From the lower limit,

$$d \leq \frac{1}{1 - \cos\theta} \tag{10.47}$$

The minimum value of d corresponds to the maximum value of $(1 - \cos\theta)$. As θ ranges from $-\infty$ to $+\infty$, $(1 - \cos\theta)$ varies between 0 and 2. Consequently, the minimum value of d is $\frac{1}{2}$. Thus, $|G| \leq 1$ for all values of $\theta = k_m \, \Delta x$ if

$$\boxed{d \leq \tfrac{1}{2}} \tag{10.48}$$

Consequently, the FTCS approximation of the diffusion equation is conditionally stable. This result explains the behavior of the FTCS method for $d = 0.6$ and $d = 1.0$ illustrated in Example 10.1.

The behavior of the amplification factor G also can be determined by graphical methods. Equation (10.45) can be written in the form

$$G = (1 - 2d) + 2d \cos \theta \tag{10.49}$$

In the complex plane, Eq. (10.49) represents an oscillation on the real axis, centered at $(1 - 2d + I0)$, with an amplitude of $2d$, as illustrated in Figure 10.14. The stability boundary, $|G| = 1$, is a circle of radius unity in the complex plane. For G to remain on or inside the unit circle, $-1 \leq |G| \leq 1$, as θ varies from $-\infty$ to $+\infty$, $2d \leq 1$. The graphical approach is very useful when G is a complex function.

10.5.3 Convergence

The proof of convergence of a finite difference method is in the domain of the mathematician. We shall not attempt to prove convergence directly. However, the convergence of a finite difference method is related to the consistency and stability of the finite difference equation. The *Lax equivalence theorem* [Lax (1954) states:

> Given a properly posed linear initial-value problem and a finite difference approximation to it that is consistent, stability is the necessary and sufficient condition for convergence.

Thus, the question of convergence of a finite difference method is answered by a study of the consistency and stability of the finite difference equation. If the finite difference equation is consistent and stable, then the finite difference method is convergent.

The Lax equivalence theorem applies to well-posed, linear, initial-value problems. Many problems in engineering and science are not linear, and nearly all problems involve boundary conditions in addition to the initial conditions. There is no equivalence theorem for such problems. Nonlinear PDEs must be linearized locally, and the FDE that approximates the linearized PDE is analysed for stability. Experience has shown that the stability criteria obtained for the FDE which approximates the linearized PDE also apply to

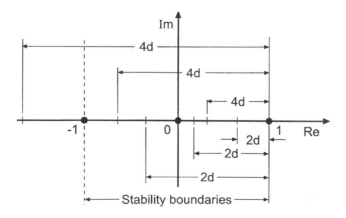

Figure 10.14   Locus of the amplification factor G for the FTCS method.

the FDE which approximates the nonlinear PDE, and that FDEs that are consistent and whose linearized equivalent is stable generally converge, even for nonlinear initial-boundary-value problems.

10.5.3 Summary

The concepts of consistency, stability, and convergence must be considered when choosing a finite difference approximation of a partial differential equation. Consistency is demonstrated by developing the modified differential equation (MDE) and letting the grid increments go to zero. The MDE also yields the order of the FDE. Stability is ascertained by developing the amplification factor, G, and determining the conditions required to ensure that $|G| \leq 1$.

Convergence is assured by the Lax equivalence theorem if the finite difference equation is consistent and stable.

10.6 THE RICHARDSON AND DUFORT-FRANKEL METHODS

The forward-time centered-space (FTCS) approximation of the diffusion equation $\bar{f}_t = \alpha \bar{f}_{xx}$ presented in Section 10.4 has several desirable features. It is an explicit, two-level, single-step method. The finite difference approximation of the spatial derivative is second order. However, the finite difference approximation of the time derivative is only first order. An obvious improvement would be to use a second-order finite difference approximation of the time derivative. The Richardson (leapfrog) and DuFort-Frankel methods are two such methods.

10.6.1 The Richardson (Leapfrog) Method

Richardson (1910) proposed approximating the diffusion equation $\bar{f}_t = \alpha \bar{f}_{xx}$ by replacing the partial derivative $\bar{f}_t$ by the three-level second-order centered-difference approximation based on time levels $n-1$, n, and $n+1$, and replacing the partial derivative $\bar{f}_{xx}$ by the second-order centered-difference approximation, Eq. (10.23). The corresponding finite difference stencil is presented in Figure 10.15. The Taylor series for f_i^{n+1} and f_i^{n-1} with base point (i, n) are given by

$$\bar{f}_i^{n+1} = \bar{f}_i^n + \bar{f}_t|_i^n \, \Delta t + \tfrac{1}{2}\bar{f}_{tt}|_i^n \, \Delta t^2 + \tfrac{1}{6}\bar{f}_{ttt}|_i^n \, \Delta t^3 + \tfrac{1}{24}\bar{f}_{tttt}|_i^n \, \Delta t^4 + \cdots \tag{10.50}$$

$$\bar{f}_i^{n-1} = \bar{f}_i^n - \bar{f}_t|_i^n \, \Delta t + \tfrac{1}{2}\bar{f}_{tt}|_i^n \, \Delta t^2 - \tfrac{1}{6}\bar{f}_{ttt}|_i^n \, \Delta t^3 + \tfrac{1}{24}\bar{f}_{tttt}|_i^n \, \Delta t^4 + \cdots \tag{10.51}$$

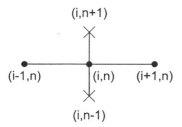

Figure 10.15 The Richardson (leapfrog) method stencil.

Adding Eqs. (10.50) and (10.51) gives

$$\bar{f}_i^{n+1} + \bar{f}_i^{n-1} = 2\bar{f}_i^n + \bar{f}_{tt}|_i^n \, \Delta t^2 + \tfrac{1}{12}\bar{f}_{tttt}|_i^n \, \Delta t^4 + \cdots \tag{10.52}$$

Solving Eq. (10.52) for $\bar{f}_t|_i^n$ gives

$$\bar{f}_t|_i^n = \frac{\bar{f}_i^{n+1} - \bar{f}_i^{n-1}}{2 \, \Delta t} - \tfrac{1}{12}\bar{f}_{tttt}(\tau) \, \Delta t^2 \tag{10.53}$$

where $t^{n-1} \leq \tau \leq t^{n+1}$. Truncating the remainder term yields the second-order centered-time approximation:

$$\boxed{f_t|_i^n = \frac{f_i^{n+1} - f_i^{n-1}}{2 \, \Delta t}} \tag{10.54}$$

Substituting Eqs. (10.54) and (10.23) into the diffusion equation gives

$$\frac{f_i^{n+1} - f_i^{n-1}}{2 \, \Delta t} = \alpha \frac{f_{i+1}^n - 2f_i^n + f_{i-1}^n}{\Delta x^2} \tag{10.55}$$

Solving Eq. (10.55) for f_i^{n+1} yields

$$\boxed{f_i^{n+1} = f_i^{n-1} + 2d(f_{i+1}^n - 2f_i^n + f_{i-1}^n)} \tag{10.56}$$

where $d = \alpha \, \Delta t / \Delta x^2$ is the diffusion number.

The Richardson method appears to be a significant improvement over the FTCS method because of the increased accuracy of the finite difference approximation of $\bar{f}_t$. However, Eq. (10.56) is unconditionally unstable. Performing a von Neumann stability analysis of Eq. (10.56) (where $f_i^n = Gf_i^{n-1}$) yields

$$G = \frac{1}{G} + 4d(\cos \theta - 1) \tag{10.57}$$

which yields

$$G^2 + bG - 1 = 0 \tag{10.58}$$

where $b = -4d(\cos \theta - 1) = 8d \sin^2(\theta/2)$. Solving Eq. (10.58) by the quadratic formula yields

$$G = \frac{-b \pm \sqrt{b^2 + 4}}{2} \tag{10.59}$$

When $b = 0$, $|G| = 1$. For all other values of b, $|G| > 1$. Consequently, the Richardson (leapfrog) method is unconditionally unstable when applied to the diffusion equation.

Since the Richardson method is unconditionally unstable when applied to the diffusion equation, it cannot be used to solve that equation, or any other parabolic PDE. This conclusion applies only to parabolic differential equations. The combination of a three-level centered-time approximation of $\bar{f}_t$ combined with a centered-space approximation of a spatial derivative may be stable when applied to hyperbolic partial differential equations. For example, when applied to the hyperbolic convection equation, where it is known simply as the leapfrog method, a conditionally stable finite difference method is obtained. However, when applied to the convection-diffusion equation, an unconditionally unstable finite difference equation is again obtained. Such occurrences of diametrically opposing results require the numerical analyst to be constantly alert when applying finite difference approximations to solve partial differential equations.

10.6.2 The DuFort-Frankel Method

DuFort and Frankel (1953) proposed a modification to the Richardson method for the diffusion equation $\bar{f}_t = \alpha \bar{f}_{xx}$ which removes the unconditional instability. In fact, the resulting FDE is unconditionally stable. The central grid point value f_i^n in the second-order centered-difference approximation of $\bar{f}_{xx}|_i^n$ is replaced by the average of f_i at time levels $n + 1$ and $n - 1$, that is, $f_i^n = (f_i^{n+1} + f_i^{n-1})/2$. Thus, Eq. (10.55) becomes

$$\frac{f_i^{n+1} - f_i^{n-1}}{2\,\Delta t} = \alpha \frac{f_{i+1}^n - (f_i^{n+1} + f_i^{n-1}) + f_{i-1}^n}{\Delta x^2} \qquad (10.60)$$

At this point, it is not obvious how the truncation error is affected by this replacement. The value f_i^{n+1} appears on both sides of Eq. (10.60). However, it appears linearly, so Eq. (10.60) can be solved explicitly for f_i^{n+1}. Thus,

$$(1 + 2d) f_i^{n+1} = (1 - 2d) f_i^{n-1} + 2d(f_{i+1}^n + f_{i-1}^n) \qquad (10.61)$$

where $d = \alpha\,\Delta t/\Delta x^2$ is the diffusion number.

The modified differential equation (MDE) corresponding to Eq. (10.61) is

$$f_t = \alpha f_{xx} - \tfrac{1}{6} f_{ttt}\,\Delta t^2 - \cdots - \alpha f_{tt}\frac{\Delta t^2}{\Delta x^2} - \tfrac{1}{12}\alpha f_{tttt}\frac{\Delta t^4}{\Delta x^2} + \cdots$$
$$+ \tfrac{1}{12}\alpha f_{xxxx}\,\Delta x^2 + \tfrac{1}{360}\alpha f_{xxxxxx}\,\Delta x^4 + \cdots \qquad (10.62)$$

As $\Delta t \to 0$ and $\Delta x \to 0$, the terms involving the ratio $(\Delta t/\Delta x)$ do not go to zero. In fact, they become indeterminate. Consequently, Eq. (10.62) is not a consistent approximation of the diffusion equation. A von Neumann stability analysis does show that $|G| \le 1$ for all values of d. Thus, Eq. (10.61) is unconditionally stable. However, due to the inconsistency illustrated in Eq. (10.62), the DuFort-Frankel method is not an acceptable method for solving the parabolic diffusion equation, or any other parabolic PDE. Consequently, it will not be considered further.

10.7 IMPLICIT METHODS

The forward-time centered-space (FTCS) method is an example of an explicit finite difference method. In explicit methods, the finite difference approximations of the individual exact partial derivatives in the partial differential equation are evaluated at the known time level n. Consequently, the solution at a point at the solution time level $n + 1$ can be expressed explicitly in terms of the known solution at time level n. Explicit finite difference methods have many desirable features. However, they share one undesirable feature: they are only conditionally stable, or as in the case of the DuFort-Frankel method, they are not consistent with the partial differential equation. Consequently, the allowable time step is generally quite small, and the amount of computational effort required to obtain the solution of some problems is quite large. A procedure for avoiding the time step limitation is obviously desirable. Implicit finite difference methods provide such a procedure.

In implicit methods, the finite difference approximations of the individual exact partial derivatives in the partial differential equation are evaluated at the solution time level $n + 1$. Fortuitously, implicit difference methods are unconditionally stable. There is no limit on the allowable time step required to achieve a numerically stable solution. There is, of course, some practical limit on the time step required to maintain the truncation errors within reasonable limits, but this is not a stability consideration; it is an accuracy consideration. Implicit methods do have some disadvantages, however. The foremost disadvantage is that the solution at a point in the solution time level $n + 1$ depends on the solution at neighboring points in the solution time level, which are also unknown. Consequently, the solution is implied in terms of unknown function values, and a system of finite difference equations must be solved to obtain the solution at each time level. Additional complexities arise when the partial differential equations are nonlinear. In that case, a system of nonlinear finite difference equations results, which must be solved by some manner of linearization and/or iteration.

In spite of their disadvantages, the advantage of unconditional stability makes implicit finite difference methods attractive. Consequently, two implicit finite difference methods are presented in this section: the backward-time centered-space (BTCS) method and the Crank-Nicolson (1947) method.

10.7.1 The Backward-Time Centered-Space (BTCS) Method

In this subsection the unsteady one-dimensional diffusion equation, $\bar{f}_t = \alpha \bar{f}_{xx}$, is solved by the *backward-time centered-space (BTCS) method*. This method is also called the *fully implicit method*. The finite difference equation which approximates the partial differential equation is obtained by replacing the exact partial derivative $\bar{f}_t$ by the first-order backward-time approximation, which is developed below, and the exact partial derivative $\bar{f}_{xx}$ by the second-order centered-space approximation, Eq. (10.23), evaluated at time level $n + 1$. The finite difference stencil is illustrated in Figure 10.16. The Taylor series for $\bar{f}_i^n$ with base point $(i, n + 1)$ is given by

$$\bar{f}_i^n = \bar{f}_i^{n+1} + \bar{f}_t|_i^{n+1}(-\Delta t) + \tfrac{1}{2}\bar{f}_{tt}|_i^{n+1}(-\Delta t)^2 + \cdots \tag{10.63}$$

Solving Eq. (10.63) for $\bar{f}_t|_i^{n+1}$ gives

$$\bar{f}_t|_i^{n+1} = \frac{\bar{f}_i^{n+1} - \bar{f}_i^n}{\Delta t} + \tfrac{1}{2}\bar{f}_{tt}(\tau)\,\Delta t \tag{10.64}$$

Truncating the remainder term yields the first-order backward-time approximation:

$$\boxed{f_t|_i^{n+1} = \frac{f_i^{n+1} - f_i^n}{\Delta t}} \tag{10.65}$$

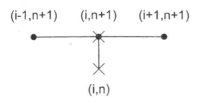

Figure 10.16 The BTCS method stencil.

Substituting Eqs. (10.65) and (10.23) into the diffusion equation, $\bar{f}_t = \alpha \bar{f}_{xx}$, yields

$$\frac{f_i^{n+1} - f_i^n}{\Delta t} = \alpha \frac{f_{i+1}^{n+1} - 2f_i^{n+1} + f_{i-1}^{n+1}}{\Delta x^2} \tag{10.66}$$

Rearranging Eq. (10.66) yields the implicit BTCS FDE:

$$\boxed{-df_{i-1}^{n+1} + (1+2)f_i^{n+1} - df_{i+1}^{n+1} = f_i^n} \tag{10.67}$$

where $d = \alpha \Delta t / \Delta x^2$ is the diffusion number.

Equation (10.67) cannot be solved explicitly for f_i^{n+1} because the two unknown neighboring values f_{i-1}^{n+1} and f_{i+1}^{n+1} also appear in the equation. The value of f_i^{n+1} is implied in Eq. (10.67), however. Finite difference equations in which the unknown solution value f_i^{n+1} is implied in terms of its unknown neighbors rather than being explicitly given in terms of known initial values are called implicit FDEs.

The modified differential equation (MDE) corresponding to Eq. (10.67) is

$$f_t = \alpha f_{xx} + \tfrac{1}{2} f_{tt} \Delta t - \tfrac{1}{6} f_{ttt} \Delta t^2 + \cdots + \tfrac{1}{12} \alpha f_{xxxx} \Delta x^2 + \tfrac{1}{360} \alpha f_{xxxxxx} \Delta x^4 + \cdots \tag{10.68}$$

As $\Delta t \to 0$ and $\Delta x \to 0$, all of the truncation error terms go to zero, and Eq. (10.68) approaches $f_t = \alpha f_{xx}$. Consequently, Eq. (10.67) is consistent with the diffusion equation. The truncation error is $0(\Delta t) + 0(\Delta x^2)$. From a von Neumann stability analysis, the amplification factor G is

$$G = \frac{1}{1 + 2d(1 - \cos\theta)} \tag{10.69}$$

The term $(1 - \cos\theta)$ is greater than or equal to zero for all values of $\theta = (k_m \Delta x)$. Consequently, the denominator of Eq. (10.69) is always ≥ 1. Thus, $|G| \leq 1$ for all positive values of d, and Eq. (10.67) is unconditionally stable. The BTCS approximation of the diffusion equation is consistent and unconditionally stable. Consequently, by the Lax Equivalence Theorem, the BTCS method is convergent.

Consider now the solution of the unsteady one-dimensional diffusion equation by the BTCS method. The finite difference grid for advancing the solution from time level n

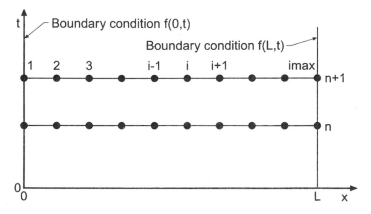

Figure 10.17 Finite difference grid for implicit methods.

to time level $n + 1$ is illustrated in Figure 10.17. For Dirichlet boundary conditions (i.e., the value of the function is specified at the boundaries), the finite difference equation must be applied only at the interior points, points 2 to imax $- 1$. At grid point 1, $f_1^{n+1} = \bar{f}(0, t)$, and at grid point imax, $f_{imax}^{n+1} = \bar{f}(L, t)$. The following set of simultaneous linear equations is obtained:

$$(1 + 2d)f_2^{n+1} - df_3^{n+1} = f_2^n + d\bar{f}(0, t) = b_2$$
$$-df_2^{n+1} + (1 + 2d)f_3^{n+1} - df_4^{n+1} = f_3^n = b_3$$
$$-df_3^{n+1} + (1 + 2d)f_4^{n+1} - df_5^{n+1} = f_4^n = b_4 \qquad (10.70)$$
$$\cdots\cdots\cdots\cdots\cdots\cdots\cdots\cdots\cdots\cdots\cdots\cdots\cdots\cdots$$
$$-df_{imax-2}^{n+1} + (1 + 2d)f_{imax-1}^{n+1} = f_{imax-1}^n + d\bar{f}(L, t) = b_{imax-1}$$

Equation (10.70) comprises a tridiagonal system of linear algebraic equations. That system of equations may be written as

$$\mathbf{A}\mathbf{f}^{n+1} = \mathbf{b} \qquad (10.71)$$

where $\mathbf{A}$ is the (imax $- 2 \times$ imax $- 2$) coefficient matrix, $\mathbf{f}^{n+1}$ is the (imax $- 2 \times 1$) solution column vector, and $\mathbf{b}$ is the (imax $- 2 \times 1$) column vector of nonhomogeneous terms. Equation (10.71) can be solved very efficiently by the Thomas algorithm presented in Section 1.5. Since the coefficient matrix $\mathbf{A}$ does not change from one time level to the next, LU factorization can be employed with the Thomas algorithm to reduce the computational effort even further.

The FTCS method and the BTCS method are both first order in time and second order in space. So what advantage, if any, does the BTCS method have over the FTCS method? The BTCS method is unconditionally stable. The time step can be much larger than the time step for the FTCS method. Consequently, the solution at a given time level can be reached with much less computational effort by taking time steps much larger than those allowed for the FTCS method. In fact, the time step is limited only by accuracy requirements.

Example 10.4.   The BTCS method applied to the diffusion equation.

Let's solve the heat diffusion problem described in Section 10.1 by the BTCS method with $\Delta x = 0.1$ cm. For the first case, let $\Delta t = 0.5$ s, so $d = \alpha \, \Delta t / \Delta x^2 = 0.5$. The results at selected time levels are presented in Figure 10.18. It is obvious that the numerical solution is a good approximation of the exact solution. The general features of the numerical solution presented in Figure 10.18 are qualitatively similar to the numerical solution obtained by the FTCS method for $\Delta t = 0.5$ s and d $= 0.5$, which is presented in Figure 10.11. Although the results obtained by the BTCS method are smoother, there is no major difference. Consequently, there is no significant advantage of the BTCS method over the FTCS method for $d = 0.5$.

The numerical solutions at $t = 10.0$ s, obtained with $\Delta t = 1.0, 2.5, 5.0$, and 10.0 s, for which $d = 1.0, 2.5, 5.0$, and 10.0, respectively, are presented in Figure 10.19. These results clearly demonstrate the unconditional stability of the BTCS method. However, the numerical solution lags the exact solution seriously for the larger values of d. The advantage of the BTCS method over explicit methods is now apparent. If the decreased accuracy associated with the larger time steps is acceptable, then the solution can be

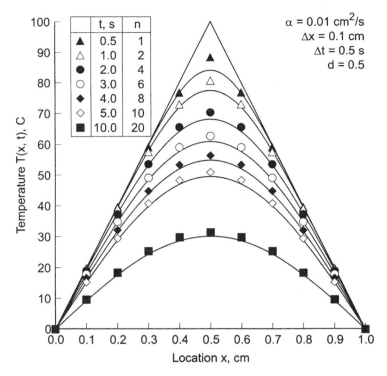

Figure 10.18 Solution by the BTCS method with $d = 0.5$.

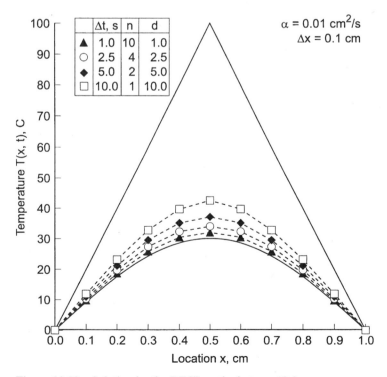

Figure 10.19 Solution by the BTCS method at $t = 10.0$ s.

obtained with less computational effort with the BTCS method than with the FTCS method. However, the results presented in Figure 10.19 suggest that large values of the diffusion number, d, lead to serious decreases in the accuracy of the solution.

The final results presented for the BTCS method are a parametric study in which the value of $T(0.4, 5.0)$ is calculated using values of $\Delta x = 0.1$, 0.05, 0.025, and 0.0125 cm, for values of $d = 0.5$, 1.0, 2.0, and 5.0. The value of Δt for each solution is determined by the specified values of Δx and d. The exact solution is $\bar{T}(0.5, 5.0) = 47.1255$ C. The results are presented in Table 10.5. The truncation error of the BTCS method is $0(\Delta t) + 0(\Delta x^2)$. For a constant value of d, $\Delta t = d \, \Delta x^2 / \alpha$. Thus, as Δx is successively halved, Δt is quartered. Consequently, both the $0(\Delta t)$ error term and the $0(\Delta x^2)$ error term, and thus the total error, should decrease by a factor of approximately 4 as Δx is halved for a constant value of d. This result is clearly evident from the results presented in Table 10.5.

The backward-time centered-space (BTCS) method has an infinite numerical information propagation speed. Numerically, information propagates throughout the entire physical space during each time step. The diffusion equation has an infinite physical information propagation speed. Consequently, the BTCS method correctly models this feature of the diffusion equation.

When a PDE is nonlinear, the corresponding FDE is nonlinear. Consequently, a system of nonlinear FDEs must be solved. For one-dimensional problems, this situation is the same as described in Section 8.7 for ordinary differential equations, and the solution procedures described there also apply here. When systems of nonlinear PDEs are considered, a corresponding system of nonlinear FDEs is obtained at each solution point, and the combined equations at all of the solution points yield block tridiagonal systems of FDEs. For multidimensional physical spaces, banded matrices result. Such problems are frequently solved by alternating-direction-implicit (ADI) methods or approximate-factorization-implicit (AFI) methods, as described by Peaceman and Rachford (1955) and Douglas (1962). The solution of nonlinear equations and multidimensional problems are discussed in Section 10.9. The solution of a coupled system of several nonlinear multidimensional PDEs by an implicit finite difference method is indeed a formidable task.

Table 10.5 Parametric Study of $T(0.4, 5.0)$ by the BTCS Method

| | $T(0.4, 5.0)$, C | | | |
| | Error $(0.4, 5.0) = [T(0.4, 5.0) - \bar{T}(0.4, 5.0)]$, C | | | |
Δx, cm	$d = 0.5$	$d = 1.0$	$d = 2.5$	$d = 5.0$
0.1	48.3810	49.0088	50.7417	53.1162
	1.2555	1.8830	3.6162	5.9907
0.05	47.4361	47.5948	48.0665	48.8320
	0.3106	0.4693	0.9410	1.7065
0.025	47.2029	47.2426	47.3614	47.5587
	0.0774	0.1171	0.2359	0.4332
0.0125	47.1448	47.1548	47.1845	47.2340
	0.0193	0.0293	0.0590	0.1085

In summary, the backward-time centered-space approximation of the diffusion equation is implicit, single step, consistent, $0(\Delta t) + 0(\Delta x^2)$, unconditionally stable, and convergent. Consequently, the time step is chosen based on accuracy requirements, not stability requirements. The BTCS method can be used to solve nonlinear PDEs, systems of PDEs, and multidimensional problems. However, in those cases, the solution procedure becomes quite complicated.

10.7.2 The Crank-Nicolson Method

The backward-time centered-space (BTCS) approximation of the diffusion equation $\bar{f}_t = \alpha \bar{f}_{xx}$, presented in the previous subsection, has a major advantage over explicit methods: It is unconditionally stable. It is an implicit single step method. The finite difference approximation of the spatial derivative is second order. However, the finite difference approximation of the time derivative is only first order. Using a second-order finite difference approximation of the time derivative would be an obvious improvement.

Crank and Nicolson (1947) proposed approximating the partial derivative $\bar{f}_t$ at grid point $(i, n + 1/2)$ by the second-order centered-time approximation obtained by combining Taylor series for $\bar{f}_i^{n+1}$ and $\bar{f}_k^n$. Thus,

$$\bar{f}_i^{n+1} = \bar{f}_i^{n+1/2} + \bar{f}_t|_i^{n+1/2}\left(\frac{\Delta t}{2}\right) + \tfrac{1}{2}\bar{f}_{tt}|_i^{n+1/2}\left(\frac{\Delta t}{2}\right)^2 + \tfrac{1}{6}\bar{f}_{ttt}|_i^{n+1/2}\left(\frac{\Delta t}{2}\right)^3 + \cdots \qquad (10.72)$$

$$\bar{f}_i^{n} = \bar{f}_i^{n+1/2} - \bar{f}_t|_i^{n+1/2}\left(\frac{\Delta t}{2}\right) + \tfrac{1}{2}\bar{f}_{tt}|_i^{n+1/2}\left(\frac{\Delta t}{2}\right)^2 - \tfrac{1}{6}\bar{f}_{ttt}|_i^{n+1/2}\left(\frac{\Delta t}{2}\right)^3 + \cdots \qquad (10.73)$$

Subtracting these two equations and solving for $\bar{f}_t|_i^{n+1/2}$ gives

$$\bar{f}_t|_i^{n+1/2} = \frac{\bar{f}_i^{n+1} - \bar{f}_i^{n}}{\Delta t} - \tfrac{1}{24}\bar{f}_{ttt}(\tau)\,\Delta t^2 \qquad (10.74)$$

where $t^n \leq \tau \leq t^{n+1}$. Truncating the remainder term in Eq. (10.74) yields the second-order centered-time approximation of $\bar{f}_t$:

$$f_t|_i^{n+1/2} = \frac{f_i^{n+1} - f_i^{n}}{\Delta t} \qquad (10.75)$$

The partial derivative $\bar{f}_{xx}$ at grid point $(i, n + \tfrac{1}{2})$ is approximated by

$$\bar{f}_{xx}|_i^{n+1/2} = \tfrac{1}{2}(f_{xx}|_i^{n+1} + \bar{f}_{xx}|_i^{n}) \qquad (10.76)$$

The order of the FDE obtained using Eqs. (10.75) and (10.76) is expected to be $0(\Delta t^2) + 0(\Delta x^2)$, but that must be proven from the MDE. The partial derivative $\bar{f}_{xx}$ at time levels n and $n + 1$ are approximated by the second-order centered-difference approximation, Eq. (10.23), applied at time levels n and $n + 1$, respectively. The finite difference stencil is illustrated in Figure 10.20. The resulting finite difference approximation of the one-dimensional diffusion equation is

$$\frac{f_i^{n+1} - f_i^{n}}{\Delta t} = \alpha\tfrac{1}{2}\left(\frac{f_{i+1}^{n+1} - 2f_i^{n+1} + f_{i-1}^{n+1}}{\Delta x^2} + \frac{f_{i+1}^{n} - 2f_i^{n} + f_{i-1}^{n}}{\Delta x^2}\right) \qquad (10.77)$$

(i-1,n+1)   (i,n+1)   (i+1,n+1)

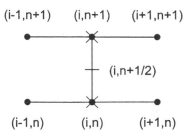

(i,n+1/2)

(i-1,n)   (i,n)   (i+1,n)

Figure 10.20   The Crank-Nicolson method stencil.

Rearranging Eq. (10.77) yields the Crank-Nicolson finite difference equation:

$$-d_{i-1}^{n+1} + 2(1+d)f_i^{n+1} - df_{i+1}^{n+1}$$
$$= df_{i-1}^n + 2(1-d)f_i^n + df_{i+1}^n \tag{10.78}$$

where $d = \alpha \, \Delta t / \Delta x^2$ is the diffusion number.

The modified differential equation (MDE) obtained by writing Taylor series for $f(x,t)$ about point $(i, n+\tfrac{1}{2})$ is

$$f_t = \alpha f_{xx} - \tfrac{1}{24} f_{ttt} \, \Delta t^2 + \cdots + \tfrac{1}{8} \alpha f_{xxtt} \, \Delta t^2$$
$$+ \tfrac{1}{12} \alpha f_{xxxx} \Delta x^2 + \tfrac{1}{360} \alpha f_{xxxxxx} \Delta x^4 + \cdots \tag{10.79}$$

As $\Delta t \to 0$ and $\Delta x \to 0$, all of the truncation error terms go to zero, and Eq. (10.79) approaches $f_t = \alpha f_{xx}$. Consequently, Eq. (10.78) is consistent with the diffusion equation. The leading truncation error terms are $0(\Delta t^2)$ and $0(\Delta x^2)$. From a von Neumann stability analysis, the amplification factor G is

$$G = \frac{1 - d(1 - \cos\theta)}{1 + d(1 - \cos\theta)} \tag{10.80}$$

The term $(1 - \cos\theta) \geq 0$ for all values of $\theta = (k_m \, \Delta x)$. Consequently $|G| \leq 1$ for all positive values of d, and Eq. (10.78) is unconditionally stable. The Crank-Nicolson approximation of the diffusion equation is consistent and unconditionally stable. Consequently, by the Lax equivalence theorem, the Crank-Nicolson approximation of the diffusion equation is convergent.

Now consider the solution of the unsteady one-dimensional diffusion equation by the Crank-Nicolson method. The finite difference grid for advancing the solution from time level n to time level $n+1$ is illustrated in Figure 10.17. For Dirichlet boundary conditions (i.e., the value of the function is specified at the boundaries), the finite difference equation must be applied only at the interior points, points 2 to imax -1. At grid point 1, $f_1^{n+1} = \bar{f}(0,t)$, and at grid point imax, $f_{imax}^{n+1} = \bar{f}(L,t)$. The following set of simultaneous linear equations is obtained:

$$2(1+d)f_2^{n+1} - df_3^{n+1} = df_1^n + 2(1-d)f_2^n + df_3^n + d\bar{f}(0,t) = b_2$$
$$-df_2^{n+1} + 2(1+d)f_3^{n+1} - df_4^{n+1} = df_2^n + 2(1-d)f_3^n + df_4^n = b_3$$
$$-df_3^{n+1} + 2(1+d)f_4^{n+1} - df_5^{n+1} = df_3^n + 2(1-d)f_4^n + df_5^n = b_4$$

$$\cdots\cdots\cdots\cdots\cdots\cdots\cdots\cdots\cdots\cdots\cdots\cdots\cdots\cdots\cdots\cdots\cdots\cdots\cdots\cdots\cdots\cdots\cdots\cdots$$

$$-df_{imax-2}^{n+1} + 2(1+d)f_{imax-1}^{n+1} = df_{imax-2}^n + 2(1-d)f_{imax-1}^n + df_{imax}^n + d\bar{f}(L,t) = b_{imax-1}$$
$$\tag{10.81}$$

Equation (10.81) comprises a tridiagonal system of linear algebraic equations, which is very similar to the system of equations developed in Section 10.7.1. for the backward-time centered-space (BTCS) method. Consequently, the present system of equations can be solved by the Thomas algorithm, as discussed in that section.

Like the backward-time centered space (BTCS) method, the Crank-Nicolson method is unconditionally stable. Consequently, the solution at a given time level can be reached with much less computational effort by taking large time steps. The time step is limited only by accuracy requirements.

Example 10.5.   The Crank-Nicolson method applied to the diffusion equation

Let's solve the heat diffusion problem described in Section 10.1 by the Crank-Nicolson method with $\Delta x = 0.1$ cm. Let $\Delta t = 0.5$ s, so $d = 0.5$. The numerical solution is presented in Figure 10.21. As expected, the results are more accurate than the corresponding results presented in Figure 10.18 for the BTCS method.

The numerical solution at $t = 10.0$ s, obtained with $\Delta t = 1.0$, 2.5, 5.0, and 10.0 s, for which $d = 1.0$, 2.5, 5.0, and 10.0, respectively, is presented in Figure 10.22. These results clearly demonstrate the unconditional stability of the Crank-Nicolson method. However, an overshoot and oscillation exists in the numerical solution for all values of d considered in Figure 10.22. These oscillations are not due to an instability. They are an inherent feature of the Crank-Nicolson method when the diffusion number becomes large. The source of these oscillations can be determined by examining the eigenvalues of the

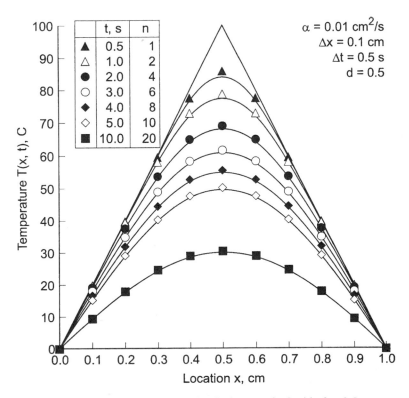

Figure 10.21   Solution by the Crank-Nicolson method with $d = 0.5$.

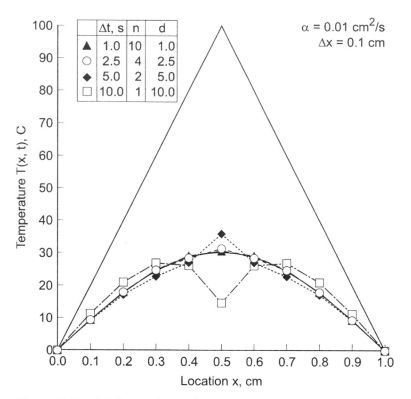

Figure 10.22 Solution by the Crank-Nicolson method at $t = 10.0$ s.

coefficient matrix for the complete system of linear equations, Eq. (10.81). The results presented in Figure 10.22 suggest that values of the diffusion number d much greater than 1.0 lead to a serious loss of accuracy in the transient solution.

The final results presented for the Crank-Nicolson method are a parametric study in which the value of $T(0.4, 5.0)$ is calculated using values of $\Delta x = 0.1$, 0.05, 0.025, and 0.0125 cm, for values of $d = 0.5$, 1.0, 2.0, and 5.0. The value of Δt for each solution is determined by the specified values of Δx and d. The exact solution is $\bar{T}(0.4, 5, 0) = 46.1255$ C. Results are presented in Table 10.6. The truncation error of the Crank-Nicolson method is $0(\Delta t^2) + 0(\Delta x^2)$. For a given value of d, $\Delta t = d \, \Delta x^2/\alpha$. Thus, as Δx is successively halved, Δt is quartered. Consequently, the $0(\Delta t^2)$ term should decrease by a factor of approximately 16 and the $0(\Delta x^2)$ term should decrease by a factor of approximately 4 as Δx is halved for a constant value of d. The results presented in Table 10.6 show that the total error decreases by a factor of approximately 4, indicating that the $0(\Delta x^2)$ term is the dominant error term.

The Crank-Nicolson method has an infinite numerical information propagation speed. Numerically, information propagates throughout the entire physical space during each time step. The diffusion equation has an infinite physical information propagation speed. Consequently, the Crank-Nicolson method correctly models this feature of the diffusion equation.

Table 10.6   Parametric Study of $T(0.4, 5.0)$ by the Crank-Nicolson Method

| | $T(0.4, 5.0)$, C | | | |
| | Error $(0.4, 5.0) = [T(0.4, 5.0) - \bar{T}(0.4, 5.0)]$, C | | | |
Δx, cm	$d = 0.5$	$d = 1.0$	$d = 2.5$	$d = 5.0$
0.1	47.7269	47.7048	46.1511	47.9236
	0.6014	0.5793	−0.9744	0.7981
0.05	47.2762	47.2744	47.2831	47.0156
	0.1507	0.1489	0.1576	−0.1099
0.025	47.1632	47.1631	47.1623	47.1584
	0.0377	0.0376	0.0368	0.0329
0.0125	47.1349	47.1349	47.1349	47.1347
	0.0094	0.0094	0.0094	0.0092

The implicit Crank-Nicolson method can be used to solve nonlinear PDEs, systems of PDEs, and multidimensional PDEs. The techniques and problems are the same as those discussed in the previous section for the BTCS method and in Section 10.9.

In summary, the Crank-Nicolson approximation of the diffusion equation is implicit, single step, consistent, $0(\Delta t^2) + 0(\Delta x^2)$, unconditionally stable, and convergent. Consequently, the time step size is chosen based on accuracy requirements, not stability requirements.

10.8   DERIVATIVE BOUNDARY CONDITIONS

All of the finite difference solutions of the unsteady one-dimensional diffusion equation presented thus far in this chapter have been for Dirichlet boundary conditions, that is, the values of the function are specified on the boundaries. In this section, a procedure for implementing derivative, or Neumann, boundary conditions is presented.

The general features of a derivative boundary condition can be illustrated by considering a modification of the heat diffusion problem presented in Section 10.1, in which the thickness of the plate is $L = 0.5$ cm and the boundary condition on the right side of the plate is

$$\bar{T}_x(0.5, t) = 0.0 \qquad\qquad\qquad (10.82)$$

The initial condition, $\bar{T}(x, 0.0)$, and the boundary condition of the left side, $\bar{T}(0.0, t)$, are the same as in the original problem. This problem is identical to the original problem due to the symmetry of the initial condition and the boundary conditions. The exact solution is given by Eq. (10.3), tabulated in Table 10.1, and illustrated in Figure 10.23. The solution smoothly approaches the asymptotic steady state solution, $\bar{T}(x, \infty) = 0.0$.

In this section, we will solve this problem numerically using the forward-time centered-space (FTCS) method at the interior points. The implementation of a derivative boundary condition does not depend on whether the problem is an equilibrium problem or a propagation problem, nor does the number of space dimensions alter the procedure. Consequently, the procedure presented in Section 8.5 for implementing a derivative boundary condition for one-dimensional equilibrium problems can be applied directly to

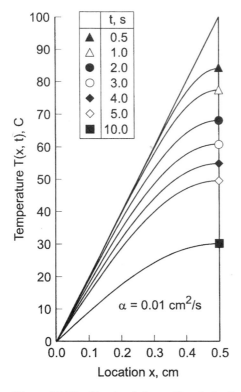

Figure 10.23   Exact solution with a derivative BC.

one-dimensional propagation problems. The finite difference grid for implementing a right-hand side derivative boundary condition is illustrated in Figure 10.24.

Let's apply the FTCS finite difference equation (FDE) at grid point I on the right-hand boundary, as illustrated in Figure 10.24. The FDE is Eq. (10.25):

$$f_I^{n+1} = f_I^n + d(f_{I-1}^n - 2f_I^n + f_{I+1}^n) \tag{10.83}$$

Grid point $I + 1$ is outside of the solution domain, so f_{I+1}^n is not defined. However, a value for f_{I+1}^n can be determined from the boundary condition on the right-hand boundary $\bar{f}_x|_I^n = $ known.

The finite difference approximation employed in Eq. (10.83) for the space derivative $\bar{f}_{xx}$ is second order. It's desirable to match this truncation error by using a second-order

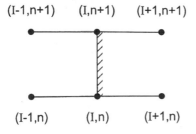

Figure 10.24   Finite difference stencil for right-hand side derivative BC.

finite difference approximation for the derivative boundary condition $\bar{f}_x|_I^n = $ known. Applying Eq. (10.21) at grid point I gives

$$\bar{f}_x|_I^n = \frac{\bar{f}_{I+1}^n - \bar{f}_{I-1}^n}{2\,\Delta x} + 0(\Delta x^2) \tag{10.84}$$

Truncating the remainder term and solving Eq. (10.84) for f_{I+1}^n gives

$$f_{I+1}^n = f_{I-1}^n + 2\bar{f}_x|_I^n\,\Delta x \tag{10.85}$$

Substituting Eq. (10.85) into Eq. (10.83) yields

$$f_I^{n+1} = f_I^n + d[\,f_{I-1}^n - 2f_I^n + (\,f_{I-1}^n + 2\bar{f}_x|_I^n\,\Delta x)] \tag{10.86}$$

Rearranging Eq. (10.86) gives the FDE applicable at the right-hand side boundary:

$$\boxed{f_I^{n+1} = f_I^n + 2d(\,f_{I-1}^n - f_I^n + \bar{f}_x|_I^n\,\Delta x)} \tag{10.87}$$

Equation (10.87) must be examined for consistency and stability. Consider the present example where $\bar{f}_x|_I^n = 0$. The modified differential equation (MDE) corresponding to Eq. (10.87) is

$$f_t = \alpha f_{xx} - \tfrac{1}{2} f_{tt}\,\Delta t - \tfrac{1}{3} f_{xxx}\Delta x + \tfrac{1}{12} f_{xxxx}\Delta x^2 + \cdots \tag{10.88}$$

As $\Delta x \to 0$ and $\Delta t \to 0$, all of the truncation error terms go to zero, and Eq. (10.88) approaches $f_t = \alpha f_{xx}$. Consequently, Eq. (10.87) is consistent with the diffusion equation. A matrix method stability analysis yields the stability criterion $d \le \tfrac{1}{2}$.

Example 10.6. Derivative boundary condition for the diffusion equation.

Let's work the example problem with $\Delta x = 0.1$ cm and $\Delta t = 0.1$ s, so $d = 0.1$, using Eq. (10.25) at the interior points and Eq. (10.87) at the right-hand boundary. The results are presented in Figure 10.25. The numerical solution is a good approximation of the exact solution. These results are identical to the results presented in Figure 10.10 for this problem which has a symmetrical initial condition and symmetrical boundary conditions.

10.9 NONLINEAR EQUATIONS AND MULTDIMENSIONAL PROBLEMS

All of the finite difference equations and examples presented so far in this chapter are for the linear unsteady one-dimensional diffusion equation. Some of the problems which arise for nonlinear partial differential equations and multidimensional problems are discussed briefly in this section.

10.9.1 Nonlinear Equations

Consider the nonlinear unsteady one-dimensional convection-diffusion equation:

$$\boxed{\bar{f}_t + u(\bar{f})\,\bar{f}_x = \alpha(\bar{f})\,\bar{f}_{xx}} \tag{10.89}$$

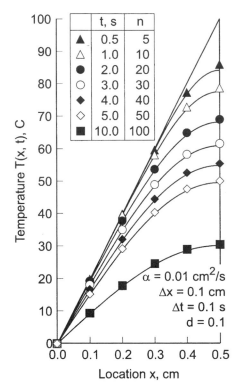

Figure 10.25   Solution with a derivative BC.

where the convection velocity u and the diffusion coefficient α depend on $\bar{f}$. The FTCS approximation of Eq. (10.89) is

$$\frac{f_i^{n+1} - f_i^n}{\Delta t} + u_i^n \frac{f_{i+1}^n - f_{i-1}^n}{2\,\Delta x} = \alpha_i^n \frac{f_{i+1}^n - 2f_i^n + f_{i-1}^n}{\Delta x^2} \tag{10.90}$$

The nonlinear coefficients are simply evaluated at the base point (i, n) where f_i^n, and hence u_i^n and α_i^n, are known. The FDE can be solved directly for f_i^{n+1}. The nonlinear coefficients cause no numerical complexities. This type of result is typical of all explicit finite difference approximations.

The BTCS approximations of Eq. (10.89) is

$$\frac{f_i^{n+1} - f_i^n}{\Delta t} + u_i^{n+1} \frac{f_{i+1}^{n+1} - f_{i-1}^{n+1}}{2\,\Delta x} = \alpha_i^{n+1} \frac{f_{i+1}^{n+1} - 2f_i^{n+1} + f_{i-1}^{n+1}}{\Delta x^2} \tag{10.91}$$

The nonlinear coefficients present a serious numerical problem. The nonlinear coefficients u_i^{n+1} and α_i^{n+1} depend on f_i^{n+1}, which is unknown. Equation (10.91), when applied at every point in the finite difference grid, yields a system of coupled nonlinear finite difference equations. The system of coupled nonlinear FDEs can be solved by simply lagging the nonlinear coefficients (i.e., letting $u_i^{n+1} = u_i^n$ and $\alpha_i^{n+1} = \alpha_i^n$), by iteration, by Newton's method, or by time linearization. Iteration and Newton's method are discussed in Section 8.7 for nonlinear one-dimensional boundary-value problems. Time linearization is presented in Section 7.11 for nonlinear one-dimensional initial-value problems. The Taylor series for $\bar{f}_i^{n+1}$ with base point (i, n) is

$$\bar{f}_i^{n+1} = \bar{f}_i^n + \bar{f}_t|_i^n \, \Delta t + 0(\Delta t^2) \tag{10.92}$$

The derivative $\bar{f}_t|_i^n$ is obtained from the PDE, which is evaluated at grid point (i, n) using the same spatial finite difference approximations used to derive the implicit FDE. Values of u_i^{n+1} and α_i^{n+1} can be evaluated for the value f_i^{n+1} obtained from Eq. (10.92). Time linearization requires a considerable amount of additional work. It also introduces additional truncation errors that depend on Δt, which reduces the accuracy and generally restricts the time step, thus reducing the advantage of unconditional stability associated with implicit finite difference equations.

10.9.2 Multidimensional Problems

All of the finite difference equations and examples presented so far in this chapter are for the linear unsteady one-dimensional diffusion equation. Some of the problems which arise for multidimensional problems are discussed in this section.

Consider the linear unsteady two-dimensional diffusion equation:

$$\boxed{\bar{f}_t = \alpha(\bar{f}_{xx} + \bar{f}_{yy})} \tag{10.93}$$

The FTCS approximation of Eq. (10.93) is

$$\frac{f_{i,j}^{n+1} - f_{i,j}^n}{\Delta t} = \alpha\left(\frac{f_{i+1,j}^n - 2f_{i,j}^n + f_{i-1,j}^n}{\Delta x^2} + \frac{f_{i,j+1}^n - 2f_{i,j}^n + f_{i,j-1}^n}{\Delta y^2}\right) \tag{10.94}$$

Equation (10.94) can be solved directly for $f_{i,j}^{n+1}$. No additional numerical complexities arise because of the second spatial derivative. For the three-dimensional diffusion equation, the additional derivative $\bar{f}_{zz}$ is present in the PDE. Its finite different approximations is simply added to Eq. (10.94) without further complications. This type of result is typical of all explicit finite difference approximations.

The BTCS approximation of Eq. (10.93) is

$$\frac{f_{i,j}^{n+1} - f_{i,j}^n}{\Delta t} = \alpha\left(\frac{f_{i+1,j}^{n+1} - 2f_{i,j}^{n+1} + f_{i-1,j}^{n+1}}{\Delta x^2} + \frac{f_{i,j+1}^{n+1} - 2f_{i,j}^{n+1} + f_{i,j-1}^{n+1}}{\Delta y^2}\right) \tag{10.95}$$

Applying Eq. (10.95) at every point in a two-dimensional finite difference grid yields a banded pentadiagonal matrix, which requires a large amount of computational effort. Successive-over-relaxation (SOR) methods can be applied for two-dimensional problems, but even that approach becomes almost prohibitive for three-dimensional problems. Alternating-direction-implicit (ADI) methods [Peaceman and Rachford (1955) and Douglas (1962)] and approximate-factorization-implicit (AFI) methods can be used to reduce the banded matrices to two (or three for three-dimensional problems) systems of tridiagonal matrices, which can be solved successively by the Thomas algorithm (see Section 1.5).

10.9.2.1 Alternating-Direction-Implicit (ADI) Method

The *alternating-direction-implicit* (*ADI*) approach consists of solving the PDE in two steps. In the first time step, the spatial derivatives in one direction, say y, are evaluated at the known time level n and the other spatial derivatives, say x, are evaluated at the unknown time level $n + 1$. On the next time step, the process is reversed. Consider the two-dimensional diffusion equation, Eq. (10.93). For the first step, the semidiscrete (i.e., time equation, discretization only) finite difference approximation yields

$$\frac{f_{i,j}^{n+1} - f_{i,j}^{n}}{\Delta t} = \alpha f_{xx}|_{i,j}^{n+1} + \alpha f_{yy}|_{i,j}^{n} \tag{10.96}$$

For the second step,

$$\frac{f_{i,j}^{n+2} - f_{i,j}^{n+1}}{\Delta t} = \alpha f_{xx}|_{i,j}^{n+1} + \alpha f_{yy}|_{i,j}^{n+2} \tag{10.97}$$

If the spatial derivatives in Eqs. (10.96) and (10.97) are replaced by second-order centered-difference approximations, Eqs. (10.96) and (10.97) both yield a tridiagonal system of FDEs, which can be solved by the Thomas algorithm (see Section 1.5). Ferziger (1981) shows that the alternating-direction-implicit method is consistent, $0(\Delta t^2) + (0(\Delta x^2) + 0(\Delta y^2)$, and unconditionally stable.

Alternating-direction-implicit (ADI) procedures can also be applied to three-dimensional problems, in which case a third permutation of Eqs. (10.96) and (10.97) involving the z-direction derivatives is required. A direct extension of the procedure presented above does not work. A modification that does work in three dimensions is presented by Douglas (1962) and Douglas and Gunn (1964).

The ADI method must be treated carefully at the $n + 1$ time step at the boundaries. No problem arises for constant BCs. However, for time dependent BCs, Eq. (10.96), which is $0(\Delta t)$, yields less accurate solutions than Eq. (10.97), which is $0(\Delta t^2)$. When accurate BCs are specified, the errors in the solution at the boundaries at time steps $n + 1$ and $n + 2$ are different orders, which introduces additional errors into the solution. Ferziger (1981) discusses techniques for minimizing this problem.

10.9.2.2   Approximate-Factorization-Implicit (AFI) Method

The *approximate-factorization-implicit* (*AFI*) approach can be illustrated for the BTCS approximation of the two-dimensional diffusion equation, Eq. (10.93), by expressing it in the semidiscrete operator form

$$\frac{f_i^{n+1} - f_i^{n}}{\Delta t} = \alpha \left(\frac{\partial^2}{\partial x^2} + \frac{\partial^2}{\partial y^2} \right) f_{i,j}^{n+1} \tag{10.98}$$

Collecting term yields the two-dimensional operator

$$\left[1 - \alpha \, \Delta t \left(\frac{\partial^2}{\partial x^2} + \frac{\partial^2}{\partial y^2} \right) \right] f_{i,j}^{n+1} = f_{i,j}^{n} \tag{10.99}$$

Equation (10.99) can be approximated by the product of two one-dimensional operators:

$$\left(1 - \alpha \, \Delta t \frac{\partial^2}{\partial x^2} \right) \left(1 - \alpha \Delta t \frac{\partial^2}{\partial y^2} \right) f_{i,j}^{n+1} = f_{i,j}^{n} \tag{10.100}$$

Equation (10.100) can be solved in two steps:

$$\left(1 - \alpha\, \Delta t\, \frac{\partial^2}{\partial y^2}\right) f^*_{i,j} = f^n_{i,j} \tag{10.101}$$

$$\left(1 - \alpha\Delta t\, \frac{\partial^2}{\partial x^2}\right) f^{n+1}_{i,j} = f^*_{i,j} \tag{10.102}$$

If the spatial derivatives in Eqs. (10.101) and (10.102) are replaced by three-point second-order centered-difference approximations, Eqs. (10.101) and (10.102), both yield a tridiagonal sytstem of FDEs, which can be solved by the Thomas algorithm (see Section 1.5).

Multiplying the two operators in Eq. (10.100) yields the single operator

$$\left[1 - \alpha\Delta t\left(\frac{\partial^2}{\partial x^2} + \frac{\partial^2}{\partial y^2}\right) + \alpha^2\, \Delta t^2\, \frac{\partial^2}{\partial x^2}\, \frac{\partial^2}{\partial y^2}\right] f^{n+1}_{i,j} = f^n_{i,j} \tag{10.103}$$

The $0(\Delta t^2)$ term in Eq. (10.103) is not present in the original finite difference equation, Eq. (10.99). Thus, the factorization has introduced a local $0(\Delta t^2)$ error term into the solution. For this reason this approach is called an *approximate factorization*. The local error of the BTCS approximation is $0(\Delta t^2)$, so the approximate factorizaton preserves the order of the BTCS approximation.

Approximate factorization can be applied to three-dimensional problems, in which case a third one-dimensional operator is added to Eq. (10.100) with a corresponding third step in Eqs. (10.101) and (10.102).

10.10 THE CONVECTION-DIFFUSION EQUATION

The solution of the parabolic unsteady diffusion equation has been discussed in Sections 10.2 to 10.9. The solution of the parabolic convection-diffusion equation is discussed in this section.

10.10.1 Introduction

Consider the unsteady one-dimensional parabolic convection-diffusion equation for the generic dependent variable $\bar{f}(x, t)$:

$$\boxed{\bar{f}_t + u\bar{f}_x = \alpha\bar{f}_{xx}} \tag{10.104}$$

where u is the convection velocity and α is the diffusion coefficient. Since the classification of a PDE is determined by the coefficients of its highest-order derivatives, the presence of the first-order convection term $u\bar{f}_x$ in the convection-diffusion equation does not affect its classification. The diffusion equation and the convection-diffusion equation are both parabolic PDEs. However, the presence of the first-order convection term has a major influence on the numerical solution procedure.

Most of the concepts, techniques, and conclusions presented in Sections 10.2 to 10.9 for solving the diffusion equation are directly applicable, sometimes with very minor modifications, for solving the convection-diffusion equation. The finite difference grids and finite difference approximations presented in Section 10.3 also apply to the convection-diffusion equation. The concepts of consistency, order, stability, and convergence presented in Section 10.5 also apply to the convection-diffusion equation. The present section is devoted to the numerical solution of the convection-diffusion equation, Eq. (10.104).

The solution to Eq. (10.104) is the function $\bar{f}(x, t)$. This function must satisfy an initial condition at $t = 0$, $\bar{f}(x, 0) = F(x)$. The time coordinate has an unspecified (i.e, open) final value. Equation (10.104) is second order in the space coordinate x. Consequently, two boundary conditions (BCs) are required. These BCs may be of the Dirichlet type (i.e., specified values of $\bar{f}$), the Neumann type (i.e., specified values of $\bar{f}_x$), or the mixed type (i.e., specified combinations of $\bar{f}$ and $\bar{f}_x$). The space coordinate x must be a closed physical domain.

The convection-diffusion equation applies to problems in mass transport, momentum transport, energy transport, etc. Most people have some physical feeling for heat transfer due to its presence in our everyday life. Consequently, the convection-diffusion equation governing heat transfer in a porous plate is considered in this chapter to demonstrate numerical methods for solving the convection-diffusion equation. That equation is presented in Section III.8, Eq. (III.101), which is repeated below:

$$T_t + \mathbf{V} \cdot \nabla T = \alpha \, \nabla^2 T \tag{10.105}$$

where T is the temperature, $\mathbf{V}$ is the vector convection velocity, and α is the thermal diffusivity. For unsteady one-dimensional heat transfer, Eq. (10.105) becomes

$$T_t + u T_x = \alpha T_{xx} \tag{10.106}$$

The following problem is considered in this chapter to illustrate the behavior of finite difference methods for solving the convection-diffusion equation. A porous plate of thickness $L = 1.0$ cm is cooled by a fluid flowing through the porous material, as illustrated in Figure 10.26. The thermal conductivity of the porous material is small compared to the thermal conductivity of the fluid, so that heat conduction through the porous material itself is negligible compared to heat transfer through the fluid by convection and diffusion (i.e., conduction). The temperatures on the two faces of the plate are

$$T(0.0, t) = 0.0 \text{ C} \qquad \text{and} \qquad T(L, t) = 100.0 \text{ C} \tag{10.107}$$

The initial fluid velocity is zero, so the initial temperature distribution is the pure diffusion distribution:

$$T(x, 0.0) = 100.0 x/L \qquad 0.0 \leq x \leq L \tag{10.108}$$

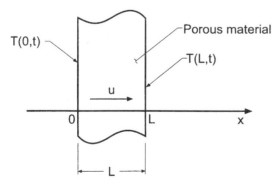

Figure 10.26 Heat convection-diffusion in a porous plate.

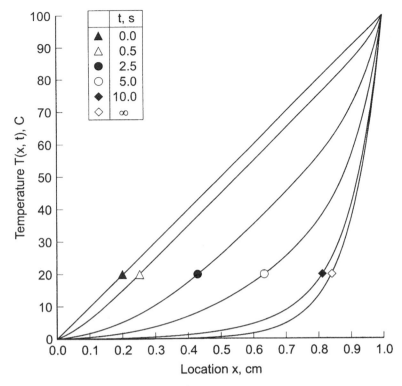

Figure 10.27 Exact solution of the heat convection-diffusion problem for $P = 10$.

This initial temperature distribution is illustrated by the top curve in Figure 10.27. The diffusion coefficient of the fluid is $\alpha = 0.01$ cm^2/s. At time $t = 0$, the fluid in the plate is instantaneously given a constant velocity $u = 0.1$ cm/s to the right. The temperature distribution $T(x, t)$ in the fluid is required.

The exact solution to this problem is obtained by replacing the original problem with two auxiliary problems. Both problems satisfy Eq. (10.106). For the first problem, $T(x, 0.0) = 0.0$, $T(0.0, t) = 0.0$, and $T(L, t) = 100.0$. For the second problem, $T(x, 0.0)$ is given by Eq. (10.108), $T(0.0, t) = 0.0$, and $T(L, t) = 0.0$. Since Eq. (10.106) is linear, the solution to the original problem is the sum of the solutions to the two auxiliary problems. The exact solution to each of the two auxiliary problems is obtained by assuming a product solution of the form $T(x, t) = X(x)\hat{T}(t)$, separating variables, integrating the resulting two ordinary differential equations, applying the boundary conditions at $x = 0$ and $x = L$, and superimposing an infinite number of harmonic functions (i.e., sines and cosines) in a Fourier series to satisfy the initial condition. The final result is

$$T(x, t) = 100 \left[\frac{\exp(Px/L) - 1}{\exp(P) - 1} \right.$$
$$\left. + \frac{4\pi \exp(Px/2L)\sinh(P/2)}{\exp(P) - 1} \sum_{m=1}^{\infty} A_m + 2\pi \ \exp(Px/2L) \sum_{m=1}^{\infty} B_m \right]$$

$$(10.109)$$

where P is the Peclet number

$$P = \frac{uL}{\alpha}$$ (10.110)

the coefficients A_m and B_m are given by

$$A_m = (-1)^m m\beta_m^{-1} \sin(m\pi x/L)e^{-\lambda_m t}$$ (10.111)

$$B_m = \left[(-1)^{m+1} m\beta_m^{-1}\left(1 + \frac{P}{\beta_m}\right)e^{-P/2} + \frac{mP}{\beta_m^2}\right]\sin\left(\frac{m\pi x}{L}\right)e^{-\lambda_m t}$$ (10.112)

and

$$\beta_m = \left(\frac{P}{2}\right)^2 + (m\pi)^2$$ (10.113)

$$\lambda_m = \frac{u^2}{4\alpha} + \frac{m^2\pi^2\alpha}{L} = \frac{\alpha\beta_m}{L^2}$$ (10.114)

The exact transient solution at selected values of time t for $L = 1.0$ cm, $u = 0.1$ cm/s, and $\alpha = 0.01$ cm^2/s, for which the Peclet number $P = uL/\alpha = 10$, is tabulated in Table 10.7 and illustrated in Figure 10.27. As expected, heat flows out of the faces of the plate to the surroundings by diffusion, and heat is convected out of the plate by convection. The temperature distribution smoothly approaches the asymptotic steady-state solution

$$T(x, \infty) = 100\frac{\exp(Px/L) - 1}{\exp(P) - 1}$$ (10.115)

From Table 10.7, it can be seen that the transient solution has reached the asymptotic steady state by $t = 50.0$ s.

Table 10.7 Exact Solution of the Heat Convection-Diffusion Problem for $P = 10$

	Temperature $T(x, t)$, C							
t, s	$x = 0.0$	$x = 0.2$	$x = 0.4$	$x = 0.6$	$x = 0.7$	$x = 0.8$	$x = 0.9$	$x = 1.0$
0.0	0.00	20.00	40.00	60.00	70.00	80.00	90.00	100.00
0.5	0.00	15.14	35.00	55.00	65.00	75.02	85.43	100.00
1.0	0.00	11.36	30.05	50.00	60.02	70.18	81.57	100.00
1.5	0.00	8.67	25.39	45.02	55.07	65.50	77.99	100.00
2.0	0.00	6.72	21.25	40.14	50.18	60.91	74.55	100.00
2.5	0.00	5.29	17.71	35.46	45.41	56.43	71.20	100.00
5.0	0.00	1.80	7.13	17.74	25.64	36.78	56.13	100.00
10.0	0.00	0.30	1.38	4.81	9.09	18.39	40.96	100.00
50.00	0.00	0.03	0.24	1.83	4.97	13.53	36.79	100.00
∞	0.00	0.03	0.24	1.83	4.97	13.53	36.79	100.00

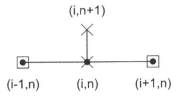

(i-1,n)   (i,n)   (i+1,n)

Figure 10.28 The FTCS stencil method.

10.10.2 The Forward-Time Centered-Space Method

In this section the convection-diffusion equation $\bar{f}_t + u\bar{f}_x = \alpha \bar{f}_{xx}$ is solved numerically by the forward-time centered-space (FTCS) method. The base point for the finite difference approximations (FDAs) of the individual exact partial derivatives is grid point (i, n). The partial derivative $\bar{f}_t$ is approximated by the first-order forward-difference FDA, Eq. (10.17), the partial derivative $\bar{f}_x$ is approximated by the second-order centered-difference FDA, Eq. (10.21), and the partial derivative $\bar{f}_{xx}$ is approximated by the second-order centered-difference FDA, Eq. (10.23). The corresponding finite difference stencil is illustrated in Figure 10.28. The resulting finite difference equation (FDE) is

$$f_i^{n+1} = f_i^n - \frac{c}{2}(f_{i+1}^n - f_{i-1}^n) + d(f_{i+1}^n - 2f_i^n + f_{i-1}^n) \qquad (10.116)$$

where $c = u\,\Delta t/\Delta x$ is the convection number and $d = \alpha\,\Delta t/\Delta x^2$ is the diffusion number. The modified differential equation (MDE) corresponding to Eq. (10.116) is

$$f_t + uf_x = \alpha f_{xx} - \tfrac{1}{2}f_{tt}\,\Delta t - \tfrac{1}{6}f_{ttt}\,\Delta t^2 - \cdots - \tfrac{1}{6}uf_{xxx}\Delta x^2 - \cdots$$
$$+ \tfrac{1}{12}\alpha f_{xxxx}\Delta x^4 + \cdots \qquad (10.117)$$

As $\Delta t \to 0$ and $\Delta x \to 0$, Eq. (10.117) approaches $f_t + uf_x = \alpha f_{xx}$. Consequently, Eq. (10.116) is a consistent approximation of the convection-diffusion equation, Eq. (10.104). The FDE is $0(\Delta t) + 0(\Delta x^2)$. The amplification factor G corresponding to Eq. (10.116) is

$$G = (1 - 2d) + 2d\cos\theta - Ic\sin\theta \qquad (10.118)$$

For $-\infty \le \theta \le \infty$, Eq. (10.118) represents an ellipse in the complex plane, as illustrated in Figure 10.29. The center of the ellipse is at $(1 - 2d + I0)$ and the axes are $2d$ and c. For stability, $|G| \le 1$, which requires that the ellipse lie on or within the unit circle $|G| = 1$. From Figure 10.29, two stability criteria are obvious. The real and imaginary axes of the ellipse must both be less than or equal to unity. From curves a and b,

$$c \le 1 \qquad \text{and} \qquad 2d \le 1 \qquad (10.119)$$

In addition, at point $(1 + I0)$ the curvature of the ellipse must be greater than the curvature of the unit circle, or the ellipse will not remain within the unit circle even though $c < 1$ and $d < 1/2$, as illustrated by curve c. This condition is satisfied if

$$c^2 \le 2d \qquad (10.120)$$

which, with $2d \le 1$, includes the condition $c \le 1$. Thus, the stability criteria for the FTCS approximation of the convection-diffusion equation are

$$c^2 \le 2d \le 1 \qquad (10.121)$$

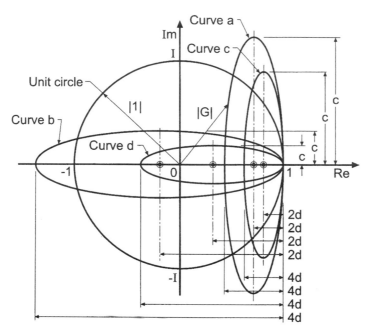

Figure 10.29 Locus of the amplification factor G for the FTCS method.

Consequently, the method is conditionally stable. Curve d in Figure 10.29 illustrates a stable condition. The FTCS approximation of the convection-diffusion equation is consistent and conditionally stable. Consequently, by the Lax equivalence theorem, it is a convergent approximation of that equation.

Example 10.7.   The FTCS method applied to the convection-diffusion equation

Let's solve the heat convection-diffusion problem using Eq. (10.116) with $\Delta x = 0.1$ cm. The exact solution at selected times is presented in Table 10.7. The numerical solution for $\Delta t = 0.5$ s,   for   which $c = u\,\Delta t/\Delta x = (0.1)(0.5)/(0.1) = 0.5$ and $d = \alpha\,\Delta t/\Delta x^2 = (0.01)(0.5)/(0.1)^2 = 0.5$ is presented in Figure 10.30, it is apparent that the numerical solution is a reasonable approximation of the exact solution. Compare these results with the solution of the diffusion equation presented in Example 10.1 and illustrated in Figure 10.11. It is apparent that the solution of the convection-diffusion equation has larger errors than the solution of the diffusion equation. These larger errors are a direct consequence of the presence of the convection term $u\bar{f}_x$. As the solution progresses in time, the numerical solution smoothly approaches the steady-state solution.

At $t = 50$ s, the exact asymptotic steady state solution has been reached. The numerical solution at $t = 50.0$ s is a reasonable approximation of the steady state solution.

In summary, the FTCS approximation of the convection-diffusion equation is explicit, single step, consistent, $0(\Delta t) + 0(\Delta x^2)$, conditionally stable, and convergent. The FTCS approximation of the convection-diffusion equation yields reasonably accurate transient solutions.

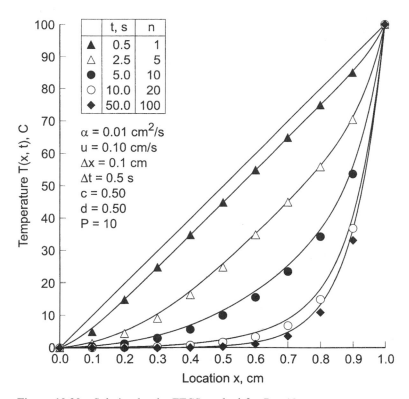

	t, s	n
▲	0.5	1
△	2.5	5
●	5.0	10
○	10.0	20
◆	50.0	100

$\alpha = 0.01$ cm^2/s
$u = 0.10$ cm/s
$\Delta x = 0.1$ cm
$\Delta t = 0.5$ s
$c = 0.50$
$d = 0.50$
$P = 10$

Figure 10.30 Solution by the FTCS method for $P = 10$.

10.10.3 The Backward-Time Centered-Space Method

The BTCS method is applied to the convection-diffusion equation $\bar{f}_t + u\bar{f}_x = \alpha\bar{f}_{xx}$ in this section. The base point for the finite difference approximation of the individual exact partial derivatives is grid point $(i, n + 1)$. The partial derivative $\bar{f}_t$ is approximated by the first-order backward-difference FDA, Eq. (10.65), the partial derivative $\bar{f}_x$ is approximated by the second-order centered-difference FDA, Eq. (10.21), and the partial derivative $\bar{f}_{xx}$ is approximated by the second-order centered-difference FDA, Eq. (10.23), both evaluated at time level $n + 1$. The corresponding finite difference stencil is illustrated in Figure 10.31. The resulting finite difference equation (FDE) is

$$\frac{f_i^{n+1} - f_i^n}{\Delta t} + u\frac{f_{i+1}^{n+1} - f_{i-1}^{n+1}}{2\,\Delta x} = \alpha\frac{f_{i+1}^{n+1} - 2f_i^{n+1} + f_{i-1}^{n+1}}{\Delta x^2} \qquad (10.122)$$

(i-1,n+1)   (i,n+1)   (i+1,n+1)

(i,n)

Figure 10.31 The BTCS method stencil.

Rearranging Eq. (10.122) yields

$$-\left(\frac{c}{2}+d\right)f_{i-1}^{n+1} + (1+2d)f_i^{n+1} + \left(\frac{c}{2}-d\right)f_{i+1}^{n+1} = f_i^n \qquad (10.123)$$

where $c = u\,\Delta t/\Delta x$ is the convection number and $d = \alpha\Delta t/\Delta x^2$ is the diffusion number. Equation (10.123) cannot be solved explicitly for f_i^{n+1} because the two unknown neighboring values f_{i-1}^{n+1} and f_{i+1}^{n+1} also appear in the equation. Consequently, an implicit system of finite difference equations results.

The modified differential equation (MDE) corresponding to Eq. (10.123) is

$$f_t + uf_x = \alpha f_{xx} + \tfrac{1}{2}f_{tt}\,\Delta t - \tfrac{1}{6}f_{ttt}\,\Delta t^2 + \cdots - \tfrac{1}{6}uf_{xxx}\,\Delta x^2 + \cdots + \tfrac{1}{12}\alpha f_{xxxx}\,\Delta x^2 + \cdots \qquad (10.124)$$

As $\Delta t \to 0$ and $\Delta x \to 0$, Eq. (10.124) approaches $f_t + uf_x = \alpha f_{xx}$. Consequently, Eq. (10.123) is a consistent approximation of the convection-diffusion equation, Eq. (10.104). From a von Neumann stability analysis, the amplification factor G is

$$G = \frac{1}{1 + 2d(1 - \cos\theta) + Ic\sin\theta} \qquad (10.125)$$

The term $(1 - \cos\theta)$ is ≥ 0 for all values of $\theta = (k_m\,\Delta x)$. Consequently, the denominator of Eq. (10.125) is always ≥ 1, $|G| \leq 1$ for all values of c and d, and Eq. (10.123) is unconditionally stable. The BTCS approximation of the convection-diffusion equation is consistent and unconditionally stable. Consequently, by the Lax Equivalence Theorem it is a convergent approximation of that equation.

Consider now the solution of the convection-diffusion equation by the BTCS method. As discussed in Section 10.7 for the diffusion equation, a tridiagonal system of equations results when Eq. (10.123) is applied at every grid point. That system of equations can be solved by the Thomas algorithm (see Section 1.5). For the linear convection-diffusion equation, LU factorization can be used.

Example 10.8. The BTCS method applied to the convection-diffusion equation

---

Let's solve the heat convection-diffusion problem using Eq. (10.123) with $\Delta x = 0.1$ cm and $\Delta t = 0.5$ s. The transient solution for $\Delta t = 0.5$ s, for which $c = d = 0.5$ s, is presented in Figure 10.32. These results are a reasonable approximation of the exact transient solution. At $t = 50.0$ s, the exact asymptotic steady state solution has been reached. The numerical solution at $t = 50.0$ s is a reasonable approximation of the steady state solution.

The implicit BTCS method becomes considerably more complicated when applied to nonlinear PDEs, systems of PDEs, and multidimensional problems. A brief discussion of these problems is presented in Section 10.9.

In summary, the BTCS approximation of the convection-diffusion equation is implicit, single step, consistent, $0(\Delta t) + 0(\Delta x^2)$, unconditionally stable, and convergent. The implicit nature of the method yields a set of finite difference equations which must be solved simultaneously. For one-dimensional problems, that can be accomplished by the Thomas algorithm. The BTCS approximation of the convection-diffusion equation yields reasonably accurate transient solutions for modest values of the convection and diffusion numbers.

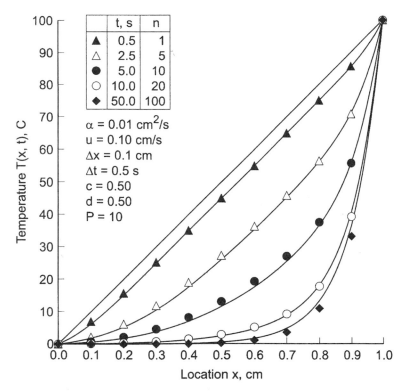

Figure 10.32 Solution by the BTCS method for $P = 10$.

10.11 ASYMPTOTIC STEADY STATE SOLUTION TO PROPAGATION PROBLEMS

Marching methods are employed for solving unsteady propagation problems, which are governed by parabolic and hyperbolic partial differential equations. The emphasis in those problems is on the transient solution itself.

Marching methods also can be used to solve steady equilibrium problems and steady mixed (i.e., elliptic-parabolic or elliptic-hyperbolic) problems as the asymptotic solution in time of an appropriate unsteady propagation problem. Steady equilibrium problems are governed by elliptic PDEs. Steady mixed problems are governed by PDEs that change classification from elliptic to parabolic or elliptic to hyperbolic in some portion of the solution domain, or by systems of PDEs which are a mixed set of elliptic and parabolic or elliptic and hyperbolic PDEs. Mixed problems present serious numerical difficulties due to the different types of solution domains (closed domains for equilibrium problems and open domains for propagation problems) and different types of auxiliary conditions (boundary conditions for equilibrium problems and boundary conditions and initial conditions for propagation problems). Consequently, it may be easier to obtain the solution of a steady mixed problem by reposing the problem as an unsteady parabolic or hyperbolic problem and using marching methods to obtain the asymptotic steady state solution. That approach to solving steady state problems is discussed in this section.

The appropriate unsteady propagation problem must be governed by a parabolic or hyperbolic PDE having the same spatial derivatives as the steady equilibrium problem or the steady mixed problem and the same boundary conditions. As an example, consider the steady convection-diffusion equation:

$$\boxed{u\hat{f}_x = \alpha\hat{f}_{xx}}$$ (10.126)

The solution to Eq. (10.126) is the function $\hat{f}(x)$, which must satisfy two boundary conditions. The boundary conditions may be of the Dirichlet type (i.e., specified values of $\hat{f}$), the Neumann type (i.e., specified values of $\hat{f}_x$), or the mixed type (i.e., specified combinations of $\hat{f}$ and $\hat{f}_x$).

An appropriate unsteady propagation problem for solving Eq. (10.126) as the asymptotic solution in time is the unsteady convection-diffusion equation:

$$\bar{f}_t + u\bar{f}_x = \alpha\bar{f}_{xx}$$ (10.127)

The solution to Eq. (10.127) is the function $\bar{f}(x, t)$, which must satisfy an initial condition, $\bar{f}(x, 0) = F(x)$, and two boundary conditions. If the boundary conditions for $\bar{f}(x, t)$ are the same as the boundary conditions for $\hat{f}(x)$, then

$$\boxed{\hat{f}(x) = \lim_{t\to\infty} \bar{f}(x, t) = \bar{f}(x, \infty)}$$ (10.128)

As long as the asymptotic solution converges, the particular choice for the initial condition, $\bar{f}(x, 0) = F(x)$, should not affect the steady state solution. However, the steady state solution may be reached in fewer time steps if the initial condition is a reasonable approximation of the general features of the steady state solution.

The steady state solution of the transient heat convection-diffusion problem presented in Section 10.10 is considered in this section to illustrate the solution of steady equilibrium problems as the asymptotic solution of unsteady propagation problems. The exact solution to that problem is

$$\hat{T}(x) = 100\frac{e^{(Px/L)} - 1}{e^P - 1}$$ (10.129)

where $P = (uL/\alpha)$ is the Peclet number. The solution for $P = 10$ is tabulated in Table 10.7 as the last row of data in the table corresponding to $t = \infty$.

As shown in Figures 10.30 and 10.32, the solution of the steady state convection-diffusion equation can be obtained as the asymptotic solution in time of the unsteady convection-diffusion equation. The solution by the FTCS method required 100 time steps. The solution by the BTCS method also required 100 time steps. However, the BTCS method is unconditionally stable, so much larger time steps can be taken if the accuracy of the transient solution is not of interest. The results of this approach are illustrated in Example 10.9.

Example 10.9.   Asymptotic steady state solution of the convection-diffusion equation

Let's solve the unsteady heat convection-diffusion problem for the asymptotic steady state solution with $\Delta x = 0.1$ cm by the BTCS method. Figure 10.33 presents seven solutions of the heat convection-diffusion equation, each one for a single time step with a different

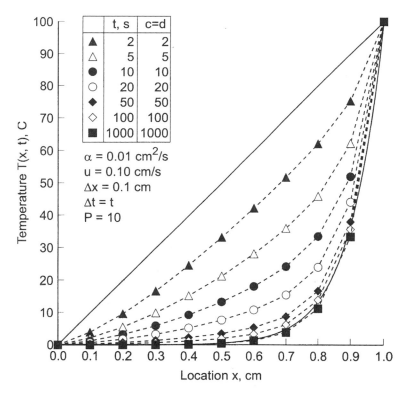

Figure 10.33 Single-step solutions by the BTCS method for $P = 10$.

value of Δt. As Δt is increased from 2.0 s to 1000.0 s, the single-step solution approaches the steady state solution more and more closely. In fact, the solution for $\Delta t = 1000.0$ s is essentially identical to the steady state solution and was obtained in a single step.

In summary, steady equilibrium problems, mixed elliptic/parabolic problems, and mixed elliptic/hyperbolic problems can be solved as the asymptotic steady state solution of an appropriate unsteady propagation problem. For linear problems, the asymptotic steady state solution can be obtained in one or two steps by the BTCS method, which is the recommended method for such problems. For nonlinear problems, the BTCS method becomes quite time consuming, since several time steps must be taken to reach the asymptotic steady-state solution. The asymptotic steady state approach is a powerful procedure for solving difficult equilibrium problems and mixed equilibrium/propagation problems.

10.12 PROGRAMS

Three FORTRAN subroutines for solving the diffusion equation are presented in this section:

1. The forward-time centered-space (FTCS) method
2. The backward-time centered-space (BTCS) method
3. The Crank-Nicolson (CN) method

The basic computational algorithms are presented as completely self-contained subroutines suitable for use in other programs. Input data and output statements are contained in a main (or driver) program written specifically to illustrate the use of each subroutine.

10.12.1 The Forward-Time Centered-Space (FTCS) Method

The diffusion equation is given by Eq. (10.4):

$$f_t = \alpha f_{xx} \tag{10.130}$$

When Dirichlet (i.e., specified f) boundary conditions are imposed, those values must be specified at the boundary points. That type of boundary condition is considered in this section. The first-order forward-time second-order centered-space (FTCS) approximation of Eq. (10.130) is given by Eq. (10.25):

$$f_i^{n+1} = f_i^n - d(f_{i+1}^n + 2.0f_i^n + f_{i-1}^n) \tag{10.131}$$

A FORTRAN subroutine, *subroutine ftcs*, for solving Eq. (131) is presented in Program 10.1. *Program main* defines the data set and prints it, calls *subroutine ftcs* to implement the solution, and prints the solution.

Program 10.1. The FTCS method for the diffusion equation program

```
      program main
c     main program to illustrate diffusion equation solvers
c     nxdim x-direction array dimension, nxdim = 11 in this program
c     ntdim t-direction array dimension, ntdim = 101 in this progra
c     imax  number of grid points in the x direction
c     nmax  number of time steps
c     iw    intermediate results output flag: 0 none, 1 all
c     ix    output increment: 1 every grid point, n every nth point
c     it    output increment: 1 every time step, n every nth step
c     f     solution array, f(i,n)
c     dx    x-direction grid increment
c     dt    time step
c     alpha diffusion coefficient
      dimension f(11,101)
      data nxdim,ntdim,imax,nmax,iw,ix,it/11,101,11,101, 0, 1, 10/
      data (f(i,1),i=1,11) / 0.0, 20.0, 40.0, 60.0, 80.0, 100.0,
     1 80.0, 60.0, 40.0, 20.0, 0.0 /
      data dx,dt,alpha,n,t / 0.1, 0.1, 0.01, 1, 0.0 /
      write (6,1000)
      call ftcs (nxdim,ntdim,imax,nmax,f,dx,dt,alpha,n,t,iw,ix)
      if (iw.eq.1) stop
      do n=1,nmax,it
```

```
        t=float(n-1)*dt
        write (6,1010) n,t,(f(i,n),i=1,imax,ix)
      end do
      stop
1000 format (' Diffusion equation solver (FTCS method)'/' '/
   1  '   n',2x,'time',3x,'f(i,n)'/' ')
1010 format (i3,f5.1,11f6.2)
      end


      subroutine ftcs(nxdim,ntdim,imax,nmax,f,dx,dt,alpha,n,t,iw,ix)
c     the FTCS method for the diffusion equation
      dimension f(nxdim,ntdim)
      d=alpha*dt/dx**2
      do n=1,nmax-1
        t=t+dt
        do i=2,imax-1
          f(i,n+1)=f(i,n)+d*(f(i+1,n)-2.0*f(i,n)+f(i-1,n))
        end do
        if (iw.eq.1) write (6,1000) n+1,t,(f(i,n+1),i=1,imax,ix)
      end do
      return
1000 format (i3,f7.3,11f6.2)
      end
```

The data set used to illustrate *subroutine ftcs* is taken from Example 10.1. The output generated by the program is presented in Output 10.1.

A Neumann (i.e., derivative) boundary condition on the right-hand side of the solution domain can be implemented by solving Eq. (10.87) at grid point imax. Example 10.6 can be solved to illustrate the application of this boundary condition.

Output 10.1.   Solution of the diffusion equation by the FTCS method

```
Diffusion equation solver (FTCS method)

  n  time   f(i,n)

  1  0.0   0.00 20.00 40.00 60.00 80.00100.00 80.00 60.00 40.00 20.00   0.00
 11  1.0   0.00 19.96 39.68 58.22 72.81 78.67 72.81 58.22 39.68 19.96   0.00
 21  2.0   0.00 19.39 37.81 53.73 64.87 68.91 64.87 53.73 37.81 19.39   0.00
 31  3.0   0.00 18.21 35.06 48.99 58.30 61.58 58.30 48.99 35.06 18.21   0.00
 41  4.0   0.00 16.79 32.12 44.51 52.63 55.45 52.63 44.51 32.12 16.79   0.00
 51  5.0   0.00 15.34 29.25 40.39 47.61 50.11 47.61 40.39 29.25 15.34   0.00
 61  6.0   0.00 13.95 26.57 36.63 43.11 45.35 43.11 36.63 26.57 13.95   0.00
 71  7.0   0.00 12.67 24.11 33.20 39.06 41.07 39.06 33.20 24.11 12.67   0.00
 81  8.0   0.00 11.49 21.86 30.10 35.39 37.21 35.39 30.10 21.86 11.49   0.00
 91  9.0   0.00 10.42 19.82 27.28 32.07 33.72 32.07 27.28 19.82 10.42   0.00
101 10.0   0.00  9.44 17.96 24.72 29.07 30.56 29.07 24.72 17.96  9.44   0.00
```

10.12.2 The Backward-Time Centered-Space (BTCS) Method

The first-order backward-time second-order centered-space (BTCS) approximation of Eq. (10.130) is given by Eq. (10.67):

$$-df_{i-1}^{n+1} + (1 + 2d)f_i^{n+1} - df_{i+1}^{n+1} = f_i^n \qquad (10.132)$$

A FORTRAN subroutine, *subroutine btcs*, for solving the system equation arising from the application of Eq. (10.132) at every interior point in a finite difference grid is presented in Program 10.2. *Subroutine thomas* presented in Section 1.8.3 is used to solve the tridiagonal system equation. Only the statements which are different from the statements in *program main* and *program ftcs* in Section 10.12.1 are presented. *Program main* defines the data set and prints it, calls *subroutine btcs* to implement the solution, and prints the solution.

Program 10.2. The BTCS method for the diffusion equation program

```
      program main
c     main program to illustrate diffusion equation solvers
      dimension f(11,101),a(11,3),b(11),w(11)
      call btcs(nxdim,ntdim,imax,nmax,f,dx,dt,alpha,n,t,iw,ix,a,b,w
 1000 format (' Diffusion equation solver (BTCS method)'/' '/
     1 '  n',2x,'time',3x,'f(i,n)'/' ')
      end

      subroutine btcs (nxdim,ntdim,imax,nmax,f,dx,dt,alpha,n,t,iw,
     1 ix,a,b,w)
c     implements the BTCS method for the diffusion equation
      dimension f(nxdim,ntdim),a(nxdim,3),b(nxdim),w(nxdim)
      d=alpha*dt/dx**2
      a(1,2)=1.0
      a(1,3)=0.0
      b(1)=0.0
      a(imax,1)=0.0
      a(imax,2)=1.0
      b(imax)=0.0
      do n=1,nmax-1
        t=t+dt
        do i=2,imax-1
          a(i,1)=-d
          a(i,2)=1.0+2.0*d
          a(i,3)=-d
          b(i)=f(i,n)
        end do
```

```
       call thomas (nxdim,imax,a,b,w)
       do i=2,imax-1
          f(i,n+1)=w(i)
       end do
       if (iw.eq.1) write (6,1000) n+1,t,(f(i,n+1),i=1,imax,ix)
     end do
     return
1000 format (i3,f5.1,11f6.2)
     end
```

The data set used to illustrate *subroutine btcs* is taken from Example 10.4. The output generated by the program is presented in Output 10.2.

Output 10.2. Solution of the diffusion equation by the BTCS method

```
Diffusion equation solver (BTCS method)

 n   time   f(i,n)

 1   0.0   0.00 20.00 40.00 60.00 80.00100.00 80.00 60.00 40.00 20.00   0.00
 3   1.0   0.00 19.79 39.25 57.66 73.06 80.76 73.06 57.66 39.25 19.79   0.00
 5   2.0   0.00 19.10 37.40 53.62 65.58 70.28 65.58 53.62 37.40 19.10   0.00
 7   3.0   0.00 18.03 34.91 49.20 59.06 62.65 59.06 49.20 34.91 18.03   0.00
 9   4.0   0.00 16.75 32.19 44.91 53.39 56.39 53.39 44.91 32.19 16.75   0.00
11   5.0   0.00 15.42 29.50 40.90 48.38 50.99 48.38 40.90 29.50 15.42   0.00
13   6.0   0.00 14.11 26.93 37.22 43.90 46.22 43.90 37.22 26.93 14.11   0.00
15   7.0   0.00 12.87 24.53 33.84 39.86 41.94 39.86 33.84 24.53 12.87   0.00
17   8.0   0.00 11.73 22.33 30.77 36.21 38.09 36.21 30.77 22.33 11.73   0.00
19   9.0   0.00 10.67 20.31 27.97 32.90 34.60 32.90 27.97 20.31 10.67   0.00
21  10.0   0.00  9.70 18.46 25.42 29.90 31.44 29.90 25.42 18.46  9.70   0.00
```

10.12.3 The Crank-Nicolson (CN) Method

The Crank-Nicolson approximation of Eq. (10.130) is given by Eq. (10.78):

$$-df_{i-1}^{n+1} + 2(1+d)f_i^{n+1} - df_{i+1}^{n+1} = df_{i-1}^n + 2(1-d)f_i^n + df_{i+1}^n \qquad (10.133)$$

A FORTRAN subroutine, *subroutine cn*, for implementing the system equation arising from the application of Eq. (10.133) at every interior point in a finite difference grid is presented in Program 10.3. Only the statements which are different from the statements in *program main* and *program btcs* in Section 10.12.2 are presented. *Program main* defines the data set and prints it, calls *subroutine cn* to implement the solution, and prints the solution.

Program 10.3.   The CN method for the diffusion equation program

```
      program main
c     main program to illustrate diffusion equation solvers
      call cn (nxdim,ntdim,imax,nmax,f,dx,dt,alpha,n,t,iw,ix,a,b,w)
 1000 format (' Diffusion equation solver (CN method)'/' '/
     1 '   n',2x,'time',3x,'f(i,n)'/' ')
      end

      subroutine cn (nxdim,ntdim,imax,nmax,f,dx,dt,alpha,n,t,iw,ix,
     1 a,b,w)
c     the CN method for the diffusion equation
          a(i,2)=2.0*(1.0+d)
          b(i)=d*f(i-1,n)+2.0*(1.0-d)*f(i,n)+d*f(i+1,n)
      end

      subroutine thomas (ndim,n,a,b,x)
c     the thomas algorithm for a tridiagonal system
      end
```

The data set used to illustrate *subroutine cn* is taken from Example 10.5. The output generated by the program is presented in Output 10.3.

Output 10.3.   Solution of the diffusion equation by the CN method

```
Diffusion equation solver (CN method)

  n  time    f(i,n)

  1  0.0   0.00 20.00 40.00 60.00 80.00100.00 80.00 60.00 40.00 20.00  0.00
  3  1.0   0.00 19.92 39.59 58.20 72.93 78.79 72.93 58.20 39.59 19.92  0.00
  5  2.0   0.00 19.34 37.76 53.75 64.97 69.06 64.97 53.75 37.76 19.34  0.00
  7  3.0   0.00 18.19 35.06 49.04 58.41 61.72 58.41 49.04 35.06 18.19  0.00
  9  4.0   0.00 16.80 32.15 44.59 52.74 55.59 52.74 44.59 32.15 16.80  0.00
 11  5.0   0.00 15.36 29.30 40.48 47.73 50.24 47.73 40.48 29.30 15.36  0.00
 13  6.0   0.00 13.98 26.64 36.73 43.23 45.48 43.23 36.73 26.64 13.98  0.00
 15  7.0   0.00 12.70 24.18 33.31 39.18 41.21 39.18 33.31 24.18 12.70  0.00
 17  8.0   0.00 11.53 21.94 30.21 35.52 37.35 35.52 30.21 21.94 11.53  0.00
 19  9.0   0.00 10.46 19.90 27.39 32.21 33.86 32.21 27.39 19.90 10.46  0.00
 21 10.0   0.00  9.49 18.04 24.84 29.20 30.70 29.20 24.84 18.04  9.49  0.00
```

10.12.4   Packages For Solving The Diffusion Equation

Numerous libraries and software packages are available for solving the diffusion equation. Many work stations and main frame computers have such libraries attached to their operating systems.

Many commercial software packages contain algorithms for integrating diffusion type (i.e., parabolic) PDEs. Due to the wide variety of parabolic PDEs governing physical problems, many parabolic PDE solvers (i.e., programs) have been developed. For this reason, no specific programs are recommended in this section.

10.13 SUMMARY

The numerical solution of parabolic partial differential equations by finite difference methods is discussed in this chapter. Parabolic PDEs govern propagation problems which have an infinite physical information propagation speed. They are solved numerically by marching methods. The unsteady one-dimensional diffusion equation $\bar{f}_t = \alpha \bar{f}_{xx}$ is considered as the model parabolic PDE in this chapter.

Explicit finite difference methods, as typified by the FTCS method, are conditionally stable and require a relatively small step size in the marching direction to satisfy the stability criteria. Implicit methods, as typified by the BTCS method, are unconditionally stable. The marching step size is restricted by accuracy requirements, not stability requirements. For accurate solutions of transient problems, the marching step size for implicit methods cannot be very much larger than the stable step size for explicit methods. Consequently, explicit methods are generally preferred for obtaining accurate transient solutions. Asymptotic steady state solutions can be obtained very efficiently by the BTCS method with a large marching step size.

Nonlinear partial differential equations can be solved directly by explicit methods. When solved by implicit methods, systems of nonlinear FDEs must be solved. Multidimensional problems can be solved directly by explicit methods. When solved by implicit methods, large banded systems of FDEs results. Alternating-direction-implicit (ADI) methods and approximate-factorization-implicit (AFI) methods can be used to solve multidimensional problems.

After studying Chapter 10, you should be able to:

1. Describe the physics of propagation problems governed by parabolic PDEs
2. Describe the general features of the unsteady diffusion equation
3. Understand the general features of pure diffusion
4. Discretize continuous physical space
5. Develop finite difference approximations of exact partial derivatives of any order
6. Develop a finite difference approximation of an exact partial differential equation
7. Understand the differences between an explicit FDE and an implicit FDE
8. Understand the theoretical concepts of consistency, order, stability, and convergence, and how to demonstrate each
9. Derive the modified differential equation (MDE) actually solved by a FDE
10. Perform a von Neumann stability analysis
11. Implement the forward-time centered-space method
12. Implement the backward-time centered-space method
13. Implement the Crank-Nicolson method

14. Describe the complications associated with nonlinear PDEs
15. Explain the difference between Dirichlet and Neumann boundary conditions and how to implement both.
16. Describe the general features of the unsteady convection-diffusion equation
17. Understand how to solve steady state problems as the asymptotic solution in time of an appropriate unsteady propagation problem
18. Choose a finite difference method for solving a parabolic PDE

EXERCISE PROBLEMS

Section 10.2 General Features of Parabolic PDEs

1. Consider the unsteady one-dimensional diffusion equation $\bar{f}_t = \alpha \bar{f}_{xx}$. Classify this PDE. Determine the characteristic curves. Discuss the significance of these results as regards domain of dependence, range of influence, signal propagation speed, auxiliary conditions, and numerical solution procedures.
2. Develop the exact solution of the heat diffusion problem presented in Section 10.1, Eq. (10.3).
3. By hand, calculate the exact solution for $T(0.5, 10.0)$.

Section 10.4 The Forward-Time Centered-Space (FTCS) Method

4. Derive the FTCS approximation of the unsteady one-dimensional diffusion equation, Eq. (10.25), including the leading truncation error terms in Δt and Δx.
5.* By hand calculation, determine the solution of the example heat diffusion problem by the FTCS method at $t = 0.5$ s for $\Delta x = 0.1$ cm and $\Delta t = 0.1$ s.
6. By hand calculation, derive the results presented in Figures 10.12 and 10.13.
7. Implement the program presented in Section 10.12.1 to reproduce Table 10.2. Compare the results with the exact solution presented in Table 10.1.
8. Solve Problem 7 with $\Delta x = 0.1$ cm and $\Delta t = 0.5$ s. Compare the results with the exact solution presented in Table 10.1.
9. Solve Problem 8 with $\Delta x = 0.05$ cm and $\Delta t = 0.125$ s. Compare the errors and the ratios of the errors for the two solutions at $t = 5.0$ s.

Section 10.5 Consistency, Order, Stability, and Convergence

Consistency and Order

10. Derive the MDE corresponding to the FTCS approximation of the diffusion equation, Eq. (10.25). Analyze consistency and order.
11. Derive the MDE corresponding to the Richardson approximation of the diffusion equation, Eq. (10.56). Analyze consistency and order.

12. Derive the MDE corresponding to the DuFort-Frankel approximation of the diffusion equation, Eq. (10.61). Analyze consistency and order.
13. Derive the MDE corresponding to the BTCS approximation of the diffusion equation, Eq. (10.67). Analyze consistency and order.
14. Derive the MDE corresponding to the Crank-Nicolson approximation of the diffusion equation, Eq. (10.78). Analyze consistency and order.

Stability

15. Perform a von Neumann stability analysis of Eq. (10.25).
16. Perform a von Neumann stability analysis of Eq. (10.56).
17. Perform a von Neumann stability analysis of Eq. (10.61).
18. Perform a von Neumann stability analysis of Eq. (10.67).
19. Perform a von Neumann stability analysis of Eq. (10.78).

Section 10.6   The Richardson and Du-Fort-Frankel methods

20. Derive the Richardson approximation of the unsteady one-dimensional diffusion equation, Eq. (10.56), including the leading truncation error terms in Δt and Δx.
21. Derive the DuFort-Frankel approximation of the unsteady one-dimensional diffusion equation, Eq. (10.61), including the leading truncation error terms in Δt and Δx.

Section 10.7   Implicit Methods

The Backward-Time Centered-Space (BTCS) Method

22. Derive the BTCS approximation of the unsteady one-dimensional diffusion equation, Eq. (10.67), including the leading truncation error terms in Δt and Δx.
23.* By hand calculation, determine the solution of the example heat diffusion problem by the BTCS method at $t = 0.5$ s for $\Delta x = 0.1$ cm and $\Delta t = 0.5$ s.
24. Implement the program presented in Section 10.12.2 to reproduce the results presented in Figure 10.18. Compare the results with the exact solution presented in Table 10.1.
25. Implement the program presented in Section 10.12.2 and repeat the calculations requested in the previous problem for $\Delta x = 0.05$ cm and $\Delta t = 0.125$ s. Compare the errors and ratios of the errors for the two solutions at $t = 10.0$ s.
25. Implement the program presented in Section 10.12.2 to reproduce the results presented in Figure 10.19.

The Crank-Nicolson Method

27. Derive the Crank-Nicolson approximation of the unsteady one-dimensional diffusion equation, Eq. (10.78), including the leading truncation error terms in Δt and Δx.
28.* By hand calculation, determine the solution of the example heat diffusion problem by the Crank-Nicolson method at $t = 0.5$ s for $\Delta x = 0.1$ cm and $\Delta t = 0.5$ s.
29. Implement the program presented in Section 10.12.3 to reproduce the results

presented in Figure 10.21. Compare the results with the exact solution presented in Table 10.1.

30. Implement the program developed in Section 10.12.3 and repeat the calculations requested in the previous problem for $\Delta x = 0.05$ cm and $\Delta t = 0.25$ s. Compare the errors and the ratios of the errors for the two solutions at $t = 10.0$ s.

31. Use the program presented in Section 10.12.3 to reproduce the results presented in Figure 10.22.

Section 10.8 Derivative Boundary Conditions

32. Derive Eq. (10.87) for a right-hand side derivative boundary condition.

33.* By hand calculation using Eq. (10.87) at the boundary point, determine the solution of the example heat diffusion problem presented in Section 10.8 at $t = 2.5$ s for $\Delta x = 0.1$ cm and $\Delta t = 0.5$ s.

34. Modify the program presented in Section 10.12.1 to incorporate a derivative boundary condition on the right-hand boundary. Check out the program by reproducing Figure 10.25.

Section 10.9 Nonlinear Equations and Multidimensional Problems

Nonlinear Equations

35. Consider the following nonlinear parabolic PDE for the generic dependent variable $f(x, y)$, which serves as a model equation in fluid mechanics:

$$ff_x = \alpha f_{yy} \tag{A}$$

where $f(x, 0) = f_1$, $f(x, Y) = f_2$, and $f(0, y) = F(y)$. (a) Derive the FTCS approximation of Eq. (A). (b) Perform a von Neumann stability analysis of the linearized FDE. (c) Derive the MDE corresponding to the linearized FDE. Investigate consistency and order. (d) Discuss a strategy for solving this problem numerically.

36. Solve the previous problem for the BTCS method. Discuss a strategy for solving this problem numerically (a) using linearization, and (b) using Newton's method.

37. Equation (A) can be written as

$$(f^2/2)_x = \alpha f_{yy} \tag{B}$$

(a) Derive the FTCS approximation for this form of the PDE. (b) Derive the BTCS approximation for this form of the PDE.

Multidimensional Problems

38. Consider the unsteady two-dimensional diffusion equation:

$$\bar{f}_t = \alpha(\bar{f}_{xx} + \bar{f}_{yy}) \tag{C}$$

(a) Derive the FTCS approximation of Eq. (C), including the leading truncation error terms in Δt, Δx, and Δy. (b) Derive the corresponding MDE. Analyze consistency and order. (c) Perform a von Neumann stability analysis of the FDE.

39. Solve Problem 38 using the BTCS method.

40. Derive the FTCS approximation of the unsteady two-dimensional convection-diffusion equation:

$$\bar{f}_t + u\bar{f}_x + v\bar{f}_y = \alpha(\bar{f}_{xx} + \bar{f}_{yy}) \tag{D}$$

41. Derive the MDE for the FDE derived in Problem 40.
42. Derive the amplification factor G for the FDE derived in Problem 40.
43. Derive the BTCS approximation of the unsteady two-dimensional convection-diffusion equation, Eq. (D).
44. Derive the MDE for the FDE derived in Problem 43.
45. Derive the amplification factor G for the FDE derived in Problem 43.

Section 10.10 The Convection-Diffusion Equation

Introduction

46. Consider the unsteady one-dimensional convection-diffusion equation:

$$\bar{f}_t + u\bar{f}_x = \alpha\bar{f}_{xx} \tag{E}$$

Classify Eq. (E). Determine the characteristic curves. Discuss the significance of these results as regards domain of dependence, range of influence, signal propagation speed, auxiliary conditions, and numerical solution procedures.

47. Develop the exact solution for the heat transfer problem presented in Section 10.10, Eqs. (10.109) and (10.115).
48. By hand calculation, evaluate the exact solution of the heat transfer problem for $P = 10$ for $T(0.8, 5.0)$ and $T(0.8, \infty)$.

The Forward-Time Centered-Space Method

49. Derive the FTCS approximation of the unsteady one-dimensional convection-diffusion equation, Eq. (10.116), including the leading truncation error terms in Δt and Δx.
50. Derive the modified differential equation (MDE) corresponding to Eq. (10.116). Analyze consistency and order.
51. Perform a von Neumann stability analysis of Eq. (10.116).
52. By hand calculation, determine the solution of the example heat transfer problem for $P = 10.0$ at $t = 1.0$ s by the FTCS method for $\Delta x = 0.1$ cm and $\Delta t = 0.5$ s. Compare the results with the exact solution in Table 10.7.
53. Modify the program presented in Section 10.12.1 to implement the numerical solution of the example convection-diffusion problem by the FTCS method. Use the program to reproduce the results presented in Figure 10.30.
54. Use the program to solve the example convection-diffusion problem with $\Delta x = 0.05$ cm and $\Delta t = 0.125$ s.

The Backward-Time Centered-Space Method

55. Derive the BTCS approximation of the unsteady one-dimensional convection-diffusion equation, Eq. (10.123), including the leading truncation error terms in Δt and Δx.
56. Derive the modified differential equation (MDE) corresponding to Eq. (10.123). Analyze consistency and order.
57. Perform a von Neumann stability analysis of Eq. (10.123).

58. By hand calculation, determine the solution of the example heat transfer problem for $P = 10.0$ at $t = 1.0$ s with $\Delta x = 0.1$ cm and $\Delta t = 1.0$ s.

59. By hand calculation, estimate the asymptotic steady state solution of the example heat transfer problem for $P = 10.0$ with $\Delta x = 0.1$ cm by letting $\Delta t = 1000.0$ s.

60. Modify the program presented in Section 10.12.2 to implement the numerical solution of the example convection-diffusion problem by the BTCS method. Use the program to reproduce the results presented in Figure 10.32.

61. Use the program to solve the convection-diffusion problem for $\Delta x = 0.05$ cm and $\Delta t = 0.25$ s. Compare the errors and the ratios of the errors for the two solutions $t = 5.0$ s.

Section 10.11 Asymptotic Steady State Solution of Propagation Problems

62. Consider steady heat transfer in a rod with an insulated end, as discussed in Section 8.6. The steady boundary-value problem is specified by

$$\hat{T}_{xx} - \alpha^2(\hat{T} - T_a) = 0 \qquad \hat{T}(0) = T_1 \text{ and } \hat{T}_x(L) = 0 \tag{F}$$

where $\alpha^2 = hP/kA$, which is defined in Section 8.6. The exact solution for $T_1 = 100.0$, $\alpha = 2.0$, and $L = 1.0$ is given by Eq. (8.70) and illustrated in Figure 8.10. This steady state problem can be solved as the asymptotic solution in time of the following unsteady problem:

$$\beta \bar{T}_t = \bar{T}_{xx} - \alpha^2(\bar{T} - T_a) \qquad \bar{T}(0) = T_1 \text{ and } \bar{T}_x(L) = 0 \tag{G}$$

with the initial temperature distribution $\bar{T}(x, 0) = F(x)$, where $\beta = \rho C/k$, ρ is the density of the rod (kg/m³), C is the specific heat (J/kg-K), and k is the thermal conductivity (J/s-m-K). Equation (G) can be derived by combining the analyses presented in Sections II.5 and II.6. (a) Derive Eq. (G). (b) Develop the FTCS approximation of Eq. (G). (c) Let $\hat{T}(0.0) = 100.0$, $\hat{T}_x(1.0) = 0.0$, $T_a = 0.0$, $L = 1.0$, $\alpha = 2.0$, $\beta = 10.0$, and the initial temperature distribution $\bar{T}(x, 0.0) = 100.0(1.0 - x)$. Solve for the steady state solution by solving Eq. (G) by the FTCS method with $\Delta x = 0.1$ cm and $\Delta t = 0.1$ s. Compare the results with the exact solution presented in Table 8.9.

63. Solve Problem 61 using the BTCS method. Try large values of Δt to reach the steady state as rapidly as possible.

Section 10.12 Programs

64. Implement the forward-time centered-space (FTCS) program for the diffusion equation presented in Section 10.12.1. Check out the program using the given data set.

65. Solve any of Problems 5 to 9 with the program.

66. Implement the backward-time centered-space (BTCS) program for the diffusion equation presented in Section 10.12.2. Check out the program using the given data set.

67. Solve any of Problem 23 to 26 with the program.

68. Implement the Crank-Nicolson program for the diffusion equation presented in Section 10.12.3. Check out the program using the given data set.

69. Solve any of Problems 28 to 31 with the program.

11

Hyperbolic Partial Differential Equations

Examples

11.1 INTRODUCTION

The constant-area tube illustrated in Figure 11.1 is filled with a stationary incompressible fluid having a very low thermal conductivity, so that heat diffusion is neglible. The fluid is heated to an initial temperature distribution, $T(x, 0)$, at which time the heat source is turned off and the fluid is instantaneously given the velocity $u = 0.1$ cm/s to the right. The

Incompressible Liquid

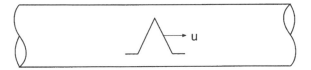

$$f_t + uf_x = 0, \quad f(x,0) = F(x), \quad f(x,t) = ?$$

Compressible Gas

$$f_{tt} = a^2 f_{xx}, \quad f(x,0) = F(x), \quad f_t(x,0) = G(x), \quad f(x,t) = ?$$

Figure 11.1   Unsteady wave propagation problems.

temperature distribution within the tube is required. The temperature distribution is governed by the unsteady one-dimensional convection equation:

$$T_t + uT_x = 0 \tag{11.1}$$

In the range $0.0 \le x \le 1.0$, the initial temperature of the fluid is given by

$$T(x, 0.0) = 200.0x \qquad\qquad 0.0 \le x \le 0.5 \tag{11.2a}$$
$$T(x, 0.0) = 200.0(1.0 - x) \qquad 0.5 \le x \le 1.0 \tag{11.2b}$$

where $T(x, t)$ is measured in degrees Celsius (C). The initial temperature is zero everywhere outside of this range. This initial temperature distribution is illustrated by the curve labelled $t = 0.0$ in Figure 11.2.

For the present problem, the temperature distribution specified by Eq. (11.2) simply moves to the right at the speed $u = 0.1$ cm/s. The exact solutions for several values of time are presented in Figure 11.2. Note that the discontinuity in slope at the peak of the temperature distribution is preserved during convection.

The lower sketch in Figure 11.1 illustrates a long duct filled with a stagnant compressible gas. The gas is initially at rest. A small triangularly shaped acoustic pressure perturbation is created in the duct. As shown in Section III.7, the acoustic motion within the duct is governed by a set of coupled first-order PDEs, Eqs. (III.89) and (III.90), where the subscript zero and the superscript prime have been dropped for clarity:

$$\rho u_t + P_x = 0 \tag{11.3}$$
$$P_t + \rho a^2 u_x = 0 \tag{11.4}$$

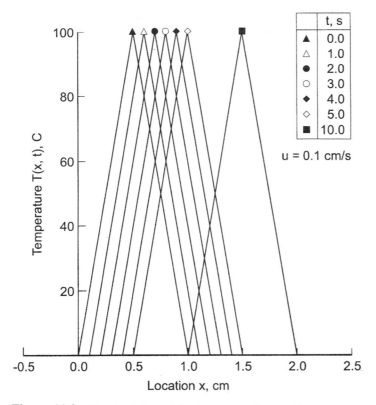

Figure 11.2 Exact solution of the heat convection problem.

As shown in Section III.7, Eqs. (11.3) and (11.4) can be combined to yield the wave equation:

$$P_{tt} = a^2 P_{xx} \tag{11.5}$$

The acoustic pressure distribution within the duct $P(x, t)$ is required. The specific problem, its exact solution, and its numerical solution are presented in Section 11.9.

Quite a few hyperbolic partial differential equations are encountered in engineering and science. Two of the more common ones are the *convection equation* and the *wave equation*, presented below for the generic dependent variable $f(x, t)$:

$$f_t + u f_x = 0 \tag{11.6}$$
$$f_{tt} = c^2 f_{xx} \tag{11.7}$$

where u is the convection velocity and c is the wave propagation speed. The convection equation applies to problems in fluid mechanics, heat transfer, etc. The wave equation applies to problems of vibrating systems, such as acoustic fields and strings.

The convection equation models a wave travelling in one direction, the direction of the velocity u. Thus, the convection equation models the essential features of the more complex wave motion governed by the wave equation, in which waves travel in both directions at the velocities $+c$ and $-c$. The general features of the numerical solution of the convection equation also apply to the numerical solution of the wave equation. Consequently, this chapter is devoted mainly to the numerical solution of the convection equation to gain insight into the numerical solution of more complicated hyperbolic PDEs such as the wave equation. Section 11.9 presents an introduction to the numerical solution of the wave equation.

The solution to Eqs. (11.6) and (11.7) is the function $f(x, t)$. For Eqs. (11.6) and (11.7), this function must satisfy an initial condition at time $t = 0$, $f(x, 0) = F(x)$. Equation (11.7) must also satisfy a second initial condition $f_t(x, 0) = G(x)$. Since Eq. (11.6) is first order in space x, only one boundary condition can be applied. Since Eq. (11.7) is second order in space, it requires two boundary conditions. In both cases, these boundary conditions may be of the Dirichlet type (i.e., specified values of f), the Neumann type (i.e., specified values of f_x), or the mixed type (i.e., specified combinations of f and f_x). The basic properties of finite difference methods for solving propagation problems governed by hyperbolic PDEs are presented in this chapter.

Figure 11.3 presents the organization of Chapter 11. After this introductory section, the general features of hyperbolic PDEs are reviewed. This discussion is followed by an

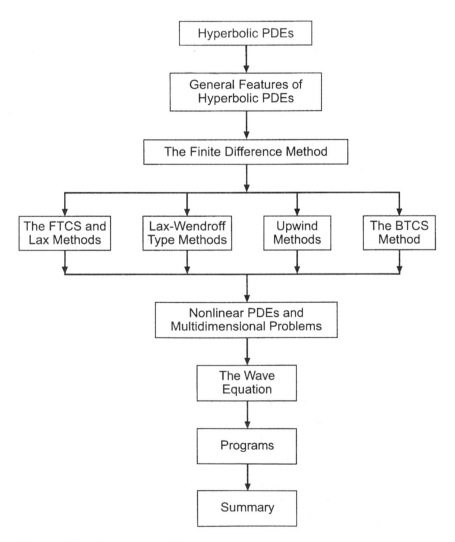

Figure 11.3 Organization of Chapter 11.

introduction to the finite difference method as it applies to hyperbolic PDEs. At this point, the presentation splits into a discussion of four major types of finite difference methods for solving hyperbolic PDEs: (1) the FTCS and Lax methods, (2) Lax-Wendroff type methods, (3) upwind methods, and (4) the BTCS method. Following these four sections, a brief discussion of nonlinear PDEs and multidimensional problems is presented. An introduction to the numerical solution of the wave equation follows. Several programs for solving the simple convection equation are then presented. The chapter ends with a summary.

11.2 GENERAL FEATURES OF HYPERBOLIC PDES

Several concepts must be considered before a propagation type PDE can be solved by a finite difference method. Most of these concepts are discussed in Section 10.2, which is concerned mainly with finite difference methods for solving parabolic PDEs. That section should be reviewed and considered relevant to finite difference methods for solving hyperbolic PDEs. In this section, the concepts which are different for hyperbolic PDEs are presented, the general features of convection are illustrated, and the concept of characteristics is discussed.

11.2.1 Fundamental Considerations

Propagation problems are *initial-boundary-value problems* in *open domains* (open with respect to time or a timelike variable) in which the solution in the domain of interest is marched forward from the initial state, guided and modified by the boundary conditions. Propagation problems are governed by parabolic or hyperbolic partial differential equations. The general features of parabolic and hyperbolic PDEs are discussed in Part III. Those features which are relevant to the finite difference solution of parabolic PDEs are summarized in Section 10.2. Those features which are relevant to the finite difference solution of hyperbolic PDEs are summarized in this section.

The general features of *hyperbolic partial differential equations (PDEs)* are discussed in Section III.7. In that section it is shown that hyperbolic PDEs govern propagation problems, which are initial-boundary-value problems in open domains. Consequently, hyperbolic PDEs are solved numerically by marching methods. From the characteristic analysis presented in Section III.7, it is known that problems governed by hyperbolic PDEs have a *finite physical information propagation speed*. As a result, the solution at a given point P at time level n depends on the solution only within a finite domain of dependence in the solution domain at times preceding time level n, and the solution at a given point P at time level n influences the solution only within a finite range of influence in the solution domain at times after time level n. Consequently, the physical information propagation speed, $c = dx/dt$, is finite. These general features of hyperbolic PDEs are illustrated in Figure 11.4.

11.2.2 General Features of Convection

Consider pure convection, which is governed by the convection equation:

$$\boxed{f_t + uf_x = 0} \tag{11.8}$$

where u is the convection velocity. The exact solution of Eq. (11.8) is given by

$$f(x, t) = F(x - ut) \tag{11.9}$$

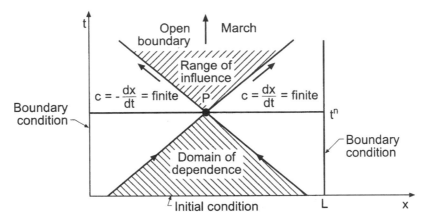

Figure 11.4   General features of hyperbolic PDEs.

which can be demonstrated by direct substitution. Equation (11.9) defines a right-traveling wave which propagates (i.e., convects) the initial property distribution to the right at the convection velocity u. The first-order (in time) convection equation requires one initial condition:

$$f(x, 0) = \phi(x) \tag{11.10}$$

Substituting Eq. (11.10) into Eq. (11.9) gives

$$F(x) = \phi(x) \tag{11.11}$$

Equation (11.11) shows that the functional form of $F(x - ut)$ is identical to the functional form of $\phi(x)$. That is, $F(x - ut) = \phi(x - ut)$. Thus, Eq. (11.9) becomes

$$\boxed{f(x, t) = \phi(x - ut)} \tag{11.12}$$

Equation (11.12) is the exact solution of the convection equation. It shows that the initial property distribution $f(x, 0) = \phi(x)$ simply propagates (i.e., convects) to the right at the constant convection velocity u unchanged in magnitude and shape.

11.2.3 Characteristic Concepts

The concept of characteristics of partial differential equations is introduced in Section III.3. In two-dimensional space, which is the case considered here (i.e., physical space x and time t), characteristics are paths (curved, in general) in the solution domain $D(x, t)$ along which physical information propagates. If a partial differential equation possesses real characteristics, then physical information propagates along the characteristic paths. The presence of characteristics has a significant impact on the solution of a partial differential equation (by both analytical and numerical methods).

Consider the unsteady one-dimensional convection equation $f_t + u f_x = 0$. It is shown in Section III.3 that the pathline is the characteristic path for the convection equation:

$$\frac{dx}{dt} = u \tag{11.13}$$

Consider the one-dimensional wave equation $f_{tt} = a^2 f_{xx}$. It is shown in Section III.7 that the wavelines are the characteristic paths for the wave equation:

$$\frac{dx}{dt} = \pm a \tag{11.14}$$

Thus, information propagates along the characteristic paths. These preferred information propagation paths should be considered when solving hyperbolic PDEs by numerical methods.

11.3 THE FINITE DIFFERENCE METHOD

The objective of a finite difference method for solving a partial differential equation (PDE) is to transform a calculus problem into an algebra problem by:

1. *Discretizing* the continuous physical domain into a discrete finite difference grid
2. *Approximating* the individual exact partial derivatives in the partial differential equation (PDE) by algebraic finite difference approximations (FDAs)
3. *Substituting* the FDAs into the PDE to obtain an algebraic finite difference equation (FDE)
4. *Solving* the resulting algebraic FDEs

These steps are discussed in detail in Section 10.3. That section should be reviewed and considered equally relevant to the finite difference solution of hyperbolic PDEs.

The objective of the numerical solution of a hyperbolic PDE is to march the solution at time level n forward in time to time level $n + 1$, as illustrated in Figure 11.5, where the physical domain of dependence of a hyperbolic PDE is illustrated. In view of the finite physical information propagation speed $c = dx/dt$ associated with hyperbolic PDEs, the solution at point P at time level $n + 1$ should not depend on the solution at any of the other points at time level $n + 1$. This requires a finite numerical information propagation speed, $c_n = \Delta x/\Delta t$.

A discussion of explicit and implicit finite difference methods is presented in Section 10.2. From that discussion, it is obvious that the numerical domain of dependence of explicit finite difference methods matches the physical domain of dependence of hyperbolic PDEs. Consequently, hyperbolic PDEs should be solved by explicit finite difference

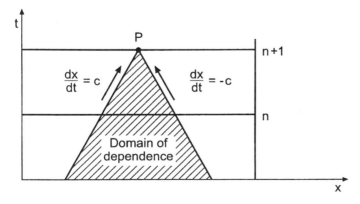

Figure 11.5 Physical domain of dependence of hyperbolic PDEs.

LIVERPOOL JOHN MOORES UNIVERSITY
LEARNING SERVICES

methods. The only exception is when solving steady state problems as the asymptotic solution in time of an appropriate unsteady propagation problem. In that case, as discussed in Section 10.11, implicit methods may have some advantages over explicit methods.

The solution domain $D(x, t)$ is discretized for the finite difference solution of a hyperbolic PDE in the same manner as done for a parabolic PDE, as illustrated in Figure 10.8.

Finite difference approximations of the individual exact partial derivatives in a PDE must be developed. As in Chapters 9 and 10, the exact solution of the PDE will be denoted by an overbar over the symbol for the dependent variable, that is, $\bar{f}(x, t)$, and the approximate solution of the PDE will be denoted by the symbol for the dependent variable without an overbar, that is $f(x, t)$. Thus,

$$
\boxed{
\begin{aligned}
\bar{f}(x, t) &= \text{exact solution} \\
f(x, t) &= \text{approximate solution}
\end{aligned}
}
$$

The exact time derivative $\bar{f}_t$ can be approximated by forward-time, backward-time, or centered-time finite difference approximations, as described in Section 10.2.

The first-order space derivative $\bar{f}_x$ is a model of physical convection. From characteristic concepts, it is known that the physical information propagation speed associated with first-order spatial derivatives is finite, and that information propagates along distinct characteristic paths. For the convection equation, the characteristic paths are the pathlines, given by $dx/dt = u$, and the physical information propagation speed is the convection velocity u. The solution at a point depends only on the information in the domain of dependence specified by the upstream characteristic paths, and the solution at a point influences the solution only in the range of influence specified by the downstream propagation paths.

These characteristic concepts suggest that first-order spatial derivatives, such as $\bar{f}_x$, should be approximated by one-sided approximations in the direction from which the physical information is being propagated. Such approximations are called *upwind approximations*. A first-order backward-space approximation of the first-order spatial derivative $\bar{f}_x$ can be obtained by writing the backward-space Taylor series for $\bar{f}_{i-1}$ and solving for $\bar{f}_x|_i$. Thus,

$$
\bar{f}_{i-1} = \bar{f}_i + \bar{f}_x|_i(-\Delta x) + \tfrac{1}{2}\bar{f}_{xx}|_i(-\Delta x^2) + \tfrac{1}{6}\bar{f}_{xxx}|_i(-\Delta x^3) + \cdots \tag{11.15}
$$

Solving Eq. (11.15) for $\bar{f}_x|_i$ gives

$$
\bar{f}_x|_i = \frac{\bar{f}_i - \bar{f}_{i-1}}{\Delta x} + \tfrac{1}{2}\bar{f}_{xx}(\xi)\,\Delta x \tag{11.16}
$$

where $x_{i-1} \le \xi \le x_i$. Truncating the remainder term yields the *first-order backward-space* (i.e., upwind) approximation of $\bar{f}_x|_i$, denoted by $f_x|_i$:

$$
\boxed{f_x|_i = \frac{f_i - f_{i-1}}{\Delta x}} \tag{11.17}
$$

Two upwind approximations of the convection equation are presented in Section 11.6.

First-order spatial derivatives can also be approximated with centered-space approximations with acceptable results. A second-order centered-space approximation of the first-order spatial derivative $\bar{f}_x$ can be obtained by combining the forward-space Taylor series for $\bar{f}_{i+1}$, presented below, with the backward-space Taylor series for $\bar{f}_{i-1}$, Eq. (11.15). Thus,

$$\bar{f}_{i+1} = \bar{f}_i + \bar{f}_x|_i \, \Delta x + \tfrac{1}{2} \, \bar{f}_{xx}|_i \, \Delta x^2 + \tfrac{1}{6} \, \bar{f}_{xxx}|_i \, \Delta x^3 + \cdots \qquad (11.18)$$

Subtracting Eq. (11.15) from Eq. (11.18) and solving for $\bar{f}_x|_i$ gives

$$\bar{f}_x|_i = \frac{\bar{f}_{i+1} - \bar{f}_{i-1}}{2\,\Delta x} - \tfrac{1}{2} \, \bar{f}_{xxx}(\xi) \, \Delta x^2 \qquad (11.19)$$

where $x_{i-1} \le \xi \le x_{i+1}$. Truncating the remainder term yields the second-order centered-space approximation of $\bar{f}_x|_i$, denoted by $f_x|_i$:

$$f_x|_i = \frac{f_{i+1} - f_{i-1}}{2\,\Delta x} \qquad (11.20)$$

Several centered-space approximations of the convection equation are presented in Sections 11.4, 11.5, and 11.7.

A characteristic analysis of the wave equation shows that information propagates along two distinct characteristic paths, $dx/dt = \pm a$, in both the positive and negative directions, with the propagation speed a. Since both directions of physical information propagation are implicit in the PDEs, upwind spatial derivative approximations cannot be employed directly. However, centered spatial derivative approximations can be employed, since both forward and backward information are used in the finite difference approximations. This approach is applied to the wave equation in Section 11.9.

11.4 THE FORWARD-TIME CENTERED-SPACE (FTCS) METHOD AND THE LAX METHOD

The most straightforward finite difference method for solving hyperbolic partial differential equations might appear to be the forward-time centered-space (FTCS) method. The FTCS method is applied to the diffusion equation in Section 10.4. It is shown there that the FTCS approximation of the diffusion equation is conditionally stable (i.e., $d = \alpha \, \Delta t / \Delta x^2 \le \tfrac{1}{2}$). However, when applied to the convection equation $\bar{f}_t + u\bar{f}_x = 0$, the FTCS method is unconditionally unstable. A modification of the FTCS method suggested by Lax (1954) removes the unconditional instability of the FTCS method, but introduces an inconsistency into the FDE, which results in excessive numerical damping (i.e., numerical diffusion). These two methods are presented in this section. An introduction to numerical diffusion and numerical dispersion is presented at the end of the section.

11.4.1 The Forward-Time Centered-Space (FTCS) Method

Introducing the first-order forward-difference approximation for $\bar{f}_t$, Eq. (10.17), and the second-order centered-difference approximation for $\bar{f}_x$, Eq. (11.20), into the convection equation $\bar{f}_t + u\bar{f}_x = 0$ yields

$$\frac{f_i^{n+1} - f_i^n}{\Delta t} + u \frac{f_{i+1}^n - f_{i-1}^n}{2\,\Delta x} = 0 \qquad (11.21)$$

Solving Eq. (11.21) for f_i^{n+1} yields the FTCS approximation of the convection equation:

$$\boxed{f_i^{n+1} = f_i^n - \frac{c}{2}(f_{i+1}^n - f_{i-1}^n)} \qquad (11.22)$$

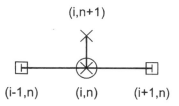

Figure 11.6   The FTCS method stencil.

where $c = u \, \Delta t / \Delta x$ is the *convection number*. The corresponding finite difference stencil is illustrated in Figure 11.6, where the circle denotes the base point for the FDAs, the crosses, $\times$, denote the points used to approximate $\bar{f}_t$, and the open squares denote the points used to approximate $\bar{f}_x$.

The modified differential equation (MDE) corresponding to Eq. (11.22) is

$$f_t + uf_x = -\tfrac{1}{2}f_{tt}\,\Delta t - \tfrac{1}{6}f_{ttt}\Delta t^2 - \cdots - \tfrac{1}{6}uf_{xxx}\Delta x^2 - \tfrac{1}{120}uf_{xxxx}\Delta x^4 - \cdots \qquad (11.23)$$

As $\Delta t \to 0$ and $\Delta x \to 0$, Eq. (11.23) approaches $f_t + uf_x = 0$, which is the convection equation. Consequently, Eq. (11.22) is a consistent approximation of that equation. The truncation error is $0(\Delta t) + 0(\Delta x^2)$. From a von Neumann stability analysis, the amplification factor G corresponding to Eq. (11.22) is

$$G = 1 - Ic \sin \theta \qquad (11.24)$$

The magnitude of G is

$$|G| = (1 + c^2 \sin^2 \theta)^{1/2} \qquad (11.25)$$

which is greater than unity for $c > 0$. Consequently, Eq. (11.22) is unconditionally unstable. From a graphical point of view, Eq. (11.24) represents a vertical line segment in the complex plane, as illustrated in Figure 11.7. Its center is at $(1 + I0)$, and the entire line segment is outside of the unit circle for all values of $c > 0$. Consequently, Eq. (11.22) is unconditionally unstable.

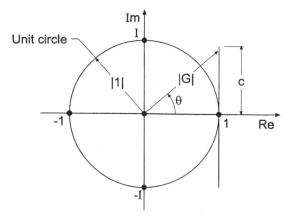

Figure 11.7   Locus of the amplification factor G for the FTCS method.

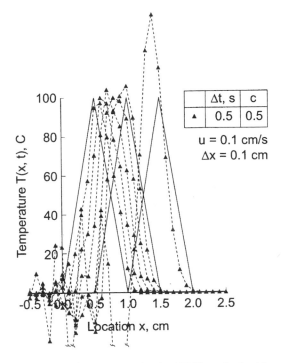

Figure 11.8   Solution by the FTCS method with $c = 0.5$.

Example 11.1.   The FTCS method applied to the convection equation

To illustrate the unstable behavior of the FTCS method applied to the convection equation, the solution to the convection problem presented in Section 11.1 is presented in Figure 11.8 for $\Delta x = 0.1$ cm with $\Delta t = 0.5$ s, for which $c = u \, \Delta t / \Delta x = (0.1)(0.5)/0.1 = 0.5$. The solution is presented at times from 1.0 to 5.0 s and at 10.0 s. The amplitude of the solution increases in an unstable manner as the wave propagates to the right. Solutions for larger values of c are not shown because they are totally unrealistic. All of the other examples in this chapter are solved with $\Delta x = 0.05$ cm and $\Delta t = 0.25$ s. For the FTCS method, that combination, for which $c = 0.5$ as in the present example, goes unstable more rapidly, so the results could not be illustrated in the figure.

In summary, the FTCS approximation of the convection equation is unconditionally unstable. Consequently, it is unsuitable for solving the convection equation, or any other hyperbolic PDE.

11.4.2   The Lax Method

Lax (1954) proposed a modification to the FTCS method that yields a conditionally stable method. In that modification, the value f_i^n in the finite difference approximation of $f_t|_i^n$ used in Eq. (11.21) is approximated by $f_i^n = (f_{i+1}^n + f_{i-1}^n)/2$. The resulting finite difference equation is

$$f_i^{n+1} = \tfrac{1}{2}(f_{i+1}^n + f_{i-1}^n) - \frac{c}{2}(f_{i+1}^n - f_{i-1}^n) \qquad (11.26)$$

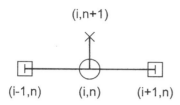

Figure 11.9 The Lax method stencil.

Equation (11.26) is the Lax approximation of the convection equation. The corresponding finite difference stencil is illustrated in Figure 11.9. The modified differential equation (MDE) corresponding to Eq. (11.26) is

$$f_t + u f_x = -\tfrac{1}{2} f_{tt}\, \Delta t - \tfrac{1}{6} f_{ttt}\, \Delta t^2 - \cdots + \tfrac{1}{2} f_{xx}\, \frac{\Delta x^2}{\Delta t} - \tfrac{1}{6} u f_{xxx}\, \Delta x^2 - \cdots \tag{11.27}$$

As $\Delta t \to 0$ and $\Delta x \to 0$, the third term on the right-hand side of Eq. (11.27) becomes indeterminate. Consequently, Eq. (11.26) is not a consistent approximation of the convection equation. From a von Neumann stability analysis, the amplification factor G corresponding to Eq. (11.26) is

$$G = \cos\theta - Ic\sin\theta \tag{11.28}$$

The magnitude of G is

$$|G| = (\cos^2\theta + c^2 \sin^2\theta)^{1/2} = [1 - \sin^2\theta(1 - c^2)]^{1/2} \tag{11.29}$$

Since $\sin^2\theta \geq 0$ for all values of $\theta = (k_m\, \Delta x)$, $|G| \leq 1$ if

$$\boxed{c = \frac{u\, \Delta t}{\Delta x} \leq 1} \tag{11.30}$$

Thus, the Lax approximation of the convection equation is conditionally stable. Equation (11.30) is the celebrated Courant-Friedrichs-Lewy (1928) stability criterion, commonly called the *CFL stability criterion*. In essence, Eq. (11.30) states that the numerical speed of information propagation $u_n = \Delta x / \Delta t$ must be greater than or equal to the physical speed of information propagation $u = dx/dt$. From a graphical point of view, Eq. (11.28) is an ellipse in the complex plane, as illustrated in Figure 11.10. The center of the ellipse is at $(0 + I0)$, and the axes are 1 and c. For stability, $|G| \leq 1$, which requires that the ellipse remain on or within the unit circle. From Figure 11.10, it is obvious that the ellipse remains on or within the unit circle for $c \leq 1$.

The Lax approximation of the convection equation may behave in a numerically consistent manner if c is held constant as Δx or Δt is changed. It is conditionally stable. Consequently, by the Lax equivalence theorem, it may behave as a convergent approximation of the convection equation.

Example 11.2. The Lax method applied to the convection equation

Let's solve the convection problem presented in Section 11.1 using Eq. (11.26) for $\Delta x = 0.05$ cm, for $\Delta t = 0.05, 0.25, 0.45$, and 0.50 s. For $u = 0.1$ cm/s and $\Delta x = 0.05$ cm, $c = u\, \Delta t/\Delta x = (0.1)\Delta t/(0.05) = 2\, \Delta t$. Thus, the value of c is twice the value of Δt for this

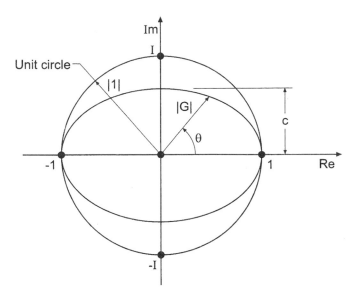

Figure 11.10 Locus of the amplification factor G for the Lax method.

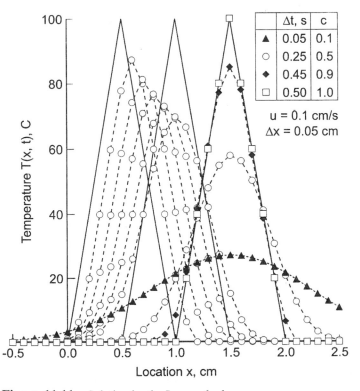

Figure 11.11 Solution by the Lax method.

choice of physical properties. The results are presented in Figure 11.11 at times from 1.0 to 5.0 s for $c = 0.5$, and at 10.0 s for $c = 0.1$, 0.5, 0.9, and 1.0. The solution is presented at every other grid point.

Several important features are illustrated in Figure 11.11. When $c = 1.0$, the numerical solution is identical to the exact solution, for the linear convection equation. This is not true for nonlinear PDEs. When $c = 0.5$, the amplitude of the solution is severely damped as the wave propagates, and the peak of the wave is rounded. The general shape of the solution is maintained, but the leading and trailing edges of the wave are quite smeared out. The result at $t = 10.0$ s for $c = 0.1$ is completely smeared out. The numerical solution does not even resemble the general shape of the exact solution. These effects are the result of the numerical damping that is present in the Lax method. In effect, the initial-data distribution is being both convected and diffused, and the effect of diffusion increases as the time step is decreased. The solution for $c = 0.9$ is much closer to the exact solution, except at the peak, which is severely damped. The presence of large amounts of numerical damping at small values of the convection number, c, is a serious problem with the Lax method.

In summary, the Lax approximation of the convection equation is explicit, single step, inconsistent, $0(\Delta t) + 0(\Delta x^2) + 0(\Delta x^2/\Delta t)$, and conditionally stable. Excessive numerical damping is present, which makes the Lax method a poor choice for solving hyperbolic PDEs.

Example 11.2 shows that the accuracy of the solution increases as the convection number is increased. In fact, the exact solution is obtained for $c = 1.0$, for the linear convection equation. For nonlinear PDEs, the exact solution is not obtained for $c = 1.0$. However, these results suggest that the most accurate solution might correspond to $c = 1.0$. This is indeed the case. The convection number c can be written as

$$c = \frac{u \, \Delta t}{\Delta x} = \frac{u}{\Delta x/\Delta t} = \frac{u}{u_n} \tag{11.31}$$

where u_n is the numerical information propagation speed. Using values of c close to 1.0 causes the numerical information propagation speed to be close to the physical information propagation speed, which accounts for the increase in accuracy when c is close to 1.0. This result is true for all explicit finite difference methods.

11.4.3 Numerical Diffusion and Dispersion

The severe numerical damping present the Lax method is a result of the leading spatial truncation error term in the MDE, Eq. (11.27). This term contains the second-order spatial derivative f_{xx}, which acts in the same manner as real physical diffusion. Hence, this effect is called *numerical diffusion*, or *numerical damping*. As shown in Section 10.2 for physical diffusion, the effect of the second spatial derivative is to diffuse, or spread out, the initial property distribution as time progresses. The same result is obtained when the second spatial derivative is present in a truncation error term. For the Lax method, the numerical diffusion coefficient $(\Delta x^2/2 \, \Delta t)$ becomes very large as Δt decreases. This effect is responsible for the severe damping present in the Lax method. In a more general sense, all even spatial derivatives in the truncation error contribute to numerical diffusion.

Although the concept is not developed in this book, it can be shown that all odd spatial derivatives in the truncation error contribute to *numerical dispersion*, which is a type of higher-order convection which tends to distort the shape of the property distribution in space and causes *wiggles* in the solution. Numerical diffusion and/or numerical dispersion are present in all numerical solutions of all PDEs.

11.5   LAX-WENDROFF TYPE METHODS

Lax and Wendroff (1960) proposed an $0(\Delta t^2) + 0(x^2)$ method based on an $0(\Delta t^2)$ forward-time Taylor series for $\bar{f}_i^{n+1}$. The first and second time derivatives in the Taylor series are expressed in terms of first and second spatial derivatives, respectively, by using the PDE to obtain $\bar{f}_t = -u\bar{f}_x$ and by differentiating the PDE to obtain $\bar{f}_{tt} = u^2\bar{f}_{xx}$. This gives a semi-discrete expression for $\bar{f}_i^{n+1}$ in terms of $\bar{f}_x$ and $\bar{f}_{xx}$. The spatial derivatives are then approximated by second-order centered-difference approximations. When first published, this approach was called the Lax-Wendroff method. Several other methods have been developed subsequently, which can be interpreted as Lax-Wendroff type methods. The original Lax-Wendroff method, which is now called the Lax-Wendroff one-step method, a two-step Lax-Wendroff type method developed by Richtmyer (1963), and a predictor-corrector Lax-Wendroff type method developed by MacCormack (1969) are presented in this section.

11.5.1   The Lax-Wendroff One-Step Method

The Lax-Wendroff (1960) one-step method is a very popular $0(\Delta t^2) + (\Delta x^2)$ explicit finite difference method. For the unsteady one-dimensional convection equation, $\bar{f}_t + u\bar{f}_x = 0$, the function to be determined is $\bar{f}(x, t)$. Expanding $\bar{f}(x, t)$ in a Taylor series in time gives

$$\boxed{\bar{f}_i^{n+1} = \bar{f}_i^n + \bar{f}_t|_i^n \, \Delta t + \tfrac{1}{2}\bar{f}_{tt}|_i^n \, \Delta t^2 + 0(\Delta t^3)} \tag{11.32}$$

The first-order time derivative $\bar{f}_t$ is determined directly from the partial differential equation:

$$\bar{f}_t = -u\bar{f}_x \tag{11.33}$$

The second-order time derivative $\bar{f}_{tt}$ is determined by differentiating the partial differential equation with respect to time. Thus,

$$\bar{f}_{tt} = (\bar{f}_t)_t = (-u\bar{f}_x)_t = -u(\bar{f}_t)_x = -u(-u\bar{f}_x)_x = u^2\bar{f}_{xx} \tag{11.34}$$

Note that this procedure does not work for a nonlinear PDE where $u = u(f)$. Substituting Eqs. (11.33) and (11.34) into Eq. (11.32) yields

$$\bar{f}_i^{n+1} = \bar{f}_i^n - u\bar{f}_x|_i^n \, \Delta t + \tfrac{1}{2}u^2\bar{f}_{xx}|_i^n \, \Delta t^2 + 0(\Delta t^3) \tag{11.35}$$

Truncating the remainder term, and approximating the two spatial derivatives $\bar{f}_x|_i^n$ and $\bar{f}_{xx}|_i^n$ by second-order centered-difference approximations, Eqs. (11.20) and (10.23), respectively, gives

$$f_i^{n+1} = f_i^n - u\left(\frac{f_{i+1}^n - f_{i-1}^n}{2\,\Delta x}\right)\Delta t + \tfrac{1}{2}u^2\left(\frac{f_{i+1}^n - 2f_i^n + f_{i-1}^n}{\Delta x^2}\right)\Delta t^2 \tag{11.36}$$

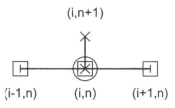

Figure 11.12 The Lax-Wendroff one-step method stencil.

Introducing the convection number, $c = u\,\Delta t/\Delta x$, yields

$$f_i^{n+1} = f_i^n - \frac{c}{2}(f_{i+1}^n - f_{i-1}^n) + \frac{c^2}{2}(f_{i+1}^n - 2f_i^n + f_{i-1}^n) \qquad (11.37)$$

Equation (11.37) is the Lax-Wendroff one-step approximation of the linear convection equation. The corresponding finite difference stencil is presented in Figure 11.12.

The modified differential equation (MDE) corresponding to Eq. (11.37) is

$$f_t + uf_x = -\tfrac{1}{2}f_{tt}\,\Delta t - \tfrac{1}{6}f_{ttt}\,\Delta t^2 - \tfrac{1}{24}f_{tttt}\,\Delta t^3 - \cdots - \tfrac{1}{6}uf_{xxx}\Delta x^2 - \cdots$$
$$+\tfrac{1}{2}u^2 f_{xx}\Delta t + \tfrac{1}{24}u^2 f_{xxxx}\Delta x^2 \Delta t + \cdots \qquad (11.38)$$

As $\Delta t \to 0$ and $\Delta x \to 0$, Eq. (11.38) approaches $f_t + uf_x$. Consequently, Eq. (11.37) is a consistent approximation of the convection equation. Equation (11.38) suggests that the FDE is $0(\Delta t) + 0(\Delta x^2)$. However, substituting $f_{tt} = u^2 f_{xx}$ into Eq. (11.38) shows that the two $0(\Delta t)$ terms cancel exactly, and the FDE is $0(\Delta t^2) + 0(\Delta x^2)$. From a von Neumann stability analysis, the amplification factor G is given by

$$G = [(1 - c^2) + c^2 \cos\theta] - Ic\sin\theta \qquad (11.39)$$

Equation (11.39) represents an ellipse in the complex plane, as illustrated in Figure 11.13. The center of the ellipse is at $(1 - c^2 + I0)$, and the axes are c and c^2. For stability,

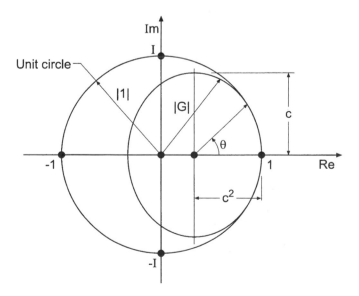

Figure 11.13 Locus of the amplification factor G for the Lax-Wendroff one-step method.

$|G| \leq 1$, which requires that the ellipse lie on or within the unit circle. From Figure 11.13, three conditions are obvious. The axes c and c^2 must both be less than or equal to unity, that is, $c \leq 1$. In addition, at point $(1 + I0)$, the curvature of the ellipse must be greater than the curvature of the unit circle. With some further analysis, it can be shown that this condition is satisfied if $c \leq 1$. All three necessary conditions are satisfied by the single sufficient condition

$$c = \frac{u\,\Delta t}{\Delta x} \leq 1 \tag{11.40}$$

Consequently, the FDE is conditionally stable. The Lax-Wendroff one-step approximation of the convection equation is consistent and conditionally stable. Consequently, by the Lax equivalence theorem, it is a convergent finite difference approximation of the convection equation.

Example 11.3. The Lax-Wendroff one-step method applied to the convection equation.

Let's solve the convection problem presented in Section 11.1 using Eq. (11.37) for $\Delta x = 0.05$ cm. The results are presented in Figure 11.14 at times from 1.0 to 5.0 s for $c = 0.5$, and at 10.0 s for $c = 0.1$, 0.5, 0.9, and 1.0.

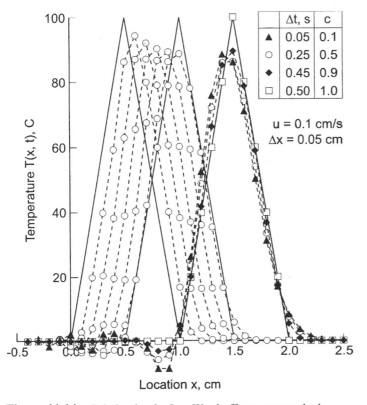

Figure 11.14   Solution by the Lax-Wendroff one-step method.

Figure 11.14 illustrates several important features of Eq. (11.37). When $c = 1.0$, the numerical solution is identical to the exact solution, for the linear convection equation. That is not the case for nonlinear PDEs. However, these results suggest that the most accurate solution of nonlinear PDEs will be obtained when $c = 1.0$. Experience shows that this is indeed the case. When $c = 0.5$, the amplitude of the solution is damped slightly, and the sharp peak becomes rounded. However, the wave shape is maintained quite well. The result at $t = 10.0$ s, for $c = 0.1$, 0.5, and 0.9, are all reasonable approximations of the exact solution. There is some decrease in the numerical convection velocity. Slight *wiggles* appear in the trailing portion of the wave. Such wiggles, which are caused by numerical dispersion, are a common feature of second-order finite difference approximations of the time derivative in convection problems. Overall, the Lax-Wendroff one-step method yields a good solution to the linear convection equation.

The Lax-Wendroff one-step method is an efficient and accurate method for solving the linear convection equation. For nonlinear PDEs and systems of PDEs, however, the method becomes quite complicated. The complications arise in the replacement of the second-order time derivative $\bar{f}_{tt}$ in terms of space derivatives by differentiating the governing partial differential equation. The simple result obtained in Eq. (11.34) no longer applies. Consequently, the Lax-Wendroff one-step method is not used very often. More efficient methods, such as the Lax-Wendroff two-step method and the MacCormack method are generally used for nonlinear equations and systems of equations. These methods have the same general features as the Lax-Wendroff one-step method, but they are considerably less complicated for nonlinear PDEs, and thus considerably more efficient.

In summary, the Lax-Wendroff one-step method applied to the convection equation is explicit, single step, consistent, $0(\Delta t^2) + 0(\Delta x^2)$, conditionally stable, and convergent. The method is quite complicated for nonlinear PDEs, systems of PDEs, and two- and three-dimensional physical spaces.

11.5.2 The Lax-Wendroff (Richtmyer) Two-Step Method

The Lax-Wendroff (1960) one-step method has many desirable features when applied to the linear convection equation. However, when applied to a nonlinear PDE or a system of PDEs, the method becomes considerably more complicated. Richtmyer (1963) proposed a three-time-level two-step method which is equivalent to the Lax-Wendroff one-step method for the linear convection equation. The first time step uses the Lax (1954) method to obtain provisional values at the second time level, and the second time step uses the leapfrog method to obtain final values at the third time level. The Richtmyer method is much simpler than the Lax-Wendroff one-step method for nonlinear PDEs and systems of PDEs. Quite commonly, any two-step method which can be interpreted as a second-order Taylor series in time is referred to as a *two-step Lax-Wendroff method* or a method of the *Lax-Wendroff type*.

For the linear convection equation $\bar{f}_t + u\bar{f}_x = 0$, the two-step method proposed by Richtmyer (1963) is as follows:

$$f_i^{n+1} = \tfrac{1}{2}(f_{i+1}^n + f_{i-1}^n) - \frac{c}{2}(f_{i+1}^n - f_{i-1}^n) \tag{11.41}$$

$$f_i^{n+2} = f_i^n - c(f_{i+1}^{n+1} - f_{i-1}^{n+1}) \tag{11.42}$$

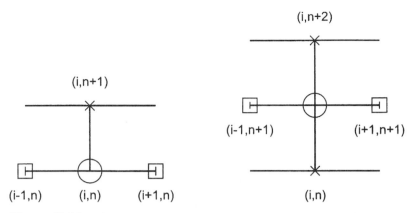

Figure 11.15   The Lax-Wendroff (Richtmyer) two-step method stencil.

where Eq. (11.41) is the Lax method, Eq. (11.26), applied from time level n to time $n + 1$, and Eq. (11.42) is the leapfrog method, which is described in Section 10.6, applied from time level $n + 1$ to time level $n + 2$. The first step (i.e., the Lax method) is a provisional step. The results of this step are used only to implement the second step. The result of the second step is the desired solution. The finite difference stencil is illustrated in Figure 11.15.

Equations (11.41) and (11.42) do not look anything like the Lax-Wendroff one-step method, Eq. (11.37). However, substituting Eq. (11.41) applied at grid points $(i - 1)$ and $(i + 1)$ into Eq. (11.42) gives Eq. (11.37), for a time step of $2\,\Delta t$ and a space increment of $2\,\Delta x$. Consequently, the two methods are equivalent for the linear convection equation. Thus, for specific values of Δt and Δx, the global error of the Lax-Wendroff two-step method is four times larger than the global error of the Lax-Wendroff one-step method. Alternatively, to obtain the same global error, Δt and Δx for the Lax-Wendroff two-step method must be one-half the values of Δt and Δx of the Lax-Wendroff one-step method. Thus, four times as much work is required to reach the same time level. For nonlinear PDEs or systems of PDEs, the two methods, while similar in behavior, are not identical.

Equations (11.41) and (11.42) comprise a Lax-Wendroff type two-step method for the linear convection equation. This Lax-Wendroff two-step method is an explicit, three-time-level, two-step, $(\Delta t^2) + 0(\Delta x^2)$, finite difference method. The third time level is not a problem because the value of f_i^{n+2} can be stored in place of f_i^n, so only two levels of computer storage are required.

Since the Lax-Wendroff two-step method proposed by Richtmyer is equivalent to the Lax-Wendroff one-step method for the linear convection equation, it follows that the consistency and stability analyses are identical. Thus, as demonstrated for the Lax-Wendroff one-step method, the method is consistent with the convection equation, $0(\Delta t^2) + 0(\Delta x^2)$, conditionally stable ($c = u\,\Delta t/\Delta x \leq 1$), and convergent.

Example 11.4.   The Lax-Wendroff (Richtmyer) two-step method applied to the convection equation

Let's solve the convection problem presented in Section 11.1 using Eqs. (11.41) and (11.42) for $\Delta x = 0.05$ cm. The results are presented in Figure 11.16 at times from 1.0 to 5.0 s for $c = 0.5$, and at 10.0 s for $c = 0.1, 0.5, 0.9$, and 1.0. The general features of the

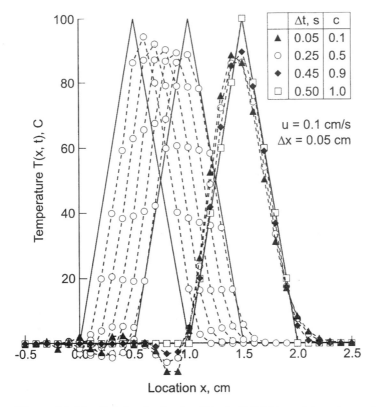

Figure 11.16 Solution by the Lax-Wendroff (Richtmyer) two-step method.

numerical solution are similar to those presented in Figure 11.14 for the Lax-Wendroff one-step method. The errors are somewhat larger, however, because the Lax-Wendroff two-step method is equivalent to the Lax-Wendroff one-step method with a time step of $2 \, \Delta t$ and a spatial grid size of $2 \, \Delta x$. In fact, applying the Lax-Wendroff two-step method with $\Delta x = 0.025$ cm, with a corresponding halving of Δt, yields the same results as presented in Figure 11.14.

In summary, the Lax-Wendroff two-step method is explicit, three-time-level, two-step, consistent, $0(\Delta t^2) + 0(\Delta x^2)$, conditionally stable, and convergent. The method is equivalent to the Lax-Wendroff one-step method for the linear convection equation. However, for nonlinear PDEs, systems of PDEs, and two- and three-dimensional physical spaces, the Lax-Wendroff two-step method is much easier to apply than the Lax-Wendroff one-step method. The explicit Lax-Wendroff two-step method can be used in a straightforward manner to solve nonlinear PDEs, systems of PDEs, and multidimensional problems, as discussed in Section 11.9.

11.5.3 The MacCormack Method

MacCormack (1969) proposed a predictor-corrector method of the Lax-Wendroff type. The MacCormack method uses the same grid spacings as the Lax-Wendroff (1960) one-step method, thus eliminating the requirement of more grid points associated with the Lax-

Wendroff (Richtmyer) two-step method. The MacCormack method can be used to solve linear partial differential equations, nonlinear PDEs, and systems of PDEs with equal ease, whereas the Lax-Wendroff one-step method becomes quite complicated for nonlinear PDEs and systems of PDEs. Consequently, the MacCormack method is a widely used method.

The basis of the Lax-Wendroff one-step method is the second-order Taylor series given in Eq. (11.32), where $\bar{f}_t$ is determined directly from the PDE and $\bar{f}_{tt}$ is determined by differentiating the PDE with respect to time. MacCormack (1969) proposed an alternate approach for evaluating $\bar{f}_{tt}|_i^n$ which employs a first-order forward-time Taylor series for $\bar{f}_t|_i^{n+1}$ with base point (i, n). Thus,

$$\bar{f}_t|_i^{n+1} = \bar{f}_t|_i^n + \bar{f}_{tt}|_i^n \, \Delta t + 0(\Delta t^2) \tag{11.43}$$

Solving Eq. (11.43) for $\bar{f}_{tt}|_i^n$ yields

$$\bar{f}_{tt}|_i^n = \frac{\bar{f}_t|_i^{n+1} - \bar{f}_t|_i^n}{\Delta t} + 0(\Delta t) \tag{11.44}$$

Substituting Eq. (11.44) into Eq. (11.32) gives

$$\bar{f}_i^{n+1} = \bar{f}_i^n + \tfrac{1}{2}(\bar{f}_t|_i^n + \bar{f}_t|_i^{n+1})(\Delta t + 0(\Delta t^3) \tag{11.45}$$

Introducing the PDE, $\bar{f}_t = -u\bar{f}_x$, into Eq. (11.45) gives

$$\bar{f}_i^{n+1} = \bar{f}_i^n - \tfrac{1}{2}(u\bar{f}_x|_i^n + u\bar{f}_x|_i^{n+1})\Delta t + 0(\Delta t^3) \tag{11.46}$$

Replacing $\bar{f}_x|_i^n$ and $\bar{f}_x|_i^{n+1}$ by second-order centered-difference approximations and truncating the remainder terms yields an $0(\Delta t^3) + 0(\Delta x^2)$ FDE, which has $0(\Delta t^2)$ global order. This replacement yields an implicit FDE, which is difficult to solve for nonlinear PDEs.

MacCormack proposed a predictor-corrector procedure which calculates provisional (i.e., predicted) values of f_i^{n+1} using first-order forward-difference approximations of $\bar{f}_t|_i^n$ and $\bar{f}_x|_i^n$ to give

$$\boxed{f_i^{\overline{n+1}} = f_i^n - c(f_{i+1}^n - f_i^n)} \tag{11.47}$$

where $c = u \, \Delta t/\Delta x$ is the convection number. In the second (i.e., corrector) step, Eq. (11.46) is solved by evaluating $\bar{f}_x|_i^n$ using the first-order forward-space approximation used in Eq. (11.47) and evaluating $\bar{f}_x|_i^{n+1}$ using the first-order backward-space approximation based on the provisional values of f_i^{n+1}. Equation (11.46) becomes

$$f_i^{n+1} = f_i^n - \tfrac{1}{2}[c(f_{i+1}^n - f_i^n) + c(f_i^{\overline{n+1}} - f_{i-1}^{\overline{n+1}})] \tag{11.48}$$

Rearranging Eq. (11.48) and introducing Eq. (11.47) yields a computationally more efficient form of the corrector equation:

$$\boxed{f_i^{n+1} = \tfrac{1}{2}[f_i^n + f_i^{\overline{n+1}} - c(f_i^{\overline{n+1}} - f_{i-1}^{\overline{n+1}})]} \tag{11.49}$$

Equations (11.47) and (11.49) comprise the MacCormack approximation of the linear convection equation. The finite difference stencils are presented in Figure 11.17.

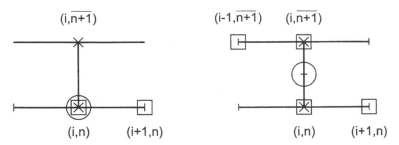

Figure 11.17 The MacCormack method stencil.

Equation (11.47) employs a forward-difference approximation of $\bar{f}_x|_i^n$ and Eq. (11.49) employs a backward-difference approximation of $\bar{f}_x|_i^{n+1}$. This differencing can be reversed. Either way, there is a slight bias in the solution due to the one-sided differences. If desired, this bias can be reduced by alternating the direction of the predictor and corrector spatial differences from one time level to the next.

The properties of the MacCormack method are not readily apparent from Eqs. (11.47) and (11.49). The time averaging of the space derivatives in the corrector, Eq. (11.48), suggest that the method may be $0(\Delta t^2)$. Since both space derivatives are one-sided first-order differences, it would appear that the overall method is $0(\Delta x)$. However, a very fortuitous cancellation of the $0(\Delta x)$ truncation error terms occurs, and the MacCormack method is $0(\Delta x^2)$.

Equations (11.47) and (11.49) do not look anything like the Lax-Wendroff one-step method, Eq. (11.37). However, substituting Eq. (11.47), applied at grid points (i, n) and $(i - 1, n)$, into Eq. (11.49) gives Eq. (11.37) identically. Consequently, the two methods are identical for the linear convection equation. For nonlinear PDEs or systems of PDEs, the two methods, while similar in behavior, are not identical.

Since the MacCormack method is identical to the Lax-Wendroff one-step method, for the linear convection equation, it follows that the consistency and stability analyses are identical. Thus, the method is consistent with the convection equation, $0(\Delta t^2) + 0(\Delta x^2)$, conditionally stable (i.e., $c = u\,\Delta t/\Delta x \leq 1$), and convergent.

Example 11.5. The MacCormack method applied to the convection equation

---

The MacCormack approximation of the linear convection equation is identical to the Lax-Wendroff one-step approximation of the linear convection equation. Consequently, the results presented in Example 11.3 also apply to the MacCormack method. The MacCormack method is an excellent method for solving convection problems.

When the partial differential equation being solved is nonlinear (i.e., the coefficient of $\bar{f}_x$ depends on $\bar{f}$), the coefficient is simply evaluated at grid point (i, n) for the predictor and at grid point $(i, \overline{n + 1})$ for the corrector. When solving systems of PDEs, Eqs. (11.47) and (11.49) are simply applied to every PDE in the system. The MacCormack method extends directly to two- and three-dimensional physical spaces simply by adding on the appropriate one-side finite difference approximations of the y and z space derivatives. The stability boundaries are more restrictive in those cases. The MacCormack method also can be used to solve the parabolic convection-diffusion equation discussed in Section 10.10. In view of these features of the MacCormack method, it is a very popular method for solving PDEs.

In summary, the MacCormack approximation of the convection equation is explicit, two step (i.e., predictor-corrector), $0(\Delta t^2) + 0(\Delta x^2)$, conditionally stable, and convergent. The method is identical to the Lax-Wendroff one-step method for the linear convection equation. However, for nonlinear equations, systems of equations, and two- and three-dimensional physical spaces, the MacCormack method is much more efficient than the Lax-Wendroff one-step method.

11.6 UPWIND METHODS

From a method of characteristics analysis of the convection equation, it is known that information propagates along the characteristic paths specified by $dx/dt = u$ [see Eq. (11.13)]. Thus, information propagates from either the left or the right side of the solution point, depending on whether $u > 0$ or $u < 0$, respectively. This type of information propagation is referred to as *upwind propagation*, since the information comes from the direction from which the convection velocity comes, that is, the upwind direction. Finite difference methods that account for the direction of information propagation are called upwind methods. Two such methods are presented in this section.

11.6.1 The First-Order Upwind Method

The simplest procedure for developing an upwind finite difference equation is to replace the time derivative $\bar{f}_t|_i^n$ by the first-order forward-difference approximation at grid point (i, n), Eq. (10.17), and to replace the space derivative $\bar{f}_x|_i^n$ by the first-order one-sided approximation in the upwind direction, Eq. (11.17), for $u > 0$. The corresponding finite difference stencil is presented in Figure 11.18. Substituting Eqs. (10.17) and (11.17) into the convection equation gives

$$\frac{f_i^{n+1} - f_i^n}{\Delta t} + u\frac{f_i^n - f_{i-1}^n}{\Delta x} = 0 \tag{11.50}$$

Solving Eq. (11.50) for f_i^{n+1} yields

$$\boxed{f_i^{n+1} = f_i^n - c(f_i^n - f_{i-1}^n)} \tag{11.51}$$

where $c = u\Delta t/\Delta x$ is the convection number.

The modified differential equation (MDE) corresponding to Eq. (11.51) is

$$f_t + uf_x = -\tfrac{1}{2}f_{tt}\,\Delta t - \tfrac{1}{6}f_{ttt}\,\Delta t^2 - \cdots + \tfrac{1}{2}uf_{xx}\,\Delta x - \tfrac{1}{6}uf_{xxx}\,\Delta x^2 + \cdots \tag{11.52}$$

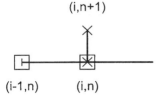

(i,n+1)

(i-1,n)   (i,n)

Figure 11.18   The first-order upwind method stencil.

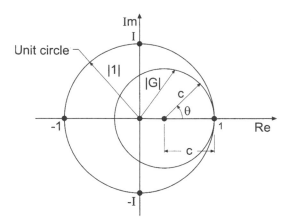

Figure 11.19 Locus of the amplification factor G for the first-order upwind method.

As $\Delta t \to 0$ and $\Delta x \to 0$, Eq. (11.52) approaches $f_t + u f_x = 0$. Consequently, Eq. (11.50) is consistent with the convection equation. The truncation error is $0(\Delta t) + 0(\Delta x)$. From a von Neumann stability analysis, the amplification factor G is given by

$$G = (1 - c) + c \cos \theta - Ic \sin \theta \tag{11.53}$$

Equation (11.53) is the equation of a circle in the complex plane, as illustrated in Figure 11.19. The center of the circle is at $(1 - c + I0)$, and its radius is c. For stability, $|G| \leq 1$, which requires the circle to be on or within the unit circle. This is guaranteed if

$$c = \frac{u \, \Delta t}{\Delta x} \leq 1 \tag{11.54}$$

Equation (11.54) is the CFL stability criterion. Consequently, the first-order upwind approximation of the convection equation is conditionally stable. It is also consistent. Consequently, by the Lax Equivalence Theorem, it is a convergent approximation of the convection equation.

Example 11.6.   The first-order upwind method applied to the convection equation

As an example of the first-order upwind method, let's solve the convection problem presented in Section 11.1 using Eq. (11.51) for $\Delta x = 0.05$ cm. The results are presented in Figure 11.20 at times from 1.0 to 5.0 s for $c = 0.5$, and at 10.0 s for $c = 0.1, 0.5, 0.9$, and 1.0.

Several important features of Eq. (11.51) are illustrated in Figure 11.20. When $c = 1.0$, the numerical solution is identical to the exact solution, for the linear convection equation. This is not true for nonlinear PDEs. When $c = 0.5$, the amplitude of the solution is damped as the wave moves to the right, and the sharp peak becomes rounded. The results at $t = 10.0$ s for $c = 0.1, 0.5, 0.9$, and 1.0 show that the amount of numerical damping (i.e., diffusion) depends on the convection number, c. The large errors associated with the numerical damping make the first-order upwind method a poor choice for solving the convection equation, or any hyperbolic PDE.

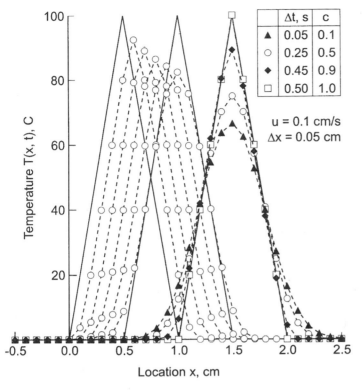

Figure 11.20 Solution by the first-order upwind method.

The first-order upwind method applied to the convection equation is explicit, single step, consistent, $0(\Delta t) + 0(\Delta x)$, conditionally stable, and convergent. However, it introduces significant amounts of numerical damping into the solution. Consequently, it is not a very accurate method for solving hyperbolic PDEs. Second-order upwind methods can be developed to give more accurate solutions of hyperbolic PDEs.

11.6.2 The Second-Order Upwind Method

An $0(\Delta t) + 0(\Delta x^2)$ finite difference approximation of the unsteady convection equation can be derived by replacing $\bar{f}_t$ by the first-order forward-difference approximation at grid point (i, n), Eq. (10.17), and replacing $\bar{f}_x$ by the second-order one-sided upwind-space approximation based on grid points $i, i-1$, and $i-2$, Eq. (5.96). Unfortunately, the resulting FDE is unconditionally unstable. It cannot be used to solve the unsteady convection equation, or any other hyperbolic PDE.

An $0(\Delta t^2) + 0(\Delta x^2)$ finite difference approximation of the unsteady convection equation is given, without derivation, by the following FDE:

$$f_i^{n+1} = f_i^n - c(f_i^n - f_{i-1}^n) - \frac{c(1-c)}{2}(f_i^n - 2f_{i-1}^n + f_{i-2}^n) \qquad (11.55)$$

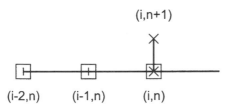

(i,n+1)

(i-2,n)   (i-1,n)   (i,n)

Figure 11.21   The second-order upwind method stencil.

where $c = u\,\Delta t/\Delta x$ is the convection number. The corresponding finite difference stencil is illustrated in Figure 11.21. The modified differential equation (MDE) corresponding to Eq. (11.55) is

$$f_t + uf_x = \tfrac{1}{2}f_{tt}\,\Delta t - \tfrac{1}{6}f_{ttt}\,\Delta t^2 + \tfrac{1}{2}u^2\,\Delta t f_{xx} + (\tfrac{1}{3}u\,\Delta x^2 - \tfrac{1}{2}u^2\,\Delta x\,\Delta t)f_{xxx} + \cdots \quad (11.56)$$

As $\Delta t \to 0$ and $\Delta x \to 0$, Eq. (11.56) approaches $f_t + uf_x = 0$. Consequently, Eq. (11.55) is consistent with the convection equation. Equation (11.56) appears to be $0(\Delta t) + 0(\Delta x^2)$. However, when $f_{tt} = u^2 f_{xx}$ is substituted into Eq. (11.56), the two $0(\Delta t)$ terms cancel, and Eq. (11.56) is seen to be $0(\Delta t^2) + 0(\Delta x^2)$, as desired.

From a von Neumann stability analysis, the amplification factor, G, is given by

$$G = \left(1 - 3\frac{c}{2} + \frac{c^2}{2} + (2c - c^2)\cos\theta + \left(c^2 - \frac{c}{2}\right)\cos 2\theta\right)$$
$$- I\left((2c - c^2)\sin\theta + \left(c^2 - \frac{c}{2}\right)\sin 2\theta\right) \quad (11.57)$$

Equation (11.57) is too complicated to solve analytically for the conditions required to ensure that $|G| \le 1$. Equation (11.57) can be solved numerically by parametrically varying θ from 0 to 2π in small increments (say 5 deg), then at each value of θ varying c parametrically from 0 to some upper value, such as 2.2, in small increments (say 0.1), and calculating $|G|$ at each combination of θ and c. Searching these results for the range of values of c which yields $|G| \le 1$ for all values of θ yields the stability range for Eq. (11.55). Performing these calculations shows that $|G| \le 1$ for $c < 2$. Thus, Eq. (11.55) is conditionally stable. It is also consistent with the convection equation. Consequently, by the Lax equivalence theorem, it is a convergent approximation of the convection equation.

Example 11.7.   The second-order upwind method applied to the convection equation

As an example of the second-order upwind method, let's solve the convection problem presented in Section 11.1 using Eq. (11.55) for $\Delta x = 0.05$ cm. The results are presented in Figure 11.22 at times from 1.0 to 5.0 s for $c = 0.5$, and at 10.0 s for $c = 0.1$, 0.5, 0.9, and 1.0.

Several important features are illustrated in Figure 11.22. When $c = 1.0$, the numerical solution is identical to the exact solution, for the linear convection equation. This is not true for nonlinear PDEs. When $c = 0.5$, the amplitude of the solution is damped only slightly as the wave propagates (i.e., convects) to the right. The results at $t = 10.0$ s for $c = 0.1$, 0.5, 0.9, and 1.0 show that the amount of numerical damping is

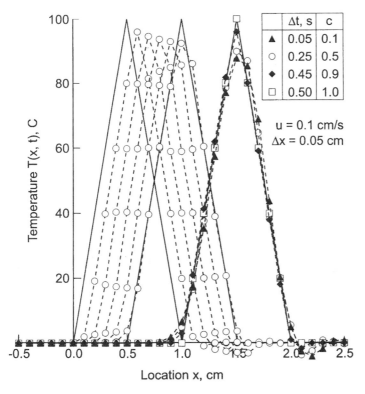

Figure 11.22 Solution by the second-order upwind method.

much less than for the first-order upwind method. The second-order upwind method is a good choice for solving the convection equation, or any hyperbolic PDE.

The second-order upwind method applied to the convection equation is explicit, single step, consistent, $0(\Delta t^2) + 0(\Delta x^2)$, conditionally stable ($c < 2$), and convergent. It is a good method for solving hyperbolic PDEs. Explicit upwind methods can be used in a straightforward manner to solve nonlinear PDEs, systems of PDEs, and multidimensional problems, as discussed in Section 11.8. Although upwind methods do not match the physical information propagation paths exactly, they do account for the direction of physical information propagation. Thus, they match the physics of hyperbolic PDEs more accurately than centered-space methods.

11.7 THE BACKWARD-TIME CENTERED-SPACE (BTCS) METHOD

The Lax method, the Lax-Wendroff type methods, and the upwind methods, are all examples of explicit finite difference methods. In explicit methods, the finite difference approximations to the individual exact partial derivatives in the partial differential equation are evaluated at grid point i at the known time level n. Consequently, the solution at grid point i at the next time level $n + 1$ can be expressed explicitly in terms of the known solution at grid points at time level n. Explicit finite difference methods have many desirable features. Foremost among these for hyperbolic PDEs is that explicit methods have a finite numerical information propagation speed, which gives rise to finite numerical domains of dependence and ranges of influence. Hyperbolic PDEs have a finite physical information propagation speed, which gives rise to finite physical domains of dependence and ranges of influence. Consequently, explicit finite difference methods closely match the physical propagation properties of hyperbolic PDEs.

However, explicit methods share one undesirable feature: they are only conditionally stable. Consequently, the allowable time step is usually quite small, and the amount of computational effort required to obtain the solution of some problems is immense. A procedure for avoiding the time step limitation would obviously be desirable. Implicit finite difference methods furnish such a procedure. Implicit finite difference methods are unconditionally stable. There is no limit on the allowable time step required to achieve a stable solution. There is, of course, some practical limit on the time step required to maintain the truncation errors within reasonable limits, but this is not a stability consideration; it is an accuracy consideration.

Implicit methods do have some disadvantages, however. The foremost disadvantage is that the solution at a point at the solution time level $n+1$ depends on the solution at neighboring points at the solution time level $n+1$, which are also unknown. Consequently, the solution is implied in terms of other unknown solutions at time level $n+1$, systems of FDEs must be solved to obtain the solution at each time level, and the numerical information propagation speed is infinite. Additional complexities arise when the partial differential equations are nonlinear. This gives rise to systems of nonlinear finite difference equations, which must be solved by some manner of linearization and/or iteration. However, the major disadvantage is the infinite numerical information propagation speed, which gives rise to infinite domains of dependence and ranges of influence. This obviously violates the finite domains of dependence and ranges of influence associated with hyperbolic PDEs. In spite of these disadvantages, the advantage of unconditional stability makes implicit finite difference methods attractive for obtaining steady state solutions as the asymptotic solution in time of an appropriate unsteady propagation problem. This concept is discussed in Section 10.11. Consequently, the backward-time centered-space (BTCS) method is presented in this section.

In this section, we will solve the unsteady one-dimensional convection equation by the *backward-time centered-space (BTCS) method*. This method is also called the *fully implicit method*. The finite difference equation (FDE) which approximates the partial differential equation is obtained by replacing the exact partial derivative $\bar{f}_t$ by the first-order backward-difference approximation, Eq. (10.65), and the exact partial derivative $\bar{f}_x$ by the second-order centered-space approximation, Eq. (11.20), evaluated at time level $n+1$. The finite difference stencil is presented in Figure 11.23. The resulting finite difference approximation of the convection equation is

$$\frac{f_i^{n+1}-f_i^n}{\Delta t}+u\frac{f_{i+1}^{n+1}-f_{i-1}^{n+1}}{2\,\Delta x}=0 \tag{11.58}$$

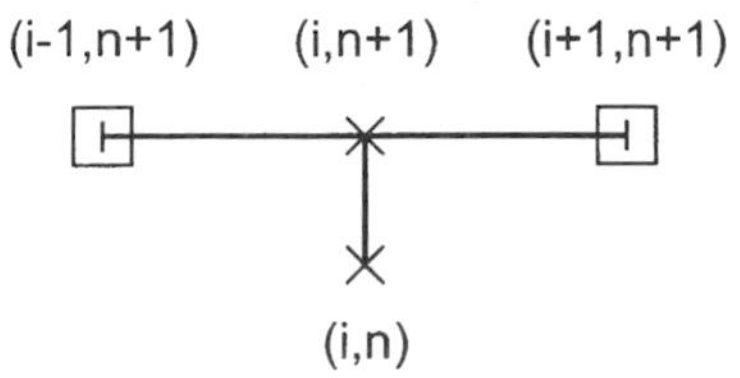

Figure 11.23 The BTCS method stencil.

Rearranging Eq. (11.58) yields

$$-\frac{c}{2}f_{i-1}^{n+1} + f_i^{n+1} + \frac{c}{2}f_{i+1}^{n+1} = f_i^n \qquad (11.59)$$

where $c = u\,\Delta t/\Delta x$ is the convection number.

Equation (11.59) cannot be solved explicitly for f_i^{n+1} because the two unknown neighboring values f_{i-1}^{n+1} and f_{i+1}^{n+1} also appear in the equation. The value of f_i^{n+1} is implied in Eq. (11.59), however. Finite difference equations in which the unknown value of f_i^{n+1} is implied in terms of its unknown neighbors, rather than being given explicitly in terms of known initial values, are called implicit finite difference equations.

The modified differential equation (MDE) corresponding to Eq. (11.59) is

$$f_t + uf_x = \tfrac{1}{2}f_{tt}\,\Delta t - \tfrac{1}{6}f_{ttt}\,\Delta t^2 - \cdots - \tfrac{1}{6}uf_{xxx}\,\Delta x^2 - \tfrac{1}{120}uf_{xxxxx}\,\Delta x^4 - \cdots \qquad (11.60)$$

As $\Delta t \to 0$ and $x \to 0$, the truncation error terms go to zero, and Eq. (11.60) approaches $f_t + uf_x = 0$. Consequently, Eq. (11.59) is consistent with the convection equation. From Eq. (11.60), the FDE is $0(\Delta t) + 0(\Delta x^2)$. From a von Neumann stability analysis, the amplification factor, G, is given by

$$G = \frac{1}{1 + Ic\sin\theta} \qquad (11.61)$$

Since $|1 + Ic\sin\theta| \ge 1$ for all values of θ and all values of c, the BTCS method is unconditionally stable when applied to the convection equation. The BTCS method applied to the convection equation is consistent and unconditionally stable. Consequently, by the Lax Equivalence Theorem, it is a convergent finite difference approximation of the convection equation.

Consider now the solution of the unsteady one-dimensional convection equation by the BTCS method. The finite difference grid for advancing the solution from time level n to time level $n+1$ by an implicit finite difference method is illustrated in Figure 11.24. There is an obvious problem with the boundary conditions for a pure initial-value problem, such as the convection problem presented in Section 11.1. Boundary conditions can be simulated in an initial-value problem by placing the open boundaries at a large distance from the region of interest and applying the initial conditions at those locations as boundary conditions.

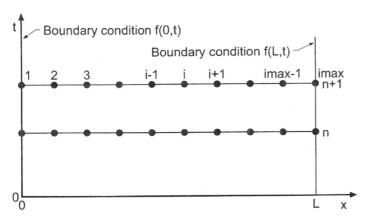

Figure 11.24   Finite difference grid for implicit methods.

Equation (11.59) applies directly at points 2 to imax − 1 in Figure 11.24. The following set of simultaneous linear equations is obtained:

$$f_2^{n+1} + \frac{c}{2}f_3^{n+1} = f_2^n + \frac{c}{2}\bar{f}(0, t) = b_2$$

$$-\frac{c}{2}f_2^{n+1} + f_3^{n+1} + \frac{c}{2}f_4^{n+1} = f_3^n = b_3$$

$$-\frac{c}{2}f_3^{n+1} + f_4^{n+1} + \frac{c}{2}f_5^{n+1} = f_4^n = b_4 \qquad (11.62)$$

$$\cdots\cdots\cdots\cdots\cdots\cdots\cdots\cdots\cdots\cdots\cdots\cdots\cdots\cdots\cdots\cdots$$

$$-\frac{c}{2}f_{imax-2}^{n+1} + f_{imax-1}^{n+1} = f_{imax-1}^n - \frac{c}{2}\bar{f}(L, t) = b_{imax-1}$$

Equation (11.62) is a tridiagonal sysem of linear equations. That system of equations may be written as

$$\mathbf{A}\mathbf{f}^{n+1} = \mathbf{b} \qquad (11.63)$$

where $\mathbf{A}$ is the (imax − 2) × (imax − 2) coefficient matrix, $\mathbf{f}^{n+1}$ is the (imax − 2) × 1 solution column vector, and $\mathbf{b}$ is the (imax − 2) × 1 column vector of nonhomogeneous terms. Equation (11.63) can be solved very efficiently by the Thomas algorithm presented in Section 1.5. Since the coefficient matrix $\mathbf{A}$ does not change from one time level to the next, LU factorization can be employed with the Thomas algorithm to reduce the computational effort ever further. As shown in Section 1.5, once the LU factorization has been performed, the number of multiplications and divisions required to solve a tridiagonal system of linear equations by the Thomas algorithm is $3n$, where $n =$ (imax − 2) is the number of equations.

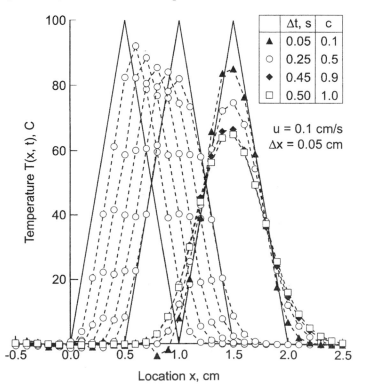

Figure 11.25   Solution by the BTCS method for $c = 0.1$ to 1.0.

Example 11.8.   The BTCS method applied to the convection equation

Let's solve the convection problem presented in Section 11.1 by the BTCS method for $\Delta x = 0.05$ cm. For this initial-value problem, numerical boundaries are located 100 grid points to the left and right of the initial triangular wave, that is, at $x = -5.0$ cm and $x = 6.0$ cm, respectively. The results are presented in Figure 11.25 at times from 1.0 to 5.0 s for $c = 0.5$, and at 10.0 s for $c = 0.1, 0.5, 0.9$, and 1.0, and in Figure 11.26 at 10.0 s for $c = 1.0, 2.5, 5.0$, and 10.0.

Several important features of the BTCS method applied to the convection equation are illustrated in Figures 11.25 and 11.26. for $c = 0.5$, the solution is severely damped as the wave propagates, and the peak of the wave is rounded. These effects are due to implicit numerical diffusion and dispersion. At $t = 10.0$ s, the best solutions are obtained for the smallest values of c. For the large values of c (i.e., $c \geq 5.0$), the solutions barely resemble the exact solution. These results demonstrate that the method is indeed stable for $c > 1$, but that the quality of the solution is very poor. The peaks in the solutions at $t = 10.0$ s for the different values of c are lagging further and further behind the peak in the exact solution, which demonstrates that the numerical information propagation speed is less than the physical information propagation speed. This effect is due to implicit numerical dispersion. Overall, the BTCS method applied to the convection equation yields rather poor transient results.

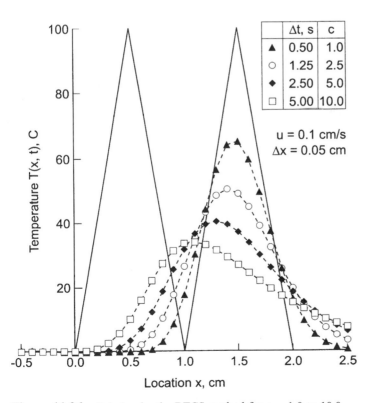

Figure 11.26   Solution by the BTCS method for $c = 1.0$ to 10.0.

The BTCS method is $0(\Delta t)$. An $0(\Delta t^2)$ implicit FDE can be developed using the Crank-Nicolson approach presented in Section 10.7.2 for the diffusion equation. The procedure is straightforward. The major use of implicit methods for solving hyperbolic PDEs is to obtain the asymptotic steady state solution of mixed elliptic/hyperbolic problems. As pointed out in Section 10.11, the BTCS method is preferred over the Crank-Nicolson method for obtaining asymptotic steady state solutions. Consequently the Crank-Nicolson method is not developed for the convection equation.

The implicit BTCS method becomes considerably more complicated when applied to nonlinear PDEs, systems of PDEs, and multidimensional problems. A discussion of these problems is presented in Section 11.8.

In summary, the BTCS approximation of the convection equation is implicit, single step, consistent, $0(\Delta t) + 0(\Delta x^2)$, unconditionally stable, and convergent. The implicit nature of the method yields a system of finite difference equations which must be solved simultaneously. For one-dimensional problems, that can be accomplished by the Thomas algorithm. The infinite numerical information propagation speed does not correctly model the finite physical information propagation speed of hyperbolic PDEs. The BTCS approximation of the convection equation yields poor results, except for very small values of the convection number, for which explicit methods are generally more efficient.

11.8 NONLINEAR EQUATIONS AND MULTIDIMENSIONAL PROBLEMS

Some of the problems associated with nonlinear equations and multdimensional problems are summarized in this section.

11.8.1 Nonlinear Equations

The finite difference equations and examples presented in this chapter are for the linear one-dimensional convection equation. In each section in this chapter, a brief paragraph is presented discussing the suitability of the method for solving nonlinear equations. The additional complexities associated with solving nonlinear equations are discussed in considerable detail in Section 10.9.1 for parabolic PDEs. The problems and solutions discussed there apply directly to finite difference methods for solving nonlinear hyperbolic PDEs.

Generally speaking, explicit methods can be extended directly to solve nonlinear hyperbolic PDEs. Implicit methods, on the other hand, yield nonlinear FDEs when applied to nonlinear PDEs. Methods for solving systems of nonlinear FDEs are discussed in Section 10.9.

11.8.2 Multidimensional Problems

The finite difference equations and examples presented in this chapter are for the linear one-dimensional convection equation. In each section a brief paragraph is presented discussing the suitability of the method for solving multidimensional problems. The additional complexities associated with solving multidimensional problems are also discussed in considerable detail in Section 10.9.2 for parabolic PDEs. The problems and solutions discussed there apply directly to finite difference methods for solving hyperbolic PDEs.

Generally speaking, explicit methods can be extended directly to solve multi-dimensional hyperbolic PDEs. When applied to multidimensional problems, implicit methods result in large banded systems of FDEs. Methods for solving these problems, such as alternating-direction-implicit (ADI) methods and approximate-factorization-implicit (AFI) methods, are discussed in Section 10.9.2.

11.9 THE WAVE EQUATION

The solution of the hyperbolic convection equation is discussed in Sections 11.4 to 11.8. The solution of the hyperbolic wave equation is discussed in this section.

11.9.1 Introduction

Consider the one-dimensional wave equation for the generic dependent variable $\bar{f}(x, t)$:

$$\boxed{\bar{f}_{tt} = c^2 \bar{f}_{xx}} \tag{11.64}$$

where c is the wave propagation speed. As shown in Section III.7, Eq. (11.64) is equivalent to the following set of two coupled first-order convection equations:

$$\boxed{\bar{f}_t + c\bar{g}_x = 0} \tag{11.65}$$
$$\boxed{\bar{g}_t + c\bar{f}_x = 0} \tag{11.66}$$

Equations (11.65) and (11.66) suggest that the wave equations can be solved by the same methods that are employed to solve the convection equation.

Sections 11.4 to 11.8 are devoted to the numerical solution of the convection equation, Eq. (11.6). Most of the concepts, techniques, and conclusions presented in Sections 11.4 to 11.8 for solving the convection equation are directly applicable, sometimes with very minor modifications, for solving the wave equation. The present section is devoted to the numerical solution of the wave equation, Eq. (11.64), expressed as a set of two coupled convection equations, Eqs. (11.65) and (11.66).

The finite difference grids and the finite difference approximations presented in Sections 10.3 and 11.3 are used to solve the wave equation. The concepts of consistency, order, stability, and convergence presented in Section 10.5 are directly applicable to the wave equation.

The exact solution of Eqs. (11.65) and (11.66) consists of the two functions $\bar{f}(x, t)$ and $\bar{g}(x, t)$. These functions must satisfy initial conditions at $t = 0$:

$$\bar{f}(x, 0) = F(x) \quad \text{and} \quad \bar{g}(x, 0) = G(x) \tag{11.67}$$

and boundary conditions at $x = 0$ or $x = L$. The boundary conditions may be of the Dirichlet type (i.e., specified $\bar{f}$ and $\bar{g}$), the Neumann type (i.e., specified derivatives of $\bar{f}$ and $\bar{g}$), or the mixed type (i.e., specified combinations of $\bar{f}$ and $\bar{g}$ and derivatives of $\bar{f}$ and $\bar{g}$).

As shown in Section 11.2, the exact solution of a single convection equation, for example, Eq. (11.6), is given by Eq. (11.12):

$$\bar{f}(x, t) = \phi(x - ut) \tag{11.68}$$

which can be demonstrated by direct substitution. Equation (11.68) defines a right-traveling wave which propagates (i.e., convects) the initial property distribution, $\bar{f}(x,0) = \phi(x)$, to the right at the velocity u, unchanged in magnitude and shape.

The exact solution of the wave equation, Eq. (11.64), is given by

$$\bar{f}(x,t) = F(x-ct) + G(x+ct) \tag{11.69}$$

which can be demonstrated by direct substitution. Equation (11.69) represents the superposition of a positive-traveling wave, $F(x-ct)$, and a negative-traveling wave, $G(x+ct)$, which propagate information to the right and left, respectively, at the wave propagation speed c, unchanged in magnitude and shape. The second-order (in time) wave equation requires two initial conditions:

$$\bar{f}(x,0) = \phi(x) \qquad \text{and} \qquad \bar{f}_t(x,0) = \theta(x) \tag{11.70}$$

Substituting Eq. (11.70) into Eq. (11.69) gives

$$\phi(x) = \bar{f}(x,0) = F(x) + G(x) \tag{11.71}$$
$$\theta(x) = \bar{f}_t(x,0) = -cF'(x) + cG'(x) \tag{11.72}$$

where the prime denotes ordinary differentiation with respect to the arguments of F and G, respectively. Integrating Eq. (11.72) yields

$$-F(x) + G(x) = \frac{1}{c}\int_{x_0}^{x} \theta(\xi)\, d\xi \tag{11.73}$$

where x_0 is a reference location and ξ is a dummy variable. Subtracting Eq. (11.73) from Eq. (11.71) gives

$$F(x) = \frac{1}{2}\left(\phi(x) - \frac{1}{c}\int_{x_0}^{x} \theta(\xi)\, d\xi\right) \tag{11.74}$$

Adding Eqs. (11.71) and (11.73) gives

$$G(x) = \frac{1}{2}\left(\phi(x) + \frac{1}{c}\int_{x_0}^{x} \theta(\xi)\, d\xi\right) \tag{11.75}$$

Equations (11.74) and (11.75) show that the functional forms of $F(x-ct)$ and $G(x+ct)$ are identical to the functional forms specified in Eqs. (11.74) and (11.75) with x replaced by $(x-ct)$ and $(x+ct)$, respectively. Substituting these values into Eqs. (11.74) and (11.75), respectively, and substituting those results into Eq. (11.69) yields

$$\boxed{\bar{f}(x,t) = \frac{1}{2}\left(\phi(x-ct) + \phi(x+ct) + \frac{1}{2c}\int_{x-ct}^{x+ct} \theta(\xi)\, d\xi\right)} \tag{11.76}$$

Equation (11.76) is the exact solution of the wave equation. It is generally called the *D'Alembert solution.*

The wave equation applies to problems of vibrating systems, such as vibrating strings and acoustic fields. Most people have some physical feeling for acoustics due to its presence in our everyday life. Consequently, the wave equation governing acoustic fields is

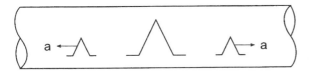

Figure 11.27   Acoustic wave propagation in an infinite duct.

considered in this section to demonstrate numerical methods for solving the wave equation. That equation is presented in Section III.7, Eq. (III.91), and repeated below:

$$P_{tt} = a^2 P_{xx} \qquad (11.77)$$

where P is the acoustic pressure perturbation (N/m^2 = Pascals = Pa) and a is the speed of sound (m/s). The superscript prime on P and the subscript 0 on a have been dropped for clarity. Equation (11.77) requires two initial conditions $P(x, 0)$ and $P_t(x, 0)$. As shown in Section III.7, Eq. (11.77) is obtained by combining Eqs. (III.89) and (III.90), which are repeated below:

$$\rho u_t + P_x = 0 \qquad (11.78)$$
$$P_t + \rho a^2 u_x = 0 \qquad (11.79)$$

where ρ is the density (kg/m^3) and u is the acoustic velocity perturbation (m/s). Equations (11.78) and (11.79) can be expressed in the form of Eqs. (11.65) and (11.66) in terms of P and the secondary variable $Q = (Pa)u$, where Pa is a constant. Thus, $Q_t + aP_x = 0$ and $P_t + aQ_x = 0$.

The following problem is considered in this section to illustrate the behavior of finite difference methods for solving the wave equation. A long duct, illustrated in Figure 11.27, is filled with a stagnant compressible gas for which the density $\rho = 1.0 \text{ kg/m}^3$ and the acoustic wave velocity $a = 1000.0$ m/s. The fluid is initially at rest, $u(x, 0) = 0.0$, and has an initial acoustic pressure distribution given by

$$P(x, 0) = 200.0(x - 1) \qquad 1.0 \le x \le 1.5 \qquad (11.80)$$
$$P(x, 0) = 200.0(2 - x) \qquad 1.5 \le x \le 2.0 \qquad (11.81)$$

where P is measured in Pa (i.e., N/m^2) and x is measured in meters. This initial pressure distribution is illustrated in Figure 11.28. For an infinitely long duct, there are no boundary conditions (except, of course, at infinity, which is not of interest in the present problem). The pressure distribution $P(x, t)$ is required.

For the acoustic problem discussed above, combining Eq. (11.79) and the initial condition $u(x, 0) = 0.0$ shows that $P_t(x, 0) = 0$, so that $\theta(x) = 0$. Combining Eqs. (11.69) and (11.76) shows that

$$\bar{f}(x, t) = F(x - at) + G(x + at) = \tfrac{1}{2}[\phi(x - at) + \phi(x + at)] \qquad (11.82)$$

Equation (11.82) must hold for all combinations of x and t. Thus,

$$F(x - at) = \tfrac{1}{2}\phi(x - at) \qquad \text{and} \qquad G(x + at) = \tfrac{1}{2}\phi(x + at) \qquad (11.83)$$

Equation (11.83) shows that at $t = 0$, $F(x) = \phi(x)/2$ and $G(x) = \phi(x)/2$. Thus, the exact solution of the acoustics problem consists of the superposition of two identical traveling waves, each having one-half the amplitude of the initial wave. One wave propagates to the right and one wave propagates to the left, both with the wave propagation speed a. Essentially, the initial distribution, which is the superposition of the two identical waves, simply decomposes into the two individual waves. The exact solution for $P(x, t)$ for several

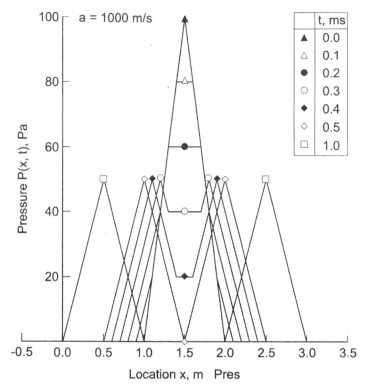

Figure 11.28   Exact solution of the wave propagation problem.

values of time t, in ms (millisec), is presented in Figure 11.28. Note that the discontinuities in the slope of the initial pressure distribution at $x = 1.0$, 1.5, and 2.0 m are preserved during the wave propagation process.

11.9.2 Characteristic Concepts

The concept of characteristics of partial differential equations is introduced in Section III.3. In two independent variables, which is the case considered here (i.e., physical space x and time t), characteristics are paths (curved, in general) in the solution domain $D(x, t)$ along which physical information propagates. If a partial differential equation possesses real characteristics, then information propagates along the characteristic paths. The presence of characteristic paths has a significant impact on the solution of a partial differential equation (by both analytical and numerical methods).

Let's apply the concepts presented in Sections III.3 and III.7 to determine the characteristics of the system of two coupled convection equations, Eqs. (11.65) and (11.66), where c has been replaced by a to model acoustic wave propagation:

$$\bar{f}_t + a\bar{g}_x = 0 \tag{11.84}$$

$$\bar{g}_t + a\bar{f}_x = 0 \tag{11.85}$$

Applying the chain rule to the continuous functions $\bar{f}(x, t)$ and $\bar{g}(x, t)$ yields

$$d\bar{f} = \bar{f}_t \, dt + \bar{f}_x \, dx \qquad \text{and} \qquad d\bar{g} = \bar{g}_t \, dt + \bar{g}_x \, dx \tag{11.86}$$

Writing Eqs. (11.84) to (11.86) in matrix form yields

$$\begin{bmatrix} 1 & 0 & 0 & a \\ 0 & a & 1 & 0 \\ dt & dx & 0 & 0 \\ 0 & 0 & dt & dx \end{bmatrix} \begin{bmatrix} \bar{f}_t \\ \bar{f}_x \\ \bar{g}_t \\ \bar{g}_x \end{bmatrix} = \begin{bmatrix} 0 \\ 0 \\ d\bar{f} \\ d\bar{g} \end{bmatrix} \tag{11.87}$$

The characteristics of Eqs. (11.84) and (11.85) are determined by setting the determinant of the coefficient matrix of Eq. (11.87) equal to zero. This gives the characteristic equation:

$$(1)[-(dx)^2] + dt(a^2\, dt) = 0 \tag{11.88}$$

Solving Eq. (11.88) for dx/dt gives

$$\boxed{\frac{dx}{dt} = \pm a} \tag{11.89}$$

Equation (11.89) shows that there are two real distinct roots associated with the characteristic equation. The physical speed of information propagation c along the characteristic curves is

$$c = \frac{dx}{dt} = \pm a \tag{11.90}$$

Consequently, information propagates in both the positive and negative x directions at the wave speed a.

11.9.3 The Lax-Wendroff One-Step Method

The one-step method developed by Lax and Wendroff (1960) is a very popular $0(\Delta t^2) + (\Delta x^2)$ explicit finite difference method for solving hyperbolic PDEs. For the pair of first-order PDEs that correspond to the linear wave equation, $\bar{f}_t + a\bar{g}_x = 0$ and $\bar{g}_t + a\bar{f}_x = 0$, the functions to be determined are $\bar{f}(x, t)$ and $\bar{g}(x, t)$. Expanding $\bar{f}(x, t)$ in a Taylor series in time gives

$$\boxed{\bar{f}_i^{n+1} = \bar{f}_i^n + \bar{f}_t|_i^n\, \Delta t + \tfrac{1}{2}\bar{f}_{tt}|_i^n\, \Delta t^2 + 0(\Delta t^3)} \tag{11.91}$$

The derivative $\bar{f}_t$ is determined directly from the PDE:

$$\bar{f}_t = -a\bar{g}_x \tag{11.92}$$

The derivative $\bar{f}_{tt}$ is determined by differentiating Eq. (11.92) with respect to time. Thus,

$$\bar{f}_{tt} = (\bar{f}_t)_t = (-a\bar{g}_x)_t = -a(\bar{g}_t)_x = -a(-a\bar{f}_x)_x = a^2\bar{f}_{xx} \tag{11.93}$$

Substituting Eqs. (11.92) and (11.93) into Eq. (11.91) yields

$$\bar{f}_i^{n+1} = \bar{f}_i^n - a\bar{g}_x|_i^n\, \Delta t + \tfrac{1}{2}a^2\bar{f}_{xx}|_i^n\, \Delta t^2 + 0(\Delta t^3) \tag{11.94}$$

Approximating the two space derivatives $\bar{g}_x|_i^n$ and $\bar{f}_{xx}|_i^n$ by second-order centered-difference approximations, Eqs. (11.20) and (10.23), respectively, gives

$$f_i^{n+1} = f_i^n - \frac{a\,\Delta t}{2\,\Delta x}(g_{i+1}^n - g_{i-1}^n) + \frac{a^2\,\Delta t^2}{2\,\Delta x^2}(f_{i+1}^n - 2f_i^n + f_{i-1}^n) \tag{11.95}$$

Introducing the convection number $c = a\,\Delta t/\Delta x$ yields

$$f_i^{n+1} = f_i^n - \frac{c}{2}(g_{i+1}^n - g_{i-1}^n) + \frac{c^2}{2}(f_{i+1}^n - 2f_i^n + f_{i-1}^n) \tag{11.96}$$

Performing the same steps for the function $\bar{g}(x, t)$ yields

$$g_i^{n+1} = g_i^n - \frac{c}{2}(f_{i+1}^n - f_{i-1}^n) + \frac{c^2}{2}(g_{i+1}^n - 2g_i^n + g_{i-1}^n) \tag{11.97}$$

Equations (11.96) and (11.97) are the Lax-Wendroff one-step approximation of the coupled convection equations that correspond to the linear wave equation.

The MDE corresponding to Eq. (11.96) is

$$f_t + ag_x = -\tfrac{1}{2}f_{tt}\,\Delta t - \tfrac{1}{6}f_{ttt}\,\Delta t^2 - \tfrac{1}{24}f_{tttt}\,\Delta t^3 - \cdots$$
$$- \tfrac{1}{6}ag_{xxx}\Delta x^2 - \cdots + \tfrac{1}{2}a^2 f_{xx}\,\Delta t + \tfrac{1}{24}a^2 f_{xxxx}\,\Delta t\,\Delta x^2 + \cdots \tag{11.98}$$

Substituting Eq. (11.93) into Eq. (11.98) gives

$$f_t + ag_x = \tfrac{1}{6}(a^3\,\Delta t^2 - a\,\Delta x^2)g_{xxx} + \tfrac{1}{8}(a^4\,\Delta t^3 - a^2\,\Delta t\,\Delta x^2)f_{xxxx} + \cdots \tag{11.99}$$

As $\Delta t \to 0$ and $\Delta x \to 0$, the remainder terms in Eq. (11.98) go to zero, and Eq. (11.98) approaches $f_t + ag_x$. Consequently, Eq. (11.96) is consistent with $\bar{f}_t + a\bar{g}_x = 0$. From Eq. (11.99), the FDE is $0(\Delta t^2) + 0(\Delta x^2)$. Similar results and conclusions apply to Eqs. (11.97).

Performing a von Neumann stability analysis of Eqs. (11.96) and (11.97) gives

$$f_i^{n+1} = f_i^n - \frac{c}{2}(g_i^n e^{I\theta} - g_i^n e^{-I\theta}) + \frac{c^2}{2}(f_i^n e^{I\theta} - 2f_i^n + f_i^n e^{-I\theta}) \tag{11.100}$$

$$g_i^{n+1} = g_i^n - \frac{c}{2}(f_i^n e^{I\theta} - f_i^n e^{-I\theta}) + \frac{c^2}{2}(g_i^n e^{I\theta} - 2g_i^n + g_i^n e^{-I\theta}) \tag{11.101}$$

Introducing the relationships between the exponential functions and the sine and cosine functions gives

$$f_i^{n+1} = f_i^n - g_i^n Ic\sin\theta + c^2 f_i^n(\cos\theta - 1) \tag{11.102}$$

$$g_i^{n+1} = g_i^n - f_i^n Ic\sin\theta + c^2 g_i^n(\cos\theta - 1) \tag{11.103}$$

Equations (11.102) and (11.103) can be written in the matrix form

$$\begin{bmatrix} f_i^{n+1} \\ g_i^{n+1} \end{bmatrix} = \mathbf{G} \begin{bmatrix} f_i^n \\ g_i^n \end{bmatrix} \tag{11.104}$$

where $\mathbf{G}$ is the *amplification matrix*:

$$\mathbf{G} = \begin{bmatrix} 1 + c^2(\cos\theta - 1) & -Ic\sin\theta \\ -Ic\sin\theta & 1 + c^2(\cos\theta - 1) \end{bmatrix} \tag{11.105}$$

For Eqs. (11.96) and (11.97) to be stable, the eigenvalues, λ, of the amplification matrix, $\mathbf{G}$, must be ≤ 1. Solving for the eigenvalues gives

$$\begin{vmatrix} [1 + c^2(\cos\theta - 1) - \lambda] & -Ic\sin\theta \\ -Ic\sin\theta & [1 + c^2(\cos\theta - 1) - \lambda] \end{vmatrix} = 0 \tag{11.106}$$

Solving Eq. (11.106) gives

$$[1 + c^2(\cos\theta - 1) - \lambda]^2 + c^2 \sin^2\theta = 0 \tag{11.107}$$

Solving Eq. (11.107) for λ gives

$$\lambda_\pm = (1 - c^2) + c^2 \cos\theta \pm Ic \sin\theta \tag{11.108}$$

Equation (11.108) represents an ellipse in the complex plane with center at $(1 - c^2 + I0)$ and axes c and c^2. For stability, $|\lambda_\pm| \leq 1$. This is guaranteed if the convection number $c = a\,\Delta t/\Delta x \leq 1$. The Lax-Wendroff one-step approximation of the wave equation is consistent and conditionally stable. Consequently, by the Lax equivalence theorem, it is a consistent approximation of that equation.

Example 11.9. The Lax-Wendroff one-step method applied to the wave equation

Now let's solve the acoustics problem presented at the beginning of this section by the Lax-Wendroff one-step method Eqs. (11.96) and (11.97), with $\Delta x = 0.05$ m. Let $f = P$ and $g = Q = (Pa)u$. Let $g(x, 0) = (Pa)u(x, 0) = 0.0$. The results are presented in Figure 11.29 at times from 0.1 to 0.5 ms for $c = 0.5$, corresponding to $\Delta t = 0.025$ ms, and at $t = 1.0$ ms for $c = 0.1, 0.5, 0.9$, and 1.0, corresponding to $\Delta t = 0.005, 0.025, 0.045$, and 0.05 ms, respectively.

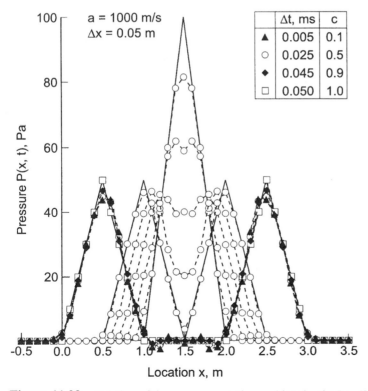

Figure 11.29   Solution of the wave propagation problem by the Lax-Wendroff one-step method.

When $c = 1.0$, the numerical solution is identical to the exact solution, for the linear wave equation. This is not true for nonlinear PDEs. For the other three values of c, the solutions are all quite good. As c decreases, the numerical solution lags the exact solution slightly due to numerical dispersion. The peak of the wave is slightly rounded. Overall, the Lax-Wendroff one-step method gives excellent results for hyperbolic PDEs.

As demonstrated in Example 11.9, the Law-Wendroff one-step method is an efficient and accurate method for solving the linear wave equation. For nonlinear PDEs and systems of PDEs, however, the method becomes quite complicated. The complications arise in the replacement of the second-order time derivatives $\bar{f}_{tt}$ and $\bar{g}_{tt}$ in terms of spatial derivatives by differentiating the governing partial differential equations. The simple result obtained in Eq. (11.93) no longer applies. Consequently, the Lax-Wendroff one-step method is not used very often. More efficient methods, such as the Lax-Wendroff two-step method presented in Section 11.5.2 and the MacCormack method presented in Section 11.5.3 are generally used for nonlinear equations and systems of equations. These methods have the same general features as the Lax-Wendroff one-step method, but they are considerably less complex for nonlinear PDEs, and thus considerably more efficient.

In summary, the Lax-Wendroff one-step method applied to the coupled convection equations and the wave equation is explicit, single step, consistent, $0(\Delta t^2) + 0(\Delta x^2)$, conditionally stable, and convergent. The method is quite complicated for nonlinear PDEs, systems of PDEs, and two- and three-dimensional problems.

11.9.4 Flux-Vector-Splitting Methods

As shown in Section 11.9.2, the coupled system of two linear convection equations, Eqs. (11.84) and (11.85), that corresponds to the linear wave equation, (Eq. (11.64), has preferred physical information propagation paths: $dx/dt = +a$ and $dx/dt = -a$. Information comes from and propagates to both the positive and negative directions at every point. When the spatial derivatives are approximated by centered-difference approximations, the preferred paths of information propagation are ignored. Upwind methods account for the preferred paths of information propagation. However, the upwind direction is not readily apparent in Eqs. (11.84) and (11.85). The *flux-vector-splitting method* identifies these preferred propagation paths in a system of hyperbolic PDES.

Consider the system of two coupled linear convection equations, Eqs. (11.84) and (11.85):

$$\boxed{\bar{f}_t + a\bar{g}_x = 0} \tag{11.109}$$

$$\boxed{\bar{g}_t + a\bar{f}_x = 0} \tag{11.110}$$

Assume $\bar{f}(t, x)$ and $\bar{g}(t, x)$ are continuous functions. Thus,

$$d\bar{f} = \bar{f}_t\, dt + \bar{f}_x\, dx \tag{11.111}$$

$$d\bar{g} = \bar{g}_t\, dt + \bar{g}_x\, dx \tag{11.112}$$

which can be written as

$$\frac{d\bar{f}}{dt} = \bar{f}_t + \frac{dx}{dt}\bar{f}_x \tag{11.113}$$

$$\frac{d\bar{g}}{dt} = \bar{g}_t + \frac{dx}{dt}\bar{g}_x \tag{11.114}$$

Along the positive-traveling wave, $dx/dt = +a$, Eqs. (11.113) and (11.114) give

$$\frac{d\bar{f}}{dt} = \bar{f}_t + a\bar{f}_x \quad \text{and} \quad \frac{d\bar{g}}{dt} = \bar{g}_t + a\bar{g}_x \tag{11.115}$$

and along the negative-traveling wave, $dx/dt = -a$, Eqs. (11.113) and (11.114) give

$$\frac{d\bar{f}}{dt} = \bar{f}_t - a\bar{f}_x \quad \text{and} \quad \frac{d\bar{g}}{dt} = \bar{g}_t - a\bar{g}_x \tag{11.116}$$

Adding Eqs. (11.115a) and (11.115b) yields

$$(\bar{f}_t + a\bar{f}_x) + (\bar{g}_t + a\bar{g}_x) = 0 \tag{11.117}$$

which applies along $dx/dt = +a$, and adding Eqs. (11.116a) and (11.116b) yields

$$(\bar{f}_t - a\bar{f}_x) - (\bar{g}_t - a\bar{g}_x) = 0 \tag{11.118}$$

which applies along $dx/dt = -a$.

The spatial flux derivatives (i.e., $\bar{f}_x$ and $\bar{g}_x$) in Eqs. (11.117) and (11.118) are associated with positive-traveling waves and negative-traveling waves, respectively. Consequently, they should be differenced in the appropriate upwind directions. Let's attach superscripts $+$ and $-$ to the spatial flux derivatives in Eqs. (11.117) and (11.118), respectively, to remind us that they are associated with positive-traveling and negative-traveling waves, respectively. Thus,

$$(\bar{f}_t + a\bar{f}_x^+) + (\bar{g}_t + a\bar{g}_x^+) = 0 \tag{11.119}$$

$$(\bar{f}_t - a\bar{f}_x^-) - (\bar{g}_t - a\bar{g}_x^-) = 0 \tag{11.120}$$

Solving Eqs. (11.119) and (11.120) explicitly for $\bar{f}_t$ and $\bar{g}_t$ yields the final form of the *flux-vector-split PDEs*:

$$\boxed{\bar{f}_t + \frac{a}{2}(\bar{f}_x^+ - \bar{f}_x^-) + \frac{a}{2}(\bar{g}_x^+ + \bar{g}_x^-)} \tag{11.121}$$

$$\boxed{\bar{g}_t + \frac{a}{2}(\bar{g}_x^+ - \bar{g}_x^-) + \frac{a}{2}(\bar{f}_x^+ + \bar{g}_x^-)} \tag{11.122}$$

Equations (11.121) and (11.122) are in a form suitable for developing upwind finite difference approximations. First-order FDEs can be developed using finite difference approximations such as Eq. (11.51), and second-order FDEs can be developed using finite difference approximations such as Eq. (11.55).

11.10 PROGRAMS

Five FORTRAN subroutines for solving the convection equation are presented in this section:

1. The Lax method
2. The Lax-Wendroff one-step method
3. The MacCormack method
4. The upwind method
5. The backward-time centered-space (BTCS) method

The basic computational algorithms are presented as completely self-contained subroutines suitable for use in other programs. Input data and output statements are contained in a main (or driver) program written specifically to illustrate the use of each subroutine.

11.10.1 The Lax Method

The convection equation is given by Eq. (11.6):

$$f_t + uf_x = 0 \tag{11.123}$$

The Lax approximation of the convection is given by Eq. (11.26):

$$f_i^{n+1} = \tfrac{1}{2}(f_{i+1}^n + f_{i-1}^n) - \frac{c}{2}(f_{i+1}^n - f_{i-1}^n) \tag{11.124}$$

A FORTRAN subroutine, *subroutine lax*, for implementing Eq. (11.124) is presented in Program 11.1. *Program main* defines the data set and prints it, calls *subroutine lax* to implement the solution, and prints the solution.

Program 11.1.   The Lax method for the convection equation program

```
      program main
c     main program to illustrate convection equation solvers
c     nxdim x-direction array dimension, nxdim = 61 in this program
c     ntdim t-direction array dimension, ntdim = 41 in this program
c     imax  number of grid points in the x direction
c     nmax  number of time steps
c     iw    intermediate results output flag: 0 none, 1 all
c     ix    grid point output interval: 1 all, n every nth point
c     it    time level output interval: 1 all, n every nth level
c     f     solution array, f(i,n)
c     dx    x-direction grid increment
c     dt    time step
c     u     convection velocity
      dimension f(61,41)
      data nxdim,ntdim,imax,nmax,iw,ix,it /61, 41, 61, 41, 0, 2, 4/
      data (f(i,1),i=1,20) / 0.,0.,0.,0.,0.,0.,0.,0.,0.,0.,0.,
     1 0.,0.,0.,0.,0.,0.,0. /
      data (f(i,1),i=21,41) / 0.,10.,20.,30.,40.,50.,60.,70.,80.,
     1 90.,100.,90.,80.,70.,60.,50.,40.,30.,20.,10.,0. /
      data (f(i,1),i=42,61) / 0.,0.,0.,0.,0.,0.,0.,0.,0.,0.,0.,
     1 0.,0.,0.,0.,0.,0.,0. /
      data dx,dt,u / 0.05, 0.25, 0.1 /
      c=u*dt/dx
      write (6,1000) c
```

```
      call lax (nxdim,ntdim,imax,nmax,f,dx,dt,u,c,iw,ix,it)
      if (iw.eq.1) stop
      do n=1,nmax,it
        t=float(n-1)*dt
        write (6,1010) n-1,t,(f(i,n),i=1,imax,ix)
      end do
      stop
1000 format (' Convection equation solver (Lax method), c=',f4.2
   1 /' '/'  n',2x,'time',3x,'f(i,n)'/' ')
1010 format (i3,f5.1,11f6.2/14x,10f6.2/14x,10f6.2/14x,10f6.2/
   1 14x,10f6.2/14x,10f6.2)
      end


      subroutine lax (nxdim,ntdim,imax,nmax,f,dx,dt,u,c,iw,ix,it)
c     implements the Lax method for the convection equation
      dimension f(nxdim,ntdim)
      do n=1,nmax-1
        t=float(n-1)*dt
        f(1,n+1)=f(1,1)
        do i=2,imax-1
         f(i,n+1)=0.5*(f(i+1,n)+f(i-1,n))-0.5*c*(f(i+1,n)-f(i-1,n))
        end do
        f(imax,n+1)=f(imax,1)
        if (iw.eq.1) write (6,1000) n+1,t,(f(i,n+1),i=1,imax,ix)
      end do
      return
1000 format (i3,f5.1,11f6.2/14x,10f6.2/14x,10f6.2/14x,10f6.2/
   1 14x,10f6.2/14x,10f6.2)
      end
```

The data set used to illustrate *subroutine lax* is taken from Example 11.2. The output generated by the program is presented in Output 11.1.

Output 11.1. Solution of the convection equation by the Lax method

```
Convection equation solver (Lax method), c = 0.50

 n  time   f(i,n)

 0  0.0  0.00  0.00  0.00  0.00  0.00  0.00  0.00  0.00  0.00  0.00  0.00
             20.00 40.00 60.00 80.00100.00 80.00 60.00 40.00 20.00  0.00
              0.00  0.00  0.00  0.00  0.00  0.00  0.00  0.00  0.00  0.00

 4  1.0  0.00  0.00  0.00  0.00  0.00  0.00  0.00  0.00  0.00  0.08  1.09
              6.33 20.00 40.00 59.84 77.81 87.34 80.00 60.00 40.08 21.09
              6.33  0.00  0.00  0.00  0.00  0.00  0.00  0.00  0.00  0.00
```

```
 8   2.0   0.00   0.00   0.00   0.00   0.00   0.00   0.00   0.00   0.01   0.09   0.64
                  2.91   9.34  21.99  39.82  58.72  74.17  81.31  76.00  60.09  40.64
                 22.91   9.34   2.00   0.00   0.00   0.00   0.00   0.00   0.00   0.00
 .............................................................
20   5.0   0.00   0.00   0.00   0.00   0.00   0.00   0.00   0.00   0.00   0.02   0.10
                  0.38   1.20   3.23   7.47  14.97  26.12  39.98  54.09  65.05  69.75
                 66.62  56.45  42.13  27.39  15.17   6.88   2.37   0.55   0.06   0.00
 .............................................................
40  10.0   0.00   0.00   0.00   0.00   0.00   0.00   0.00   0.00   0.00   0.00   0.00
                  0.02   0.05   0.14   0.37   0.89   1.96   3.95   7.35  12.58  19.86
                 28.90  38.81  48.10  55.01  58.03  56.38  50.34  41.10  29.32   0.00
```

11.10.2   The Lax-Wendroff One-Step Method

The Lax-Wendroff one-step approximation of the convection equation is given by Eq. (11.37):

$$f_i^{n+1} = f_i^n - \frac{c}{2}(f_{i+1}^n - f_{i-1}^n) + \frac{c^2}{2}(f_{i+1}^n - 2f_i^n + f_{i-1}^n) \qquad (11.125)$$

A FORTRAN subroutine, *subroutine lw1*, for implementing Eq. (11.125) is presented in Program 11.2. Only the statements which are different from the statements in *program main* and *program lax* in Section 11.10.1 are presented. *Program main* defines the data set and prints it, calls *subroutine lw1* to implement the solution, and prints the solution.

Program 11.2.   The Lax-Wendroff method for the convection equation program

```
      program main
c     main program to illustrate convection equation solvers
      call lw1 (nxdim,ntdim,imax,nmax,f,dx,dt,u,c,iw,ix,it)
 1000 format (' Convection equation solver (Lax-Wendroff method),'
     1 ' c = ',f4.2/' '/'  n',2x,'time',3x,'f(i,n)'/' ')
      end

      subroutine lw1 (nxdim,ntdim,imax,nmax,f,dx,dt,u,c,iw,ix,it)
c     the Lax-Wendroff method for the convection equation
            f(i,n+1)=f(i,n)-0.5*c*(f(i+1,n)-f(i-1,n))+0.5*c**2
     1             *(f(i+1,n)-2.0*f(i,n)+f(i-1,n))
      end
```

The data set used to illustrate *subroutine lw1* is taken from Example 11.3. The output generated by the program is presented in Output 11.2.

Output 11.2. **Solution of the convection equation by the Lax-Wendroff method**

```
Convection equation solver (Lax-Wendroff method), c = 0.50

 n  time   f(i,n)

 0  0.0  0.00  0.00  0.00  0.00  0.00  0.00  0.00  0.00  0.00  0.00  0.00
               20.00 40.00 60.00 80.00100.00 80.00 60.00 40.00 20.00  0.00
                0.00  0.00  0.00  0.00  0.00  0.00  0.00  0.00  0.00  0.00

 4  1.0  0.00  0.00  0.00  0.00  0.00  0.00  0.00  0.00  0.00 -0.05 -0.75
                1.98 20.00 40.00 60.11 81.50 96.04 80.00 60.00 39.95 19.25
                1.98  0.00  0.00  0.00  0.00  0.00  0.00  0.00  0.00  0.00

 8  2.0  0.00  0.00  0.00  0.00  0.00  0.00  0.00  0.00 -0.01 -0.09  0.47
               -1.87  2.66 20.08 40.18 59.05 83.74 94.67 79.85 59.91 40.47
               18.13  2.66  0.07  0.00  0.00  0.00  0.00  0.00  0.00  0.00
.......................................................................
20  5.0  0.00  0.00  0.00  0.00  0.00  0.00  0.00  0.00  0.00 -0.01  0.06
               -0.20  0.49 -0.27 -3.02  3.68 20.80 39.04 60.54 86.08 92.45
               78.99 60.45 39.73 16.95  3.80  0.41  0.02  0.00  0.00  0.00
.......................................................................
40 10.0  0.00  0.00  0.00  0.00  0.00  0.00  0.00  0.00  0.00  0.00  0.00
                0.00 -0.01 -0.01  0.10 -0.33  0.35  0.93 -2.32 -3.62  5.56
               20.23 38.28 64.68 86.92 89.88 78.47 60.68 37.64 16.48  0.00
```

11.10.3 The MacCormack Method

The MacCormack approximation of the convection equation is given by Eqs. (11.47) and (11.49):

$$f_i^{\overline{n+1}} = f_i^n - c(f_{i+1}^n - f_i^n) \tag{11.126}$$

$$f_i^{n+1} = \tfrac{1}{2}[f_i^n + f_i^{\overline{n+1}} - c(f_i^{\overline{n+1}} - f_{i-1}^{\overline{n+1}})] \tag{11.127}$$

A FORTRAN subroutine, *subroutine mac*, for implementing Eqs. (11.126) and (11.127) is presented in Program 11.3. Only the statements which are different from the statements in *program main* and *program lax* in Section 11.10.1 are presented. *Program main* defines the data set and prints it, calls *subroutine mac* to implement the solution, and prints the solution.

Program 11.3. The MacCormack method for the convection equation program

```
      program main
c     main program to illustrate convection equation solvers
      dimension f(61,41),g(61,41)
      call mac (nxdim,ntdim,imax,nmax,f,g,dx,dt,u,c,iw,ix,it)
 1000 format (' Convection equation solver (MacCormack method),'
     1 ' c = ',f4.2/' '/'  n',2x,'time',3x,'f(i,n)'/' ')
      end


      subroutine mac (nxdim,ntdim,imax,nmax,f,g,dx,dt,u,c,iw,ix,it)
c     the MacCormack method for the convection equation
      dimension f(nxdim,ntdim),g(nxdim,ntdim)
        do i=1,imax-1
          g(i,n+1)=f(i,n)-c*(f(i+1,n)-f(i,n))
        end do
        do i=2,imax-1
          f(i,n+1)=0.5*(f(i,n)+g(i,n+1)-c*(g(i,n+1)-g(i-1,n+1)))
        end do
      end
```

The data set used to illustrate *subroutine mac* is taken from Example 10.5. The output generated by the program is presented in Output 11.3.

Output 11.3. Solution of the convection equation by the MacCormack method

```
Convection equation solver (MacCormack method), c = 0.50

 n  time   f(i,n)

 0  0.0  0.00  0.00  0.00  0.00  0.00  0.00  0.00  0.00  0.00  0.00  0.00
              20.00 40.00 60.00 80.00100.00 80.00 60.00 40.00 20.00  0.00
               0.00  0.00  0.00  0.00  0.00  0.00  0.00  0.00  0.00  0.00

 4  1.0  0.00  0.00  0.00  0.00  0.00  0.00  0.00  0.00  0.00 -0.05 -0.75
               1.98 20.00 40.00 60.11 81.50 96.04 80.00 60.00 39.95 19.25
               1.98  0.00  0.00  0.00  0.00  0.00  0.00  0.00  0.00  0.00

 8  2.0  0.00  0.00  0.00  0.00  0.00  0.00  0.00  0.00 -0.01 -0.09  0.47
              -1.87  2.66 20.08 40.18 59.05 83.74 94.67 79.85 59.91 40.47
              18.13  2.66  0.07  0.00  0.00  0.00  0.00  0.00  0.00  0.00
```

```
. . . . . . . . . . . . . . . . . . . . . . . . . . . . . . . . . . . . . . . . . . . . . . . . . . . .
20  5.0  0.00   0.00   0.00   0.00   0.00   0.00   0.00   0.00   0.00 -0.01   0.06
             -0.20   0.49  -0.27  -3.02   3.68  20.80  39.04  60.54  86.08  92.45
             78.99  60.45  39.73  16.95   3.80   0.41   0.02   0.00   0.00   0.00

. . . . . . . . . . . . . . . . . . . . . . . . . . . . . . . . . . . . . . . . . . . . . . . . . . . .
40 10.0  0.00   0.00   0.00   0.00   0.00   0.00   0.00   0.00   0.00   0.00   0.00
              0.00  -0.01  -0.01   0.10  -0.33   0.35   0.93  -2.32  -3.62   5.56
             20.23  38.28  64.68  86.92  89.88  78.47  60.68  37.64  16.48   0.00
```

11.10.4 The Upwind Method

The first-order upwind approximation of the convection equation is given by Eq. (11.51):

$$f_i^{n+1} = f_i^n - c(f_i^n - f_{i-1}^n) \tag{11.128}$$

The second-order upwind approximation of the convection equation is given by Eq. (11.55):

$$f_i^{n+1} = f_i^n - c(f_i^n - f_{i-1}^n) - \frac{c(1-c)}{2}(f_i^n - 2f_{i-1}^n + f_{i-2}^n) \tag{11.129}$$

A FORTRAN subroutine, *subroutine up*, for implementing Eqs. (11.128) and (11.129) is presented in Program 11.4. When $iu = 1$, the first-order upwind method is implemented. When $iu = 2$, the second-order upwind method is implemented. Only the statements which are different from the statements in *program main* and *program lax* in Section 11.10.1 are presented. *Program main* defines the data set and prints it, calls *subroutine up* to implement the solution, and prints the solution.

Program 11.4. The upwind method for the convection equation program

```
      program main
c     main program to illustrate convection equation solvers
c     iu    upwind method selector: 1 first-order, 2 second-order
      data dx,dt,u,iu / 0.05, 0.25, 0.1, 2 /
      write (6,1000) iu,c
      call up (nxdim,ntdim,imax,nmax,f,g,dx,dt,u,c,iw,ix,it,iu)
```

```
1000 format (' Convection equation solver (Upwind method), iu = ',
    1 i1,' and c = ',f4.2/' '/'  n',3x,'time',3x,'f(i,n)'/' ')
      end

      subroutine up(nxdim,ntdim,imax,nmax,f,g,dx,dt,u,c,iw,ix,it,iu)
c     the upwind method for the convection equation
        do i=2,imax-1
          f(i,n+1)=f(i,n)-c*(f(i,n)-f(i-1,n))
          if (iu.eq.2) f(i,n+1)=f(i,n+1)-0.5*c*(c-1.0)*(f(i,n)
    1                            -2.0*f(i-1,n)+f(i-2,n))
        end do
      end
```

The data set used to illustrate *subroutine up* is taken from Example 11.7. The output generated by the program is presented in Output 11.4.

Output 11.4. Solution of the convection equation by the second-order upwind method

```
Convection equation solver (Upwind method), iu = 2 and c = 0.50

 n   time    f(i,n)

 0  0.0  0.00   0.00   0.00   0.00   0.00   0.00   0.00   0.00   0.00   0.00   0.00
                20.00  40.00  60.00  80.00 100.00  80.00  60.00  40.00  20.00   0.00
                 0.00   0.00   0.00   0.00   0.00   0.00   0.00   0.00   0.00   0.00

 4  1.0  0.00   0.00   0.00   0.00   0.00   0.00   0.00   0.00   0.00   0.00   0.00
                 5.49  20.65  40.02  60.00  80.00  89.01  78.69  59.95  40.00  20.00
                 5.49   0.65   0.02   0.00   0.00   0.00   0.00   0.00   0.00   0.00

 8  2.0  0.00   0.00   0.00   0.00   0.00   0.00   0.00   0.00   0.00   0.00   0.00
                 1.21   7.88  21.87  40.27  60.02  77.58  84.25  76.26  59.47  39.96
                21.21   7.88   1.87   0.27   0.02   0.00   0.00   0.00   0.00   0.00
      ...............................................................
20  5.0  0.00   0.00   0.00   0.00   0.00   0.00   0.00   0.00   0.00   0.00   0.00
                 0.01   0.17   1.19   4.71  12.55  25.23  41.44  58.10  70.67  74.92
                69.52  56.62  40.23  24.51  12.51   5.24   1.77   0.48   0.10   0.00
      ...............................................................
40 10.0  0.00   0.00   0.00   0.00   0.00   0.00   0.00   0.00   0.00   0.00   0.00
                 0.00   0.00   0.00   0.02   0.10   0.45   1.54   4.18   9.31  17.60
                28.85  41.70  53.74  62.15  64.79  61.04  51.99  40.00  27.72   0.00
```

Example 11.6, which illustrates the first-order upwind method, can be solved by setting $iu = 1$ in *program main*.

11.10.5 The BTCS Method

The BTCS approximation of the convection equation is given by Eq. (11.59):

$$-\frac{c}{2}f_{i-1}^{n+1} + f_i^{n+1} + \frac{c}{2}f_{i+1}^{n+1} = f_i^n \qquad (11.130)$$

A FORTRAN subroutine, *subroutine btcs*, for solving the system equation arising from the application of Eq. (11.130) at every interior grid point is presented in Program 11.5. Only the statements which are different from the statements in *program main* and *program lax* in Section 11.10.1 are presented. To close the implicit solution grid, 100 grid points are added to the left and right sides of the triangular initial-value data. *Program main* defines the data set and prints it, calls *subroutine btcs* to implement the solution, and prints the solution. *Subroutine thomas* from Section 1.8.4 is used to solve the tridiagonal system equation.

Program 11.5. The BTCS method for the convection equation program

```
      program main
c     main program to illustrate convection equation solvers
      dimension f(221,41),a(221,3),b(221),w(221)
      data nxdim,ntdim,imax,nmax,iw,ix,it/221,41, 221, 41, 0, 2, 4/
      data (f(i,1),i=101,121) / 0.,10.,20.,30.,40.,50.,60.,70.,80.,
     1 90.,100.,90.,80.,70.,60.,50.,40.,30.,20.,10.,0. /
      data dx,dt,u, / 0.05, 0.25, 0.1 /
      do i=1,100
        f(i,1)=0.0
      end do
      do i=122,221
        f(i,1)=0.0
      end do
      call btcs(nxdim,ntdim,imax,nmax,f,dx,dt,u,c,n,t,iw,ix,it,a,b,w)
 1000 format (' Convection equation solver (BTCS method),'
     1 ' c = ',f4.2/' '/' n',2x,'time',3x,'f(i,n)'/' ')
      end

      subroutine btcs (nxdim,ntdim,imax,nmax,f,dx,dt,u,c,n,t,iw,ix,
     1 it,a,b,w)
c     the BTCS method for the diffusion equation
      dimension f(nxdim,ntdim),a(nxdim,3),b(nxdim),w(nxdim)
      d=alpha*dt/dx**2
      a(1,2)=1.0
      a(1,3)=0.0
      b(1)=0.0
```

```
a(imax,1)=0.0
a(imax,2)=1.0
b(imax)=0.0
do n=1,nmax-1
   t=t+dt
   do i=2,imax-1
      a(i,1)=-0.5*c
      a(i,2)=1.0
      a(i,3)=0.5*c
      b(i)=f(i,n)
   end do
   call thomas (nxdim,imax,a,b,w)
   do i=2,imax-1
      f(i,n+1)=w(i)
   end do
   if (iw.eq.1) write (6,1000) n+1,t,(f(i,n+1),i=81,141,ix)
end do
return
```

The data set used to illustrate *subroutine btcs* is taken from Example 11.8. The output generated by the program is presented in Output 11.5.

Output 11.5. Solution of the convection equation by the BTCS method

```
Convection equation solver (BTCS method), c = 0.50

 n   time    f(i,n)

 0   0.0   0.00   0.00   0.00   0.00   0.00   0.00   0.00   0.00   0.00   0.00   0.00
                 20.00  40.00  60.00  80.00 100.00  80.00  60.00  40.00  20.00   0.00
                  0.00   0.00   0.00   0.00   0.00   0.00   0.00   0.00   0.00   0.00

 4   1.0   0.00   0.00   0.00   0.00   0.00   0.00   0.00  -0.02  -0.13  -0.68  -1.23
                  3.94  20.59  40.32  61.37  82.47  92.13  78.88  59.75  39.31  18.76
                  3.94   0.55   0.06   0.01   0.00   0.00   0.00   0.00   0.00   0.00
 8   2.0   0.00   0.00   0.00   0.00   0.00  -0.01  -0.04  -0.18  -0.56  -0.76   0.61
                 -1.11   6.00  22.36  41.73  58.83  82.35  88.53  76.96  58.81  40.53
                 18.80   5.64   1.24   0.22   0.03   0.00   0.00   0.00   0.00   0.00
   ..............................................................................
20   5.0   0.00  -0.01  -0.03  -0.09  -0.21  -0.39  -0.44  -0.11   0.32   0.35   0.83
                  0.84   0.51  -1.10   0.79   8.48  22.85  40.36  61.82  78.27  82.14
                 73.57  58.30  38.79  20.74   8.93   3.18   0.96   0.25   0.06   0.01
   ..............................................................................
40  10.0  -0.28  -0.27  -0.11   0.18   0.45   0.59   0.60   0.40  -0.03  -0.29  -0.30
                 -0.28  -0.25  -0.01   0.00   0.06   0.01  -0.36   0.07   4.05  12.55
                 25.87  43.26  60.72  72.35  74.84  68.20  54.65  38.24  23.32  12.45
```

11.10.6 Packages For Solving The Convection Equation

Numerous libraries and software packages are available for solving the convection equation. Many work stations and mainframe computers have such libraries attached to their operating systems.

Many commercial software packages contain algorithms for integrating convection type (i.e., hyperbolic) PDEs. Due to the wide variety of hyperbolic PDEs governing physical problems, many hyperbolic PDE solvers (i.e., programs) have been developed. For this reason, no specific programs are recommended in this section.

11.11 SUMMARY

The numerical solution of hyperbolic partial differential equations by finite difference methods is discussed in this chapter. Hyperbolic PDEs govern propagation problems, which have a finite physical information propagation speed. They are solved numerically by marching methods. The unsteady one-dimensional convection equation $\bar{f}_t + u\bar{f}_x = 0$ is considered as the model hyperbolic PDE in this chapter.

Explicit finite difference methods, as typified by the Lax-Wendroff type methods and the upwind methods, are conditionally stable and require a relatively small step size in the marching direction to satisfy the stability criteria. Implicit methods, as typified by the BTCS method, are unconditionally stable. The marching step size is restricted by accuracy requirements, not stability requirements. For accurate solutions of transient problems, explicit methods are recommended. When steady state solutions are to be obtained as the asymptotic solution in time of an appropriate unsteady propagation problem, the BTCS method with a large step size is recommended.

In all the examples solved by explicit FDEs in this chapter, the most accurate solutions are obtained for the largest value of the convection number, $c = u \, \Delta t/\Delta x$. In fact, for linear PDEs, the exact solution is obtained for $c = 1$. Although this is not the case for nonlinear PDEs, these results suggest that all explicit finite difference methods applied to hyperbolic PDEs should march forward in time with the largest possible time step allowed by stability considerations.

Nonlinear partial differential equations can be solved directly by explicit methods. When solved by implicit methods, systems of nonlinear FDEs must be solved. Multi-dimensional problems can be solved directly by explicit methods. When solved by implicit methods, large banded systems of FDEs result. As discussed in Section 10.9.2, alternating direction implicit (ADI) methods and approximate factorization implicit (AFI) methods can be used to solve multidimensional problems.

After studying Chapter 11, you should be able to:

1. Describe the physics of propagation problems governed by hyperbolic PDEs
2. Describe the general features of the unsteady convection equation
3. Understand the general features of pure convection
4. Discretize continuous physical space
5. Understand the differences between an explicit FDE and an implicit FDE
6. Understand the theoretical concepts of consistency, order, stability, and convergence

7. Derive the modified differential equation (MDE) actually solved by a FDE
8. Perform a von Neumann stability analysis
9. Implement the Lax method
10. Describe the concepts underlying Lax-Wendroff type methods
11. Implement the Lax-Wendroff one-step method
12. Implement the Law-Wendroff (Ritchtmyer) two-step method
13. Implement the MacCormack method
14. Describe the concepts underlying upwind methods
15. Implement the first-order upwind method
16. Implement the second-order upwind method
17. Implement the backward-time centered-space (BTCS) method
18. Describe the complications associated with nonlinear PDEs
19. Describe the complications associated with multidimensional problems
20. Describe the differences between explicit and implicit FDEs
21. Describe the general features of the wave equation
22. Describe the similarities and differences between the convection equation and the wave equation
23. Solve linear wave propagation problems by the Law-Wendroff one-step method
24. Describe the concepts underlying flux-vector splitting
25. Choose a finite difference method for solving a hyperbolic PDE

EXERCISE PROBLEMS

Section 11.2   General Features of Hyperbolic PDEs

1. Consider the unsteady one-dimensional convection equation $\bar{f}_t + u\bar{f}_x = 0$. Classify this PDE. Determine the characteristic curves. Discuss the significance of these results as regards domain of dependence, range of influence, physical information propagation speed, auxiliary conditions, and numerical solution procedures.
2. Discuss the general feaures of hyperbolic PDEs.
3. Discuss the major similarities of parabolic and hyberboic PDEs.
4. Discuss the major differences between parabolic and hyperbolic PDEs.
5. Develop the exact solution of the convection equation, Eq. (11.12).
6. Discuss the significance of Eq. (11.12) as regards the general behavior of convection problems.

Section 11.4   The Forward-Time Centered-Space (FTCS) Method and The Lax Method

The Forward-Time Centered Space Method

7. Derive the forward-time centered-space (FTCS) approximation of the unsteady one-dimensional convection equation, (Eq. (11.22)), including the leading truncation error terms in Δt and Δx.

8. Perform a von Neumann stability analysis of Eq. (11.22).
9.* By hand calculation, solve the convection problem presented in Section 11.1 by the FTCS method with $\Delta x = 0.1$ cm and $\Delta t = 1.0$ s for $t = 2.0$ s.

The Lax method

10. Derive the Lax approximation of the unsteady one-dimensional convection equation, Eq. (11.26), including the leading truncation error terms in Δt and Δx.
11. Derive the modified differential equation (MDE) corresponding to Eq. (11.26). Analyze consistency and order.
12. Perform a von Neumann stability analysis of Eq. (11.26).
13.* By hand calculation, solve the convection problem presented in Section 11.1 by the Lax method with $\Delta x = 0.1$ cm and $\Delta t = 0.5$ s for $t = 1.0$ s. Compare the results with the exact solution.
14. Implement the program presented in Section 11.10.1 to solve the example convection problem by the Lax method. Use the program to solve the example convection problem with $\Delta x = 0.1$ cm and $\Delta t = 0.5$ s for $t = 10.0$ s. Compare the results with the exact solution and the results of Problem 13.
15. Use the program to reproduce the results presented in Figure 11.11, where $\Delta x = 0.05$ cm. Compare the errors with the errors in Problem 14 at selected locations and times.

Section 11.5   Lax-Wendroff-Type Methods

The Lax-Wendroff One-Step Method

16. Derive the Lax-Wendroff one-step approximation of the unsteady one-dimensional convection equation, Eq. (11.37), including the leading truncation error terms in Δt and Δx.
17. Derive the MDE corresponding to Eq. (11.37). Analyze consistency and order.
18. Perform a von Neumann stability analysis of Eq. (11.37).
19. By hand calculation, solve the convection problem presented in Section 11.1 by the Lax-Wendroff one-step metehod with $\Delta x = 0.1$ cm and $\Delta t = 0.5$ s for $t = 1.0$ s. Compare the results with the exact solution.
20. Implement the program presented in Section 11.10.2 to solve the example convection problem by the Lax-Wendroff one-step method. Use the program to solve the example convection problem with $\Delta x = 0.1$ cm and $\Delta t = 0.5$ s for $t = 10.0$ s. Compare the results with the exact solution and the results of Problem 19.
21. Use the program to reproduce the results presented in Figure 11.14, where $\Delta x = 0.05$ cm. Compare the errors with the errors in Problem 20 at selected locations and times.

The Lax-Wendroff (Richtmyer) Two-Step Method

22. Discuss the Lax-Wendroff (Richtmyer) two-step approximation of the unsteady one-dimensional convection equation, Eqs. (11.41) and (11.42). Show that, for

the linear convection equation, the two-step method is equivalent to the Lax-Wendroff one-step method for 2 Δt and 2 Δx.

23. By hand calculation, solve the convection problem presented in Section 11.1 by the Lax-Wendroff two-step method with $\Delta x = 0.1$ cm and $\Delta t = 0.5$ s for $t = 1.0$ s. Compare the results with the exact solution.

24. Modify the program presented in Section 11.10.2 to solve the example convection problem by the Lax-Wendroff (Richtmyer) two-step method. Use the program to solve the example convection problem with $\Delta x = 0.1$ cm and $\Delta t = 0.5$ s for $t = 10.0$ s. Compare the results with the exact solution and the results of Problem 23.

25. Use the program to reproduce the results presented in Figure 11.16, where $\Delta x = 0.05$ cm. Compare the errors with the errors in Problem 24 at selected locations and times.

The MacCormack Method

26. Develop the MacCormack approximation of the unsteady one-dimensional convection equation, Eqs. (11.47) and (11.49), including the leading truncation error terms. Show that, for the linear convection equation, the two-step method is identical to the Lax-Wendroff one-step method.

27.* By hand calculation, solve the convection problem presented in Section 11.1 by the MacCormack method with $\Delta x = 0.1$ cm and $\Delta t = 0.5$ s for $t = 1.0$ s. Compare the results with the exact solution and the results of Problem 19.

28. Implement the program presented in Section 11.10.3 to solve the example convection problem by the MacCormack method. Use the program to solve the example convection problem with $\Delta x = 0.1$ cm and $\Delta t = 0.5$ s for $t = 10.0$ s. Compare the results with the exact solution and the results of Problem 27.

29. Use the program to reproduce the results presented in Figure 11.14, where $\Delta x = 0.05$ cm. Compare the errors with the errors of Problems 21 and 28.

Section 11.6 Upwind Methods

The First-Order Upwind Method

30. Derive the first-order upwind approximation of the unsteady one-dimensional convection equation for $u > 0$, Eq. (11.51), including the leading truncation error terms in Δt and Δx.

31. Derive the MDE corresponding to Eq. (11.51). Analyze consistency and order.

32. Perform a von Neumann stability analysis of Eq. (11.51).

33.* By hand calculation, solve the convection problem presented in Section 11.1 by the first-order upwind method with $\Delta x = 0.1$ cm and $\Delta t = 0.5$ s for $t = 1.0$ s. Compare the results with the exact solution.

34. Implement the program presented in Section 11.10.4 to solve the example convection problem by the first-order upwind method. Use the program to solve the example convection problem with $\Delta x = 0.1$ cm and $\Delta t = 0.5$ s for $t = 10.0$ s. Compare the results with the exact solution and the results of Problem 33.

35. Use the program to reproduce the results presented in Figure 11.20, where $\Delta x = 0.05$ cm. Compare the errors with the errors in Problem 34 at selected locations and times.

The Second-Order Upwind Method

36. A second-order upwind approximation of the unsteady one-dimensional convection equation can be developed by using the second-order backward difference approximation for $\bar{f}_x$ specified by Eq. (5.101). (a) Derive the FDE, including the leading truncation error terms in Δt and Δx. (b) Perform a von Neumann stability analysis of this FDE.

37. Derive the MDE corresponding to Eq. (11.55). Analyze consistency and order.

38. Perform a von Neumann stability analysis of Eq. (11.55). This is best accomplished numerically.

39. By hand calculation, solve the convection problem presented in Section 11.1 by the second-order upwind method with $\Delta x = 0.1$ cm and $\Delta t = 0.5$ s for $t = 1.0$ s. Compare the results with the exact solution.

40. Implement the program presented in Section 11.10.4 to solve the example convection problem by the second-order upwind method. Use the program to solve the example convection problem with $\Delta x = 0.1$ cm and $\Delta t = 0.5$ s for $t = 10.0$ s. Compare the results with the exact solution and the results of Problem 39.

41. Use the program to reproduce the results presented in Figure 11.22, where $\Delta x = 0.05$ cm. Compare the errors with the errors of Problem 40 at selected locations and times.

Section 11.7 The Backward-Time Centered-Space Method

42. Derive the BTCS approximation of the unsteady one-dimensional convection equation, Eq. (11.59), including the leading truncation error terms in Δt and Δx.

43. Derive the MDE corresponding to Eq. (11.59). Analyze consistency and order.

44. Peform a von Neumann stability analysis of Eq. (11.59).

45.* By hand calculation, determine the solution of the example convection problem by the BTCS method for $t = 1.0$ s for $\Delta x = 0.25$ cm and $\Delta t = 1.0$ s. Apply the initial conditions as boundary conditions at $x = -0.5$ and 1.5 cm. Compare the results with the exact solution.

46. Implement the program presented in Section 11.10.5 to solve the example convection problem by the BTCS method. Use the program to solve the example convection problem with $\Delta x = 0.1$ cm and $\Delta t = 0.5$ s for $t = 10.0$ s. Apply the initial conditions as boundary conditions 100 grid points to the left and right of the initial triangular wave. Compare the results with the exact solution.

47. Use the program to reproduce the results presented in Figure 11.25, where $\Delta x = 0.05$ cm. Apply the initial conditions as boundary conditions 100 grid

points to the left and right of the initial triangular wave. Compare the errors with the errors in Problem 46 at selected locations and times.

48. Use the program to reproduce the results presented in Figure 11.26. Discuss these results.

Section 11.8 Nonlinear Equations and Multidimensional Problems

Nonlinear Equations

49. Consider the following hyperbolic PDE for the generic dependent variable $\bar{f}(x,t)$, which serves as a model equation in fluid dynamics:

$$\bar{f}_t + \bar{f}\bar{f}_x = 0 \tag{A}$$

where $\bar{f}(x,0) = F(x)$. (a) Develop the Lax approximation of Eq. (A). (b) Discuss a strategy for solving this problem numerically.

50. Solve Problem 49 by the MacCormack method.

51. Solve Problem 49 by the BTCS method. Discuss a strategy for solving this problem numerically by (a) linearization, (b) iteration, and (c) Newton's method.

52. Equation (A) can be written as

$$\bar{f}_t + (\bar{f}^2/2)_x = 0 \tag{B}$$

which is the conservation form of the nonlinear PDE. (a) Develop the Lax approximation of Eq. (B). (b) Discuss a strategy for solving this problem numerically.

53. Solve Problem 52 by the MacCormack method. Develop the MacCormack approximation of Eq. (B). Discuss a strategy for solving this problem numerically.

54. Solve Problem 52 by the BTCS method. Develop the BTCS approximation of Eq. (B). Discuss a strategy for solving this problem numerically by (a) linearization, (b) iteration, and (c) Newton's method.

55. Equation (B) can be written in the form

$$Q_t + E_x = 0 \tag{C}$$

where $Q = \bar{f}$ and $E = (\bar{f}^2/2)$. Solving Eq. (C) by the BTCS method yields a nonlinear FDE:

$$Q_i^{n+1} - Q_i^n + \frac{\Delta t}{2\,\Delta x}(E_{i+1}^{n+1} - E_{i-1}^{n+1}) = 0 \tag{D}$$

Equation (D) can be time linearized as follows:

$$E^{n+1} = E^n + \frac{\partial E}{\partial Q}|^n(Q^{n+1} - Q^n) = E^n + A^n(Q^{n+1} - Q_n) \tag{E}$$

where $A^n = (\partial E/\partial Q)^n$. Combining Eqs. (D) and (E) and letting $\Delta Q = (Q^{n+1} - Q^n)$ yields the delta form of the FDE, which is linear in ΔQ:

$$\Delta Q_i + \frac{\Delta t}{2\,\Delta x}(A^n_{i+1}\Delta Q_{i+1} - A^n_{i-1}\Delta Q_{i-1}) = -\frac{\Delta t}{2\,\Delta x}(E^n_{i+1} - E^n_{i-1}) \qquad (F)$$

Apply this procedure to develop a strategy for solving Eq. (B).

56. Write a program to solve Problem 55 numerically for $F(x) = 200.0x$ for $0.0 \leq x \leq 0.5$ and $F(x) = 200.0(1.0 - x)$ for $0.5 \leq x \leq 1.0$. March from $t = 0.0$ to $t = 10.0$ s with $\Delta x = 0.1$ cm and $\Delta t = 1.0$ s.

Multidimensional Problems

57. Consider the unsteady two-dimensional convection equation:

$$\bar{f}_t + u\bar{f}_x + v\bar{f}_y = 0 \qquad (G)$$

(a) Derive the Lax-Wendroff one-step approximation of Eq. (G), including the leading truncation error terms in Δt, Δx, and Δy. (b) Derive the corresponding MDE. Analyse consistency and order. (c) Perform a von Neumann stability analysis of the FDE.

58. Solve Problem 57 by the BTCS method. (a) Derive the backward-time centered-space (BTCS) approximation of Eq. (G), including the leading truncation error terms in Δt, Δx, and Δy. (b) Derive the corresponding MDE. Analyze consistency and order. (c) Peforma a von Neumann stability analysis of the FDE.

Section 11.9   The Wave Equation

Introduction

59. Consider the set of two coupled unsteady one-dimensional convection equations:

$$\bar{f}_t + a\bar{g}_x = 0 \qquad \text{and} \qquad \bar{g}_t + a\bar{f}_x = 0 \qquad (H)$$

Classify this set of PDEs. Determine the characteristic curves. Discuss the significance of these results as regards domain of dependence, range of influence, physical information propagation speed, and numerical solution procedures.

60. Develop the exact solution for the acoustics problem presented in Section 11.9.1 and discuss its significance.

Characteristic Concepts

61. Develop the method of characteristics analysis of the two coupled unsteady one-dimensional convection equations presented in Section 11.9.2. Discuss the effects of nonlinearities on the results.

The Lax-Wendroff One-Step Method

62. Derive the Lax-Wendroff one-step approximation of the coupled convection equations, Eq. (H), including the leading truncation error terms in Δt and Δx.

63. Derive the MDE corresponding to the finite difference approximation of Eq. (H). Analyze consistency and order.

64. Perform a von Neumann stability analysis of the finite difference approximation of Eq. (H).

65. By hand calculation, determine the solution of the example acoustics problem at $t = 0.1$ ms by the Lax-Wendroff one-step method with $\Delta x = 0.1$ m and $\Delta t = 0.05$ ms. Compare the results with the exact solution.

66. Modify the program presented in Section 11.10.2 to solve the example acoustics problem by the Lax-Wendroff one-step method with $\Delta x = 0.1$ m and $\Delta t = 0.05$ ms for $t = 1.0$ ms. Compare the results with the results of Problem 65.

67. Use the program to reproduce the results in Figure 11.29 where $\Delta x = 0.05$ m. Compare the errors with the errors in Problem 66.

Flux-Vector-Splitting Methods

68. Develop the flux-vector-splitting approximation of Eq. (C), $Q_t + E_x = 0$.

69. Substitute the first-order upwind finite difference approximation, Eq. (11.51), into Eqs. (11.121) and (11.122) to derive the first-order flux-vector-split FDEs. Derive the corresponding MDEs. Investigate consistency and order. Perform a von Neumann stability analysis of the FDEs.

70. By hand calculation, determine the solution of the example acoustics problem by the first-order flux-vector-splitting method with $\Delta x = 0.1$ m and $\Delta t = 0.05$ ms for $t = 0.1$ ms. Compare the results with the exact solution.

71. Modify the program presented in Section 11.10.2 to solve the example acoustics problem by the first-order flux-vector-splitting method with $\Delta x = 0.1$ m and $\Delta t = 0.05$ ms for $t = 1.0$ ms. Compare the results with the results of Problem 70.

72. Use the program to solve the example acoustics problem with $\Delta x = 0.05$ m and $\Delta t = 0.01, 0.025, 0.045$, and 0.05 ms for $t = 1.0$ ms. Compare the errors with the errors in Problem 71.

73. Substitute the second-order finite difference approximation, Eq. (11.55), into Eqs. (11.121) and (11.122) to derive the second-order flux-vector-split FDEs. Derive the corresponding MDEs. Investigate consistency and order. Perform a von Neumann stability analysis of the FDEs.

74. By hand calculation, determine the solution of the example acoustics problem by the second-order flux-vector-splitting method with $\Delta x = 0.1$ m and $\Delta t = 0.05$ ms for $t = 0.1$ ms. Compare the results with the exact solution and the results of Problem 65.

75. Modify the program presented in Section 11.10.2 to solve the example acoustics problem by the second-order flux-vector-splitting method with $\Delta x = 0.1$ m and $\Delta t = 0.05$ ms for $t = 1.0$ ms. Compare the results with the results of Problem 74.

76. Use the program to solve the example acoustics problem with $\Delta x = 0.05$ m and $\Delta t = 0.025$ ms for $t = 1.0$ ms. Compare the errors with the errors in Problem 75.

Section 11.10   Programs

77. Implement the Lax method program presented in Section 11.10.1. Check out the program using the given data set.

78. Solve any of Problems 13 to 15 with the program.
79. Implement the Lax-Wendroff method program presented in Section 11.10.2. Check out the program using the given data set.
80. Solve any of Problems 19 to 21 with the program.
81. Implement the MacCormack method program presented in Section 11.10.3. Check out the program using the given data set.
82. Solve any of Problems 27 to 29 with the program.
83. Implement the upwind method program presented in Section 11.10.4. Check out the program using the given data set.
84. Solve any of the Problems 33 to 35 and 38 to 40 with the program.
85. Implement the BTCS method program presented in Section 11.10.5. Check out the program using the given data set.
86. Solve any of Problems 44 to 46 with the program.

12

The Finite Element Method

12.1. Introduction
12.2. The Rayleigh-Ritz, Collocation, and Galerkin Methods
12.3. The Finite Element Method for Boundary-Value Problems
12.4. The Finite Element Method for the Laplace (Poisson) Equation
12.5. The Finite Element Method for the Diffusion Equation
12.6. Programs
12.7. Summary
Problems

Examples

12.1. The Rayleigh-Ritz method
12.2. The collocation method
12.3. The FEM on a one-dimensional uniform grid
12.4. The FEM on a one-dimensional nonuniform grid
12.5. The FEM with a derivative boundary condition
12.6. The FEM for the Laplace equation
12.7. The FEM for the Poisson equation
12.8. The FEM for the diffusion equation

12.1. INTRODUCTION

All the methods for solving differential equations presented in Chapters 7 to 11 are based on the *finite difference approach*. In that approach, all of the derivatives in a differential equation are replaced by algebraic finite difference approximations, which changes the differential equation into an algebraic equation that can be solved by simple arithmetic. Another approach for solving differential equations is based on approximating the exact solution by an approximate solution, which is a linear combination of specific trial functions, which are typically polynominals. The trial functions are linearly independent functions that satisfy the boundary conditions. The unknown coefficients in the trial functions are then determined in some manner.

To illustrate this approach, consider the one-dimensional boundary-value problem:

$$\bar{y}'' + Q\bar{y} = F \qquad \text{with appropriate boundary conditions} \qquad (12.1)$$

where $Q = Q(x)$ and $F = F(x)$. Let's approximate the exact solution $\bar{y}(x)$ by an approximate solution $y(x)$, which is a linear combination of specific trial functions $y_i(x)(i = 1, 2, \ldots, I)$:

$$\bar{y}(x) \approx y(x) = \sum_{i=1}^{I} C_i y_i(x) \tag{12.2}$$

This approach can be applied to the global solution domain $D(x)$. The Rayleigh-Ritz method, the collocation method, and the Galerkin weighted residual method for determining the coefficients, C_i $(i = 1, 2, \ldots, I)$ for the global solution domain are presented in Section 12.2. The heat transfer problem illustrated in Figure 12.1a is solved by these methods in Section 12.2.

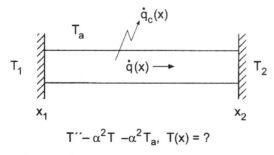

$$T'' - \alpha^2 T - \alpha^2 T_a, \quad T(x) = ?$$

(a) One-dimensional boundary-value problem.

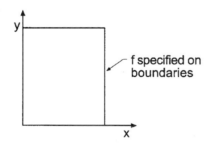

$$f_{xx} + f_{yy} = F(x,y), \quad f(x,y) = ?$$

(b) The Laplace (Poisson) equation.

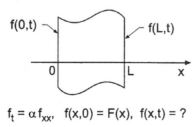

$$f_t = \alpha f_{xx}, \quad f(x,0) = F(x), \quad f(x,t) = ?$$

(c) The diffusion equation.

Figure 12.1.   Finite element problems.

Another approach is based on applying Eq. (12.2) to a subdomain of the global solution domain $D_i(x)$, which is called an *element* of the global solution domain. The solutions for the individual elements are assembled to obtain the global solution. This approach is called the *finite element method*. The finite element method for solving Eq. (12.1) is presented in Section 12.3. The heat transfer problem illustrated in Figure 12.1a is solved by the finite element method in Section 12.3.

The finite element method also can be applied to solve partial differential equations. Consider the two-dimensional Poisson equation:

$$f_{xx} + f_{yy} = F(x, y) \qquad \text{with appropriate boundary conditions} \tag{12.3}$$

The finite element method for solving Eq. (12.3) is presented in Section 12.4. Figure 12.1b illustrates the heat transfer problem presented in Chapter 9 to illustrate finite difference methods for solving elliptic partial differential equations. That problem is solved by the finite element method in Section 12.4.

Consider the one-dimensional diffusion equation:

$$f_t = \alpha f_{xx} \qquad \text{with appropriate initial and boundary conditions} \tag{12.4}$$

Figure 12.1c illustrates the heat transfer problem presented in Chapter 10 to illustrate finite difference methods for solving parabolic partial differential equations. That problem is solved in Section 12.5 to illustrate the application of the finite element method for solving unsteady time marching partial differential equations.

The treatment of the finite element method presented in this chapter is rather superficial. A detailed treatment of the method requires a complete book devoted entirely to the finite element method. The books by Rao (1982), Reddy (1993), Strang and Fix (1973), and Zienkiewicz and Taylor (1989 and 1991) are good examples of such books. The objective of this chapter is simply to introduce this important approach for solving differential equations.

The organization of Chapter 12 is illustrated in Figure 12.2. After the general introduction presented in this section, the Rayleigh-Ritz, collocation, and Galerkin methods are presented for one-dimensional boundary-value problems. That presentation is followed by a discussion of the finite element method applied to one-dimensional boundary-value problems. Brief introductions to the application of the finite element method to the Laplace (Poisson) equation and the diffusion equation follow. The chapter closes with a Summary, which discusses the advantages and disadvantages of the finite element method, and lists the things you should be able to do after studying Chapter 12.

12.2. THE RAYLEIGH-RITZ, COLLOCATION, AND GALERKIN METHODS

Consider the one-dimensional boundary-value problem specified by Eq. (12.1):

$$\bar{y}'' + Q\bar{y} = F \qquad \text{with appropriate boundary conditions} \tag{12.5}$$

where $Q = Q(x)$ and $F = F(x)$. The Rayleigh-Ritz, collocation, and Galerkin methods for solving Eq. (12.5) are presented in this section.

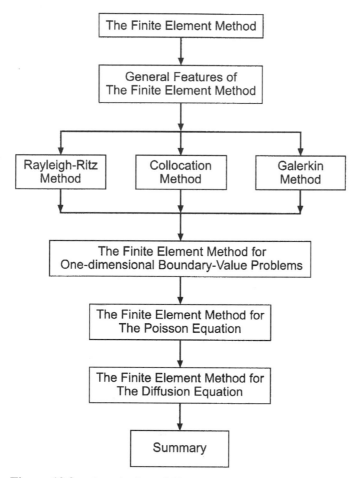

Figure 12.2. Organization of Chapter 12.

12.2.1. The Rayleigh-Ritz Method

The *Rayleigh-Ritz method* is based on the branch of mathematics known as the *calculus of variations*. The objective of the calculus of variations is to extremize (i.e., minimize or maximize) a special type of function, called a *functional*, which depends on other unknown functions. The simplest problem of the calculus of variations in one independent variable (i.e., x) is concerned with the extremization of the following integral:

$$I[\bar{y}(x)] = \int_a^b G(x, \bar{y}, \bar{y}')dx \qquad\qquad (12.6)$$

where $G(x, \bar{y}, \bar{y}')$, which is called the *fundamental function*, is a function of the independent variable x, the unknown function $\bar{y}(x)$, and its first derivative $\bar{y}'(x)$. The end points a and b are fixed. The square bracket notation, $I[\bar{y}(x)]$, is used to emphasize that I is not a function of x; it is a function of the function $\bar{y}(x)$. In fact, $I[\bar{y}(x)]$ does not depend on x at all since x is not present in the result when the definite integral is evaluated for a specific function $\bar{y}(x)$. The objective of the calculus of variations is to determine the particular function $\bar{y}(x)$ which extremizes (i.e., minimizes or maximizes) the functional $I[\bar{y}(x)]$.

Functionals are extremized (i.e., minimized or maximized) in a manner analogous to extremizing ordinary functions, which is accomplished by setting the first derivative of the ordinary function equal to zero. The derivative of a functional is called a *variation* and is denoted by the symbol δ to distinguish it from the derivative of an ordinary function, which is denoted by the symbol d. The first variation of Eq. (12.6) (for fixed end points a and b) is given by

$$\delta I = \int_a^b \left(\frac{\partial G}{\partial y} \delta y + \frac{\partial G}{\partial y'} \delta y' \right) dx = \int_a^b \left(\frac{\delta G}{\delta y} \delta y + \frac{\partial G}{\partial y'} \frac{d(\delta y)}{dx} \right) dx \qquad (12.7)$$

where $\delta y' = \delta(dy/dx) = d(\delta y)/dx$ from continuity requirements. Integrating the last term in Eq. (12.7) by parts yields

$$\int_a^b \frac{\partial G}{\partial y'} \frac{d(\delta y)}{dx} dx = -\int_a^b \frac{d}{dx}\left(\frac{\partial G}{\partial y'} \right) \delta y \, dx + \frac{\partial G}{\partial y'} \delta y \Big|_a^b \qquad (12.8)$$

where the last term in Eq. (12.8) is zero since $\delta y = 0$ at the boundaries for fixed end points. Substituting Eq. (12.8) into Eq. (12.7) and setting $\delta I = 0$ gives

$$\delta I = \int_a^b \left[\frac{\partial G}{\partial y} - \frac{d}{dx}\left(\frac{\partial G}{\partial y'} \right) \right] \delta y \, dx = 0 \qquad (12.9)$$

Equation (12.9) must be satisfied for arbitrary distributions of δy, which requires that

$$\boxed{\frac{\partial G}{\partial y} - \frac{d}{dx}\left(\frac{\partial G}{\partial y'} \right) = 0} \qquad (12.10)$$

Equation (12.10) is known as the *Euler equation* of the calculus of variations.

How does the calculus of variations relate to the solution of a boundary-value ordinary differential equation? To answer this question, consider the following simple linear boundary-value problem with Dirichlet boundary conditions:

$$\boxed{\bar{y}'' + Q\bar{y} = F \qquad \bar{y}(x_1) = \bar{y}_1 \text{ and } \bar{y}(x_2) = \bar{y}_2} \qquad (12.11)$$

where $Q = Q(x)$ and $F = F(x)$. The problem is to determine a functional $I[\bar{y}(x)]$ whose extremum (i.e., minimum or maximum) is precisely Eq. (12.11). If such a functional can be found, extremizing that functional yields the solution to Eq. (12.11).

The particular functional whose extremization yields Eq. (12.11) is given by

$$I[\bar{y}(x)] = \int_a^b [(\bar{y}')^2 - Q\bar{y}^2 + 2F\bar{y}] \, dx \qquad (12.12)$$

where the fundamental function $G(x, \bar{y}, \bar{y}')$ is defined as

$$G(x, \bar{y}, \bar{y}') = (\bar{y}')^2 - Q\bar{y}^2 + 2F\bar{y} \qquad (12.13)$$

Applying the Euler equation, Eq. (12.10), to the fundamental function given by Eq. (12.13) gives

$$\frac{\partial}{\partial \bar{y}}[(\bar{y}')^2 - Q\bar{y}^2 + 2F\bar{y}] = \frac{d}{dx}\left\{ \frac{\partial}{\partial \bar{y}'}[(\bar{y}')^2 - Q\bar{y}^2 + 2F\bar{y}] \right\} \qquad (12.14)$$

Performing the differentiations gives

$$-2Q\bar{y} + 2F = \frac{d}{dx}(2\bar{y}') = 2\frac{d^2\bar{y}}{dx^2} \tag{12.15}$$

which yields the result

$$\bar{y}'' + Q\bar{y} = F \tag{12.16}$$

which is identically Eq. (12.11). Thus, the function $\bar{y}(x)$ which extremizes the functional $I[\bar{y}(x)]$ given by Eq. (12.12) also satisfies the boundary-value ODE, Eq. (12.11).

The Rayleigh-Ritz method is based on approximating the exact solution $\bar{y}(x)$ of the variational problem by an approximate solution $y(x)$, which depends on a number of unspecified parameters $a, b, \ldots$. That is, $\bar{y}(x) \approx y(x) = y(x, a, b, \ldots)$. Thus, Eq. (12.12) becomes

$$I[\bar{y}(x)] \approx I[y(x)] = I[y(x, a, b, \ldots)] \tag{12.17}$$

Taking the first variation of Eq. (12.17) with respect to the parameters a, b, etc., yields

$$\delta I[y(x, a, b, \ldots)] = \frac{\partial I}{\partial a}\delta a + \frac{\partial I}{\partial b}\delta b + \cdots \tag{12.18}$$

which is satisfied only if

$$\frac{\partial I}{\partial a} = \frac{\partial I}{\partial b} = \cdots = 0 \tag{12.19}$$

Equation (12.19) yields exactly the number of equations required to solve for the parameters $a, b, \ldots$, which determines the function $y(x)$ that extremizes the functional $I[y(x)]$. The function $y(x)$ is also the solution of the differential equation, Eq. (12.11).

In summary, the steps in the Rayleigh-Ritz method are as follows:

1. Determine the functional $I[\bar{y}(x)]$ that yields the boundary-value ODE when the Euler equation s applied.
2. Assume that the functional form of the approximate solution $y(x)$ is given by

$$\bar{y}(x) \approx y(x) = \sum_{i=1}^{I} C_i y_i(x) \tag{12.20}$$

Choose the functional forms of the trial functions $y_i(x)$, and ensure that they are linearly independent and satisfy the boundary conditions.
3. Substitute the approximate solution, Eq. (12.20), into the functional $I[\bar{y}(x)]$ to obtain $I[C_i]$.
4. Form the partial derivatives of $I[C_i]$ with respect to C_i, and set them equal to zero:

$$\frac{\partial I}{\partial C_i} = 0 \qquad (i = 1, 2, \ldots, I) \tag{12.21}$$

5. Solve Eq. (12.21) for the coefficients C_i $(i = 1, 2, \ldots, I)$.

Let's illustrate the Rayleigh-Ritz method by applying it to solve the boundary-value problem specified by Eq. (12.11):

$$\boxed{\bar{y}'' + Q\bar{y} = F \qquad \bar{y}(x_1) = \bar{y}_1 \text{ and } \bar{y}(x_2) = \bar{y}_2} \tag{12.22}$$

As a specific example, let the boundary conditions be $\bar{y}(0.0) = 0.0$ and $\bar{y}(1.0) = Y$. Thus, Eq. (12.22) becomes

$$\bar{y}'' + Q\bar{y} = F \qquad \bar{y}(0.0) = 0.0 \text{ and } \bar{y}(1.0) = Y \tag{12.23}$$

Step 1. The functional $I[\bar{y}(x)]$ corresponding to Eq. (12.23) is given by Eq. (12.12):

$$I[\bar{y}(x)] = \int_0^1 [(\bar{y}')^2 - Q\bar{y}^2 + 2F\bar{y}]\,dx \tag{12.24}$$

where the fundamental function $G(x, \bar{y}, \bar{y}'^{-1})$ is defined as

$$G(x, \bar{y}, \bar{y}') = (\bar{y}')^2 - Q\bar{y}^2 + 2F\bar{y} \tag{12.25}$$

As shown by Eqs. (12.14) to (12.16), the function $\bar{y}(x)$ which extremizes the functional $I[\bar{y}(x)]$ given by Eq. (12.24) also satisfies the boundary-value ordinary differential equation, Eq. (12.23).

Step 2. Assume that the functional form of the approximate solution $y(x)$ is given by

$$y(x) = C_1 y_1(x) + C_2 y_2(x) + C_3 y_3(x) = C_1 x + C_2 x(x-1) + C_3 x^2(x-1) \tag{12.26}$$

The three trial functions in Eq. (12.26) are linearly independent. Applying the boundary conditions yields $C_1 = Y$. Thus, the approximate solution is given by

$$y(x) = Yx + C_2 x(x-1) + C_3 x^2(x-1) = y(x, C_2, C_3) \tag{12.27}$$

Step 3. Substituting the approximate solution, Eq. (12.27), into Eq. (12.24) gives

$$I[y(x)] = \int_0^1 [(y')^2 - Qy^2 + 2Fy]\,dx = I[C_2, C_3] \tag{12.28}$$

Step 4. Form the partial derivatives of Eq. (12.28) with respect to C_2 and C_3:

$$\frac{\partial I}{\partial C_2} = \int_0^1 \frac{\partial}{\partial C_2}[(y')^2]\,dx - \int_0^1 \frac{\partial}{\partial C_2}[Qy^2]\,dx + \int_0^1 \frac{\partial}{\partial C_2}[2Fy]\,dx = 0 \tag{12.29a}$$

$$\frac{\partial I}{\partial C_3} = \int_0^1 \frac{\partial}{\partial C_3}[(y')^2]\,dx - \int_0^1 \frac{\partial}{\partial C_3}[Qy^2]\,dx + \int_0^1 \frac{\partial}{\partial C_3}[2Fy]\,dx = 0 \tag{12.29b}$$

Evaluating Eq. (12.29) yields

$$\frac{\partial I}{\partial C_2} = \int_0^1 2y'\frac{\partial y'}{\partial C_2}\,dx - \int_0^1 2Qy\frac{\partial y}{\partial C_2}\,dx + \int_0^1 2F\frac{\partial y}{\partial C_2}\,dx = 0 \tag{12.30a}$$

$$\frac{\partial I}{\partial C_3} = \int_0^1 2y'\frac{\partial y'}{\partial C_3}\,dx - \int_0^1 2Qy\frac{\partial y}{\partial C_3}\,dx + \int_0^1 2F\frac{\partial y}{\partial C_3}\,dx = 0 \tag{12.30b}$$

Step 5. Solve Eq. (12.30) for C_2 and C_3. Equation (12.30) requires the functions $y(x)$, $\partial y/\partial C_2$, $\partial y/\partial C_3$, $y'(x)$, $\partial y'/\partial C_2$, and $\partial y'/\partial C_3$. Recall Eq. (12.27):

$$y(x) = Yx + C_2 x(x-1) + C_3 x^2(x-1) \tag{12.31}$$

Differentiating Eq. (12.31) with respect to x, C_2, and C_3 gives

$$y'(x) = Y + C_2(2x-1) + C_3(3x^2 - 2x) \tag{12.32}$$

$$\frac{\partial y}{\partial C_2} = (x^2 - x) \qquad \text{and} \qquad \frac{\partial y}{\partial C_3} = (x^3 - x^2) \tag{12.33}$$

Differentiating Eq. (12.32) with respect to C_2 and C_3 gives

$$\frac{\partial y'}{\partial C_2} = (2x - 1) \qquad \text{and} \qquad \frac{\partial y'}{\partial C_3} = (3x^2 - 2x) \tag{12.34}$$

Substituting Eqs. (12.31) to (12.34) into Eq. (12.30) and dividing through by 2 gives

$$\int_0^1 [Y + C_2(2x - 1) + C_3(3x^2 - 2x)](2x - 1)\,dx$$

$$-\int_0^1 Q[Yx + C_2(x^2 - x) + C_3(x^3 - x^2)](x^2 - x)\,dx + \int_0^1 F(x^2 - x)\,dx = 0 \tag{12.35a}$$

$$\int_0^1 [Y + C_2(2x - 1) + C_3(3x^2 - 2x)](3x^2 - 2x)\,dx$$

$$-\int_0^1 Q[Yx + C_2(x^2 - x) + C_3(x^3 - x^2)](x^3 - x^2)\,dx + \int_0^1 F(x^3 - x^2)\,dx = 0 \tag{12.35b}$$

The functions $Q = Q(x)$ and $F = F(x)$ must be substituted into Eq. (12.35) before integration.

At this point, all that remains is a considerable amount of simple algebra, integration, evaluation of the integrals at the limits of integration, and simplification of the results. Integrate Eq. (12.35) for $Q =$ constant and $F =$ constant and evaluate the results. The final result is:

$$C_2\left(\frac{1}{3} - \frac{Q}{30}\right) + C_3\left(\frac{1}{6} - \frac{Q}{60}\right) = -\frac{QY}{12} + \frac{F}{6} \tag{12.36a}$$

$$C_2\left(\frac{1}{6} - \frac{Q}{60}\right) + C_3\left(\frac{2}{15} - \frac{Q}{105}\right) = -\frac{QY}{20} + \frac{F}{12} \tag{12.36b}$$

Solving Eq. (12.36) for C_2 and C_3 and substituting the results into Eq. (12.27) yields the approximate solution $y(x)$.

Example 12.1.   The Rayleigh-Ritz method.

Let's apply the Rayleigh-Ritz method to solve the heat transfer problem presented in Section 8.1. The boundary-value ODE is [see Eq. (8.1)]

$$T'' - \alpha^2 T = -\alpha^2 T_a \qquad T(0.0) = 0.0 \text{ and } T(1.0) = 100.0 \tag{12.37}$$

Let $Q = -\alpha^2 = -16.0$ cm^{-2}, $T_a = 0.0$ (which gives $F = 0.0$), and $Y = 100.0$. For these values, Eq. (12.36) becomes

$$C_2\left(\frac{1}{3} + \frac{16}{30}\right) + C_3\left(\frac{1}{6} + \frac{16}{60}\right) = \frac{(16)(100)}{12} \tag{12.38a}$$

$$C_2\left(\frac{1}{6} + \frac{16}{60}\right) + C_3\left(\frac{2}{15} + \frac{16}{105}\right) = \frac{(16)(100)}{20} \tag{12.38b}$$

Solving Eq. (12.38) gives $C_2 = 57.294430$ and $C_3 = 193.103448$. Substituting these results into Eq. (12.27) gives the approximate solution $T(x)$:

$$T(x) = 100x + 57.294430(x^2 - x) + 193.103448(x^3 - x^2) \tag{12.39}$$

Table 12.1 Solution by the Rayleigh-Ritz Method

x, cm	$T(x)$, C	$\bar{T}(x)$, C	Error(x), C
0.00	0.000000	0.000000	
0.25	5.205570	4.306357	0.899214
0.50	11.538462	13.290111	−1.751650
0.75	37.102122	36.709070	0.393052
1.00	100.000000	100.000000	

Simplifying Eq. (12.39) yields the final solution:

$$T(x) = 42.705570x - 135.809109x^2 + 193.103448x^3 \qquad (12.40)$$

Table 12.1 presents values from Eq. (12.40) at five equally spaced points (i.e., $\Delta x = 0.25$ cm). The solution is most accurate near the boundaries and least accurate in the middle of the physical space. The Euclidean norm of the errors in Table 12.1 is 2.007822 C, which is comparable to the Euclidean norm of 1.766412 C for the errors obtained by the second-order equilibrium method presented in Table 8.8.

12.2.2. The Collocation Method

The *collocation method* is a member of a family of methods known as *residual methods*. In residual methods, an approximate form of the solution $y(x)$ is assumed, and the residual $R(x)$ is defined by substituting the approximate solution into the exact differential equation. The approximate solution is generally chosen as the sum of a number of linearly independent trial functions, as done in the Rayleigh-Ritz method. The coefficients are then chosen to minimize the residual in some sense. In the collocation method, the residual itself is set equal to zero at selected locations. The number of locations is the same as the number of unknown coefficients in the approximate solution $y(x)$.

In summary, the steps in the collocation method are as follows:

1. Determine the differential equation which is to be solved, for example, Eq. (12.5).
2. Assume that the functional form of the approximate solution $y(x)$ is given by

$$\bar{y}(x) \approx y(x) = \sum_{i=1}^{I} C_i y_i(x) \qquad (12.41)$$

Choose the functional form of the trial functions $y_i(x)$ and ensure that they are linearly independent and satisfy the boundary conditions.
3. Substitute the approximate solution $y(x)$ into the differential equation and define the residual $R(x)$:

$$R(x) = y'' + Qy - F = R(x, C_1, C_2, \ldots, C_I). \qquad (12.42)$$

4. Set $R(x, C_1, C_2, \ldots, C_I) = 0$ at I values of x.
5. Solve the system of residual equations for the coefficients C_i $(i = 1, 2, \ldots, I)$.

Let's illustrate the collocation method by applying it to solve the boundary-value problem specified by Eq. (12.11). Consider the specific example given by Eq. (12.23):

$$\boxed{\bar{y}'' + Q\bar{y} = F \qquad \bar{y}(0.0) = 0.0 \text{ and } \bar{y}(1.0) = Y} \tag{12.43}$$

Step 1. The differential equation to be solved is given by Eq. (12.43).
Step 2. Assume that the functional form of the approximate solution $y(x)$ is given by Eq. (12.27):

$$y(x) = Yx + C_2 x(x-1) + C_3 x^2(x-1) \tag{12.44}$$

Step 3. Define the residual $R(x)$:

$$R(x) = y'' + Qy - F \tag{12.45}$$

From Eq. (12.44):

$$y''(x) = 2C_2 + C_3(6x - 2) \tag{12.46}$$

Substituting Eqs. (12.44) and (12.46) into Eq. (12.45) gives

$$R(x) = 2C_2 + C_3(6x - 2) + Q[Yx + C_2(x^2 - x) + C_3(x^3 - x^2)] - F \tag{12.47}$$

Step 4. Since there are two unknown coefficients in Eq. (12.47), the residual can be set equal to zero at two arbitrary locations. Choose $x = 1/3$ and $2/3$. Thus,

$$R(1/3) = 2C_2 + C_3\left(\frac{6}{3} - 2\right) + Q\left[\frac{Y}{3} + C_2\left(\frac{1}{9} - \frac{1}{3}\right) + C_3\left(\frac{1}{27} - \frac{1}{9}\right)\right] - F = 0 \tag{12.48a}$$

$$R(2/3) = 2C_2 + C_3\left(\frac{12}{3} - 2\right) + Q\left[\frac{2Y}{3} + C_2\left(\frac{4}{9} - \frac{2}{3}\right) + C_3\left(\frac{8}{27} - \frac{4}{9}\right)\right] - F = 0 \tag{12.48b}$$

Step 5. Solve Eq. (12.48) for C_2 and C_3. The final result is:

$$\left(2 - \frac{2Q}{9}\right)C_2 - \left(\frac{2Q}{27}\right)C_3 = -\frac{QY}{3} + F \tag{12.49a}$$

$$\left(2 - \frac{2Q}{9}\right)C_2 + \left(2 - \frac{4Q}{27}\right)C_3 = -\frac{2QY}{3} + F \tag{12.49b}$$

Solving Eq. (12.49) for C_2 and C_3 and substituting the results into Eq. (12.44) yields the approximate solution $y(x)$.

Example 12.2. The collocation method.

To illustrate the collocation method, let's solve the heat transfer problem presented in Section 8.1 [see Eq. (8.1)]:

$$T'' - \alpha^2 T = -\alpha^2 T_a \qquad T(0.0) = 0.0 \text{ and } T(1.0) = 100.00 \tag{12.50}$$

Table 12.2 Solution by the Collocation Method

x, cm	$T(x)$, C	$\bar{T}(x)$, C	Error(x), C
0.00	0.000000	0.000000	
0.25	5.848837	4.306357	1.542480
0.50	14.000000	13.290111	0.709888
0.75	40.151163	36.709070	3.442092
1.00	100.000000	100.000000	

Let $Q = -\alpha^2 = -16.0\,\text{cm}^{-2}$, $T_a = 0.0\,\text{C}$ (which gives $F = 0.0$), and $Y = 100.0$. Equation (12.49) becomes

$$\left(2 + \frac{32}{9}\right)C_2 + \frac{32}{27}C_3 = \frac{1600}{3} \tag{12.51a}$$

$$\left(2 + \frac{32}{9}\right)C_2 + \left(2 + \frac{64}{27}\right)C_3 = \frac{3200}{3} \tag{12.51b}$$

Solving Eq. (12.51) gives $C_2 = 60.279070$ and $C_3 = 167.441861$. Substituting these results into Eq. (12.44) gives the approximate solution $T(x)$:

$$T(x) = 100x + 60.279070(x^2 - x) + 167.441861(x^3 - x^2) \tag{12.52}$$

Simplifying Eq. (12.52) yields the final solution:

$$\boxed{T(x) = 39.720930x - 107.162791x^2 + 167.441861x^3} \tag{12.53}$$

Table 12.2 presents values from Eq. (12.53) at five equally spaced points (i.e., $\Delta x = 0.25$ cm). The Euclidean norm of the errors is 3.838122 C, which is 91 percent larger than the Euclidean norm of the errors for the Rayleigh-Ritz method presented in Example 12.1.

12.2.3. The Galerkin Weighted Residual Method

The *Galerkin weighted residual method*, like the collocation method, is a *residual method*. Unlike the collocation method, however, the Galerkin weighting residual method is based on the *integral* of the residual over the domain of interest. In fact, the residual $R(x)$ is weighted over the domain of interest by multiplying $R(x)$ by weighting functions $W_j(x)$ ($j = 1, 2, \ldots$), integrating the weighted residuals over the range of integration, and setting the integrals of the weighted residuals equal to zero to give equations for the evaluation of the coefficients C_i of the trial functions $y_i(x)$.

In principle, any functions can be used as the weighting functions $W_j(x)$. For example, letting $W_j(x)$ be the Dirac delta function yields the collocation method presented in Section 12.2.2. Galerkin showed that basing the weighting functions $W_j(x)$ on the trial functions $y_i(x)$ of the approximate solution $y(x)$ yields exceptionally good results. That choice is presented in the following analysis.

In summary, the steps in the Galerkin weighted residual method are as follows:

1. Determine the differential equation which is to be solved, for example Eq. (12.5).

2. Assume that the functional form of the approximate solution $y(x)$ is given by

$$\bar{y}(x) \approx y(x) = \sum_{i=1}^{I} C_i y_i(x) \tag{12.54}$$

Choose the functional form of the trial functions $y_i(x)$, and ensure that they are linearly independent and satisfy the boundary conditions.

3. Introduce the approximate solution $y(x)$ into the differential equation and define the residual $R(x)$:

$$R(x) = y'' + Qy - F \tag{12.55}$$

4. Choose the weighting functions $W_j(x)$ ($j = 1, 2, \ldots$).
5. Set the integrals of the weighted residuals $W_j(x)R(x)$ equal to zero:

$$\int_{x_1}^{x_2} W_j(x)R(x)\,dx = 0 \qquad (j = 1, 2, \ldots) \tag{12.56}$$

6. Integrate Eq. (12.56) and solve the system of weighted residual integrals for the coefficients C_i ($i = 1, 2, \ldots, I$).

To illustrate the Galerkin weighted residual method, let's apply it to solve the boundary-value problem specified by Eq. (12.11). Consider the specific example given by Eq. (12.23):

$$\bar{y}'' + Qy - F \qquad \bar{y}(0.0) = 0.0 \text{ and } \bar{y}(1.0) = Y \tag{12.57}$$

Step 1. The differential equation to be solved is given by Eq. (12.57).
Step 2. Assume the functional form of the approximate solution $y(x)$ given by Eq. (12.27):

$$y(x) = Yx + C_2 x(x-1) + C_3 x^2(x-1) \tag{12.58}$$

Step 3. Define the residual $R(x)$:

$$R(x) = y'' + Qy - F \tag{12.59}$$

From Eq. (12.58):

$$y'' = 2C_2 + C_3(6x - 2) \tag{12.60}$$

Substituting Eqs. (12.58) and (12.60) into Eq. (12.59) gives

$$R(x) = 2C_2 + C_3(6x - 2) + Q[Yx + C_2(x^2 - x) + C_3(x^3 - x^2)] - F \tag{12.61}$$

which is the same as the residual given by Eq. (12.47) for the collocation method.
Step 4. Choose two weighting functions $W_2(x)$ and $W_3(x)$. Let $W_2(x) = y_2(x)$ and $W_3(x) = y_3(x)$ from Eq. (12.26). Thus,

$$W_2(x) = x^2 - x \qquad \text{and} \qquad W_3(x) = x^3 - x^2 \tag{12.62}$$

Step 5. Set the integrals of the weighted residuals equal to zero. Thus,

$$\int_0^1 (x^2 - x)\{2C_2 + C_3(6x - 2) + Q[Yx + C_2(x^2 - x) + C_3(x^3 - x^2)] - F\}\, dx = 0$$

(12.63a)

$$\int_0^1 (x^3 - x^2)\{2C_2 + C_3(6x - 2) + Q[Yx + C_2(x^2 - x) + C_3(x^3 - x^2)] - F\}\, dx = 0$$

(12.63b)

The functions $Q = Q(x)$ and $F = F(x)$ must be substituted into Eq. (12.63) before integrating.

Step 6. Integrate Eq. (12.63), for $Q =$ constant and $F =$ constant, evaluate the results, and collect terms. The final result is

$$C_2\left(\frac{1}{3} - \frac{Q}{30}\right) + C_3\left(\frac{1}{6} - \frac{Q}{60}\right) = -\frac{QY}{12} + \frac{F}{6}$$

(12.64a)

$$C_3\left(\frac{1}{6} - \frac{Q}{60}\right) + C_3\left(\frac{2}{15} - \frac{Q}{105}\right) = -\frac{QY}{20} + \frac{F}{12}$$

(12.64b)

Equation (12.64) is identical to the result obtained by the Rayleigh-Ritz method, Eq. (12.36). This correspondence always occurs when the weighting functions $W_j(x)$ ($j = 1, 2, \ldots$), are chosen as the trial functions, $y_i(x)$.

12.2.4. Summary

The Rayleigh-Ritz method, the collocation method, and the Galerkin weighted residual method are based on approximating the global solution to a boundary-value problem by a linear combination of specific trial functions. The Rayleigh-Ritz method is based on the calculus of variations. It requires a functional whose extremum (i.e., minimum or maximum) is also a solution to the boundary-value ODE. The collocation method is a residual method in which an approximate solution is assumed, the residual of the differential equation is defined, and the residual is set equal to zero at selected points. The collocation method is generally not as accurate as the Rayleigh-Ritz method and the Galerkin weighted residual method, so it is seldom used. Its main utility lies in the introduction of the concept of a residual, which leads to the Galerkin weighted residual method in which the integral of a weighted residual over the domain of interest is set equal to zero. The most common choices for the weighting functions are the trial functions of the approximate solution $y(x)$. In that case, the Rayleigh-Ritz method and the Galerkin method yield identical results.

There are problems in which the Rayleigh-Ritz approach is preferred, and there are problems in which the Galerkin weighted residual approach is preferred. If the variational functional is known, then it is logical to apply the Rayleigh-Ritz approach directly to the functional rather than to develop the corresponding differential equation and then apply the Galerkin weighted residual approach. This situation arises often in solid mechanics problems where Hamilton's principle (a variational approach on an energy principle) can be employed. If the governing differential equation is known, then it is logical to apply the Galerkin weighted residual approach rather than look for the functional corresponding to the differential equation. This situation arises often in fluid mechanics and heat transfer problems.

12.3. THE FINITE ELEMENT METHOD FOR BOUNDARY-VALUE PROBLEMS

The Rayleigh-Ritz method presented in Section 12.2.1 and the Galerkin weighted residual method presented in Section 12.2.3 are based on approximating the exact solution of a boundary-value ordinary differential equation $\bar{y}(x)$ by an approximate solution $y(x)$, which is a combination of linearly independent trial functions $y_i(x)$ $(i = 1, 2, \ldots)$ that apply over the *global solution domain* $D(x)$. The trial functions are typically polynominals. To increase the accuracy of either of these two methods, the degree of the polynominal trial functions must be increased. This leads to rapidly increasing complexity. As discussed in Chapter 4, increased accuracy of polynominal approximations can be obtained more easily by applying low degree polynominals to subdomains of the global domain. That is the fundamental idea of the *finite element method*. The finite element method (FEM) discretizes the global solution domain $D(x)$ into a number of subdomains $D_i(x)$ $(i = 1, 2, \ldots)$, called *elements*, and applies either the Rayleigh-Ritz method or the Galerkin weighted residual method to the discretized global solution domain.

The finite element method is developed in this section by applying it to solve the following simple linear boundary-value problem with appropriate boundary conditions (BCs):

$$\bar{y}'' + Q\bar{y} = F \qquad \text{with appropriate boundary conditions} \tag{12.65}$$

where $Q = Q(x)$ and $F = F(x)$.

The concept underlying the extension of the basic Rayleigh-Ritz approach or the Galerkin weighted residual approach to the finite element approach is illustrated in Figure 12.3. Figure 12.3a illustrates the global solution domain $D(x)$. The functional $I[C_i]$ from the Rayleigh-Ritz approach, or the weighted residual integral $I(C_i)$ from the Galerkin weighted residual approach, applies over the entire global solution domain $D(x)$. Let the symbol I denote either $I[C_i]$ or $I(C_i)$. Figure 12.3b illustrates the discretized global solution domain $D(x)$ which is discretized into I nodes and $I - 1$ elements. Note that the symbol I is being used for the functional $I[C_i]$, the weighted residual integral $I(C_i)$, and the number of nodes. The subscript i denotes the grid points, or nodes, and the superscript (i) denotes the elements. Element (i) starts at node i and ends at node $i + 1$. The element lengths (i.e., grid increments) are $\Delta x_i = x_{i+1} - x_i$. Figure 12.3c illustrates the discretization of the global integral I into the sum of the discretized integrals $I^{(i)}(i = 1, 2, \ldots, I - 1)$. Each discretized integral $I^{(i)}$ in Figure 12.3c is evaluated exactly as the global integral I in Figure 12.3a. This process yields a set of equations relating the nodal values within each element, which are called the *nodal equations*.

The global integral $I = \sum I^{(i)}$ could be differentiated directly with respect to C_i in one step by differentiating all of the individual element integrals (i.e., $\partial I^{(i)}/\partial C_i$) and summing the results. This approach would immediately yield I equations for the I nodal values C_i. However, the algebra is simplified considerably by differentiating a single generic discretized integral $I^{(i)}$ with respect to every C_i present in $I^{(i)}$, to obtain a generic set of equations involving those values of C_i. These equations are called the *element equations*. This generic set of element equations is then applied to all of the discretized elements to obtain a complete set of I equations for the nodal values C_i. This complete set of element equations is called the *system equation*. The system equation is adjusted to

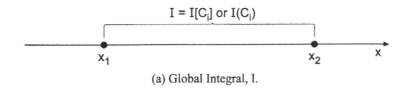

(a) Global Integral, I.

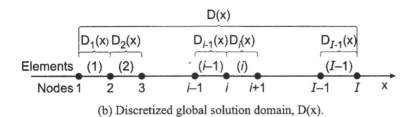

(b) Discretized global solution domain, D(x).

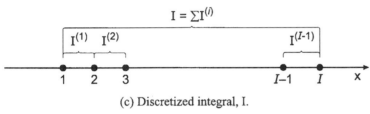

(c) Discretized integral, I.

Figure 12.3.   Finite element discretization.

account for the boundary conditions, and the *adjusted system equation* is solved for the nodal values C_i $(i = 1, 2, \ldots, I)$.

In summary, the steps in the finite element approach are as follows:

1. *Formulate* the problem. If the Rayleigh-Ritz approach is to be used, find the *functional I* to be extremized. If the Galerkin weighted residual approach is to be used, determine the *differential equation* to be solved.

2. *Discretize* the global solution domain $D(x)$ into subdomains (i.e., elements) $D_i(x)$ $(i = 1, 2, \ldots, I)$. Specify the type of element to be used (i.e., linear, quadratic, etc.).

3. Assume the functional form of the *approximate solution* $y^{(i)}(x)$ within each element, and choose the *interpolating functions* for the elements.

4. For the Rayleigh-Ritz approach, substitute the approximate solution $y(x)$ into the functional I to determine $I[C_i]$. For the Galerkin weighed residual approach, substitute the approximate solution $y(x)$ into the differential equation to determine the residual $R(x)$, weight the residual with the weighting functions $W_j(x)$, and form the weighted residual integral $I(C_i)$.

5. Determine the *element equations*. For the Rayleigh-Ritz approach, evaluate the partial derivatives of the functional $I[C_i]$ with respect to the nodal values C_i, and equate them to zero. For the Galerkin weighted residual approach, evaluate the partial derivatives of the weighted residual integral $I(C_i)$ with respect to the nodal values C_i, and equate them to zero.

6.   *Assemble* the element equations to determine the *system equation.*
7.   *Adjust* the system equation to account for the *boundary conditions.*
8.   Solve the *adjusted system equation* for the nodal values C_i.

Discretization of the global solution domain and specification of the interpolating polynominals are accomplished in the same manner for both the Rayleigh-Ritz approach and the Galerkin weighted residual approach. Consequently, those steps are considered in the next section before proceeding to the Rayleigh-Ritz approach and the Galerkin weighted residual approach developments in Sections 12.3.2 and 12.3.3, respectively.

12.3.1.   Domain Discretization and the Interpolating Polynominals

Let's discretize the global solution domain $D(x)$ into I nodes and $I - 1$ elements, as illustrated in Figure 12.4, where the subscript i denotes the grid points, or nodes, and the superscript (i) denotes the elements. Element (i) starts at node i and ends at node $i + 1$. The element lengths (i.e., grid increments) are $\Delta x_i = x_{i+1} - x_i$.

Let the global exact solution $\bar{y}(x)$ be approximated by the global approximate solution $y(x)$, which is the sum of a series of local interpolating polynominals $y^{(i)}(x)$ ($i = 1, 2, \ldots, I - 1$) that are valid within each element.

$$y(x) = y^{(1)}(x) + y^{(2)}(x) + \cdots + y^{(i)}(x) + \cdots + y^{(I-1)}(x) = \sum_{i=1}^{I-1} y^{(i)}(x) \qquad (12.66)$$

The local interpolating polynominals $y^{(i)}(x)$ are defined as follows:

$$y^{(i)}(x) = y_i N_i^{(i)}(x) + y_{i+1} N_{i+1}^{(i)}(x) \qquad (12.67)$$

where y_i and y_{i+1} are the values of $y(x)$ at nodes i and $i + 1$, respectively, and $N_i^{(i)}(x)$ and $N_{i+1}^{(i)}(x)$ are linear interpolating polynominals within element (i). The subscript i denotes the grid point where $N_i^{(i)}(x) = 1.0$, and the superscript (i) denotes the element within which $N_i^{(i)}(x)$ applies. The interpolating polynominals are generally called *shape functions* in the finite element literature. The shape functions are defined to be unity at their respective nodes, zero at the other nodes, and zero everywhere outside of their element. Thus, $y^{(i)}(x_i) = y_i$, that is, the to-be-determined coefficients y_i represent the solution at the nodes. Figure 12.5 illustrates the linear shape functions for element (i). From Figure 12.4,

$$N_i^{(i)}(x) = -\frac{x - x_{i+1}}{x_{i+1} - x_i} = -\frac{x - x_{i+1}}{\Delta x_i} \qquad (12.68)$$

$$N_{i+1}^{(i)}(x) = \frac{x - x_i}{x_{i+1} - x_i} = \frac{x - x_i}{\Delta x_i} \qquad (12.69)$$

Substituting Eqs. (12.68) and (12.69) into Eq. (12.67) gives

$$y^{(i)}(x) = y_i \left(-\frac{x - x_{i+1}}{\Delta x_1} \right) + y_{i+1} \left(\frac{x - x_i}{\Delta x_i} \right) \qquad (12.70)$$

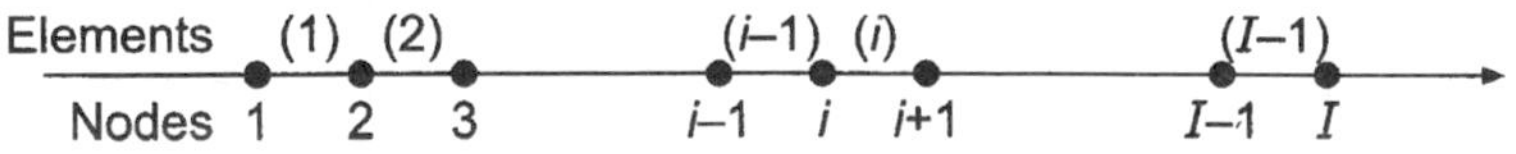

Figure 12.4.   Discretized global solution domain.

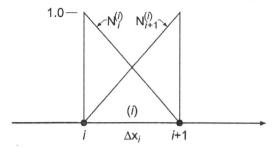

Figure 12.5. Linear shape functions for element (i).

Equation (12.70) is actually a linear Lagrange polynominal applied to element (i). Since there are $I - 1$ elements, there are $2(I - 1)$ shape functions in the global solution domain $D(x)$. The $2(I - 1)$ shape functions in Eqs. (12.68) and (12.69) form a linearly independent set.

In the interpolating polynominal presented in Eq. (12.70) is a linear polynominal. The corresponding element is called a *linear element*. Higher-order interpolating polynominals can be developed by placing additional nodes within each element. Thus, quadratic, cubic, etc., elements can be defined. All of the results presented in this chapter are based on linear elements.

12.3.2. The Rayleigh-Ritz Approach

As discussed in Section 12.2.1, the Rayleigh-Ritz approach is based on extremizing (i.e., minimizing or maximizing) the following functional [see Eq. (12.6)]:

$$I[\bar{y}(x)] = \int_a^b G(x, \bar{y}, \bar{y}')\, dx \qquad (12.71)$$

where the functional $I[\bar{y}(x)]$ yields the boundary-value ODE, Eq. (12.65), when the Euler equation, Eq. (12.10), is applied.

In terms of the global approximate solution $y(x)$ and the discretized global solution domain illustrated in Figure 12.4, Eq. (12.71) can be written as follows:

$$I[y(x)] = \int_{x_1}^{x_2} G\, dx + \int_{x_2}^{x_3} G\, dx + \cdots + \int_{x_{i-1}}^{x_i} G\, dx + \int_{x_i}^{x_{i+1}} G\, dx + \cdots + \int_{x_{I-1}}^{x_I} G\, dx \qquad (12.72)$$

$$I[y(x)] = I^{(1)}[y(x)] + I^{(2)}[y(x)] + \cdots + I^{(i-1)}[y(x)] + I^{(i)}[y(x)] + \cdots + I^{(I-1)}[y(x)] \qquad (12.73)$$

where $G(x, y, y')$ within each element (i) depends on the interpolating polynominal within that element $y^{(i)}(x)$. Consider element (i). Substituting Eq. (12.67) for $y^{(i)}(x)$ into the integral in Eq. (12.73) corresponding to element (i) gives

$$I^{(i)}[y(x)] = \int_{x_i}^{x_{i+1}} G(x, y, y')\, dx = I^{(i)}[y^{(i)}(x)] = I^{(i)}[y_i, y_{i+1}] \qquad (12.74)$$

Thus, Eq. (12.73) can be expressed in terms of the nodal values y_i ($i = 1, 2, \ldots, I$), as follows:

$$I[y(x)] = I^{(1)}[y_1, y_2] + I^{(2)}[y_2, y_3] + \cdots + I^{(i-1)}[y_{i-1}, y_i]$$
$$+ I^{(i)}[y_i, y_{i+1}] + \cdots + I^{(I-1)}[y_{I-1}, y_I] \tag{12.75}$$

Extremizing (i.e., minimizing or maximizing) Eq. (12.75) is accomplished by setting the first variation δI of $I[y(x)]$ equal to zero. This gives

$$\delta I[y(x)] = \frac{\partial I}{\partial y_1} \delta y_1 + \frac{\partial I}{\partial y_2} \delta y_2 + \cdots + \frac{\partial I}{\partial y_{i-1}} \delta y_{i-1} + \frac{\partial I}{\partial y_i} \delta y_i + \cdots + \frac{\partial I}{\partial y_I} \delta y_I = 0$$
$$\tag{12.76}$$

Since the individual variations δy_i ($i = 1, 2, \ldots, I$) are arbitrary, Eq. (12.76) is satisfied only if

$$\frac{\partial I}{\partial y_1} = \frac{\partial I}{\partial y_2} = \cdots = \frac{\partial I}{\partial y_i} = \cdots \frac{\partial I}{\partial y_I} = 0 \tag{12.77}$$

Equation (12.77) yields I equations for the determination of the I nodal values y_i ($i = 1, 2, \ldots, I$).

Consider the general nodal equation corresponding to $\partial I / \partial y_i$. From Eq. (12.75), y_i appears only in $I^{(i-1)}[y_{i-1}, y_i]$ and $I^{(i)}[y_i, y_{i+1}]$. Thus,

$$\frac{\partial I}{\partial y_i} = \frac{\partial I^{(i-1)}}{\partial y_i} + \frac{\partial I^{(i)}}{\partial y_i} = 0 \tag{12.78}$$

which yields

$$\frac{\partial I}{\partial y_i} = \frac{\partial}{\partial y_i} \int_{x_{i-1}}^{x_i} G\left[x, y^{(i-1)}(x), \frac{d[y^{(i-1)}(x)]}{dx}\right] dx + \frac{\partial}{\partial y_i} \int_{x_i}^{x_{i+1}} G\left[x, y^{(i)}(x), \frac{d[y^{(i)}(x)]}{dx}\right] dx = 0$$
$$\tag{12.79}$$

The result of evaluating Eq. (12.79) is the *nodal equation* corresponding to node i. A similar nodal equation is obtained at all the other nodes. For Dirichlet boundary conditions, $y_1 = \bar{y}_1$ and $y_I = \bar{y}_I$, so $\delta y_1 = \delta y_I = 0$, and nodal equations are not needed at the boundaries. For Neumann boundary conditions, $y_1' = \bar{y}_1'$ and $y_I' = \bar{y}_I'$, and $\bar{y}_1$ and $\bar{y}_I$ are not specified. Thus, δy_1 and δy_I are arbitrary. In that case, the Euler equation must be supplemented by equations arising from the variations of the boundary points, which yields nodal equations at the boundary points. That process is not developed in this analysis. The results presented in the remainder of this section are based on Dirichlet boundary conditions. Neumann boundary conditions are considered in Section 12.3.3, which is based on the Galerkin weighted residual approach. Gathering all the nodal equations into a matrix equation yields the *system equation*, which can be solved for the nodal values, $y_i(i = 2, 3, \ldots, I - 1)$.

Let's develop the finite element method using the Rayleigh-Ritz approach to solve Eq. (12.65). As shown in Section 12.2.1, the functional whose extremum (i.e., minimum or maximum) is equivalent to Eq. (12.65) is [see Eq. (12.12)]

$$I[\bar{y}(x)] = \int_a^b [(\bar{y}')^2 - Q\bar{y}^2 + 2F\bar{y}] \, dx \tag{12.80}$$

where the fundamental function $G(x, \bar{y}, \bar{y}')$ is defined as

$$G(x, \bar{y}, \bar{y}') = (\bar{y}')^2 - Q\bar{y}^2 + 2F\bar{y} \tag{12.81}$$

Let's evaluate the second integral in Eq. (12.79), $\partial I^{(i)}/\partial y_i$, to illustrate the procedure. The corresponding result for the first integral in Eq. (12.79), $\partial I^{(i-1)}/\partial y_i$, will be presented later. Substituting Eq. (12.81), with $\bar{y}(x)$ approximated by $y(x)$, into the second integral in Eq. (12.79) and differentiating with respect to y_i gives

$$\frac{\partial I^{(i)}}{\partial y_i} = \frac{\partial}{\partial y_i} \int_{x_i}^{x_{i+1}} G(x, y, y')\, dx = \frac{\partial}{\partial y_i} \int_{x_i}^{x_{i+1}} [(y')^2 - Qy^2 + 2Fy]\, dx \tag{12.82}$$

$$\frac{\partial I^{(i)}}{\partial y_i} = 2 \int_{x_1}^{x_{i+1}} y' \frac{\partial y'}{\partial y_i}\, dx - 2 \int_{x_i}^{x_{i+1}} Qy \frac{\partial y}{\partial y_i}\, dx + 2 \int_{x_i}^{x_{i+1}} F \frac{\partial y}{\partial y_i}\, dx \tag{12.83}$$

Equation (12.83) requires the functions $y^{(i)}(x)$, $\partial[y^{(i)}(x)]/\partial y_i$, $y'(x) = d[y^{(i)}(x)]/dx$, and $\partial[y'(x)]/\partial y_i$. Recall Eq. (12.70) for $y^{(i)}(x)$:

$$y^{(i)}(x) = y_i \left(-\frac{x - x_{i+1}}{\Delta x_i} \right) + y_{i+1} \left(\frac{x - x_i}{\Delta x_i} \right) \tag{12.84}$$

Differentiating Eq. (12.84) with respect to y_i yields

$$\frac{\partial[y^{(i)}(x)]}{\partial y_i} = -\frac{x - x_{i+1}}{\Delta x_i} \tag{12.85}$$

Differentiating Eq. (12.84) with respect to x gives

$$y'(x) = \frac{d[y^{(i)}(x)]}{dx} = -\frac{y_i}{\Delta x_i} + \frac{y_{i+1}}{\Delta x_i} = \frac{y_{i+1} - y_i}{\Delta x_i} \tag{12.86}$$

Differentiating Eq. (12.86) with respect to y_i gives

$$\frac{\partial[y'(x)]}{\partial y_i} = -\frac{1}{\Delta x_i} \tag{12.87}$$

Substituting Eqs. (12.84) to (12.87) into Eq. (12.83) yields

$$\frac{\partial I^{(i)}}{\partial y_i} = 2 \int_{x_i}^{x_{i+1}} \left(\frac{y_{i+1} - y_i}{\Delta x_i} \right) \left(-\frac{1}{\Delta x_i} \right) dx + 2 \int_{x_i}^{x_{i+1}} F \left(-\frac{x - x_{i+1}}{\Delta x_i} \right) dx$$
$$- 2 \int_{x_i}^{x_{i+1}} Q \left[y_i \left(-\frac{x - x_{i+1}}{\Delta x_i} \right) + y_{i+1} \left(\frac{x - x_i}{\Delta x_i} \right) \right] \left(-\frac{x - x_{i+1}}{\Delta x_i} \right) dx \tag{12.88}$$

where the order of the second and third terms in Eq. (12.83) has been interchanged. Simplifying the integrals in Eq. (12.88) gives

$$\frac{\partial I^{(i)}}{\partial y_i} = -2 \int_{x_i}^{x_{i+1}} \frac{(y_{i+1} - y_i)}{\Delta x_i^2}\, dx - 2 \int_{x_i}^{x_{i+1}} \frac{F(x - x_{i+1})}{\Delta x_i}\, dx$$
$$- 2 \int_{x_i}^{x_{i+1}} \frac{Qy_i}{\Delta x_i^2} (x^2 - 2x_{i+1}x + x_{i+1}^2)\, dx$$
$$+ 2 \int_{x_i}^{x_{i+1}} \frac{Qy_{i+1}}{\Delta x_i^2} (x^2 - x_i x - x_{i+1}x + x_i x_{i+1})\, dx \tag{12.89}$$

Now let's evaluate the integrals in Eq. (12.89). Let the values of Q and F in Eq. (12.89) be average values, so they can be taken out of the integrals. Thus,

$$\bar{Q}^{(i)} = \frac{(Q_i + Q_{i+1})}{2} \tag{12.90}$$

$$\bar{F}^{(i)} = \frac{(F_i + F_{i+1})}{2} \tag{12.91}$$

Integrating Eq. (12.89) yields

$$\frac{\partial I^{(i)}}{\partial y_i} = -\frac{2(y_{i+1} - y_i)}{\Delta x_i^2} x \Big|_{x_i}^{x_{i+1}} - \frac{2\bar{F}^{(i)}}{\Delta x_i} \left(\frac{x^2}{2} - x_{i+1}x \right) \Big|_{x_i}^{x_{i+1}}$$

$$- \frac{2\bar{Q}^{(i)}y_i}{\Delta x_i^2} \left(\frac{x^3}{3} - x_{i+1}x^2 + x_{i+1}^2 x \right) \Big|_{x_i}^{x_{i+1}}$$

$$+ \frac{2\bar{Q}^{(i)}y_{i+1}}{\Delta x_i^2} \left(\frac{x^3}{3} - x_i\frac{x^2}{2} - x_{i+1}\frac{x^2}{2} + x_i x_{i+1}x \right) \Big|_{x_i}^{x_{i+1}} \tag{12.92}$$

Evaluating Eq. (12.92) gives

$$\frac{\partial I^{(i)}}{\partial y_i} = -\frac{2(y_{i+1} - y_i)}{\Delta x_i^2}(x_{i+1} - x_i) - \frac{2\bar{F}^{(i)}}{\Delta x_i}\left(\frac{x_{i+1}^2}{2} - x_{i+1}^2 - \frac{x_i^2}{2} + x_{i+1}x_i \right)$$

$$- \frac{2\bar{Q}^{(i)}y_i}{\Delta x_i^2}\left(\frac{x_{i+1}^3}{3} - x_{i+1}^3 + x_{i+1}^3 - \frac{x_i^3}{3} + x_{i+1}x_i^2 - x_{i+1}^2 x_i \right)$$

$$+ \frac{2\bar{Q}^{(i)}}{\Delta x_i^2}y_{i+1}\left(\frac{x_{i+1}^3}{3} - \frac{x_i x_{i+1}^2}{2} - \frac{x_{i+1}^3}{2} + x_{i+1}^2 x_i - \frac{x_i^3}{3} + \frac{x_i^3}{2} + \frac{x_i^2 x_{i+1}}{2} - x_i^2 x_{i+1} \right) \tag{12.93}$$

The four terms in parentheses involving x_i and x_{i+1} on the right-hand side of Eq. (12.93) reduce as follows:

Term 1: Δx_i

Term 2: $-\Delta x_i^2/2$

Term 3: $\Delta x_i^3/3$

Term 4: $-\Delta x_i^3/6$ (12.94)

Substituting Eq. (12.94) into Eq. (12.93) and dividing through by 2 yields

$$\frac{\partial I^{(i)}}{\partial y_i} = -\frac{(y_{i-1} + y_i)}{\Delta x_i} + \frac{\bar{F}^{(i)} \Delta x_i}{2} - \frac{\bar{Q}^{(i)} y_i \Delta x_i}{3} - \frac{\bar{Q}^{(i)} y_{i+1} \Delta x_i}{6} \tag{12.95}$$

The first integral in Eq. (12.79), $\partial I^{(i-1)}/\partial y_i$, is evaluated by repeating the steps presented above. The result is

$$\frac{\partial I^{(i-1)}}{\partial y_i} = \frac{(y_i - y_{i-1})}{\Delta x_{i-1}} + \frac{\bar{F}^{(i-1)} \Delta x_{i-1}}{2} - \frac{\bar{Q}^{(i-1)} y_{i-1} \Delta x_{i-1}}{6} - \frac{\bar{Q}^{(i-1)} y_i \Delta x_{i-1}}{3} \tag{12.96}$$

Substituting Eqs. (12.95) and (12.96) into Eq. (12.79), collecting terms, and multiplying through by -1 yields the nodal equation for node i for a nonuniform grid:

$$y_{i-1}\left(\frac{1}{\Delta x_{i-1}}+\frac{\bar{Q}^{(i-1)}\Delta x_{i-1}}{6}\right)-y_i\left(\frac{1}{\Delta x_{i-1}}+\frac{1}{\Delta x_i}-\frac{\bar{Q}^{(i-1)}\Delta x_{i-1}}{3}-\frac{\bar{Q}^{(i)}\Delta x_i}{3}\right)$$

$$+y_{i+1}\left(\frac{1}{\Delta x_i}+\frac{\bar{Q}_i^{(i)}\Delta x_i}{6}\right)$$

$$=\frac{(\bar{F}^{(i-1)}\Delta x_{i-1}+\bar{F}^{(i)}\Delta x_i)}{2} \qquad (12.97)$$

Equation (12.97) is valid for a nonuniform grid. Letting $\Delta x_{i-1}=\Delta x_i=\Delta x$ and multiplying through by Δx yields the nodal equation for node i for a uniform grid:

$$y_{i-1}\left(1+\frac{\bar{Q}^{(i-1)}\Delta x^2}{6}\right)-2y_i\left[1-\frac{(\bar{Q}^{(i-1)}+\bar{Q}^{(i)})\Delta x^2}{6}\right]+y_{i+1}\left(1+\frac{\bar{Q}^{(i)}\Delta x^2}{6}\right)$$

$$=\left(\frac{\bar{F}^{(i-1)}+\bar{F}^{(i)}}{2}\right)\Delta x^2 \qquad (12.98)$$

Example 12.3.   The FEM on a one-dimensional uniform grid.

Let's apply the results obtained in this section to solve the heat transfer problem presented in Section 8.1. The boundary-value ODE is [see Eq. (8.1)]

$$T''-\alpha^2 T=-\alpha^2 T_a \qquad T(0.0)=0.0 \text{ and } T(1.0)=100.00 \qquad (12.99)$$

Let $Q=-\alpha^2=-16.0\,\text{cm}^{-2}$, $T_a=0.0\,\text{C}$ (which gives $F=0.0$), and $\Delta x=0.25\,\text{cm}$, corresponding to the physical domain discretization illustrated in Figure 12.6. For these values, Eq. (12.98) becomes

$$\tfrac{5}{6}T_{i-1}-\tfrac{8}{3}T_i+\tfrac{5}{6}T_{i+1}=0 \qquad (12.100)$$

Applying Eq. (12.100) at nodes 2 to 4 in Figure 12.5 gives

Node 2 : $\tfrac{5}{6}T_1-\tfrac{8}{3}T_2+\tfrac{5}{6}T_3=0$ \qquad (12.101a)

Node 3 : $\tfrac{5}{6}T_2-\tfrac{8}{3}T_3+\tfrac{5}{6}T_4=0$ \qquad (12.101b)

Node 4 : $\tfrac{5}{6}T_3-\tfrac{8}{3}T_4+\tfrac{5}{6}T_5=0$ \qquad (12.101c)

Setting $T_1=0.0$ and $T_5=100.0$ yields the following system of linear algebraic equations:

$$\begin{bmatrix} -2.666667 & 0.833333 & 0.000000 \\ 0.833333 & -2.666667 & 0.833333 \\ 0.000000 & 0.833333 & -2.666667 \end{bmatrix}\begin{bmatrix} T_2 \\ T_3 \\ T_4 \end{bmatrix}=\begin{bmatrix} 0.000000 \\ 0.000000 \\ -83.333333 \end{bmatrix} \qquad (12.102)$$

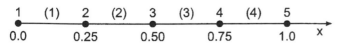

Figure 12.6.   Uniform physical domain discretization.

Table 12.3 Solution by the FEM on a Uniform Grid

x, cm	$T(x)$, C	$\bar{T}(x)$, C	Error(x), C
0.00	0.000000	0.000000	
0.25	3.792476	4.306357	−0.513881
0.50	12.135922	13.290111	−1.154189
0.75	35.042476	36.709070	−1.666594
1.00	100.000000	100.000000	

Solving Eq. (12.102) using the Thomas algorithm yields the results presented in Table 12.3. The Euclidean norm of the errors in Table 12.3 is 2.091354 C, which is 18 percent larger than the Euclidean norm of 1.766412 C for the errors for the second-order equilibrium finite difference method presented in Table 8.8.

Let's illustrate the variable Δx capability of the finite element method by reworking Example 12.3 on a nonuniform grid.

Example 12.4. The FEM on a one-dimensional nonuniform grid.

Let's solve the heat transfer problem in Example 12.3, Eq. (12.99), by applying Eq. (12.97) on the nonuniform grid generated in Example 8.12 in Section 8.8 and illustrated in Figures 8.22 and 12.7. Let $Q = -\alpha^2 = -16.0 \text{ cm}^{-2}$ and $T_a = 0.0 \text{ C}$ (which gives $F = 0.0$). The geometric data from Table 8.25 are tabulated in Table 12.4. Those results and the coefficients of T_{i-1}, T_i, and T_{i+1} in Eq. (12.97) are presented in Table 12.4.
    Applying Eq. (12.97) at nodes 2 to 4 gives:

$$\text{Node 2}: \quad 1.666667T_1 - 9.650794T_2 + 2.650794T_3 = 0 \tag{12.103a}$$

$$\text{Node 3}: \quad 2.650794T_2 - 10.895238T_3 + 4.244445T_4 = 0 \tag{12.103b}$$

$$\text{Node 4}: \quad 4.244445T_3 - 14.577778T_4 + 7.666667T_5 = 0 \tag{12.103c}$$

Setting $T_1 = 0.0$ and $T_5 = 100.0$ yields the following tridiagonal system of FDEs:

$$\begin{bmatrix} -9.650794 & 2.650794 & 0.000000 \\ 2.650794 & -10.895238 & 4.244445 \\ 0.000000 & 4.244445 & -14.577778 \end{bmatrix} \begin{bmatrix} T_2 \\ T_3 \\ T_4 \end{bmatrix} = \begin{bmatrix} 0.000 \\ 0.000 \\ -766.667 \end{bmatrix} \tag{12.104}$$

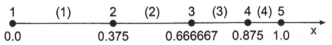

Figure 12.7. Nonuniform physical domain discretization.

Table 12.4 Geometric Data and Coefficients for the Nonuniform Grid

Node	x, cm	Δx_-, cm	Δx_+, cm	$(\cdots)T_{i-1}$	$(\cdots)T_i$	$(\cdots)T_{i+1}$
1	0.000					
2	0.375	0.375000	0.291667	1.666667	−9.650794	2.650794
3	0.666667	0.291667	0.208333	2.650794	−10.895238	4.244445
4	0.875	0.208333	0.125000	4.244445	−14.577778	7.666667
5	1.000					

Table 12.5   Solution by the FEM on a Nonuniform Grid

Node	x, cm	$T(x)$, C	$\bar{T}(x)$, C	Error(x), C
1	0.0	0.000000	0.000000	
2	0.375	6.864874	7.802440	−0.937566
3	0.666667	24.993076	26.241253	−1.248179
4	0.875	59.868411	60.618093	−0.749682
5	1.0	100.000000	100.000000	

Solving Eq. (12.104) using the Thomas algorithm yields the results presented in Table 12.5.

Comparing these results with the results presented in Table 12.3 for a uniform grid shows that the errors are a little larger at the left end of the rod and a little smaller at the right end of the rod. The Euclidean norm of the errors in Table 12.5 is 1.731760 C, which is smaller than the Euclidean norm of 2.091354 C for the uniform grid results in Table 12.3, and which is comparable to the Euclidean norm of 1.766412 C for the errors presented in Table 12.7 for the second-order equilibrium finite difference method.

12.3.3.   The Galerkin Weighted Residual Approach

The Rayleigh-Ritz approach is applied in Section 12.3.2 to develop the finite element method. As discussed in Section 12.2.4, the Galerkin weighted residual approach is generally more straightforward than the Rayleigh-Ritz approach, since there is no need to look for the functional corresponding to the boundary-value ODE. The finite element method based on the Galerkin weighted residual approach is illustrated in this section by applying it to solve the following simple linear boundary-value problem:

$$\boxed{\bar{y}'' + Q\bar{y} = F \qquad \text{with appropriate boundary conditions}} \tag{12.105}$$

where $Q = Q(x)$ and $F = F(x)$.

As discussed in Section 12.2.3, the Galerkin weighted residual method is based on the residual obtained when the exact solution $\bar{y}(x)$ of the boundary-value ODE, Eq. (12.105), is approximated by an approximate solution $y(x)$. The resulting residual $R(x)$ is then

$$R(x) = y'' + Qy - F \tag{12.106}$$

The residual $R(x)$ is multiplied by a set of weighting factors $W_j(x)$ $(j = 1, 2, \ldots)$ and integrated over the global solution domain $D(x)$ to obtain the weighted residual integral:

$$I(y(x)) = \int_a^b W_j(x)R(x)\,dx = 0 \tag{12.107}$$

Substituting Eq. (12.106) into Eq. (12.107) gives

$$I(y(x)) = \int_a^b W_j(x)(y'' + Qy - F)dx = 0 \tag{12.108}$$

Integrating the first term in Eq. (12.108) by parts gives

$$\int_a^b W_j y'' \, dx = -\int_a^b y' W_j' \, dx + y' W_j \Big|_a^b = -\int_a^b y' W_j' \, dx + y_b' W_j(b) - y_a' W_j(a) \quad (12.109)$$

The last two terms in Eq. (12.109) involve the derivative at the boundary points. For Dirichlet boundary conditions, these terms are not needed. For Neumann boundary conditions, these two terms introduce the derivative boundary conditions at the boundaries of the global solution domain. Substituting Eq. (12.109) into Eq. (12.108) yields

$$I(y(x)) = \int_a^b (-y' W_j' + Qy W_j - F W_j) \, dx + y_b' W_j(b) - y_a' W_j(a) = 0 \quad (12.110)$$

In terms of the global approximate solution $y(x)$ and the discretized global solution domain illustrated in Figure 12.4, Eq. (12.110) can be written as follows:

$$I(y(x)) = I^{(1)}(y(x)) + I^{(2)}(y(x)) + \cdots + I^{(i-1)}(y(x)) + I^{(i)}(y(x)) + \cdots$$
$$+ I^{(I-1)}(y(x)) + y_b' W_I(b) - y_a' W_1(a)$$
$$= 0 \quad (12.111)$$

where $I^{(i)}(y(x))$ is given by

$$I^{(i)}(y(x)) = \int_{x_i}^{x_{i+1}} \left(-y' \frac{d[W_j(x)]}{dx} + Qy W_j - F W_j \right) dx \quad (12.112)$$

where $y^{(i)}(x)$ is the interpolating polynominal and $W_j(x)$ denotes the weighting factors applicable to element (i). The interpolating polynominal $y^{(i)}(x)$ is given by Eq. (12.67):

$$y^{(i)}(x) = y_i N_i^{(i)}(x) + y_{i+1} N_{i+1}^{(i)}(x) \quad (12.113)$$

where the shape functions $N_i^{(i)}(x)$ and $N_{i+1}^{(i)}(x)$ are given by Eqs. (12.68) and (12.69):

$$N_i^{(i)}(x) = -\frac{x - x_{i+1}}{\Delta x_i} \quad (12.114)$$

$$N_{i+1}^{(i)}(x) = \frac{x - x_i}{\Delta x_i} \quad (12.115)$$

In the Galerkin weighted residual approach, the weighting factors $W_j(x)$ are chosen to be the shape functions $N_i^{(i)}(x)$ and $N_{i+1}^{(i)}(x)$. Recall that $N_i^{(i)}(x)$ and $N_{i+1}^{(i)}(x)$ are defined to be zero everywhere outside of element (i). Letting $W_j = N_i^{(i)}$ in Eq. (12.112) gives

$$I(y(x)) = \int_{x_i}^{x_{i+1}} \left(-y' \frac{d[N_i^{(i)}(x)]}{dx} + Qy N_i^{(i)} - F N_i^{(i)} \right) dx = 0 \quad (12.116)$$

Equation (12.116) is equal to zero since $N_i^{(i)}(x)$ is zero in all the integrals except $I^{(i)}(y(x))$ and $N_i^{(i)}(a) = N_i^{(i)}(b) = 0.0$. Letting $W_j(x) = N_{i+1}^{(i)}(x)$ in Eq. (12.112) gives

$$I(y(x)) = \int_{x_i}^{x_{i+1}} \left(-y' \frac{d[N_{i+1}^{(i)}(x)]}{dx} + Qy N_{i+1}^{(i)} - F N_{i+1}^{(i)} \right) dx = 0 \quad (12.117)$$

Equations (12.116) and (12.117) are the *element equations* for element (i).

An alternate approach is based on the function $N_i(x)$ illustrated in Figure 12.8. Thus,

$$N_i(x) = N_i^{(i-1)}(x) + N_i^{(i)}(x) \quad (12.118)$$

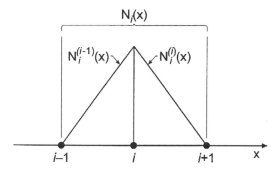

Figure 12.8. Shape function for node i.

Equation (12.118) simply expresses the fact that $N_i(x) = N_i^{(i-1)}(x)$ in element $(i-1)$ and $N_i(x) = N_i^{(i)}(x)$ in element (i). Letting $W_j(x) = N_i(x)$ in Eq. (12.110) gives

$$I(y(x)) = \int_{x_{i-1}}^{x_i} \left[-y' \frac{d[N_i^{(i-1)}(x)]}{dx} + QyN_i^{(i-1)} - FN_i^{(i-1)} \right] dx$$

$$+ \int_{x_i}^{x_{i+1}} \left[-y' \frac{d[N_i^{(i)}(x)]}{dx} + QyN_i^{(i)} - FN_i^{(i)} \right] dx = 0 \qquad (12.119)$$

Equation (12.119) is the *nodal equation* for node i.

Note that Eq. (12.116), which is the first element equation for element (i), is identical to the second integral in Eq. (12.119). When the element equations are developed for element $(i-1)$, it is found that the second element equation for element $(i-1)$, which corresponds to Eq. (12.117) for element (i), is identical to the first integral in Eq. (12.119). Thus, Eq. (12.119) can be obtained by combining the proper element equations from elements $(i-1)$ and (i). This process is called *assemblying* the element equations. Thus, the element equation approach and the nodal equation approach yield identical results.

Which of two approaches is preferable? For one-dimensional domains, there is no appreciable difference in the amount of effort involved in the two approaches. However, for two- and three-dimensional domains, the element approach is considerably simpler. Thus, the element approach is used in the remainder of this section to illustrate that approach.

Let's illustrate the Galerkin weighted residual approach by applying it to solve Eq. (12.105). Steps 1 and 2, discretizing the solution domain and choosing the interpolating polynominals, are discussed in Section 12.3.1 and illustrated in Figures 12.4 and 12.5. For element (i), the shape functions and the linear interpolating polynominal are given by Eqs. (12.68) to (12.70). Thus,

$$N_i^{(i)}(x) - \frac{x - x_{i+1}}{\Delta x_i} \qquad \text{and} \qquad N_{i+1}^{(i)}(x) = \frac{x - x_i}{\Delta x_i} \qquad (12.120)$$

$$y^{(i)}(x) = y_i \left(-\frac{x - x_{i+1}}{\Delta x_i} \right) + y_{i+1} \left(\frac{x - x_i}{\Delta x_i} \right) \qquad (12.121)$$

Note that Eq. (12.121) is simply a linear Lagrange interpolating polynominal for element (*i*). The element equations for element (*i*) are given by Eqs. (12.116) and (12.117). Thus,

$$I(y(x)) = \int_{x_i}^{x_{i+1}} \left[-y' \frac{d[N_i^{(i)}(x)]}{dx} + QyN_i^{(i)} - FN_i^{(i)} \right] dx = 0 \qquad (12.122)$$

$$I(y(x)) = \int_{x_i}^{x_{i+1}} \left[-y' \frac{d[N_{i+1}^{(i)}]}{dx} + QyN_{i+1}^{(i)} - FN_{i+1}^{(i)} \right] dx = 0 \qquad (12.123)$$

From Eq. (12.120),

$$\frac{d[N_i^{(i)}(x)]}{dx} = -\frac{1}{\Delta x_i} \qquad (12.124)$$

$$\frac{d[N_{i+1}^{(i)}(x)]}{dx} = \frac{1}{\Delta x_i} \qquad (12.125)$$

Substituting Eqs. (12.124) and (12.125) into Eqs. (12.122) and (12.123), respectively, gives

$$\frac{1}{\Delta x_i} \int_{x_i}^{x_{i+1}} y' \, dx + \int_{x_i}^{x_{i+1}} N_i^{(i)} Qy \, dx - \int_{x_i}^{x_{i+1}} N_i^{(i)} F \, dx = 0 \qquad (12.126a)$$

$$-\frac{1}{\Delta x_i} \int_{x_i}^{x_{i+1}} y' \, dx + \int_{x_i}^{x_{i+1}} N_{i+1}^{(i)} Qy \, dx - \int_{x_i}^{x_{i+1}} N_{i+1}^{(i)} F \, dx = 0 \qquad (12.126b)$$

Equation (12.126) requires the functions $y(x)$, $y'(x)$, $N_i^{(i)}(x)$, and $N_{i+1}^{(i)}(x)$, which are given by Eqs. (12.121), (12.86), and (12.120). Substituting all of these expressions into Eq. (12.126), evaluating $Q(x)$ and $F(x)$ as average values for each element as done in Eqs. (12.90) and (12.191), integrating, and evaluating the results at the limits of integration yields the two element equations:

$$-y_i \left(\frac{1}{\Delta x_i} - \frac{\bar{Q}^{(i)} \Delta x_i}{3} \right) + y_{i+1} \left(\frac{1}{\Delta x_i} + \frac{\bar{Q}^{(i)} \Delta x_i}{6} \right) - \frac{\bar{F}^{(i)} \Delta x_i}{2} = 0 \qquad (12.127a)$$

$$y_i \left(\frac{1}{\Delta x_i} + \frac{\bar{Q}^{(i)} \Delta x_i}{6} \right) - y_{i+1} \left(\frac{1}{\Delta x_i} - \frac{\bar{Q}^{(i)} \Delta x_i}{3} \right) - \frac{\bar{F}^{(i)} \Delta x_i}{2} = 0 \qquad (12.127b)$$

Equation (12.127) is valid for nonuniform Δx. Letting $\Delta x_i = \Delta x = \text{constant}$ and multiplying through by Δx yields

$$-y_i \left(1 - \frac{\bar{Q}^{(i)} \Delta x^2}{3} \right) + y_{i+1} \left(1 + \frac{\bar{Q}^{(i)} \Delta x^2}{6} \right) - \frac{\bar{F}^{(i)} \Delta x^2}{2} = 0 \qquad (12.128a)$$

$$y_i \left(1 + \frac{\bar{Q}^{(i)} \Delta x^2}{6} \right) - y_{i+1} \left(1 - \frac{\bar{Q}^{(i)} \Delta x^2}{3} \right) - \frac{\bar{F}^{(i)} \Delta x^2}{2} = 0 \qquad (12.128b)$$

Equation (12.128) is valid for uniform Δx.

Let's assemble the element equations for a uniform grid, Eq. (12.128). Applying Eq. (12.128) for element $(i-1)$ gives:

$$-y_{i-1}\left(1 - \frac{\bar{Q}^{(i-1)}\,\Delta x^2}{3}\right) + y_i\left(1 + \frac{\bar{Q}^{(i-1)}\,\Delta x^2}{6}\right) - \frac{\bar{F}^{(i-1)}\Delta x^2}{2} = 0 \qquad (12.129a)$$

$$y_{i-1}\left(1 + \frac{\bar{Q}^{(i-1)}\,\Delta x^2}{6}\right) - y_i\left(1 - \frac{\bar{Q}^{(i-1)}\,\Delta x^2}{3}\right) - \frac{\bar{F}^{(i-1)}\,\Delta x^2}{2} = 0 \qquad (12.129b)$$

Applying Eq. (12.128) for element (i) gives:

$$-y_i\left(1 - \frac{\bar{Q}^{(i)}\,\Delta x^2}{3}\right) + y_{i+1}\left(1 + \frac{\bar{Q}^{(i)}\,\Delta x^2}{6}\right) - \frac{\bar{F}^{(i)}\,\Delta x^2}{2} = 0 \qquad (12.130a)$$

$$y_i\left(1 + \frac{\bar{Q}^{(i)}\,\Delta x^2}{6}\right) - y_{i+1}\left(1 - \frac{\bar{Q}^{(i)}\,\Delta x^2}{3}\right) - \frac{\bar{F}^{(i)}\,\Delta x^2}{2} = 0 \qquad (12.130b)$$

Adding Eqs. (12.129b) and (12.130a) yields the nodal equation for node i. Thus,

$$\boxed{\begin{aligned} &y_{i-1}\left(1 + \frac{\bar{Q}^{(i-1)}\,\Delta x^2}{6}\right) - 2y_i\left(1 - \frac{(\bar{Q}^{(i-1)} + \bar{Q}^{(i)})\,\Delta x^2}{6}\right) + y_{i+1}\left(1 + \frac{\bar{Q}^{(i)}\,\Delta x^2}{6}\right) \\ &= \left(\frac{\bar{F}^{(i-1)} + \bar{F}^{(i)}}{2}\right)\Delta x^2 \end{aligned}}$$

$$(12.131)$$

Equation (12.131) is identical to Eq. (12.98), which was obtained by the Rayleigh-Ritz nodal approach. Applying the assembly step to Eq. (12.127), which is valid for $\Delta x_{i-1} \neq \Delta x_i$, yields Eq. (12.97). These results demonstrate that the nodal approach and the element approach yield identical results, and that the Rayleigh-Ritz approach and the Galerkin weighted residual approach yield the same results when the weighting factors $W_j(x)$ are the shape functions of the interpolating polynominals.

At the left and right boundaries of the global solution domain, elements (1) and (I), respectively, Eq. (12.110) shows that $y_a' W_1(a)$ and $y_b' W_I(b)$, respectively, must be added to the element equations corresponding to $W_1 = N_1^{(1)}(x)$ and $W_I = N_1^{(I-1)}(x)$. Note that $W_1(a) = 1.0$ and $W_I(b) = 1.0$. Subtracting $y_1' = y'(a)$ from Eq. (12.128a) and adding $y_I' = y'(b)$ to Eq. (12.128b) yields

$$-y_1\left(1 - \frac{\bar{Q}^{(1)}\,\Delta x^2}{3}\right) + y_2\left(1 + \frac{\bar{Q}^{(1)}\,\Delta x^2}{6}\right) = \frac{\bar{F}^{(1)}\,\Delta x^2}{2} + \Delta x\,y_1' \qquad (12.132)$$

$$y_{I-1}\left(1 + \frac{\bar{Q}^{(I)}\,\Delta x^2}{6}\right) - y_I\left(1 - \frac{\bar{Q}^{(I)}\,\Delta x^2}{3}\right) = \frac{\bar{F}^{(I)}\,\Delta x^2}{2} - \Delta x\,y_I' \qquad (12.133)$$

For Dirichlet boundary conditions, $\bar{y}(x_1) = \bar{y}_1$ and $\bar{y}(x_I) = \bar{y}_I$, and Eqs. (12.132) and (12.133) are not needed. However, for Neumann boundary conditions, $\bar{y}'(x_1) = \bar{y}_1'$ and $\bar{y}'(x_I) = \bar{y}_I'$, Eqs. (12.132) and (12.133) are included as the first and last equations, respectively, in the system equation.

Example 12.5.   The FEM with a derivative boundary condition.

Let's apply the Galerkin finite element method to solve the heat transfer problem presented in Section 8.5. The boundary-value ODE is [see Eq. (8.75)]

$$T'' - \alpha^2 T = -\alpha^2 T_a \qquad T(0.0) = 100.0 \text{ and } T'(1.0) = 0.0 \qquad (12.134)$$

Let $Q = -\alpha^2 = -16.0 \text{ cm}^{-2}$, $T_a = 0.0 \text{ C}$ (which gives $F = 0.0$), and $\Delta x = 0.25 \text{ cm}$, corresponding to the physical domain discretization illustrated in Figure 12.6. For interior nodes 2 to 4, the nodal equations are the same as Eq. (12.101) in Example 12.3:

$$\text{Node 2}: \qquad \tfrac{5}{6}T_1 - \tfrac{8}{3}T_2 + \tfrac{5}{6}T_3 = 0 \qquad (12.135a)$$

$$\text{Node 3}: \qquad \tfrac{5}{6}T_2 - \tfrac{8}{3}T_3 + \tfrac{5}{6}T_4 = 0 \qquad (12.135b)$$

$$\text{Node 4}: \qquad \tfrac{5}{6}T_3 - \tfrac{8}{3}T_4 + \tfrac{5}{6}T_5 = 0 \qquad (12.135c)$$

In Eq. (12.135a), $T_1 = 100.0$. The boundary condition at $x = 1.0 \text{ cm}$ is $T_5' = 0.0$. Applying Eq. (12.133) at node 5 yields

$$\text{Node 5}: \qquad \tfrac{5}{6}T_4 - \tfrac{4}{3}T_5 = \frac{(0.0)(0.25)^2}{2} - (0.25)(0.0) = 0 \qquad (12.135d)$$

Equation (12.135) yields the following system of linear algebraic equations:

$$\begin{bmatrix} -2.666667 & 0.833333 & 0.000000 & 0.000000 \\ 0.833333 & -2.666667 & 0.833333 & 0.000000 \\ 0.000000 & 0.833333 & -2.666667 & 0.833333 \\ 0.000000 & 0.000000 & 0.833333 & -1.333333 \end{bmatrix} \begin{bmatrix} T_2 \\ T_3 \\ T_4 \\ T_5 \end{bmatrix}$$

$$= \begin{bmatrix} -83.333333 \\ 0.000000 \\ 0.000000 \\ 0.000000 \end{bmatrix} \qquad (12.136)$$

Solving Eq. (12.136) using the Thomas algorithm yields the results presented in Table 12.6. The Euclidean norm of the errors in Table 12.6 is 2.359047 C, which is 15 percent larger than the Euclidean norm of 2.045460 C for the errors for the second-order equilibrium finite difference method results presented in Table 8.17.

Table 12.6   Solution by the FEM with a Derivative
Boundary Condition

x, cm	$T(x)$, C	$\bar{T}(x)$, C	Error(x), C
0.00	100.000000	100.000000	
0.25	35.157568	36.866765	−1.709197
0.50	12.504237	13.776782	−1.272545
0.75	4.856011	5.650606	−0.794595
1.00	3.035006	3.661899	−0.626893

12.3.4. Summary

The finite element method is an extremely important and popular method for solving boundary-value problems. It is one of the most popular methods for solving boundary-value problems in two and three dimensions, which are elliptic PDEs. The application of the finite element method to elliptic PDEs is discussed in Section 12.4.

The finite element method breaks the global solution domain into a number of subdomains, called elements, and applies either the Rayleigh-Ritz approach or the Galerkin weighted residual approach to the individual elements. The global solution is obtained by assemblying the results for all of the elements. As discussed in Section 12.2.4, the choice between the Rayleigh-Ritz approach and the Galerkin weighted residual approach generally depends on whether the variational principle is known or the governing differential equation is known.

Two approaches can be taken to the finite element method: the *nodal approach* and the *element approach*. The ultimate objective of both approaches is to develop a system of nodal equations, called the *system equation*, for the global solution. The nodal approach yields a set of nodal equations directly, which gives the system equation directly. The element approach yields a set of element equations, which must be assembled to obtain the nodal equations and the system equation. For one-dimensional problems, the two approaches are comparable in effort since each node belongs only to the two elements lying on either side of the node. However, in two- and three-dimensional problems, each node can belong to many elements, thus, the element approach is generally simpler and preferred.

One of the major advantages of the finite element method is that the element sizes do not have to be uniform. Thus, many small elements can be placed in regions of large gradients, and fewer large elements can be placed in regions of small gradients. This feature is extremely useful in two- and three-dimensional problems. The finite element method is a very popular method for solving boundary-value problems.

12.4. THE FINITE ELEMENT METHOD FOR THE LAPLACE (POISSON) EQUATION

The finite element method is applied to one-dimensional boundary-value problems in Section 12.3. In this section, the finite element method is applied to the two-dimensional Laplace (Poisson) equation:

$$\boxed{\bar{f}_{xx} + \bar{f}_{yy} = F(x, y) \qquad \text{with appropriate boundary conditions}} \qquad (12.137)$$

The steps in the finite element approach presented in Section 12.3 also apply to multidimensional problems. The Galerkin weighted residual approach presented in Section 12.3.3 is applied to develop the element equations for a rectangular element. The element equations are assembled to develop the system equation for a rectangular physical space.

12.4.1. Domain Discretization and the Interpolating Polynominals

Consider the rectangular global solution domain $D(x, y)$ illustrated in Figure 12.9a. The global solution domain $D(x, y)$ can be discretized in a number of ways. Figure 12.9b illustrates discretization into rectangular elements, and Figure 12.9c illustrates discretization into right triangles.

Triangular elements and quadrilateral elements are the two most common forms of two-dimensional elements. Figure 12.10a illustrates a general triangular element, and Figure 12.10b illustrates a set of right triangular elements. Figure 12.11a illustrates a general quadrilateral element, and Figure 12.11b illustrates a rectangular quadrilateral element.

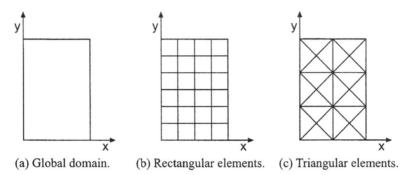

(a) Global domain.   (b) Rectangular elements.   (c) Triangular elements.

Figure 12.9.   Rectangular solution domain $D(x, y)$.

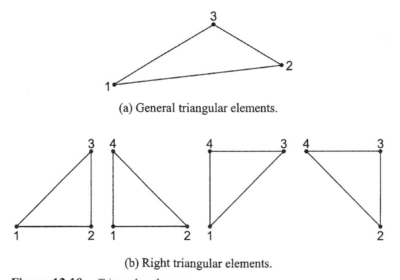

(a) General triangular elements.

(b) Right triangular elements.

Figure 12.10.   Triangular elements.

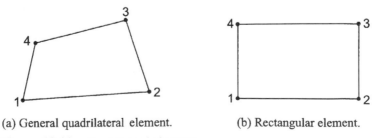

(a) General quadrilateral element.   (b) Rectangular element.

Figure 12.11.   Quadrilateral elements.

In this section, we'll discretize the rectangular global solution domain $D(x, y)$ illustrated in Figure 12.9a into rectangular elements, as illustrated in Figure 12.12. The global solution domain $D(x, y)$ is covered by a two-dimensional grid of lines. There are I lines perpendicular to the x axis, which are denoted by the subscript i. There are J lines perpendicular to the y axis, which are denoted by the subscript j. There are $(I - 1) \times (J - 1)$ elements, which are denoted by the superscript (i, j). Element (i, j) starts at node i, j and ends at node $i + 1, j + 1$. The grid increments are $\Delta x_i = x_{i+1} - x_i$ and $\Delta y_j = y_{j+1} - y_j$.

Let the global exact solution $\bar{f}(x, y)$ be approximated by the global approximate solution $f(x, y)$, which is the sum of a series of local interpolating polynominals $f^{(i,j)}(x, y)$ $(i = 1, 2, \ldots, I - 1, j = 1, 2, \ldots, J - 1)$ that are valid within each element. Thus,

$$f(x, y) = \sum_{i=1}^{I-1} \sum_{j=1}^{J-1} f^{(i,j)}(x, y) \tag{12.138}$$

Let's define the local interpolating polynominal $f^{(i,j)}(x, y)$ as a linear bivariate polynominal. Element (i, j) is illustrated in Figure 12.13. Let's use a local coordinate system, where node i, j is at (0.0), node $i + 1, j$ is at $(\Delta x, 0)$, etc. Denote the grid points as 1, 2, 3, and 4. The linear interpolating polynominal, $f^{(i,j)}(x, y)$, corresponding to element (i, j) is given by

$$f^{(i,j)}(x, y) = f_1 N_1(x, y) + f_2 N_2(x, y) + f_3 N_3(x, y) + f_4 N_4(x, y) \tag{12.139}$$

where f_1, f_2, etc., are the values of $f(x, y)$ at nodes 1, 2, etc., respectively, and $N_1(x, y)$, $N_2(x, y)$, etc., are linear interpolating polynominals within element (i, j). The interpolating polynominals, $N_1(x, y)$, $N_2(x, y)$, etc., are called *shape functions* in the finite element literature. The subscripts of the shape functions denote the node at which the corresponding shape function is equal to unity. The shape function is defined to be zero at the other

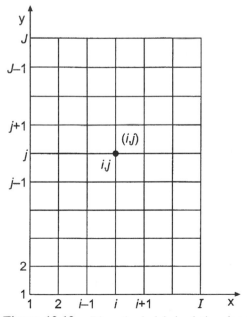

Figure 12.12.   Discretized global solution domain $D(x, y)$.

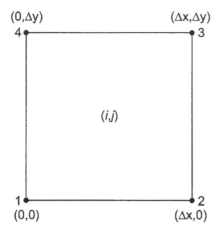

Figure 12.13. Rectangular element (i, j).

three nodes and zero everywhere outside of the element. Since the element approach is being used, only one element is involved. Thus, the superscript (i, j) identifying the element will be omitted for clarity. Figure 12.14 illustrates the shape functions $N_1(x, y)$, $N_2(x, y)$, etc.

Next, let's develop the expressions for the shape functions $N_1(x, y)$, $N_2(x, y)$, etc. First, consider $N_1(x, y)$:

$$N_1(x, y) = a_0 + a_1\bar{x} + a_2\bar{y} + a_3\bar{x}\bar{y} \tag{12.140}$$

where $\bar{x}$ and $\bar{y}$ are normalized values of x and y, respectively, that is, $\bar{x} = x/\Delta x$ and $\bar{y} = y/\Delta y$. Introducing the values of $N_1(x, y)$ at the four nodes into Eq. (12.140) gives

$$N_1(0, 0) = 1.0 = a_0 + a_1(0) + a_2(0) + a_3(0)(0) = a_0 \tag{12.141a}$$
$$N_1(1, 0) = 0.0 = a_0 + a_1(1) + a_2(0) + a_3(1)(0) = a_0 + a_1 \tag{12.141b}$$
$$N_1(0, 1) = 0.0 = a_0 + a_1(0) + a_2(1) + a_3(0)(1) = a_0 + a_2 \tag{12.141c}$$
$$N_1(1, 1) = 0.0 = a_0 + a_1(1) + a_2(1) + a_3(1)(1) = a_0 + a_1 + a_2 + a_3 \tag{12.141d}$$

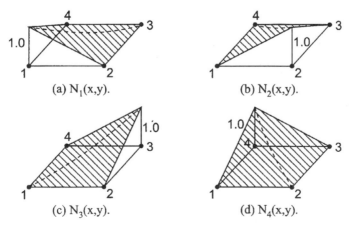

Figure 12.14. Rectangular element shape functions.

Let $\mathbf{a}^T = [a_0 \ \ a_1 \ \ a_2 \ \ a_3]$. Solving Eq. (12.141) by Gauss elimination yields $\mathbf{a}^T = [1.0 \ \ -1.0 \ \ -1.0 \ \ 0.0]$. Thus, $N_1(x, y)$ is given by

$$N_1(x, y) = 1.0 - \bar{x} - \bar{y} + \bar{x}\bar{y} \tag{12.142a}$$

In a similar manner, it is found that

$$N_2(x, y) = \bar{x} - \bar{x}\bar{y} \tag{12.142b}$$

$$N_3(x, y) = \bar{x}\bar{y} \tag{12.142c}$$

$$N_4(x, y) = \bar{y} - \bar{x}\bar{y} \tag{12.142d}$$

Equation (12.142) comprises the shape functions for a rectangular element. Substituting Eq. (12.142) into Eq. (12.139) yields

$$\boxed{f(x, y) = f_1(1.0 - \bar{x} - \bar{y} + \bar{x}\bar{y}) + f_2(\bar{x} - \bar{x}\bar{y}) + f_3(\bar{x}\bar{y}) + f_4(\bar{y} - \bar{x}\bar{y})} \tag{12.143}$$

Equation (12.143) is the interpolating polynominal for a rectangular element.

12.4.2. The Galerkin Weighted Residual Approach

The Galerkin weighted residual approach is applied in this section to develop a finite element approximation of the Laplace (Poisson) equation, Eq. (12.137):

$$\boxed{\bar{f}_{xx} + \bar{f}_{yy} = F(x, y) \qquad \text{with appropriate boundary conditions}} \tag{12.144}$$

Let's approximate the exact solution $\bar{f}(x, y)$ by the approximate solution $f(x, y)$ given by Eq. (12.138). Substituting $\bar{f}(x, y)$ into Eq. (12.144) gives the residual $R(x, y)$:

$$R(x, y) = f_{xx} + f_{yy} - F \tag{12.145}$$

The residual $R(x, y)$ is multiplied by a set of weighting functions $W_k(x, y)$ $(k = 1, 2, \ldots)$ and integrated over the global solution domain $D(x, y)$ to obtain the weighted residual integral $I(f(x, y))$, which is equated to zero. Consider the general weighting function $W(x, y)$. Then

$$I(f(x, y)) = \iint_D W(f_{xx} + f_{yy} - F) \, dx \, dy \tag{12.146}$$

The first two terms in Eq. (12.146) can be integrated by parts. Thus,

$$Wf_{xx} = W(f_x)_x = (Wf_x)_x - W_x f_x \tag{12.147a}$$

$$Wf_{yy} = W(f_y)_y = (Wf_y)_y - W_y f_y \tag{12.147b}$$

Substituting Eq. (12.147) into Eq. (12.146) gives

$$I(f(x, y)) = \iint_D ((Wf_x)_x + (Wf_y)_y - W_x f_x - W_y f_y - WF) \, dx \, dy \tag{12.148}$$

The first two terms in Eq. (12.148) can be transformed by Stokes' theorem to give

$$\iint_D (Wf_x)_x \, dx \, dy = \oint_B Wf_x n_x \, ds \tag{12.149a}$$

$$\iint_D (Wf_y)_y \, dx \, dy = \oint_B Wf_y n_y \, ds \tag{12.149b}$$

where the line integrals in Eq. (12.149) are evaluated around the outer boundary B of the global solution domain $D(x, y)$ and n_x and n_y are the components of the unit normal vector to the outer boundary $\mathbf{n}$. Note that the flux of $f(x, y)$ crossing the outer boundary B of the global solution domain $D(x, y)$ is given by

$$q_n = \mathbf{n} \cdot \nabla f = n_x f_x + n_y f_y \tag{12.150}$$

Substituting Eqs. (12.149) and (12.150) into Eq. (12.148) yields

$$I(f(x, y)) = -\iint_D (W_x f_x + W_y f_y + WF)\, dx\, dy + \oint_B W q_n\, ds \tag{12.151}$$

The line integral in Eq. (12.151) specifies the flux q_n normal to the outer boundary B of the global solution domain $D(x, y)$. For all interior elements which do not coincide with a portion of the outer boundary, these fluxes cancel out when all the interior elements are assembled. For any element which has a side coincident with a portion of the outer boundary, the line integral expresses the boundary condition on that side. For Dirichlet boundary conditions, $f(x, y)$ is specified on the boundary, and the line integral is not needed. For Neumann boundary conditions, the line integral is used to apply the derivative boundary conditions.

In terms of the global approximate solution $f(x, y)$ and the discretized global solution domain illustrated in Figure 12.12, Eq. (12.151) can be written as follows:

$$I(f(x, y)) = I^{(1,1)}(f(x, y)) + \cdots + I^{(i,j)}(f(x, y)) + \cdots + I^{(I-1, J-1)}(f(x, y))$$
$$+ \oint_B W q_n\, ds = 0$$

where $I^{(i,j)}(f(x, y))$ is given by

$$I^{(i,j)}(f)(x, y)) = -\iint_D (W_x f_x + W_y f_y + WF)\, dx\, dy \tag{12.153}$$

where $f(x, y)$ is the approximate solution given by Eq. (12.143) and $W(x, y)$ is an as yet unspecified weighting function.

The evaluation of the weighed residual integral, Eq. (12.153), requires the function $f(x, y)$ and its partial derivatives with respect to x and y. From Eqs. (12.143),

$$f(x, y) = f_1(1 - \bar{x} - \bar{y} + \bar{x}\bar{y}) + f_2(\bar{x} - \bar{x}\bar{y}) + f_3\bar{x}\bar{y} + f_4(\bar{y} - \bar{x}\bar{y}) \tag{12.154}$$

Differentiating Eq. (12.154) with respect to x and y gives

$$f_x = f_1\left(-\frac{1}{\Delta x} + \frac{\bar{y}}{\Delta x}\right) + f_2\left(\frac{1}{\Delta x} - \frac{\bar{y}}{\Delta x}\right) + f_3\left(\frac{\bar{y}}{\Delta x}\right) + f_4\left(-\frac{\bar{y}}{\Delta x}\right) \tag{12.155a}$$

$$f_y = f_1\left(-\frac{1}{\Delta y} + \frac{\bar{x}}{\Delta y}\right) + f_2\left(-\frac{\bar{x}}{\Delta y}\right) + f_3\left(\frac{\bar{x}}{\Delta y}\right) + f_4\left(\frac{1}{\Delta y} - \frac{\bar{x}}{\Delta y}\right) \tag{12.155b}$$

Substituting Eq. (12.155) into Eq. (12.153) yields

$$I(f(x, y)) = \iint_D W_x \frac{1}{\Delta x}[f_1(-1+\bar{y})+f_2(1-\bar{y})+f_3(\bar{y})+f_4(-\bar{y})]\,dx\,dy$$
$$- \iint_D W_y \frac{1}{\Delta y}[f_1(-1+\bar{x})+f_2(-\bar{x})+f_3(\bar{x})+f_4(1-\bar{x})]\,dx\,dy$$
$$- \iint_D WF\,dx\,dy = 0 \tag{12.156}$$

where the superscript (i, j) has been dropped from I for simplicity.

In the Galerkin weighed residual approach, the weighting factors $W_k(x, y)$ $(k = 1, 2, \ldots)$ are chosen to be the shape functions $N_1(x, y)$, $N_2(x, y)$, etc. specified by Eq. (12.142). Let's evaluate Eq. (12.156) for $W(x, y) = N_1(x, y)$. From Eq. (12.142a),

$$W_1(x, y) = N_1(x, y) = 1 - \bar{x} - \bar{y} - \bar{x}\bar{y} \tag{12.157}$$

Differentiating Eq. (12.157) with respect to x and y gives

$$(W_1)_x = \left(-\frac{1}{\Delta x} + \frac{\bar{y}}{\Delta x}\right) \quad \text{and} \quad (W_1)_y = \left(-\frac{1}{\Delta y} + \frac{\bar{x}}{\Delta y}\right) \tag{12.158}$$

Substituting Eqs. (12.157) and (12.158) into Eq. (12.156) yields

$$I(f(x, y)) = -\frac{1}{\Delta x^2}\iint_D (-1+\bar{y})[f_1(-1+\bar{y})+f_2(1-\bar{y})+f_3(\bar{y})+f_4(-\bar{y})]\,dx\,dy$$
$$-\frac{1}{\Delta y^2}\iint_D (-1+\bar{x})[f_1(-1+\bar{x})+f_2(-\bar{x})+f_3(\bar{x})+f_4(1-\bar{x})]\,dx\,dy$$
$$-\iint_D (1-\bar{x}-\bar{y}+\bar{x}\bar{y})F\,dx\,dy = 0 \tag{12.159}$$

Recall that $\bar{x} = x/\Delta x$, and $\bar{y} = y/\Delta y$. Thus, $dx = \Delta x\,d\bar{x}$ and $dy = \Delta y\,d\bar{y}$. Thus, Eq. (12.159) can be written as

$$I(f(x, y)) = \int_0^1 \left[\int_0^1 (\cdots)\,\Delta x\,d\bar{x}\right] \Delta y\,d\bar{y} \tag{12.160}$$

where the integrand of the inner integral, denoted as $(\ldots)$, is obtained from Eq. (12.159). Evaluating the inner integral in Eq. (12.160) gives

$$(\cdots) = -\Delta x \int_0^1 \frac{1}{\Delta x^2}[f_1(1-2\bar{y}+\bar{y}^2)+f_2(-1+2\bar{y}-\bar{y}^2)$$
$$+f_3(-\bar{y}+\bar{y}^2)+f_4(\bar{y}-\bar{y}^2)]\,d\bar{x}$$
$$-\Delta x \int_0^1 \frac{1}{\Delta y^2}[f_1(1-2\bar{x}+\bar{x}^2)+f_2(\bar{x}-\bar{x}^2)$$
$$+f_3(-\bar{x}+\bar{x}^2)+f_4(-1+2\bar{x}-\bar{x}^2)]\,d\bar{x}$$
$$-\Delta x \int_0^1 (1-\bar{x}-\bar{y}+\bar{x}\bar{y})F\,dx \tag{12.161}$$

Let $\bar{F}$ denote the average value of $F(x, y)$ in the element. Integrating Eq. (12.161) gives

$$(\cdots) = -\frac{\Delta x}{\Delta x^2}[f_1(1 - 2\bar{y} + \bar{y}^2) + f_2(-1 + 2\bar{y} - \bar{y}^2) + f_3(-\bar{y} + \bar{y}^2) + f_4(\bar{y} - \bar{y}^2)]\bar{x}\Big|_0^1$$

$$-\frac{\Delta x}{\Delta y^2}\left[f_1\left(\bar{x} - \bar{x}^2 + \frac{\bar{x}^3}{3}\right) + f_2\left(\frac{\bar{x}^2}{2} - \frac{\bar{x}^3}{3}\right) + f_3\left(-\frac{\bar{x}^2}{2} + \frac{\bar{x}^3}{3}\right)\right.$$

$$\left. + f_4\left(-\bar{x} + \bar{x}^2 - \frac{\bar{x}^3}{3}\right)\right]\Big|_0^1$$

$$-\Delta x \bar{F}\left(\bar{x} - \frac{\bar{x}^2}{2} - \bar{x}\bar{y} + \frac{\bar{x}^2\bar{y}}{2}\right)\Big|_0^1 \qquad (12.162)$$

Evaluating Eq. (12.162) yields

$$(\cdots) = -\frac{\Delta x}{\Delta x^2}[f_1(1 - 2\bar{y} + \bar{y}^2) + f_2(-1 + 2\bar{y} - \bar{y}^2) + f_3(-\bar{y} + \bar{y}^2) + f_4(\bar{y} - \bar{y}^2)]$$

$$-\frac{\Delta x}{\Delta y^2}\left(\frac{1}{3}f_1 + \frac{1}{6}f_2 - \frac{1}{6}f_3 - \frac{1}{3}f_4\right) - \frac{\Delta \bar{x}\bar{F}}{2}(1 - \bar{y}) \qquad (12.163)$$

Substituting Eq. (12.163) into Eq. (12.160) gives

$$I(f(x, y)) = -\frac{\Delta x \Delta y}{\Delta x^2}\int_0^1 [f_1(1 - 2\bar{y} + \bar{y}^2) + f_2(-1 + 2\bar{y} - \bar{y}^2)$$

$$+ f_3(-\bar{y} + \bar{y}^2) + f_4(\bar{y} - \bar{y}^2)]\, d\bar{y}$$

$$-\frac{\Delta x \Delta y}{\Delta y^2}\int_0^1\left(\frac{1}{3}f_1 + \frac{1}{6}f_2 - \frac{1}{6}f_3 - \frac{1}{3}f_4\right) d\bar{y}$$

$$-\Delta x \Delta y\int_0^1 \frac{\bar{F}}{2}(1 - \bar{y})\, d\bar{y} \qquad (12.164)$$

Integrating Eq. (12.164) gives

$$I(f(x, y)) = -\frac{\Delta x \Delta y}{\Delta x^2}\left[f_1\left(\bar{y} - \bar{y}^2 + \frac{\bar{y}^3}{3}\right) + f_2\left(-\bar{y} + \bar{y}^2 - \frac{\bar{y}^3}{3}\right) + f_3\left(-\frac{\bar{y}^2}{2} + \frac{\bar{y}^3}{3}\right)\right.$$

$$\left. + f_4\left(\frac{\bar{y}^2}{2} - \frac{\bar{y}^3}{3}\right)\right]\Big|_0^1 - \frac{\Delta x \Delta y}{\Delta y^2}\left(\frac{1}{3}f_1 + \frac{1}{6}f_2 - \frac{1}{6}f_3 - \frac{1}{3}f_4\right)\bar{y}\Big|_0^1$$

$$-\frac{\Delta x \Delta y \bar{F}}{2}\left(\bar{y} - \frac{\bar{y}^2}{2}\right)\Big|_0^1 = 0 \qquad (12.165)$$

Evaluating Eq. (12.165) and collecting terms yields

$$-\frac{1}{3}\left(\frac{\Delta y}{\Delta x} + \frac{\Delta x}{\Delta y}\right)f_1 + \frac{1}{6}\left(\frac{2\Delta y}{\Delta x} - \frac{\Delta x}{\Delta y}\right)f_2 + \frac{1}{6}\left(\frac{\Delta y}{\Delta x} + \frac{\Delta x}{\Delta y}\right)f_3 + \frac{1}{6}\left(-\frac{\Delta y}{\Delta x} + \frac{2\Delta x}{\Delta y}\right)f_4$$

$$-\frac{\Delta x \Delta y \bar{F}}{4} = 0 \qquad (12.166a)$$

Repeating the steps in Eqs. (12.156) to (12.166) for $W_2 = N_2$, $W_3 = N_3$, and $W_4 = N_4$ yields the following results:

$$\frac{1}{6}\left(\frac{2\,\Delta y}{\Delta x} - \frac{\Delta x}{\Delta y}\right)f_1 - \frac{1}{3}\left(\frac{\Delta y}{\Delta x} + \frac{\Delta x}{\Delta y}\right)f_2 + \frac{1}{6}\left(-\frac{\Delta y}{\Delta x} + \frac{2\,\Delta x}{\Delta y}\right)f_3$$
$$+ \frac{1}{6}\left(\frac{\Delta y}{\Delta x} + \frac{\Delta x}{\Delta y}\right)f_4 - \frac{\Delta x\,\Delta y\bar{F}}{4} = 0 \tag{12.166b}$$

$$\frac{1}{6}\left(\frac{\Delta y}{\Delta x} + \frac{\Delta x}{\Delta y}\right)f_1 + \frac{1}{6}\left(-\frac{\Delta y}{\Delta x} + \frac{2\,\Delta x}{\Delta y}\right)f_2 - \frac{1}{3}\left(\frac{\Delta y}{\Delta x} + \frac{\Delta x}{\Delta y}\right)f_3$$
$$+ \frac{1}{6}\left(\frac{2\,\Delta y}{\Delta x} - \frac{\Delta x}{\Delta y}\right)f_4 - \frac{\Delta x\,\Delta y\bar{F}}{4} = 0 \tag{12.166c}$$

$$\frac{1}{6}\left(-\frac{\Delta y}{\Delta x} + \frac{2\,\Delta x}{\Delta y}\right)f_1 + \frac{1}{6}\left(\frac{\Delta y}{\Delta x} + \frac{\Delta x}{\Delta y}\right)f_2 + \frac{1}{6}\left(\frac{2\,\Delta y}{\Delta x} - \frac{\Delta x}{\Delta y}\right)f_3$$
$$- \frac{1}{3}\left(\frac{\Delta y}{\Delta x} + \frac{\Delta x}{\Delta y}\right)f_4 - \frac{\Delta x\,\Delta y\bar{F}}{4} = 0 \tag{12.166d}$$

Equations (12.166a) to (12.166d) are the element equations for element (i, j) for $\Delta x \neq \Delta y$.

The next step is to assemble the element equations, Eqs. (12.166a) to (12.166d), to obtain the nodal equation for node i, j. This process is considerably more complicated for two- and three-dimensional problems than for one-dimensional problems because each node belongs to several elements. Consider the portion of the discretized global solution domain which surrounds node i, j, which is illustrated in Figure 12.15. Note that $\Delta x_- = (x_i - x_{i-1}) \neq \Delta x_+ = (x_{i+1} - x_i)$, and $\Delta y_- = (y_i - y_{i-1}) \neq \Delta y_+ = (y_{i+1} - y_i)$. These differences in the grid increments must be accounted for while assembling the equations for four different elements using the element equations derived for a single element. Also note that the average value of $F(x, y)$, denoted by $\bar{F}$, can be different in the four elements surrounding node i, j. Local node 0 is surrounded by local elements (1), (2), (3), and (4). The assembled nodal equation for node 0 is obtained by combining all of the element equations for elements (1), (2), (3), and (4), respectively, which correspond to the shape functions associated with node 0.

Figure 12.16 illustrates the process. Figure 12.16a illustrates the basic element used to derive Eqs. (12.166a) to (12.166d). Figures 12.16b to 12.16e illustrate elements (1) to (4), respectively, from Figure 12.15. Consider element (1) illustrated in Figure 12.16b.

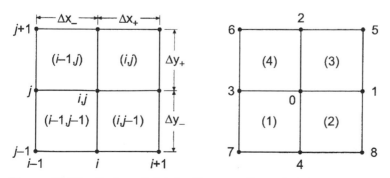

Figure 12.15. Portion of global grid surrounding node i, j.

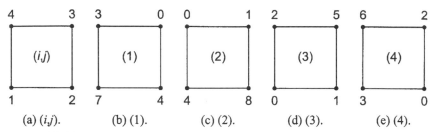

Figure 12.16. Element correspondence.

Node 0 in element (1) corresponds to node 3 in the basic element. Thus, the element equation corresponding to node 3, Eq. (12.166c), is part of the nodal equation for node 0. Renumbering the function values in Eq. (12.166c) to correspond to the nodes in Figure 12.16b yields:

Element (1) Equation (12.166c) with $\Delta x = \Delta x_-$ and $\Delta y = \Delta y_-$.

$$\frac{1}{6}\left(\frac{\Delta y_-}{\Delta x_-}+\frac{\Delta x_-}{\Delta y_-}\right)f_7 +\frac{1}{6}\left(-\frac{\Delta y_-}{\Delta x_-}+\frac{2\,\Delta x_-}{\Delta y_-}\right)f_4 -\frac{1}{3}\left(\frac{\Delta y_-}{\Delta x_-}+\frac{\Delta x_-}{\Delta y_-}\right)f_0$$
$$+\frac{1}{6}\left(\frac{2\,\Delta y_-}{\Delta x_-}-\frac{\Delta x_-}{\Delta y_-}\right)f_3 -\frac{\Delta x_-\,\Delta y_-\bar{F}^{(1)}}{4}=0 \tag{12.167a}$$

Repeating the process for elements (2) to (4) show that Eq. (12.166d) corresponds to element (2), Eq. (12.166a) corresponds to element (3), and Eq. (12.166b) corresponds to element (4). Renumbering the function values in those equations to agree with the node numbers in Figure 12.16 yields the remaining three element equations corresponding to node 0.

Element (2) Equation (12.166d) with $\Delta x = \Delta x_+$ and $\Delta y = \Delta y_-$.

$$\frac{1}{6}\left(-\frac{\Delta y_-}{\Delta x_+}+\frac{2\,\Delta x_+}{\Delta y_-}\right)f_4 +\frac{1}{6}\left(\frac{\Delta y_-}{\Delta x_+}+\frac{\Delta x_+}{\Delta y_-}\right)f_8 +\frac{1}{6}\left(\frac{2\,\Delta y_-}{\Delta x_+}-\frac{\Delta x_+}{\Delta y_-}\right)f_1$$
$$-\frac{1}{3}\left(\frac{\Delta y_-}{\Delta x_+}+\frac{\Delta x_+}{\Delta y_-}\right)f_0 -\frac{\Delta x_+\,\Delta y_-\bar{F}^{(2)}}{4}=0 \tag{12.167b}$$

Element (3) Equation (12.166a) with $\Delta x = \Delta x_+$ and $\Delta y = \Delta y_+$.

$$-\frac{1}{3}\left(\frac{\Delta y_+}{\Delta x_+}+\frac{\Delta x_+}{\Delta y_+}\right)f_0 +\frac{1}{6}\left(\frac{2\,\Delta y_+}{\Delta x_+}-\frac{\Delta x_+}{\Delta y_+}\right)f_1 +\frac{1}{6}\left(\frac{\Delta y_+}{\Delta x_+}+\frac{\Delta x_+}{\Delta y_+}\right)f_5$$
$$+\frac{1}{6}\left(-\frac{\Delta y_+}{\Delta x_+}+\frac{2\,\Delta x_+}{\Delta y_+}\right)f_2 -\frac{\Delta x_+\,\Delta y_+\bar{F}^{(3)}}{4}=0 \tag{12.167c}$$

Element (4) Equation (12.166b) with $\Delta x = \Delta x_-$ and $\Delta y = \Delta y_+$.

$$\frac{1}{6}\left(\frac{2\,\Delta y_+}{\Delta x_-}-\frac{\Delta x_-}{\Delta y_+}\right)f_3 -\frac{1}{3}\left(\frac{\Delta y_+}{\Delta x_-}+\frac{\Delta x_-}{\Delta y_+}\right)f_0 +\frac{1}{6}\left(-\frac{\Delta y_+}{\Delta x_-}+\frac{2\,\Delta x_-}{\Delta y_+}\right)f_2$$
$$+\frac{1}{6}\left(\frac{\Delta y_+}{\Delta x_-}+\frac{\Delta x_-}{\Delta y_+}\right)f_6 -\frac{\Delta x_-\,\Delta y_+\bar{F}^{(4)}}{4}=0 \tag{12.167d}$$

Summing Eqs. (12.167a) to (12.167d) yields the nodal equation for node 0:

$$f_0 \left[-\frac{1}{3}\left(\frac{\Delta y_-}{\Delta x_-} + \frac{\Delta x_-}{\Delta y_-} \right) - \frac{1}{3}\left(\frac{\Delta y_-}{\Delta x_+} + \frac{\Delta x_+}{\Delta y_-} \right) - \frac{1}{3}\left(\frac{\Delta y_+}{\Delta x_+} + \frac{\Delta x_+}{\Delta y_+} \right) - \frac{1}{3}\left(\frac{\Delta y_+}{\Delta x_-} + \frac{\Delta x_-}{\Delta y_+} \right) \right]$$

$$+ f_1 \left[\frac{1}{6}\left(\frac{2\,\Delta y_-}{\Delta x_+} - \frac{\Delta x_+}{\Delta y_-} \right) + \frac{1}{6}\left(\frac{2\,\Delta y_+}{\Delta x_+} - \frac{\Delta x_+}{\Delta y_+} \right) \right]$$

$$+ f_2 \left[\frac{1}{6}\left(-\frac{\Delta y_+}{\Delta x_+} + \frac{2\,\Delta x_+}{\Delta y_+} \right) + \frac{1}{6}\left(-\frac{\Delta y_+}{\Delta x_-} + \frac{2\,\Delta x_-}{\Delta y_+} \right) \right]$$

$$+ f_3 \left[\frac{1}{6}\left(\frac{2\,\Delta y_-}{\Delta x_-} - \frac{\Delta x_-}{\Delta y_-} \right) + \frac{1}{6}\left(\frac{2\,\Delta y_+}{\Delta x_-} - \frac{\Delta x_-}{\Delta y_+} \right) \right]$$

$$+ f_4 \left[\frac{1}{6}\left(-\frac{\Delta y_-}{\Delta x_-} + \frac{2\,\Delta x_-}{\Delta y_-} \right) + \frac{1}{6}\left(-\frac{\Delta y_-}{\Delta x_+} + \frac{2\,\Delta x_+}{\Delta y_-} \right) \right]$$

$$+ f_5 \frac{1}{6}\left(\frac{\Delta y_+}{\Delta x_+} + \frac{\Delta x_+}{\Delta y_+} \right) + f_6 \frac{1}{6}\left(\frac{\Delta y_+}{\Delta x_-} + \frac{\Delta x_-}{\Delta y_+} \right)$$

$$+ f_7 \frac{1}{6}\left(\frac{\Delta y_-}{\Delta x_-} + \frac{\Delta x_-}{\Delta y_-} \right) + f_8 \frac{1}{6}\left(\frac{\Delta y_-}{\Delta x_+} + \frac{\Delta x_+}{\Delta y_-} \right)$$

$$- \frac{1}{4}(\Delta x_- \,\Delta y_- \bar{F}^{(1)} + \Delta x_+ \,\Delta y_- \bar{F}^{(2)} + \Delta x_+ \,\Delta y_+ \bar{F}^{(3)} + \Delta x_- \,\Delta y_+ \bar{F}^{(4)})$$

$$= 0 \tag{12.168}$$

Let $\Delta x_- = \Delta x_+ = \Delta x$ and $\Delta y_- = \Delta y_+ = \Delta y$, multiply through by 3, and collect terms.

$$-4\left(\frac{\Delta y}{\Delta x} + \frac{\Delta x}{\Delta y} \right)f_0 + \left(\frac{2\,\Delta y}{\Delta x} - \frac{\Delta x}{\Delta y} \right)f_1 + \left(\frac{2\,\Delta x}{\Delta y} - \frac{\Delta y}{\Delta x} \right)f_2 + \left(\frac{2\,\Delta y}{\Delta x} - \frac{\Delta x}{\Delta y} \right)f_3$$

$$+ \left(\frac{2\,\Delta x}{\Delta y} - \frac{\Delta y}{\Delta x} \right)f_4 + \frac{1}{2}\left(\frac{\Delta y}{\Delta x} + \frac{\Delta x}{\Delta y} \right)(f_5 + f_6 + f_7 + f_8)$$

$$- \frac{\Delta x\,\Delta y(\bar{F}^{(1)} + \bar{F}^{(2)} + \bar{F}^{(3)} + \bar{F}^{(4)})}{4} = 0 \tag{12.169}$$

For $\Delta x = \Delta y = \Delta L$ and $F = \text{constant}$, Eq. (12.169) yields

$$-8f_0 + (f_1 + f_2 + f_3 + f_4 + f_5 + f_6 + f_7 + f_8) - \Delta L^2 F = 0 \tag{12.170}$$

Example 12.6. The FEM for the Laplace equation.

Let's apply the results obtained in this section to solve the heat transfer problem presented in Section 9.1. The elliptic partial differential equation is [see Eq. (9.1)]

$$T_{xx} + T_{yy} = 0 \tag{12.171}$$

with $T(x, 15.0) = 100.0\sin(\pi x/10.0)$ and $T(x, 0.0) = T(0.0, y) = T(10.0, y) = 0.0$. The exact solution to this problem is presented in Section 9.1. Let $\Delta x = \Delta y = 2.5\,\text{cm}$. The discretized solution domain is illustrated in Figure 12.17.

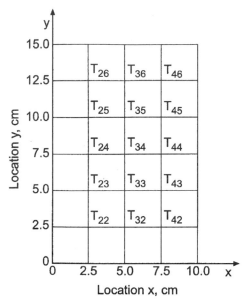

Figure 12.17.   Discretized solution domain $D(x, y)$.

In terms of the i, j notation, Eq. (12.170) becomes

$$-8T_{i,j} + (T_{i+1,j+1} + T_{i+1,j} + T_{i+1,j-1} + T_{i,j+1} + T_{i,j-1} + T_{i-1,j+1}$$
$$+ T_{i-1,j} + T_{i-1,j-1}) = 0 \qquad (12.172)$$

Solving Eq. (12.172) for $T_{i,j}$, adding the term, $\pm T_{i,j}$, to the result, and applying the over-relaxation factor, ω, yields

$$T_{i,j}^{k+1} = T_{i,j}^{k} + \omega \Delta T_{i,j}^{k+1} \qquad (12.173a)$$

$$\Delta T_{i,j}^{k+1} = \frac{\begin{aligned} T_{i+1,j+1} + T_{i+1,j} + T_{i+1,j-1} + T_{i,j+1} + T_{i,j-1} \\ + T_{i-1,j+1} + T_{i-1,j} + T_{i-1,j-1} - 8T_{i,j} \end{aligned}}{8}$$

$$(12.173b)$$

where the most recent values of the terms in the numerator of Eq. (12.173b) are used.

Let $T_{i,j}^{(0)} = 0.0$ at all the interior nodes and let $\omega = 1.23647138$, which is the optimum value of ω for the five point finite difference method. Solving Eq. (12.173) yields the results presented in Table 12.7.

The solution for both a 5×7 grid and a 9×13 grid are presented in Table 12.7. Comparing these results with the results obtained by the five point finite difference method in Section 9.4, which are presented in Table 9.2, shows that the finite element method has slightly larger errors. The Euclidean norms of the errors in Table 9.2 for the two grid sizes are 3.3075 C and 0.8503 C, respectively. The ratio of the norms is 3.89, which shows that the five-point method is second order. The Euclidean norm of the errors in Table 12.7 for the two grid sizes is 3.5889 C and 0.8679 C, respectively, which are slightly larger than the norms obtained by the five-point method. The ratio of the norms is 4.14, which shows that the finite element method is second order.

Table 12.7 Solution of the Laplace Equation by the FEM

	$T(x, y)$, C			
	Error $(x, y) = [T(x, y) - \bar{T}(x, y)]$, C			
	$\Delta x = \Delta y = 2.5$ cm, 5×7 grid		$\Delta x = \Delta y = 1.25$ cm, 9×13 grid	
y, cm	$x = 2.5$ cm	$d = 5.0$ cm	$x = 2.5$ cm	$x = 5.0$ cm
12.5	30.8511	43.6301	31.9007	45.1144
	−1.3787	−1.9497	−0.3291	−0.4654
10.0	13.4487	19.0194	14.3761	20.3308
	−1.2243	−1.7314	−0.2969	−0.4200
7.5	5.8360	8.2534	6.4438	9.1129
	−0.8063	−1.1402	−0.1985	−0.2807
5.0	2.4715	3.4952	2.8110	3.9754
	−0.4524	−0.6398	−0.1129	−0.1596
2.5	0.9060	1.2813	1.0539	1.4904
	−0.1977	−0.2795	−0.0498	−0.0704

Example 12.6 illustrates the application of the finite element method to the Laplace equation. Let's demonstrate the application of the finite element method to the Poisson equation by solving the heat diffusion problem presented in Example 9.6.

Example 12.7. The FEM for the Poisson equation.

Let's apply the FEM to solve the heat diffusion problem presented in Section 9.8. The elliptic partial differential equation is [see Eq. (9.58)]

$$T_{xx} + T_{yy} = -\frac{\dot{Q}}{k} \tag{12.174}$$

with $T(x, y) = 0.0$ C on all the boundaries and $\dot{Q}/k = 1000.0$ C/cm^2. The width of the solution domain is 1.0 cm and the height of the solution domain is 1.5 cm. The exact solution to this problem is presented in Section 9.8. Let $\Delta x = \Delta y = 0.25$ cm. The discretized solution domain is presented in Figure 12.17.

Equations (12.173a) and (12.173b) also apply to this problem, with the addition of the source term, $\Delta x^2 F(x, y) = -\Delta x^2 \dot{Q}/k = 1000.0 \, \Delta x^2$. Thus,

$$T_{i,j}^{k+1} = T_{i,j}^k + \omega \Delta T_{i,j}^{k+1} \tag{12.175a}$$

$$\Delta T_{i,j}^{k+1} = \frac{\begin{array}{c} T_{i+1,j+1} + T_{i+1,j} + T_{i+1,j-1} + T_{i,j+1} + T_{i,j-1} \\ + T_{i-1,j+1} + T_{i-1,j} + T_{i-1,j-1} - 8T_{i,j} + \Delta x^2 F_{i,j} \end{array}}{8}$$

$$\tag{12.175b}$$

where the most recent values of the terms in the numerator of Eq. (12.175b) are used.

Let $T_{i,j}^{(0)} = 0.0$ at all the interior nodes and $\omega = 1.23647138$, which is the optimum value of ω for the five-point finite difference method. Solving Eq. (12.175) gives the results presented in Table 12.8.

Table 12.8 presents the solution for both a 5×7 grid and a 9×13 grid. Comparing these results with the results presented in Table 9.8 for the five-point method shows that

Table 12.8 Solution of the Poisson Equation by the FEM

	$T(x, y)$, C $T(x, y)$, C Error $(x, y) = [T(x, y) - \bar{T}(x, y)]$, C			
	$\Delta x = \Delta y = 0.25$ cm, 5×7 grid		$\Delta x = \Delta y = 0.125$ cm, $9\times$ grid	
y, cm	$x = 0.25$ cm	$x = 0.50$ cm	$x = 0.25$ cm	$x = 0.50$ cm
1.25	52.9678	66.9910	51.0150	65.2054
	50.4429	64.6197	50.4429	64.6197
	2.5249	2.3713	0.5721	0.5857
1.00	73.2409	96.0105	71.5643	93.6686
	71.0186	92.9387	71.0186	92.9387
	2.2223	3.0718	0.5457	0.7299
0.75	78.7179	103.7400	77.1238	101.4930
	76.6063	10.7714	76.6063	100.7714
	2.1116	2.9686	0.5175	0.7216

the finite element method has about 10 percent larger errors than the five-point method. The Euclidean norms of the errors in Table 9.8 are 5.6663 C and 1.4712 C, respectively. The ratio of the norms is 3.85, which shows that the five-point method is second order. The Euclidean norms of the errors in Table 12.8 are 6.2964 C and 1.5131 C, respectively. The ratio of the norms is 4.16, which shows that the finite element method is second order.

12.5. THE FINITE ELEMENT METHOD FOR THE DIFFUSION EQUATION

Section 12.4 presents the application of the finite element method to the two-dimensional Laplace (Poisson) equation. In this section, the finite element method is applied to the one-dimensional diffusion equation:

$$\bar{f}_t = \alpha \bar{f}_{xx} + Q\bar{f} - F \qquad \text{with appropriate auxiliary conditions} \qquad (12.176)$$

where $Q = Q(x)$ and $F = F(x)$. The steps in the finite element approach presented in Section 12.3 also apply to initial-boundary-value problems, with modifications to account for the time derivative. The Galerkin weighted residual method is applied in this section to develop the element equations for the one-dimensional diffusion equation.

12.5.1. Domain Discretization and the Interpolating Polynominals

Consider the global solution domain $D(x, t)$ illustrated in Figure 12.18. The physical space is discretized into I nodes and $I - 1$ elements. The subscript i denotes the nodes and the superscript (i) denotes the elements. Element (i) starts at node i and ends at node $i + 1$. The element lengths (i.e., grid increments) are $\Delta x_i = x_{i+1} - x_i$. The time axis is discretized into time steps $\Delta t^n = t^{n+1} - t^n$. The time steps Δt^n can be variable, that is, $\Delta t^{n-1} \neq \Delta t^n$, or constant, that is, $\Delta t^{n-1} = \Delta t^n = \Delta t = $ constant.

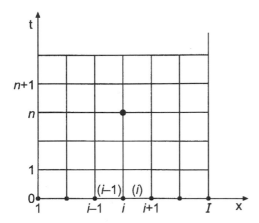

Figure 12.18. Finite element discretization.

Let the global exact solution $\bar{f}(x, y)$ be approximated by the global approximate solution $f(x, t)$, which is the sum of a series of local interpolating polynominals $f^{(i)}(x, t)$ $(i = 1, 2, \ldots, I - 1)$ that are valid within each element. Thus,

$$f(x, t) = f^{(1)}(x, t) + \cdots f^{(i)}(x, t) + \cdots f^{(I-1)}(x, t) = \sum_{i=1}^{I-1} f^{(i)}(x, t) \tag{12.177}$$

The local interpolating polynominals, $f^{(i)}(x, t)$, are defined by Eqs. (12.67) to (12.70), where the nodal values, f_i $(i = 1, 2, \ldots, I)$, are functions of time. Thus,

$$f^{(i)}(x, t) = f_i(t)N_i^{(i)}(x) + f_{i+1}(t)N_{i+1}^{(i)}(x) \tag{12.178}$$

$$N_i^{(i)}(x) = -\frac{x - x_{i+1}}{x_{i+1} - x_i} = -\frac{x - x_{i+1}}{\Delta x_i} \tag{12.179}$$

$$N_{i+1}^{(i)}(x) = \frac{x - x_i}{x_{i+1} - x_i} = \frac{x - x_i}{\Delta x_i} \tag{12.180}$$

Substituting Eqs. (12.179) and (12.180) into Eq. (12.178) yields

$$f^{(i)}(x, t) = f_i(t)\left(-\frac{x - x_{i+1}}{\Delta x_i}\right) + f_{i+1}(t)\left(\frac{x - x_i}{\Delta x_i}\right) \tag{12.181}$$

Equation (12.181) is a linear Lagrange polynominal applied to element (i). Since there are $I - 1$ elements, there are $2(I - 1)$ shape functions in the global physical space. The $2(I - 1)$ shape functions specified by Eqs. (12.179) and (12.180) form a linearly independent set.

12.5.2. The Galerkin Weighted Residual Approach

The Galerkin weighted residual approach is applied in this section to develop a finite element approximation of the one-dimensional diffusion equation, Eq. (12.176):

$$\boxed{\bar{f}_t = \alpha \bar{f}_{xx} + Q\bar{f} - F \qquad \text{with appropriate auxiliary conditions}} \tag{12.182}$$

where $Q = Q(x)$ and $F = F(x)$. Substituting the approximate solution $f(x, t)$ into Eq. (12.182) yields the residual $R(x, t)$:

$$R(x, t) = f_t - \alpha f_{xx} - Qf + F \tag{12.183}$$

The residual $R(x, t)$ is multiplied by a set of weighting functions $W_k(x)$ ($k = 1, 2, \ldots$) and integrated over the global physical domain $D(x)$ to obtain the weighted residual integral, which is equated to zero. Consider the general weighting function $W(x)$. Then,

$$I(f(x, t)) = \int_a^b W(f_t - \alpha f_{xx} - Qf + F) \, dx = 0 \tag{12.184}$$

where $I(f(x, t))$ denotes the weighted residual integral.

The second term on the right-hand size of Eq. (12.184) can be integrated by parts. Thus,

$$-\int_a^b W \alpha f_{xx} \, dx = \int_a^b \alpha W_x f_x \, dx - (W \alpha f_x)_a^b \tag{12.185}$$

The last term in Eq. (12.185) cancels out at all the interior nodes when the element equations are assembled. It is applicable only at nodes 1 and I when derivative boundary conditions are applied. Consequently, that term will be dropped from further consideration except when a derivative boundary condition is present. Substituting Eq. (12.185), without the last term, into Eq. (12.184) gives

$$I(f(x, t)) = \int_a^b W f_t \, dx + \int_a^b \alpha W_x f_x \, dx - \int_a^b W Q f \, dx + \int_a^b W F \, dx = 0 \tag{12.186}$$

In terms of the global approximate solution, $f(x, t)$, and the discretized global physical domain illustrated in Figure 12.18, Eq. (12.186) can be written as follows:

$$I(f(x, t)) = I^{(1)}(f(x, t)) + \cdots + I^{(i)}(f(x, t)) + \cdots + I^{(I-1)}(f(x, t)) = 0 \tag{12.187}$$

where $I^{(i)}(f(x, t))$ is given by

$$I^{(i)}(f(x, t)) = \int_{x_i}^{x_{i+1}} W f_t^{(i)} \, dx + \int_{x_i}^{x_{i+1}} \alpha W_x f_x^{(i)} \, dx - \int_{x_i}^{x_{i+1}} W Q f^{(i)} \, dx + \int_{x_i}^{x_{i+1}} W F \, dx = 0 \tag{12.188}$$

where $f^{(i)}(x, t)$ is given by Eq. (12.181) and $W(x)$ is an as yet unspecified weighting function. Since the shape functions $N_i^{(i)}(x)$ and $N_{i+1}^{(i)}(x)$ are defined to be zero everywhere outside of element (i), each individual weighted residual integral $I^{(i)}(f(x, t))$ must be zero to satisfy Eq. (12.187).

The evaluation of Eq. (12.188) requires the function $f^{(i)}(x, t)$ and its partial derivatives with respect to t and x. From Eqs. (12.178) to (12.80),

$$f^{(i)}(x, t) = f_i(t) N_i^{(i)}(x) + f_{i+1}(t) N_{i+1}^{(i)}(x) \tag{12.189}$$

$$N_i^{(i)}(x) = -\frac{x - x_{i+1}}{\Delta x_i} \quad \text{and} \quad N_{i+1}^{(i)}(x) = \frac{x - x_i}{\Delta x_i} \tag{12.190}$$

Differentiating Eq. (12.189) with respect to t and x gives

$$f_t^{(i)} = \dot{f}_i \left(-\frac{x - x_{i+1}}{\Delta x_i} \right) + \dot{f}_{i+1} \left(\frac{x - x_i}{\Delta x_i} \right) \tag{12.191}$$

$$f_x^{(i)} = f_i \left(-\frac{1}{\Delta x_i} \right) + f_{i+1} \left(\frac{1}{\Delta x_i} \right) = \frac{f_{i+1} - f_i}{\Delta x_i} \tag{12.192}$$

where $\dot{f}_i = d[f_i(t)]/dt$ and $\dot{f}_{i+1} = d[f_{i+1}(t)/dt]$. Substituting Eqs. (12.189) to (12.192) into Eq. (12.188) yields

$$I^{(i)}(f(x,\,t)) = \int_{x_i}^{x_{i+1}} W(\dot{f}_i N_i^{(i)} + \dot{f}_{i+1} N_{i+1}^{(i)})\,dx + \int_{x_i}^{x_{i+1}} \alpha W_x \frac{f_{i+1} - f_i}{\Delta x_i}\,dx$$

$$- \int_{x_i}^{x_{i+1}} WQ - f_i N_i^{(i)} + f_{i+1} N_{i+1}^{(i)})\,dx + \int_{x_i}^{x_{i+1}} WF\,dx = 0 \tag{12.193}$$

Let's denote $I^{(i)}(f(x,\,y))$ symbolically as

$$I^{(i)}(f(x,\,y)) = A + B + C + D \tag{12.194}$$

where A, B, etc., denote the four integrals in Eq. (12.193).

In the Galerkin weighted residual approach, the weighting factors W_k ($k = 1, 2, \ldots$) are chosen to be the shape functions $N_i^{(i)}(x)$ and $N_{i+1}^{(i)}(x)$ specified by Eq. (12.190). Let $W(x) = N_i^{(i)}(x)$. Then,

$$W(x) = N_i^{(i)}(x) = -\frac{x - x_{i+1}}{\Delta x_i} \quad \text{and} \quad W_x = -\frac{1}{\Delta x_i} \tag{12.195}$$

Substitute $W(x)$ and W_x into Eq. (12.193) and evaluate integrals A, B, C, and D.

$$A = \int_{x_i}^{x_{i+1}} -\left(\frac{x - x_{i+1}}{\Delta x_i} \right) \left[\dot{f}_i \left(-\frac{x - x_{i+1}}{\Delta x_i} \right) + \dot{f}_{i+1} \left(\frac{x - x_i}{\Delta x_i} \right) \right] dx \tag{12.196}$$

$$A = \frac{1}{\Delta x_i^2} \int_{x_i}^{x_{i+1}} [\dot{f}_i(x^2 - 2x_{i+1}x + x_{i+1}^2) - \dot{f}_{i+1}(x^2 - x_{i+1}x - x_i x + x_{i+1}x_i)]\,dx \tag{12.197}$$

$$A = \frac{1}{\Delta x_i^2} \left[\dot{f}_i \left(\frac{x^3}{3} - x_{i+1}x^2 + x_{i+1}^2 x \right) - \dot{f}_{i+1} \left(\frac{x^3}{3} - \frac{x_{i+1}x}{2} - \frac{x_i x^2}{2} + x_{i+1}x_i x \right) \right] \Bigg|_{x_i}^{x_{i+1}}$$

$$\tag{12.198}$$

Introducing the limits of integration and simplifying Eq. (12.198) gives

$$A = \frac{\Delta x_i}{6}(2\dot{f}_i + \dot{f}_{i+1}) \tag{12.199}$$

Substituting Eq. (12.195) into integral B and evaluating gives

$$B = \int_{x_i}^{x_{i+1}} \alpha \left(-\frac{1}{\Delta x_i} \right) \frac{f_{i+1} - f_i}{\Delta x_i}\,dx = -\frac{\alpha(f_{i+1} - f_i)}{\Delta x_i^2} x \Bigg|_{x_i}^{x_{i+1}} = -\frac{\alpha(f_{i+1} - f_i)}{\Delta x_i} \tag{12.200}$$

Substituting Eq. (12.195) into integral C gives

$$C = -\int_{x_i}^{x_{i+1}} \left(-\frac{x - x_{i+1}}{\Delta x_i} \right) Q \left[f_i \left(-\frac{x - x_{i+1}}{\Delta x_i} \right) + f_{i+1} \left(\frac{x - x_i}{\Delta x_i} \right) \right] dx \tag{12.201}$$

Let $\bar{Q}$ denote the average value of $Q(x)$ over element (i). Then,

$$C = -\frac{\bar{Q}}{\Delta x_i^2} \int_{x_i}^{x_{i+1}} [f_i(x^2 - 2x_{i+1}x + x_{i+1}^2) - f_{i+1}(x^2 - x_{i+1}x - x_i x + x_{i+1}x_i)]\, dx$$

(12.202)

Integrating Eq. (12.202) and evaluating the result yields

$$C = -\frac{\bar{Q}\Delta x_i}{6}(2f_i + f_{i+1})$$

(12.203)

Finally, substituting Eq. (12.195) into integral D, integrating, and evaluating the result yields

$$D = \int_{x_i}^{x_{i+1}} \left(-\frac{x - x_{i+1}}{\Delta x_i}\right) F\, dx = -\frac{\bar{F}}{\Delta x_i}\left(\frac{x^2}{2} - x_{i+1}x\right)\Big|_{x_i}^{x_{i+1}}\, dx = \frac{\Delta x_i \bar{F}}{2}$$

(12.204)

where $\bar{F}$ denotes the average value of $F(x)$ over element (i). Substituting the results for A, B, C, and D into Eq. (12.194) yields the first element equation for element (i). Thus,

$$\boxed{I^{(i)}(f(x, y)) = \frac{\Delta x_i}{6}(2\dot{f}_i + \dot{f}_{i+1}) - \frac{\alpha(f_{i+1} - f_i)}{\Delta x_i} - \frac{\bar{Q}\Delta x_i}{6}(2f_i + f_{i+1}) + \frac{\Delta x_i \bar{F}}{2} = 0}$$

(12.205)

Next, let $W(x) = N_{i+1}^{(i)}(x)$. Then,

$$W(x) = N_{i+1}^{(i)}(x) = \frac{x - x_i}{\Delta x_i} \qquad \text{and} \qquad W_x = \frac{1}{\Delta x_i}$$

(12.206)

Substituting Eq. (12.206) into Eq. (12.193) and evaluating integrals A, B, C, and D yields the second element equation for element (i). Thus,

$$\boxed{I^{(i)}(f(x, y)) = \frac{\Delta x_i}{6}(\dot{f}_i + 2\dot{f}_{i+1}) + \frac{\alpha(f_{i+1} - f_i)}{\Delta x_i} - \frac{\bar{Q}\Delta x_i}{6}(f_i + 2f_{i+1}) + \frac{\Delta x_i \bar{F}}{2} = 0}$$

(12.207)

Equations (12.205) and (12.207) are the element equations for element (i).

Next let's assemble the element equations to obtain the nodal equation for node i. Figure 12.19a illustrates the portion of the discretized global physical domain surrounding node i. Note that $\Delta x_{i-1} = x_i - x_{i-1} \neq \Delta x_i = x_{i+1} - x_i$. Consider element $(i-1)$ in Figure 12.19a. Node i in element $(i-1)$ corresponds to node $(i+1)$ in the general element illustrated in Figure 12.19b. Thus, the element equation corresponding to node i in element $(i-1)$ is Eq. (12.207) with i replaced by $i-1$. Thus

$$\frac{\Delta x_-}{6}(\dot{f}_{i-1} + 2\dot{f}_i) + \frac{\alpha(f_i - f_{i-1})}{\Delta x_-} - \frac{\bar{Q}^{(i-1)}\Delta x_-}{6}(f_{i-1} + 2f_i) + \frac{\Delta x_- \bar{F}^{(i-1)}}{2} = 0 \quad (12.208)$$

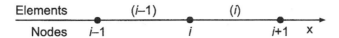

Elements   (*i*–1)   (*i*)

Nodes   *i*–1   *i*   *i*+1   x

(a) Portion of global grid surrounding node *i*.

(*i*)   (*i*–1)   (*i*)

i   *i*+1   *i*–1   *i*   *i*   *i*+1

(b) General element.   (c) Element *i*–1.   (d) Elememt *i*.

Figure 12.19.   Element correspondence.

Consider element (*i*) in Figure 12.19c. Node *i* in element (*i*) corresponds to node *i* in the general element illustrated in Figure 12.19a. Thus, the element equation corresponding to node *i* in element (*i*) is Eq. (12.205). Thus,

$$\frac{\Delta x_+}{6}(2\dot{f}_i + \dot{f}_{i+1}) - \frac{\alpha(f_{i+1} - f_i)}{\Delta x_+} - \frac{\bar{Q}^{(i)}\Delta x_+}{6}(2f_i + f_{i+1}) + \frac{\Delta x_+ \bar{F}^{(i)}}{2} = 0 \qquad (12.209)$$

Multiplying Eq. (12.208) by $6/\Delta x_-$ and Eq. (12.209) by $6/\Delta x_+$ and adding yields the nodal equation for node *i*:

$$\dot{f}_{i-1} + 4\dot{f}_i + \dot{f}_{i+1} + \frac{6\alpha(f_i - f_{i-1})}{\Delta x_-^2} - \frac{6\alpha(f_{i+1} - f_i)}{\Delta x_+^2} - \bar{Q}^{(i-1)}(f_{i-1} + 2f_i)$$
$$- \bar{Q}^{(i)}(2f_i + f_{i+1}) + 3(\bar{F}^{(i-1)} + \bar{F}^{(i)}) = 0 \qquad (12.210)$$

Next, let's develop a finite difference approximation for $\dot{f}$. Several possibilities exist. For example,

$$\dot{f}^n = \frac{f^{n+1} - f^n}{\Delta t} \qquad \dot{f}^{n+1} = \frac{f^{n+1} - f^n}{\Delta t} \qquad \dot{f}^{n+1/2} = \frac{f^{n+1} - f^n}{\Delta t} \qquad (12.211)$$

The first expression is a first-order forward-time approximation, the second expression is a first-order backward-time approximation, and the third expression is a second-order centered-time approximation. When using any of these finite difference approximations, the function values in Eq. (12.210) must be evaluated at the corresponding time level.

Let's develop the forward-time approximation. Substituting the first expression in Eq. (12.211) into Eq. (12.210), evaluating all the function values in Eq. (12.210) at time level *n*, and multiplying through by Δt yields

$$f_{i-1}^{n+1} + 4f_i^{n+1} + f_{i+1}^{n+1} = f_{i-1}^n + 4f_i^n + f_{i+1}^n$$
$$- \frac{6\alpha\,\Delta t\,(f_i^n - f_{i-1}^n)}{\Delta x_-^2} + \frac{6\alpha\,\Delta t\,(f_{i+1}^n - f_i^n)}{\Delta x_+^2}$$
$$+ \Delta t\,\bar{Q}^{(i-1)}(f_{i-1}^n + 2f_i^n) + \Delta t\,\bar{Q}^{(i)}(2f_i^n + f_{i+1}^n)$$
$$- 3\,\Delta t\,(\bar{F}^{(i-1)} + \bar{F}^{(i)}) = 0 \qquad (12.212)$$

Let $\Delta x_- = \Delta x_+ = \Delta x$, $Q = $ constant, $F = $ constant, and $d = \alpha \Delta t / \Delta x^2$. Equation (12.212) becomes

$$f_{i-1}^{n+1} + 4f_i^{n+1} + f_{i+1}^{n+1} = (f_{i-1}^n + 4f_i^n + f_{i+1}^n) + 6d(f_{i-1}^n - 2f_i^n + f_{i+1}^n)$$
$$+ \Delta t\, Q(f_{i-1}^n + 4f_i^n + f_{i+1}^n) - 6\Delta t\, F \qquad (12.213)$$

Equation (12.212) is the nodal equation for a nonuniform grid, and Eq. (12.213) is the nodal equation for a uniform grid with $Q = $ constant and $F = $ constant.

Example 12.8.   The FEM for the diffusion equation.

Let's apply the results obtained in this section to obtain the solution of the steady heat transfer problem presented in Section 8.1 as the asymptotic steady-state solution of the one-dimensional diffusion equation at large time. The boundary-value ODE is [see Eq. (8.1)]:

$$T'' - \alpha^2 T = -\alpha^2 T_a \qquad T(0.0) = 0.0 \qquad T(1.0) = 100.0 \qquad (12.214)$$

where $\alpha^2 = 16.0\,\text{cm}^{-2}$ and $T_a = 0.0\,\text{C}$. The exact solution to this steady-state problem is presented in Section 8.1. Applying Eq. (12.176) to this heat transfer problem gives

$$T_t = \alpha T_{xx} + QT - F \qquad (12.215)$$

Note that α^2 in Eq. (12.214) is not the same α as the α in Eq. (12.215). For the present problem, let $\alpha = 0.01\,\text{cm}^2/\text{s}$, $Q = -0.16\,\text{s}^{-1}$, $T_a = 0.0$ (for which $F = 0.0$), and $\Delta x = 0.25\,\text{cm}$. The discretized physical space is illustrated in Figure 12.20. Let the initial temperature distribution be $T(x, 0.0) = 100.0x$. Let $\Delta t = 1.0\,\text{s}$, and march 50 time steps to approach the asymptotic steady-state solution.

For these data, $6d = 6\alpha\,\Delta t/\Delta x^2 = 6(0.01)(1.0)/(0.25)^2 = 0.96$ and $(1 + \Delta t\, Q) = (1 + 1.0(-0.16)) = 0.84$. Equation (12.213) gives

$$f_{i-1}^{n+1} + 4f_i^{n+1} + f_{i+1}^{n+1} = 0.84(f_{i-1}^n + 4f_i^n + f_{i+1}^n) + 0.96(f_{i-1}^n - 2f_i^n + f_{i+1}^n) \qquad (12.216)$$

Applying Eq. (12.216) at nodes 2 to 4 gives

Node 2:	$T_1^{n+1} + 4T_2^{n+1} + T_3^{n+1} = b_2$	(12.217a)
Node 3:	$T_2^{n+1} + 4T_3^{n+1} + T_4^{n+1} = b_3$	(12.217b)
Node 4:	$T_3^{n+1} + 4T_4^{n+1} + T_5^{n+1} = b_4$	(12.217c)

where

$$b_i = 0.84(f_{i-1}^n + 4f_i^n + f_{i+1}^n) + 0.96(f_{i-1}^n - 2f_i^n + f_{i+1}^n) \qquad (12.218)$$

Setting $T_1(0.0) = 0.0$, $T_2(0.0) = 25.0$, $T_3 = (0.0) = 50.0$, $T_4(0.0) = 75.0$, and $T_5(0.0) = 100.0$ and applying the Thomas algorithm to solve Eq. (12.217) yields the solution presented in line 2 of Table 12.9. The results for subsequent time steps are also

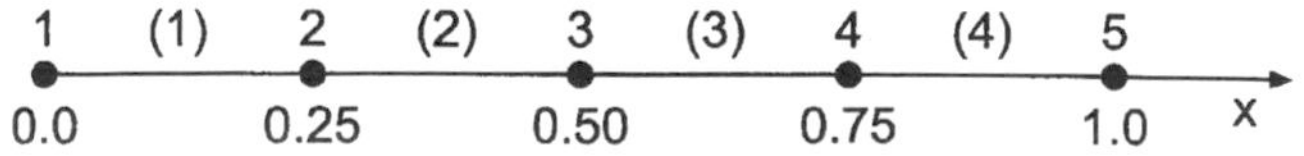

Figure 12.20.   Discretized physical space.

Table 12.9 Solution of the Diffusion Equation by the FEM

t, s	x, cm				
	0.00	0.25	0.50	0.75	1.00
0.0	0.0	25.000000	50.000000	75.000000	100.0
1.0	0.0	20.714286	43.142857	58.714286	100.0
2.0	0.0	18.466122	33.621224	52.146122	100.0
3.0	0.0	14.646134	28.524884	46.770934	100.0
4.0	0.0	12.119948	23.955433	43.684876	100.0
5.0	0.0	9.907778	20.941394	41.271152	100.0
10.0	0.0	5.132565	14.030479	36.383250	100.0
20.0	0.0	3.855092	12.224475	35.105092	100.0
30.0	0.0	3.795402	12.140060	35.045402	100.0
40.0	0.0	3.792612	12.136116	35.042612	100.0
50.0	0.0	3.792482	12.135931	35.042482	100.0
Steady-state	0.0	4.761905	14.285714	38.095238	100.0
Exact	0.0	4.306357	13.290111	36.709070	100.0

presented in Table 12.9. The next to the last line presents the solution obtained by the second-order equilibrium method in Example 8.4, and the last line presents the exact solution.

The Euclidean norm of the errors in the steady-state solution obtained in Example 8.4 is 1.766412 C. The Euclidean norm of the errors in Table 12.9 at $t = 50.0$ s is 2.091343 C, which is 18 percent larger.

12.6. PROGRAMS

Three FORTRAN programs for implementing the finite element method are presented in this section:

1. Boundary-value ordinary differential equations
2. The two-dimensional Laplace (Poisson) equation
3. The one-dimensional diffusion equation

The basic computational algorithms are presented as completely self-contained subroutines suitable for use in other programs. Input data and output statements are contained in a main (or driver) program written specifically to illustrate the use of each subroutine.

12.6.1. Boundary-Value Ordinary Differential Equations

The boundary-value ordinary differential equation considered in Section 12.3 is given by Eq. (12.65):

$$\bar{y}'' + Q\bar{y} = F \qquad \text{with appropriate boundary conditions} \qquad (12.218)$$

The finite element method applied to Eq. (12.218) yields Eq. (12.97) for a nonuniform grid. For a uniform grid, the corresponding result is given by Eq. (12.98). These equations are applied at every interior point in a finite difference grid. The resulting system of FDEs, which is called the *system equation*, is solved by the Thomas algorithm. An initial approximation $y(x)^{(0)}$ must be specified. If the ODE is linear, the solution is obtained in one pass. If the ODE is nonlinear, the solution is obtained iteratively.

A FORTRAN subroutine, *subroutine fem1*, for implementing Eqs. (12.97) and Eq. (12.98) is presented below in Program 12.1. *Program main* defines the data set and prints it, calls *subroutine fem1* to set up and solve the system of FDEs, and prints the solution. A first guess for the solution $y(i)^{(0)}$, must be supplied in a *data* statement. *Subroutine thomas*, Section 1.8.3, is used to solve the system equation.

Program 12.1.   The boundary-value ordinary differential equation FEM program.

```
      program main
c     main program to illustrate the FEM for ODEs
c     nd     array dimension, nd = 9 in this program
c     imax   number of grid points in the x direction
c     x      x direction grid points, x(i)
c     dxm    dx- grid increment
c     dxp    dx+ grid increment
c     y      solution array, y(i)
c     q      coefficient of y in the ODE
c     fx     nonhomogeneous term
c     bc     right side boundary condition, 1.0 y, 2.0 y'
c     yp2    right side derivative boundary condition, y'
c     iter   maximum number of iterations
c     tol    convergence tolerance
c     iw     intermediate results output flag: 0 none, 1 all
c     ix     output increment: 1 all points, n every nth point
      dimension x(9),dxm(9),dxp(9),y(9),a(9,3),b(9),z(9)
      data nd,imax,iter,tol,ix,iw /9, 5, 1, 1.0e-06, 1, 1/
      data (x(i),i=1,5) /0.0, 0.25, 0.50, 0.75, 1.00/
c2    data (x(i),i=1,5) /0.0, 0.375, 0.66666667, 0.875, 1.0/
      data (y(i),i=1,5) /0.00, 25.0, 50.0, 75.0, 100.0/
c2    data (y(i),i=1,5) /0.0, 37.5, 66.666667, 87.5, 100.0/
c3    data (y(i),i=1,5) /100.0, 75.0, 50.0, 25.0, 0.0/
      data bc,yp2 /1.0, 0.0/
c3    data bc,yp2 /2.0, 0.0/
      data fx,q /0.0, -16.0/
      write (6,1000)
      if (iw.eq.1) write (6,1010) (i,x(i),y(i),i=1,imax,ix)
      do i=2,imax-1
         dxm(i)=x(i)-x(i-1)
         dxp(i)=x(i+1)-x(i)
      end do
      call fem1 (nd,imax,x,dxm,dxp,y,q,fx,bc,yp2,a,b,z,iter,
     1 tol,ix,iw)
      if (iw.eq.0) write (6,1010) (i,x(i),y(i),i=1,imax,ix)
      stop
```

```
1000 format (' Finite element method for ODEs'/' '/'  i',7x,
    1 'x',12x,'f'/' ')
1010 format (i3,2f13.6)
     end


     subroutine fem1 (nd,imax,x,dxm,dxp,y,q,fx,bc,yp2,a,b,z,
    1 iter,tol,ix,iw)
c    implements the FEM for a second-order ODE
     dimension x(nd),dxm(nd),dxp(nd),y(nd),a(nd,3),b(nd),
    1 z(nd)
     a(1,2)=1.0
     a(1,3)=0.0
     b(1)=y(1)
     if (bc.eq.1.0) then
        a(imax,1)=0.0
        a(imax,2)=1.0
        b(imax)=y(imax)
     else
        a(imax,1)=1.0+q*dxp(imax-1)**2/6.0
        a(imax,2)=-(1.0-q*dxp(imax-1)**2/3.0)
        b(imax)=0.5*fx*dxp(imax-1)**2-dxp(imax-1)*yp2
     end if
     do it=1,iter
        do i=2,imax-1
           a(i,1)=1.0/dxm(i)+q*dxm(i)/6.0
           a(i,2)=-(1.0/dxm(i)+1.0/dxp(i)-q*dxm(i)/3.0
    1         -q*dxp(i)/3.0)
           a(i,3)=1.0/dxp(i)+q*dxp(i)/6.0
           b(i)=0.5*(fx*dxm(i)+fx*dxp(i))
        end do
        call thomas (nd,imax,a,b,z)
        dymax=0.0
        do i=1,imax
           dy=abs(y(i)-z(i))
           if (dy.gt.dymax) dymax=dy
           y(i)=z(i)
        end do
        if (iw.eq.1) write (6,1000)
        if (iw.eq.1) write (6,1010)(i,x(i),y(i),i=1,imax,ix)
        if (dymax.le.tol) return
     end do
     if (iter.gt.1) write (6,1020) iter
     return
1000 format (' ')
1010 format (i3,2f13.6)
1020 format (' '/' Solution failed to converge, it = ',i3)
     end


     subroutine thomas (ndim,n,a,b,x)
c    the Thomas algorithm for a tridiagonal system
     end
```

The data set used to illustrate *subroutine fem1* for a uniform grid is taken from Example 12.3. The uniform grid is defined in the *data* statements. The output generated by the program is presented in Output 12.1.

Output 12.1.   Solution of a boundary-value ODE with a uniform grid by the FEM.

Finite element method for ODEs

i	x	y
1	0.000000	0.000000
2	0.250000	25.000000
3	0.500000	50.000000
4	0.750000	75.000000
5	1.000000	100.000000
1	0.000000	0.000000
2	0.250000	3.792476
3	0.500000	12.135922
4	0.750000	35.042476
5	1.000000	100.000000

The solution for a nonuniform grid also can be obtained by *subroute fem1*. All that is required is to define the nonuniform grid in a *data* statement. The data set used to illustrate this option is taken from Example 12.4. The required *data* statements are included in *program main* as *comment statements c2*. The output generated by the program for a nonuniform grid is illustrated in Output 12.2.

Output 12.2.   Solution of a boundary-value ODE with a nonuniform grid by the FEM.

Finite element method for ODEs

i	x	y
1	0.000000	0.000000
2	0.375000	37.500000
3	0.666667	66.666667
4	0.875000	87.500000
5	1.000000	100.000000
1	0.000000	0.000000
2	0.375000	6.864874
3	0.666667	24.993076
4	0.875000	59.868411
5	1.000000	100.000000

Lastly, the solution of a boundary-value ODE with a derivative boundary condition also can be obtained by *subroutine fem1*. The data set used to illustrate *subroutine fem1* for a uniform grid with a derivative boundary condition is taken from Example 12.5. The required *data* statements are included in *program main* as *comment statements c3*. The output generated by the program is presented in Output 12.3.

Output 12.3. Solution of a boundary-value ODE with a derivative boundary condition.

Finite element method for ODEs

i	x	y
1	0.000000	100.000000
2	0.250000	75.000000
3	0.500000	50.000000
4	0.750000	25.000000
5	1.000000	0.000000
1	0.000000	100.000000
2	0.250000	35.157578
3	0.500000	12.504249
4	0.750000	4.856019
5	1.000000	3.035012

12.6.2. The Laplace (Poisson) Equation

The Laplace (Poisson) equation is given by Eq. (12.137):

$$\bar{f}_{xx} + \bar{f}_{yy} = F(x, y) \qquad \text{with appropriate boundary conditions} \qquad (12.221)$$

The finite element algorithm for solving Eq. (12.221) for a rectangular global domain with rectangular elements for uniform Δx and Δy is presented in Eq. (12.168). The corresponding algorithm for $\Delta x =$ constant and $\Delta y =$ constant, but $\Delta x \neq \Delta y$, is presented in Eq, (12.169). For $\Delta x = \Delta y =$ constant, the corresponding algorithm is given by Eq. (12.170). A FORTRAN subroutine, *subroutine fem2*, for implementing Eq. (12.170) is presented in Program 12.2. *Program main* defines the data set and prints it, calls *subroutine fem2* to implement the solution, and prints the solution.

Program 12.2. The Laplace (Poisson) equation FEM program

```
      program main
c     main program to illustrate the FEM for PDEs
c     nxd x-direction array dimension, nxd = 9
c     nyd y-direction array dimension, nyd = 13
c     imax  number of grid points in the x direction
c     jmax  number of grid points in the y direction
c     iw    intermediate results output flag: 0 none, 1 all
c     ix    output increment: 1 all points, n every nth point
c     x     x direction array, x(i,j)
c     y     y direction array, y(i,j)
c     f     solution array, f(i,j)
c     fx    right-hand side derivative boundary condition
```

```
c      fxy    nonhomogeneous term in the Poisson equation
c      dx,dy x-direction and y-direction grid increments
c      iter   maximum number of iterations
c      tol    convergence tolerance
c      omega sor overrelaxation factor
       dimension x(9,13),y(9,13),f(9,13)
       data nxd,nyd,imax,jmax,iw,ix / 9, 13, 5, 7, 0, 1 /
       data (f(i,1),i=1,5) /0.0, 100.0, 70.71067812,
      1 70.71067812, 0.0/
c2     data (f(i,1),i=1,5) /0.0, 0.0, 0.0, 0.0, 0.0/
       data (f(i,7),i=1,5) /0.0, 0.0, 0.0, 0.0, 0.0/
       data (f(1,j),j=2,6) /0.0, 0.0, 0.0, 0.0, 0.0/
       data (f(5,j),j=2,6) /0.0, 0.0, 0.0, 0.0, 0.0/
       data fx,fxy /0.0, 0.0/
c2     data fx,fxy /0.0, 1000.0/
       data dx,dy,iter,tol,omega /2.5, 2.5, 25, 1.0e-06,
      1 1.23647138/
c2     data dx,dy,iter,tol,omega /0.25, 0.25, 25, 1.0e-06,
c2    1 1.23647138/
       do i=2,imax-1
          do j=2,jmax-1
             f(i,j)=0.0
          end do
       end do
       write (6,1000)
       if (iw.eq.1) then
          do j=1,jmax,ix
             write (6,1010) (f(i,j),i=1,imax,ix)
          end do
       end if
       call fem2 (nxd,nyd,imax,jmax,x,y,f,fx,fxy,dx,dy,iter,
      1 tol,omega,iw,ix)
       if (iw.eq.0) then
          do j=1,jmax,ix
             write (6,1010) (f(i,j),i=1,imax,ix)
          end do
       end if
       stop
 1000 format (' FEM Laplace (Poisson) equation solver'/' ')
 1010 format (5f12.6)
       end


       subroutine fem2 (nxd,nyd,imax,jmax,x,y,f,fx,fxy,dx,dy,
      1 iter,tol,omega,iw,ix)
c      Laplace (Poisson) equation solver with Dirichlet BCs
  dimension x(nxd,nyd),y(nxd,nyd),f(nxd,nyd)
  do it=1,iter
     dfmax=0.0
     do j=2,jmax-1
        do i=2,imax-1
           df=(f(i+1,j+1)+f(i+1,j)+f(i+1,j-1)+f(i,j+1)
      1       +f(i,j-1)+f(i-1,j+1)+f(i-1,j)+f(i-1,j-1)
      2       -8.0*f(i,j)+3.0*dx**2*fxy)/8.0
           if (abs(df).gt.dfmax) dfmax=abs(df)
```

```
        if (abs(df).gt.dfmax) dfmax=abs(df)
        f(i,j)=f(i,j)+omega*df
      end do
      end do
      if (iw.eq.1) then
        do j=1,jmax,ix
          write (6,1000) (f(i,j),i=1,imax,ix)
        end do
      end if
      if (dfmax.le.tol) then
        write (6,1010) it,dfmax
        return
      end if
    end do
    write (6,1020) iter
    return
000 format (5f12.6)
010 format (' Solution converged, it =',i3,
  1 ',   dfmax =',e12.6/' ')
020 format (' Solution failed to converge, iter =',i3/' ')
    end
```

The data set used to illustrate *subroutine fem2* for the Laplace equation is taken from Example 12.6. The output generated by the program is presented in Output 12.4.

Output 12.4. Solution of the Laplace equation by the FEM.

```
FEM Laplace (Poisson) equation solver

The solution converged, it = 14,   dfmax =0.361700E-06

 0.000000   70.710678  100.000000   70.710678    0.000000
 0.000000   30.851112   43.630061   30.851112    0.000000
 0.000000   13.448751   19.019405   13.448751    0.000000
 0.000000    5.836032    8.253395    5.836032    0.000000
 0.000000    2.471489    3.495213    2.471489    0.000000
 0.000000    0.905997    1.281273    0.905997    0.000000
 0.000000    0.000000    0.000000    0.000000    0.000000
```

The solution of the Poisson equation also can be obtained with *subroutine fem2*. The only additional data required are the boundary values and the value of the nonhomogeneous term $F(x, y)$. The data set used to illustrate this option is taken from Example 12.7. The present *subroutine fem* is limited to a constant value of $F(x, y)$. The necessary *data* statements are included in *program main* as *comment statements c2*. The output generated by the Poisson equation program is presented in Output 12.5.

Output 12.5. Solution of the Poisson equation by the FEM.

```
FEM Laplace (Poisson) equation solver

The solution converged, it = 16,   dfmax =0.300418E-06

 0.000000    0.000000    0.000000    0.000000    0.000000
 0.000000   52.967802   66.990992   52.967802    0.000000
```

0.000000	73.240903	96.010522	73.240903	0.000000
0.000000	78.717862	103.740048	78.717862	0.000000
0.000000	73.240903	96.010522	73.240903	0.000000
0.000000	52.967802	66.990991	52.967802	0.000000
0.000000	0.000000	0.000000	0.000000	0.000000

12.6.3. The Diffusion Equation

The diffusion equation is given by Eq. (12.176):

$$\bar{f}_t = \alpha \bar{f}_{xx} + Q\bar{f} - F(x) \qquad \text{with appropriate auxiliary conditions} \qquad (12.222)$$

The finite element algorithm for solving Eq. (12.222) for a nonuniform grid is given by Eq. (12.211). The corresponding algorithm for a uniform grid is given by Eq. (12.212).

A FORTRAN subroutine, *subroutine fem3*, for implementing Eq. (12.211) is presented in Program 12.3. *Program main* defines the data set and prints it, calls *subroutine fem3* to implement the solution, and prints the solution.

Program 12.3.   The diffusion equation FEM program.

```
      program main
c     main program to illustrate the FEM for PDEs
c     nxd x-direction array dimension, nxd = 9
c     ntd t-direction array dimension, ntd = 101
c     imax  number of grid points in the x direction
c     nmax  number of time steps
c     iw    intermediate results output flag: 0 none, 1 all
c     ix,it output increment: 1 all points, n every nth point
c     f     solution array, f(i,n)
c     q     coefficient of f in differential equation
c     fx    nonhomogeneous term
c     x     x axis grid points, x(i)
c     dxm   dx- grid increment
c     dxp   dx+ grid increment
c     dt    time step
c     alpha diffusion coefficient
      dimension f(9,101),x(9),dxm(9),dxp(9),a(9,3),b(9),z(9)
      data nxd,ntd,imax,nmax,iw,ix,it /9,101,5,50,0,1,1/
c2    data nxd,ntd,imax,nmax,iw,ix,it /9,101,5,101,0,1,2/
      data (x(i),i=1,5) /0.0, 0.25, 0.50, 0.75, 1.0/
c2    data (x(i),i=1,5) /0.0, 0.375, 0.66666667, 0.875, 1.0/
      data (f(i,1),i=1,5) /0.0, 25.0, 50.0, 75.0, 100.0/
c2    data (f(i,1),i=1,5) /0.0, 37.5, 66.666667, 87.5,100.0/
      data dt,alpha,n,t,q,fx /1.0, 0.01, 1, 0.0, -0.16, 0.0/
c2    data dt,alpha,n,t,q,fx /0.5, 0.01, 1, 0.0, -0.16, 0.0/
      do i=2,imax-1
         dxm(i)=x(i)-x(i-1)
         dxp(i)=x(i+1)-x(i)
      end do
      write (6,1000)
      write (6,1010) n,t,(f(i,1),i=1,imax,ix)
```

```
      call fem3 (nxd,ntd,imax,nmax,f,q,fx,dxm,dxp,dt,alpha,n,
     1 t,iw,ix,a,b,z)
      if (iw.eq.1) stop
      do n=it+1,nmax,it
         t=float(ix*(n-1))*dt
         write (6,1010) n,t,(f(i,n),i=1,imax,ix)
      end do
      stop
 1000 format (' FEM diffusion equation solver'/' '/
     1 '  n',1x,'time',18x,'f(i,n)'/' ')
 1010 format (i3,f5.1,9f8.3)
      end


      subroutine fem3 (nxd,ntd,imax,nmax,f,q,fx,dxm,dxp,dt,
     1 alpha,n,t,iw,ix,a,b,z)
c     implements the FEM for the diffusion equation
      dimension f(nxd,ntd),dxm(nxd),dxp(nxd),a(nxd,3),
     1 b(nxd),z(nxd)
      if (iw.eq.1) write (6,1000) n,t,(f(i,1),i=1,imax,ix)
      a(1,2)=1.0
      a(1,3)=0.0
      b(1)=f(1,1)
      a(imax,1)=0.0
      a(imax,2)=1.0
      b(imax)=f(imax,1)
      do n=1,nmax-1
         t=t+dt
         do i=2,imax-1
            a(i,1)=dxm(i)
            a(i,2)=2.0*(dxm(i)+dxp(i))
            a(i,3)=dxp(i)
            b(i)=dxm(i)*f(i-1,n)+2.0*(dxm(i)+dxp(i))*f(i,n)
     1         +dxp(i)*f(i+1,n)
     2         +6.0*alpha*dt*((f(i+1,n)-f(i,n))/dxp(i)
     3         -(f(i,n)-f(i-1,n))/dxm(i))
     4         +q*dt*dxp(i)*(2.0*f(i,n)+f(i+1,n))
     5         +q*dt*dxm(i)*(f(i-1,n)+2.0*f(i,n))
     6         -3.0*dt*(dxm(i)*fx+dxp(i)*fx)
         end do
         call thomas (nxd,imax,a,b,z)
         do i=1,imax
            f(i,n+1)=z(i)
         end do
         if (iw.eq.1)write(6,1000)n,t,(f(i,n+1),i=1,imax,ix)
      end do
      return
 1000 format (i3,f5.1,9f8.3)
      end


      subroutine thomas (ndim,n,a,b,x)
c     the Thomas algorithm for a tridiagonal system
      end
```

The data set used to illustrate *subroutine fem3* for a uniform grid is taken from Example 12.8. The output generated by the diffusion equation program is presented in Output 12.6.

Output 12.6. Solution of the diffusion equation by the FEM for a uniform grid.

```
FEM diffusion equation solver

 n time                   f(i,n)

 0  0.0   0.000  25.000  50.000  75.000 100.000
 1  1.0   0.000  20.714  43.143  58.714 100.000
 2  2.0   0.000  18.466  33.621  52.146 100.000
 3  3.0   0.000  14.646  28.525  46.771 100.000
 4  4.0   0.000  12.120  23.955  43.685 100.000
 5  5.0   0.000   9.908  20.941  41.271 100.000
        ........................................
50 50.0   0.000   3.792  12.136  35.042 100.000
```

Subroutine fem3 also can implement the solution for a nonuniform physical grid. All that is required is to define the nonuniform physical grid in a *data* statement. The data set used to illustrate *subroutine fem* for a nonuniform grid is taken from Example 12.9. The necessary *data* statements are included in *program main* as *comment statements c2*. The output generated by the diffusion equation program is presented in Output 12.7.

Output 12.7. Solution of the diffusion equation by the FEM for a nonuniform grid.

```
FEM diffusion equation solver

 n time                   f(i,n)

  1  0.0   0.000  37.500  66.667  87.500 100.000
  2  0.5   0.000  34.422  61.692  78.888 100.000
  3  1.0   0.000  31.916  55.796  75.263 100.000
  4  1.5   0.000  29.365  50.990  72.549 100.000
  5  2.0   0.000  26.889  47.062  70.424 100.000
  6  2.5   0.000  24.567  43.815  68.729 100.000
         .......................................
101 50.0   0.000   6.865  24.993  59.868 100.000
```

12.6.4. Packages for the Finite Element Method

Numerous libraries and software packages are available for implementing the finite element method for a wide variety of differential equations, both ODEs and PDEs. Many work stations and mainframe computers have such libraries attached to their operating systems.

Several large commercial programs are available for solid mechanics problems, acoustic problems, fluid mechanics problems, heat transfer problems, and combustion problems. These programs generally have one-, two-, and in some cases three-dimensional capabilities. Many of the programs consider both steady and unsteady problems. They contain rather sophisticated discretization procedures and graphical output capabilities. Generally speaking, the use of these programs requires an experienced user.

12.7. SUMMARY

The Rayleigh-Ritz method, the collocation method, and the Galerkin weighted residual method for solving boundary-value ordinary differential equations are introduced in this chapter. The finite element method, based on the Galerkin weighted residual approach, is developed for a boundary-value ODE, the Laplace (Poisson) equation, and the diffusion equation. The examples presented in this chapter are rather simple, in that they all involve a linear differential equation and linear elements. Extension of the finite element method to more complicated differential equations and higher-order elements is conceptually straightforward, although it can be quite tedious. The objective of this chapter is to introduce the finite element method for solving differential equations so that the reader is prepared to study more advanced treatments of the subject.

After studying Chapter 12, you should be able to:

1. Describe the basic concepts underlying the calculus of variations
2. Describe the general features of the Rayleigh-Ritz method
3. Apply the Rayleigh-Ritz method to solve simple linear one-dimensional boundary-value problems
4. Describe the general features of residual methods
5. Describe the general features of the collocation method
6. Apply the collocation method to solve simple linear one-dimensional boundary-value problems
7. Describe the general features of the Galerkin weighted residual method
8. Apply the Galerkin weighted residual method to solve simple linear one-dimensional boundary-value problems
9. Describe the general features of the finite element method for solving differential equations
10. Disceretize a one-dimensional space into nodes and elements
11. Develop and apply the shape functions for a linear one-dimensional element
12. Apply the Galerkin weighted residual approach to develop a finite element solution of simple linear one-dimensional boundary-value differential equations
13. Discretize a two-dimensional rectangular space into nodes and elements
14. Develop and apply the shape functions for a linear two-dimensional rectangular element
15. Apply the Galerkin weighted residual approach to develop a finite element solution of the Laplace equation and the Poisson equation
16. Describe how the time derivative is approximated in a finite element solution of a partial differential equation
17. Apply the Galerkin weighted residual approach to develop a finite element solution of the one-dimensional diffusion equation

EXERCISE PROBLEMS

12.2. The Rayleigh-Ritz, Collocation, and Galerkin Methods

The Rayleigh-Ritz Method

1. Derive the Rayleigh-Ritz algorithm, Eq. (12.36), for the boundary-value ODE, Eq. (12.5).
2. Solve Example 12.1 with $T(0.0) = 0.0\,C$, $T(1.0) = 200.0\,C$, and $T_a = 100.0\,C$. Evaluate the solution at increments of $\Delta x = 0.25$ cm. Compare the results with the results obtained in Example 12.1.
3. Solve Example 12.1 with $T(0.0) = 0.0\,C$, $T(1.0) = 200.0\,C$, and $T_a = 0.0\,C$.
4. Apply the Rayleigh-Ritz approach to solve the following boundary-value ODE:

$$\bar{y}'' + P\bar{y}' + Q\bar{y} = F \qquad \bar{y}(0.0) = 0.0 \text{ and } \bar{y}(1.0) = Y \qquad (A)$$

   where P, Q, and F are constants. Let $P = 5.0$, $Q = 4.0$, $F = 1.0$, and $y(1.0) = 1.0$. Evaluate the resulting algorithm for these values. Calculate the solution at increments of $\Delta x = 0.25$. Compare the results with the exact solution.
5. Solve Problem 4 with $P = 4.0$, $Q = 6.25$, and $F = 1.0$.
6. Solve Problem 4 with $P = 5.0$, $Q = 4.0$, and $F(x) = -1.0$.
7. Solve Problem 4 with $P = 4.0$, $Q = 6.25$, and $F(x) = -1.0$.

The Collocation Method

8. Derive the collocation algorithm, Eq. (12.49), for the boundary-value ODE, Eq. (12.5).
9. Solve Example 12.2 with $T(0.0) = 0.0\,C$, $T(1.0) = 200.0\,C$, and $T_a = 100.0\,C$. Compare the results with the results obtained in Example 12.2.
10. Solve Example 12.2 with $T(0.0) = 0.0\,C$, $T(1.0) = 200.0\,C$, and $T_a = 0.0\,C$.
11. Apply the collocation approach to solve boundary-value ODE (A). Let $P = 5.0$, $Q = 4.0$, $F = 1.0$, and $y(1.0) = 1.0$. Evaluate the resulting algorithm for these values. Calculate the solution at increments of $\Delta x = 0.25$. Compare the results with the exact solution.
12. Solve Problem 11 with $P = 4.0$, $Q = 6.25$, and $F = 1.0$.
13. Solve Problem 11 with $P = 5.0$, $Q = 4.0$, and $F(x) = -1.0$.
14. Solve Problem 11 with $P = 4.0$, $Q = 6.25$, and $F(x) = -1.0$.

The Galerkin Weighted Residual Method

15. Derive the Galerkin weighted residual algorithm, Eq. (12.64), for the boundary-value ODE, Eq. (12.5).
16. Solve Example 12.1 with $T(0.0) = 0.0\,C$, $T(1.0) = 200.0\,C$, and $T_a = 100.0\,C$. Compare the results with the results obtained in Example 12.1.
17. Solve Example 12.1 with $T(0.0) = 0.0\,C$, $T(1.0) = 200.0\,C$, and $T_a = 0.0\,C$.
18. Apply the Galerkin weighted residual approach to solve boundary-value ODE (A). Let $P = 5.0$, $Q = 4.0$, $F = 1.0$, $y(0.0) = 0.0$, and $y(1.0) = 1.0$. Evaluate the resulting algorithm for these values. Calculate the solution at increments of $\Delta x = 0.25$. Compare the results with the exact solution.

19. Solve Problem 18 with $P = 4.0$, $Q = 6.25$, and $F = 1.0$.
20. Solve Problem 18 with $P = 5.0$, $Q = 4.0$, and $F(x) = -1.0$.
21. Solve Problem 18 with $P = 4.0$, $Q = 6.25$, and $F(x) = -1.0$.

12.3. The Finite Element Method for Boundary-Value Problems

22. Derive the finite element algorithm Eqs. (12.97) and (12.98), for the boundary-value ODE, Eq. (12.65).
23. Solve Example 12.3 with $T(0.0) = 0.0\,C$, $T(1.0) = 200.0\,C$, and $T_a = 100.0\,C$, using Eq. (12.98). Compare the results with the results obtained in Example 12.3.
24. Solve Example 12.3 with $T(0.0) = 0.0\,C$, $T(1.0) = 200.0\,C$, and $T_a = 0.0\,C$ using Eq. (12.98).
25. Solve Example 12.4 with $T(0.0) = 0.0\,C$, $T(1.0) = 200.0\,C$, and $T_a = 100.0\,C$, using Eq. (12.97) for a nonuniform grid. Compare the results with the results obtained in Example 12.4.
26. Solve Example 12.4 with $T(0.0) = 0.0\,C$, $T(1.0) = 200.0\,C$, and $T_a = 0.0\,C$ using Eq. (12.97) for a nonuniform grid.
27. Apply the finite element approach to solve boundary-value ODE (A), where P, Q, and F are constants. Apply the Galerkin weighted residual approach. Let $P = 5.0$, $Q = 4.0$, $F = 1.0$, and $y(1.0) = 1.0$. Apply the resulting algorithm for these values. Evaluate the solution for $\Delta x = 0.25$. Compare the results with the exact solution.
28. Solve Problem 27 with $P = 4.0$, $Q = 6.25$, and $F = 1.0$.
29. Solve Problem 27 with $P = 5.0$, $Q = 4.0$, and $F(x) = -1.0$.
30. Solve Problem 27 with $P = 4.0$, $Q = 6.25$, and $F(x) = -1.0$.
31. Apply the finite element method to solve boundary-value ODE (A), where P, Q, and F are constants. Let $P = 5.0$, $Q = 4.0$, $F = 1.0$, and $y(1.0) = 1.0$. Apply the resulting algorithm for these values. Evaluate the solution for $\Delta x = 0.25$. Compare the results with the exact solution.
32. Solve Problem 31 with $P = 4.0$, $Q = 6.25$, and $F = 1.0$.
33. Solve Problem 31 with $P = 5.0$, $Q = 4.0$, and $F(x) = -1.0$.
34. Solve Problem 31 with $P = 4.0$, $Q = 6.25$, and $F(x) = -1.0$.

12.4. The Finite Element Method for the Laplace (Poisson) Equation

35. Derive the finite element algorithm, Eqs. (12.168), (12.169), and (12.170), for solving Laplace (Poisson) equation.
36. Implement the program presented in Section 12.6.2 and solve the problem presented in Example 12.6.
37. Solve Example 12.6 with $\Delta x = \Delta y = 5.0\,cm$.
38. Solve Example 12.6 with $\Delta x = \Delta y = 1.25\,cm$. Let variable $ix = 2$ to print every other point.
39. Solve Example 12.6 with $T(x,\ 15.0) = 200.0\sin(\pi x/10.0)$.
40. Modify the problem presented in Example 12.6 by letting $T = 0.0\,C$ on the top boundary and $T = 10.0\sin(\pi y/15.0)\,C$ on the right boundary. Solve this problem by the FEM program for $\Delta x = \Delta y = 2.5\,cm$.
41. Consider steady heat diffusion in the unit square, $0.0 \le x \le 1.0$ and $0.0 \le y \le 1.0$. Let $T(0.0,\ y) = T(x,\ 0.0) = 100.0\,C$ and $T(1.0,\ y) =$

$T(x, 1.0) = 0.0\,C$. Solve this problem using the FEM program with $\Delta x = \Delta y = 0.25$.

42. Solve Problem 41 with $T(0.0, y) = T(x, 0.0) = 0.0\,C$ and $T(1.0, y) = T(x, 1.0) = 100.0\,C$.

43. Implement the program presented in Section 12.6.2 and solve the problem presented in Example 12.7.

44. Work Example 12.7 with $\Delta x = \Delta y = 5.0\,cm$.

45. Solve Example 12.7 with $\Delta x = \Delta y = 1.25\,cm$. Let variable $ix = 2$ to print every other point.

46. Solve Example 12.7 with $\dot{Q}/k = -2000.0\,C/cm^2$.

12.5. The Finite Element Method for the Diffusion Equation

47. Derive the finite element algorithm, Eqs. (12.212) and (12.213), for the one-dimensional diffusion equation.

48. Implement the program presented in Section 12.6.3 and solve the problem presented in Example 12.8.

49. Solve Example 12.8 using the nonuniform grid presented in Example 12.4.

50. Work Example 12.8 with $\Delta x = \Delta y = 0.125\,cm$.

51. Develop the finite element algorithm for the one-dimensional diffusion equation, Eqs. (12.212) and (12.213), using the backward-time approximation for $\dot{f}$. Modify the program presented in Section 12.6.3 to implement the algorithm. Solve Example 12.8 using the modified finite element program. Compare the results with the results obtained in Examples 12.3 and 12.8.

52. Develop the finite element algorithm for the one-dimensional diffusion equation, Eqs. (12.212) and (12.213), by using the centered-time approximation for $\dot{f}$. Modify the program presented in Section 12.6.3 to implement the algorithm. Solve Example 12.8 using the modified finite element program. Compare the results with the results obtained in Example 12.3 and 12.8.

12.6. Programs

53. Implement the boundary-value ordinary differential equation FEM program presented in Section 12.6.1. Check out the program using the given data.

54. Solve any of Problems 23 to 26 or 28 to 30 using the boundary-value ODE FEM program.

55. Implement the Laplace (Poisson) equation FEM program presented in Section 12.6.2. Check out the program using the given data.

56. Solve any of Problems 37 to 39 or 41 to 44 with the Laplace (Poisson) equation FEM program.

57. Modify the Laplace (Poisson) equation FEM program to account for variable Δx and Δy in the manner in which variable Δx is accounted for in the boundary-value ordinary differential equation FEM program presented in Section 12.6.1. The algorithm is given by Eq. (12.169). Solve Example 12.6 using the quadratic transformation presented in Example 8.12 to pack constant x lines on the left and right sides of the physical domain and constant y lines at the top of the physical domain.

58. Implement the diffusion equation FEM program presented in Section 12.6.3. Check out the program using the given data.

59. Solve any of Problems 48 to 52 using the diffusion equation FEM program.

APPLIED PROBLEMS

60. The deflection of a simply supported and uniformly loaded beam is governed by the ordinary differential equation (for small defections)

$$EI\frac{d^2y}{dx^2} = -\frac{qLx}{2} + \frac{qx^2}{2} \qquad y(0) = 0 \text{ and } y(L) = 0 \qquad \text{(B)}$$

where q is the uniform load per unit length, L is the length of the beam, I is the moment of inertia of the beam cross section, and E is the modulus of elasticity. For a rectangular beam, $I = wh^3/12$, where w is the width and h is the height. Consider a beam $(E = 200\,\text{GN/m}^2)$ 5.0 m long, 5.0 cm wide, and 10.0 cm high, which is subjected to the uniform load $q = -1,500\,\text{N/m}$ on the 5.0 cm face. Evaluate $\bar{F}$ for each element by integrating $F(x)$ for each element by the trapezoid rule. Solve for the deflection $y(x)$. Compare the results with the results obtained in Example 8.6 and Problem 8.119.

References

Abramowitz, M., and Stegun, I. A. (1964), *Handbook of Mathematical Functions, Applied Mathematics Series No. 55*, National Bureau of Standards, Washington, D.C.

Acton, F. S. (1970), *Numerical Methods That Work*, Harper and Row, New York.

Bashforth, F. and Adams, J. C. (1883), *An Attempt to Test the Theories of Capillary Action...with an Explanation of the Method of Integration Employed*, Cambridge University Press, Cambridge.

Brandt, A. (1977), Multi-Level Adaptive Solutions to Boundary-Value Problems, *Mathematics of Computation*, vol. 31, no. 138, pp. 333–390.

Brent, R. P. (1973), *Algorithms for Minimization without Derivatives*, Chaps. 3 and 4, Prentice Hall, Englewood Cliffs, NJ.

Chapra, S. C., and Canale, R. P. (1998), *Numerical Methods for Engineers*, 3rd ed., McGraw-Hill, New York.

Colebrook, C. F. (1939), Turbulent Flow in Pipes with Particular Reference to the Transition between Smooth and Rough Pipes, *Journal of the Institute of Civil Engineers*, London.

Courant, R., Friedricks, K. O., and Lewy, H. (1928), *Uber die Partiellen Differenzengleichungen der Mathematischen Physik*, vol. 100, pp. 32–74.

Crank, J., and Nicolson, P. (1947), A Practical Method for Numerical Evaluation of Solutions of Partial Differential Equations of the Heat-Conduction Type, *Proceedings of the Cambridge Philosophical Society*, vol. 43, no. 50, pp. 50–67.

Dennis, J. E., and Schnabel, R. B. (1983), *Numerical Methods of Unconstrained Optimization and Nonlinear Equations*, Prentice Hall, Englewood Cliffs, NJ.

Douglas, J. (1962), Alternating Direction Methods for Three Space Variables, *Numerische Mathematik*, vol. 4, pp. 41–63.

Douglas, J., and Gunn, J. E. (1964), A General Formulation of Alternating Direction Implicit Methods, Part I: Parabolic and Hyperbolic Problems, *Numerische Mathematik*, vol. 6, pp. 428–453.

DuFort, E. C., and Frankel, S. P. (1953), Stability Conditions in the Numerical Treatment of Parabolic Differential Equations, *Mathematical Tables and Other Aids to Computation*, vol. 7, pp. 135–152.

Fadeev, D. K., and Fadeeva, V. N. (1963), *Computational Methods of Linear Algebra*, Freeman, San Francisco, CA.

Fehlberg, E. (1966), New High-Order Runge-Kutta Formulas with an Arbitrary Small Truncation Error, *Zeitschrift fur Angewandte Mathematik und Mechanik*, vol. 46, pp. 1–16.

Ferziger, J. H. (1981), *Numerical Methods for Engineering Application*, John Wiley & Sons, New York.

Fox, R. W., and McDonald, A. T. (1999), *Introduction to Fluid Mechanics*, 6th ed., John Wiley & Sons, New York.

Forsythe, G. E., and Wasow, W. (1960), *Finite Difference Methods for Partial Differential Equations*, John Wiley & Sons, New York.

Frankel, S. P. (1950), Convergence Rates of Iterative Treatments of Partial Differential Equations, *Mathematical Tables and Other Aids to Computation*, vol. 4, pp. 65–75.

Freudenstein, F. (1955), Approximate Synthesis of Four-Bar Linkages, *Transactions of the American Society of Mechanical Engineers*, vol. 77, pp. 853–861.

Gear, C. W. (1971), *Numerical Initial Value Problems in Ordinary Differential Equations*, Prentice-Hall, Englewood Cliffs, NJ.

Genereaux, R. P. (1939), Fluid Flow Design Methods, *Industrial Engineering Chemistry*, vol. 29, no. 4, pp. 385–388.

Gerald, C. F., and Wheatley, P. O. (1999), *Applied Numerical Analysis*, 6th ed., Addison-Wesley, Reading, MA.

Gragg, W. (1965), On Extrapolation Algorithms for Ordinary Initial Value Problems, *Journal Soc. Ind. Appl. Math., Numer. Anal. Ser. B2*, pp. 384–403.

Hackbusch, W. (1980), *Multi-Grid Methods and Applications*, Springer-Verlag, Berlin, Heidelberg.

Hadamard, J. (1923), *Lectures on Cauchy's Problem in Linear Partial Differential Equations*, Yale University Press, New Haven, CT.

Henrici, P. K. (1964), *Elements of Numerical Analysis*, John Wiley & Sons, New York.

Hildebrand, F. B. (1956), *Introduction to Numerical Analysis*, McGraw-Hill, New York.

Householder, A. S. (1964), *The Theory of Matrices in Numerical Analysis*, Blaisdell, New York.

Householder, A. S. (1970), *The Numerical Treatment of a Single Nonlinear Equation*, McGraw-Hill, New York.

Jeeves, T. A. (1958) Secant Modification of Newton's Method, *Communications of the Association of Computing Machinery*, vol. 1, no. 8, pp. 9–10.

Lax, P. D. (1954), Weak Solutions of Nonlinear Hyperbolic Equations and their Numerical Computation, *Comm. Pure and Appl. Math.*, vol. 2, pp. 159–193.

Lax, P. D., and Wendroff, B. (1960), Systems of Conservation Laws, *Comm. Pure and Appl. Math.*, vol. 13, pp. 217–237.

MacCormack, R. W. (1969), The Effect of Viscosity in Hypervelocity Impact Cratering, *American Institute of Aeronautics and Astronautics*, Paper 69-354.

Muller, D. E. (1956), A Method of Solving Algebraic Equations Using an Automatic Computer, *Mathematical Tables and Other Aids to Computation (MTAC)*, vol. 10, pp. 208–215.

Peaceman, D. W., and Rachford, H. H. (1955), The Numerical Solution of Parabolic and Elliptic Differential Equations, *Journal Soc. Ind. Appl. Math.*, vol. 3, pp. 28–41.

Press, W. H., Flannery, B. P., Teukolsky, S. A., and Vetterling, W. T. (1989), *Numerical Recipes—The Art of Scientific Computing*, Cambridge University Press, Cambridge, England.

Ralston, A., and Rabinowitz, P. (1978), *A First-Course in Numerical Analysis*, 2nd ed., McGraw-Hill, New York.

Rao, S. S. (1982), *The Finite Element Method in Engineering*, Pergamon Press, New York.

Reddy, J. N. (1993), *An Introduction to the Finite Element Method*, 2nd ed., McGraw-Hill, New York.

Rice, J. R. (1983), *Numerical Methods, Software and Analysis*, McGraw-Hill, New York.

Richardson, L. F. (1970), The Appropriate Arithmetical Solution by Finite Differences of Physical Problems with an Application to the Stresses in a Masonary Dam, *Phil. Trans. Roy. Soc.*, London Series A, vol. 210, pp. 307–357.

Richtmyer, R. D. (1963), A Survey of Difference Methods for Nonsteady Fluid Dynamics, *NCAR Technical Note 63-2*, National Center for Atmospheric Research, Boulder, CO.

Smith, B. T., Boyle, J. M., Dongerra, J. J., Garbow, B. S., Ikebe, Y., Klema, V. C., and Moler, C. B. (1976), Matrix Eigensystem Routines—EISPACK Guide, *Lecture Notes in Computer Science*, vol. 6, Springer-Verlag, Heidelberg.

Southwell, R. V. (1940), *Relaxation Methods in Engineering Science*, Oxford University Press, London.

Stewart, G. W. (1973), *Introduction to Matrix Computation*, Academic Press, New York.

Stoer, J., and Bulirsch, R. (1980), *Introduction to Numerical Analysis*, Chapter 7, Springer-Verlag, New York.

Strang, G. (1988), *Linear Algebra and Its Applications*, 3rd ed., pp. 172–176, 376–378, Harcourt Brace Jovanovich, San Diego, CA.

Strong, G. and Fix, G. J. (1973), *An Analysis of the Finite Element Method*, Prentice Hall, Englewood Cliffs, NJ.

Thomas, L. H. (1949), Elliptic Problems in Linear Difference Equations Over a Network, *Watson Scientific Computing Laboratory Report*, Columbia University Press, New York.

Thompson, J. F., Warsi, Z. U. A., and Mastin, C. W. (1985), *Numerical Grid Generation, Foundations and Applications*, North-Holland, New York.

Timoschenko, S. (1955), *Strength of Materials*, D. Van Nostrand Company, New York.

Warming, R. F., and Hyett, B. J. (1974), The Modified Equation Approach to the Stability and Accuracy Analysis of Finite-Difference Methods, *Journal of Computational Physics*, vol. 14, pp. 159–179.

Wilkinson, J. H. (1965), *The Algebraic Eigenvalue Problem*, Clarendon Press, Oxford, England.

Zienkiewicz, O. C. and Taylor, R. L. (1991), *The Finite Element Method*, vols. 1 and 2, McGraw-Hill, New York.

Zucrow, M. J., and Hoffman, J. D. (1976), *Gas Dynamics*, vols. I and II, John Wiley & Sons, New York.

Answers to Selected Problems

All of the problems for which answers are given in this section are denoted in the individual chapters by an asterisk appearing before the corresponding problem numbers.

Chapter 1. Systems of Linear Algebraic Equations

3. (a) $\begin{bmatrix} 11 & 20 \\ 21 & 22 \\ 15 & 19 \end{bmatrix}$ (b) $\begin{bmatrix} 23 & 39 \\ 5 & -3 \\ 19 & 33 \end{bmatrix}$ (e) $\begin{bmatrix} 5 & 13 & 19 \\ 9 & 17 & 19 \\ 6 & 14 & 16 \end{bmatrix}$ (f) $\begin{bmatrix} 10 & 26 & 36 \\ 2 & 0 & -4 \\ 8 & 22 & 30 \end{bmatrix}$

(g) $\begin{bmatrix} 21 & 17 & 11 \\ 31 & 28 & 12 \end{bmatrix}$ (h) $\begin{bmatrix} 15 & 15 & 12 \\ 19 & 22 & 17 \end{bmatrix}$ (i) $\begin{bmatrix} 7 & 17 & 21 \\ 10 & 28 & 38 \end{bmatrix}$

(j) $\begin{bmatrix} 11 & 11 & 8 \\ 11 & 35 & 20 \\ 8 & 20 & 14 \end{bmatrix}$ (k) $\begin{bmatrix} 38 & -1 & 31 \\ -1 & 14 & -2 \\ 31 & -2 & 26 \end{bmatrix}$ (l) $\begin{bmatrix} 3 & 7 & 9 \\ 7 & 19 & 25 \\ 9 & 25 & 35 \end{bmatrix}$

8. (a) 24 (b) 18 (d) 4 (e) 432 (f) 96 (g) 432 (h) 96
9. (a) 24 (b) 18 (d) 4 (e) 432 (f) 96 (g) 432 (h) 96
13. $\mathbf{x}^T = \begin{bmatrix} -1 & 2 & 1 \end{bmatrix}$
15. $\mathbf{x}^T = \begin{bmatrix} 1 & 2 & 3 & 4 \end{bmatrix}$
21. $\mathbf{x}^T = \begin{bmatrix} -1 & 2 & 1 \end{bmatrix}$
23. $\mathbf{x}^T = \begin{bmatrix} 1 & 2 & 3 & 4 \end{bmatrix}$
29. $\mathbf{x}^T = \begin{bmatrix} -1 & 2 & 1 \end{bmatrix}$
31. $\mathbf{x}^T = \begin{bmatrix} 1 & 2 & 3 & 4 \end{bmatrix}$

37. $\mathbf{A}^{-1} = \begin{bmatrix} 0.272727 & 0.227273 & 0.863636 \\ 0.363636 & 0.136364 & 0.318182 \\ 0.454545 & 0.045455 & 0.772727 \end{bmatrix}, \quad \mathbf{x} = \begin{bmatrix} -1 \\ 2 \\ 1 \end{bmatrix}$

39. $\mathbf{A}^{-1} = \begin{bmatrix} 164 & -107 & 84 & -13 \\ 14 & -9 & 7 & -1 \\ -127 & 83 & -65 & 10 \\ -49 & 32 & -25 & 4 \end{bmatrix}, \quad \mathbf{x} = \begin{bmatrix} 1 \\ 2 \\ 3 \\ 4 \end{bmatrix}$

45. $\mathbf{LU} = \begin{bmatrix} -2.000000 & 3.000000 & 1.000000 \\ -1.500000 & 8.500000 & -3.500000 \\ -0.500000 & -0.058824 & 1.294118 \end{bmatrix}$,

$\mathbf{b'} = \begin{bmatrix} 9.000000 \\ 0.000000 \\ -4.000000 \end{bmatrix}$, $\quad \mathbf{x} = \begin{bmatrix} -1 \\ 2 \\ 1 \end{bmatrix}$

47. $\mathbf{LU} = \begin{bmatrix} 1.00 & 3.00 & 2.00 & -1.00 \\ 4.00 & -10.00 & -3.00 & 5.00 \\ 3.00 & 1.20 & -0.40 & 1.00 \\ -1.00 & -0.50 & 6.25 & 0.25 \end{bmatrix}$,

$\mathbf{b'} = \begin{bmatrix} 9.000000 \\ -9.000000 \\ 2.800000 \\ 1.000000 \end{bmatrix}$, $\quad \mathbf{x} = \begin{bmatrix} 1 \\ 2 \\ 3 \\ 4 \end{bmatrix}$

53. $\mathbf{x}^T = \begin{bmatrix} 1 & 2 & 3 & 4 \end{bmatrix}$

59. $k = 10, \mathbf{x}^T = [0.805664 \quad 1.611328 \quad 2.685547 \quad 3.759766]$

$\quad\, k = 75, \mathbf{x}^T = [1.000000 \quad 2.000000 \quad 3.000000 \quad 4.000000]$

64. $k = 10, \mathbf{x}^T = [1.017937 \quad 1.976521 \quad 3.018995 \quad 3.990502]$

$\quad\, k = 33, \mathbf{x}^T = [1.000001 \quad 1.999999 \quad 3.000001 \quad 3.999999]$

69. $k = 10, \mathbf{x}^T = [0.999989 \quad 1.999969 \quad 2.999997 \quad 4.000004]$

$\quad\, k = 14, \mathbf{x}^T = [1.000000 \quad 2.000000 \quad 3.000000 \quad 4.000000]$

Chapter 2. Eigenproblems

9. $k = 5, \lambda = 4.506849, \mathbf{x}^T = [1.000000 \quad 0.936778 \quad 1.285106]$

15. $k = 7, \lambda = 6.634483, \mathbf{x}^T = [1.000000 \quad 0.902441 \quad 1.437482 \quad 1.647300]$

25. The inverse power method fails when x_1 is the unity component. let x_2 be the unity component. $k = 12$, $\lambda = -0.285157$, $\mathbf{x}^T = [-0.578414 \quad 1.000000 \quad -0.128329]$

31. The inverse power method fails when x_1 or x_2 is the unity component. Let x_3 be the unity component. $k = 15$, $\lambda = -0.407585$, $\mathbf{x}^T = [0.874085 \quad -1.799727 \quad -1.000000 \quad -0.215193]$

46. $\mathbf{A}_s = \begin{bmatrix} -3.5 & 1.0 & 2.0 \\ 2.0 & -3.5 & 1.0 \\ 1.0 & 1.0 & -1.5 \end{bmatrix}$

$\quad k = 30, \lambda = -0.284702, \mathbf{x}^T = [1.000000 \quad -1.726150 \quad 0.220675]$

51. $\mathbf{D}_s = \begin{bmatrix} 0.2 & 1.0 & 2.0 \\ 2.0 & 0.2 & 1.0 \\ 1.0 & 1.0 & 2.2 \end{bmatrix}$, $\quad \mathbf{LU} = \begin{bmatrix} 0.200000 & 1.000000 & 2.000000 \\ 10.000000 & -9.800000 & -19.000000 \\ 5.000000 & 0.408163 & -0.044898 \end{bmatrix}$

$\quad k = 5, \lambda = 0.778124, \mathbf{x}^T = [1.000000 \quad 6.792161 \quad -3.507019]$

54. $\mathbf{D}_s = \begin{bmatrix} -0.5 & 1.0 & 1.0 & 2.0 \\ 2.0 & -0.5 & 1.0 & 1.0 \\ 3.0 & 2.0 & -0.5 & 2.0 \\ 2.0 & 1.0 & 1.0 & 2.5 \end{bmatrix}$,

$\mathbf{LU} = \begin{bmatrix} -0.500000 & 1.000000 & 1.000000 & 2.000000 \\ -4.000000 & 3.500000 & 5.000000 & 9.000000 \\ -6.000000 & 2.285714 & -5.928571 & -6.571429 \\ -4.000000 & 1.428571 & 0.361446 & 0.018072 \end{bmatrix}$

$k = 5, \lambda = 1.508563, \mathbf{x}^T = [1.000000 \quad 5.046599 \quad 5.671066 \quad -5.104551]$

$\mathbf{D}_s = \begin{bmatrix} 1.5 & 1.0 & 1.0 & 2.0 \\ 2.0 & 1.5 & 1.0 & 1.0 \\ 3.0 & 2.0 & 1.5 & 2.0 \\ 2.0 & 1.0 & 1.0 & 4.5 \end{bmatrix}$,

$\mathbf{LU} = \begin{bmatrix} 1.500000 & 1.000000 & 1.000000 & 2.000000 \\ 1.333333 & 0.166667 & -0.333333 & -1.666667 \\ 2.000000 & 0.000000 & -0.500000 & -2.000000 \\ 1.333333 & -2.000000 & 2.000000 & 2.500000 \end{bmatrix}$

$k = 14, \lambda = -0.407454, \mathbf{x}^T = [1.000000 \quad -2.058322 \quad 1.143127 \quad -0.246129]$

56. $\lambda = \lambda_s^{(10)} + 4.5 = -4.722050 + 4.5 = -0.222050$

$\mathbf{D}_s = \begin{bmatrix} 1.222050 & 1.000000 & 2.000000 \\ 2.000000 & 1.222050 & 1.000000 \\ 1.000000 & 1.000000 & 3.222050 \end{bmatrix}$,

$\mathbf{LU} = \begin{bmatrix} 1.222050 & 1.000000 & 2.000000 \\ 1.636594 & -0.414544 & -2.273188 \\ 0.818297 & -0.438320 & 0.589073 \end{bmatrix}$

$k = 5, \lambda_{s,i} = -15.836736, \lambda_s = -0.063093, \lambda = \lambda_s - 0.222050$

$\lambda = -0.285143, \mathbf{x}^T = [1.000000 \quad -1.728895 \quad 0.221876]$

61. $k = 7, \lambda = 4.507019, \det(\mathbf{D}) = -0.000001$

66. $k = 7, \lambda = -0.285142, \det(\mathbf{D}) = 0.000000$

74. $\mathbf{Q}^{(10)} = \begin{bmatrix} 1.000000 & 0.000000 & 0.000000 \\ 0.000000 & 1.000000 & 0.000456 \\ 0.000000 & 0.000456 & -1.000000 \end{bmatrix}$,

$\mathbf{R}^{(10)} = \begin{bmatrix} 4.507018 & 0.749863 & 0.197048 \\ 0.000000 & 0.777730 & 1.182967 \\ 0.000000 & 0.000000 & 0.285287 \end{bmatrix}$

$\mathbf{A}^{(10)} = \begin{bmatrix} 4.507019 & 0.749951 & -0.196706 \\ 0.000000 & 0.778268 & -1.182613 \\ 0.000000 & 0.000130 & -0.285287 \end{bmatrix}$

$k = 9, \lambda_1 = 4.507019, \lambda_2 = 0.778268, \lambda_3 = -0.285287$

82. $\lambda_1 = 4.507019, \mathbf{x}_1^T = [1.000000 \quad 0.936734 \quad 1.285143]$

Chapter 3. Roots of Nonlinear Equations

1. $i = 13, x = 0.739075, f(x) - 0.00001745$
6. $i = 15, x = -0.815582, f(x) = -0.00007538$
   $i = 16, x = 1.429611, f(x) = 0.00000465$
8. $i = 4, x = 0.739083, f(x) = -0.00000288$
13. $i = 8, x = -0.815544, f(x) = 0.00002565$
   $i = 16, x = 1.429600, f(x) = 0.00008531$
15. $x = \cos(x) = g(x), g'(x) = -\sin(x)$
   $i = 22, x = 0.739050, g'(x) = -0.673586$
24. $x = \exp(x)^{0.25} = g(x), g'(x) = 0.25/\exp(x)^{0.75}$
   $i = 12, x = 1.429605, g'(x) = 0.085563$
29. $i = 4, x = 0.739085, f(x) = 0.00000000$
33. $i = 4, x = 1.000000, f(x) = 0.00000000$
37. $i = 4, x = 0.739085, f(x) = 0.00000000$
41. $i = 7, x = 1.000000, f(x) = 0.00000011$
45. (a) $x_0 = 1.5, i = 12, x = 1.000088, f(x) = -0.00000002$
   $x_0 = 2.5, i = 6, x = 3.000000, f(x) = 0.00000002$
   $x_0 = -3.0, i = 17, x = 0.999923, f(x) = -0.00000001$
46. (a) $x_0 = 1.0 + I1.0, i = 11$
   $x_0 = 1.0 - I1.0, i = 11$
   $x_0 = 2.0 + I0.0, i = 6$
   $x_0 = 5.0 + I1.0, i = 9$
47. $(x_0, y_0) = (1.0, 1.0), i = 6, (x, y) = (1.906287, 0.523977)$
   $(x_0, y_0) = (0.0, 1.0), i = 5, (x, y) = (-0.691051, 1.625373)$

Chapter 4. Polynomial Approximation and Interpolation

1. (a) $P_3(1.5) = -1.8750,$ (b) $P_3'(1.5) = 5.7500,$ (c) $Q_2(x) = x^2 - 7x + 12$
13. (a) $f(2.0, 0.8) = 5.201650, \text{Error} = 0.201650, \text{Ratio} = 4.02$
   $f(2.0, 0.4) = 5.050100, \text{Error} = 0.050100$
16. (a) $P_2(x) = 10.8336 - 18.7510x + 11.4175x^2, P_2(0.9) = 3.205875$
   (b) $P_2(x) = 6.16700 - 6.25075x + 3.08375x^2, P_2(0.9) = 3.039163$
   (c) $P_3(x) = 12.8340 - 29.5865x + 30.1713x^2 - 10.4187x^3, P_3(0.9) = 3.049594$
   (d) $P_3(x) = 9.49900 - 16.5244x + 13.4963x^2 - 3.47083x^3, P_3(0.9) = 3.028750$
   (e) $P_4(x) = 14.5015 - 40.2863x + 54.8364x^2 - 34.7365x^3 + 8.68490x^4,$
   $P_4(0.9) = 3.036566$

21. (a) 3.205875 (b) 3.039163 (c) 3.049594 (d) 3.028750 (e) 3.036566
26. (a) 3.205875 (b) 3.039163 (c) 3.049594 (d) 3.028750 (e) 3.036566
43. (a) 3.205875 (b) 3.039163 (c) 3.049594 (d) 3.028750 (e) 3.036566
48. (a) 3.205875 (b) 3.039163 (c) 3.049594 (d) 3.028750 (e) 3.036566
65. $v(9000, T) = -0.0165440 + 0.0777550 \times 10^{-3}T - 0.0120500 \times 10^{-6}T^2$,

 $v(9000, 750) = 0.034994$

 $v(10000, T) = -0.0171970 + 0.0740200 \times 10^{-3}T - 0.0128000 \times 10^{-6}T^2$,

 $v(10000, 750) = 0.031118$

 $v(11000, T) = -0.0172360 + 0.0697700 \times 10^{-3}T - 0.0127000 \times 10^{-6}T^2$,

 $v(11000, 750) = 0.027948$

 $v(P, 750) = 0.101644 - 0.0105819 \times 10^{-3}P + 0.353000 \times 10^{-9}P^2$,

 $v(9500, 750) = 0.032968$

69. $v(P, T) = -7.69900 \times 10^{-3} + 0.103420 \times 10^{-3}T - 0.233000 \times 10^{-6}P - 4.86000$

 $\times 10^{-9}PT$

 $v(9500, 750) = 0.033025$

77. $C_p(T) = 0.999420 + 0.142900 \times 10^{-3}T$

82. $C_p(T) = 0.853364 + 0.454921 \times 10^{-3}T - 0.228902 \times 10^{-6}T^2 + 0.0729798$

 $\times 10^{-9}T^3 - 0.0113636 \times 10^{-}$

105. $f = 15.9146/\mathrm{Re}^{0.999125}$

Chapter 5. Numerical Differentiation and Difference Formulas

1. (a) $f(x) = -0.01363717 + 2.73191900x$

 $f'(x) = 2.73191900, f'(1.0) = 2.73191900$, Error $= 0.01363717$

 $f''(x) = 0, f''(1.0)$ cannot be calculated

 (b) $f(x) = 1.37284033 - 0.027308500x + 1.37275000x^2$

 $f'(x) = -0.02730850 + 2.745500x, f'(1.0) = 2.71819150$, Error $= -0.00009033$

 $f''(x) = 2.74550000, f''(1.0) = 2.74550000$, Error $= 0.02721817$

 (c) $f(x) = 0.08937973 + 1.39568450x - 0.03620000x^2 + 0.46500000x^3$

 $f'(x) = 1.39568450 - 0.07240000x + 1.39500000x^2$

 $f'(1.0) = 2.71828450$, Error $= 0.00000267$

 $f''(x) = -0.0724000 + 2.79000000x$,

 $f''(1.0) = 2.71760000$, Error $= -0.00068183$

8. (a) $f'(x) = \dfrac{(x - 1.01)}{-0.01}(2.71828183) + \dfrac{(x - 1.00)}{0.01}(2.74560102)$

 $f'(1.0) = 2.73191900$, Error $= 0.01363717$

 $f''(x)$ cannot be calculated

(b) $f'(x) = \dfrac{(2x - 2.03)}{0.0002}(2.71828183) + \dfrac{(2x - 2.02)}{-0.0001}(2.74560102)$

$+ \dfrac{(2x - 2.01)}{0.0002}(2.77319476)$

$f'(1.0) = 2.71819150$, Error $= -0.00009033$

$f''(1.0) = 2.74550000$, Error $= 0.02711817$

(c) $f'(1.0) = 2.71828450$, Error $= 0.00000267$

$f''(1.0) = 2.71760000$, Error $= -0.00068183$

13.

x_i	$f_i^{(0)}$	$f_i^{(1)}$	$f_i^{(2)}$	$f_i^{(3)}$
1.00	2.71828183			
		2.73191900		
1.01	2.74560102		1.37275000	
		2.75937400		0.46500000
1.02	2.77319476		1.38670000	
		2.78710800		
1.03	2.80106584			

$f'(x) = 2.73191900 + [2x - (x_0 + x_1)](1.37275000)$

$+ [3x^2 - 2(x_0 + x_1 + x_2)x + (x_0x_1 + x_0x_2 + x_1x_2)](0.46500000)$

(a) $f'(1.0) = 2.731919000$, Error $= 0.01363717$

$f''(x)$ cannot be calculated

(b) $f'(1.0) = 2.71829150$, Error $= -0.00009033$

$f''(1.0) = 2.74550000$, Error $= 0.02721817$

(c) $f'(1.0) = 2.71828450$, Error $= 0.00000267$

$f''(1.0) = 2.71760000$, Error $= -0.00068183$

17.

x	$f(x)$	$\Delta f(x)$	$\Delta^2 f(x)$	$\Delta^3 f(x)$
1.00	2.71828183			
		0.02731919		
1.01	2.74560102		0.00027455	
		0.02759374		0.00000279
1.02	2.77319476		0.00027734	
		0.02787108		
1.03	2.80106584			

(a) $f'(1.0) = \dfrac{1}{0.01}\left[0.02731919 - \dfrac{1}{2}(0.00027455) + \dfrac{1}{3}(0.00000279)\right]$

$f'(1.0) = 2.73191900, 2.71819150, 2.71828450$

Error $= 0.01363717, -0.00009033, -0.00000267$

$f''(1.0) = \dfrac{1}{(0.01)^2}[0.00027455 - 0.00000279] = 2.74550000, 2.71760000$

Error $= 0.02721817, -0.00068183$

(b) $f'(1.0) = \dfrac{1}{0.02}\left[0.05491293 - \dfrac{1}{2}(0.00110932) + \dfrac{1}{3}(0.00002241)\right]$

$f'(1.0) = 2.74564650, 2.71791350, 2.71828700$

Error $= 0.02736467, -0.00036833, 0.00000517$

$f''(1.0) = \dfrac{1}{(0.02)^2}[0.00110932 - 0.00002241]$

$= 2.77330000, 2.71727500$

Error $= 0.05501817, -0.00100683$

33. $f'(1.0) = 2.71900675, 2.71846300, 2.71832750$

(a) Error $= \dfrac{1}{2^2 - 1}(2.71846300 - 2.71900675) = -0.00018125$

(b) Error $= \dfrac{1}{2^2 - 1}(2.71832750 - 2.71846300) = -0.00004517$

Extrapolated value $= 2.71846300 - 0.00018125 = 2.71828175$

Extrapolated value $= 2.71832750 - 0.00004517 = 2.71828233$

51. (b) $u'(0.0) = \dfrac{1}{1.0}\left[55.5600 - \dfrac{1}{2}(-22.2300) + \dfrac{1}{3}(0.0100)\right]$,

$= 55.5600, 66.6750, 66.6783$

Chapter 6. Numerical Integration

5. (D)

n	h	I	Error	Ratio
1	π	15.70796327	−2.00000000	4.66
2	$\pi/2$	17.27875959	−0.42920368	4.13
4	$\pi/4$	17.60408217	−0.10388110	4.03
8	$\pi/8$	17.68219487	−0.02576840	
	Exact	17.70796327		

7.

n	h	I	Error	Ratio
1	1.6	8.12800000	2.26379084	3.30
2	0.8	6.54936000	0.68515084	3.60
4	0.4	6.05468000	0.19047084	3.83
8	0.2	5.91394000	0.04973084	
	Exact	5.86420916		

12. (D)

n	h	I	Error	Ratio
1	$\pi/2$	17.80235837	0.09439510	20.70
2	$\pi/4$	17.71252302	0.00455975	16.94
4	$\pi/8$	17.70823244	0.00026917	
	Exact	17.70796327		

19. (D)

n	h	I	Error	Ratio
1	$\pi/3$	17.74848755	0.04052428	20.16
2	$\pi/6$	17.70997311	0.00200984	16.84
4	$\pi/12$	17.70808265	0.00011938	
	Exact	17.70796327		

35. (D)

n	h	$0(h^2)$	$0(h^4)$	$0(h^6)$	$0(h^8)$
1	π	15.70796327			
2	$\pi/2$	17.27875959	17.80235837		
4	$\pi/4$	17.60408217	17.71252302	17.70653400	
8	$\pi/8$	17.68219487	17.70823244	17.70794640	17.70796882
	Exact	17.70796327			

43. (D)

n	h	I	Error	Ratio
1	π	17.64378284	-0.06418043	21.01
2	$\pi/2$	17.70490849	-0.00305478	17.00
4	$\pi/4$	17.70778360	-0.00017967	
	Exact	17.70796327		

46. (D)

n	h	I	Error	Ratio
1	π	17.70935218	0.00138891	85.52
2	$\pi/2$	17.70797951	0.00001624	67.67
4	$\pi/4$	17.70796351	0.00000024	
	Exact	17.70796327		

49. (D)

n	h	I	Error	Ratio
1	π	17.70794750	-0.00001577	315.4
2	$\pi/2$	17.70796322	-0.00000005	
4	$\pi/4$	17.70796327	-0.00000000	
	Exact	17.70796327		

Chapter 7. One-Dimensional Initial-Value Problems

15. $\Delta t = 0.2, y(1.0) = 0.672320, E(0.2) = 0.040199$, Ratio $= 2.09$
    $\Delta t = 0.1, y(1.0) = 0.651322, E(0.1) = 0.019201$

21. $\Delta t = 0.2, y(1.0) = 2.102601, E(0.2) = -0.345453$, Ratio $= 1.69$
    $\Delta t = 0.1, y(1.0) = 2.243841, E(0.1) = -0.204213$

26. $\Delta t = 0.2, y(1.0) = 0.598122, E(0.2) = -0.033998$, Ratio $= 1.92$
    $\Delta t = 0.1, y(1.0) = 0.614457, E(0.1) = -0.017664$

39. $\Delta t = 0.2, y(1.0) = 2.102601, E(0.2) = -0.232621$, Ratio $= 2.54$
    $\Delta t = 0.1, y(1.0) = 2.243841, E(0.1) = -0.091381$
    $\Delta t = 0.05, y(1.0) = 2.335222$

40. $\Delta t = 0.20, y(1.0) = 2.426707, E(0.2) = -0.019843$, Ratio $= 4.59$
    $\Delta t = 0.10, y(1.0) = 2.442230, E(0.1) = -0.004320$
    $\Delta t = 0.05, y(1.0) = 2.446550$

59. $\Delta t = 0.2, y(1.0) = 0.629260, E(0.2) = -0.002860$, Ratio $= 4.32$
    $\Delta t = 0.1, y(1.0) = 0.631459, E(0.1) = -0.000662$

65. $\Delta t = 0.2, y(1.0) = 2.401971, E(0.2) = -0.046084$, Ratio $= 3.39$
    $\Delta t = 0.1, y(1.0) = 2.434473, E(0.1) = -0.013581$

70. $\Delta t = 0.2, y(1.0) = 0.629260, E(0.2) = -0.002860$, Ratio $= 4.32$
    $\Delta t = 0.1, y(1.0) = 0.631459, E(0.1) = -0.000662$

76. $\Delta t = 0.2, y(1.0) = 2.426707, E(0.2) = -0.021348$, Ratio $= 3.66$
    $\Delta t = 0.1, y(1.0) = 2.442230, E(0.1) = -0.005825$

84. $\Delta t = 0.2, y(1.0) = 0.63211476, E(0.2) = -0.00000580$, Ratio $= 17.6$
    $\Delta t = 0.1, y(1.0) = 0.63212023, E(0.1) = -0.00000033$

90. $\Delta t = 0.2, y(1.0) = 2.44802522, E(0.2) = -0.00002910$, Ratio $= 35.9$
    $\Delta t = 0.1, y(1.0) = 2.44805351, E(0.1) = -0.00000081$

94. $\Delta t = 0.2, y(1.0) = 2.00007303, E(0.2) = -0.00007303$, Ratio $= 26.0$
    $\Delta t = 0.1, y(1.0) = 2.00000281, E(0.1) = -0.00000281$

100.

M	$y(0.2)$, $0(h^2)$	$\dfrac{4 \text{ MAV} - \text{LAV}}{3}$	$\dfrac{16 \text{ MAV} - \text{LAV}}{15}$	$\dfrac{64 \text{ MAV} - \text{LAV}}{63}$
2	0.18100000	0.18126833	0.18126925	0.18126925
4	0.18120125	0.18126919	0.18126925	
8	0.18125220	0.18126924		
16	0.18126498			

t_n	y_n	$\bar{y}_n$	Error
0.0	0.00000000	0.00000000	
0.2	0.18126925	0.18126925	0.00000000
0.4	0.32967995	0.32967995	0.00000000
0.6	0.45118836	0.45118836	0.00000000
0.8	0.55067104	0.55067104	0.00000000
1.0	0.63212056	0.63212056	0.00000000

106.

M	$y(0.2)$, $0(h^2)$	$\dfrac{4\,MAV - LAV}{3}$	$\dfrac{16\,MAV - LAV}{15}$	$\dfrac{64\,MAV - LAV}{63}$
2	0.73860604	0.73851857	0.73851815	0.73851815
4	0.73854044	0.73851818	0.73851815	
8	0.73852374	0.73851815		
16	0.73851955			

t_n	y_n	$\bar{y}_n$	Error
0.0	0.00000000	0.00000000	
0.2	0.73851815	0.73851815	0.00000000
0.4	1.01535175	1.01535175	0.00000000
0.6	1.35522343	1.35522343	0.00000000
0.8	1.80248634	1.80248634	0.00000000
1.0	2.44805432	2.44805432	0.00000000

116. $\Delta t = 0.2, y(1.0) = 0.63213778, E(0.2) = 0.00001722$, Ratio $= 14.7$

     $\Delta t = 0.1, y(1.0) = 0.63212173, E(0.1) = 0.00000117$

122. $\Delta t = 0.2, y(1.0) = 2.44802270, E(0.2) = -0.00003162$, Ratio is meaningless

     $\Delta t = 0.1, y(1.0) = 2.44821893, E(0.1) = 0.00016461$

127. $\Delta t = 0.2, y(1.0) = 0.63211820, E(0.2) = -0.00000236$, Ratio $= 26.2$

     $\Delta t = 0.1, y(1.0) = 0.63212047, E(0.1) = -0.00000009$

133. $\Delta t = 0.2, y(1.0) = 2.44835945, E(0.2) = 0.00030513$, Ratio $= 8.54$

     $\Delta t = 0.1, y(1.0) = 2.44809007, E(0.1) = 0.00003575$

135. $\Delta t = 0.2, y(1.0) = 0.598122, E(0.2) = -0.033998$, Ratio $= 1.92$

     $\Delta t = 0.1, y(1.0) = 0.614457, E(0.1) = -0.017664$

159. $\Delta t = 0.2, y(1.0) = 3.039281, E(0.2) = 0.591227$, Ratio $= 2.13$

     $\Delta t = 0.1, y(1.0) = 2.725422, E(0.1) = 0.277367$

164. $\Delta t = 0.2, y(1.0) = 0.598122, E(0.2) = -0.033998$, Ratio $= 1.92$

     $\Delta t = 0.1, y(1.0) = 0.614457, E(0.1) = -0.017664$

170. $\Delta t = 0.2, y(1.0) = 3.837679, E(0.2) = 1.389625$, Ratio $= 4.02$

     $\Delta t = 0.1, y(1.0) = 2.793804, E(0.1) = 0.345749$

185. $\Delta t = 0.2, y(1.0) = 12.391680, z(1.0) = 8.655360$

     $\Delta t = 0.1, y(1.0) = 15.654961, z(1.0) = 10.685748$

187. $\Delta t = 0.2, y(1.0) = 21.25212768, z(1.0) = 14.15934151$
    $\Delta t = 0.1, y(1.0) = 21.27386712, 1.0) = 14.17277742$

210. $\Delta t = 0.1, y(1.0) = 2.100000, E(0.01) = 0.000000$
    $\Delta t = 0.2, y(0.1) = 2.099918, E(0.05) = -0.000082$

220. $\Delta t = 0.010, y(0.990) = 0.369730, z(0.99) = -62.027036$
    $y(1.000) = 0.366032, z(1.00) = -62.396766$
    $\Delta t = 0.020, y(0.980) = 0.371602, z(0.98) = -61.839829$
    $y(1.000) = 0.364170, z(1.000000) = -62.583032$
    $\Delta t = 0.025, y(0.975) = 0.372546, z(0.975) = -61.745391$
    $y(1.000) = 0.363232, z(1.000) = -62.676756$

227. $\Delta t = 0.0010, y(0.0990) = 3.622791, z(0.0990) = -1.811396$
    $y(0.1000) = 3.619169, z(0.1000) = -1.809584$
    $\Delta t = 0.0020, y(0.0980) = 6.626240, z(0.0980) = -4.813120$
    $y(0.1000) = 0.618987, z(0.1000) = 1.190506$
    $\Delta t = 0.0025, y(0.0025) = 8.490000, z(0.0025) = -6.495000$
    $y(0.0050) = -2.769975, z(0.0050) = 4.759988$

234. $\Delta t = 1.0, y(99.0) = 0.905698, z(99.0) = 0.905698$
    $y(100.0) = 0.904792, z(100.0) = 0.904792$
    $\Delta t = 2.0, y(98.0) = -0.093440, z(98.0) = 0.906560$
    $y(100.0) = 1.904747, z(100.0) = 0.904747$
    $\Delta t = 2.5, y(2.5) = -0.502500, z(2.5) = 0.997500$
    $y(5.0) = 3.245006, z(5.0) = 0.995006$

Chapter 8. One-Dimensional Boundary-Value Problems

4.

		$\Delta x = 0.25$		$\Delta x = 0.125$		$\Delta x = 0.0625$	
x	$\bar{y}(x)$	$y(x)$	Error	$y(x)$	Error	$y(x)$	Error
0.00	0.000000	0.000000		0.000000		0.000000	
0.25	1.045057	2.027778	0.982721	1.327386	0.282329	1.162644	0.117587
0.50	1.233303	1.583333	0.350031	1.392072	0.158769	1.304867	0.071564
0.75	1.149746	1.250000	0.100254	1.203698	0.053952	1.175560	0.028515
1.00	1.000000	1.000000		1.000000		1.000000	

5.

		$\Delta x = 0.25$		$\Delta x = 0.125$		$\Delta x = 0.0625$	
x	$\bar{y}(x)$	$y(x)$	Error	$y(x)$	Error	$y(x)$	Error
0.00	0.000000	0.000000		0.000000		0.000000	
0.25	1.045057	0.819554	-0.225503	1.002372	-0.042685	1.036161	-0.008896
0.50	1.233303	1.077398	-0.155905	1.206372	-0.026931	1.227779	-0.005523
0.75	1.149746	1.087621	-0.062125	1.139889	-0.009857	1.147752	-0.001994
1.00	1.000000	1.000000		1.000000		1.000000	

6.

	$\Delta x = 0.25$			$\Delta x = 0.125$		$\Delta x = 0.0625$	
x	$\bar{y}(x)$	$y(x)$	Error	$y(x)$	Error	$y(x)$	Error
0.00	0.000000	0.000000		0.000000		0.000000	
0.25	1.045057	1.031965	−0.013092	1.044519	−0.000539	1.045030	−0.000027
0.50	1.233303	1.225251	−0.008052	1.232974	−0.000329	1.233286	−0.000017
0.75	1.149746	1.146878	−0.002868	1.149630	−0.000116	1.149740	−0.000006
1.00	1.000000	1.000000		1.000000		1.000000	

37.

	$\Delta x = 0.25$	$\Delta x = 0.125$	$\Delta x = 0.0625$
x	$y(x)$	$y(x)$	$y(x)$
0.00	0.000000	0.000000	0.000000
0.25	0.318463	0.371144	0.394892
0.50	0.619809	0.683537	0.708655
0.75	0.858085	0.896714	0.909746
1.00	0.999622	0.999981	0.999956

43.

	$\Delta x = 0.25$			$\Delta x = 0.125$		$\Delta x = 0.0625$	
x	$\bar{y}(x)$	$y(x)$	Error	$y(x)$	Error	$y(x)$	Error
0.00	0.000000	0.000000		0.000000		0.000000	
0.25	1.045057	1.176150	0.131093	1.072659	0.027601	1.051705	0.006648
0.50	1.233303	1.305085	0.071782	1.249448	0.016146	1.237241	0.003939
0.75	1.149764	1.172518	0.022772	1.155177	0.005431	1.151086	0.001340
1.00	1.000000	1.000000		1.000000		1.000000	

54.

x	$\bar{y}(x)$	$y(x)$	Error
0.000	1.000000	1.000000	
0.125	0.653429	0.653457	0.000027
0.250	0.449272	0.444518	−0.004753
0.375	0.330781	0.320687	−0.010093
0.500	0.263622	0.249256	−0.014366
0.625	0.227045	0.209876	−0.017169
0.750	0.208528	0.189911	−0.018617
0.875	0.200534	0.181536	−0.018998
1.000	0.198542	0.179923	−0.018619

60.

x	$\bar{y}(x)$	$y(x)$	Error
0.000	1.000000	1.000000	
0.125	0.653429	0.646682	−0.006748
0.250	0.449272	0.442720	−0.006551
0.375	0.330781	0.326706	−0.004074
0.500	0.263622	0.262284	−0.001338
0.625	0.227045	0.227955	0.000910
0.750	0.208528	0.211022	0.002494
0.875	0.200534	0.204009	0.003475
1.000	0.198542	0.202525	0.003983

66.

	$\Delta x = 0.25$		$\Delta x = 0.125$	
x	$y(x)$	$y'(x)$	$y(x)$	$y'(x)$
0.00	0.000000	2.580247	0.000000	2.506447
0.25	0.645062	−0.395062	0.446421	0.367691
0.50	0.546296	−0.296296	0.497341	−0.106313
0.75	0.472222	−0.222222	0.463610	−0.178353
1.00	0.416667	−0.166667	0.419605	−0.160791

74.

	$y(2.0) = 0.0$		$y(5.0) = 0.0$		$y(10.0) = 0.0$	
	$\Delta x = 0.25$	$\Delta x = 0.125$	$\Delta x = 0.25$	$\Delta x = 0.125$	$\Delta x = 0.25$	$\Delta x = 0.125$
x	$y(x)$	$y(x)$	$y(x)$	$y(x)$	$y(x)$	$y(x)$
0.00	1.000000	1.000000	1.000000	1.000000	1.000000	1.000000
0.25	0.741458	0.756494	0.749982	0.765603	0.750000	0.765625
0.50	0.545417	0.567634	0.562463	0.586138	0.562500	0.586182
0.75	0.395717	0.419969	0.421819	0.448727	0.421875	0.448795
1.00	0.280105	0.303028	0.316329	0.343513	0.316406	0.343609

86.

		$\Delta x = 0.250$		$\Delta x = 0.125$	
x	$\bar{y}(x)$	$y(x)$	Error	$y(x)$	Error
0.00	0.00000000	0.00000000		0.00000000	
0.25	0.18994383	0.18994276	−0.00000108	0.18994377	−0.00000007
0.50	0.43025304	0.43025117	−0.00000127	0.43025296	−0.00000008
0.75	0.70598635	0.70598554	−0.00000081	0.70598630	−0.00000005
1.00	1.00000000	1.00000000		1.00000000	

90.

	$\Delta x = 0.25$	$\Delta x = 0.125$
x	$y(x)$	$y(x)$
0.00	0.000000	0.000000
0.25	0.420208	0.416310
0.50	0.735142	0.729703
0.75	0.923685	0.919964
1.00	1.000000	1.000000

Chapter 9. Elliptic Partial Differential Equations

7.

		x	
y	0.0	5.0	10.0
15.0	100.000000	100.000000	100.000000
10.0	0.000000	26.666667	0.000000
5.0	0.000000	6.666667	0.000000
0.0	0.000000	0.000000	0.000000

9.

		x	
y	0.0	5.0	10.0
15.0	0.000000	0.000000	100.000000
10.0	0.000000	33.333333	100.000000
5.0	0.000000	33.333333	100.000000
0.0	0.000000	0.000000	100.000000

11.

			x		
y	0.0	0.25	0.50	0.75	1.00
1.00	100.000000	0.000000	0.000000	0.000000	0.000000
0.75	100.000000	50.000000	28.571429	14.285714	0.000000
0.50	100.000000	71.428571	50.000000	28.571429	0.000000
0.25	100.000000	85.714286	71.428571	50.000000	0.000000
0.00	100.000000	100.000000	100.000000	100.000000	100.000000

26.

		x	
y	0.0	5.0	10.0
15.0	0.000000	0.000000	0.000000
10.0	0.000000	83.333333	0.000000
5.0	0.000000	83.333333	0.000000
0.0	0.000000	0.000000	0.000000

Chapter 10. Parabolic Partial Differential Equations

5.

			x, cm			
t, s	0.0	0.1	0.2	0.3	0.4	0.5
0.0	0.0000	20.0000	40.0000	60.0000	80.0000	100.0000
0.1	0.0000	20.0000	40.0000	60.0000	80.0000	96.0000
0.2	0.0000	20.0000	40.0000	60.0000	79.6000	92.8000
0.3	0.0000	20.0000	40.0000	59.9600	78.9600	90.1600
0.4	0.0000	20.0000	39.9960	59.8640	78.1800	87.9200
0.5	0.0000	19.9996	39.9832	59.7088	77.3224	85.9720

23.

			x, cm			
t, s	0.0	0.1	0.2	0.3	0.4	0.5
0.0	0.0000	20.0000	40.0000	60.0000	80.0000	100.0000
0.5	0.0000	19.9448	39.7790	59.1713	76.9061	88.4530

28.

			x, cm			
t, s	0.0	0.1	0.2	0.3	0.4	0.5
0.0	0.0000	20.0000	40.0000	60.0000	80.0000	100.0000
0.5	0.0000	19.9881	39.9286	59.5837	77.5736	85.8579

33.

			x, cm			
t, s	0.0	0.1	0.2	0.3	0.4	0.5
0.0	0.0000	20.0000	40.0000	60.0000	80.0000	100.0000
0.5	0.0000	20.0000	40.0000	60.0000	80.0000	80.0000
1.0	0.0000	20.0000	40.0000	60.0000	70.0000	80.0000
1.5	0.0000	20.0000	40.0000	55.0000	70.0000	70.0000
2.0	0.0000	20.0000	37.5000	55.0000	62.5000	70.0000
2.5	0.0000	18.7500	37.5000	50.0000	62.5000	62.5000

Chapter 11. Hyperbolic Partial Differential Equations

9.

						x, cm							
t, s	0.0	0.1	0.2	0.3	0.4	0.5	0.6	0.7	0.8	0.9	1.0	1.1	1.2
0.0	0.00	20.00	40.00	60.00	80.00	100.00	80.00	60.00	40.00	20.00	0.00	0.00	0.00
1.0	−10.00	0.00	20.00	40.00	60.00	100.00	100.00	80.00	60.00	40.00	10.00	0.00	0.00
2.0	−10.00	−15.00	0.00	20.00	30.00	80.00	110.00	100.00	80.00	65.00	30.00	5.00	0.00

13.

	x, cm												
t, s	0.0	0.1	0.2	0.3	0.4	0.5	0.6	0.7	0.8	0.9	1.0	1.1	1.2
0.0	0.00	20.00	40.00	60.00	80.00	100.00	80.00	60.00	40.00	20.00	0.00	0.00	0.00
0.5	5.00	10.00	30.00	50.00	70.00	80.00	90.00	70.00	50.00	30.00	15.00	0.00	0.00
1.0	2.50	11.25	20.00	40.00	57.50	75.00	77.50	80.00	60.00	41.25	22.50	11.25	0.00

27.

	x, cm												
t, s	0.0	0.1	0.2	0.3	0.4	0.5	0.6	0.7	0.8	0.9	1.0	1.1	1.2
0.0	0.00	20.00	40.00	60.00	80.00	100.00	80.00	60.00	40.00	20.00	0.00	0.00	0.00
0.5	−2.50	10.00	30.00	50.00	70.00	95.00	90.00	70.00	50.00	30.00	7.50	0.00	0.00
1.0	−3.12	2.81	20.00	40.00	59.38	86.25	94.38	80.00	60.00	40.31	16.88	2.81	0.00

33.

	x, cm												
t, s	0.0	0.1	0.2	0.3	0.4	0.5	0.6	0.7	0.8	0.9	1.0	1.1	1.2
0.0	0.00	20.00	40.00	60.00	80.00	100.00	80.00	60.00	40.00	20.00	0.00	0.00	0.00
0.5	0.00	10.00	30.00	50.00	70.00	90.00	90.00	70.00	50.00	30.00	10.00	0.00	0.00
1.0	0.00	5.00	20.00	40.00	60.00	80.00	90.00	80.00	60.00	40.00	20.00	5.00	0.00

45.

	x, cm								
t, s	−0.5	−0.25	0.0	0.25	0.50	0.75	1.00	1.25	1.50
0.0	0.00	0.00	0.00	50.00	100.00	50.00	0.00	0.00	0.00
1.0	0.00	1.16	−5.82	30.27	92.85	66.03	12.70	2.54	0.00

Index